ENVIRONMENTAL & POLLUTION SCIENCE

SECOND EDITION

D1116062

ENVIRONMENTAL & POLLUTION SCIENCE

SECOND EDITION

EDITORS

Ian L. Pepper, Ph.D.

Professor of Environmental Microbiology
The University of Arizona, Tucson, Arizona

Charles P. Gerba, Ph.D.

Professor of Environmental Microbiology
The University of Arizona, Tucson, Arizona

Mark L. Brusseau, Ph.D.

Professor of Subsurface Hydrology
and Environmental Chemistry
The University of Arizona, Tucson, Arizona

Photography & Technical Editor: K.L. Josephson
Copy Editor: E.R. Loya

ELSEVIER

AMSTERDAM • BOSTON • HEIDELBERG • LONDON • NEW YORK • OXFORD
PARIS • SAN DIEGO • SAN FRANCISCO • SINGAPORE • SYDNEY • TOKYO
Academic Press is an imprint of ELSEVIER

Acquisitions Editor:	Jeremy Hayhurst
Project Manager:	Heather Furrow
Marketing Manager:	Linda Beattie
Cover Design:	Julio Esperas
Production Services:	Dovetail Content Solutions
Composition:	Maryland Composition
Cover Printer:	Transcontinental Interglobe
Interior Printer:	Transcontinental Interglobe

Academic Press is an imprint of Elsevier
30 Corporate Drive, Suite 400, Burlington, MA 01803, USA
525 B Street, Suite 1900, San Diego, California 92101-4495, USA
84 Theobald's Road, London WC1X 8RR, UK

This book is printed on acid-free paper. ⊗

Copyright © 2006, 1996, Elsevier Inc. All rights reserved.

No part of this publication may be reproduced or transmitted in any form
or by any means, electronic or mechanical, including photocopy,
recording, or any information storage and retrieval system, without
permission in writing from the publisher.

Permissions may be sought directly from Elsevier's Science &
Technology Rights Department in Oxford, UK: phone: (+44) 1865
843830, fax: (+44) 1865 853333, E-mail: permissions@elsevier.com.
You may also complete your request on-line via the Elsevier homepage
(http://elsevier.com), by selecting "Support & Contact" then "Copyright
and Permission" and then "Obtaining Permissions."

Library of Congress Cataloging-in-Publication Data
Environmental and pollution science / Ian L. Pepper, Charles P. Gerba,
Mark L. Brusseau [editors].--2nd ed.
  p. cm.
  Rev. ed. of Pollution science. 1996.
  Includes bibliographical references and index.
  ISBN-13: 978-0-12-551503-0 (casebound : alk. paper)
  ISBN-10: 0-12-551503-0 (casebound : alk. paper) 1. Pollution.
2. Environmental sciences. I. Pepper, Ian L. II. Gerba, Charles P.,
1945- III. Brusseau, Mark L. IV. Pollution science.

TD174.P68 2006
628—dc22

2005035422

British Library Cataloguing-in-Publication Data
A catalogue record for this book is available from the British Library.

ISBN 13: 978-0-12-551503-0
ISBN 10: 0-12-551503-0

For information on all Academic Press Publications
visit our Web site at www.books.elsevier.com

Printed in Canada
06 07 08 09 10   9 8 7 6 5 4 3 2 1

DEDICATION

"To all of my graduate students—you have taught me more than you will ever know."

Ian L. Pepper

"To Peggy, Peter, and Phillip for putting up with me all these years."

Charles P. Gerba

"I dedicate this to my mother, Jeannie—thanks Mom!"

Mark L. Brusseau

TABLE OF CONTENTS

CHAPTER 4
Physical-Chemical Characteristics of the Atmosphere   46

A.D. Matthias, S.A. Musil, and H.L. Bohn

CHAPTER 5
Biotic Characteristics of the Environment   58

I.L. Pepper and K.L. Josephson

CHAPTER 6
Physical Processes Affecting Contaminant Transport and Fate   78

M.L. Brusseau

CHAPTER 7
Chemical Processes Affecting Contaminant Transport and Fate   89

M.L. Brusseau and J. Chorover

CHAPTER 8
Biological Processes Affecting Contaminant Transport and Fate   105

R.M. Maier

PART 2

MONITORING, ASSESSMENT, AND REGULATION OF ENVIRONMENTAL POLLUTION

CHAPTER 9

Physical Contaminants   *123*

J. Walworth and I.L. Pepper

CHAPTER 10

Chemical Contaminants   *132*

M.L. Brusseau, C.M. McColl, G. Famisan, and J.F. Artiola

CHAPTER 11

Microbial Contaminants   *144*

C.P. Gerba and I.L. Pepper

CHAPTER 12

The Role of Environmental Monitoring in Pollution Science   *170*

J.F. Artiola and M.L. Brusseau

CHAPTER 13

Environmental Toxicology   *183*

C.P. Gerba

PART 5

WASTE AND WATER TREATMENT AND MANAGEMENT

CHAPTER 25

Industrial and Municipal Solid Waste Treatment and Disposal   *415*

J.F. Artiola

CHAPTER 26

Municipal Wastewater Treatment   *429*

C.P. Gerba and I.L. Pepper

CHAPTER 27

Land Application of Biosolids and Animal Wastes   *451*

I.L. Pepper and C.P. Gerba

PREFACE

This textbook focuses on: (i) the continuum of the environment—namely the earth, water, and atmosphere, epitomized by the term "Environmental Science"; and (ii) science-based aspects of the pollution of the environment. Thus, this textbook is designed to provide a scientific basis that will allow for prudent decisions to be made to manage and mitigate pollution throughout the environment. In general, pollution can be defined as the accumulation and adverse interaction of contaminants within the environment. Pollution is ubiquitous and can occur on or within land, oceans, or the atmosphere. Contaminants can consist of chemical compounds/elements, biological entities, particulate matter, or energy, and they may be of natural or anthropogenic origin. Given these complexities, multidisciplinary approaches are needed to address environmental pollution issues. Therefore, this text provides a rigorous science-based integration of the physical, chemical, and biological properties and processes that influence the environment and that also affect contaminants in the environment.

Following the foundation chapters that describe these factors, the text deals with numerous facets critical to pollution analysis and management:

- characterization
- risk assessment
- regulation
- fate and transport
- remediation and/or restoration

These components are described in detail for well-developed disciplines such as waste management and water treatment. However, newer areas of concern are also introduced such as global change, pathogenic contamination, and indoor air quality. The ultimate in this line of thought is presented as "Emerging Issues in Pollution Science." While scientific and technical matters are the focus of the text, social, economic, and political issues are considered as well, to provide a holistic coverage. This is accomplished primarily through the use of case studies.

This text is designed for a science-based junior/senior-level undergraduate course. Students that will benefit from this text originate from a variety of backgrounds, including: environmental science, microbiology, hydrology, earth science, and environmental engineering. The text will also serve as an introductory text for graduate students originating from traditional disciplines such as chemistry, physics, and biology, and for those with non-science backgrounds.

Naturally, the second edition is an evolution of "Pollution Science," which was first published in 1996. Time does indeed fly, and the authors are now older and (hopefully) wiser. All original chapters have been significantly revised or modified. In addition, we have added several new chapters. Other changes include an enhanced focus on providing problems and calculations, which are designed to provide a quantitative as well as qualitative aspect to the book. Ultimately, we hope this text will illustrate that the integration of physical, chemical, earth, and biological sciences with engineering, health, and social sciences can be successfully accomplished, thereby inspiring readers to implement inter- and multi-disciplinary solutions for past, present, and future environmental pollution problems.

Ian L. Pepper
Charles P. Gerba
Mark L. Brusseau
January 2006

THE EDITORS

Charles P. Gerba
*Professor of Environmental
Microbiology*

Ian L. Pepper
*Professor of Environmental
Microbiology*

Mark L. Brusseau
*Professor of Subsurface Hydrology
and Environmental Chemistry*

CONTRIBUTING AUTHORS

Robert G. Arnold
Professor of Environmental Engineering

Janick F. Artiola
Associate Professor in Environmental Chemistry

Donald J. Baumgartner
Professor Emeritus, Environmental Engineering

Jon Chorover
Professor of Environmental Chemistry

Andrew C. Comrie
Professor of Geography, Professor of Atmospheric Sciences

Michael A. Crimmins
Assistant Professor and Climate Science Extension Specialist

Gale Famisan
Assistant Research Scientist

Kevin M. Fitzsimmons
Professor of Environmental Science

Gregor Grass
Assistant Professor of Molecular Microbiology

Karen L. Josephson
Research Specialist, Principal

Raina M. Maier

Professor of Environmental Microbiology

Allan D. Matthias

Associate Professor of Soil, Water and Environmental Science

Colleen M. McColl

Environmental Scientist

Sheri A. Musil

Research Specialist, Senior

David M. Quanrud

Assistant Research Scientist in Arid Lands Studies

Christopher Rensing

Assistant Professor of Molecular Microbiology

Kelly A. Reynolds

Assistant Research Scientist

Geoffrey R. Tick

Assistant Professor of Hydrogeology

David B. Walker

Assistant Research Scientist

James L. Walworth

Associate Professor of Soil Science

The Department of Soil, Water and Environmental Science (SWES)

SWES is part of the College of Agriculture and Life Sciences within the University of Arizona. The roots of this department were in Soil Science, but over the past fifteen years, we have added exceptional environmental science programs which are nationally recognized. Faculty members of the department form an interdisciplinary team of microbiologists, chemists, physicists, hydrologists, and earth scientists, that can address environmental and pollution science problems.

The department is also home to the Environmental Research Laboratory (ERL). This off-campus facility is an established leader of environmental research and education in arid regions. The overall goal of ERL is to improve the health, welfare and living standards of communities in desert areas through the development of new sustainable technologies. ERL also houses the University of Arizona, National Science Foundation Water Quality Center (WQC), the only such Center in the United States.

CHAPTER 1

THE EXTENT OF GLOBAL POLLUTION

I.L. Pepper, C.P. Gerba, and M.L. Brusseau

Pollution is ubiquitous, and can even cause beautiful sunsets. *Photo courtesy Ian Pepper.*

1.1 SCIENCE AND POLLUTION

Pollution is ubiquitous and takes many forms and shapes. For example, the beautiful sunsets that we may see in the evening are often due to the interaction of light and atmospheric contaminants, as illustrated above.

Pollution can be defined as the accumulation and adverse affects of contaminants or pollutants on human health and welfare, and/or the environment. But in order to truly understand pollution, we must define the identity and nature of potential contaminants. Contaminants can result from waste materials produced from the activity of living organisms,

INFORMATION BOX 1.1

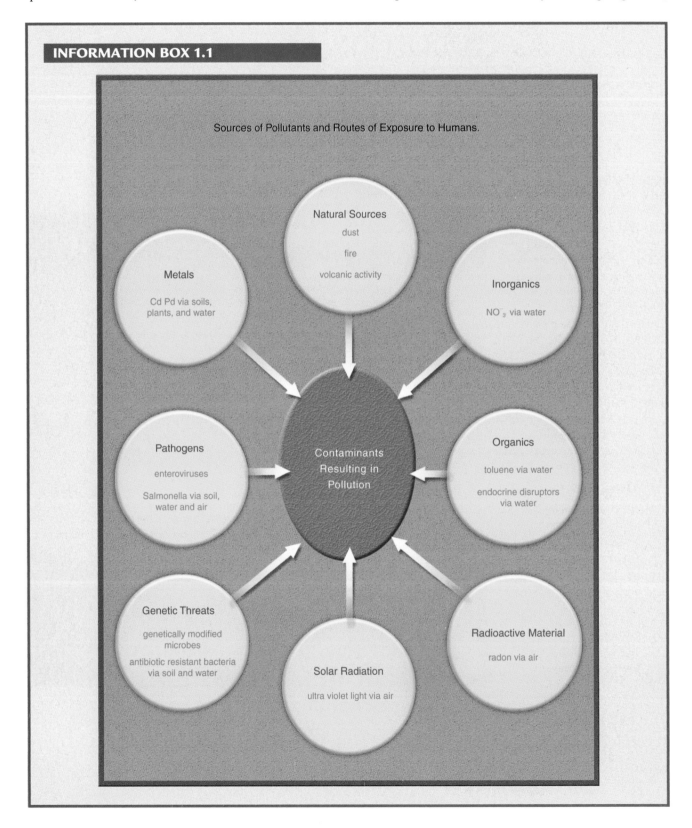

Sources of Pollutants and Routes of Exposure to Humans.

Natural Sources
dust
fire
volcanic activity

Metals
Cd Pd via soils, plants, and water

Inorganics
NO_3^- via water

Pathogens
enteroviruses
Salmonella via soil, water and air

Contaminants Resulting in Pollution

Organics
toluene via water
endocrine disruptors via water

Genetic Threats
genetically modified microbes
antibiotic resistant bacteria via soil and water

Solar Radiation
ultra violet light via air

Radioactive Material
radon via air

TABLE 1.1 Recently discovered microbes that have had a significant impact on human health.

AGENT	MODE OF TRANSMISSION	DISEASE / SYMPTOMS
Rotavirus	Waterborne	Diarrhea
Legionella	Waterborne	Legionnaire's disease
Escherichia coli O157:H7	Foodborne Waterborne	Enterohemorrhagic fever, kidney failure
Hepatitis E virus	Waterborne	Hepatitis
Cryptosporidium	Waterborne Foodborne	Diarrhea
Calicivirus	Waterborne Foodborne	Diarrhea
Helicobacter pylori	Foodborne Waterborne	Stomach ulcers
Cyclospora	Foodborne Waterborne	Diarrhea

especially humans. However, contamination can also occur from natural processes such as arsenic dissolution from bedrock into groundwater, or air pollution from smoke that results from natural fires. Pollutants are also ubiquitous in that they can be in the solid, liquid, or gaseous state. Information Box 1.1 presents the major categories of pollutants and their predominant routes of human exposure. Clearly, many of the agents identified in Information Box 1.1 occur directly through activities such as mining or agriculture. But in addition, pollution is also produced as an indirect result of human activity. For example, fossil fuel burning increases atmospheric carbon dioxide levels and increases global warming. Other classes of pollutants can occur due to poor waste management or disposal, which can lead to the presence of pathogenic microorganisms in water. Some examples of microbial pathogens and associated diseases are shown in Table 1.1. Another example of pollution due to human activity is accidental spillage of organics that can be toxic, such as chlorinated solvents or petroleum hydrocarbons that contaminate

TABLE 1.2 Common organic and inorganic contaminants found in the environment.

CHEMICAL CLASS	FREQUENCY OF OCCURRENCE
Gasoline, fuel oil	Very frequent
Polycyclic aromatic hydrocarbons	Common
Creosote	Infrequent
Alcohols, ketones, esters	Common
Ethers	Common
Chlorinated organics	Very frequent
Polybrominated diphenyl ethers (PBDEs)	
Polychlorinated biphenyls (PCBs)	Infrequent
Nitroaromatics (TNT)	Common
Metals (Cd, Cr, Cu, Hg, Ni, Pb, Zn)	Common
Nitrate	Common

From *Environmental Microbiology* © 2000, Academic Press, San Diego, CA.

groundwater. Some common contaminants that find their way into the environment, with the potential to adversely affect human health and welfare, are shown in Table 1.2.

In this textbook, we will discuss these major sources of pollution in a science-based context, hence the name: *Environmental and Pollution Science* (Information Box 1.2).

The focus of the text will be to identify the basic scientific processes that control the transport and fate of pollutants in the environment. We will also try to define the potential for adverse effects to human health and welfare, and the environment using a risk-based approach. Finally, we will present real world "case studies." The diverse nature of the scientific disciplines needed to study pollution science are shown in Information Box 1.3. It is the holistic integration of these diverse and complex entities that presents the major challenge to understanding both "Environmental and Pollution Science."

1.2 GLOBAL PERSPECTIVE OF THE ENVIRONMENT

The environment plays a key role in the ultimate fate of pollutants. The environment consists of land, water, and the atmosphere. All sources of pollution are initially released or dumped into one of these phases of the environment. As pollutants interact with the environment, they undergo physical and chemical changes, and are ultimately incorporated into the environment. The environment thus acts as a continuum into which all waste materials are placed. The pollutants, in

INFORMATION BOX 1.2

Environmental and Pollution Science is the study of the physical, chemical, and biological processes fundamental to the transport, fate, and mitigation of contaminants that arise from human activities as well as natural processes.

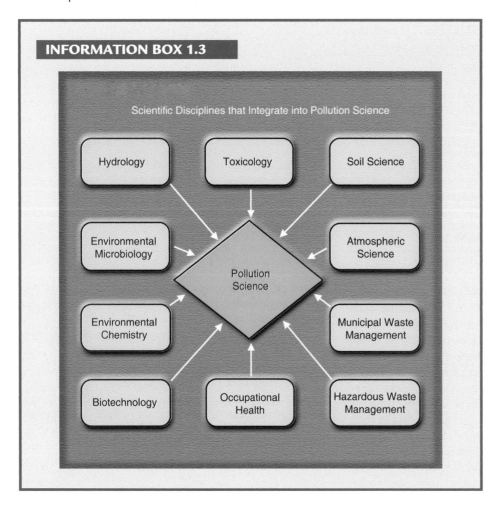

INFORMATION BOX 1.3

Scientific Disciplines that Integrate into Pollution Science

Hydrology

Toxicology

Soil Science

Environmental Microbiology

Atmospheric Science

Pollution Science

Environmental Chemistry

Municipal Waste Management

Biotechnology

Occupational Health

Hazardous Waste Management

turn, obey the second law of thermodynamics: matter cannot be destroyed; it is merely converted from one form to another. Thus, taken together, the way in which substances are added to the environment, the rate at which these wastes are added, and the subsequent changes that occur determine the impact of the waste on the environment. It is important to recognize the concept of the environment as a continuum, because many physical, chemical, and biological processes occur not within one of these phases, such as the air alone, but rather at the interface between two phases such as the soil/water interface.

The concept of the continuum relies on the premise that resources are utilized at a rate at which they can be replaced or renewed, and that wastes are added to the environment at a rate at which they can be assimilated without disturbing the environment. Historically, natural wastes were generated that could easily be broken down or transformed into beneficial, or at least benign, compounds. However, post-industrial contamination has resulted in the formation of **xenobiotic waste**—compounds that are foreign to natural ecosystems and that are less subject to degradation. In some cases, natural processes can actually enhance the toxicity of the pollutants. For example, organic compounds that are not themselves carcinogenic can be microbially converted into carcinogenic substances. Other compounds, even those not normally considered pollutants, can cause pollution if they are added to the

environment in quantities that result in high concentrations of these substances. An excellent example here is nitrate fertilizer, which is often added to soil at high levels. Such nitrates can end up in drinking water supplies and cause methemoglobinemia (blue baby disease) in newborn infants (see Chapter 16).

Some pollutants, such as microbial pathogens, are entirely natural and may be present in the environment at very low concentrations. Even so, they are still capable of causing pathogenic diseases in humans or animals. Such natural microorganisms are also classified as pollutants, and their occurrence within the environment needs to be carefully controlled.

1.3 POLLUTION AND POPULATION PRESSURES

To understand the relationship between population and pollution, let us examine a typical curve for the growth of a pure culture of bacteria in a liquid medium (Figure 1.1). Early on, the bacteria growing in the medium do not increase significantly in number, due to low population densities, which results in organisms operating as separate entities. This initial low-growth phase is known as the **lag period**. Next, the

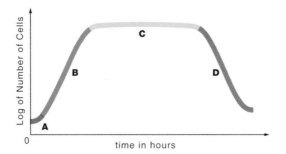

Figure 1.1 Typical growth curve for a pure culture of bacteria. A = lag period, B = exponential phase, C = stationary phase, D = death phase. From *Pollution Science* © 1996, Academic Press, San Diego, CA.

number of organisms increases exponentially for a finite period of time. This phase of growth is known as the **exponential phase** or **log phase**. After this exponential phase of growth, a **stationary phase** occurs, during which the total number of organisms remains constant as new organisms are constantly being produced while other organisms are dying. Finally, we observe the **death phase**, in which the total number of organisms decreases. We know that bacteria reproduce by binary fission, so it is easy to see how a doubling of bacteria occurs during exponential growth. But what causes the stationary and death phases of growth?

Two mechanisms prevent the number of organisms from increasing *ad infinitum*: first, the organisms begin to run out of nutrients; and second, waste products build up within the growth medium and become toxic to the organisms. An analogous situation exists for humans. Initially, in

prehistoric times, population densities were low and population numbers did not increase significantly or rapidly (Figure 1.2.). During this time resources were plentiful; thus, the environment could easily accommodate the amount of wastes produced. Later, populations began to increase very rapidly. Although not exponential, this phase of growth was comparable to the log phase of microbial growth. During this period then, large amounts of resources were utilized, and wastes were produced in ever-greater quantities. This period of growth is still under way. However, we seem to be approaching a period in which lack of resources or buildup of wastes (*i.e.,* pollution) will limit continued growth—hence the renewed interest in recycling materials as well as in controlling, managing, and cleaning up waste materials. To do this, we must arrive at an understanding of the predominant biotic and abiotic characteristics of the environment.

Currently the world population is 6.3 billion and increasing rapidly. This population pressure has caused intense industrial and agricultural activities that produce hazardous contaminants in their own right. In addition, increased populations result in the production of wastes that at low concentrations are not hazardous, but which at high concentrations become hazardous. Hence, concentrated animal feedlot operations (CAFOs), where large numbers of animals are kept in close proximity, require special attention to minimize potential pollution (see Chapter 27). Finally, note that as the world population increases, people tend to relocate from sparsely populated rural areas to more congested urban centers or "mega-cities." Typically, urbanized areas consume more natural resources and produce more waste per

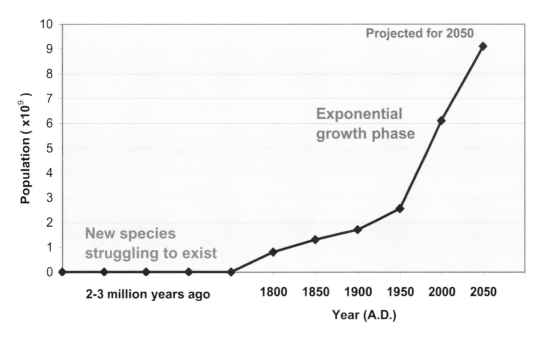

Figure 1.2 World population increases from the inception of the human species. From Population Reference Bureau, Inc., 1990. Adapted from *Pollution Science* © 1996, Academic Press, San Diego, CA.

TABLE 1.3 Collection and storage specifications for a Quality Assurance Project Plan (QAPP). [a]

Sampling strategies: Number and type of samples, locations, depths, times, intervals
Sampling methods: Specific techniques and equipment to be used
Sample storage: Types of containers, preservation methods, maximum holding times

[a]The QAPP normally also includes details of the proposed microbial analysis to be conducted on the soil samples.

From *Environmental Microbiology* © 2000, Academic Press, San Diego, CA.

capita than rural areas. The trend towards mega-cities will intensify this problem.

1.4 OVERVIEW OF ENVIRONMENTAL CHARACTERIZATION

In order to ascertain the potential or actual extent of pollution that has occurred, it is necessary to undertake environmental monitoring of the polluted site (see Chapter 12; also Artiola et al., 2004). This frequently involves site characterization, which involves identifying the area and/or volume of the environment which has been polluted. It can also involve comparisons with nonpolluted control sites to evaluate normal background levels of contaminants. In order to undertake site characterization, it is important to establish proper sampling regimens for the particular environmental sample, be it soil, water, or air. Here we provide an overview of the basic strategies for environmental sampling. Because so many choices are available, it is important to ensure that quality assurance is addressed by developing a quality assurance project plan (QAPP), as shown in Table 1.3.

1.4.1 Soil and the Subsurface

Physically, surface soil samples are easy to obtain using inexpensive equipment such as a shovel or a soil auger (Figure 1.3).

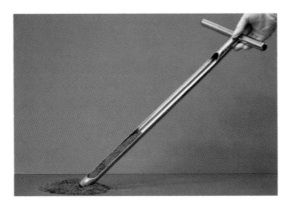

Figure 1.3 Hand auger. From *Environmental Microbiology* © 2000, Academic Press, San Diego, CA.

Augers are useful in that they allow samples to be taken at exactly the same depth on every occasion. Augers are available that can take soil cores to a depth of 2 meters, in 1-foot increments. Typically a soil sample consists of about 2 kilograms. Because soils are heterogeneous, it is frequently better to collect multiple cores that are mixed together to give a composite sample. Soil samples that are collected for microbial analysis should be kept on ice while transported to the laboratory. Microbial analyses should be performed as soon as possible to minimize the effects of storage on microbial populations and should not be air dried prior to analyses. Soils sampled for chemical analyses should be air dried and can then be kept indefinitely pending analysis.

For subsurface sampling, mechanical drill rigs, such as rotary mud drilling (Figure 1.4) or hollow-stem augers (Figure 1.5), are necessary. Subsurface sampling is more complex and more expensive than surface soil sampling, particularly when deep subsurface sampling is attempted.

1.4.2 Water

Collecting water samples tends to be somewhat easier than sampling soils. First of all, water at a given site tends to be more homogenous than soils, with less site-to-site variability between two samples collected within the same vicinity. Secondly, it is often physically easier to collect water sam-

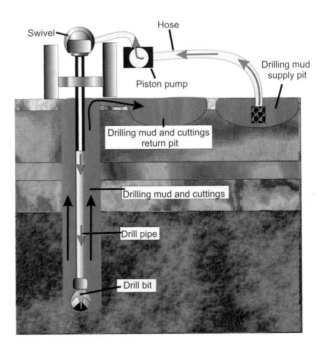

Figure 1.4 Rotary mud drilling. With rotary drilling the mechanical rotation of a drilling tool is used to create a borehole. Either air (air rotary drilling) or a fluid often called a drilling mud (mud rotary drilling) is forced down the drill stem to displace the borehole cuttings to the outside of the drill and upward to the surface. From *Environmental Microbiology* © 2000, Academic Press, San Diego, CA.

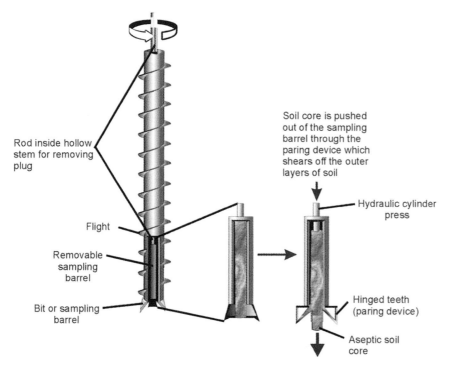

Figure 1.5 Diagram of a hollow-stem auger. Note the reverse threading on the outside of the auger. This is used to displace the borehole cuttings upward to the surface. A subcore of each core collected is taken using a split spoon sampler or a push tube. In either case, the outside of the core must be regarded as contaminated. Therefore, the outside of the core is shaved off with a sterile spatula or a subcore can be taken using a sterile plastic syringe. Alternatively, as shown in this figure, intact cores are automatically pared to remove the outer contaminated material, leaving an inner sterile core. From *Environmental Microbiology* © 2000, Academic Press, San Diego, CA.

ples. Surface water samples can be collected in wide-mouth polyethylene jars or with a bucket. Subsurface water samples can be collected through the use of bailers or garden hose lines submerged to specific depths and attached to a pump. The amount of water collected can be a few milliliters, such as when routine analyses such as pH are to be done. In other cases, large volumes need to be collected (1000 liters), as in the case of determining the presence of enteric viruses in marine waters. Normally water samples are kept as cool as possible in sealed containers to prevent microbial and chemical activity and preclude evaporation.

1.4.3 Air

The collection of air samples for analysis can be done in a variety of ways. In some cases samples of air are diverted automatically into instruments for continuous measurement of pollutant concentrations. In yet other applications, air samples are collected in sample bags for later laboratory chemical analysis.

Aerosolized biological particles including microorganisms are known as **bioaerosols**. Many devices have been designed for the collection of bioaerosols, including impingement and impaction devices. **Impingement** is the trapping of airborne particles in a liquid matrix. In contrast,

impaction is the forced deposition of airborne particles on a solid surface. Two of the most commonly used devices for microbial air sampling are the SKC biosamplers (SKC-West Incorporated, Fullerton, CA) (Figure 1.6) and the Andersen Six Stage Impaction Sampler (Anderson Instruments Incorporated, Atlanta, GA) (Figure 1.7) (see also Chapter 27).

Figure 1.6 SKC Biosamplers. Photo courtesy J. Brooks.

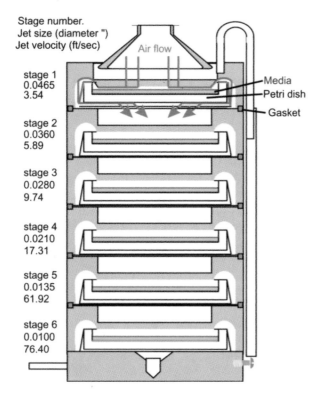

Stage number.
Jet size (diameter ")
Jet velocity (ft/sec)

Air flow

stage 1
0.0465
3.54

Media
Petri dish
Gasket

stage 2
0.0360
5.89

stage 3
0.0280
9.74

stage 4
0.0210
17.31

stage 5
0.0135
61.92

stage 6
0.0100
76.40

Figure 1.7 Schematic representation of the Andersen six-stage impaction air sampler. Air enters through the top of the sampler and larger particles are impacted upon the surface of the petri dish on stage 1. Smaller particles, which lack sufficient impaction potential, follow the air stream to the subsequent levels. As the air stream passes through each stage, the air velocity increases, thus increasing the impaction potential, so that particles are trapped on each level based upon their size. Therefore, larger particles are trapped efficiently on stage 1 and slightly smaller particles on stage 2 and so on until even very small particles are trapped on stage 6. The Andersen six-stage sampler thus separates particles based upon their size. From *Environmental Microbiology* © 2000, Academic Press, San Diego, CA.

1.5 ADVANCES IN ANALYTICAL DETECTION TECHNOLOGY

When evaluating environmental quality, the question is frequently asked, "How clean is clean?" The answer to the question depends on the technologies available for analytical detection of contaminants. As our technologies improve, they are continually redefining our understanding of the term "clean." Using new instruments and innovative techniques, we are increasingly capable of measuring environmental parameters with greater sensitivity and accuracy.

1.5.1 Advances in Chemical Analysis

An example of enhanced chemical detection technology is illustrated by the determination of heavy metal concentrations in water. Thirty years ago, atomic absorption (AA) spectroscopy was utilized, which gave measurements at the

level of milligrams per liter. A newer improved flameless AA technique using a graphite furnace improved detection limits to the level of micrograms per liter. The latest technology utilizes **inductively coupled plasma (ICP)** spectroscopy, which has detection limits of nanograms per liter.

Advanced methods have been developed that allow investigation of physical, chemical, and microbial processes at the molecular scale. For example, atomic force microscopy is being used to examine the distribution of atoms and molecules at solid surfaces. X-ray absorption fine structure spectroscopy is being used to determine the geometry, composition, and mode of attachment of ions at mineral-water interfaces. Nuclear magnetic resonance methods are being used to study the interaction of contaminants with soil organic matter. Recent advances in imaging methods, such as synchrotron x-ray microtomography, have allowed us to begin to directly measure the pore-scale distribution of fluids in porous media. This will provide a better understanding of how water and organic liquids move through the subsurface.

1.5.2 Advances in Biological Analysis

Great progress has been made towards new innovative technology for characterizing microbial properties and activities. State-of-the-art approaches are shown in Table 1.4.

The use of molecular technology in particular has revolutionized biological detection capabilities. The **polymerase chain reaction** based technique allows for detection of an organism's DNA at the nanogram level (Figure 1.8). Sequence analysis using PCR and computer searches allows for enhanced identification of new microbes.

Overall, the advent of these new supersensitive technologies allows us to reexamine the question of "How clean is clean?" Environmental samples that were analyzed decades ago and found to contain no detectable heavy metals were considered pristine. Using today's technology allows for quantification of metal concentrations, albeit at extremely low levels. Perhaps the real question is not "Are the samples pristine," but "Are they pristine enough?"

1.6 THE RISK BASED APPROACH TO POLLUTION SCIENCE

Risk assessment is an integral part of pollution science and is covered in detail in Chapter 14. Its importance lies in the fact that it provides a quantifiable answer to the question "Is this polluted site safe?" Throughout this text where appropriate, we evaluate two types of risk assessment: health-based risks and ecological risks. The former focuses on human health, whereas the latter focuses on potential detrimental effects to parts of the environment or the entire environment.

Regardless of the focus, the **risk assessment** process consists of four basic steps (Information Box 1.4).

TABLE 1.4 State-of-the-art approaches to the monitoring of microbial properties and activities.

TECHNIQUE	FUNCTION
I. Microbial	
Epifluorescent microscopy	Detection of specific microbes
Electron microscopy	Magnification up to 10^6
Confocal scanning microscopy	3-D images
II. Physiological	
Carbon respiration	Heterotrophic activity
Radiolabeled tracers	Degradation of specific compounds
Enzyme assays	Specific microbial transformations
III. Immunological	
Immunoassays	Detection of specific microbes
Immunocytochemical assays	Location of specific antigens
Immunoprecipitation assays	Semi-quantitative determination of antigens
IV. Nucleic Acid Based	
Polymerase chain reaction (PCR)	Detection of genes or specific microbes
Reverse transcriptase (RT-PCR)	Detection of mRNA or viruses
Gene probes	Detection of genes
Denaturing gradient gel electrophoresis	Microbial diversity changes
Plasmid profile analysis	Unique microbial functions

From *Environmental Monitoring and Characterization* © 2004. Elsevier Academic Press, San Diego.

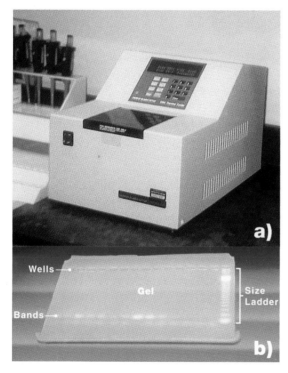

Figure 1.8 (a) An automated PCR thermalcycler that is used to amplify target DNA. (b) A gel stained with ethidium bromide. From *Pollution Science* © 1996, Academic Press, San Diego, CA.

INFORMATION BOX 1.4

The Risk Assessment Process

Hazard identification—Defining the hazard and nature of the harm; for example, identifying a chemical contaminant, such as lead or carbon tetrachloride, and documenting its toxic effects on humans.

Exposure assessment—Determining the concentration of a contaminating agent in the environment and estimating its rate of intake in target organisms; for example, finding the concentration of aflatoxin in peanut butter and determining the dose an "average" person would receive.

Dose-response assessment—Quantifying the adverse effects arising from exposure to a hazardous agent based on the degree of exposure. This assessment is usually expressed mathematically as a plot showing the response in living organisms to increasing doses of the agent.

Risk characterization—Estimating the potential impact of a hazard based on the severity of its effects and the amount of exposure.

Once a given risk has been calculated, informed decisions can be made with respect to the severity of the pollution, and what should be done about it.

1.7 WASTE MANAGEMENT, SITE REMEDIATION, AND ECOSYSTEM RESTORATION

As noted above, human activities produce enormous volumes of waste, much of which ends up in the environment and has the potential to cause pollution. Improper management of this waste exacerbates the pollution problem. Thus, significant resources are expended to control and treat this waste. These methods will be discussed in Part 5 of the book.

Once a site becomes contaminated with hazardous pollutants and is judged to pose a risk to human health or the environment, it must be cleaned up or **remediated**. Remediation of contaminated sites has become a major activity since the advent of Superfund in 1980, a federal program designed to support such activities. The numerous methods available for remediation of hazardous waste sites will be reviewed in Chapter 19. Over the past two decades we have learned that once contaminated, sites can not be completely cleaned up, and also that site remediation is often very expensive. This leads to the axiom of pollution prevention: that preventing pollution from occurring in the first place is much preferred to the alternative. This in turn leads back to the use of best management practices for pollution control.

Ecosystems damaged through human activity or natural processes may lose productivity or sustainability. Examples of this issue include the loss of native vegetation via deforestation and loss of soil fertility due to salinization, the buildup of salts in soil. These damaged ecosystems need to be restored through a process analogous to hazardous-waste site remediation. The basis of, and methods used for ecosystem restoration are discussed in Chapter 20.

REFERENCES AND ADDITIONAL READING

Artiola J.F., Brusseau M.L., and Pepper I.L. (2004) *Environmental Monitoring and Characterization*. Academic Press, San Diego, California.

Maier R.M., Pepper I.L., and Gerba C.P. (2000) *A Textbook of Environmental Microbiology*. Academic Press, San Diego, California.

Pepper I.L., Gerba C.P., and Brusseau M.L. (1996) *Pollution Science*. Academic Press, San Diego, California.

CHAPTER 2

PHYSICAL-CHEMICAL CHARACTERISTICS OF SOILS AND THE SUBSURFACE

I.L. Pepper and M.L. Brusseau

Soil is a heterogeneous medium. *Photo courtesy K.L. Josephson.*

2.1 SOIL AND SUBSURFACE ENVIRONMENTS

The human environment is located at the earth's surface and is heavily dependent on the soil/water/atmosphere continuum. Ultimately this continuum moderates all of our activities, and the physical, chemical, and biological properties of each component are interactive. The geological zone between the land surface and subsurface groundwater consists of unsaturated material and is known as the **vadose zone**. A subset of the vadose zone is the near-surface **soil** environment, which is in direct contact with both surface water and the atmosphere. Since pollutants are often disposed of into surface soils, the transport of these contaminants into both the atmosphere and groundwater is influenced by the properties of soil and the vadose zone. In addition, since plants are grown in surface soils, the potential for uptake of contaminants such as heavy metals is also controlled by soil properties.

Soil is an intricate, yet durable entity that directly and indirectly influences our quality of life. Colloquially known as dirt, soil is taken for granted by most people and yet it is essential to our daily existence. It is responsible for plant growth, for the cycling of all nutrients through microbial transformations, and for maintaining the oxygen/carbon dioxide balance of the atmosphere; it is also the ultimate site of disposal for most waste products. Soil is a complex mixture of weathered rock particles, organic residues, water, and billions of living organisms. It can be as thin as 6 inches, or it may be hundreds of feet thick. Because soils are derived from unique sources of parent material under specific environmental conditions, no two soils are exactly alike. Hence there are literally thousands of different kinds of soils just within the United States. These soils have different properties that influence the way soils are used optimally.

Soil is the weathered end product of the action of climate and living organisms on soil parent material with a particular topography over time. We refer to these factors as the five **soil-forming factors** (Information Box 2.1). The biotic component consists of both microorganisms and plants. The **vadose zone** is the water-unsaturated and generally unweathered material between ground water and the land surface.

The major difference between a surface soil and a vadose zone is the fact that the vadose zone parent material has generally not been modified by climate. A model of a

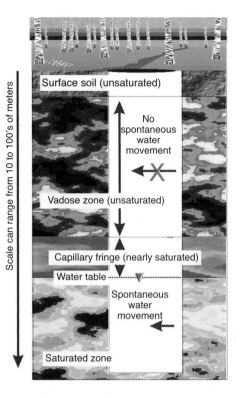

Figure 2.1 Cross section of the subsurface showing surface soil, vadose zone, and saturated zone. Adapted from *Environmental Microbiology* © 2000, Academic Press, San Diego, CA.

cross section of a typical subsurface environment is shown in Figure 2.1.

There are several parameters of soil that vitally affect the transport and fate of environmental pollutants. We will now discuss these parameters while providing an overview of soil as a natural body as it affects pollution.

2.2 SOLID PHASE

2.2.1 Soil Profiles

The process of soil formation generates different horizontal layers, or **soil horizons**, that are characteristic of that particular soil. It is the number, nature, and extent of these horizons that give a particular soil its unique character. A typical soil profile is illustrated in Figure 2.2. Generally, soils contain a dark organic-rich layer, designed as the O horizon, then a lighter colored layer, designated as the A horizon, where some humified organic matter accumulates. The layer that underlies the A horizon is called the E horizon because it is characterized by **eluviation**, which is the process of removal or transport of nutrients and inorganics out of the A horizon. Beneath the E horizon is the B horizon, which is characterized by **illuviation**. Illuviation is the deposition of the substances from the E horizon into the B horizon. Beneath the B horizon is the C horizon, which contains the parent material from which the soil was derived. The C horizon is generally unweathered parent material. Although certain

---

INFORMATION BOX 2.1

The Five Soil-forming Factors

· Parent material
· Climate
· Organisms (plants and microbes)
· Topography
· Time

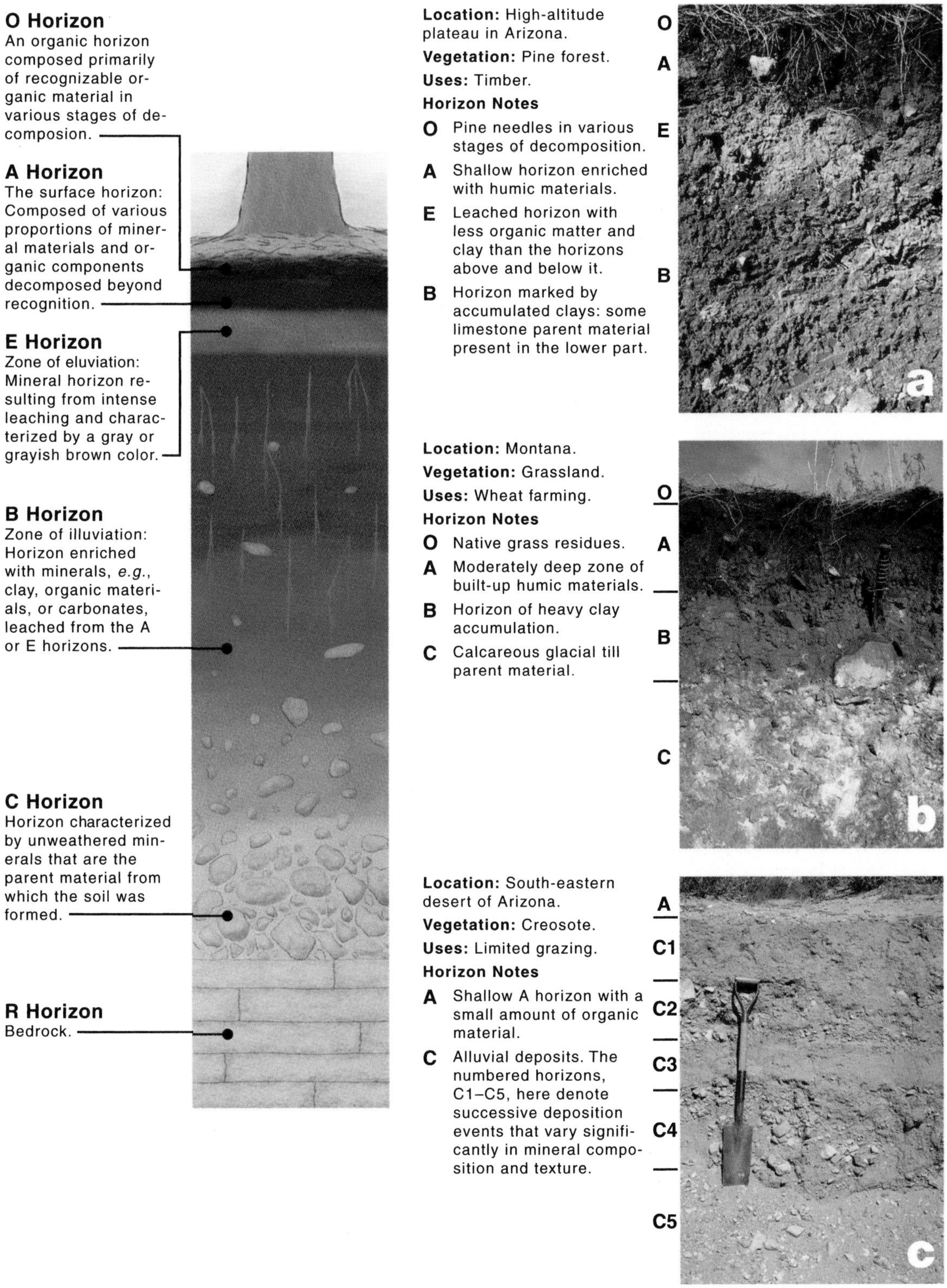

O Horizon
An organic horizon composed primarily of recognizable organic material in various stages of decomposition.

A Horizon
The surface horizon: Composed of various proportions of mineral materials and organic components decomposed beyond recognition.

E Horizon
Zone of eluviation: Mineral horizon resulting from intense leaching and characterized by a gray or grayish brown color.

B Horizon
Zone of illuviation: Horizon enriched with minerals, *e.g.*, clay, organic materials, or carbonates, leached from the A or E horizons.

C Horizon
Horizon characterized by unweathered minerals that are the parent material from which the soil was formed.

R Horizon
Bedrock.

Location: High-altitude plateau in Arizona.
Vegetation: Pine forest.
Uses: Timber.
Horizon Notes

O Pine needles in various stages of decomposition.
A Shallow horizon enriched with humic materials.
E Leached horizon with less organic matter and clay than the horizons above and below it.
B Horizon marked by accumulated clays: some limestone parent material present in the lower part.

Location: Montana.
Vegetation: Grassland.
Uses: Wheat farming.
Horizon Notes

O Native grass residues.
A Moderately deep zone of built-up humic materials.
B Horizon of heavy clay accumulation.
C Calcareous glacial till parent material.

Location: South-eastern desert of Arizona.
Vegetation: Creosote.
Uses: Limited grazing.
Horizon Notes

A Shallow A horizon with a small amount of organic material.
C Alluvial deposits. The numbered horizons, C1–C5, here denote successive deposition events that vary significantly in mineral composition and texture.

Figure 2.2 Typical soil profiles illustrating different soil horizons. These horizons develop under the influence of the five soil-forming factors and result in unique soils. Adapted from *Pollution Science* © 1996, Academic Press, San Diego, CA.

INFORMATION BOX 2.2

Size of the Primary Particle

Sand = 2 mm−0.05 mm
Silt = 0.05−0.002 mm
Clay = <0.002 mm (2 μ)

---

diagnostic horizons are common to most soils, not all soils contain each of these horizons.

2.2.2 Primary Particles and Soil Texture

Soil normally consists of about 95–99% inorganic and 1–5% organic matter. The primary inorganic material is in turn composed of three primary particles—sand, silt, and clay—which are delineated on a size basis (Information Box 2.2).

The differences in the size of the particles are due to the weathering of the parent rock. Table 2.1 illustrates the size fractionation of soil constituents including mineral, organic, and biological constituents. This table also illustrates the effect of size on specific surface area.

The percentage of sand, silt, and clay in a particular soil determines its **soil texture**, which affects many of the physical and chemical properties of the soil. Various

mixtures of the three primary components result in different textural classes (Figure 2.3). Of the three primary particles, clay is by far the dominant factor in determining a soil's properties. This is because there are more particles of clay per unit weight, than sand or silt, due to the smaller size of the clay particles. In addition, the clay particles are the primary soil particles that have an associated electric charge (see Chapter 7). The predominance of clay particles explains why any soil with greater than 35% clay has the term *clay* in its textural class. In addition, because increases in soil clay concentrations results in increased surface area, this also increases the chemical reactivity of the soil (see Chapter 7).

2.2.3 Soil Structure

The three primary particles do not normally remain as individual entities. Rather, they aggregate to form secondary structures, which occur because microbial gums, polysaccharides, and other microbial metabolites bind the primary particles together. In addition, particles can be held together physically by fungal hyphae and plant roots. These secondary aggregates, which are known as **peds**, can be of different sizes and shapes, depending on the particular soil. Soils with even modest amounts of clay usually have well defined peds and hence a well-defined **soil structure**. These aggregates of primary particles usually remain intact as long as the soil is not disturbed, for example, by plowing. In

TABLE 2.1 Size fractionation of soil constituents.

SPECIFIC SURFACE AREA	SOIL		
	Mineral constituents	Size	Organic and biologic constituents
0.0003 m^2/g	**Sand** Primary minerals: quartz, silicates, carbonates	2 mm	**Organic debris**
0.12 m^2/g	**Silt** Primary minerals: quartz, silicates, carbonates	50 μm	**Organic debris, large microorganisms** Fungi Actionomycetes Bacterial colonies

3 m^2/g	**Granulometric clay**
	Microcrystals of primary minerals
	Phyllosilicates
	   Inherited; illite, mica
	   Transformed: vermiculite,
	      high-charge smectite
	   Neoformed; kaolinite, smectite
	Oxides and hydroxides
		2 μm	**Amorphous organic matter**
			Humic substances
			Biopolymers

30 m^2/g	**Fine clay** Swelling clay minerals Interstratified clay minerals Low range order crystalline compounds	0.2 μm	**Small viruses**

From *Environmental Microbiology* © 2000, Academic Press, San Diego, CA.

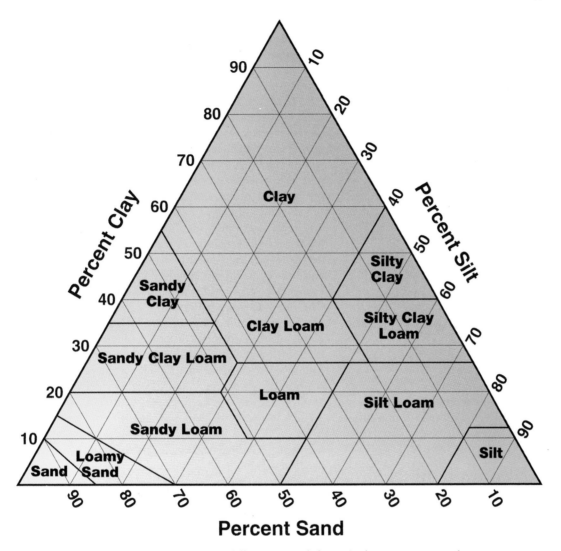

Figure 2.3 A soil textural triangle showing different textural classes in the USDA system. These textural classes characterize soil with respect to many of their physical properties. *From Pollution Science* © 1996, Academic Press, San Diego, CA.

contrast, sandy soils with low amounts of clay generally have less well defined soil structure.

The phenomenon of soil structure has a profound influence on the physical properties of the soil. Because its particles are arranged in secondary aggregates, a certain volume of the soil include voids that are filled with either soil air or soil water. Soils in which the structure has many voids within and between the peds offer favorable environments for soil organisms and plant roots, both of which require oxygen and water. Soils with no structure, that is, those consisting of individual primary particles are characterized as **massive**. Massive soils have very few (and very small) void spaces and therefore little room for air or water.

Void spaces are collectively known as **pore space,** which is made up of individual pores. These pores allow movement of air, water, and microorganisms through the soil. Pores that exist between aggregates are called **interaggregate pores**, whereas those within the aggregates are termed **intraggregate pores** (Figure 2.4). Although the average pore size is smaller in a clay soil, there are many more pores than in a sandy soil, and as a result, the total amount of pore space is larger in a fine-textured (clay) soil than in a coarse-textured (sandy) soil (Figure 2.5). However, because small pores do not transmit water as fast as larger pores, a fine-textured soil will slow the movement of any material moving through it, including air, water, and microorganisms (see Chapter 6). Sometimes fine-textured layers of clay known as **clay lenses** can be found within volumes of coarser materials resulting in heterogeneous environments. In this case, water will move through the coarser material and flow around the clay lens. This has implications for remediative strategies such as "pump and treat," which is used to remove contaminants from the saturated zone.

Together texture and structure are important factors that control the movement of water, contaminants and microbes through soils, and hence affect contaminant transport and fate.

Pore space may also be increased by plant roots, worms, and small mammals, whose root channels, worm

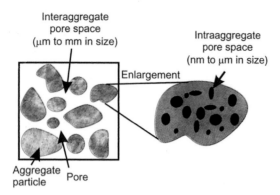

Interaggregate
pore space
(μm to mm in size)

Intraaggregate
pore space
(nm to μm in size)

Enlargement

Aggregate   Pore
particle

Figure 2.4 Pore space. In surface soils, mineral particles are tightly packed together and even cemented in some cases with microbial polymers forming soil aggregates. The pore spaces between individual aggregates are called interaggregate pores and vary in size from micrometers to millimeters. Aggregates also contain pores within aggregates that are smaller in size, ranging from nanometers to micrometers. These are called intraggregate pores. From *Environmental Microbiology*© 2000, Academic Press, San Diego, CA.

holes, and burrows create macro openings. These larger openings can result in significant aeration of surface and subsurface soils and sediments, as well as **preferential flow** of water through the soil.

2.2.4 Cation-Exchange Capacity

The parameter known as cation-exchange capacity (CEC) arises because of the charge associated with clay particles. Normally, this is a negative charge that occurs for one of two reasons:

1. **Isomorphic substitution:** Clay particles exist as inorganic lattices composed of silicon and aluminum oxides. Substitution of a divalent magnesium cation (Mg^{2+}) for a trivalent aluminum cation (Al^{3+}) can result in the loss of one positive charge, which is equivalent to a gain of one negative charge. Other substitutions can also lead to increases in negative charge.

2. **Ionization:** Hydroxyl groups (OH) at the edge of the lattice can ionize, resulting in the formation of negative charge:

$$Al - OH = Al - O^- + H^+$$

These are also known as **broken-edge bonds.** Ionizations such as these usually increase as the pH increases, and are therefore known as **pH-dependent charge**. The functional groups of organic matter, such as carboxyl moieties, are also subject to ionization and can contribute to the total pH-dependent charge. The total amount of negative charge is usually measured in terms of equivalents of negative charge per 100 g of soil and is a measure of the potential CEC of the soil. A milliequivalent (meq) is 1,000[th] of an equivalent weight. Equivalents of chemicals are related to hydrogen, which has a defined equivalent weight of 1. The equivalent weight of a chemical is the atomic weight divided by its valence. For example, the equivalent weight of calcium is 40/2 = 20 g. A CEC of 15–20 meq per 100 g of soil is considered to be average, whereas a CEC > 30 is considered high. Note that it is the clays and organic particles that are negatively charged. Due to their small particle size, they are collectively called the **soil colloids.** The existence of CEC allows the phenomenon of cation exchange to occur (see Chapter 7).

2.2.5 Soil pH

We define pH as the negative logarithm of the hydrogen ion concentration:

$$pH = - \log[H^+]$$

Usually, water ionizes to H^+ and OH^-:

$$HOH \leftrightarrow H^+ + OH^-$$

The **dissociation constant** (K_{eq}) is defined as

$$K_{eq} = \frac{[H^+][OH]}{[HOH]} = 10^{-14} \, mol \, L^{-1}$$

Since the concentration of HOH is large relative to that of H^+ or OH^-, it is normally given the value of 1; therefore

$$[H][OH^-] = 10^{-14} \, mol \, L^{-1}$$

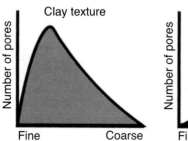

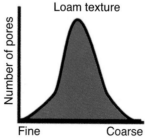

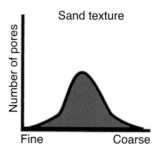

Figure 2.5 Typical pore size distributions for clay, loam, and sand-textured horizons. Note that the clay-textured material has the smallest average pore size, but the greatest total volume of pore space. From *Environmental Microbiology* © 2000, Academic Press, San Diego, CA.

TABLE 2.2 Soil pH regimes.

SOIL	pH REGIME
Acidic	>5.5
Neutral	6–8
Alkaline	>8.5

From *Pollution Science* © 1996, Academic Press, San Diego, CA.

TABLE 2.3 Major constituents of plant residues.

CONSTITUENT	% DRY WEIGHT
Cellulose	15–60
Hemicellulose	10–30
Lignin	5–30
Protein and nucleic acids	2–15
Soluble substances, *e.g.*, sugars	10

From *Pollution Science* © 1996, Academic Press, San Diego, CA.

For a neutral solution

$$[H^+] = [OH^-] = 1 \times 10^{-7}$$

and

$$pH = -\log[H^+] = -(-7) = 7$$

A pH value of less than 7 indicates acidity, whereas a pH value greater than 7 indicates alkalinity (or basicity) (Table 2.2).

In areas with high rainfall, basic cations tend to leach out of the soil profile; moreover, soils developed in these areas have higher concentrations of organic matter, which contain acidic components and residues. Thus, such soils tend to have decreased pH values and are acidic in nature. Soils in arid areas do not undergo such basic leaching, and the concentrations of organic matter are lower. In addition, water tends to evaporate in such areas, allowing salts to accumulate. These soils are therefore alkaline, with higher pH values.

Soil pH affects the solubility of chemicals in soils by influencing the degree of ionization of compounds and their subsequent overall charge. The extent of ionization is a function of the pH of the environment and the dissociation constant (pK) of the compound. Thus, soil pH may be critical in affecting transport of potential pollutants through the soil and vadose zone.

2.2.6 Organic Matter

Organic compounds are incorporated into soil at the surface via plant residues such as leaves or grassy material. These organic residues are degraded microbially by soil microorganisms, which utilize the organics as food or microbial substrate. The main plant constituents, shown in Table 2.3, vary in degree of complexity and ease of breakdown by microbes. In general, soluble constituents are easily metabolized and break down rapidly, whereas lignin, for example, is very resistant to microbial decomposition. The net result of microbial decomposition is the release of nutrients for microbial or plant metabolism, as well as the particle breakdown of complex plant residues. These microbial modified complex residues are ultimately incorporated into large macromolecules that form the stable basis of soil organic matter. This stable organic matrix is slowly metabolized by indigenous soil organisms, a process that results in about 2% breakdown of the complex materials annually. Owing to the slow but constant decomposition of the organic matrix and annual fresh additions of plant residues, an equilibrium is achieved in which the overall amount of soil organic matter remains constant. In humid areas with high rainfall, soil organic matter contents can be as high as 5% on a dry-weight basis. In arid areas with high rates of decomposition and low inputs of plant residues, values are usually less than 1%. The formation of soil organic matter is illustrated in Figure 2.6, and terms used to define soil organic matter are shown in Table 2.4.

The release of nutrients that occurs as plant residues degrade has several effects on soil. The enhanced microbial activity causes an increase in soil structure, which affects most of the physical properties of soil, such as aeration and infiltration. The stable humic substances contain many moi-

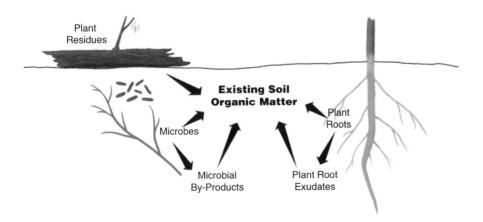

Figure 2.6 Schematic representation of the formation of soil organic matter. From *Pollution Science* © 1996, Academic Press, San Diego, CA.

TABLE 2.4 Terms used to define soil organic matter.

TERM	DEFINITION
Organic residues	Undecayed plant and microbial biomass and their partial decomposition products.
Soil biomass	Live microbial biomass.
Soil organic matter or humus	All soil organic matter, except organic residues and soil biomass.
Humic substances	High-molecular-weight complex stable macromolecules with no distinct physical or chemical properties. These substances are never exactly the same in any two soils because of variable inputs and environments. This is the stable backbone of soil organic matter and is degraded only slowly (2% per year).
Nonhumic substances	Known chemical materials such as amino acids, organic acids, carbohydrates, or fats. They include all known biochemical compounds and have distinct physical and chemical properties. They are normally easily degraded by microbes.

From *Pollution Science* © 1996, Academic Press, San Diego, CA.

eties that contribute to the pH-dependent CEC of the soil. In addition, many of the humic and nonhumic substances can complex or chelate heavy metals, and sorb organic contaminants. This retention affects their availability to plants and soil microbes as well as their potential for transport into the subsurface (see Chapter 6).

2.2.7 Vadose Zone—Solid Phase

The **vadose zone** is defined as the unsaturated environment that lies between the surface soil and the saturated zone. Physically, the vadose zone parent material may be very similar to that of the surface soil above it, except that it is less weathered and has very low organic matter content. In terms of texture, vadose zones normally contain larger rocks and cobbles than surface soils, but still have high amounts of sand, silt, and clay. The low organic content is due to the fact that organic material added to surface soils as vegetative leaf litter is usually degraded within the surface soil. Therefore, the organic carbon content of vadose zones is usually very low. This leads to **oligotrophoic** (low nutrient) conditions. Hence, microbial activity in vadose zones is normally lower than in surface soils (see also Chapter 5). This may affect the fate of subsurface organic contaminants, since there may be decreased rates of biodegradation (see also Chapter 8).

2.3 GASEOUS PHASE

2.3.1 Constituents of Soil Atmosphere

Soil and the atmosphere are in direct contact; therefore, most of the gases found in the atmosphere are also found in the air phase within the soil (called the soil atmosphere), but at different concentrations. The main gaseous constituents are oxygen, carbon dioxide, nitrogen, and other volatile compounds such as hydrogen sulfide or ethylene. The concentrations of oxygen and carbon dioxide in the soil atmosphere are normally different than in the atmosphere (Table 2.5). This variable reflects the use of oxygen by aerobic soil organisms and subsequent release of carbon dioxide. In addition, the gaseous concentrations in soil are normally regulated by diffusion of oxygen into soil and of carbon dioxide from soil.

2.3.2 Availability of Oxygen and Soil Respiration

The oxygen content of soil is vital for **aerobic microorganisms**, which utilize oxygen as a terminal electron acceptor during degradation of organic compounds (see Chapter 8). **Facultative anaerobes** can utilize oxygen or combined forms of oxygen (such as nitrate) as a terminal electron acceptor. **Anaerobes** cannot utilize oxygen as an acceptor. Strict anaerobes are lethally affected by oxygen because they

TABLE 2.5 Characteristics of the soil atmosphere.

LOCATION	COMPOSITION (% VOLUME BASIS)		
	Nitrogen (N_2)	Oxygen (O_2)	Carbon dioxide (CO_2)
Atmosphere	78.1	20.9	0.03
Well-aerated soil surface	78.1	18–20.5	0.3–3
Fine clay or saturated soil	>79	~0–10	Up to 10

From *Environmental Microbiology* © 2000, Academic Press, San Diego, CA.

do not contain enzymes that can degrade toxic peroxide radicals. Since microbial degradation of many organic compounds in soil, including xenobiotics, is carried out by aerobic organisms, the presence of oxygen in soil is necessary for such decomposition. Oxygen is found dissolved either in the soil solution or in the soil atmosphere, but soil oxygen concentrations in solution are much lower than in the soil atmosphere.

The total amount of pore space depends on soil texture and soil structure. Soils high in clays have more total pore space, but smaller pore sizes. In contrast, sandy soils have larger pore sizes, allowing more rapid water and air movement. In any soil, as the amount of soil structure increases, the total pore space of the soil increases. Aerobic soil microbes require both water and oxygen, which are both found within the pore space. Therefore, the soil moisture content controls the amount of available oxygen in a soil. In soils saturated with water, all pores are full of water and the oxygen content is very low. In dry soils, all pores are essentially full of air, so the soil moisture content is very low. In soils at **field capacity**, that is, soils having moderate soil moisture, both air (oxygen) and moisture are readily available to soil microbes. In such situations, soil respiration via aerobic microbial metabolism is normally at a maximum. It is important to note, however, that low-oxygen concentrations may exist in certain isolated pore regions, allowing anaerobic microsites to exist even in aerobic soils, thereby supporting transformation processes carried out by facultative anaerobes and strict anaerobes. This is an excellent example of how soil can function as a discontinuous environment of great diversity.

2.3.3 Gaseous Phase Within the Vadose Zone

Vadose zones generally are primarily aerobic regions. However, due to the heterogeneous nature of the subsurface, anaerobic zones can occur, particularly in clay lenses. Thus, both aerobic and anaerobic microbial processes may occur.

At contaminated sites, volatile organic compounds can be found in the gaseous phase of the vadose zone. For example, chlorinated solvents, which are ubiquitous organic contaminants (see Chapter 10), are volatile and are often found in the vadose-zone gaseous phase below hazardous waste sites. In such cases, soil venting is often used to remove the contamination (see Chapter 19). The porosity, structure, and water content of the vadose zone are critical to effective application of soil venting.

2.4 LIQUID PHASE

Water is, of course, essential for all biological forms of life, in part because of the unique nature of its structure. The fact that the oxygen moiety of the molecule is slightly more electronegative than the hydrogen counterparts results in a polar molecule. This polarity, in turn, allows water to hydrogen bond both to other water molecules and to other polar molecules. This capacity to bond with almost anything has a profound influence on biological systems, and it explains why water is a near-universal solvent. It also explains the hydration of cations and the adsorption of water to soil colloids (see Chapter 7).

By definition, the vadose zone is unsaturated and contains low moisture content. However, whenever rainfall or irrigation events occur at the soil surface, some moisture leaches into the vadose zone. Other avenues by which moisture can reach the subsurface are through burrowing animal holes or worm holes, which result in preferential flow. Even so, significant moisture in the vadose zone is the exception rather than the rule. Basic properties of water in both surface and subsurface environments are discussed in Chapter 3.

2.5 BASIC PHYSICAL PROPERTIES

2.5.1 Bulk Density

Soil bulk density is defined as the ratio of dry mass of solids to bulk volume of the soil sample:

$$\rho_b = \frac{M_s}{V_T} = \frac{M_s}{V_s + V_w + V_a}$$

where:

ρ_b = Soil bulk density [M L^{-3}]
M_s = Dry mass of solid [M]
V_s = Volume of solids [L^3]
V_w = Volume of water [L]
V_a = Volume of air [L^3]
V_T = Bulk volume of soil [L^3]

The bulk volume of soil represents the combined volume of solids and pore space. In SI units, bulk density is usually expressed in g cm^{-3} or kg m^{-3}. Bulk density is used as a measure of soil structure. It varies with a change in soil structure, particularly due to differences in packing. In addition, in swelling soils, bulk density varies with the water content. Therefore, it is not a fixed quantity for such soils.

2.5.2 Porosity

Porosity (n) is defined as the ratio of void volume (pore space) to bulk volume of a soil sample:

$$n = \frac{V_v}{V_T} = \frac{V_v}{V_w + V_s + V_a}$$

where:

n is the total porosity [n]
V_v is the volume of voids [L^3]
V_T is the bulk volume of sample [L^3]

TABLE 2.6 Porosity (n) values of selected porous media.

TYPE OF MATERIAL	n (%)
Unconsolidated media	
Gravel	20–40
Sand	20–40
Silt	25–50
Clay	30–60
Rocks	
Karst Limestone	5–30
Sandstone	5–30
Shale	0–10
Fractured crystalline rock	0–20
Dense crystalline rock	0–10

From *Environmental Monitoring* © 2004, Academic Press, San Diego, CA.

It is dimensionless and described either in percentages with values ranging from 0 to 100%, or as a fraction where values range from 0 to 1. The general range of porosity that can be expected for some typical materials is listed in Table 2.6.

Porosity of a soil sample is determined largely by the packing arrangement of grains and the grain-size distribution. Cubic arrangements of uniform spherical grains provide the ideal porosity with a value of 47.65 %. Rhombohedral packing of similar grains presents the least porosity with a value of 25.95 %. Because both packings have uniformly sized grains, porosity is independent of grain size. If grain size varies, porosity is dependent on grain size as well as distribution. Total porosity can be separated into two types, primary and secondary, as discussed in Section 2.2.3. The porosity of a soil sample or unconsolidated sediment is determined as follows. First the bulk volume of the soil sample is calculated from the size of the sample container. Next, the soil sample is placed into a beaker containing a known volume of water. After becoming saturated, the volume of water displaced by the soil sample is equal to the volume of solids in the soil sample. The volume of voids is calculated by subtracting the volume of water displaced from the bulk volume of the bulk soil sample.

In a saturated soil, porosity is equal to water content, since all pore spaces are filled with water. In such cases, total porosity can also be calculated by weighing the saturated sample, drying it, and then weighing it again. The difference in mass is equal to the mass of water, which, using a water density of 1 g cm^{-3}, can be used to calculate the volume of void spaces. Porosity is then calculated as the ratio of void volume and total sample volume.

Porosity can also be estimated using the following equation:

$$n = 1 - \frac{\rho_b}{\rho_d}$$

where:

ρ_b is the bulk density of soil [M L^{-3}]

and

ρ_d is the particle density of soil [M L^{-3}]

A value of 2.65 g cm^{-3} is often used for the latter, based on silica sand as a primary soil component. Void ratio (e), which is used in engineering, is the ratio of volume of voids to volume of solids:

$$e = \frac{V_v}{V_s}$$

The relationship between porosity and void ratio is described as:

$$e = \frac{n}{1 - n}$$

It is dimensionless. Values of void ratios are typically less than 1.

2.5.3 Soil Water Content

Soil-water content can be expressed in terms of mass (θ_g) or volume (θ_v). Gravimetric (mass) water content is the ratio of water mass to soil mass, usually expressed as a percentage. Typically, the mass of dry soil material is considered as the reference state; thus:

$$\theta_g \% = [(\text{mass wet soil} - \text{mass dry soil})/\text{mass dry soil}] \times 100$$

Volumetric water content expresses the volume (or mass, assuming a water density, ρ_w, of 1 g cm^3) of water per volume of soil, where the soil volume is comprised of the solid grains and the pore spaces between the grains. When the soil is completely saturated with water, θ_v should generally equal the porosity. The relationship between gravimetric and volumetric water contents is given by:

$$\theta_v = \theta_g \left[\frac{\rho_b}{\rho_w} \right]$$

A related term that is often used to quantify the amount of water associated with a sample of soil is "saturation", S_w, which describes the fraction of the pore volume (void space) filled with water:

$$S_w = \frac{\theta_v}{n}$$

2.5.4 Soil Temperature

Soil temperature is often a significant factor especially in agriculture and land treatment of organic wastes, since growth of biological systems is influenced by soil temperature. In addition, soil temperature influences the physical, chemical, and microbiological processes that take place in soil. These processes may control the transport and fate of contaminants in the subsurface environment. The temperature of the soil zone fluctuates throughout the year in accordance with the above-ground temperature. Conversely, the temperature below the upper few meters of the subsurface remains relatively constant throughout the year.

QUESTIONS AND PROBLEMS

1. The hydrogen ion concentration of the soil solution from a particular soil is 3×10^{-6} mol L^{-1}. What is the pH of the soil solution?

2. What is the soil textural class of a soil with 20% sand, 60% silt, and 20% clay?

3. A 100-g sample of a moist soil initially has a moisture content of 15% on a dry weight basis. What is the new moisture content if 10 g of water is uniformly mixed into the soil?

4. Which factors within this chapter affect the cation-exchange capacity (CEC) of a soil? Explain why.

5. Which factors can potentially affect the transport of contaminants through soil and vadose zone? Explain why.

6. How does soil moisture content affect the activity of aerobic and anaerobic soil microorganisms?

7. Compare and contrast surface soils with the vadose zone.

REFERENCES AND ADDITIONAL READING

Maier R.M., Pepper I.L., and Gerba C.P. 2000. *Environmental Microbiology*. Academic Press, San Diego, California.

CHAPTER **3**

PHYSICAL-CHEMICAL
CHARACTERISTICS OF WATER

D.B. Walker, M.L. Brusseau, and K. Fitzsimmons

Apache Reservoir, Arizona. *Photo courtesy D. Walker.*

3.1 THE WATERY PLANET

3.1.1 Distribution

Ninety seven per cent of water on the Earth is marine (saltwater), while only 3% is freshwater (Figure 3.1). With regard to the freshwater, 79% is stored in polar ice caps and mountain glaciers, 20% is stored in aquifers or soil moisture, and 1% is surface water (primarily lakes and rivers). An estimated 110,000 km^3 of rain, snow, and ice falls annually on land surfaces, and this is what replenishes fresh water resources. Possible effects of global warming, combined with continued increases in human population and economic development are resulting in critical concern for the future sustainability of freshwater resources.

The limited supplies of surface waters and groundwater receive significant amounts of the pollutants generated by humans. Lakes across the planet have an average retention time of 100 years, meaning it takes 100 years to replace that volume of water. Rivers, on the other hand, have a much shorter retention time. The relatively long retention time in lakes highlights the danger of introducing pollutants that will be present for a long time (*i.e.*, they are "environmentally persistent"). The short retention time in rivers means that pollutants are transferred rapidly to other areas such as groundwater or oceans. The retention time of groundwater is measured in hundreds if not thousands of years. In the groundwater environment, persistent pollutants may remain intact for extremely long periods because of constraints to transformation. The characteristics of groundwater are described in Section 3.10. Pollution of groundwater and surface water is discussed in Chapters 17 and 18, respectively.

Pollutants in the ocean may be introduced into the food chain by filter-feeding organisms or possibly may be sequestered in cold, deep basins where they are resistant to degradation by natural processes. Much of the world's population inhabits coastal areas, making oceans especially

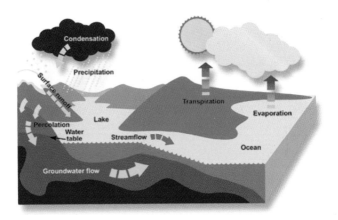

Figure 3.2 The hydrologic cycle. Courtesy Environment Canada. (http://www.ec.gc.ca/water/en/nature/prop/e_cycle.htm)

vulnerable to pollutants introduced directly or from surface water and groundwater drainage.

3.1.2 The Hydrologic Cycle

Water covers much more of earth's surface than does land. The continual movement of water across the earth due to evaporation, condensation, or precipitation is called the hydrologic cycle (Figure 3.2). The consistency of this cycle has taken millennia to establish, but can be greatly altered by human activities including global warming, desertification, or excessive groundwater pumping. Water, in its constantly changing and various forms, has been and continues to be an important factor driving evolutionary processes in all living things.

Evaporating water moderates temperature; clouds and water vapor protect us from various forms of radiation; and precipitation spreads water to all regions of the globe, allowing life to flourish from the highest peaks to the deepest caves. Solar energy drives evaporation from open water surfaces as well as soil and plants. Air currents distribute this vaporized water around the globe. Cloud formation, condensation, and precipitation are functions of cooling. When vaporized, water cools to a certain temperature, condensation occurs, and often results in precipitation to the earth's surface. Once back on the surface of the earth, whether on land or water, solar energy then continues the cycle. The latent heat of water (the energy that is required or released as water changes states) serves to moderate global temperatures, maintaining them in a range suitable for humans and other living organisms.

Some processes involved with the hydrologic cycle aid in purifying water of the various contaminants accumulated during its cycling. For instance, precipitation reaching the soil will allow weak acids absorbed from air to react with various minerals and neutralize the acids. Suspended sediments entrained through erosion and runoff will settle out as the water loses velocity in ponds or lakes. Other solids will be filtered out as water percolates through soil and vadose zones and ultimately to an aquifer. Many organic compounds will be degraded by bacteria in soil or sediments.

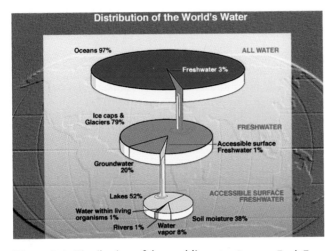

Figure 3.1 Distribution of the world's water Courtesy Earth Forum, Houston Museum of Natural Science (http://earth.rice. edu/mtpe/hydro/hydrosphere/hot/freshwater/0water_chart.html)

Salts and other dissolved solids will be left behind as water evaporates and returns to a gaseous phase or freezes into a solid phase (ice). These processes maintained water quality of varying degrees before human impacts on the environment; however, the current scale of these impacts often tends to overwhelm the ability of natural systems to cleanse water through the hydrologic cycle. Further, we have introduced many compounds that are resistant to normal removal or degradation processes (Chapters 16–18).

3.2 UNIQUE PROPERTIES OF WATER

3.2.1 Structure and Polarity

Water is an unusual molecule in that the structure of two hydrogen atoms and one oxygen atom provides several characteristics that make it a universal solvent. First is the fact that the two hydrogen atoms, situated on one side of the oxygen atom, carry positive charges, while the oxygen atom retains a negative charge (Figure 3.3).

This induced polarity allows water molecules to attract both positive and negative ions to the respective poles of the molecule. It also causes water molecules to attract one another. This contributes to the viscosity of water and to the alignment that water molecules will take when temperatures decrease to the point of ice formation. The fact that water becomes less dense in its solid state, compared to its liquid state, is yet another unusual characteristic. Because of this, ice floats and insulates deeper water. This is critical to maintaining deep bodies of liquid waters on Earth rather than a thin layer of water on top of an increasingly deep bed of solid ice.

The bipolar nature of water and its attraction to other polar compounds makes it an easy conduit for the dissolution and transport for any number of pollutants. Because so many materials dissolve so completely in water, their removal from water is often difficult.

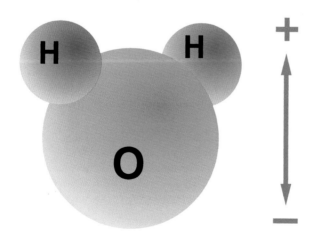

Figure 3.3 Structure and charge distribution of water. (http://faculty.uca.edu/~benw/biol1400/notes32.htm)

3.2.2 Thermal Properties

Water has unique thermal properties that enable it to exist in three different states: vapor; solid; and liquid under environmentally relevant conditions. Changes in each phase have certain terminology, depending upon state changes, as described below:

Condensation: vapor → liquid

Evaporation: liquid → vapor

Freezing: liquid → solid

Melting: solid → liquid

Sublimation: solid → vapor

Frost Formation: vapor → solid

Most liquids contract with decreasing temperature. This contraction also makes these liquids denser (*i.e.,* "heavier) as temperature decreases. Water is unique because its density increases only down to approximately 4°C, at which point it starts to become less dense (Figure 3.4). This is important because without this unique property, icebergs and other solid forms of water would sink to the bottom of the ocean, displacing liquid water as they did so. Also, lakes and ponds would freeze from the bottom up with the same effect.

The **specific heat** of water is the amount of energy required to raise one gram of water, one degree C, and is usually expressed as *joules per gram-degree Celsius* ($J g^{-1} °C^{-1}$). Specific heat values for the different phases of water are given below.

PHASE	$J G^{-1} °C^{-1}$
Vapor	2.02
Liquid	4.18
Solid	2.06

The **latent heat of fusion** is the amount of energy required to change 1 gram of ice, at its melting point temperature, to liquid. It is considered "latent" because there is no temperature change associated with this energy transfer, only a change in phase. The heat of fusion for water is -333 $J g^{-1} °C^{-1}$

The energy required for the phase changes of water are given in Table 3.1.

Earth is unique because it contains the necessary temperatures and pressures for all three states of water to exist. Water, under the correct combination of temperature and pressure, is capable of existing in all three states (solid, liquid, and vapor) simultaneously and in equilibrium. This is referred to as the *triple point*, where infinitesimally small increases or decreases in either pressure or temperature will cause water to be either a liquid, solid, or gas. Specifically, the triple point of water exists at a temperature and pressure of 273.16 Kelvin (0.0098°C) and 611.73 pascals (0.00603 atm) respectively. Figure 3.5 shows that decreasing temperature and increasing pressure causes water to pass directly from a

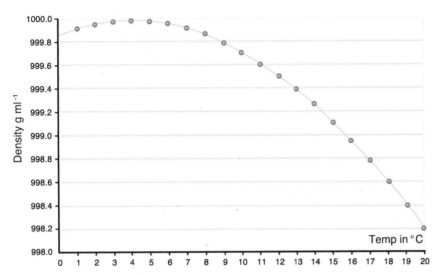

Figure 3.4 The density of water at varying temperatures. (http://www.cyberlaboratory.com/library/basicsofdensity/whatisdensity.htm)

gas to a solid. At pressures higher than the triple point, increasing temperature causes solid water (ice) to transform into liquid and eventually gas (vapor). Liquid water cannot exist in pressures lower than the triple point and ice instantaneously becomes steam with increasing temperature. This process is known as sublimation.

3.3 MECHANICAL PROPERTIES

3.3.1 Interception, Evaporation, Infiltration, Runoff

Precipitation in a nonpolluted environment provides a fairly pure form of water. However, today precipitation may absorb pollutant gases in the environment to form acid rain (see also Chapter 23). Precipitation can also pick up fine particulates that were suspended in the air. As the forms of precipitation reach the surface, they are likely to fall upon and be intercepted by various types of vegetation. In many regions, much of the precipitation may settle in or on trees, shrubs, or grasses and never actually reach the ground. In others, the plants may slow the rate of fall of raindrops, break them into smaller drops, or channel them more gently to the surface. Interception leads to several factors that impact the water and its role with later pollution events. First, the water

may evaporate directly from the plant, never reaching the soil. Second, it may entrain materials settled on the plant surfaces. Third, by slowing the momentum and reducing the energy of falling rain, physical impacts on the soil and resulting erosion may be reduced.

Certain anthropogenic land use practices or natural events can lead to decreases in interception and subsequent increases in sediment suspended within water. Often, sediment may have other pollutants attached to it thereby polluting the water as well. Certain mining practices, if not re-vegetated, can result in increased erosion resulting in contamination of streams. Natural events, such as wildfires, can also result in substantial erosion and contamination of downstream areas (see also Chapter 16).

Evaporation of water is another crucial part of the hydrologic cycle. The rate of evaporation from a body of

TABLE 3.1 Phase changes of water.

PROCESS	FROM	TO	ENERGY GAINED OR LOST ($J\ g^{-1}\ °C^{-1}$)
Condensation	Vapor	Liquid	2500
Deposition	Vapor	Ice	2833
Evaporation	Liquid	Vapor	−2500
Freezing	Liquid	Ice	333
Melting	Ice	Liquid	−333
Sublimation	Ice	Vapor	2833

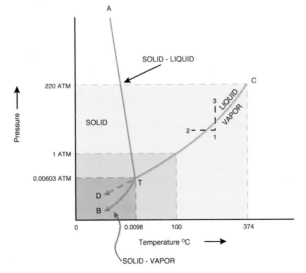

Figure 3.5 Phase diagram of the triple point of water. (http://www.sv.vt.edu/classes/MSE2094_NoteBook/96ClassProj/pics/trip_pt1)

Figure 3.6 This satellite image is of the Great Salt Lake in Utah. This is the largest lake in the U.S. west of the Mississippi River covering some 1,700 square miles. It is also 3 to 5 times more saline than the world's oceans. It is a fishless lake with only the most saline-tolerant ("halophytic") organisms capable of surviving. The largest organisms inhabiting its waters are species of brine shrimp and brine flies. (http://ut.water.usgs.gov/greatsaltlake/)

water, or mass of soil, is a function of the relative humidity, temperature, and wind speed. An important subcomponent of evaporation is transpiration, the active transport and evaporation of water from plants. Plants transport nutrients in an aqueous solution and then dispose of the water through their leaves by evaporation. As water evaporates, it leaves a concentrated amount of compounds that were formerly dissolved in that water. This applies to nutrients left in plants, as well as to pollutants that were introduced with the water.

Water that is not contained in oceans is often referred to as "freshwater," implying that it is not saline. This is not always the case, and some inland waters can be much more saline than the world's oceans. This is especially true in arid environments or enclosed basins that have limited or no drainage. Often, salinity in inland waters reaches such high levels that it supports little, if any, life. Salinity in inland waters, and in the world's oceans, is largely a result of evaporation. As water is vaporized and once again enters the hydrologic cycle, salts accumulate on the earth's surface, and in lieu of adequate dilution and flushing, can often make water increasingly saline (Figure 3.6).

Precipitation that reaches the soil surface either infiltrates the ground or runs off the surface. Human uses of water also deliver enormous amounts of water onto soils or human-made structures that can either infiltrate or contribute to run-off.

This is a major source of pollutants introduced into the environment. The infiltration rate of water into the ground is an important measure used to determine how foundations and sewer systems are designed, how irrigation water should be applied, and how pollutants may migrate to a water supply.

How water runs off of surfaces is also a matter of interest to hydrology, fisheries, aquatic biology, and pollution science. Not only are pollutants entrained in flowing water, but erosion and flooding can also occur. Studies of run off and surface flow focus upon the amounts of soil and pollutants that are transported and their eventual fate as they arrive into lakes or streams.

3.4 THE UNIVERSAL SOLVENT

One of the most unique properties of water is its ability to dissolve other substances. It is this ability that can lead to large-scale landscape transformations (Figure 3.7), and the

Figure 3.7 The Grand Canyon of the Colorado River was formed by the dissolution and erosion of material over eons. Historically, most of this material was deposited in the Gulf of California. With the construction of large dams along the course of the Colorado River, most of this material is now deposited in storage reservoirs. (http://www.kaibab.org/tr961/lg961110.jpg)

TABLE 3.2 Examples of typical concentrations of solutes in water.

Percent		parts per hundred	10^2
Milligram	$mg\,L^{-1}$	parts per million	10^6
Microgram	$\mu g\,L^{-1}$	parts per billion	10^9
Nanogram	$ng\,L^{-1}$	parts per trillion	10^{12}

ability to carry contaminants relatively long distances. If it were not for the various substances dissolved in water, an organism's cells would quickly be deprived of essential nutrients, salts, and gasses, leading to eventual death. The dissolution of materials in water has shaped the nature of all living creatures on the planet.

3.4.1 Concentration Terminology

It is important to quantify the amount of material dissolved in water. Quantification require a range of values so that we can determine high versus low concentrations for a given constituent. The values are always expressed as a ratio of solute to water (Table 3.2). The importance of very small concentrations should never be underestimated. This is especially true in toxicological studies where very small concentrations can lead to toxic impacts on organisms (Chapter 13).

There are two major expressions in concentration terminology.

– Mass/mass. An example would be parts per million (ppm), which equals parts of solution/parts of material $\times 10^6$.
– Mass/volume. An example would be milligrams/liter ($mg\,L^{-1}$), which equals milligrams of dissolved solid(s)/liter of solution.

Most of the time, $mg\,L^{-1}$ and ppm will be the same number. Their relationship is that the specific gravity of solution $\times$ ppm = $mg\,L^{-1}$. Note that this same relationship holds true when using other concentrations such as parts per billion (ppb) and $\mu g\,L^{-1}$ or parts per trillion and $ng\,L^{-1}$.

To calculate this we assume the following:

• Water weighs 8.33 pounds^{-1} gallon.
• There are 7.5 gallons ft^{-3} and this therefore weighs 62.43 pounds.
• One acre foot = 43,560 ft^3
• So, 43,560 ft^3 × 62.42 pounds ft^{-3} = 2,718,144 pounds of solution.
• Therefore, 2,718,144 × 0.000023 (or 23 ppm) = 62.5 pounds of Na and 2,718,144 × 0.000035 (or 35 ppm) = 95.1 pounds of Cl.
• 62.5 (Na) + 95.1 (Cl) = 157.7 pounds of NaCl.

Now that the farmer knows how much is in one acre foot, if the rate of water flow onto his crops is known, he can calculate an accumulation rate. For instance, let's say the farmer wants to know how many tons day^{-1} and tons year^{-1} of NaCl flow onto his crops and into his soil if the flow is held constant at 2 ft^3 second^{-1} (commonly written as cfs for "cubic feet per second").

• 2.0 cfs 3600 sec hour^{-1} × 24 hr day^{-1} × 62.4 pounds ft^{-3} × 0.000058 (58 ppm) = 625 pounds NaCl day^{-1}.
• 2000 pounds ton^{-1} divided by 625 pounds = 0.313 tons NaCl day^{-1}.
• 0.313 tons/day × 365 days/year = 114 tons NaCl year^{-1}.

3.4.2. Oxygen and Other Gases in Water

Just like terrestrial counterparts, aquatic organisms (other than anaerobic microbes) need dissolved oxygen and other gases in order to survive. Additionally, the world's oceans "absorb" an estimated $1/4$ to $1/3$ of carbon dioxide emitted by human activity. If it were not for the ocean's ability to absorb carbon dioxide, an important greenhouse gas, global warming would proceed at an unprecedented rate (see also Chapter 24). The amount of gas that an aqueous solution can hold is dependent upon several variables, the most important of which is atmospheric pressure. Simply stated, increasing

---

EXAMPLE CALCULATION 3.1

Using Concentration

Knowing the concentrations of constituents in water has many utilitarian uses. For example, a farmer may want to know how much salt will accumulate in the soil on his property when using water where both sodium (Na) and chloride (Cl) concentrations are known.

• Suppose the water contained 35 ppm Cl and 23 pm Na (*i.e.*, 58 ppm NaCl).
• How many pounds of Na, Cl, and NaCl are contained in an acre foot of water? (1 acre foot = 1 acre of land with a water depth of 1 foot).

INFORMATION BOX 3.1

Examples of Why Small Numbers are Important

• An AIDS virus is only 10^{-8} meters in size, or 0.00001 mm, yet it only takes one virus to have potentially devastating effects on the human immune system.

• From *Science*, 20 February 1991:

"In the end, after all the antibaryons had been consumed, one odd baryon out of 10 billion was left over. It was this tiny remnant that gave rise to all the planets, stars, and galaxies."

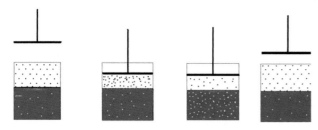

Figure 3.8 The solubility of oxygen in water under different atmospheric pressures.

atmospheric pressure causes a greater amount of gas to go into solution at a given temperature (Figure 3.8). Generally, increasing water temperature will result in an increased solubility of gas. This constant is otherwise known as "Henry's Law" and is written as:

$$\rho = K'_c c \qquad \text{(Eq. 3.1)}$$

ρ = partial pressure of the gas in mmHg

c = concentration of gas in mmoles, mL, or mg L^{-1} at a constant temp

K'_c = the solubility factor, different for each gas

The constant K'_c is specific for every gas and solute at a given temperature (see also Chapter 7). There is a direct, linear relationship between the partial pressure and the concentration of gas in solution. For example, if the partial pressure is increased by $^1/_4$, the concentration of gas in solution is increased by $^1/_4$ and so on. This is because the number of collisions of gas molecules on the surface of the solute (water in this case) is directly proportional to increases or decreases in partial pressure. Since the concentration: pressure ratio remains the same, we can predict the concentration of gas in water under differing partial pressures. This relationship can be written as:

$$\frac{\text{Concentration1}}{\text{Pressure1}} = \frac{\text{Concentration2}}{\text{Pressure2}}$$

For example, 1 liter of water under 1 atmosphere of pressure, will contain 0.0404 grams of oxygen. What will the concentration of oxygen be if the partial pressure is increased to 15 atmospheres?

$$\text{C1} = 0.0404 \text{ g O}_2/1 \text{ liter solution}$$
$$\text{P1} = 1 \text{ atm}$$
$$\text{P2} = 15 \text{ atm}$$
$$\text{C2} = ?$$
$$\frac{.0404 \text{ g O}_2}{1 \text{ atm}} = \frac{\text{C2}}{15 \text{ atm}}$$
$$\text{C2} = (15 \text{ atm}) (0.0404 \text{ g O}_2 \text{ per 1 liter/1 atm})$$
$$\text{C2} = 0.606 \text{ g O}_2$$

In any body of water, there are sources and sinks of dissolved oxygen. Sources include atmospheric re-aeration through turbulence; ripples and waves; and dams and waterfalls. Another potential source of dissolved oxygen is photosynthesis primarily by algae or submersed aquatic vegetation. During photosynthesis, plants convert CO_2 into oxygen in the process outlined below.

$$6CO_2 + 12 \text{ H}_2O + \text{Light Energy}$$
$$\rightarrow C_6H_{12}O_6 + 6O_2 + 6H_2O \quad \text{(Eq. 3.2)}$$

All natural waters also have sinks of dissolved oxygen, which include:

Sediment Oxygen Demand (SOD): Due to decomposition of organic material deposited on bottom sediments.

Biological Oxygen Demand (BOD): The oxygen required for cellular respiration by microorganisms.

Chemical Oxygen Demand (COD): The oxygen required for *all* organic compounds. Note that BOD is a subset of COD.

Respiration is the metabolic process by which organic carbon is oxidized to carbon dioxide and water with a net release of energy (see also Chapter 5). Aerobic respiration requires, and therefore consumes, oxygen.

$$C_6H_{12}O_6 + 6O_2 \rightarrow 6CO_2 + 6H_2O + \text{energy} \quad \text{(Eq. 3.3)}$$

This is, essentially, the opposite of photosynthesis. In the absence of light, the CO_2 collected by plants via photosynthesis during the day, is released back into the water at night, resulting in a net loss of dissolved oxygen. Depending upon the amount of nutrients, algae, and available light, this often results in large daily fluctuations in dissolved oxygen levels known as Diel patterns (Figure 3.9).

The implications of dissolved oxygen sinks and sources on aquatic organisms and overall water quality are crucial in determining whether or not a river, lake, or stream is polluted and to what degree. If dissolved oxygen sinks are greater than sources for extended periods of time, it is safe to assume some degree of contamination has occurred. Examples of anthropogenic wastes that can cause dissolved oxygen impairment of receiving waters are sewage (raw and treated, human and nonhuman), agricultural runoff, slaughterhouses, and pulp mills.

3.4.3 Carbon Dioxide in Water

Carbon dioxide only accounts for approximately 0.033% of the gases in earth's atmosphere, yet is abundant in surface water. The biggest reason for the abundance of carbon dioxide in water is due to its relatively high solubility; almost 30 times that of oxygen. In the atmosphere, carbon dioxide is released when fossil fuels are burned for human uses, and as a result of large worldwide increases in the use of fossil fuels during the last century or so, the amount of carbon dioxide in the atmosphere has steadily increased. Carbon dioxide is currently rising at a rate of approximately 1 mg L^{-1} year^{-1} or about 40% since the beginning of the Industrial Revolution. Since carbon dioxide is a major greenhouse gas, changes in global climate may have long-term environmental consequences (Chapter 24).

At room temperature, carbon dioxide has a solubility in water of 90 ml^3 of carbon dioxide per 100 ml^3 of water.

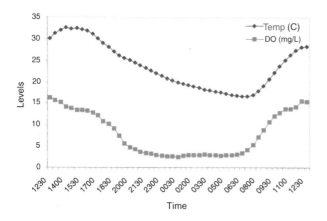

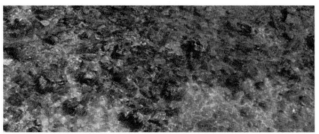

Figure 3.9 Diel pattern of temperature and dissolved oxygen in Rio de Flag, an effluent-dominated stream in Flagstaff, Arizona. (A) Data was collected every 30 minutes over a 24-hour period on 08/12/03. (B) Profuse growth of attached algae ("periphyton") growing in the stream at the time. Photosynthesis and respiration by these algae likely contributed to the large swings in dissolved oxygen levels within the water over the 24-hour period. Photo courtesy D. Walker.

Carbon dioxide dissociates and exists in several forms in water. First, carbon dioxide can simply dissolve into water going from a gas to an aqueous form. A very small portion of carbon dioxide (less than 1%) dissolved in water is hydrated to form carbonic acid, (H_2CO_3). Equilibrium is then established between the dissolved carbon dioxide and carbonic acid.

$$CO_2 + H_2O \leftrightarrow H_2CO_3 \qquad (Eq.\ 3.4)$$

Carbonic acid, a very weak acid, is then dissociated in two steps.

$$H_2CO_3 \leftrightarrow H^+ + HCO_3^- \qquad (Eq.\ 3.5)$$
$$HCO_3^- \leftrightarrow H^+ + CO_3^{2-} \qquad (Eq.\ 3.6)$$

As carbon dioxide is dissolved in water, equilibrium is eventually established with the carbonate ion (CO_3^{2-}). Carbonate, being a largely insoluble anion, then reacts with cations in the water, causing these cations to precipitate out of solution. As a result, Ca^+ and Mg^2 often precipitate as carbonates. Calcium carbonate ($CaCO_3$), otherwise known as limestone, has resulted in large deposits as a result of this process. As limestone is once again dissolved, carbon dioxide is released back into the atmosphere. In addition, several aquatic organisms, such as corals and shelled creatures such as clams, oysters, and scallops, are capable of converting the carbon dioxide in water into calcium carbonate.

Diel fluctuations in oxygen and pH levels can occur during the day in waters where photosynthesis is taking place (see Figure 3.9). Algae and plants convert carbon dioxide into carbohydrates to be used in metabolic processes. In very productive waters, this process can leave bicarbonate or carbonate in excess, leading to increased pH levels. In the absence of adequate light for photosynthesis, respiration predominates, resulting in carbon dioxide once again being restored to the water resulting in decreased pH levels.

Calcium carbonate, while insoluble at neutral to basic pH levels, readily dissolves in acidic conditions. In the initial step, carbonate acts as a base resulting in calcium ions and carbonic acid. In the next step, carbonic acid is dissociated releasing carbon dioxide as a gas.

$$CaCO_3 + 2\,H^+ \rightarrow Ca^{2+} + H_2CO_3 \qquad (Eq.\ 3.7)$$
$$H_2CO_3 \leftrightarrow H_2O + CO_2 \qquad (Eq.\ 3.8)$$

Rain is often slightly acidic due to the dissolution of atmospheric carbon dioxide. Recently, due to the burning of fossil fuels, other gases can also be dissolved in rain resulting in "acid rain." Atmospheric pollutants responsible for acid rain include sulfur dioxide (SO^2) and nitrous oxides (NO_x). More than 2/3 of these pollutants come from burning fossil fuels for electrical power generation, and prevailing winds can result in acid rain being deposited far from original source. Acid rain has far-reaching environmental consequences including acidification of lakes and streams, making them uninhabitable by aquatic life; extensive damage to forests, plants, and soil; damage to building materials and automotive finishes; and human health concerns. However an amendment to the Clean Air Act, the **Acid Rain Program**, whose goal is to lower electrical power emissions of the pollutants causing acid rain, shows recent evidence of success, and lakes, rivers, and streams have responded favorably (see also Chapters 4 and 23).

In lieu of any anthropogenic acidification of rain or surface water, conditions often exist that can result in the dissolution of limestone:

$$CO_2 + H_2O + CaCO_3 \leftrightarrow Ca^{2+} + 2\,HCO_3^- \qquad (Eq.\ 3.9)$$

The remaining reaction is a 3-step process:

$$CaCO_3 \leftrightarrow Ca^{2+} + CO_3^{2-} \qquad (Eq.\ 3.10)$$
$$CO_2 + H_2O \leftrightarrow H_2CO_3 \qquad (Eq.\ 3.11)$$
$$H_2CO_3 + CO_3^{2-} \leftrightarrow 2\,HCO_3^- \qquad (Eq.\ 3.12)$$

This reaction can result in the formation of caves when naturally acidic rainwater reacts with a subterranean layer of limestone, dissolving the calcium carbonate and forming openings. As slightly acidic water reaches the cave ceiling, the water evaporates and carbon dioxide escapes. It is this reaction that is responsible for the many elaborate formations in cave ecosystems (Figure 3.10).

Total alkalinity is the total concentration of bases, usually carbonate and bicarbonate, in water and is expressed as mg/L of calcium carbonate. Analytically, total alkalinity is expressed as the amount of sulfuric acid needed to bring a so-

Figure 3.10 Cave formation in the Big Room, Carlsbad Caverns National Park, New Mexico. (http://www.nps.gov/cave/home.htm)

lution to a pH of 4.2. At this pH, the alkalinity in the solution is "used up," and any further addition of an acid results in drastic decreases in pH levels. Total alkalinity, by definition, is the ability of a water body to neutralize acids. In other words, it is the "buffering capacity" of a water body, and it is influenced by the minerals in local soils. In areas of the northeastern U.S., where parent material contributes little to the total alkalinity in the water, the cumulative effects of acid rain have been most devastating and have extirpated aquatic life from several streams. Mining activity and pulp mills can also add to reductions in total alkalinity and subsequent decreases in pH. Stopgap measures in watersheds, lakes, or streams where alkalinity has been depleted include such drastic actions as dropping lime from helicopters to increase the buffering capacity for aquatic life.

3.5 OXIDATION-REDUCTION REACTIONS

Oxidation-reduction reactions involve the transfer of electrons from one atom to another. Oxidation is defined as the loss of an electron from an atom, and reduction is the gain of an electron from an atom.

$$Zn(s) + 2 H^+ (aq) \leftrightarrow Zn^{2+}(aq) + H_2(g) \qquad \text{(Eq. 3.13)}$$

In the above example, the oxidation number of Zn has changed from 0 to +2, producing Zn^{2+}, and the oxidation number of H^+ has changed from +1 to 0, producing H_2. In this reaction, Zn has been oxidized and H+ has been reduced. Since Zn(s) was oxidized, it caused the reduction of H^+(aq) and is therefore the reducing agent. Likewise H^+(aq) caused the oxidation of Zn(s), making H^+ the oxidizing agent.

Loss of electrons from one substance must simultaneously be accompanied by the gain of electrons from another. Electrons are neither created nor destroyed in chemical reactions, and we can envision oxidation-reduction pairs. These

oxidation-reduction reactions are often referred to as **redox** reactions (see also Chapter 7). Just as the transfer of hydrogen ions determines the pH of a solution, the transfer of electrons between species determines the redox potential of an aqueous solution. Redox potential is also referred to as "ORP" for oxidation-reduction potential, and is measured in volts or *Eh* (1 volt = 1 *Eh*). ORP specifically measures the tendency for a solution to either gain or lose electrons when it is subject to change by the introduction of a new species. A solution with a higher ORP will have a tendency to gain electrons (*i.e.*, oxidize them) and a solution with a lower ORP will have a tendency to lose electrons to new species (*i.e.*, reduce them).

Whether a chemical species in solution is oxidized or reduced has a profound influence on biogeochemical cycling of metals, nutrients, salts, organic compounds, and so on. Examples of redox couplings of interest to water quality include the following (after Cole 1994):

REDOX COUPLE	VOLTS	DISSOLVED O_2 (MG/L)
NO_3^- to NO_2^-	0.45–0.40	4.0
NO_2^- to NO_3^-	0.40–0.35	0.4
Fe^{+++} to Fe^{++}	0.30–0.20	0.1
SO_4^{--} to S^{--}	0.10–0.06	0.0

Redox reactions have an effect on bioavailability of nutrients. For example, iron exists either as particulate and oxidized ferric (Fe^{+++}), or the reduced and soluble ferrous (Fe^{++}). Phosphorous is an essential nutrient for plant and animal growth and under oxidizing conditions, is bound to ferric iron forming a ferro-phosphate complex that is biologically unavailable. If pollutants enter into a water body, dissolved oxygen may be depleted and reducing conditions prevail. Under these reducing conditions, iron loses its normally close association with phosphorous, with the latter becoming biologically available for algal growth sometimes forming noxious, and potentially toxic, blooms. Reducing conditions often prevail in the bottom of thermally stratified lakes and reservoirs, and phosphorous can accumulate leading to large growths of algae when the lake de-stratifies.

Methylation of mercury (addition of CH_3) also relies upon physical-chemical characteristics such as dissolved oxygen, pH, and ORP. Mercury is a metal that occurs naturally in the environment and has many different chemical forms. Methylmercury is the form of mercury that bioaccumulates (increases from one trophic level to the next by ingestion or absorption) and can cause mercury poisoning in wildlife and humans. Methylation of mercury can occur both abiotically and biotically by sulfate-reducing bacteria (SRB's) of the genera *Desulfovibrio* and *Desulfotomaculum*. SRB's require anaerobic and reducing conditions in their environment in order to reduce sulfate to sulfide. Specifically, they use sulfate as a terminal electron acceptor to break down molecular hydrogen or organic matter for metabolism through the following processes (see also Chapter 5).

$$3H_2SO_4 + C_6H_{12}O_6 \rightarrow 6CO_2 + 6H_2O + 3S^{-2} \quad \text{(Eq. 3.14)}$$

Mercury (Hg) normally binds very tightly to organic matter. The above process is the method in which mercury can become de-coupled from organic matter and subsequently methylated via methyl transfer from cobalamin (vitamin B12) to Hg^{2+}.

3.6 LIGHT IN AQUATIC ENVIRONMENTS

Just as in terrestrial systems, light at the water's surface marks the beginning of photosynthesis or "primary production." Light is the driving force behind almost all metabolic processes in aquatic ecosystems. Light carries heat energy to be used in many chemical and biological processes, and can simultaneously regulate and/or damage, aquatic biota. Pollutants, especially suspended and dissolved substances, can have profound effects on both the amount of light available for photosynthesis and the heat energy needed for these processes.

3.6.1 Light Energy

Light contains differing amounts of energy, depending upon frequency and wavelength. Quantum theory states that electromagnetic energy, such as light, is transmitted in discrete amounts or "packets" called **quanta**. A single quanta (*i.e.,* "quantum") of electromagnetic energy is also referred to as a **photon**, and the energy carried by each photon is proportional to its frequency. The quantity of electromagnetic energy flow over time is measured as a rate, *i.e.,* quanta second^{-1} known as the **radiant flux of light**.

The arrangement of light based upon differing wavelengths, frequencies, and energies is described by *spectra*. For example, the spectrum formed by white light contains all colors and is therefore said to be continuous (Figure 3.11). Certain biological, chemical, and physical processes occur only at specific frequencies of spectra, some of which can be seen with the human eye and several of which cannot. Light that is divided over a certain range of spectra is divided by color and measured by its frequency in nanometers.

Light travels at 299,792 km second^{-1}. It takes approximately 400 trillion waves of red light at 750 nm to span the distance light travels every second. It takes almost twice as many violet waves, at 380 nm, to fill the same volume of a light second. The amount of radiation emitted by the sun that reaches the earth's outer atmosphere is 1.94 calories cm^{-2} minute^{-1} and is known as the **solar constant**. The most common wavelength that makes it to our outer atmosphere is ~480nm.

Light wave frequency and energy are interrelated as explained by **Planck's equation,** which is expressed as:

$$E = h\nu \qquad \text{(Eq. 3.15)}$$

where:

E is the energy in a photon of light
h is Planck's constant of 6.6255×10^{-34} joules second^{-1}
ν is the wave frequency

Frequency of a light wave is given by $\nu = c/\lambda$
where:

c is the speed of light (3×10^8 m/second), and
λ is the wavelength

For example, the frequency of red light (750 nm) is calculated from

$$\nu = \frac{3 \times 10^8}{7.5 \times 10^{-7}} = \frac{4.00 \times 10^{14}}{\text{second}^{-1} \text{ wavelengths light}}$$

Substituting this number into Plancks equation ($E = h\nu$), 26.5×10^{-20} joules are contained in a single photon of light. Compare this with the fact that it takes 6.024×10^{23} photons just to initiate a photosynthetic reaction (otherwise known as **Avogadro's number**). Thus, it can be seen that it takes a large amount of light energy to perform what we would consider a simple biological process. Photosynthetically active radiation, the spectrum of light needed for photosynthesis by most plants, is approximately 400 to 700 nm. Specific types of chlorophylls and accessory pigments in plants have narrowed the requirements of wavelength ranges. For example, chlorophyll *a*, a photosynthesizing pigment common to all algae, absorbs light in two peaks, 670–680 nm and again at 435 nm.

3.6.2 Light at and Below the Waters Surface

Several processes can affect both the intensity and quality of light reaching the earth's surface. One such process is simple scattering of light by particles in the atmosphere, including water vapor. Refraction of light occurs when the speed of light changes going from one medium, such as air, into another, such as water. Light can also be reflected off of a water surface due to several factors such as the incident angle of light, wave height and frequency, or the presence of ice. Another factor affecting light intensity is absorption due to the decrease in light energy by its transformation into heat. Both atmospheric gases and water can cause absorption.

Absorption (*i.e.,* "quenching") of light entering a body of water can be quantified using a **vertical absorption coefficient** expressed as:

$$k = \frac{\ln I_o - \ln I_z}{z}$$

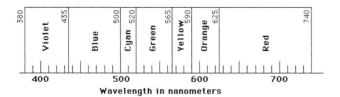

Figure 3.11 Wavelengths of light and associated spectra. (http://hyperphysics.phy-astr.gsu.edu/hbase/vision/specol.html)

where:

I_o is the natural log of the initial amount of light entering the water

I_z is the natural log of light remaining at any given depth

z is the thickness of the water in meters

The vertical absorption coefficient is somewhat analogous to the coefficient of extinction, except that the latter uses the base10 logarithm. The vertical coefficient of absorption is therefore 2.3 times the coefficient of extinction. Another important variable when considering the fate of light in water is **the total coefficient of absorption**, which is the sum of all factors leading to the intensity (or "extinction") of light at any given depth. The total coefficient of absorption can be expressed as:

$$I_z = I_{o^-}{}^{-kw} + I_{o^-}{}^{-kp} + I_{o^-}{}^{-kc}$$

where:

kw = the coefficient of absorption in pure water

kp = suspended particulate matter

kc = dissolved substances

Note that kp or kc can only be determined after filtration or centrifugation. The total coefficient of absorption is different for each body of water and is dependent upon the amount of dissolved or suspended material in the water. Dissolved substances are normally humic or fulvic acids, tannins, lignins, or anything that constitutes colored, dissolved, organic matter absorbing light strongly at relatively short wavelengths (*e.g.*, blues and ultraviolet radiation < 500 nm). Suspended material includes fine clays, and phytoplankton, which absorb light evenly over the entire spectrum.

In standing water, vertical light penetration can be roughly estimated using a **secchi disk** which is standardized, 20-cm diameter, black and white, weighted disk lowered into the water using a calibrated line. The depth at which the disk almost, but not quite, disappears is recorded. This is known as the **secchi disk transparency** expressed as Z_{sd}. To be comparable, this has to be done between 10 am and 2 pm on any given day. Secchi disk depth is often mistakenly used as a proxy for primary production (the amount of standing algal biomass) in a water body. In reality there are many other mitigating factors besides algae that can cause either increases or decreases in transparency. The **photic zone** is the volume of water from the surface to where 99% of the light needed for photosynthesis has been extinguished. A very rough estimate of photic zone depth is anywhere from 2.7 to 3 times Z_{sd}.

3.7 OCEANS

3.7.1 Salts

The oceans are saline due to the constant input of dissolved salts leached from rocks and soils on land surfaces. These dissolved solids consist of many salts including sodium, calcium, and magnesium salts. There are several common methods of determining total dissolved solids and/or salinity including electrical conductivity, density, light refraction, silver titration, and simple evaporation of a known volume. Each of these methods will provide a result that can be converted to a percentage, or more commonly parts per thousand, of salinity. Open ocean seawater will vary from 33 to 35 parts per thousand (ppt), while coastal waters may have less than 1 ppt.

3.7.2 Transport and Accumulation of Pollutants

The oceans tend to become the repository of many pollutants. Air pollution and water pollution often transport the compounds to the ocean through rain events or runoff. Historically, one of the most common mindsets was that the "solution to pollution is dilution." Over time we have discovered that even the oceans are not vast enough to handle the volume of pollutants that can be discharged by human activities. In many cases, the effects of pollutants are so toxic that even vast dilution is not effective. In other instances we have found that filter-feeding organisms bioaccumulate toxic compounds that ascend food chains and can affect grazers and top carnivores, as well as humans consuming various seafoods. Mercury in swordfish and certain sharks is such an example.

3.7.3 Wave Morphology and Currents

Waves and currents move pollutants within surface water. This is especially critical in marine systems as we attempt to control and/or track pollution movement. Waves are typically the result of winds blowing across the surface of a body of water. As the air friction pushes against the water, small ripples form. Continued breezes push against the sides of each ripple providing additional energy. The tops of the ripples may blow off, forming whitecaps. This releases some of the energy, but more will continue with the bulk of the water below. Continued wind energy transferred to the waves can store enormous amounts of energy. The size of a wave is a function of the average velocity of the wind, the period of time it blows, and the distance of open water across which it blows. The distance over which blowing winds creates waves is called the **fetch**. Waves are usually described by the period (the time between two crests passing the same point), the wavelength (distance between two crests), and the wave height (vertical distance between a trough and the next crest).

Waves are important factors in the dispersion of pollutants, especially oil spills. Calm waters facilitate the recovery of oil and other floating pollutants. However, in cases where it cannot be collected, wave action can break up the thick mats of oil and spread the material so that bacterial degradation can break down the organic molecules. It will also allow the lighter fractions to volatilize.

Water motion in the oceans is a function of waves, which are wind driven, and currents, which are driven by a number of factors. The most important is the Coriolis effect caused by the spinning of the earth (see Chapter 4). However, wind, runoff from rivers, density differences from temperature or salinity extremes, and tidal fluctuations can all drive currents.

Currents are often compared to the circulatory system of a living organism. Trade winds will power currents that transport water and its constituents across vast distances along the surface. Cooling of surface waters in the high latitudes causes cold, dense water to sink into the depths, where it flows along the bottom until it upwells in lower latitudes to replace warmer surface waters that are blown away from coastlines. Tidal action will also drive currents in local situations.

Currents and waves effectively mix surface waters on a short time scale. Deep-water currents mix water on a much longer time scale. Together they effectively spread pollutants to every corner of the ocean. Water motion can be effectively measured in two ways. If the motion is fairly consistent in one direction, a current meter can be used to determine velocity. A more common situation is when water motion is not consistent, but varies in direction and speed in three dimensions over short periods of time. In this case, a better measure is to use a clod card. Clod cards consist of a block of calcium sulfate that slowly dissolves in water. The block is mounted on a card stock for easy attachment and handling. Its rate of dissolution can be measured by recording initial and final dry weights. These are normally used to compare two or more environments. The clod cards can be calibrated, if necessary, by placing controls in waters or tanks with known velocities of water motion.

3.8 LAKES AND RESERVOIRS—THE LENTIC SYSTEM

Lentic systems are closed ecosystems such as lakes. However, while lakes and reservoirs are relatively more "closed" than rivers and streams, they are far from isolated. Although some lakes and reservoirs have subterranean groundwater inputs, the majority of water entering them is a result of overland flow; therefore, lakes and reservoirs are reflections of all processes that have occurred in the watershed up to that point. Both natural and anthropogenic watershed influences can have profound effects on both water quality for human use and aquatic communities living within lentic systems.

3.8.1 Lentic Typology

There are several ways to classify lakes and reservoirs: by origin, ecoregion, shape and size, regimen of mixing, and stratification. Detail about every different type of lake or reservoir is beyond the scope of this chapter, and the reader is instead referred to any of several available limnological texts. Rather, this section will discuss the differences between two main types of lentic systems: lakes and reservoirs.

The main difference between lakes and reservoirs is that the former have natural origins, while the latter are manufactured by humans for anthropogenic needs. Both are lentic systems and therefore share some common attributes. Lakes are dominant where glaciers have scoured the landscape, as, for example, around the Great Lakes. In other cases, tectonic activity has formed rifts, allowing for the African Rift Lakes,

or some other depression has been made through natural causes and there is adequate ground or surface water inputs to fill these depressions. Reservoirs, on the other hand, are constructed where lakes are not in abundance and water is needed for human use. Reservoirs are often found in greater abundance in arid and semi-arid regions such as in the western U.S., where large natural lakes are not abundant. The large number of reservoirs built in arid regions often means sacrificing lotic habitats through either direct impoundment or some change in water chemistry caused by impoundment. Endemic aquatic organisms living in streams and rivers of the western U.S. are among the most endangered species on the planet due to impoundment of habitat and changes in environmental conditions below large dams.

Some of the major differences between lakes and reservoirs are given in Table 3.3. The arrows are either increasing or decreasing as they relate to either lakes or reservoirs. For example, lakes generally have a much smaller watershed area than reservoirs.

3.8.2 Trophic State

Material within lentic systems are generally classified as either **autochthonous** or **allochthonous** in origin. Allochthonous (from the Greek, meaning "other than from the earth or land itself") material is everything that has been imported to the lentic system from somewhere else in the watershed. This material can be thought of in terms of *loading*. Autochthonous (from the Greek, meaning "of or from the earth or land itself") material is that which is recycled within the lake or reservoir. Both sources play a role in a lentic system's **trophic state**.

The trophic status of a lake or reservoir is largely a means to communicate the ecological condition of a water body. Trophic state is based upon the total weight of living biological material or biomass in a water body at a specific location and time. Time and location-specific measurements can be aggregated to produce waterbody-level estimations of trophic state. Trophic status is not equivalent to **primary production**, which is the rate of carbon fixed (usually expressed as g of C fixed $day^{-1} m^{-3}$). Trophic state, being a multidimensional phenomenon, has no single trophic indicator that adequately measures its underlying concept. Combining the major physical, chemical, and biological expressions of trophic state into a single index reduces the

TABLE 3.3 Comparisons of natural lakes and reservoirs.

VARIABLE	NATURAL LAKES	RESERVOIRS
Watershed Area	↓	↑
Maximum Depth	↓	↑
Mean Depth	↑	↓
Resident Time	↑	↓

TABLE 3.4 Trophic categories for lakes and reservoirs.

TROPHIC STATE	CHLOROPHYLL A (μg/L)	SECCHI DISK DEPTH (m)	TOTAL P (μg/L)	ATTRIBUTES
<30	<0.95	>8	<6	**Oligotrophy:** Clear water, dissolved oxygen throughout the year in the hypolimnion
30–40	0.95–2.6	8–4	6–12	Hypolimnia of shallower lakes may become anoxic
40–50	2.9–7.3	4–2	12–24	**Mesotrophy:** Water moderately clear; increasing probability of hypolimnetic anoxia during the summer.
60–60	7.3–20	2–1	24–48	**Eutrophy:** Problems with excessive primary production begin. Anoxic hypolimnia in stratified lakes/ reservoirs.
60–70	20–56	0.5–1	48–96	Cyanobacteria dominate the phytoplankton. Increasing problems with anoxia.
>70	56–155	0.25–0.5	96–192	**Hyper-eutrophy:** Primary production limited only by light. Dense growths of algae and/or aquatic plants. Increasing prevalence of anoxia throughout the water column. Fish kills possible.
>80	>155	<0.25	192–384	Few aquatic plants or other forms of life. Sustained periods of anoxia.

variability associated with individual indicators and provides a reasonable composite measure of trophic conditions in a water body.

Several trophic state indices have been devised. The selection of which one to use depends upon several different chemical and physical parameters. One of the most-used trophic state indices for lakes is the **Carlson's TSI** (Carlson, 1977). Especially important are the nutrients phosphorous and nitrogen, both essential macronutrients for algal growth and primary production. Other variables used include measures of chlorophyll *a* and secchi disk depth. Chlorophyll *a* is a pigment common to all algae and gives a measure of standing biomass. Phosphorous is often the nutrient that is most "limiting" in natural waters (although nitrogen limitation also does occur). Carlson's TSI relies upon three variables—chlorophyll *a*, secchi disk depth, and total phosphorous—to determine trophic status of any water body that is phosphorous-limited. In the broadest sense, trophic status of a lake or reservoir is often divided into the categories presented in Table 3.4.

The advantage of a fixed boundary system is its easy application by managers and technical personnel with only limited limnological training. However, trophic terminology has a history of being mis-used. For example, deeming a lake as eutrophic does not automatically mean it has poor water quality. Although the concepts of water quality and trophic state are related, they should not be used interchangeably. Trophic state is an absolute scale that describes the biological condition of a body of water. The trophic scale is a division of variables used in the definition of trophic state and is not subject to change because of the attitude or biases of the observer. An oligotrophic or a eutrophic lake has attributes of production that remain constant no matter what the use of the water or where the lake is located. For the trophic state terms to have meaning, they must be applicable in any situation and location, while keeping in mind that trophic status is just one of several aspects of the biology of the water body in question. Water quality, on the other hand, is a term used to describe the condition of a water body in relation to human needs or values. Quality is not an absolute; the terms "good" or "poor" water quality only have meaning relative to the attitude of the user. An oligotrophic lake might have "good" water quality for swimming, but have "poor" water quality for fishing. Confusion can ensue when trophic state is used to infer water quality.

3.8.3 Density and Layering

As light enters the water, different wavelengths are quenched exponentially (see 3.6.2). Wavelengths in the 620- to 740-nm range are absorbed first. This range also contains those wavelengths of light that carry the most amount of heat energy. The relative density of water is temperature-dependent. Water becomes increasingly dense down to about 4°C, at which point it becomes less dense (see 3.2.2). Thus, differential heating leads to differential vertical layering of the water column usually beginning in spring and early summer. The definitions of the various layers are given in Information Box 3.2. An example of thermal stratification in a lake is presented in Figure 3.12.

Interaction between the epilimnion and hypolimnion often results in the formation of autochthonous feedback mechanisms. In almost every case, anoxia within the hypolimnion mirrors epilimnetic production so that increases in trophic state result in increased hypolimnetic anoxia. Prolonged anoxia within the hypolimnion often results in the prevalence of reducing conditions (see Section 3.5). Under these reducing conditions, nutrients that would otherwise be bound to material within sediments become unbound and once again available for biological uptake. The density

INFORMATION BOX 3.2

Stratification of Lakes and Reservoirs

The uppermost layer is called the **epilimnion**, and is characterized by relatively warm water where most photosynthesis occurs. Depending upon environmental conditions, it is more oxygenated than layers below it. The middle layer is called the **metalimnion** and contains an area known as the *thermocline*. The thermocline is that area within the water column where the temperature gradient is the steepest. The metalimnion is that region surrounding the thermocline where the temperature gradient is steep compared to the upper and bottom layers. Due to the temperature gradient becoming increasingly steep within a correspondingly smaller volume of water, the thermocline becomes an infinitesimally small plane, whereas the metalimnion is a larger region encompassing the mean of the greatest rate of change. The **hypolimnion** is the bottom layer and is colder and denser than either the epilimnion or metalimnion. When a lake or reservoir is thermally stratified, the hypolimnion becomes largely isolated from atmospheric conditions and is often referred to as being stagnant. Additionally, the hypolimnion receives organic debris from the epilimnion, and as respiring bacteria begin the process of decomposition of this received material, consumption of dissolved oxygen (*e.g.*, respiration) usually exceeds either production of oxygen from photosynthesis or atmospheric re-aeration. The eplimnion is often referred to as the **trophogenic** area of lentic systems, where mixing through wind and wave action as well as photosynthesis exceeds respiration, whereas the hypolimnion is referred to as the tropholytic region, where organic material is synthesized and mineralization by bacteria occurs.

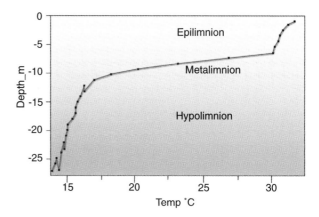

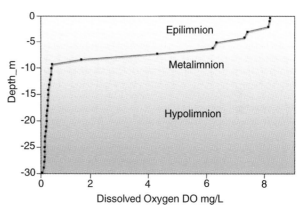

Figure 3.12 Thermal stratification in Lake Pleasant, Arizona. Data collected on 08/05/2003. The upper graph plots depth and temperature and the bottom graph depth and dissolved oxygen.

differences between the epilimnion and hypolimnion means that relatively few of these nutrients are available for uptake by phytoplankton until the lake or reservoir de-stratifies or "turns over," which occurred when the epilimnion cools to a temperature similar to the hypolimnion (usually during the fall). Algal "blooms" or sudden increases in biomass of algae are common during this time.

Recycling of nutrients through the thermocline from the hypolimnion to the trophogenic epilimnion does occur, although this is relatively small compared to overall nutrient levels within the hypolimnion. Increases in algal biomass means that more organic material is available for transport back through the thermocline into the hypolimnion, adding to anoxia and the potential release of more nutrients from sediments, so that a positive feedback loop is established. This autochthonous cycling of nutrients can keep a lake or reservoir locked into a eutrophic state even after other sources of pollutants from the watershed have been reduced.

3.9 STREAMS AND RIVERS—THE LOTIC SYSTEM

Lotic systems consist of running water such as rivers or streams and are the great transporter of material to oceans, lakes, and reservoirs. Rivers and streams are also vulnerable to both natural and anthropogenic sources of pollution, and have a history of being used and misused.

3.9.1 Stream Morphometry

Stream morphometry was initiated by R.E. Horton and A.E. Strahler in the 1940s and 50s to find suites of holistic stream properties from the measurement of various attributes. This was designed to allow some type of classification system that could be used as a communicative tool for hydrologists (Figure 3.13).

The original idea was to develop a hierarchical classification system of stream segments. These segments were ordered numerically from headwaters so that individual tributaries at the headwaters were given the order of "1." The joining of two 1st order streams were given the order of

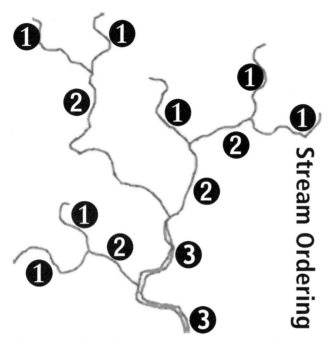

Stream Ordering

Figure 3.13 Example of stream ordering. Notice that it takes 2 stream orders of the same magnitude to increase downstream ranking. For example, the confluence of a 2nd order and a 1st order stream does not equal a third order, but the joining of two 2nd order streams does. (http://www.cotf.edu/ete/modules/waterq/streamorder.html)

"2" and the joining of two 2nd order streams, the order of "3," and so on.

Horton found that the ratio between number of stream segments in one order and the next was consistently around three. This is called the **bifurcation ratio**. This ratio has also been discovered in the rooting system of plants, the branching structure of woody plants, leaf venation, and the human circulatory system.

ORDER	# OF SEGMENTS	BIFURCATION RATIO
1	30	3.0
2	10	3.3
3	3	3.0

Horton called this association the **Law of Stream Numbers**, which is defined as the "morphometric relationship observed in the number of stream segments of a particular classification order in stream order branching." Horton combined the information that he obtained to define the **Laws of Stream Lengths** and **Basin Areas**. The Law of Stream Lengths states that a geometric relationship exists between the numbers of stream segments in successive stream orders, whereas the Law of Basin Areas indicates that the mean basin area of successively ordered streams form a linear relationship.

A quantifiable measure of the morphometry of drainage networks is **drainage density**, which is the length of stream channel per unit area of drainage basin expressed as:

$$\text{Drainage density (Dd)} = \frac{\text{stream length}}{\text{basin area}}$$

Drainage density is useful numerical measure of landscape dissection and runoff potential.

Rivers are often divided into relatively homogenous units or **reaches**. A river has distinct chemical, physical, and biological attributes, depending upon stream order and overall size of the channel. General physical characteristics based upon channel size are given in Table 3.5.

3.9.2 Stream Hydraulics

The flow of fluids is generally classified into two types: **laminar** and **turbulent**. For laminar flow, which occurs at lower velocities, the individual fluid (*e.g.*, water) molecules move uniformly in the direction of the mean gradient, with minimal mixing. Conversely, at higher velocities, the fluid molecules do not always move uniformly; instead, they may also cross the paths of other molecules, mixing and forming eddies. This is refereed to as turbulent flow. The Reynolds number, a dimensionless parameter, is used to characterize conditions for which flow will be laminar or turbulent.

The volume of water flowing in a stream at a given time is referred to as the stream discharge. Discharge has units of volume per time (*e.g.*, cubic meters per second). It is calculated as: $Q = A * v$, where Q is discharge, A is cross-sectional area and v is velocity. The cross-sectional area of the stream, which has units of length squared, is controlled by the shape and size of the channel and the height of water in the channel. This latter term is often referred to as the stream stage. The velocity (units of length per time) of water in a stream is controlled by the gradient (slope) of the stream and roughness of the stream channel surfaces. Water velocities are generally not uniform across the stream cross-section. Rather, the highest velocities typically occur in the center of the channel just below the surface, and the lowest velocities occur along the channel surfaces (where friction is greatest).

3.10 GROUNDWATER—WATER IN THE SUBSURFACE

Water in the subsurface serves as a critical resource for human consumption, both directly and indirectly (see Chapter 17). Groundwater resources serve as one of the two primary sources of potable water supply in the world (the other being surface water). In addition, water in the soil profile supports plant life, upon which humans are dependent in several ways. Water is also central to the transport and fate of contaminants in the subsurface. We will examine the distribution and movement of groundwater in this section. The impact of water flow on transport of contaminants in the subsurface is discussed in Chapters 6 and 17.

TABLE 3.5 Properties of stream channels.

CHANNEL SIZE	PHYSICAL CHARACTERISTICS	ORDER
Small	Cobbles (>64 mm dia.) And boulders (>256 mm dia.) Dominate the substrate.	Low
	Pools form behind rocks or logs (step-pool formations)	
	Relatively steep gradients (2° to 20°)	
	Banks composed of bedrock, boulders, and roots.	
	Highly erosional areas.	
Intermediate	20–30 m max. width	Intermediate
	Dominated by pool-riffle-bar units	
	Riffles are zones of relatively shallow, rapid flow	
	Rapids, cascades, and glides (extended riffles) may be prominent.	
	Major pool types include *backwater* (formed from either obstructions in the main channel or from periodic flooding of banks), *dammed* (found upstream of boulder lies and gravel bars), and *scour* (where flow converges past an obstruction).	
Large	Dominated by pool-riffle sequences, bar formations, and meanders.	High
	Reach gradient largely determined by valley gradient.	
	Transport of sediment may increase *sinuosity*.	
	Sinuosity is defined as river length/valley length.	
	Braided channels may form when the river can no longer carry its sediment load.	
	Increased deposition of sediment in large channels.	

Data from: Church, M. (1992). Channel morphology and typology. *The Rivers Handbook: Hydrological and Ecological Principles.* P. Callow and G.E. Petts. Oxford, Blackwell Scientific Publications. Vol 1, 126-143.

3.10.1 Water in the Subsurface

We can observe a cross-section of water distribution in the subsurface by drilling a borehole or excavating a pit. A schematic of a typical subsurface profile was presented in Figure 2.1. The vadose zone, also known as the zone of aeration or unsaturated zone, represents a region extending from near the ground surface to a water table. The water table is defined as a water surface that is at atmospheric pressure. In the soil and vadose zones, all pores are usually not filled with water; many pores will also contain air. In such cases, the porous medium is considered to be unsaturated. Water pressure in the soil and vadose zones is less than atmospheric pressure (P<0). The thickness of the vadose zone varies from a meter or less in tropical regions to a few hundred meters in arid regions, depending upon the climate (*e.g.*, precipitation), soil texture, and vegetation (see also Chapter 2).

Water stored in the soil and vadose zones is retained by surface and capillary forces acting against gravitational forces. Molecular forces hold water in a thin film around soil grains. Capillary forces hold water in the small pores between soil grains. Gravity forces are not sufficient to force this water to percolate downward. Thus, there is very little movement of water in the vadose zone when water contents are relatively low. For higher water-content conditions,

some of the water is free to move under the influence of gravity. When this occurs, water movement would generally be vertically downward. If this water moves all the way to the water table, it serves to replenish (or recharge) groundwater. The **capillary fringe** is the region above the water table where water is pulled from the water table by capillary forces. This zone is also called the **tension-saturated zone**. The thickness of this zone is a function of grain-size distribution and varies from a few centimeters in coarse-grained soils to a few meters in fine-grained soils. The water content in this zone ranges from saturated to partially saturated, but fluid pressure acting on the water is less than atmospheric pressure (P<0).

The region beneath the water table is called the saturated or phreatic zone. In this zone, all pores are saturated with water and the water is held under positive pressure. Because all pores are filled with water, soil-water content is equal to porosity, except when liquid organic contaminants are also present in the pore spaces (see Chapter 17). The water in the saturated zone is usually referred to as groundwater. Water movement in the saturated zone is generally horizontal. Specific sections of the saturated zone, particularly those comprised of sands and gravels, are called aquifers, which are geologic units that store and transmit significant quantities of groundwater.

3.10.2 Principles of Subsurface Water Flow

Water at any point in the subsurface possesses energy in mechanical, thermal, and chemical forms. The energy status of water, for example the effort required to move water from one point to another, is a critical aspect of quantifying water flow. For groundwater flow, the contributions of chemical and thermal energies to the total energy of water are generally relatively minor and thus are usually ignored. Therefore, we consider water flow through porous media to be primarily a mechanical process. Fluid flow through porous media always occurs from regions where energy per unit mass of fluid (fluid potential) is higher, to regions where it is lower.

From fluid mechanics, the mechanical energy of water at any point is composed of the kinetic energy of the fluid, the potential (or elevation) energy, and the energy of fluid pressure. For water flow in porous media, kinetic energy is generally negligible because pore-water velocities are usually small. Thus, total energy of water is considered to consist of potential and pressure energies.

The potential or elevation energy results from the force of gravity acting on the water. In the absence of pressure energy considerations, water always flows from regions of higher elevation potential to lower. This is why surface water flows "downhill." Similarly, groundwater usually flows downward from higher elevations (underneath mountain peaks) to lower elevations (underneath the valley floor).

In the vadose zone, the pressure potential is negative, indicating that energy is required to "pull" water away from the soil surfaces and small pores. In the saturated zone, the pressure potential is positive due to the pressure exerted by overlying water. The water pressure increases with depth in the saturated zone. This condition occurs in all water bodies—including swimming pools—which is manifested by the increasing pressure one feels on their sinuses and eardrums, as one swims deeper below the surface.

The energy potentials are commonly expressed in terms of length to simplify their use. In length terms, the energy potentials are referred to as "heads." The total energy potential head for water is called the **hydraulic head**; sometimes, particularly for vadose-zone applications, it is called the **soil-water potential head**. The equation we use to relate the hydraulic head (h) to its two parts, elevation head (z) and pressure head (Ψ), is: $h = \Psi + z$. Each head has a dimension of length (L) and is generally expressed in meter or feet.

Hydraulic head measurements are essential pieces of information that are required for characterizing groundwater flow systems (*i.e.*, direction and magnitude of flow), determining hydraulic properties of aquifers, and evaluating the influence of pumping on water levels in a region. Piezometers are used to measure the hydraulic head at distinct points in saturated regions of the subsurface. A piezometer is a hollow tube or pipe drilled or forced into a profile to a specific depth. Water rises inside the tube to a level corresponding to the pressure head at the terminus. The level to which water rises in the piezometer with reference to a datum such as sea level is the hydraulic head (Figure 3.14).

The relationship between the three head components is illustrated in a piezometer in Figure 3.14. The value of z represents the distance between the measurement point in the profile and a reference datum. Sea level is often taken as

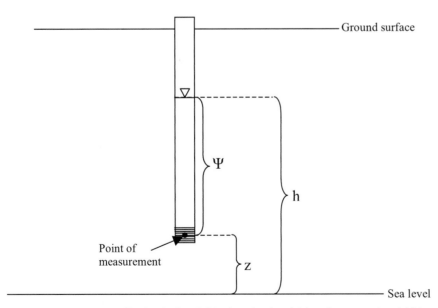

Figure 3.14 Concept of hydraulic head (h), elevation head (z) and pressure head (Ψ) in a piezometer. The cross-hatched section at the bottoms of the tube (terminus) represents the screened interval, which allows water to flow into the piezometer. Water level is denoted by the "$\triangledown$" symbol. From Yolcubal et al., 2004.

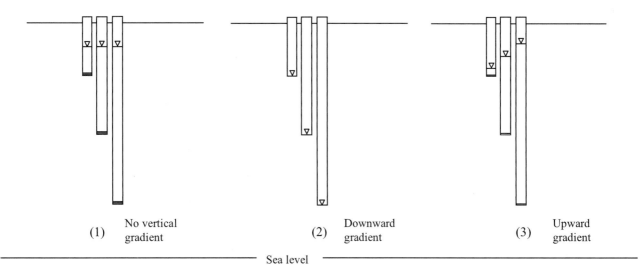

Figure 3.15 Three hypothetical situations for vertical hydraulic gradient water level is denoted by the "∇" symbol. From Yolcubal et al., 2004.

the reference point where z = 0, although some people use the elevation of land surface as the reference datum. The value of Ψ represents the distance between the measurement point and the water level in the well. The value of h represents the elevation of water from the reference datum. This basic hydraulic head relationship is essential to an understanding of groundwater flow. Water flow in porous media always occurs from regions in which hydraulic head is higher to regions in which it is lower.

Water level measurements in a network of production or monitor wells are used to define the potentiometric (or water table) surface of regional ground water. Knowledge of this surface is required to define hydraulic gradients and flow directions. Data from piezometers terminating in depth-wise increments provide information about the vertical flow direction of water in saturated regions (Figure 3.15). In the first (left side) diagram, the water level is uniform, indicating no vertical flow. In the second (middle) diagram, the water level is highest in the shallowest piezometer and lowest in the deepest one, indicating downward flow. The reverse is true for the third (right side) diagram.

3.10.3 Darcy's Law

In 1856, a French hydraulic engineer, Henry Darcy, established a relationship that bears his name to this day. The relationship is based on studies of water flow through columns of sand, similar to the schematic shown in Figure 3.16. In Darcy's experiment, the column is packed with sand and plugged on both ends with stoppers. Water is introduced into the column under pressure through an inlet in the stopper and allowed to flow through it until all the pores are fully saturated with water and inflow and outflow rates are equal. Water pressures along the flow path are measured by the manometers installed at the ends of the column.

In his series of experiments, Darcy studied the relationship between flow rate and the head loss between the inlet and outlet of the column. He found that:

1. The flow rate is proportional to the head loss between the inlet and outlet of the column:

$$Q \propto (H_a - h_b)$$

The flow rate is inversely proportional to the length of flow path:

$$Q \propto \frac{1}{dl}$$

2. The flow rate is proportional to the cross-sectional area of the column:

$$Q \propto A$$

Mathematically, these experimental results can be written as:

$$Q = KA \left[\frac{h_a - h_b}{\Delta l} \right] = KA \left[\frac{\Delta h}{\Delta l} \right]$$

where:

Q = flow rate or discharge [$L^3 T^{-1}$]
A = cross-sectional area of the column [L^2]
$h_{a,b}$ = hydraulic head [L]
dh = head loss between two measurement points [L]
dl = the distance between the measurement locations [L]
dh/dl = hydraulic gradient []
K = proportionality constant or hydraulic conductivity [$L T^{-1}$]

We can rewrite Darcy's Law as:

$$\frac{Q}{A} = q = \frac{K \, dh}{dl}$$

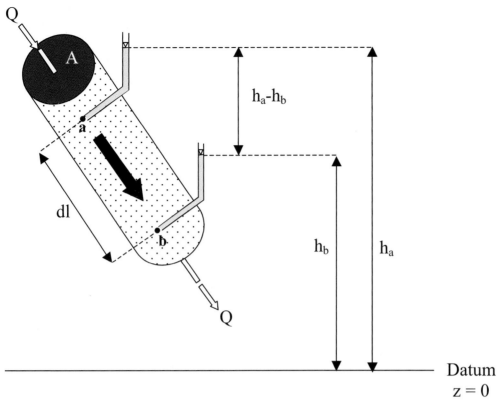

Figure 3.16 Schematic of Darcy's experimental apparatus (original apparatus was vertically oriented). From Yolcubal et al., 2004.

where q is called specific discharge or Darcy velocity with units designated as [L T^{-1}]. This is an apparent velocity because Darcy velocity represents the total discharge over a cross-sectional area of the porous medium. Cross-sectional area includes both void and solid spaces; however, water flow occurs only in the connected pore spaces of the cross sectional area. Therefore, to determine the actual mean water velocity, specific discharge is divided by the porosity of the porous medium: v = q/n, where v is the pore-water velocity or average linear velocity. Pore-water velocity is always greater than Darcy velocity.

3.10.4 Hydraulic Conductivity

The proportionality constant in Darcy's law, which is called **hydraulic conductivity (K)** or **coefficient of permeability**, is a measure of the fluid transmitting capacity of a porous medium and is expressed as:

$$K = \frac{k\rho g}{\mu}$$

where:

   k is the intrinsic or specific permeability [L^2]
   ρ is the fluid density [M L^{-3}]
   g is the acceleration due to gravity [L T^{-2}]
   μ is the dynamic viscosity of the fluid [M T^{-1} L^{-1}]

Hydraulic conductivity has a dimension of velocity [L T^{-1}], and is usually expressed in m s^{-1}, cm s^{-1}, or m day^{-1} in SI units, or ft s^{-1}, ft day^{-1}, or gal day^{-1} ft^{-2}.

As indicated by the equation above, hydraulic conductivity depends on properties of both the fluid and porous medium. The two fluid properties are density and dynamic viscosity. Intrinsic permeability is a property that in most cases depends solely on the physical properties of the porous medium. This relationship can be illustrated using the expression called the **Hazen approximation**:

$$k = C(d_{10})^2$$

where:
   C is a shape factor [dimensionless]
   d_{10} is the effective grain diameter [L]

C is a constant that represents the packing geometry, grain morphology (size and shape), and grain-size distribution of the porous medium. The value of C ranges between 45 for clays and 140 for sand. A value of C = 100 is often used as an average. d_{10} is the diameter for which 10% (by weight) of the sample has grain diameters smaller than that diameter, as determined by sieve analysis. The Hazen approximation is applicable to sand with an effective mean diameter between 0.1 and 3.0 mm. Intrinsic permeability (k) has dimensions of square feet (ft^2), square meter (m^2), or square centimeter (cm^2).

As noted above, porous-media properties that control K include pore size, grain-size distribution, grain geometry, and packing of grains. Among those properties, the influence of grain size on K is dramatic, since K is linearly proportional to the square of grain diameter. The larger the grain diameter, the larger is the hydraulic conductivity. For example, hydraulic conductivity of sands ranges from 10^{-4} to 10^{-1} cm s^{-1}, whereas the hydraulic conductivity of clays ranges from 10^{-9} to 10^{-7} cm s^{-1}. The values of saturated hydraulic conductivity vary by several orders of magnitude, depending on the material. The range of values of hydraulic conductivity and intrinsic permeability for different media is illustrated in Figure 3.17.

In the vadose zone, hydraulic conductivity is not only a function of fluid and media properties, but also the soil-water content (θ), and is described by the following equation:

$$K(\theta) = K \, k_r(\theta)$$

where:

K(θ) is the unsaturated hydraulic conductivity
K is the saturated hydraulic conductivity
$k_r(\theta)$ is the relative permeability or relative hydraulic conductivity

Relative permeability is a dimensionless number that ranges between 0 and 1. The $k_r(\theta)$ term equals 1 when all the pores are fully saturated with water, and equals 0 when the porous medium is dry. Unsaturated hydraulic conductivity is always lower than saturated hydraulic conductivity.

Unsaturated hydraulic conductivity is a function of soil-water content. As the soil-water content decreases, so does K(θ). In fact, a small drop in the soil-water content of a porous medium, depending upon its texture, may result in a dramatic decrease (*e.g.*, 10^3, 10^6) in the unsaturated hydraulic conductivity. As we discussed earlier, the hydraulic conductivity of sands is always greater than that of clays for saturated porous media. However, in the vadose zone, this relationship may not always hold true. For example, during drainage of a soil, larger pores drain first and the residual water remains in the smaller pores. Since sand has larger pores than clay, it will lose a greater proportion of water for a given suction. Consequently, at relatively low soil-water contents (high suctions), most of the pores of a sand will be drained, while many for a clay will remain saturated. Therefore, the unsaturated hydraulic conductivity of a clay unit may become greater than that of a sand unit at lower soil-water contents.

Hydraulic conductivity is a critical piece of information required for evaluating aquifer performance, characterizing contaminated sites for remediation, and determining the fate and transport of contaminant plumes in subsurface environments. For example, for water management issues, one needs to know the hydraulic conductivity to calculate the water-transmitting and storage capacities of the aquifers. For remediation applications, knowledge of K distribution of contaminated soils is necessary for calculating plume velocity and travel time, to determine if the plume may reach a downgradient location of concern. Hydraulic conductivity can be measured in the laboratory as well as in the field. Laboratory measurements are performed on either disturbed or undisturbed samples that are collected in the field. Laboratory measurements are relatively inexpensive, quick, and easy to make compared to field measurements. They are often used to obtain an initial characterization of a site before on-site characterization is initiated. However, measurements made for a sample represent that specific volume of media. A single sample will rarely provide an accurate representation of the field because of the heterogeneity inherent to the subsurface. Thus, a large number of samples would usually be required to characterize the hydraulic conductivity distribution

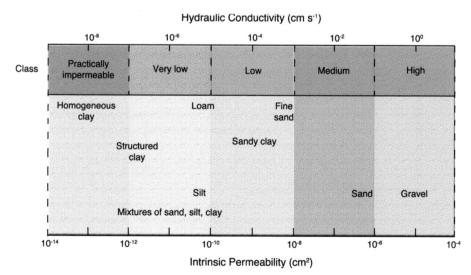

Figure 3.17 Hydraulic conductivity and permeability at saturation. (Adapted from Klute, A., and Dirksen, C. [1982] Methods of Soil Analysis. Part I— Physical and mineralogical methods. Soil Science Society of America, Madison, WI).

present at the site. Thus, field tests, while more expensive, are generally preferred to characterize large sites.

3.11 A WATERSHED APPROACH

Strictly defined, a watershed is a bound hydrologic unit where all drainages flow to a common water source. Referring back to the section on stream order, we can see that a watershed encompasses all stream orders from beginning to end. No matter where one goes on the earth (excluding the oceans), that person would be in a distinct watershed. Watersheds are usually delineated by sharp gradients and precipitation or snowmelt flows along one side of the gradient or the other, depending upon local topography and natural hydrology.

Increasingly, resource managers are using watersheds as the unit of measure for management and delineation purposes on a landscape scale. In hydrologic terms, it makes perfect sense to focus management efforts in these terms, because a land use activity in one part of a watershed often has an effect on downstream areas. The more traditional approach of addressing specific sources of pollution in a specific area has been successful in addressing problems that are readily noticeable; however, this approach has met with limited success in addressing the chronic and subtle stressors that often contribute to impairment within a watershed. A watershed framework is better able to capture these subtle changes over time.

It's important to understand that watersheds, and the organisms they contain, pay no attention to human-imposed political boundaries. Natural physical, chemical, and biological processes within watersheds provide civilization-sustaining benefits when functioning properly. Unfortunately, the same civilization to which watersheds provide these benefits is often the source of impairment and disruption of natural processes within them. Acknowledging this fact while understanding that political boundaries do exist is a necessary prerequisite toward cooperative management of the world's most precious natural resource.

While water within watersheds follows the physicochemical "laws" previously described in this chapter, watershed management is equal parts hydrology and social science. Watershed management acknowledges that humans are a part of, not separated from, watershed processes and functioning. Watershed management, as a component of ecosystem management, tries to unify communities, managers, water quality experts, and as many stakeholders as possible, with the overall goal of increasing or sustaining water quality over large geographic areas for the common good of not only affected communities, but also of the watershed itself.

This ideology is far from revolutionary, and perhaps the best definition of a watershed was by the geologist, explorer, and teacher John Wesley Powell, who stated over 100 years ago that:

"A watershed is that area of land, a bounded hydrologic system within which all living things are inextricably linked by their common water course and where, as humans settled, simple logic demanded that they become part of a community."

QUESTIONS AND PROBLEMS

1. A. Determine pressure head (Ψ), elevation head (z), and hydraulic head (h) for both wells.
   B. Calculate the hydraulic gradient.
   C. Does groundwater flow from well #1 to #2, or from #2 to #1?

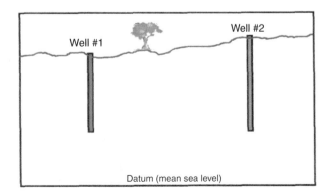

Datum (mean sea level)

The horizontal distance between the wells is 250 meters.
Well 1: The elevation at the ground surface is 100 m; the measured water level in the well is 40 meters below the ground surface. The well casing length is 60 m.

Well 2. The elevation at the ground surface is 115 m; the measured water level in the well is 50 meters below the ground surface. The well casing length is 80 m.

2. Why is hydraulic conductivity a function of the fluid as well as the porous medium? Is the conductivity for water greater than or less than that for air (for a given porous medium)?

3. Water has been described as a universal solvent. Describe three properties of water that account for its ability to dissolve so many types of compounds.

4. Describe some physical, chemical, and biological characteristics within the epilimnion and hypolimnion of a thermally stratified lake.

5. For the temperatures and pressures given below, determine if water will be in a solid, liquid, or gaseous state (or some combination thereof).
   a. 219 ATM and 0.0000098°C
   b. 100 ATM and 100°C
   c. 1 ATM and 373°C
   d. 0.006 ATM and 0.01°C

REFERENCES AND ADDITIONAL READING

Birge E.A. (1898) Plankton studies on Lake Mendota II. The crustacea of the plankton from July, 1894, to December, 1896. *Trans. Wis. Acad. Sci. Arts Lett.* **11**:274–448.

Brezonik P.L. (1994) *Chemical Kinetics and Process Dynamics in Aquatic Systems*. Lewis-CRC Press, Boca Raton, Florida. 754 pp.

Callow P. and Petts G.E. (1992) *The Rivers Handbook*. Vol. 1. Blackwell Sci. Publ., Cambridge. 526 p.

Carlson R.E. (1977) A trophic state index for lakes. *J. Limnol Ocean.* **22**:361–369.

Clesceri L.S., Greenberg A.E. and Eaton A.D. eds. (1999) *Standard Methods for the Examination of Water and Wastewater*. 20th Ed. American Public Health Association, Washington, DC.

Cole G.A. (1994) *Textbook of Limnology*. 4th ed. Waveland Press Inc., Prospect Heights, Illinois.

Edmondson W.T. (1994) What is limnology? In *Limnology Now: A Paradigm of Planetary Problems*, R. Margalef, ed. Elsevier, New York, New York.

Frey D.G. (1963) Wisconsin: The Birge and Juday years. Chapter 1 in *Limnology in North America*, D.G. Frey, ed. University of Wisconsin Press, Madison, Wisconsin. 734 pp.

Horton R.E. (1931) The field, scope, and status of the science of hydrology. pp. 189–202 in *Reports and Papers, Hydrology, Trans. AGU*. National Research Council, Washington, DC.

Hutchinson G.E. (1957) *A Treatise on Limnology*, Vol. I. John Wiley & Sons, New York.

Hynes H.B.N. (1970) The Ecology of Running Waters. University of Toronto Press, Toronto.

Kelly C.A., Rudd J.W.M., Hesslein R.H., Schindler D.W., Dillon P.J., Driscoll C.T., Gherini S.A. and Hecky R.E. (1988) Prediction of biological acid neutralization in acid sensitive lakes. *Biogeochemistry* **3**:129–140.

Klute A. and Dirkson C. (1982) *Methods of Soil Analysis. Part I—Physical and mineralogical methods*. Soil Science Society of America, Madison, Wisconsin.

Mitsch W., and J.G. Gosselink. (1993) *Wetlands*, 2nd ed. Van Nostrand Reinhold, New York, New York. 722 pp.

Reynolds O. (1883) An experimental investigation of the circumstances which determine whether the motion of water shall be direct or sinuous, and of the law of resistance in parallel channels. *Royal Society, Phil. Trans.*

Schlesinger W.H. (1991) *Biogeochemistry: An Analysis of Global Change*. Academic Press, Orlando, Florida.

Streeter H.W. and Phelps E.B. (1925) A study of the pollution and natural purification of the Ohio River. III. Factors concerned in the phenomena of oxidation and reaeration. *U.S. Public Health Serv. Public Health Bull.* 146.

Swain E.B., Engstrom D.R., Brigham M.E., Henning T.A. and Brezonik P.L. (1992) Increasing rates of atmospheric mercury deposition in midcontinental North America. *Science* **257**:784–787.

van der Leeden F., Troise F.L. and Todd D.K. (199) *The Water Encyclopedia*. Lewis Pubs. Chelsea, Michigan.

Vannote R.L., Minshall G.W., Cummins K.W., Sedell J.R. and Cushing C.E. (1980) The river continuum concept. *Can. J. Fish. Aquat. Sci.* **37**:130–137.

Vollenweider R.A. (1975) Input-output models with special reference to the phosphorous loading concept in limnology. *Schweiz. Z. Hydrol.* **37**:53–84.

Wetzel R.G. and Likens G.E. (1991) *Limnological Analyses*. Springer-Verlag, New York, New York.

Williams J.E., Wood C.A. and Dombeck M.P. eds. (1997) *Watershed Restoration: Principles and Practices*. American Fisheries Society, Bethesda, Maryland.

Yolcubal I., Brusseau M.L., Artiola J.F., Wierenga P. and Wilson L.G. (2004). Environmental physical properties and processes. In *Environmental Monitoring and Characterization* (J.F. Artiola, I.L. Pepper, and M. Brusseau, eds.). Elsevier Academic Press, San Diego, California. pp. 207–237.

CHAPTER 4

PHYSICAL-CHEMICAL CHARACTERISTICS OF THE ATMOSPHERE

A.D. Matthias, S.A. Musil, and H.L. Bohn

Wildland fires, such as this one outside of Tucson, Arizona, can contribute particulate pollutants to the atmosphere. *Photo courtesy Janick F. Artiola.*

4.1 CHEMICAL COMPOSITION

By mass and by volume, more than 99% of the atmosphere is made up of nitrogen (N_2), oxygen (O_2), and argon (Ar) gases (Table 4.1). The concentrations of these atmospheric gases, together with neon (Ne), helium (He), and krypton (Kr), have probably been constant for many millions of years and are unlikely to change markedly, either by natural or anthropogenic means.

The trace gas concentrations, on the other hand, are variable. These gases, which are the subject of considerable concern, are listed in Table 4.2. [*Note:* Although water vapor is listed among the variable components, it will not be discussed here.] We know that all of these gases are affected by human activities as well as by reactions with the soil, biosphere, and oceans. But how much change in trace gas concentrations is due to human activity and how much to lesser-known natural causes is still unclear.

Soils serve as both source and sink for virtually all of the gases in Table 4.2. Whether soils function as a source or sink can vary between day and night, with the season, with water content, with cultivation, with fertilization, and with the gas being considered, as can the strength of that function. The oceans and biosphere also fluctuate in their source/sink behavior. This chapter emphasizes the role of soils in controlling trace gas concentrations within the atmosphere.

The concentration[1] of carbon dioxide (CO_2) is about 380 μL L^{-1} as of 2005, which is about 100 μL L^{-1} larger than pre- Industrial Revolution (18th century) concentrations. The CO_2 concentration decreases by a few parts per million each summer because of increased photosynthesis by terrestrial plants. From fall through spring, CO_2 rises because of microbial decomposition of organic matter and plant respiration. The heights of the annual peaks and valleys are buffered by the less seasonal photosynthesis-degradation cycle in the oceans. In the southern hemisphere, the CO_2 peaks and valleys are six months out of phase with those of the northern hemisphere. The amplitude of the annual CO_2 variation in the southern hemisphere is also much smaller because the southern land areas are much smaller and a larger percentage is arid.

We usually attribute the CO_2 increase to combustion of fossil fuels. Fossil-fuel carbon burning results in only about 10% of the CO_2 annually released by the decomposition of organic matter in soils and oceans. Other, less quantifiable CO_2 sources are (1) cultivation of native soils,

which converts some of the soil organic matter into CO_2; (2) agricultural plants, which accumulate less biomass carbon because they are annuals, are less dense, and have a shorter growing season than do native plants; and (3) clearing of forests, which decreases biomass carbon. Because the increase in atmospheric CO_2 amounts to only about half of the total amount of CO_2 produced by fuel consumption alone, we can infer that the capacity of the biosphere, soils, and oceans to buffer changes in atmospheric CO_2 is quite large. The sinks include increased photosynthesis and biomass, increased CO_2 absorption by the oceans, and probably increased soil organic matter.

Another carbon-based gas is methane (CH_4), whose major sources are swamps, natural gas seepage, and termite activity. Methane concentrations in the atmosphere have been increasing over the past several decades. Hydrogen is also liberated in small amounts from wetlands.

Nitrous oxide (N_2O) and ammonia (NH_3) are released from and absorbed by soils naturally, and releases of these gases are higher after fertilization. Nitrous oxide, whose concentration has also been increasing slowly, is a rather unreactive gas that has a long residence time ($\sim$150 years) in the atmosphere. Atmospheric ammonia has also been observed in higher concentrations, especially in industrial regions, where it often takes the form of ammonium sulfate. Ammonia reacts rapidly with soils, plants, and the ocean.

TABLE 4.1 Constant atmospheric components.

GAS	PERCENT BY VOLUME OF DRY AIR	CONCENTRATION (μL L^{-1})
Nitrogen (N_2)	78.1	780,840
Oxygen (O_2)	20.9	209,460
Argon (Ar)	0.9	9,340
Neon + helium + krypton (Ne + He + Kr)	0.002	24

From *Pollution Science*, © 1996, Academic Press, San Diego, CA.

TABLE 4.2 Variable gas concentrations in the atmosphere.

GAS	CONCENTRATION (μL L^{-1})
Water vapor (H_2O)	< 10,000
Carbon dioxide (CO_2)	380
Methane (CH_4)	1.5
Hydrogen (H_2)	0.50
Nitrous oxide (N_2O)	0.31
Ozone (O_3)	0.02
Carbon monoxide (CO)	< 0.05
Ammonia (NH_3)	0.004
Nitrogen dioxide (NO_2)	0.001
Sulfur dioxide (SO_2)	0.001
Nitric oxide (NO)	0.0005
Hydrogen sulfide (H_2S)	0.00005

Adapted from *Pollution Science*, © 1996, Academic Press, San Diego, CA.

---

[1] Various units are used to express gas concentrations in the atmosphere. Here, μL L^{-1} and percentages by mass and volume will be used. The unit μL L^{-1} corresponds to the commonly used unit of ppmv (parts per million by volume). We can also interpret both μL L^{-1} and ppmv as a mol fraction $\times 10^6$, or the number of gas molecules (or atoms) of a component gas in one million air molecules (or atoms) (as we can show through derivation). For air pollutants and regulatory purposes, a mass per volume unit, mg m^{-3}, is also used.

Nitrogen dioxide (NO_2), its dimer N_2O_4, and nitric oxide (NO) (which are often combined and written as NO_x) are produced by combustion of coal and petroleum, as is sulfur dioxide (SO_2). While sulfur is a constituent of coal and oil, the nitrogen oxides are by-products of high-temperature furnaces and internal-combustion engines. In addition, lightning produces NO_x. These gases are highly reactive in air: they rapidly oxidize to nitric and sulfuric acid, which quickly dissolve in water and wash out as **acid rain**. These gases are also absorbed directly from the air by plants and calcareous soils.

Rain is naturally acidic (pH 5–6) because it absorbs atmospheric carbon dioxide to form carbonic acid. Rain becomes even more acidic as it absorbs SO_2 and NO_x and forms nitric and sulfuric acids. The effect of this pollutant absorption is particularly evident in the low pH of rain downwind of the major industrial centers of North America and Europe (see Information Box 4.1). The low pH of rain can have a number of deleterious effects on living organisms in freshwater and terrestrial ecosystems, as most organisms do best in a rather narrow range of pH levels (near neutral, pH 5 to 9). At its worst, acid rain can gradually decrease the pH of the local water and soil to the point of indirectly killing many organisms. The strength of the effect of acid rain is dependent on the overall amount of acid deposited and the buffering capacity (acid-neutralizing capacity) of the soil and the underlying bedrock. For instance, areas with limestone tend to be less sensitive to acid rain because the limestone reacts with the acids to keep pH levels more neutral. Large regions of eastern Canada are strongly affected by acid rain, in part because the parent soil material is granite, which has very little buffering capacity.

Air pollutants, including SO_2 and NO_x, can be transported long distances from their sources by wind. Reducing acid rain requires a regional approach, often including multiple countries. The U.S. and Canada have joint agreements on pollutant controls and emissions trading programs. The United Nations also has developed protocols that are followed by many European countries. As a result, recent studies show that acid deposition has decreased in many parts of

Figure 4.1 Limestone pillar from an old (circa 1200 A.D.) church in central Paris. Acid rain accelerates weathering of building materials such as marble and limestone. Photo courtesy M.A. Crimmins.

North America and Europe due in part to new emissions regulations over the last 20 years. With this decrease, some regions are showing signs of relatively rapid recovery. Other areas continue to acidify, perhaps in part because the local buffering capacity is poor. However, most of the reduced emissions are in sulfur compounds, with NO_x continuing to contribute to acid deposition. Over 85% of SO_2 production in the U.S. in 2002 (see Figure 4.2) was from stationary fuel combustion (power and industrial plants). About 54% of NO_x production in the U.S. was from transportation, which is more difficult to control due to the large number of sources. As newer vehicles with improved catalytic converters become common, there should be some reduction of NO_x emissions. Finally, some areas of the world, such as China, are currently undergoing rapid industrialization and are experiencing an increase in acid rain.

On the global scale, carbon monoxide (CO) is not considered an air pollutant because soil microorganisms adsorb it relatively rapidly and oxidize it to CO_2. In urban areas, however, carbon monoxide can accumulate during rush hour traffic. Long-term exposure to low levels of carbon monoxide can affect cardiovascular health. Exposure to high levels of CO is toxic.

To decrease air pollution, regulatory agencies worldwide have put increasing restrictions on SO_2, NO_x, and organic chemical emissions. London, Pittsburgh, Salt Lake City, Los Angeles, and many other North American and European cities have already shown obvious improvements. Figure 4.3 illustrates smog in Salt Lake City, Utah, during the early 1970s, before strict emissions standards were promulgated. Visibility and air quality in Salt Lake City has improved because of regulatory restrictions on emissions. Unfortunately, in other cities, such as Mexico City, the situation will probably worsen before it improves.

INFORMATION BOX 4.1

Acid Rain Effects on Built Environments

Acid rain can cause significant surface damage to man-made items such as buildings, metals, and glass (see Figure 4.1). The Acropolis in Athens, Greece, is a good example of an ancient structure that has survived earthquakes, wars, and time, only to be severely damaged by acid deposition. SO_2 and sulfuric acid also attack medieval stained glass, which is causing damage at the Canterbury Cathedral in England and the Chartres Cathedral in France.

Ozone (O_3) is considered an air pollutant in the lower atmosphere because it is harmful to plants and humans. Ozone is produced by the action of ultraviolet (UV) sunlight on polluted air that contains nitrogen oxides (NO and NO_2) and organic gases such as industrial solvents, fuels, and partially oxidized hydrocarbons. The ozone concentration has therefore been adopted by the U.S. Environmental Protection Agency (EPA) as an index of air pollution.

In the upper atmosphere, ozone is beneficial to life on Earth in that it absorbs much of the UV fraction of sunlight. The UV fraction of sunlight can cause skin cancer in humans and animals and stunt plant growth. A distinctive pollution problem that increases harmful UV light at the Earth's surface is the **ozone hole**, a seasonal (springtime) decrease in stratospheric ozone measured over the South Pole. The hole was discovered by British scientists in the 1980s. During the 1980s and 1990s the size of the hole generally increased, but recent observations indicates that the size is decreasing (see Figure 23.10).

At higher altitudes, ozone is attacked by chlorine (Cl) and to a lesser extent by NO_x. In the more intense UV of that region, chlorofluorocarbons (CFCs, such as CCl_2F_2) decompose to Cl_2, which degrades O_3 to oxygen. NO_x degrades O_3 as well, but to a lesser extent; N_2O at higher altitudes is converted by UV light to NO_x (see Section 23.2.4.2).

Figure 4.3 Smog covers Salt Lake City in July 1972 prior to increased emissions restrictions. Visibility and air quality have improved because of regulatory restrictions on emissions. Photo courtesy U.S. Environmental Protection Agency.

Hydrogen sulfide is a colorless gas that smells like rotten eggs. There are numerous natural sources of hydrogen sulfide, including thermal springs and swamps. Anthropogenic sources include oil production, pulp and paper mills, municipal sewer plants and large livestock operations. Since the odor can be perceived at levels as low as 10 ppb, in low quantities it can be a nuisance gas. High exposures (>300 ppm) can cause severe respiratory distress, with exposures over 700–800 ppm usually resulting in death. Hydrogen sulfide gas reacts with water vapor to form sulfuric acid, which can contribute to acid rain.

4.2 PHYSICAL PROPERTIES AND STRUCTURE

The ability of the atmosphere to accept, disperse, and remove pollutants is strongly related to its various physical and dynamic properties. Atmospheric winds, for example, determine the pathways and speeds at which pollutants are transported away from sources such as cars and smoke stacks. Another physical process, the condensation of water vapor into rain and fog droplets, scavenges water-soluble pollutants from the atmosphere, ultimately determining the rate of their removal. In addition, the vertical variation of temperature greatly influences atmospheric stability and hence the turbulent mixing of polluted air with clean air. Temperature also affects reaction rates between chemical species, such as those involved in ozone formation in polluted urban environments.

The following sections provide a brief introduction to the physical and dynamic properties of the atmosphere that are most relevant to our understanding of air pollution processes. The purpose here is to gain an overall understanding of air density, pressure, wind, water vapor, precipitation, ra-

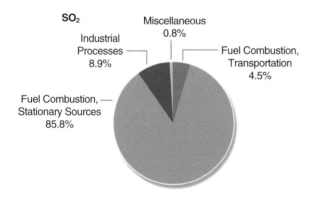

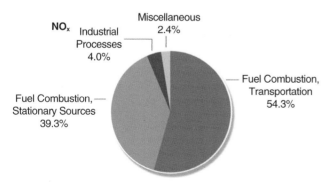

Figure 4.2 Sources of SO_2 and NO_x in the U.S. in 2002. Data: U.S. Environmental Protection Agency.

diation transfer, and temperature. We will not cover several important topics concerning the atmosphere, such as large-scale weather disturbances (e.g., hurricanes) or forecasting weather conditions for air pollution advisories. Interested readers should consult more comprehensive textbooks in atmospheric science for detailed information. Suggested references are listed at the end of this chapter.

4.2.1 Density, Pressure, and Wind

Air is a multicomponent mixture of gaseous molecules and atoms, which are constantly moving about and undergoing frequent collisions. The mass and kinetic energy of each of these moving molecules imparts a force upon collision, which gives rise to **atmospheric pressure.** The horizontal variations of pressure, which result in air flow (winds) across the earth's surface, are an important factor in air pollution dispersal.

Pressure is the force per unit area exerted by air molecules. Usually expressed in units of Newtons per square meter ($N\,m^{-2}$) or Pascals (Pa), pressure is exerted equally in all directions because molecular scale motion is uniformly distributed in all directions. Thus, at any height in the atmosphere, pressure is the cumulative force (weight) per unit area exerted by all molecules above that height. Under static equilibrium conditions, the weight of the atmosphere pushing down on an air parcel at any height is exactly balanced by a pressure gradient force pushing upward. The weight of the atmosphere compresses air molecules near the earth's surface. In fact, nearly two-thirds of all atmospheric molecules are contained within a one scale-height distance of about 8.4 km above the surface. Air density (measured in $kg\,m^{-3}$) and pressure are both highest at sea level, with values of about 1.2 kg m^{-3} and 101.3 kPa (1013 mbar) on average, respectively. From sea level upward, both decrease exponentially with height, as illustrated in Figure 4.4. At heights greater than about 60 km, so few molecules (and atoms and ions) are present that both density and pressure become almost negligible.

Sea-level pressure varies both temporally and spatially across the earth's surface. For example, at any time of day, the pressure at Seattle, Washington may be several millibars higher or lower than the pressure at Miami, Florida, even though both cities are at sea level.

Surface pressure differences (gradients) can result from several factors, such as variations in the heating of air molecules by the sun, fluctuations in atmospheric water vapor and cloud cover, and rotation of the earth. Belts of semipermanent high (H) and low (L) surface pressure circle the earth at various latitudes. These belts are a result of the general circulation of the atmosphere, as illustrated simplistically in Figure 4.5a. Large-scale circulation is composed, on average, of six main convective cells (three each in the northern and the southern hemispheres). The two main large equatorial cells shown in Figure 4.5a are known as **Hadley cells**. The six convective cells result primarily from the heating of the earth's surface by sunlight. More sunlight is absorbed per unit area over the equator than at higher latitudes, thus heated air ascends over the equator and cooler air

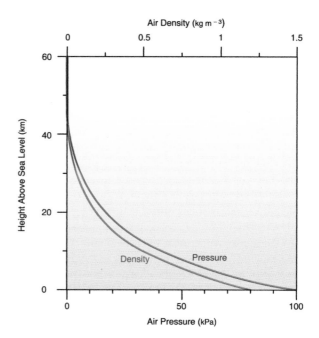

Figure 4.4 Approximate density and pressure variations with altitude in the earth's atmosphere.

descends at higher latitudes. If the earth did not spin on its axis of rotation, there would likely be only one large convective cell in the northern hemisphere and one large cell in the southern hemisphere—extending between the equator and the poles. There are six cells instead of two cells, however, because of the earth's spin on its axis, which deflects the winds. The six convective cells shown in Figure 4.5a are important because they carry heat away from the warm equatorial region toward the poles and they transport air pollutants long distances over the earth's surface.

Low surface pressure results when warm, moist, buoyant air, such as that over the equator, ascends from the earth's surface. Moist air rises above the equator because the temperature is relatively high and because moist air, containing perhaps 2–4% water vapor, is less dense (and lighter) than dry air. The relatively low molecular weight of water vapor (18) relative to dry air (29) lowers the average molecular weight of moist air. The lighter air moves upward and exerts relatively less pressure at the surface.

Surface flows of moist air associated with the trade winds converge at the **intertropical convergence zone (ITCZ)** near the equator (see Figure 4.5a). As the flows converge from the northern and southern hemispheres, the air is heated by the equatorial sun and thus rises. As the air rises, it cools and loses its moisture by condensation and precipitation. At high altitudes the rising air current diverges northward and southward. At subtropical latitudes (about 30 degrees north and south), the dry upper-level air flow subsides (sinks) toward the (mostly) ocean surfaces where it again becomes moist and flows back to the ITCZ. The subsiding air compresses (and thus warms) the atmosphere and increases pressure. Subsidence may occur at a rate of about 1 km per day at subtropical latitudes. High pressure is therefore associated with relatively warm, dry, subsiding air. This process has important implications for dispersal of air pollutants.

Surface wind patterns associated with atmospheric circulation are more complex than the simple idealized flow patterns shown in Figure 4.5. In general, surface winds are influenced by several forces acting on air masses, including pressure-gradient (flow from high to low pressure), Coriolis (deflection of air flow to the right in the northern hemisphere due to the rotation of the earth), frictional, and centrifugal forces. In the northern hemisphere, these forces combine to cause air to flow counterclockwise around low pressure and clockwise around high pressure.

Flow around low pressure is called **cyclonic flow.** Flow around high pressure is termed **anti-cyclonic flow.** At low latitudes in the northern hemisphere, prevailing surface **trade winds** are generally from northeast to southwest. At midlatitudes in the northern hemisphere, prevailing surface winds are generally from southwest to northeast—the **westerlies** shown in Figure 4.5a. At high latitudes over the Arctic, air flow is generally northeast to southwest. Because of convergence at the ITCZ, winds tend to be light over the equator.

The surface wind patterns shown in Figure 4.5a are also illustrated by the black arrows on the world map in Figure 4.5b. Frictional drag between the winds and the ocean surface tends to cause the ocean currents (red arrows in Figure 4.5b) to generally follow the wind patterns. Note for example in Figure 4.5b how the westerly winds off the east coast of North America coincide with the direction of the Gulf Stream, which brings warm tropical ocean water to northern Europe. Although the British Islands, for example, are at a relatively high latitude range ($\sim$ 50 to 60 degrees north), the Gulf Stream helps moderate the climate there.

The general circulation and resultant spatial pattern of high and low pressure influence long-range pollutant transport such as dust (see Figure 4.6). In regions with semipermanent high pressure features (such as those within the subtropical high-pressure belt at about 30 degrees latitude), calm, stagnant conditions often persist for long periods, thereby amplifying air pollutant concentrations near the surface. For example, air subsidence associated with high pressure over the eastern Pacific Ocean (see Figure 4.5a) markedly influences air quality in the coastal cities of California. Similarly, high pressure over the southwestern United States also adversely affects air quality in the region, particularly over large urban areas such as Phoenix, Arizona. Similar high pressures can cause air quality problems over Cairo, Egypt.

4.2.2 Temperature

Temperature is a measure of the kinetic energy (heat content) of molecules and atoms. Air temperature affects nearly all physical, chemical, and biological processes within the earth-atmosphere system. A good example is the influence it has on the atmosphere of polluted urban environments, where high temperature greatly increases the rate of photochemical smog formation. Furthermore, once smog is formed, it may be dispersed upward and downward by buoyancy-generated atmospheric turbulence resulting from temperature (and hence density) differences between individual air parcels and their surrounding environment. We can see the effects of buoyancy on air motion by watching the erratic motion of a helium-filled balloon once it is released into the atmosphere.

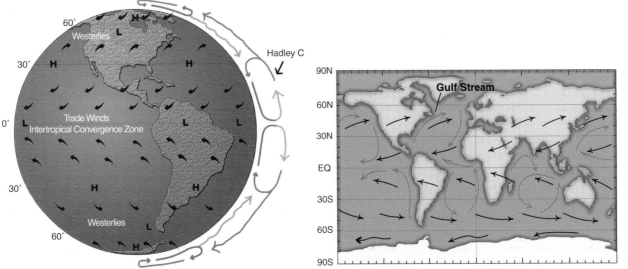

Figure 4.5 (A) General circulation of the earth's atmosphere showing large-scale convective cells, predominant surface wind directions, and long-term surface pressure features. H = high pressure, L = low pressure. Black arrows indicate the direction of surface winds. (B) General surface wind patterns (black arrows) and ocean currents (red arrows). From *Pollution Science,* © 1996, Academic Press, San Diego, CA.

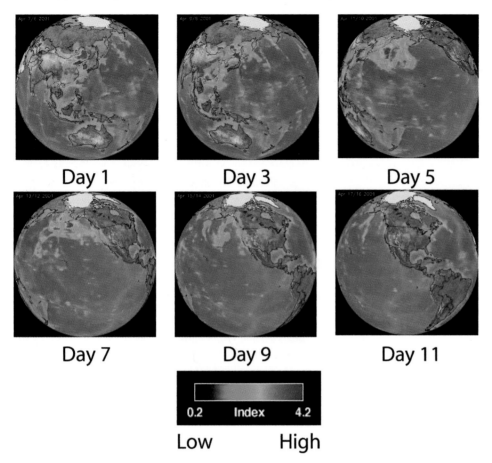

Figure 4.6 A large dust storm moving from east Asia across North America in April 2001. Note the storm is intense enough to carry aerosol particles from China to northeastern North America in 11 days. Atmospheric circulation can move air pollutants on a global scale. Data from the Earth Probe TOMS (Total Ozone Mapping Spectrometer). (Images: U.S. National Aeronautics and Space Administration)

Air temperature near the earth's surface varies markedly over different time scales, ranging from seconds to years. By midsummer, for example, air temperature at 2 m above the Sonoran Desert floor in Arizona may vary diurnally from about 45°C maximum (at midafternoon) to 20°C minimum (at dawn). In midwinter the daily variation in the desert may range about 10–30°C. Temporal variations of air temperature are caused mainly by varying solar energy input to the surface.

Air temperature at a given height, say, 2 m, also varies markedly across the earth's surface owing to spatial variations in energy input. Obviously, the lowest temperatures occur in the polar regions where solar energy input per unit surface area is small. The highest temperatures occur in low-latitude deserts, such as the Sahara in Africa, where solar energy input per unit surface area is very large and little water is available for evaporative cooling of the ground.

The question of how air temperature changes with increasing height above the ground is important when considering how air pollution is dispersed near the ground. To answer this question, we must recognize that energy exchange takes place almost continuously between the surface and the atmosphere. Some heat exchange occurs by **conduction** through a very thin layer of air over the surface; however, most heat exchange occurs by means of **convection,** which is the turbulent exchange caused by buoyancy and shear stress. Convection becomes increasingly more efficient with increasing height. This increase is due to the decreased effect of surface frictional drag at greater heights. At midday the change in temperature with height above the surface is often very large. Temperature gradients within the first few millimeters above a hot desert soil surface may be as high as -1°C per millimeter. Note that a negative temperature gradient means that air temperature decreases with increasing height from the ground surface. Because convection quickly becomes very efficient in mixing air with increased height, the temperature gradients within turbulent air rapidly decrease in magnitude with increasing height. It is important to remember that air in contact with the earth's surface during daylight hours is generally warmer than air aloft because of strong surface heating. Thus, air temperature generally decreases with increased height within the lower part of the atmosphere during the daytime.

Atmospheric scientists use the term **lapse rate** to describe the observed change (generally a decrease) of air temperature with height $\Delta T/\Delta z$). The lapse rate at a given height (z, in meters) and location may vary greatly throughout the day in response to changes in heat flow between the surface and the atmosphere. At a height of 2 m at midday above a hot desert soil surface, for example, the lapse rate may range from about -0.01 to $-0.2°C$ per meter. Within the first few kilometers of the lower atmosphere however the lapse rate is, on average, about $-0.0065°C$ per meter.

During the night, however, the situation is generally the reverse of daytime conditions. The ground surface may quickly lose energy to space by infrared radiation emission (see Section 4.2.4) and become relatively cool. This cooling process also cools the air in direct contact with the surface. Thus at night, air temperature often increases with increasing height, typically on the order of 0.1 to 1°C per meter. An **air temperature inversion** occurs when temperature increases with height up to a level (called the **inversion height**) of maximum air temperature. Above the inversion height, the temperature decreases with height. Radiation inversions are particularly common in the dry desert environment of the southwestern United States and northeastern Africa, where nocturnal loss of radiant energy from the ground to space causes cooling of the air in contact with the ground. Inversions can also occur as a result of subsidence associated with anti-cyclonic flow, which is also common over the southwestern United States. (These and other causes of inversions are discussed further in Section 23.4.1.) As discussed in the following paragraphs, the stable atmospheric conditions associated with inversions tend to trap pollutants near their source. The stability of air defines its ability to mix and disperse pollutants. Air can be unstable, stable, or neutral.

Unstable air results in turbulent motion associated with free convection due to buoyancy within the atmosphere. Buoyant motion enhances upward penetration of air parcels into the atmosphere, thus helping to disperse pollutants. Under unstable conditions, an air parcel that is displaced adiabatically (without heat exchange with its surroundings) upward or downward a short distance is accelerated away from its initial position by buoyancy. Air is unstable because the net buoyancy force acting on the parcel accelerates it either upward or downward, depending upon the temperature (density) difference between the parcel and its surrounding environment. Unstable conditions are prevalent during daytime when convection carries heat upward from the soil surface.

In **stable air**, turbulence is suppressed or even absent. In stable conditions, buoyancy tends to restore an adiabatically displaced parcel to its original height. In other words, the buoyancy force acts in the direction opposite to the motion of the displaced parcel. Stable conditions occur most often at night, when convective heat flow is downward from the atmosphere to the soil surface.

Neutral stability means that the buoyancy force is zero and that a balance exists between gravity (acting downward) and the pressure gradient force (acting upward) on the parcel. The pressure gradient force is the difference between the pressures at the top and bottom of the parcel divided by the distance between top and bottom. Thus, under neutral conditions an air parcel displaced upward or downward from its initial height remains at its new height unless acted upon by an external force. Neutral conditions often occur briefly after sunrise, and before sunset, when convective heat flow is zero. Cloudy, windy days are also favorable for neutral stability.

We base the assessment of the pollutant-dispersal ability of the atmosphere on quantification of the stability of the atmosphere. Stability is largely determined by the value of the measured lapse rate $\Delta T/\Delta z$ relative to the adiabatic lapse rate (Γ). The constant Γ is defined as the change in the temperature of the air parcel when the parcel is displaced upward or downward adiabatically from a base height z_b (see Figure 4.7). This change in temperature results from a change in pressure, as described by the ideal gas law. When the atmosphere is relatively dry, Γ is equal to $-g/c_p$

where:
  g is the acceleration due to gravity,
  c_p is the specific heat of air at constant pressure

The Γ thus has a value of about $-0.01°C$ per meter. This means that the temperature (T_b) of an air parcel adiabatically lifted from height z_b will decrease by 0.01°C per meter of displacement. Likewise, if the parcel is lowered from z_b its temperature will increase by 0.01°C per meter of displacement.

- When $\Delta T/\Delta z < \Gamma$ (e.g., $-0.05°C$ per meter is less than $-0.01°C$ per meter), the atmosphere is unstable, as shown in Figure 4.7 *(center)*. It is unstable because a parcel adiabatically displaced upward from its initial position at z_b is always warmer than its surroundings. When it is adiabatically displaced downward, it is cooler than its surroundings. Thus, it may be accelerated up or down from z_b by the buoyancy force, causing turbulence.

- When $\Delta T/\Delta z > \Gamma$, conditions are said to be stable (Figure 4.7, *bottom*). During stable conditions, a parcel adiabatically displaced upward from z_b becomes cooler than its surroundings. When it is displaced downward, it becomes warmer than its surroundings. Thus, buoyancy restores the parcel to z_b, suppressing turbulent motion.

- When $\Delta T/\Delta z = \Gamma$, neutral conditions are present (Figure 4.7, *top*), and the buoyancy force acting on the parcel is zero at any height.

The criteria for characterizing stability are summarized as follows:

  unstable conditions: $\Delta T/\Delta z < \Gamma$
  stable conditions: $\Delta T/\Delta z > \Gamma$
  neutral conditions: $\Delta T/\Delta z = \Gamma$.

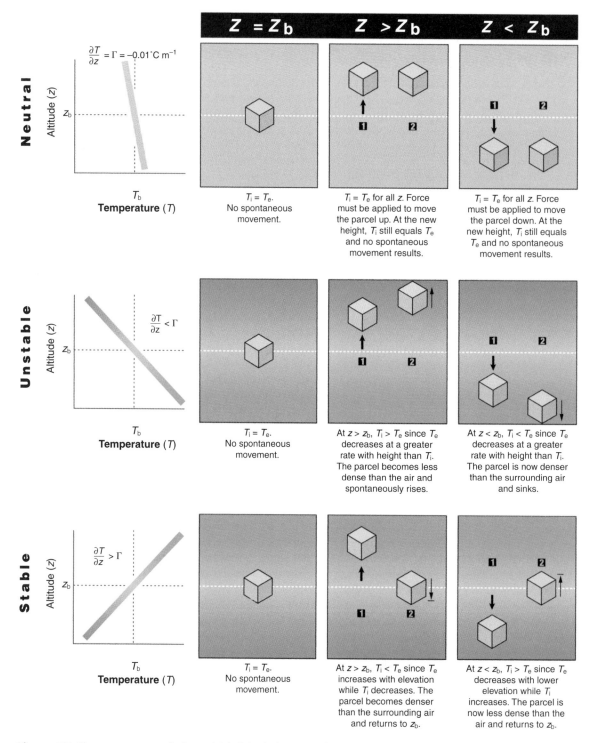

Figure 4.7 Air-temperature variations with height during neutral, unstable, and stable conditions.
From *Pollution Science*, © 1996, Academic Press, San Diego, CA.

4.2.3 Water Vapor and Precipitation

Water vapor is a highly variable part of the atmosphere. In warm, humid, tropical rain forests, high rates of evaporation of water from the earth's surface keep the lower atmosphere almost continuously saturated. On the other hand, in dry, hot deserts, there is usually little water to evaporate, and the amount of water vapor in the atmosphere is almost negligible.

Atmospheric water vapor is characterized by various parameters, including vapor pressure, relative humidity, dew point temperature, water vapor density, and specific humidity. **Relative humidity** is probably the most familiar. It is defined as the ratio of the actual vapor pressure to the saturation vapor pressure of the air, which is solely a function of air temperature.

Condensation of water vapor into cloud and fog droplets occurs when air is cooled to saturation at the **dew point temperature.** Cooling occurs by various processes, such as the radiational cooling of the surface at night, the upward convective movement, advective motion of the atmosphere (in which a cold air front displaces warm moist air upward), and orographic lifting (in which air rises over mountain ranges).

Precipitation in the form of rain or snow rids the air of many types of particulate matter and gaseous pollutants. This removal by scavenging is known as **wet deposition.** Pollutants may dissolve directly in the water droplets, or they may be adsorbed on the droplets. Soluble pollutants include, among other compounds, dioxides of sulfur and nitrogen. Removal of these pollutants increases the acidity of precipitation, producing acid rain.

4.2.4 Radiative Transfer

The transfer of energy within the earth-atmosphere system by electromagnetic radiation (light) is very important for maintaining the climate of the earth. The energy associated with electromagnetic radiation is proportional to the wavelength of the light waves. Shorter wavelength light, such as the visible light produced by the sun, has higher energy per unit wavelength than longer wavelength light, such as infrared radiation emitted by the ground and the earth's atmosphere.

The radiation environment of the earth is largely a function of three main radiative transfer processes:

1. The flux of radiant energy reaching the earth's surface from the sun,
2. The redistribution of the radiant energy between the ground and the atmosphere,
3. The loss of radiant energy to space.

The overall radiation balance of the earth-atmosphere system involves absorption, scattering, transmission, and emission processes, which are described briefly in this section. Two wavelength intervals of the electromagnetic spectrum are of primary importance to the overall radiation balance: shortwave (solar) and longwave radiation. Shortwave consists of the wavelength interval from about 0.15–3.0 μm. Within this portion are the components of **ultraviolet** (0.15–0.36 μm), **visible** (0.36–0.75 μm), and **near infrared** (0.75–3 μm) radiation. Shortwave energy is emitted by the sun, which is an almost perfect blackbody radiator with a temperature of about 6000 K. A perfect blackbody radiator emits the maximum possible radiant energy per unit wavelength at a given temperature. Ultraviolet and visible light are the shortwave components most significant to the global environment.

On a per-unit-wavelength basis, ultraviolet (UV) is very high-energy radiation. Fortunately, most of the high-energy UV wavelengths are selectively absorbed by ozone, oxygen, and other constituents in the earth's upper atmosphere. Some UV, however, reaches the earth's surface, where it can be harmful to life and contributes to the production of photochemical smog in the lower atmosphere.

Most shortwave radiation is within the visible portion of the spectrum. In fact, the wavelength of maximum energy flux from the sun is at 0.48 μm, which is visible to the human eye as green light. Most visible light from the sun passes through clear air without significant loss. Scattering of visible light by atmospheric molecules, clouds, and aerosols does occur, however. Visible light reaching the land surface is either absorbed (about 75%) or reflected (about 25%) by surface matter (*e.g.*, plants, water, and soil). The absorbed radiant energy heats the soil and air and evaporates water. Some of the absorbed energy is re-emitted back to the atmosphere in the form of longwave radiation.

Longwave radiation encompasses the spectrum from about 3.0–100 μm and is emitted by matter within the earth-atmosphere system. Since terrestrial absolute temperatures are about 288 K (kelvin) (~15°C), the wavelength of maximum longwave emission is about 10 μm. Long wave radiation is commonly referred to as terrestrial or **infrared** radiation.

The earth's atmosphere is largely opaque to most of the longwave spectrum. A window exists, however, between about 8 and 11 μm that permits escape of a portion of longwave energy to space. Absorption and re-emission of longwave energy within the atmosphere occur within the vibrational energy mode of various molecular species, including water vapor, carbon dioxide, nitrous oxide, and methane. These and a few other species are the well-known "greenhouse" gases, and are responsible for the earth's **greenhouse effect.**

Without the warming by the natural greenhouse effect, the earth surface would be about 33°C colder than its present mean temperature of about 15°C. There is, however, much public and scientific concern that an additional 1.4–5.8°C global warming will occur by the end of this century owing to gradual accumulation of atmospheric greenhouse gases from anthropogenic sources.

The major greenhouse gas is water vapor because of its high concentration relative to the other greenhouse gases. Dust and aerosol clouds also affect radiative transfer. The enormous Tambora volcanic explosion April 10–11, 1815, on the island of Sumbawa in Indonesia caused the "year without a summer" in 1816 and crop failures in the northern hemisphere due to cooling by dust in

---

INFORMATION BOX 4.2

Effects of a Large Volcanic Explosion

The Tambora explosion of 1815 is an example of how dust and aerosols from a volcano can affect the earth's climate. The eruption of the Tambora volcano on the island of Sumbawa in Indonesia during April 10–11, 1815, took the lives of about 92,000 people. It was a "supercolossal" eruption that rated a 7 out of 8 (8 being the largest) on the Volcanic Explosive Index scale. It was the largest volcanic eruption in the past 500 years. The energy of the eruption (equivalent to about 20,000 MTons of TNT, which is far larger than the largest nuclear bomb ever detonated (58 MTons TNT)), injected about 150 km^3 (~36 $miles^3$) of ash and dust presumably more than 25 km into the stratosphere that darkened the earth for many days. Dust and aerosols from the volcano remained suspended in the atmosphere for several months and ultimately significantly cooled the earth on average by ~0.7°C by reflecting sunlight back to space. The cooling led to "the year without a summer," causing crop failure throughout North America and Northern Europe in 1816. The anomalous cold weather brought snow every month of the year to New England. The crop failures there may also have caused an increased migration of people from the New England states to the new territories in the Midwest.

---

shear forces (forced convection). Within the first few meters above the ground surface, vertical gradients of air temperature, wind speed, humidity, and other scalar quantities are often large and variable with time. These gradients are due mainly to the temporal variability of energy and mass exchanges (e.g., evaporation of water) between the surface and the atmosphere. The depth of the boundary layer varies over the course of the day. By midafternoon, rising air from the heated ground may extend the boundary layer up to the 1 km height. This height is often referred to as the **mixing depth or mixed layer.** The turbulent air parcels, often referred to as **eddies,** undergo eddying motion. By night however, the atmosphere cools, and the boundary may shrink to a thickness of only about 0.1 km. Most of the important atmospheric pollutant transport and transformations occur within the boundary layer. However, some chemically stable gases are dispersed upward throughout much of the troposphere. Some, such as N_2O and CFCs, eventually diffuse upward into the stratosphere.

At the very bottom of the boundary layer, directly above the earth's surface, is a sublayer known as the **surface**

the atmosphere (see Information Box 4.2). However, some volcanos can cause cooling by injecting large amounts of SO_2, in addition to or in place of dust, into the upper atmosphere. The 1991 eruption of Mt. Pinatubo is a good example (see Figure 4.8).

4.2.5 Lower Atmosphere

Temperature variation with height defines the various layers of the atmosphere. The major atmospheric layers are shown in Figure 4.9. The **troposphere** is the lowest major layer. Within the troposphere, vertical variation of temperature is characterized by lapse-rate conditions; thus the troposphere is generally unstable and well mixed. The troposphere extends upward from the surface to a height of about 10–15 km, depending upon latitude and season of the year. The troposphere is certainly familiar to us, since it is tropospheric air that we breathe. Also, most of our weather occurs in the troposphere, including cloud formation, rain, winds, and other meteorological processes. The **tropopause** is the upper limit of the troposphere, which separates the troposphere from the stratosphere above.

The **atmospheric boundary layer,** which is an important sublayer at the bottom of the troposphere, forms the atmospheric interface between the troposphere and the ground surface. In this region of the atmosphere, airflow patterns are strongly affected by buoyancy (free convection) and surface

Figure 4.8 A major eruption of Mount Pinatubo, Philippines, in June 1991. The eruption sent a cloud of ash and gases into the stratosphere that circled the world multiple times. It is estimated that the cloud cooled annual temperatures in some regions by as much as 0.5 °C. Photograph: Karin Jackson, courtesy of the U.S. Geological Survey.

layer. This sublayer generally extends upward to about one-tenth of the boundary-layer depth. The properties of the surface layer are most directly affected by surface roughness and surface heat exchange. Energy and mass fluxes are nearly constant with height in the surface layer; thus this sublayer is sometimes called the **constant flux layer.**

4.2.6 Upper Atmosphere

The **stratosphere** is the stable (stratified) layer of atmosphere extending from the tropopause upward to a height of about 50 km (Figure 4.9). The stratosphere is highly stable because the air temperature increases with height up to the stratopause, which is the height of the temperature inversion. The increased temperature in this layer is due mainly to UV absorption by various chemical species, including ozone and molecular oxygen present in the stratosphere. Maximum heating takes place in the upper part of the stratosphere. Because of the stable air, pollutant mixing is suppressed within this layer. Thus, natural (*e.g.,* N_2O) and synthetic (*e.g.,* CFC) chemicals that reach the stratosphere from the troposphere tend to diffuse upward very slowly within the stratosphere.

Ozone is formed naturally and photochemically within the stratosphere. Ozone is considered a pollutant in the troposphere, but in the stratosphere it is essential to life on earth because it absorbs biologically harmful UV radiation.

The **mesosphere** and the **thermosphere** are two additional atmospheric layers above the stratosphere. These layers are largely decoupled from the stratosphere and troposphere below; therefore, they exert little influence on our weather and on pollutant transport processes. Likewise, pollution has little or no effect upon these two upper layers.

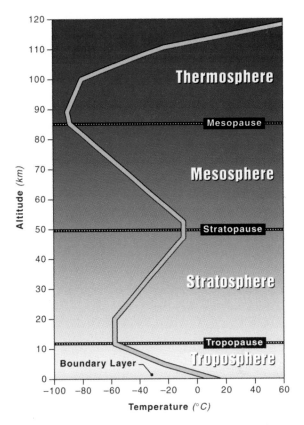

Figure 4.9 The structure of the atmosphere as defined by average variation of temperature with altitude. From *Pollution Science,* © 1996, Academic Press, San Diego, CA.

QUESTIONS AND PROBLEMS

1. What causes wind? How does wind affect the movement of heat, water vapor, and pollution in the atmosphere?

2. How can surface pressures differ between Venice and New York even though both cities are at sea level?

3. Describe how and why air temperature varies with increasing height in the troposphere and stratosphere.

4. Why is air generally well-mixed in the troposphere but not in the stratosphere?

5. The dry adiabatic lapse rate of the atmosphere is given by $\Gamma = g/c_p$ where g is acceleration due to gravity (9.8 m s^{-2}) and c_p is the specific heat of air (1010 J kg^{-1} K^{-1}). Calculate the actual numerical value of Γ and show that its units are K m^{-1}. [Hint: Recall that Joules (*J*) of energy can be represented in terms of kinetic energy (kg m^2 s^{-2})].

6. Suppose a small parcel of air is lifted adiabatically (*i.e.,* it is insulated from its surroundings) from the ground upward to a height of 100 m. How much cooler or warmer will the parcel be at 100 m than at the ground surface?

7. Suppose the small parcel of air in question #6 is 1 degree Celsius warmer than the air surrounding it at 100 m height. Will the parcel move upward, downward, or remain stationary at 100 m? Is the atmosphere unstable or stable at 100 m height? Explain.

8. The concentration of atmospheric carbon dioxide is increasing about 1.5 µL L^{-1} per year. If the current rate of increase continues, when (at about what year) will its concentration be double the pre-Industrial Revolution concentration of 280 µL L^{-1}?

9. Why is there environmental concern about the increasing concentration of carbon dioxide in the atmosphere? What is causing the increase?

REFERENCES AND ADDITIONAL READING

Ahrens C.D. (2003) *Meteorology Today An Introduction to Weather, Climate and the Environment.* 7th Edition. Brooks/Cole-Thomson Learning Inc. Pacific Grove, California.

Mitchell J.F.B. (1989) The greenhouse effect and climate change. *Rev. Geophys.* **27**, 115–139.

CHAPTER 5

BIOTIC CHARACTERISTICS OF THE ENVIRONMENT

I.L. Pepper and K.L. Josephson

The environment can be the whole world. However, even a few grams of soil on a microscope slide contains billions of organisms. *Photo courtesy K.L. Josephson.*

5.1 MAJOR GROUPS OF MICROBES

Microorganisms in the environment are ubiquitous and diverse in origin. Microorganisms in the environment are also fundamentally different from laboratory-maintained or clinical isolates of microbes, because they are adapted to harsh and often widely fluctuating environments. The smallest organisms are the viruses, which do not carry out metabolic reactions and thus require a host for self-replication. Viruses are unique in that they consist solely of nucleic acids and proteins, and are not technically viable living organisms. Although new entities are currently being discovered, such as infectious proteins known as prions, generally all biological entities are characterized by the presence of nucleic acids. In environmental microbiology, we can also categorize microbes as prokaryotes or eukaryotes, both of which clearly affect human health and welfare, and are essential for maintaining life as we know it (Figure 5.1). Bacteria and actinomycetes are prokaryotic. Larger and more complex organisms include the eukaryotic fungi, algae, and protozoa.

Viruses are significant because of their ability to infect other living organisms and cause disease. Bacteria can also cause infections, but are also important because of their ability to transform biochemicals. Fungi are also involved in biochemical transformations. Algae affect surface water pollution, and can also produce microbial toxins, but their overall impact in the environmental microbiology arena is not as significant as that of the bacteria and fungi. Finally, note that the protozoa are significant sources of pathogens that directly affect human health. A classic example of this would be the outbreak of *Cryptosporidium* contamination in potable water supplies that resulted in several deaths in 1993 in Milwaukee, Wisconsin. Protozoa are also key players in surface environments as grazers of bacteria, helping to maintain a steady ecological balance.

In this chapter the basic structure and function of key environmental microbes will be described. We will focus primarily on viruses, bacteria, and fungi because of their importance and relevance to environmental issues. The objective of this chapter is to enable a reader without a microbiological background to understand the characteristics of these microbes, their roles in the environment, and their requirements to function successfully in the environment. Special emphasis is placed on the bacteria because of their multiple roles in environmental microbiology. The intent of the chapter is not to establish a treatise on each organism but rather to provide a basic framework of information relevant to subsequent chapters that deal with the effects of these organisms in environmental microbiology. A massive amount of detailed information is available in the literature on each of the types of organisms covered in this chapter.

Microorganisms in the environment are represented by diverse populations and communities. Microscopic organisms, including viruses, bacteria, fungi, algae, and protozoa, are too small to be seen with the naked eye but large enough to be studied under a microscope. However, there are certain genera of fungi, algae, and protozoa that are macroscopic in nature, and therefore only viruses and bacteria are totally microscopic (Figure 5.2).

Viruses are the smallest microorganisms, and are considered to be ultramicroscopic. **Phage** are viruses that infect bacteria. As defined, viruses are submicroscopic, parasitic, nonfilterable agents consisting primarily and simply of nucleic acid material surrounded by a protein coat. Electron microscopes are required to visualize viruses because their size is below the resolution capacity of light microscopes. Ranging

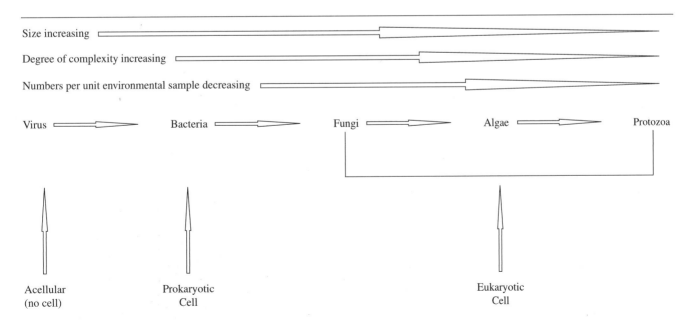

Figure 5.1 Scope and diversity of microbes found in the environment.

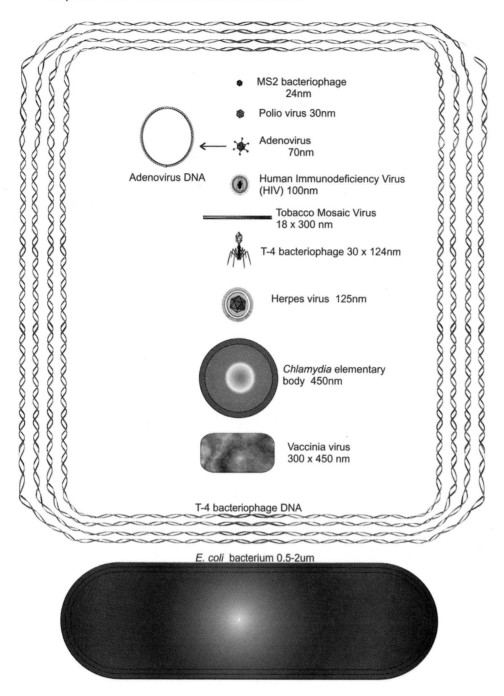

MS2 bacteriophage
24nm

Polio virus 30nm

Adenovirus
70nm

Adenovirus DNA

Human Immunodeficiency Virus
(HIV) 100nm

Tobacco Mosaic Virus
18 x 300 nm

T-4 bacteriophage 30 x 124nm

Herpes virus 125nm

Chlamydia elementary
body 450nm

Vaccinia virus
300 x 450 nm

T-4 bacteriophage DNA

E. coli bacterium 0.5-2um

Figure 5.2 Comparative sizes of selected bacteria, viruses, and nucleic acids. From *Environmental Microbiology* © 2000, Academic Press, San Diego, CA.

in size from 20 to several hundred nanometers, virus particles are very simple in organization, and normally consist of only an inner nucleic acid genome, an outer protein capsid, and sometimes an additional membrane envelope (Figure 5.3). Unlike other microorganisms, viruses do not necessarily fulfill the requirements for being alive as they have no ribosomes and do not metabolize. Viruses must infect and replicate inside living cells. For example, noroviruses can be transmitted via ingestion of contaminated water. When a human ingests contaminated water, noroviruses enter the intestines, attach,

infect, and replicate in intestinal cells, ultimately causing symptoms of gastroenteritis.

Bacteria are very small, relatively simple, single-celled organisms whose genetic material is not enclosed in a nuclear membrane (Figures 5.4A and 5.4B). Based upon this cellular organization bacteria are classified as prokaryotes and include the eubacteria and the Archaea. Although they are classified as bacteria, special mention should be made of the **actinomycetes**. These prokaryotic microbes consist of long chains of single cells, which allow them to exist in a

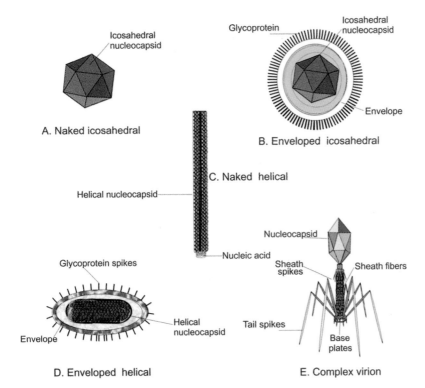

Figure 5.3 Simple forms of viruses and their components. The naked icosahedral viruses (A) resemble small crystals; the enveloped icosahedral viruses (B) are made up of icosahedral nucleocapsids surrounded by the envelope; naked helical viruses (C) resemble rods with a fine, regular helical pattern in their surface; enveloped helical viruses (D) are helical nucleocapsids surrounded by the envelope; and complex viruses (E) are mixtures of helical and icosahedral and other structural shapes. From *Environmental Microbiology* © 2000, Academic Press, San Diego, CA.

filamentous form. Structurally, from the exterior, actinomycetes resemble miniature fungi, and this is the reason why they are often reported as a specialized subgroup of bacteria. Actinomycetes are important producers of antibiotics such as streptomycin, are important in the biodegradation of complex organics, and also produce geosmin, which is the compound that gives soil its characteristic odor. Geosmin can also result in taste and odor problems in potable waters (see also Chapter 28). Prokaryotes lack developed internal structures, especially a membrane-enclosed nucleus, and

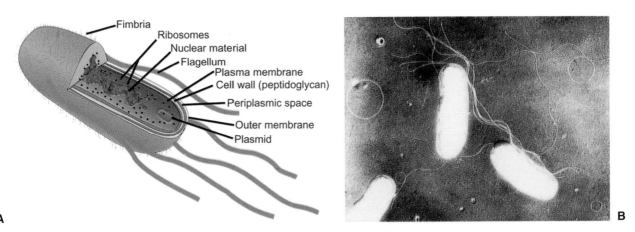

Figure 5.4 (A) Schematic representation of a typical bacterial cell. (B) Scanning electron micrograph of a soil bacterium with multiple flagella. The circles are detached flagella that have spontaneously assumed the shape of a circle. From *Environmental Microbiology* © 2000, Academic Press, San Diego, CA.

thus are distinguished from more complex microorganisms such as fungi. Also lacking are internal cell membranes and complex internal cell organelles involved in growth, nutrition, or metabolism. Bacteria are the most ubiquitous organisms and can be found in even the most extreme environments, as they have evolved a capacity for rapid growth, metabolism, and reproduction, as well as the ability to use a diverse range of organic and inorganic substances as carbon and energy sources. Bacteria are considered to be the simplest form of life, because viruses are not considered to be alive. A single gram of soil can contain up to 10^{10} bacteria. Bacteria are especially vital to life on earth because of their functioning in the major environments. They play an important role in biogeochemical processes, nutrient cycling in soils, bioremediation, human and plant diseases, plant-microbe interactions, municipal waste treatment, and the production of important drug agents including antibiotics (see Chapter 30).

In contrast to bacteria, **protozoa** are unicellular eukaryotes, meaning that they have characteristic organelles (mitochondria, plasma membrane, nuclear envelope, eukaryotic ribosomal RNA, endoplasmic membranes, chloroplast, and flagella). Protozoa can be large and even be visible to the naked eye (Figure 5.5). Their sizes can range from 2 μm to several cm. In addition, they have independent metabolic pathways. Although they are single-celled organisms, they are by no means simple in structure, and many diverse forms can be observed among the more than 65,000 named species. Morphological variability, evolved over hundreds of millions of years, has enabled protozoan adaptation to a wide variety of environments. Protozoa can be found in nearly all terrestrial and aquatic environments and are thought to play a valuable role in ecological cycles by in part controlling bacterial populations. Many species are able to exist in extreme environments from polar-regions to hot springs and desert soils. In recent years, protozoan pathogens such as *Giardia, Cyclospora, Cryptosporidium,* and the *Microsporidia,* have emerged to become threats to safe drink-

ing water (see Chapter 28). Protozoa may be free living, capable of growth and reproduction outside any host, or parasitic, meaning that they colonize host cell tissues. Some are opportunists, adapting either a free-living or parasitic existence as their environment dictates.

Fungi, like protozoa, are also eukaryotic organisms (Figure 5.6). They are very ubiquitous in the environment and also critically affect human health and welfare. For example, the fungi can be both beneficial and harmful to plants, animals, and humans. Certain species of fungi promote the health of many plants through mycorrhizal associations, while other species are phytopathogenic and capable of destroying plant tissue and even whole crops. Fungi are also important in the cycling of organics and in bioremediation. One of the most important fungi are the yeasts, which are utilized in the fermentation of sugars to alcohol in the brewing and wine industries. Fungi range from microscopic, with a single cell, to macroscopic, with filaments of a single fungal cell being several cm in length. Overall, fungi are heterotrophic in nature, with different genera metabolizing everything from simple sugars to complex aromatic hydrocarbons. Fungi are also important in the degradation of the plant polymers cellulose and lignin. For the most part, fungi are aerobic, though some such as the yeasts are capable of fermentation. Fungi are distinguished from algae by their lack of photosynthetic ability.

Algae are a group of photosynthetic organisms that can be macroscopic, as in the case of seaweeds and kelps, or microscopic in size. Algae are aerobic eukaryotic organisms and as such exhibit structural similarities to fungi with the exception of the algal chloroplast (Figure 5.7). Chloroplasts are photosynthetic cell organelles found in algae that are capable of converting the energy of sunlight into chemical energy through a photosynthetic process. Algae are abundant in fresh and salt water, in soil, and in association with plants. When excessive nutrient conditions occur in aquatic environments, they can cause eutrophication. The capacity of algae to photosynthesize is critical because it forms the basis

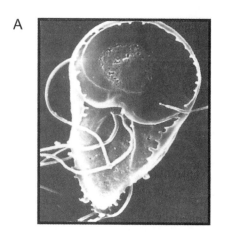
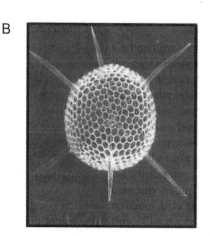
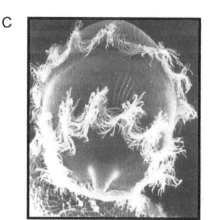

Figure 5.5 Basic morphology of protozoa. (A) Scanning electron micrograph of a flagella protozoa, *Giardia*. (B) Scanning electron micrograph of a testate cilia, *Heliosoma*. (C) Electron micrograph of ciliated *Didinium*. From *Environmental Microbiology* © 2000, Academic Press, San Diego, CA.

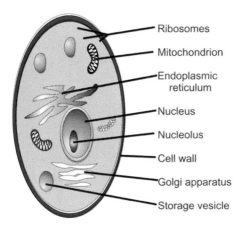

Figure 5.6 Structure of a typical fungal cell. From *Environmental Microbiology* © 2000, Academic Press, San Diego, CA.

of aquatic food chains, and can also be important in the initial colonization of disturbed terrestrial environments. Overall, the algae are sometimes known by common names such as green algae, brown algae, or red algae based on their predominant color.

Blue-green algae are actually classified as bacteria known as *Cyanobacteria*. The *Cyanobacteria* are photosynthetic and some can also fix atmospheric nitrogen. Large blooms of freshwater *Cyanobacteria* may produce toxins that are harmful to animals and humans. Conversely, Spirulina is a blue-green algae grown commercially and sold as a natural dietary supplement.

Most environmental monitoring of beneficial microbes focuses on bacteria, actinomycetes and fungi, particularly in

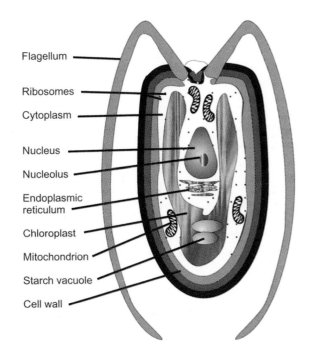

Figure 5.7 Structure of a typical algal cell. From *Environmental Microbiology* © 2000, Academic Press, San Diego, CA.

TABLE 5.1 Relative estimates of the abundance of soil microbes in the environment.

MICROBE	NUMBER (PER GRAM OF SOIL)	BIOMASS WITHIN ROOT ZONE (KG/HA[1])
Bacteria	10^8	500
Actinomycetes	10^7	500
Fungi	10^6	1500

From *Pollution Science* © 1996 Academic Press, San Diego, CA.

the soil environment. Estimates of typical soil microbial numbers and their biomass are shown in Table 5.1. A comparison of their microbial properties is shown in Table 5.2.

5.2 MICROORGANISMS IN SURFACE SOILS

Surface soils are predominantly occupied by indigenous populations of bacteria (including actinomycetes), fungi, algae, and protozoa. In general, as the size of these organisms increases from bacteria to protozoa, the number present decreases (see Figure 5.1). It is also known that potentially there may be phage or viruses present that can infect each class of organism, but information on the extent of these infectious agents in surface soils is limited. In addition to these indigenous populations, specific microbes can be introduced into soil by human or animal activity. Human examples include the deliberate direct introduction of bacteria as biological control agents or as biodegradative agents. Microbes are also introduced indirectly as a result of application of biosolids to agricultural fields (see Chapter 27). Animals introduce microbes through bird droppings and animal excrement. Regardless of the source, introduced organisms rarely significantly affect the abundance and distribution of indigenous populations, which usually out-compete introduced organisms.

The following discussion is an overview of the dominant types of microbes found in surface soils, including their occurrence, distribution, and function.

5.2.1 Bacteria

Bacteria, which are the most numerous organisms in soil or anywhere else on earth, are prokaryotic organisms lacking a nuclear membrane. They are characterized by a complex cell envelope, which contains cytoplasm but no cell organelles. Bacteria are capable of rapid growth and reproduction, both of which occur by binary fission. Genetic exchange occurs predominantly by **conjugation** (cell-to-cell contact) or **transduction** (exchange via viruses), although **transformation** (transfer of naked DNA) also occurs. The size of bacteria generally ranges from 0.1 to 2 μm. Soil bacteria can be rod-shaped, coccoidal, helical, or pleomorphic (Figure 5.8). Soil bacteria also exhibit great diversity with respect to colony morphology (Figure 5.9).

TABLE 5.2 Characteristics of bacteria, actinomycetes, and fungi.

CHARACTERISTIC	BACTERIA	ACTINOMYCETES	FUNGI
Population	Most numerous	Intermediate	Least numerous
Biomass	Bacteria and actinomycetes have similar biomass		Largest biomass
Degree of branching	Slight	Filamentous, but some fragment to individual cells	Extensive filamentous forms
Aerial mycelium	Absent	Present	Present
Growth in liquid culture	Yes—turbidity	Yes—pellets	Yes—pellets
Growth rate	Exponential	Cubic	Cubic
Cell wall	Murein, teichoic acid, and lipopolysaccharide	Murein, teichoic acid, and lipopolysaccharide	Chitin or cellulose
Complex fruiting bodies	Absent	Simple	Complex
Competitiveness for simple organics	Most competitive	Least competitive	Intermediate
Fix N	Yes	Yes	No
Aerobic	Aerobic, anaerobic	Mostly aerobic	Aerobic except yeast
Moisture stress	Least tolerant	Intermediate	Most tolerant
Optimum pH	6–8	6–8	6–8
Competitive pH	6–8	>8	<5
Competitiveness in soil	All soils	Dominate dry, high-pH soils	Dominate low-pH soils

From *Pollution Science* © 1996, Academic Press, San Diego, CA.

5.2.1.1 Mode of nutrition

We can classify bacteria according to their mode of nutrition:

1. **Autotrophic mode:** Strict soil autotrophs obtain energy from inorganic sources, and carbon from carbon dioxide. These kinds of organisms generally have few growth-factor requirements. Chemoautotrophs obtain energy from the oxidation of inorganic substances, whereas photoautotrophs obtain energy from photosynthesis.

2. **Heterotrophic mode:** Heterotrophs obtain energy and carbon from organic substances. Chemoheterotrophs obtain energy from oxidations; however, photoheterotrophs obtain energy from photosynthesis but require organic electron donors.

In soil, chemoheterotrophs and chemoautotrophs predominate; phototrophs of either variety are not as numerous because soil is not permeable to sunlight.

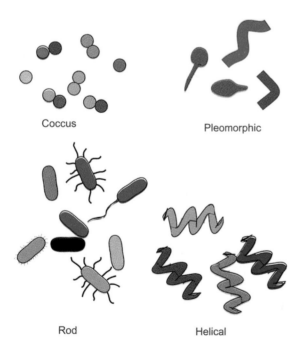

Coccus

Pleomorphic

Rod

Helical

Figure 5.8 Typical shapes of representative bacteria. From *Environmental Microbiology* © 2000, Academic Press, San Diego, CA.

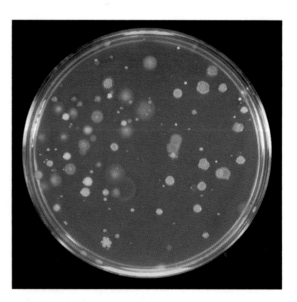

Figure 5.9 Here, a variety of bacteria with some actinomycetes are isolated from a field soil on a Petri dish on a general heterotrophic medium. From *Pollution Science*, First Edition © 1996, Academic Press, San Diego, CA.

5.2.1.2 Type of electron acceptor

Aerobic bacteria utilize oxygen as a terminal electron acceptor and possess superoxide dismutase or catalase enzymes that are capable of degrading peroxide radicals. **Anaerobic bacteria** do not utilize oxygen as a terminal electron acceptor. Strict anaerobes do not possess superoxide dismutase or catalase enzymes, and are thus poisoned by the presence of oxygen. Although other kinds of anaerobes do possess these enzymes, they utilize terminal electron acceptors other than oxygen, such as nitrate or sulfate. **Facultative anaerobes** can use oxygen or combined forms of oxygen as terminal electron acceptors.

5.2.1.3 Ecological classification

Bacteria can be classified based on the concept of *r* **and** *K* **selection**. Organisms adapted to living under conditions in which substrate is plentiful are designated as *K*-selected. **Rhizosphere** organisms living off root exudates are examples of *K*-selected organisms. Organisms that are *r*-selected live in environments in which substrate is the limiting factor except for occasional flushes of substrate. *r*-Selected organisms rely on rapid growth rates when substrate is available, and generally occur in uncrowded environments. In contrast, *K*-selected organisms exist in crowded environments and are highly competitive.

5.2.1.4 Dominant culturable soil bacteria

1. **Arthrobacter:** The most numerous bacteria in soil, as determined by plating procedures, arthrobacters represent as much as 40% of the culturable soil bacteria. These autochthonous organisms are pleomorphic and Gram-variable. Young cells are Gram-negative rods, which later become Gram-positive cocci.

2. **Streptomyces:** These organisms are actually actinomycetes (which are discussed later). They are Gram-positive, chemoheterotrophic organisms that can comprise 5–20% of the bacterial count in soil. These organisms produce antibiotics, including streptomycin (discovered by Selman Waksman, who was awarded the Nobel Prize in Medicine in 1942). They are also involved in nutrient cycling and biodegradation.

3. **Pseudomonas:** These Gram-negative organisms, which are also known as pseudomonads, are ubiquitous and diverse in nature. They are generally heterotrophic and aerobic, but some are facultative autotrophs. As a group, they possess many different enzyme systems and are capable of degrading a wide variety of organic compounds, including recalcitrant compounds. They are also used as biocontrol agents. These organisms can comprise 10–20% of the bacterial population.

4. **Bacillus:** About as prevalent as the pseudomonads, *Bacilli* are characterized as Gram-positive aerobic organisms that produce endospores. This genus is heterotrophic and diverse. *Bacilli* often constitute 10% of the bacterial population. They are important in nutrient cycling and biodegradation. *Bacillus anthracis* is the causative agent of anthrax.

5.2.1.5 Distribution and function

Bacteria are the most abundant soil organisms, with a biomass of about 500 kg per ha to the depth of the root zone. Generally, aerobes are more prevalent than anaerobes, particularly in the A horizon. Dominant culturable genera include *Streptomyces, Arthrobacter, Bacillus, Pseudomonas, Alcaligenes, Agrobacteria, Corynebacterium, and Flavobacterium.*

Bacteria are almost always the most abundant organisms found in surface soils in terms of numbers. Culturable numbers vary depending on specific environmental conditions, particularly soil moisture and temperature (Roszak and Colwell, 1987). Culturable bacteria can be as numerous as 10^7 to 10^8 cells per gram of soil, whereas total populations (including viable but nonculturable organisms) can exceed 10^{10} cells per gram. In unsaturated soils, aerobic bacteria usually outnumber anaerobes by two or three orders of magnitude. Anaerobic populations increase with increasing soil depth, but rarely predominate unless soils are saturated with water.

Overall estimates have indicated that surface soils can contain up to 10,000 individual species (Turco and Sadowsky, 1995). Not all of these are culturable, and information on the nonculturable species is limited, but more recently, information has increased in terms of diversity. Most studies of soil bacteria have focused on culturable organisms, and a massive amount of information is available on key genera.

Tables 5.3 and 5.4 identify some of the culturable bacteria genera that are known to be critical to environmental microbiology. Of course, the lists are by no means all inclusive.

TABLE 5.3 Examples of important autotrophic soil bacteria.

ORGANISM	CHARACTERISTICS	FUNCTION
Nitrosomonas	Gram negative, aerobe	Converts $NH_4^+ \rightarrow NO_2^-$ (first step of nitrification)
Nitrobacter	Gram negative, aerobe	Converts $NO_2^- \rightarrow NO_3^-$ (second step of nitrification)
Thiobacillus	Gram negative, aerobe	Oxidizes $S \rightarrow SO_4^{2-}$ (sulfur oxidation)
Thiobacillus denitrificans	Gram negative, facultative anaerobe	Oxidizes $S \rightarrow SO_4^{2-}$; functions as a denitrifier
Thiobacillus ferrooxidans	Gram negative, aerobe	Oxidizes $Fe^{2+} \rightarrow Fe^{3+}$

TABLE 5.4 Examples of important heterotrophic soil bacteria.

ORGANISM	CHARACTERISTICS	FUNCTION
Actinomycetes, *e.g.*, *Streptomyces*	Gram positive, aerobic, filamentous	Produce geosmins, "earthy odor," and antibiotics
Bacillus	Gram positive, aerobic, spore former	Carbon cycling, production of insecticides and antibiotics
Clostridium	Gram positive, anaerobic, spore former	Carbon cycling (fermentation), toxin production
Methanotrophs, *e.g.*, *Methylosinus*	Aerobic	Methane oxidizers that can cometabolize trichloroethene (TCE) using methane monooxygenase
Alcaligenes eutrophus	Gram negative, aerobic	2,4-D degradation via plasmid pJP4
Rhizobium	Gram negative, aerobic	Fixes nitrogen symbiotically with legumes
Frankia	Gram positive, aerobic	Fixes nitrogen symbiotically with nonlegumes
Agrobacterium	Gram negative, aerobic	Important plant pathogen, causes crown gall disease

Bacteria are critically involved in almost all soil biochemical transformations, including the metabolism of both organics and inorganics. The importance of soil bacteria in the fate and mitigation of pollutants cannot be overestimated. Because of their prevalence and diversity, as well as fast growth rates and adaptability, they have an almost unlimited ability to degrade most natural products and many xenobiotics. We will examine, in detail, the influence of bacteria on waste disposal and pollution mitigation in the succeeding chapters of this text (Chapters 8, 19, and 26).

Actinomycetes are organisms that technically are classified as bacteria, but are unique enough to be discussed as an individual group. They have some characteristics in common with bacteria, but are also similar in some respects to fungi. For the most part they are aerobic chemoheterotrophic organisms consisting of elongated single cells. They display a tendency to branch into filaments, or hyphae, that resemble fungal mycelia; these hyphae are morphologically similar to those of fungi, but are smaller in diameter (about 0.5–2 μm). The diversity of soil bacteria including actinomycetes are shown in Figure 5.9. Finally, the spatial relationship of soil bacteria, actionomycetes and fungi are illustrated in Figure 5.10. The total number of actinomycetes in soils is often about 10^7 per gram of soil. Generally, the population of actinomycetes is 1 to 2 orders of magnitude less than that of other bacteria in soil. They are not known to reproduce sexually, but all produce asexual spores called **conidia**. The genus *Streptomyces* dominates the actinomycete population, and these Gram-positive organisms may represent 90% of the total actinomycete population.

Actinomycetes are an important component of bacterial populations, especially under conditions of high pH, high temperature, or water stress. One distinguishing feature of this group of bacteria is that they are able to utilize a great variety of substrates found in soil, especially some of the less degradable insect and plant polymers such as chitin, cellulose, and hemicellulose.

Like all bacteria, actinomycetes are prokaryotic organisms. In addition, the adenine-thymine and guanine-cytosine contents of bacteria and actinomycetes are similar, as are the cell wall constituents of both types of organisms. Actinomycete filaments are also about the same size as those of bacteria.

Like fungi, however, actinomycetes display extensive mycelial branching, and both types of organisms form aerial mycelia and conidia. Moreover, growth of actinomycetes in liquid culture tends to produce fungus-like clumps or pellets rather than the uniform turbidity produced by bacteria. Finally, growth rates in fungi and actinomycetes are not exponential as they are in bacteria; rather, they are cubic.

Actinomycetes can metabolize a wide variety of organic substrates including organic compounds that are not usually metabolized such as phenols and steroids. They are also important in the metabolism of heterocyclic compounds such as complex nitrogen compounds and pyrimidines. The breakdown products of their metabolites are frequently aromatic, and these metabolites are important in the formation of humic substances and soil humus. The earthy odor associated with most soils is due to **geosmin**, a compound produced by actinomycetes.

Actinomycetes often comprise about 10% of the total bacterial population, with a biomass of about 500 kg per ha to the depth of the root zone. More tolerant of alkaline soils (pH > 7.5) and less tolerant of acidic soils (pH < 5.5), actinomycetes are also more tolerant to low soil moisture contents than other bacteria. Because of this and their tolerance of alkaline soils, actinomycete populations tend to be larger in desert soils, such as the southwestern U.S.

These organisms can be important in the control of other soil organisms. Many actinomycetes interact with plants in symbiotic and pathogenic associations. For example, the genus *Frankia* initiates root nodules with nonleguminous, nitrogen-fixing plants, whereas the species *Streptomyces scabies* is the causative agent of potato scab. On the other hand, many streptomycetes produce antibiotics, including streptomycin, chloramphenicol, tetracycline, and cycloheximide (see Chapter 30).

5.2.2 Fungi

Fungi other than yeasts are aerobic and are abundant in most surface soils. Numbers of fungi usually range from 10^5 to 10^6 per gram of soil. Despite their lower numbers compared with bacteria, fungi usually contribute a higher proportion of the total soil microbial biomass. This is due to their comparatively large size; a fungal hypha can range from 2 to 10 μm in diameter. Figure 5.11 shows an

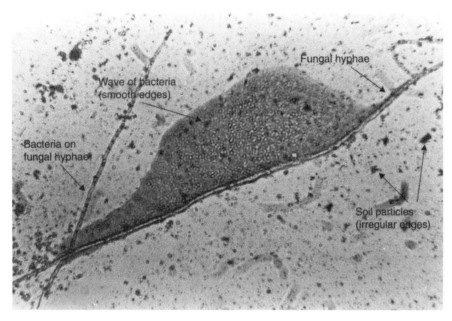

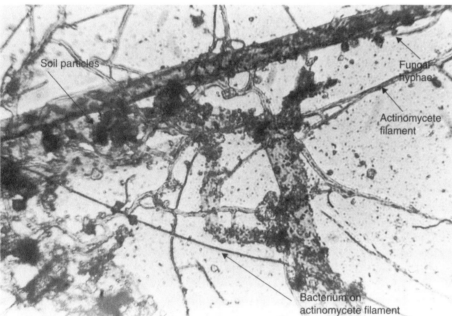

Figure 5.10 Comparison of soil bacteria, actinomycetes and fungi viewed under a light. From *Environmental Microbiology Laboratory Manual, Second Edition* © 2004, Academic Press, San Diego, CA.

example of the diverse fungal population that can be isolated from surface soil. Because of their large size, fungi are more or less restricted to the interaggregate regions of the soil matrix. Yeasts can metabolize anaerobically (fermentation) and are less numerous than aerobic mycelium-forming fungi. Generally, yeasts are found at populations of up to 10^3 per gram of soil. Because of their reliance on organic sources for substrates, fungal populations are greatest in the surface O and A horizons, and numbers decrease rapidly with increasing soil depth. As with bacteria, soil fungi are normally found associated with soil particles or within plant rhizospheres.

Fungi are important components of the soil with respect to nutrient cycling, especially the decomposition of organic matter. They decompose both simple sugars and complex polymers such as cellulose and lignin. The role of fungi in decomposition is increasingly important when the soil pH declines because fungi tend to be more tolerant of acidic conditions than bacteria. Some of the common genera of soil fungi involved in nutrient cycling are *Penicillium* and *Aspergillus*. These organisms are also important in the development of soil structure because they physically entrap soil particles with fungal hyphae. Fungi are critical in the degradation of complex plant polymers such as cellulose and

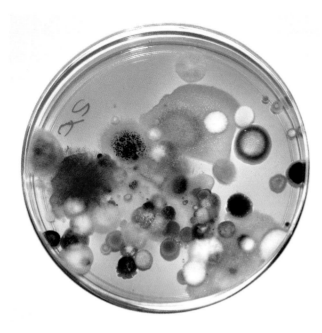

Figure 5.11 Soil fungi isolated from a surface soil grown in a Petri dish containing Rose Bengal Agar. Photo courtesy K.L. Josephson.

lignin, and some fungi can also degrade a variety of pollutant molecules. The best-known example of such a fungus is the white rot fungus *Phanerochaete chrysosporium*. Other fungi such as *Fusarium* spp., *Pythium* spp., and *Rhizoctonia* spp., are important plant pathogens. Still, others cause disease; for example, *Coccidioides immitis* causes a chronic human pulmonary disease known as valley fever in the southwestern deserts of the United States. Finally, note that mycorrhizal fungi are critical for establishing plant-fungal interactions that act as an extension of the root system of almost all higher plants. Without these mycorrhizal associations, plant growth as we know it would be impossible.

5.2.3 Algae

Algae are typically phototrophic and thus would be expected to survive and metabolize in the presence of a light-energy source and CO_2 carbon source. Therefore, one would expect to find algal cells predominantly in areas where sunlight can penetrate the surface of the soil. One can actually find algae to a depth of 1 m because some algae, including the green algae and diatoms, can grow heterotrophically as well as photoautotrophically. In general, though, algal populations are highest in the surface 10 cm of soil. Typical algal populations close to the soil surface range from 5,000 to 10,000 per gram of soil. Note that a surface soil where a visible algal bloom has developed can contain millions of algal cells per gram of soil.

Algae are often the first to colonize surfaces in a soil that are devoid of preformed organic matter. Colonization by this group of microbes is important in establishing soil formation processes, especially in barren volcanic areas, desert soils, and rock faces. Algal metabolism is critical to soil formation in two ways: algae provide a carbon input through photosynthesis, and as they metabolize they produce and release carbonic acid, which aids in weathering the surrounding mineral particles. Further, algae produce large amounts of extracellular polysaccharides, which also aid in soil formation by causing aggregation of soil particles (Killham, 1994).

Populations of soil algae generally exhibit seasonal variations with numbers being highest in the spring and fall. This is because desiccation caused by water stress tends to suppress growth in the summer and cold stress affects growth in the winter. Four major groups of algae are found in soil. The green algae or the Chlorophyta (for example, the *Chlamydomonas*) are the most common algae found in acidic soils. Also widely distributed are diatoms such as *Navicula*, which are members of the Bacillariophyta. Diatoms are found primarily in neutral and alkaline soils. Diatoms are characterized by the presence of a silicon dioxide cell wall. Less numerous are the golden or yellow algae (Chrysophyta), such as *Botrydiopsis*, and the red algae (Rhodophyta), such as *Prophyridium*). In addition to these algal groups, there are the cyanobacteria, often called blue-green algae, (*e.g.*, *Nostoc* and *Anabaena*), which are actually classified as bacteria but have many characteristics in common with algae. The cyanobacteria participate in the soil-forming process discussed in the previous paragraph, and some cyanobacteria also have the capacity to fix nitrogen, a nutrient that is usually limiting in a barren environment. In temperate soils the relative abundance of the major algal groups follows the order green algae > diatoms > cyanobacteria > yellow-green algae. In tropical soils the cyanobacteria predominate.

5.2.4 Protozoa

Protozoa are unicellular, eukaryotic organisms that range up to 5.5 mm in length, although most are much smaller. Most protozoa are heterotrophic and survive by consuming bacteria, yeast, fungi, and algae. There is evidence that they may also be involved, to some extent, in the decomposition of soil organic matter. Because of their large size and requirement for large numbers of smaller microbes as a food source, protozoa are found mainly in the top 15 to 20 cm of the soil. Protozoa are usually concentrated near root surfaces that have high densities of bacteria or other prey. Soil protozoa are flatter and more flexible than aquatic protozoa, which makes it easier to move around in the thin films of water that surround soil particle surfaces as well as to move into small soil pores.

There are three major categories of protozoa: the flagellates, the amoebae, and the ciliates. The flagellates are the smallest of the protozoa and move by means of one to several flagella. Some flagellates (*e.g.*, *Euglena*), contain chlorophyll, although most do not. The amoebae, also called rhizopods, move by protoplasmic flow, either with extensions called pseudopodia or by whole-body flow. Amoebae are usually the most numerous types of protozoan found in a given soil environment. Ciliates are protozoa that move by beating short cilia that cover the surface of the cell. The protozoan population of a soil is often correlated with the bacterial population, which is the protozoan's major food source. Numbers of protozoa can range from 30,000 to 10^6 per gram of soil.

5.3 MICROORGANISMS IN THE SUBSURFACE

Although the microorganisms of surface soils have been studied extensively, the study of subsurface microorganisms is relatively new, beginning in earnest in the 1980s. Complicating the study of subsurface life are the facts that sterile sampling is problematic and many subsurface microorganisms are difficult to culture.

Because subsurface microbiology is still a developing field, information is limited in comparison with that for surface microorganisms. Yet there is enough information available to document that subsurface environments, once thought to contain very few if any microorganisms, actually have a significant and diverse population of microorganisms. In particular, shallow subsurface zones, specifically those with a relatively rapid rate of water recharge, have high numbers of microorganisms. The majority of these organisms are bacteria, but protozoa and fungi are also present. As a general rule, total numbers of bacteria, as measured by direct counts, remain fairly constant, ranging between 10^5 and 10^7 cells per gram throughout the profile of a shallow subsurface system. For comparison, numbers in surface soils range from 10^9 to 10^{10} cells per gram. This decrease in numbers is directly correlated with the low amounts of inorganic nutrients and organic matter in subsurface materials. Subsurface eukaryotic counts are also lower than surface counts by several orders of magnitude. Low eukaryotic counts are a result of low organic matter content but, perhaps more important, result from removal by physical straining by small soil pores as they move downward. A final point to be made is that both prokaryotic and eukaryotic counts are highest in portions of the subsurface containing sandy sediments. This does not mean that clayey regions are not populated, but the numbers tend to be lower. This may also be due to exclusion and physical straining of microorganisms by small pores in clay-rich media.

Numbers of culturable bacteria in subsurface environments show more variability than direct counts, ranging from zero to almost equal numbers of direct counts. Thus, in general, the difference between direct and viable counts in the subsurface is greater than the difference in surface soils, which is normally one to two orders of magnitude. The larger difference between direct and viable counts in the subsurface may be because nutrients are much more limiting in the subsurface, and therefore a greater proportion of the population may be in a nonculturable state.

The dominant populations are aerobic, heterotrophic bacteria, although there are small populations of eukaryotic organisms as well as anaerobic and autotrophic organisms. Subsurface microbes are diverse in type, although not as diverse as surface organisms. Great diversity in heterotrophic activity has been found, and subsurface microbes have shown the capacity to degrade simple substrates such as glucose as well as more complex substrates such as aromatic compounds, surfactants, and pesticides. It has also been shown that subsurface bacteria are capable of denitrification activity (Artiola and Pepper, 1992).

5.4 BIOLOGICAL GENERATION OF ENERGY

Biological activity requires energy, and all microorganisms generate energy. This energy is subsequently stored as **adenosine triphosphate** (ATP), which can then be utilized for growth and metabolism as needed, subject to the second law of thermodynamics.

The Second Law of Thermodynamics

In a chemical reaction, only part of the energy is used to do work. The rest of the energy is lost as entropy.

Gibbs free energy ΔG is the amount of energy available for work for any chemical reaction. For the reaction

$$A + B \leftrightarrow C + D$$

the thermodynamic equilibrium constant is defined as

$$K_{eq} = \frac{[C]\,[D]}{[A]\,[B]}$$

Case 1: If product formation is favored:
That is, if

$$[C]\,[D] > [A]\,[B]$$

then $K_{eq} > 1$ and $\ln K_{eq}$ is positive.
For example, if $K_{eq} = 2.0$, then

$$\ln K_{eq} = 0.69$$

Case 2: If product formation is *not* favored:

$$[C]\,[D] < [A]\,[B]$$

then $K_{eq} < 1$ and $\ln K_{eq}$ is negative.
For example, if $K_{eq} = 0.20$, then

$$\ln K_{eq} = -1.61$$

The relationship between the equilibrium coefficient and the free energy ΔG is given by

$$\Delta G = -RT \log K_{eq}$$

where:

R is the universal gas constant and T = absolute temperature (K).

If $\underline{\Delta G \text{ is negative}}$, energy is released from the reaction due to a spontaneous reaction. This is because if ΔG is negative, log K_{eq} must be positive. Therefore $K_{eq} > 1$, which means energy must be released.

If $\underline{\Delta G \text{ is positive}}$, energy is needed to make reaction proceed. This is because if ΔG is positive, log K_{eq} must be negative. Therefore $K_{eq} < 1$, which means energy must be added to promote the reaction.

Thus we can use ΔG values for any biochemical reaction mediated by microbes to determine whether energy is liberated for work, and how much energy is liberated. Soil organisms can generate energy via several mechanisms, which can be divided into two main categories.

1. **Photosynthesis:**

$$2 H_2O + CO_2 \xrightarrow{\text{light}} CH_2O + O_2 + H_2O$$
$$\text{biomass}$$
$$\Delta G = + 115 \text{ K cal mol}^{-1}$$

For this reaction, energy supplied by sunlight is necessary. The fixed organic carbon is then used to generate energy via respiration. Examples of soil organisms that undergo photosynthesis are *Rhodospirillum*, *Chromatium*, and *Chlorobium*.

2. **Respiration**

(a) Aerobic heterotrophic respiration: Many soil organisms undergo aerobic, heterotrophic respiration, *e.g.*, *Pseudomonas* and *Bacillus*.

$$C_6 H_{12} O_6 + 6 O_2 \rightarrow 6 CO_2 + 6 H_2O$$
$$\Delta G = -686 \text{ K cal mol}^{-1}$$

(b) Aerobic autotrophic respiration: The reactions carried out by *Nitrosomonas* and *Nitrobacter* are known as **nitrification**:

$$NH_3 + 1\tfrac{1}{2} O_2 \rightarrow HNO_2 + H_2O \qquad (Nitrosomonas)$$
$$\Delta G = -66 \text{ K cal mol}^{-1}$$

$$KNO_2 + \tfrac{1}{2} O_2 \rightarrow KNO_3 \qquad (Nitrobacter)$$
$$\Delta G = -17.5 \text{ K cal mol}^{-1}$$

The following two reactions are examples of **sulfur oxidation**:

$$2 H_2S + O_2 \rightarrow 2 H_2O + 2 S \qquad (Beggiatoa)$$
$$\Delta G = -83 \text{ K cal mol}^{-1}$$

$$2 S + 3 O_2 \rightarrow 2 H_2SO_4 \qquad (Thiobacillus\ thiooxidans)$$
$$\Delta G = -237 \text{ K cal mol}^{-1}$$

The next reaction involves the **degradation of cyanide**:

$$\overset{(Streptomyces)}{2 KCN + 4 H_2O + O_2 \rightarrow 2 KOH + 2NH_3 + 2CO_2}$$
$$\Delta G = -56 \text{ K cal mol}^{-1}$$

All of the above reactions illustrate how soil organisms mediate reactions that can cause or negate pollution. For example, nitrification and sulfur oxidation can result in the production of specific pollutants, *i.e.*, nitrate and sulfuric acid, whereas the destruction of cyanide is obviously beneficial with respect to the mitigation of pollution.

(c) Facultative anaerobic, heterotrophic respiration:

Pseudomonas denitrificans can achieve this kind of metabolism by utilizing nitrate as a terminal electron acceptor rather than oxygen. Note that these organisms can use oxygen as a terminal electron acceptor if it is available, and that aerobic respiration is more efficient than anaerobic respiration.

$$5 C_6H_{12}O_6 + 24 KNO_3 \rightarrow 30 CO_2 + 18 H_2O$$
$$+ 24 KOH + 12 N_2$$
$$\Delta G = -36 \text{ K cal mol}^{-1}$$

(d) Facultative anaerobic autotrophic respiration:

$$S + 2 KNO_3 \rightarrow K_2 SO_4 + N_2 + O2 \quad \overset{(Thiobacillus}{\underset{denitrificans)}{}}$$
$$\Delta G = -66 \text{ K cal mol}^{-1}$$

(e) Anaerobic heterotrophic respiration: *Desulfovibrio* is an example of an organism that carries out this type of metabolism.

$$\overset{Lactic\ acid}{CH_3CHOHCOOH} + SO_4^{2-} \rightarrow \overset{Acetic\ acid}{2CH_3COOH} + HS^-$$
$$+ H_2CO_3 + HCO_3^-$$
$$\Delta G = -170 \text{ kJ mol}^{-1}$$

Soil organisms can also undergo fermentation, which is also an anaerobic process. But fermentation is not widespread in soil. Overall, there are many ways in which soil organisms can and do generate energy. The above-listed mechanisms illustrate the diversity of soil organisms and explain the ability of the soil community to break down or transform almost any natural substance. In addition, enzyme systems have evolved to metabolize complex molecules, be they organic or inorganic. These enzymes can also be used to degrade xenobiotics with similar chemical structures. Xenobiotics, which do not degrade easily in soil, are normally chemically different from any known natural substance; hence soil organisms have not evolved enzyme systems capable of metabolizing such compounds. There are many other factors that influence the breakdown or degradation of chemical compounds. These will be discussed in Chapter 8.

5.5 SOIL AS AN ENVIRONMENT FOR MICROBES

5.5.1 Biotic Stress

Since indigenous soil microbes are in competition with one another, the presence of large numbers of organisms results in biotic stress factors. Competition can be for substrate, water, or growth factors. In addition, microbes can secrete inhibitory or toxic substances, including antibiotics, that harm neighboring organisms. Finally, many organisms are predatory or parasitic on neighboring microbes. For example, phages infect both bacteria and fungi. Because of biotic stress, nonindigenous organisms that are introduced into a soil environment often survive for fairly short periods of time (days to several weeks). This effect has important consequences for pathogens (see Chapter 11) and for other organisms introduced to aid biodegradation (see Chapter 8). This process of adding organisms to assist biodegradation is called **bioaugmentation**.

5.5.2 Abiotic Stress

Light: Soil is impermeable to light, that is, no sunlight penetrates beyond the top few centimeters of the soil surface. Phototrophic organisms are therefore limited to the top few centimeters of soil. At the surface of the soil, however, such physical parameters as temperature and moisture fluctuate significantly throughout the day and also seasonally. Hence most soils tend to provide a harsh environment for photosynthesizing organisms. A few phototrophic organisms, including algae, have the ability to switch to a heterotrophic respiratory mode of nutrition in the absence of light. Such "switch-hitters" can be found at significant depths within soils. Normally, these organisms do not compete with other indigenous heterotrophic organisms for organic substrates.

Soil Moisture: Typically, soil moisture content varies considerably in any soil, and soil organisms must adapt to a wide range of soil moisture contents. Soil aeration is dependent on soil moisture: saturated soils tend to be anaerobic, whereas dry soils are usually aerobic. But soil is a heterogeneous environment; even saturated soils contain pockets of aerobic regimes, and dry soils harbor anaerobic microsites that exist within the centers of secondary aggregates. Although the bacteria are the least tolerant of low soil moisture, as a group they are the most flexible with respect to soil aeration. They include aerobes, anaerobes, and facultative anaerobes, whereas the actinomycetes and fungi are predominantly aerobic.

Soil Temperature: Soil temperatures vary widely, particularly near the soil surface. Most soil populations are resistant to wide fluctuations in soil temperature although soil populations can be psychrophilic, mesophilic, or thermophilic, depending on the geographic location of the soil. Most soil organisms are mesophilic because of the buffering effect of soil on soil temperature, particularly at depths beneath the soil surface.

Soil pH: Undisturbed soils usually have fairly stable soil pH values within the range of 6–8, and most soil organisms have pH optima within this range. There are, of course, exceptions to this rule, as exemplified by *Thiobacillus thiooxidans*, an organism that oxidizes sulfur to sulfuric acid and has an optimum pH of 2 to 3. Microsite variations of soil pH can also

occur owing, perhaps, to local decomposition of an organic residue to organic acids. Here again, we see that soil behaves as a heterogeneous or discontinuous environment, allowing organisms with differing pH optima to coexist in close proximity. Normally, soil organisms are not adversely affected by soil pH unless drastic changes occur. Drastic change can happen, for example, when lime is added to soil to increase the pH or when sulfur is added to decrease the pH.

Soil Texture: Almost all soils contain populations of soil organisms regardless of the soil texture. Even soils whose textures are extreme, such as pure sands or clays, usually contain populations of microbes, albeit in lower numbers than in soils with less extreme textures. Most nutrients are associated with clay or silt particles, which also retain soil moisture efficiently. Thus, soils with at least some silt or clay particles offer a more favorable habitat for organisms than do soils without these materials.

Soil Carbon and Nitrogen: Carbon and nitrogen are both nutrients that are found in soils. Since both of these nutrients are present in low concentrations, growth and activity of soil organisms are limited. In fact, many organisms exist in soil under limited starvation conditions and hence are dormant. Without added substrate or amendment, soil organisms generally metabolize at low rates. The major exception to this is the rhizosphere, whose root exudates maintain a high population level. In most cases, all available soil nutrients are immediately utilized. Soil humus represents a source of organic nutrients that is mineralized slowly by autochthonous organisms. Similarly, specific microbial populations can utilize xenobiotics as a substrate, even though the rate of degradation is sometimes quite low.

Soil Redox Potential: Redox potential (E_h) is the measurement of the tendency of an environment to oxidize or reduce substrates. In a sense, we can think of it as the availability of different terminal electron acceptors that are necessary for specific organisms. Such electron acceptors exist only at specific redox potentials, which are measured in millivolts (mV). An aerobic soil, which is an oxidizing environment, has a redox potential or E_h of $+800$ mV; an anaerobic soil, which is a reducing environment, has an E_h of about to -300 mV. Oxygen is found in soils at a redox potential of about $+800$ mV. When soil is placed in a closed container, aerobic organisms use oxygen as a terminal electron acceptor until all of it is depleted. As this process occurs, the redox potential of the soil decreases, and other compounds can be used as terminal electron acceptors. Table 5.5 illustrates the redox potential at which various substrates are reduced, and the activity of different types of organisms in a soil.

The fact that different terminal electron acceptors are available for various organisms having diverse pH requirements means that some soil environments are more suitable than others for various groups of organisms. Figure 5.12, which illustrates optimal E_h/pH relationships for various groups of organisms, shows that redox potential affects the activity of all organisms.

TABLE 5.5 **Redox potential at which soil substrates are reduced.**

REDOX POTENTIAL (mV)	REACTION	TYPE OF ORGANISM
+800	$O_2 \rightarrow H_2O$	Aerobes
+740	$NO_3 \rightarrow N_2, N_2O$	Facultative anaerobes
220	$SO_4 \rightarrow S^{2-}$	Anaerobes
300	$CO_2 \rightarrow CH_4$	Anaerobes

From *Pollution Science* © 1996, Academic Press, San Diego, CA.

5.6 ACTIVITY AND PHYSIOLOGICAL STATE OF MICROBES IN SOIL

Is the soil environment a favorable one for soil microbes? There are actually two possible answers to this question. On the one hand, soil is a very harsh environment; on the other hand, soil contains very large populations of microbes. How can this be? It appears that soil organisms are well adapted to this harsh environment, so they must have mechanisms to exploit the resources available. These mechanisms also allow organisms to survive long periods of time when resources are not available. Viewing soil as a community having large numbers of organisms and great microbial diversity, we can infer that each species has a habitat and a niche, rather like a home and a job. Because soil is a heterogeneous environment, many different kinds of organisms can coexist. Diversity also ensures that all available nutrients are utilized in soil.

Owing to the harsh physical soil environment and the fact that nutrients are usually limiting, soil organisms do not

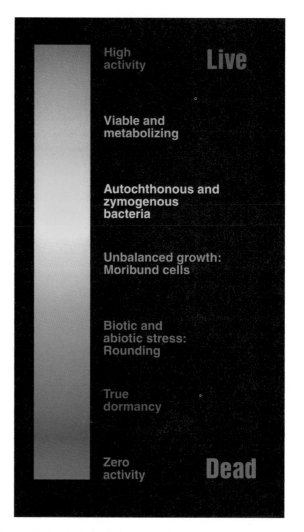

Figure 5.13 Physiological states of soil organisms. From *Pollution Science* © 1996, Academic Press, San Diego, CA.

actively metabolize most of the time. In fact, they exist under stress, so they may exist injured or even be killed. Figure 5.13 shows the various potential states of soil organisms, with the two extremes being live microbes and dead ones. Between these two extremes, other physiological states are possible, including metabolically active and dormant states. Because many organisms are in fact injured in soil, and because they have diverse specific nutritional needs, many soil organisms cannot be cultured by conventional methods. These are the so-called **viable but nonculturable organisms**. In practice, perhaps 99% of all soil organisms may be nonculturable (Roszak and Colwell, 1987). Thus, any methodology that relies on obtaining soil organisms via a culturable procedure may in fact be sampling a very small subsection of the soil population.

Based on our discussion of the soil as an environment for microbes, we may reasonably infer that indigenous organisms within a particular soil are selected by the specific environment in that soil. Indigenous organisms are often capable of surviving wide variations in particular environmental parameters. Introduced organisms, however, are

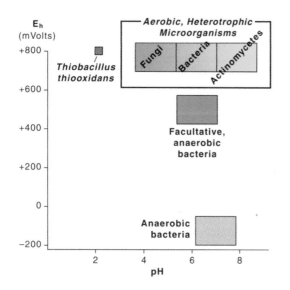

Figure 5.12 Optimal E_h/pH relationships for various soil organisms. From *Pollution Science* © 1996, Academic Press, San Diego, CA.

unlikely to be as well adapted and cannot be expected to compete with indigenous organisms unless a specific niche is available.

5.7 ENUMERATION OF SOIL BACTERIA VIA DILUTION AND PLATING

Because of the importance of soil bacteria, there is extensive interest in estimating soil bacterial numbers. Perhaps the most common method of bacterial enumeration is the culturable technique known as "dilution and plating." The following section provides an overview of this technique and an example calculation.

Since soils generally contain millions of bacteria per gram, normally a dilution series of the soil is made by suspending a given amount of soil in a dispersing solution (often a dilute buffered peptone or saline solution), and transferring aliquots of the suspensions to fresh solution until the suspension is diluted sufficiently to allow individual discrete bacterial colonies to grow on agar plates.

After inoculation on several replicate agar plates, the plates are incubated at an appropriate temperature, and counted after they have formed macroscopic fungal colonies. Because the assumption is that one colony is derived from one organism, the term **colony forming units (CFUs)** is used in the final analysis, with the results expressed in terms of CFUs per gram of dry soil.

Figure 5.14 describes a dilution and plating protocol procedure. Beginning at step 1, a 10-fold dilution series is performed. A 10-fold series is very common as the calculations for the determination of the organism count is very simple. Here, 10 g of moist soil is added to 95 mL (solution A) of deionized water and shaken well to disperse the organisms. The reason that 10g of soil are used is that 10g of soil occupies approximately 5 mL. Thus, we have 10g of soil in 100 mL total volume, thereby forming a 1:10 w/v dilution.

Next, 1.0 mL of suspension is removed from the bottle and added to a tube (B) containing 9.0 mL of the same dispersion solution as in A. The tube is capped and vortexed. Working diligently, the dilution series is continued to the highest desired dilution (tubes C, D, and E). The three most dilute suspensions are plated. Three different dilutions (tubes C, D, and E) are plated so as to increase the chance of obtaining a dilution that will result in a countable number of organisms (Step 2).

Here, pour plates are utilized for the plating procedure. The dilution of interest is vortexed and 1.0 mL of suspension is removed from the tube and added to each of two sterile Petri dishes. Before the soil particles in the inoculum can settle, pour plates are made (step 3a). Here, a suitable agar such as R_2A medium poured into the plate with the one mL of inoculum. The agar is at a temperature warm enough to keep the agar fluid, but cool enough not to kill the organisms.

Then the plate is *gently* swirled (step 3b) to distribute the agar and inoculum across the bottom of the plate *(without splashing agar on the sides or lid of the dish)*. Finally, the agar is allowed to solidify, and the plates are incubated upside down to prevent condensation from falling on the growing surface of the agar (step 4). Counting takes place after an incubation period suitable for the organism(s) of interest (often 5–7 days). An example calculation is shown in Calculation Box 5.1.

5.8 MICROORGANISMS IN AIR

Microbes that are found in air are generally known as **bioaerosols** and the process that moves microbes into air is known as **aerosolization**. Bioaerosols can be whole microorganisms including viruses, bacteria, fungi, or spores, or they can be biological remnants such as endotoxin and cell wall constituents. Bioaerosol sizes range typically from 0.5 to 30 μm in diameter and are normally surrounded by a thin film of water. Biological particles can also be associated with particulate matter such as soil or wastes such as biosolids, depending on the origin of the bioaerosol. Biological particles in the lower spectrum of sizes (0.5 to 5 μm) are of most concern since these are most easily inhaled or ingested (Stetzenbach, 2001).

5.8.1 Incidence of Bioaerosols

Microbes that are aerosolized can be pathogenic or non pathogenic depending on the source of microorganisms. Soil itself can be a large source of non pathogenic microbes that are known as **heterotrophic plate count (HPC)** bacteria. These are indigenous soil microbial inhabitants that do not normally pose adverse human health threats.

Numbers of HPC bacteria are frequently a function of wind duration and speed, as well as processes involving agriculture, mining, or road construction. They can range from essentially zero to 10^5 to 10^6 per m³ of air. Numbers of HPC can be estimated by use of specialized air samplers (Figure 5.15). In this technique, vacuum pumps are used to deliver known volumes of air to a sampler containing a microbial trapping solution. After air collection, microbes trapped in the liquid are identified and enumerated.

There are also numerous airborne pathogenic microorganisms that can infect plants, animals, or humans (Information Box 5.1). Clearly disease caused by airborne microorganisms can be devastating, as in the case of foot and mouth disease of animals, or common cold infections in humans. Generally these pathogens are released from infected animals or humans, or they can be of soil borne origin as in the case of valley fever, which is particularly prevalent in the southwest U.S. Bioaerosols have also been of concern to residents close to biosolid land application sites. A case study on this issue is presented in Chapter 27. Indoor air quality within homes and buildings is also an issue as discussed in Chapter 22.

Step 1. Make a 10-fold dilution series.

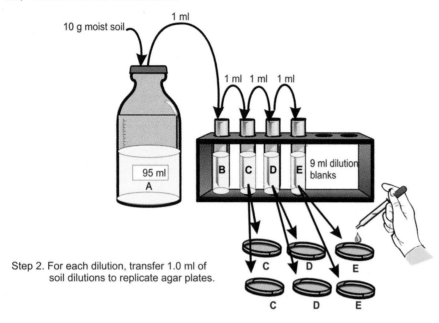

Step 2. For each dilution, transfer 1.0 ml of
soil dilutions to replicate agar plates.

Step 3a. Add molten agar cooled to 45°C
to the dish containing the soil
suspension.

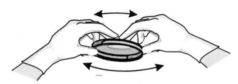

Step 3b. After pouring each plate, replace
the lid on the dish and gently swirl
the agar to mix in the inoculum and
completely cover the bottom of the
plate.

Step 4. Incubate plates under specified conditions.

Step 5. Count dilutions yielding 30-300 colonies
per plate. Express counts as CFUs per
g dry soil.

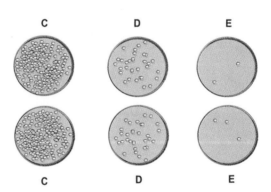

Figure 5.14 Dilution and pour plating technique. Here, the diluted soil suspension is incorporated directly in the agar medium rather than being surface applied as in the case of spread plating. From *Environmental Microbiology* © 2000, Academic Press, San Diego, CA.

5.8.2 Fate and Transport of Bioaerosols

Air is a harsh environment for airborne microbes, and generally contains no nutrients to sustain microorganisms. Particularly in the case of pathogens, there is considerable interest in how far bioaerosols will travel prior to death or inactiva-tion. Plant pathogens such as spores of wheat rust can be spread several hundred or even a thousand kilometers through airborne transmission and remain viable. In contrast, bacterial and viral pathogens are generally inactivated over shorter distances during air borne transmission. Environmental factors clearly affect transport and fate. Desiccation

---

EXAMPLE CALCULATION BOX 5.1

Dilution and Plating Calculations

A 10-gram sample of soil with a moisture content of 20% on a dry weight basis is analyzed for viable culturable bacteria via dilution and plating techniques. The dilutions were made as follows:

Process			Dilution
10 g soil	→	95 ml saline (solution A)	10^{-1} (weight/volume)
1 ml solution A	→	9 ml saline (solution B)	10^{-2} (volume/volume)
1 ml solution B	→	9 ml saline (solution C)	10^{-3} (volume/volume)
1 ml solution C	→	9 ml saline (solution D)	10^{-4} (volume/volume)
1 ml solution D	→	9 ml saline (solution E)	10^{-5} (volume/volume)

1 ml of solution E is pour plated onto an appropriate medium and results in 200 bacterial colonies.

$$\textbf{Number of CFU} = \frac{1}{\text{dilution factor}} \times \text{number of colonies}$$

$$= \frac{1}{10^{-5}} \times 200 \text{ CFU/g moist soil}$$

$$= 2.00 \times 10^7 \text{ CFU/g moist soil}$$

But, for 10 g of moist soil,

$$\text{Moisture content} = \frac{\text{Moist weight} - \text{dry weight (D)}}{\text{dry weight (D)}}$$

Therefore,

$$0.20 = \frac{10 - D}{D} \quad \text{and}$$

$$D = 8.33 \text{ g}$$

Number of CFU per g dry soil $= 2.00 \times 10^7 \times \dfrac{1}{8.33} = 2.4 \times 10^7$

From *Environmental Microbiology* © 2000, Academic Press, San Diego, CA.

---

due to increased air temperature and low relative humidity can kill or inactivate bacteria and viruses. Ultraviolet radiation from sunlight can also damage microbes. With respect to transport, wind speed is the greatest factor that influences movement of bioaerosols.

Figure 5.15 Air samplers used to estimate bioaerosol concentrations. From *Environmental Microbiology Laboratory Manual, Second Edition* © 2004, Academic Pres, San Diego, CA.

INFORMATION BOX 5.1

Examples of Important Airborne Pathogens

Plant Disease	Fungal Pathogen
Dutch Elm Disease	*Ceratocystis ulmi*
Stem rust of wheat	*Puccinia graminis*
Potato blight	*Phytophthora infestans*

Animal Disease	Pathogen
Brucellosis	*Brucella* spp. (bacteria)
Aspergillosis	*Aspergillus* spp. (fungi)
Rabies	*Rhabdoviridae* (virus)
Foot and mouth disease	Aphthovirus (virus)

Human Disease	Pathogen
Pulmonary anthrax	*Bacillus anthracis* (bacteria)
Typhoid fever	*Salmonella typhi* (bacteria)
Valley fever	*Coccidioides immitis* (fungus)
Common cold	Picornavirus
Hepatitis	Hepatitis (virus)

5.9 MICROORGANISMS IN SURFACE WATERS

The study of microorganisms in water is known as aquatic microbiology. Seventy percent of the earth's surface is covered with water in the form of oceans, estuaries, lakes, wetlands, and streams. Clearly the diversity of microbial communities within these environments is enormous, and an in-depth review is beyond the scope of this book. However, microbes that are known to inhabit surface waters include viruses, bacteria, fungi, and protozoa. With respect to pollution we are often concerned with microbes that are either pathogenic or produce toxins that adversely affect human health and welfare (see Chapter 11). In other situations we are concerned with the transport and fate of pathogens introduced into surface waters via, for example, waste disposal, or contamination of surface waters with animal wastes. Contamination of surface waters with animal wastes resulted in a large outbreak of water-borne disease in Milwaukee in 1993. In this outbreak, more than 100 individuals died of infection with *Cryptosporidium*—a protozoan parasite (see Case Study, Chapter 11).

5.9.1 General Characteristics of Microbial Communities in Aquatic Environments

Inland Surface Waters (Lakes, Rivers, Streams): These are essentially freshwater environments not directly affected by marine waters. The study of freshwater microbiology is known as **microlimnology**. If the freshwater is standing water, such as in lakes or ponds, the environment is known as a **lentic habitat**. In contrast, running water is known as a **lotic habitat**, exemplified by streams or rivers.

Marine Waters and Estuaries: True marine waters such as those found in oceans are characterized by salinity of 33–37%. In contrast transitional areas between fresh and marine waters such as estuaries are less saline than marine waters.

TABLE 5.6 Types and numbers of dominant microbes in surface waters.

ENVIRONMENT	TYPE OF MICROBE	NUMBERS PER ml
Springs	Primary Producers	10^6–10^9
	Heterotrophic Bacteria	Up to 10^6
Rivers	Primary Producers	Up to 10^8
	Heterotrophic Bacteria	Up to 10^9
Lakes	Primary Producers	Highly variable
	Heterotrophic Bacteria	Highly variable
Estuary	Primary Producers	Up to 10^7
	Heterotrophic Bacteria	10^6–10^8
Marine	Primary Producers	Up to 10^8
	Heterotrophic Bacteria	Up to 10^8

From *Environmental Microbiology* © 2000, Academic Press, San Diego, CA.

In all of these different types of surface waters there can be vast differences in any specific environment due to variations in environmental conditions (Information Box 5.2). These differences can result in tremendous ranges of microbial numbers in any given environment (Table 5.6). There are two major types of microbial populations in surface waters: **primary producers** and heterotrophic bacteria. The primary producers are phototrophic algae or bacteria, which utilize light and carbon dioxide as a source of energy and carbon. In contrast, heterotrophic populations of bacteria require dissolved organic matter as a carbon and energy source. As can be imagined, the relative numbers of these two groups in any aquatic environment are variable and site specific (Information Box 5.2).

Apart from bacterial and algal populations, streams, rivers, and lakes also contain fungal, protozoan and viral populations. The protozoa and viruses are thought to be important in controlling the bacterial and algal populations. Overall these microbial communities are important in controlling the oxygen content of surface waters, and the degradation of organics (see Chapter 18). Aquatic microorganisms are also important in urban settings through the formation of **biofilms**, in potable water distribution lines (see Chapter 28).

INFORMATION BOX 5.2

Environmental Influences on Aquatic Microbial Communities

Parameter	Microbial Community Influence	Effect
Depth of water	Primary producers	As depth increases photosynthesis and primary producer population decreases
Proximity to habitat	Heterotrophic populations	The greater the proximity to terrestrial habitat, the more dissolved organic matter and heterotrophic activity
Size of stream	Primary producers and heterotrophic populations	As the size increases, more dissolved organic matter is acquired. Heterotrophic activity increases, while light penetration and phototrophic activity decreases.

QUESTIONS AND PROBLEMS

1. Briefly discuss the statement "Soil is a favorable environment as a habitat for microorganisms."

2. For the chemical reaction

$$A + B \leftrightarrow C + D$$

the thermodynamic equilibrium constant $K_{eq} = 0.38$.

Deduce whether ΔG for the reaction is negative or positive. Is energy liberated from the reaction, or must energy be added to promote the reaction?

3. A 10-gram sample of soil with a moisture content of 25% on a dry weight basis is analyzed for viable culturable bacteria via dilution and plating techniques. A 10^{-5} dilution of the soil suspension resulted in 421 bacterial colonies. How many culturable bacteria are there per gram of dry soil?

4. For the following organisms identify the substrate that can be oxidized as well as the terminal electron acceptor utilized by the organism.

   a. *Nitrosomonas* spp.

   b. *Thiobacillus thiooxidans*

   c. *Thiobacillus denitrificans*

   d. *Desulfovibrio* spp.

5. Discuss the likely microbial populations for the following scenarios. Identify modes of nutrition and whether or not populations are likely to be aerobic or anaerobic.

   a. peat bog

   b. pristine alpine lake

   c. waters in the middle of the Pacific Ocean, far away from any island

   d. the air in a city on a rainy day

   e. the air surrounding a farm during a windy day when a dry bare field is being plowed

REFERENCES AND ADDITIONAL READING

Artiola J.F. and I.L. Pepper. (1992) Denitrification activity in the root zone of a sludge amended desert soil. *Biol. Fert. Soils.* **13**, 200–205.

Fuhrman J. (1992) Bacterioplankton roles in cycling of organic matter. The microbial food web. In *Primary Productivity and Biogeochemical Cycles in the Sea* (ed. P.G. Falkowski and A.D. Woodhead) Plenum, New York, pp. 361–383.

Killham K. (1994) "Soil Ecology." Cambridge University Press, Cambridge.

Maier R.M., Pepper I.L. and Gerba C.P. (2000) *A Textbook of Environmental Microbiology.* Academic Press, San Diego, California.

Roszak D.B. and Colwell R.R. (1987) Survival strategies of bacteria in the natural environment. *Microbiol. Rev.* **51**, 365–379.

Stetzenbach L.D. (2001) Introduction to aerobiology. In *Manual of Environmental Microbiology*, 2nd ed. (ed. C.J. Hurst, R.L. Crawford, G.R. Knudsen, M.J. McInerney, and L.D. Stetzenbach (ed.). ASM Press, Washington, DC.

Turco R.F. and Sadowsky M. (1995) The microflora of bioremediation. In *Bioremediation: Science and Applications* (ed. H.D. Skipper and R.F. Turco). Special Publ. No. 43, Soil Science Society of America, Madison, Wisconsin, pp. 87–102.

CHAPTER **6**

PHYSICAL PROCESSES AFFECTING CONTAMINANT TRANSPORT AND FATE

M.L. Brusseau

A trench site in New Mexico used in the study of water and solute movement.
From Pollution Science © 1996, Academic Press, San Diego, CA.

6.1 CONTAMINANT TRANSPORT AND FATE IN THE ENVIRONMENT

An understanding of the transport and fate of contaminants in the environment is required to evaluate the potential impact of contaminants on human health and the environment. For example, such knowledge is needed to conduct risk assessments, such as evaluating the probability that a contaminant spill would result in groundwater pollution. Such knowledge is also required to develop and evaluate methods for remediating environmental contamination. Just as important, knowledge of contaminant transport and fate is necessary to help design chemicals and processes that minimize adverse impacts on human health and the environment to enhance pollution prevention.

The four general processes that control the transport and fate of contaminants in the environment are advection, dispersion, interphase mass transfer, and transformation reactions. These are defined in Information Box 6.1.

Many of the processes influencing contaminant transport and fate are illustrated in Figure 6.1, which shows the disposition of an organic liquid contaminant spilled into the subsurface (DNAPL is a "denser than water" nonaqueous phase liquid).

The fate of a specific contaminant in the environment is a function of the combined influences of these four general processes. The combined impact of the four processes determines the "pollution potential" and "persistence" of a

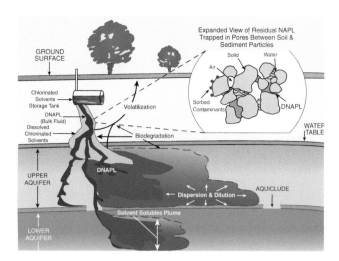

Figure 6.1 Disposition of organic liquid spilled into the subsurface. Schematic of chlorinated solvent pollution: dense nonaqueous phase liquids migrate downward in the subsurface, and serve as a source of contamination for groundwater. Also shown are natural attenuation processes. *DNAPL*, Dense nonaqueous phase liquid; *NAPL*, nonaqueous phase liquid. From U.S. EPA, 1999.

contaminant in the environment. The **pollution potential** characterizes, in essence, the "ability" of the chemical to contaminate the medium of interest (soil, water, air). Compounds that are transported readily (*e.g.*, high aqueous solubility, low sorption), and that are not transformed to any great extent (*i.e.*, are persistent) generally have larger pollution potentials. Large transport rates mean that the contaminant readily spreads from the site where it first entered the environment. A low transformation potential means that the chemical will persist in the environment, and thus remain hazardous, for longer times.

The health risk posed by a specific contaminant to humans or other organisms is, of course, a function of its toxicological characteristics (as discussed in Chapter 13), as well as its pollution potential. Thus, it is important to understand both types of properties. For example, the greatest potential health risk will generally be associated with contaminants that are persistent and highly toxic. However, actual harmful effects will occur only if the organism is exposed to the contaminant. Thus, the pathways of exposure are critical to assessing risks posed by contaminants. The processes influencing the transport and fate of contaminants in the environment have a significant impact on pathways and levels of exposure.

Once a chemical is applied to (or spilled onto) the land surface, it may remain in place or it may transfer to the air, surface runoff, or the subsurface. Chemicals with moderate to large vapor pressures may evaporate or volatilize into the gas phase, thus becoming subject to atmospheric transport and fate processes including advection (carried along by wind), dispersion, and transformation reactions. Breathing contaminated air is one potential source of exposure to hazardous contaminants. However, atmospheric concentrations of most toxic chemicals are relatively low due to dilution effects. Transfer of contami-

---

INFORMATION BOX 6.1

Processes Influencing Contaminant Transport and Fate

Advection is the transport of matter via the movement of a fluid; as discussed in Chapter 3, a fluid moves in response to a gradient of fluid potential. For example, contaminant molecules dissolved in water will be carried along by the water as it flows through (infiltration) or above (runoff) the soil. Similarly, contaminant molecules residing in air will be carried along as the air flows. **Dispersion** represents spreading of matter about the center of the contaminant mass. Spreading is caused by molecular diffusion and nonuniform flow fields. Contaminant molecules can reside in several phases in the environment, such as in air (atmosphere and soil gas phase), in water, and associated with soil particles. So-called **mass transfer** processes, such as sorption, evaporation, and volatilization, involve the transfer of matter between phases in response to gradients of chemical potential or, more simply, concentration gradients. **Transformation reactions** include any process by which the physicochemical nature of a chemical is altered. Examples include biotransformation (metabolism by organisms), hydrolysis (interaction with water molecules), and radioactive decay.

nants to surface runoff during precipitation or irrigation events, is a major concern associated with the non-point source pollution issue. Once entrained into surface runoff, the contaminant may then be transported to surface water bodies. Consumption of contaminated surface water is one potential route of human or animal exposure to toxic chemicals. Another route of toxic chemical exposure for humans is consumption of contaminated groundwater. Once applied to a land surface, a contaminant can partition to the soil-pore water. The contaminant then has the potential to move downward to a saturated zone (aquifer), thereby contaminating groundwater. Whether or not this will occur, as well as the time it will take, and the resulting magnitude of contamination, depends on numerous factors. These include: the magnitude and rate of infiltration or recharge; the soil type; the depth to the aquifer; and the quantity of contaminant and its physicochemical properties (*e.g.*, solubility, degree of sorption, and transformation potential). Volatile contaminants can move by advection and diffusion in the soil-gas phase in addition to the pore-water phase, which provides an additional means to travel to an aquifer.

6.2 CONTAMINANT PROPERTIES

The physicochemical properties of the contaminant have influence in its transport and fate behavior. For example, as noted above, chemicals with moderate to large vapor pressures may evaporate or volatilize into the gas phase, thus becoming subject to atmospheric transport and fate processes. Such chemicals can also undergo transport in the gaseous phase in the vadose zone. As another example, chemicals with larger aqueous solubilities will more readily transfer to water, and thus be subject to transport by water flow. The physicochemical proper-

ties of contaminants are controlled by their molecular structure (see Chapter 7). The biodegradability of contaminants is also dependent upon their molecular structure (see Chapter 8).

A critical property to consider when evaluating transport and fate behavior is the phase state of the contaminant. Under "natural" conditions (temperature $T = 25°C$, pressure $P = 1$ atm), chemicals in their pure form exist as solids, liquids, or gases (see Table 6.1). Clearly, the mobility of a chemical in the environment will depend in part on the phase in which it occurs, with gases generally being most mobile and solids least mobile.

Many of the organic contaminants of greatest concern happen to exist as liquids in their pure state under natural conditions. These organic compounds are referred to as immiscible or **nonaqueous phase liquids (NAPLs)**. Examples of NAPLs include fuels (gasoline, aviation fuel), chlorinated solvents, and polychlorinated biphenyls. The presence of NAPLs in the subsurface at a contaminated site greatly complicates remediation efforts (see Chapter 19). Once released into the subsurface, the NAPL becomes trapped in pore spaces, after which it is very difficult to physically remove. Hence, they serve as long-term sources of contamination as the molecules transfer to other phases (see Chapter 17). An additional complicating factor is that many NAPLs comprise multiple constituents. Examples of such multi-component NAPLs include fuels (gasoline, diesel fuel, and aviation fuel), coal tar, and creosote, all of which contain hundreds of organic compounds. These multi-component NAPLs can contain individual compounds, such as naphthalene and anthracene, that normally occur as solids but which are "dissolved" in the organic liquids.

Most inorganic contaminants of concern occur as solids in their elemental state. One notable exception is mercury,

TABLE 6.1 Contaminant properties.

	REPRESENTATIVE CONTAMINANTS	SOLUBILITY	VAPOR PRESSURE	VOLATILITY	SORPTION POTENTIAL	BIODEGRADATION RATE
Solids						
Organic	naphthalene	low	medium	medium	medium	medium
	pentachlorophenol	low	medium	low	high	medium
	DDT	low	low	low	high	low
Inorganic	lead	low	low	low	medium	non degradable
	chromium	high	low	low	low	non degradable
	arsenic	medium	low	low	low	non degradable
	cadmium	low	low	low	medium	non degradable
Liquids						
Organic	trichloroethene	medium	high	medium	low	low
	benzene	medium	high	medium	low	medium
Inorganic	mercury	low	medium	low	medium	non degradable
Gases						
Organic	methane	medium	very high	very high	low	low
Inorganic	carbon dioxide	medium	very high	very high	low	non degradable
	carbon monoxide	low	very high	very high	low	non degradable
	sulfur dioxide	medium	very high	very high	low	non degradable

From *Environmental Monitoring and Characterization* © 2004, Elsevier Academic Press, San Diego, CA.

which is a liquid under standard conditions. An important factor for inorganic contaminants is their "speciation." For example, many inorganics occur primarily in ionic form in the environment (*e.g.*, Pb^{+2}, Cd^{+2}, NO_3^-). Speciation can greatly influence aqueous solubility and sorption potential. In addition, many inorganics may combine with other inorganics, forming complexes whose transport behavior differs from that of the parent ions. These concepts are discussed further in Chapter 7.

6.3 ADVECTION

As introduced above, advection is the transport of matter by the movement of a fluid. Any contaminant dissolved in water (which we call a "solute") will be carried along by the water as it flows. Similarly, contaminant molecules residing in air (vapor phase) will be carried along as the air flows. Advection is generally the single most important mode of transport of contaminants in the environment. Advective transport by water or air is the primary reason for large-scale movement of contaminants in the environment.

Transport of contaminant mass by advection is proportional to the rate of fluid movement. Thus, characterizing the advective transport potential of a contaminant at a specific location requires one to determine the direction and rate of fluid flow at that location. Monitoring air flow in the atmosphere was discussed in Chapter 4. Determining the direction and rate of surface water flow is relatively straightforward—*e.g.*, for streams, the direction is defined by the stream channel and the rate is related to the water level (see Chapter 3). Conversely, characterizing the direction and rate of water or air movement in the subsurface is much more complex.

Characterizing groundwater flow in an aquifer system generally requires constructing groundwater maps. These maps are basic tools of hydrogeological interpretation from which one can derive the magnitude of hydraulic gradient and the direction of groundwater flow (see Figure 6.2). To construct these maps, one needs hydraulic head measurements from a number of wells or piezometers that are installed at various depths in an aquifer of interest. Groundwater maps are prepared by plotting hydraulic head values of all wells on a base map showing the well locations, and drawing a line connecting the points of equal hydraulic heads (equipotential line). Equipotential lines are somewhat similar to elevation contours on a topographic map.

In three dimensions, the points with equivalent head values form an equipotential surface called the piezometric or potentiometric surface. The potentiometric surface is simply the map of hydraulic head, or the level to which water will rise in a well that is screened within the aquifer. If an aquifer is characterized by primarily horizontal flow, one can draw a potentiometric surface for the entire aquifer. In an unconfined aquifer, the water table represents the potentiometric surface of the aquifer. However, in a confined aquifer

Figure 6.2 Example of a groundwater contour map. Image courtesy of the Arizona Department of Water Resources and the University of Arizona Water Resources Research Center.

(one where a low-permeability unit resides on top of the aquifer), the potentiometric surface is an imaginary surface connecting the water levels in wells. Groundwater maps are actually two-dimensional representations of three-dimensional potentiometric surfaces, and are denoted as water-table maps for unconfined aquifers, and potentiometric-surface maps for confined aquifers.

Measurements of groundwater velocity are important for hydrogeological studies and environmental monitoring applications including predicting the rate of contaminant movement. There are several techniques available for determining groundwater flow velocities in the field. These include estimation based on Darcy's Law, borehole flowmeters, and tracer tests. An estimation of groundwater velocity can be obtained using **Darcy's Law**, rewritten as:

$$v = q/n = (K\,\Delta h/\Delta l)/n$$

where:

v is average pore-water velocity (L/T)
q is Darcy velocity (L/T)
n is porosity $(-)$
K is hydraulic conductivity (L/T)
$\Delta h/\Delta l$ is hydraulic gradient $(-)$

This approach requires knowledge of hydraulic conductivity and the hydraulic gradient, which can be obtained as described in Chapter 3. Borehole flowmeters are used for many applications including well-screen positioning, recharge-zone determination; and estimation of hydraulic-conductivity distribution. In addition, they can be used for measurements of horizontal and vertical flow characteristics in a cased well or borehole. Tracers may also be used in field studies to determine velocities of groundwater. A tracer test is conducted by injecting a tracer solution into the aquifer and monitoring tracer concentrations at downgradient locations. The time required for the tracer to travel from the injection point to the monitoring point can be used to calculate groundwater velocity. A suitable tracer should not react physically or chemically with groundwater or aquifer material, and must not undergo transformation reactions. These types of substances are generally called conservative tracers.

6.4 DISPERSION

Dispersion represents spreading of matter about the center of the contaminant mass. In essence, as a mass of contamination (called a "pulse" or "plume") moves by advection, the size of the plume increases due to dispersion. In other words, the plume "grows" as it moves. This spreading is caused by molecular diffusion and nonuniform flow fields. It is important to note that spreading occurs in both water flow and air flow systems.

Molecular diffusion is the result of random motion of individual molecules. Every molecule vibrates and moves due to its individual kinetic energy. At any given time, each molecule may move in any given direction; thus, we call this random motion. The net result of this series of individual movements is that molecules will spread from regions containing greater numbers of molecules (*i.e.*, higher concentrations) to regions with fewer numbers (lower concentrations).

The effect of diffusion is readily seen by adding a drop of food coloring dye to a beaker of water—the dye will spread and eventually color the entire volume of water. The contribution of molecular diffusion to overall transport and spreading of a contaminant is generally small. It becomes significant primarily only in systems where advection is minimal, such as clay units in the saturated zone of the subsurface.

The major cause of dispersion is nonuniform flow fields. When a fluid flows through the environment, it does not move as one uniform body. Rather, sections of the fluid move at different rates or velocities. For example, in a river, water flowing along the channel walls moves slower due to friction, than water located in the center of the channel. In the subsurface, the dispersion effect occurs at a range of spatial scales. For smaller scales ($<$ 1 meter), dispersion is caused by non-uniform flow in the porous-medium pores, and occurs in three major ways (Figure 6.3a). First, fluid flow in a single pore is faster in the center of the pore because friction slows the fluid near the pore walls. Second, fluid flow is faster in larger-diameter pores than it is in smaller-diameter pores (a smaller proportion of fluid is influenced by friction for larger pores). Third, the time it takes a fluid or contaminant molecule to travel from one location to another is less

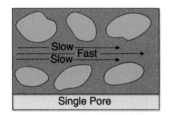
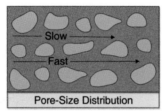
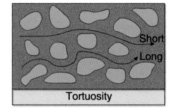

A

Single Pore   Pore-Size Distribution   Tortuosity

Cross-Sectional View

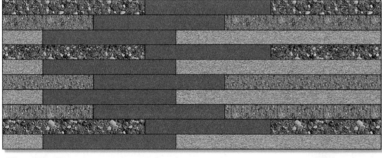

Flow Direction ⟶

	Gravel
	Sand
	Silt

Map View

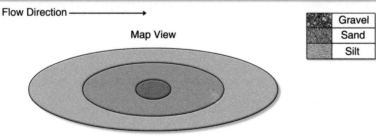

B

Figure 6.3 Processes causing dispersion (spreading of a pulse or plume) as a result of non-uniform fluid flow: (A) pore scale; (B) field scale.

for pore sequences that have fewer twists and turns (less tortuous flow path). For larger scales (field scale), fluid moves faster in larger-permeability units such as sand and slower in lower-permeability units such as clay (see Chapter 3). As a result, the rates of advection differ for different portions of the fluid. Thus, as a plume of dissolved contaminant moves, different sections of the plume will be moving at different velocities. This results in spreading or "growth" of the plume in the direction of travel. This dispersion effect is illustrated in Figure 6.3b.

6.5 MASS TRANSFER

As noted in a previous section, a contaminant may reside in the air, in water, or associated with a solid phase. The transfer of contaminants from their original phase to other phases is an important aspect of contaminant behavior. The primary mass-transfer processes of interest for contaminant transport in the environment are: dissolution, evaporation, volatilization, and sorption. These four processes will be discussed briefly below; a more detailed discussion is presented in Chapter 7.

The transfer of molecules from their pure state to water is called **dissolution**. For example, placing salt crystals in water will result in dissolution—the salt will dissolve in the water. Similarly, placing an immiscible organic liquid (such as a cleaning solvent) in contact with water will result in the transfer of some of the organic-liquid molecules to the water. This is how water becomes polluted by a contaminant. The extent to which contaminant molecules or ions will dissolve in water is governed by their aqueous solubility (see Chapter 7). The solubilities of different contaminants vary by orders of magnitude. A critical aspect of solubility is how it compares to pollution action levels for the contaminant. For example, the aqueous solubility of trichloroethene (a chlorinated cleaning solvent) is approximately 1 g/L. This is very low compared to many other compounds. However, the maximum contaminant level for trichloroethene is 5 μg/L, 200,000 times smaller. Thus, it does not take very much trichloroethene to pollute water.

Evaporation of a compound involves transfer from the pure liquid or pure solid phase to the gas phase. The vapor pressure of a contaminant, then, is the pressure of its gas phase in equilibrium with the solid or liquid phase, and is an index of the degree to which the compound will evaporate. In other words, we can think of the vapor pressure of a compound as its "solubility" in air. Evaporation can be an important transfer process when pure-phase contaminant is present in the vadose zone, such that contaminant molecules evaporate into the soil gas.

Volatilization is the transfer of contaminant between water and gas phases. Volatilization is different from evaporation, which specifies a transfer of molecules from their pure phase to the gas phase. For example, the transfer of benzene molecules from a pool of gasoline to the atmosphere is evaporation, whereas the transfer of benzene molecules from water (where they are dissolved) to the atmosphere is volatilization. The vapor pressure of a compound gives us a rough idea of the extent to which a compound will volatilize, but volatilization also depends on the solubility of the compound as well as environmental factors. Volatilization is an important component of the transport and remediation of volatile organic contaminants in the vadose zone (see Chapters 17 and 19).

Sorption also influences the transport of many contaminants in the environment. The broadest definition of sorption (or retention) is the association of contaminant molecules with the solid phase of the porous medium (*e.g.*, soil particles). Sorption can occur by numerous mechanisms, depending on the properties of the contaminant and of the sorbent (see Chapter 7). A critical impact of sorption is that it slows or retards the rate of movement of contaminants (referred to as "retardation"). As long as the soil grains are immobile, contaminant molecules will also be immobile when they are sorbed to the grains. Thus, contaminants that sorb will move more slowly than the mean water velocity. Conversely, contaminants that do not sorb will generally move at the same velocity as water.

6.6 TRANSFORMATION REACTIONS

The transport and fate of many contaminants in the environment is influenced by transformation reactions, such as biodegradation, hydrolysis, and radioactive decay. The susceptibility of a contaminant to such transformation reactions is very dependent on their molecular structure (organics) or speciation (inorganics). For example, chlorinated hydrocarbons are generally more resistant to biodegradation than are nonchlorinated hydrocarbons. Abiotic and biotic transformation reactions are discussed in Chapters 7 and 8, respectively.

A critical impact of transformation reactions on transport and fate is that the mass of original contaminant is reduced. This is usually a positive result, as it leads to reducing the amount of potential pollution present in the environment. However, in some cases it may produce a negative result, such as when the transformation reaction produces a more hazardous compound.

6.7 CHARACTERIZING SPATIAL AND TEMPORAL DISTRIBUTIONS OF CONTAMINANTS

Characterizing the transport and fate behavior of contaminants in the environment is based on evaluating their distribution in the environment in terms of phase distribution (are they in groundwater, sorbed to soil, etc.), spatial distribution, and temporal distribution (*i.e.*, how concentrations change with time). The information needed to assess these distributions is obtained from sampling and monitoring programs, as

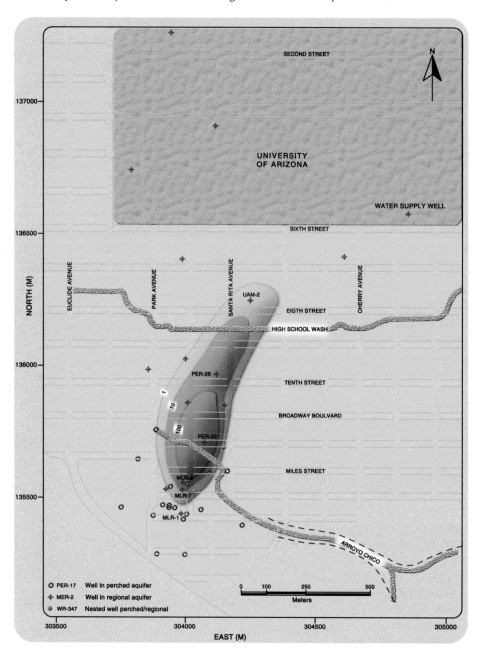

Figure 6.4 Groundwater contaminant plume contour map for a site in Tucson, Arizona. The contours represent tetrachloroethene concentrations in Mg/L. (Drawn by Concepción Carreón Diazconti.)

described in Chapter 12 (detailed information on environmental monitoring is available in Artiola et al., 2004).

Examining spatial distributions of contamination is based on developing contaminant contour maps. This is done by plotting contaminant concentrations, obtained from analysis of samples collected from discrete points in space, on a map of the site. The points of equal concentrations are connected by contour lines. This procedure produces a diagram of the contaminant plume, which provides a means to visualize the size and distribution of contamination at the site (see Figure 6.4). These contaminant plume maps can be produced periodically (at different times), which allows exami-

nation of the transport behavior of the contamination—is the contaminant plume moving or staying in place?

Simple examples of spatial distributions of contaminant plumes as affected by the four transport and fate processes are presented in Figure 6.5. The impact of fluid velocity on advective transport is illustrated by comparing the locations of plumes 1 and 2—plume 1 has traveled a greater distance downgradient from the point of origin because of the greater fluid velocity for that case. The impact of dispersion is shown for plume 3, which is longer than plume 2 (which has no dispersion). The impact of sorption and retardation on transport is observed for plume 4, which has not traveled as

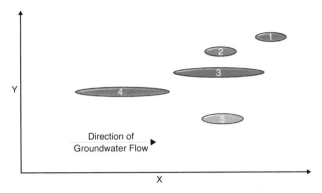

Figure 6.5 Simple examples of spatial distributions of contaminant plumes as affected by the four transport and fate processes. Case 1: Advection (larger fluid velocity, v); Case 2: Advection (small v); Case 3: Advection (smaller v) and Dispersion; Case 4: Advection (smaller v), Dispersion, and Retardation; Case 5: Advection (smaller v), Dispersion, and Transformation. The plumes are drawn with a single contour representing the detectable concentration. The plumes all originated at the $x = 0$ plane.

far as plume 3. The impact of a transformation reaction that causes loss of contaminant mass is illustrated for plume 5, which is smaller than plume 3.

The change in contaminant concentrations at a site over time is an important aspect of the risk posed by the contamination. The temporal variability of contaminant concentrations is characterized by plotting concentration histories, in which concentrations of samples collected from one location at different times are plotted as a function of time. A special case of these concentration histories is obtained when we are monitoring a location at which contamination is just beginning to arrive. These plots are called breakthrough curves.

Simple examples of breakthrough curves as affected by the four transport and fate processes are presented in Figure 6.6. The impact of fluid velocity on advective transport is illustrated by comparing the locations of breakthrough curves 1 and 2; breakthrough curve 1 appears earlier than breakthrough curve 2 because of the greater fluid velocity for case 1. The impact of dispersion is shown for breakthrough curve 3, which is more spread out (rotated clockwise) than breakthrough curve 2 (which has no dispersion). The impact of sorption and retardation on transport is observed for breakthrough curve 4, which appears later than breakthrough curve 3. The impact of a transformation reaction that causes loss of contaminant mass is illustrated for breakthrough curve 5, which peaks at a lower concentration than breakthrough curve 3.

6.8 ESTIMATING PHASE DISTRIBUTIONS OF CONTAMINANTS

As discussed in the beginning of this chapter, contaminants can reside in multiple phases of the environment—as vapor in air, dissolved in water, and sorbed to porous-medium particles. It is often important to know how a contaminant will distribute among these phases. For example, conducting risk assessments requires an evaluation of potential routes of exposure. To do this, we would need to know if and how much contaminant is present in the various phases. In addition, we would like to know how much contaminant is present in the various phases when we design systems to clean up contaminated sites.

A simple way to estimate contaminant distributions is to use phase-distribution coefficients. These coefficients provide information about the distribution of contaminant between two phases, and are presented as ratios of concen-

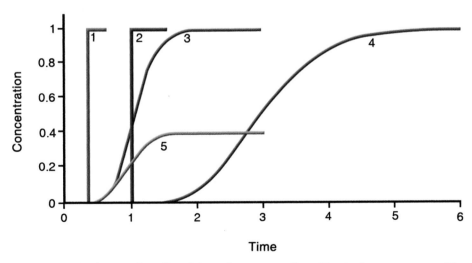

Figure 6.6 Simple examples of breakthrough curves as affected by the four transport and fate processes. Case 1: Advection (larger fluid velocity, v); Case 2: Advection (smaller v); Case 3: Advection (smaller v) and Dispersion; Case 4: Advection (smaller v), Dispersion, and Retardation; Case 5: Advection (smaller v), Dispersion, and Transformation. The input concentration is equal to 1 concentration unit.

trations in the two phases. Two key distribution coefficients are the sorption coefficient K_d ($K_d = S/C_w$), which describes the distribution of a contaminant between porous-medium particles and water (*i.e.*, sorption), and Henry's Constant, H ($H = C_g/C_w$), which describes the distribution between gas (air) and water phases (*i.e.*, volatilization).

In these equations,

> C_w is the concentration of contaminant in water
> C_g is the concentration of contaminant in air
> S is the concentration of contaminant in the sorbed phase

The first piece of information of interest is the total amount of contaminant in the system. For example, the total mass of contaminant contained in a given volume of soil can be defined as:

$$M = V_w C_w + M_s S + V_g C_g \qquad \text{(Eq. 6.1)}$$

where:

> M is total contaminant mass
> V_w is the volume of the water phase
> M_s is the mass of soil particles
> V_g is the volume of soil gas

We can define a unit-mass of contaminant (M^*) by dividing equation 1 by V_T, the total volume of the soil system. We can also define M^* in terms of one concentration by substituting the distribution equations [$S = K_d C_w$, $C_g = H C_w$] into Equation 6.1. We use C_w as the key concentration since it is usually the concentration measured in subsurface monitoring programs. The modified equation for estimating contaminant mass is:

$$M^* = [\theta_w + \rho_b K_d + \theta_g H] C_w \qquad \text{(Eq. 6.2)}$$

where:

> θ_w is the soil-water content (volume of water per volume of soil)
> θ is the soil-gas content (volume of soil gas per volume of soil)
> ρ_b is soil bulk density (mass of soil per volume of soil)

Remember that M^* is mass of contaminant per soil volume element; thus, total contaminant mass (M) is calculated by $M^* \times V_T$.

The fraction of contaminant residing in each phase can be calculated by the following equations:

$$\text{Fraction in Water} = \frac{\theta_w C_w}{M^*} \qquad \text{(Eq. 6.3)}$$

$$\text{Fraction Retained by Soil Particles} = \frac{\rho_b K_d C_w}{M^*} \qquad \text{(Eq. 6.4)}$$

$$\text{Fraction in Soil Gas} = \frac{\theta_g H C_w}{M^*} \qquad \text{(Eq. 6.5)}$$

It is important to remember that this approach is based on assuming that the distribution processes have reached equilibrium. If this is not true, the estimates obtained with Equations 6.1–6.5 can be erroneous.

EXAMPLE CALCULATION 6.1

The calculation of contaminant phase distributions will be illustrated with the following example. We will assume the following values for the soil properties: $\theta_w = 0.25$, $\theta_g = 0.25$, and $\rho_b = 1.5$ g/cm^3. To simplify the calculations, we will assume

$$K_d = 1 \frac{mL}{g}$$

$$H = 1$$

$$C_w = 1 \text{ mg/L}$$

Using these values, the unit-mass of contaminant is calculated to be:

$$M^* = [0.25 + 1.5 \frac{g}{mL} \times 1 \frac{mL}{g} + 0.25 \times 1]$$

$$\times 1 \frac{mg}{L} = 2 \frac{mg}{L}$$

The fraction of pollutant residing in each phase is:

$$\text{Fraction in water} = \frac{0.25 \times 1 \frac{mg}{L}}{2 \frac{mg}{L}} = 0.125$$

$$\text{Fraction in Soil Gas} = \frac{0.25 \times 1 \times 1 \frac{mg}{L}}{2 \frac{mg}{L}} = 0.125$$

Fraction Retained by Soil Particles

$$= \frac{1.5 \frac{g}{ml} \times 1 \frac{mL}{g} \times 1 \frac{mg}{L}}{2 \frac{mg}{L}} = 0.75$$

Thus, 75% of the contaminant mass is associated with the soil particles, while 12.5% is associated each with water and with air.

6.9 QUANTIFYING CONTAMINANT TRANSPORT AND FATE

The transport and fate of contaminants in the environment is quantified by developing a set of governing equations. These equations are used to represent the various processes that influence transport and fate, as discussed above. The set of equations are developed based upon a conceptual model of the transport system (*i.e.*, an idea of what is happening in the system). This set of equations constitutes a mathematical model of the system. It is important to recognize that a mathematical model is an inexact representation of reality, built upon a suite of assumptions and simplifications. The accuracy of a mathematical model will depend on the validity of the assumptions and representativeness of the simplifications.

The most widely used model to represent the transport and fate of contaminants in the environment is the advection-dispersion equation. A simplified version of this equation for solute transport in porous media is given by:

$$R \frac{\partial C}{\partial t} = -v \frac{\partial C}{\partial x} + D \frac{\partial^2 C}{\partial x^2} - T \qquad \text{(Eq. 6.6)}$$

where:

C is contaminant concentration
D is the dispersion coefficient
R is the retardation factor
T represents possible transformation reactions
v is mean fluid velocity
t is time, and x is distance

This equation is developed with numerous assumptions, including uniform, one-dimensional fluid flow, homogeneous conditions (for example, a homogeneous porous medium), and that sorption is the only mass-transfer process.

The term on the left-hand side of the equation represents the change in the amount of contaminant present at a given location. Changes in the amount of contaminant present at a given location are caused by transport of contaminant to and from that location, which occurs by advection and dispersion, and by mass transfer and transformation processes. The first term on the right-hand side of the equation represents advection, while the second term represents dispersion. The term T represents potential transformation reactions. A specific equation would be used for each given transformation process. One widely used equation is the "first-order" reaction equation, which is discussed in Chapter 7.

The effect of sorption on transport, namely retardation, is represented by the retardation factor, R. The retardation factor is defined as:

$$R = 1 + \frac{\rho_b}{\theta} K_d = \frac{v_w}{v_p} = \frac{d_w}{d_p} \qquad \text{(Eq. 6.7)}$$

where:

v_w is the velocity of the fluid
v_p is velocity of the contaminant
d_w is the distance traveled by the fluid
d_p is the distanced traveled by the contaminant

Inspection of Equation 6.7 shows that as sorption increases (*i.e.*, K_d increases), the magnitude of the retardation factor increases. When the contaminant is not sorbed by the soil (*i.e.*, $K_d = 0$), the retardation factor is equal to 1. This means that the contaminant will move at the same velocity as the mean velocity of the fluid [$v_p = v_w$]. When $R = 2$, the contaminant moves at an effective velocity that is half that of the fluid. In other words, the contaminant moves only half as fast as the fluid. When $R = 10$, the contaminant moves at one-tenth the velocity of water. Contaminants that have small retardation factors (< 10) are considered to be relatively mobile. They can move rapidly from a spill site and thus quickly contaminant a large area. Conversely, contaminants that have very large retardation factors ($> 1,000$) move very slowly with respect to fluid flow. Thus, they probably will not create a contaminated zone as extensive as that created by mobile contaminants.

A special case exists for some systems wherein the retardation factor is less than 1, which would mean that the solute moves *faster* than the fluid by which it is being carried. A major example of this behavior is the transport of anionic substances such as Cl^- in the subsurface. As discussed in Chapter 2, the surfaces of many soils have a net-negative charge. This leads to a repulsion of negatively charged solutes. For soil domains characterized by very small pores, this expulsion could prevent the solute from entering the water residing in the pores. Thus, the solute would travel through only a portion of the soil, which results in transport that appears more rapid than the rate of water movement. This effect is termed anion exclusion. Of course, the individual solute molecules are not actually moving faster than the water molecules; they only appear to do so because they travel through a smaller portion of the porous medium than do the water molecules. Another cause of $R < 1$ behavior is called size exclusion. In this case, extremely large molecules may be too large to pass through the smallest pores comprising a soil. Thus, just like the case above, the solute would travel through only a portion of the soil, resulting in transport that appears more rapid than the rate of water movement. Size exclusion has been observed in transport experiments for large colloids such as bacteria and protozoa.

Contaminant transport in the environment is usually much more complicated than what is represented by the simple advection-dispersion equation presented above. More complicated models have been developed to attempt to quantify transport and fate in real systems. A major factor limiting our ability to use these complex models is the difficulty in determining values for all of the unknown parameters in the model. This requires extensive characterization of the site, which is expensive and time consuming.

QUESTIONS AND PROBLEMS

1. A tanker truck containing a large volume of a liquid contaminant has overturned off the side of the freeway, and the contents of the tank have spilled onto the ground surface.

   a. How will the compound move through the environment?

   b. Identify and define the major processes that will control the transport and fate of the contaminant.

   c. What contaminant properties will affect the transport of the compound in the environment?

2. a. Calculate the total contaminant mass given the following data:

   $V_T = 1 \text{ m}^3$ $H = 0.5$

   $\theta_w = 0.3$ $K_d = 2 \text{ mL/g}$

   $\theta_g = 0.1$

   $\rho_b = 1.5 \text{ g/cm}^3$

   $C_w = 0.1 \text{ mg/L}$

   b. Calculate the fraction of contaminant present in water, in soil atmosphere, and sorbed to the soil.

3. a. Calculate retardation factors for the following chemicals, given the following data:

CHEMICAL	K_d (ml/g)	R
Benzene	0.1	
Trichloroethane	0.2	
Chlorobenzene	0.4	
Naphthalene	0.6	
PCB	10	

[$\rho_b = 1.5 \text{ g/cm}^3$, $\theta_w = 0.3$]

   b. Calculate the distances traveled by each chemical given the following information: the velocity of water moving through the soil is 1 cm/day; the elapsed time is 300 days.

REFERENCES AND ADDITIONAL READING

Artiola J.F., Pepper I.L., and Brusseau M.L. (2004). *Environmental Monitoring and Characterization*. Academic Press, San Diego, California.

CHAPTER **7**

CHEMICAL PROCESSES AFFECTING CONTAMINANT TRANSPORT AND FATE

M.L. Brusseau and J. Chorover

The use of permanganate solution (dark red) to oxidize trichloroethene (light red) trapped in sand. The zone of black discoloration in the "after" photo results from the formation of MnO_2 solids associated with the degradation of trichloroethene when it reacts with permanganate.
Photo courtesy Justin Marble.

7.1 INTRODUCTION

From the perspective of chemistry, the environment is a heterogeneous system, meaning that it contains solid, liquid, and gaseous phases. As discussed in Chapter 6, the transport behavior of contaminants in the environment is influenced by their partitioning or transfer among these phases and also by transformation reactions. Major mass-transfer processes will be discussed in this chapter, including precipitation-dissolution and sorption-desorption. We will examine physical and chemical properties of contaminants that influence their speciation and solubility in water, and the magnitude of their sorption by porous media. The mass-transfer behavior of inorganic and organic contaminants can be quite different; therefore, we will discuss them separately. In both cases, however, we rely on the principles of thermodynamic equilibrium to assist in the prediction of phase-transfer processes. This will be followed by a brief discussion of selected transformation reactions that alter the physical and chemical properties of contaminants.

7.2 BASIC PROPERTIES OF INORGANIC CONTAMINANTS

7.2.1 Speciation of Inorganic Pollutants

The precise chemical form of an element at a given point in time and space is defined as its speciation. Aqueous-phase speciation, which can be predicted on the basis of thermodynamics, is strongly affected by the environmental redox status and pH. Whereas redox status affects an element's most stable oxidation state, the solution-phase acidity (pH) affects its charge and degree of hydrolysis.

For example, the trace element arsenic (As) occurs in two different principal oxidation states [As(III) and As(V)], and each of these oxidation states is represented by various species with differing reactivities. The reduced form—As(III) —is favored under oxygen-depleted conditions, such as might be found in wetland sediments, and it occurs principally as the neutral species of arsenite, $As(OH)_3^0$, in the pH range (2–9) of most natural waters. The oxidized form—As(V), known as arsenate—is favored under well-oxygenated conditions, and occurs dominantly as the monovalent anion $H_2AsO_4^-$ between pH 3 and 6, and as the bivalent anion $HAsO_4^{2-}$ at pH > 7. Aqueous-phase speciation is also affected by the availability of complexing agents that can form **soluble complexes** with the ion or molecule of interest.

In addition to influencing the aqueous-phase speciation of solutes, pH and redox status also affect the solubility and charge of mineral solids that serve to sequester inorganic contaminants in soils. Solid-phase speciation of a contaminant can be diverse, as it is for aqueous systems. Contaminants become part of the solid phase via various mechanisms of **adsorption** to the surfaces of natural particles, and/or **precipitation** within existing or newly formed

solids. Thus, solution-phase speciation controls the total amount of a contaminant element in equilibrium with these adsorbed or solid phases. The various inter-related chemical processes affecting the disposition of inorganic compounds in heterogeneous environmental systems are depicted in Figure 7.1. They are discussed in more detail in the following sections.

7.2.2 Aqueous Phase Activities and Concentrations

Prediction of contaminant behavior requires working with balanced chemical reactions that represent changes in speciation both within a phase (*e.g.*, within the aqueous solution), and between phases (*e.g.*, between the solution and solid phases). A **balanced chemical reaction** is one that contains an equal number of moles of every element (*mass balance*) and an equal number of moles of charge (*charge balance*) on both sides of the equation. Since water is ubiquitous in the environment, many chemical reactions of inorganic contaminants involve aqueous species as reactants or products.

When we use thermodynamic equilibrium constants to predict speciation, the reactions must be written in terms of *effective concentration* or **activities** of dissolved species, not their concentrations. The activity and concentration of a species, *i*, are related through the **activity coefficient** (γ_i):

$$(i) = \gamma_i\,[i] \qquad \text{(Eq. 7.1)}$$

where the term in parentheses on the left side represents activity and the term in square brackets on the right side denotes aqueous-phase concentration in either *molal* (mol kg^{-1}) or *molar* (mol L^{-1}) concentration units. *Molal* is used when a mass-based concentration is desired and *molar* is

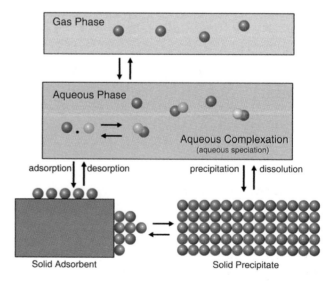

Figure 7.1 Disposition of inorganic contaminants in the environment is controlled by aqueous-phase complexation reactions, precipitation-dissolution reactions, and sorption-desorption reactions.

used when a volume-based concentration is needed. Since activity itself is unitless, effective units for γ_i are inverse to those of concentration.

In very dilute solutions similar to pure water, the activity coefficients of all dissolved species approach the value of 1.0 and therefore concentration is equal to activity. As the concentration of electrolytes increases, however, the behavior of individual ions is affected by the presence of others that are in close proximity. This results in a decrease in the value of γ_i with increasing electrolyte concentration over the range observed for fresh waters, *and a corresponding decrease in activity relative to concentration* of ion i. In essence, an activity coefficient that deviates from 1.0 indicates that the impact of that particular species on the aqueous system (*i.e.*, its activity) is not directly proportional to the amount of the species present (*i.e.*, its concentration). An activity coefficient less than 1, as is observed for inorganic species, represents a case wherein the activity is reduced relative to the amount of the species present. Conversely, as will be discussed below, many organic compounds have activity coefficients greater than 1. Knowledge of activity coefficients is important for evaluating the behavior of contaminants in aqueous systems.

The reduced-activity effects observed for inorganics arise from mutual electrostatic interactions that are proportional to the charges of the ions involved. They are embodied in the definition of the **ionic strength** (I) of an aqueous solution, which is given by

$$I = \tfrac{1}{2} \sum_i [i]\, Z_i^2 \qquad \text{(Eq. 7.2)}$$

where Z_i is the valence of species i and the sum is over all charged species in solution. For example, the ionic strength of a 0.05 M NaCl solution is given by:

$$I\,(0.05\ \text{M NaCl}) = \tfrac{1}{2}\,[(0.05)1^2\ (\text{From Na})$$
$$+ (0.05)1^2]\,)\ (\text{From Cl}) = 0.05\ \text{M} \quad \text{(Eq. 7.3a)}$$

whereas the ionic strength of a 0.05 M $CaCl_2$ solution is substantially higher:

$$I\,(0.05\ \text{M CaCl}_2) = 1/2\,[(0.05)2^2\ (\text{From Ca})$$
$$+ 2(0.05)1^2]\ (\text{From Cl}_2) = 0.15\ \text{M} \quad \text{(Eq. 7.3b)}$$

Activities of individual, charged aqueous species are then estimated from empirical relationships between γ_i and I as given by the **Davies equation**:

$$\log \gamma_i = -0.512 Z_i^2 \left[\frac{\sqrt{I}}{1 + \sqrt{I}} - 0.3 I \right] \qquad \text{(Eq. 7.4)}$$

This equation is valid for natural waters with ionic strengths approaching that of seawater ($\simeq 0.7$ M). The Davies equation shows that the influence of ionic strength on γ_i increases with charge (Z) of the dissolved species (Figure 7.2). This reflects the fact that ions of higher charge interact more strongly with each other, as is expected for Coulombic attraction (or repulsion) between ions of opposite (or like) charge. It is important to understand the influence of ionic strength on speciation of inorganics in aqueous solution, because for example, the transport behavior of metals will be

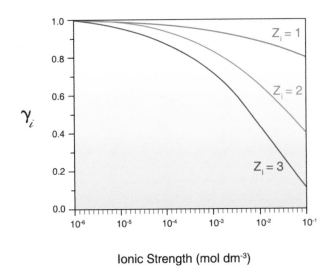

Figure 7.2 The influence of ionic strength on activity coefficients of charged ($Z_i = 1$, 2, or 3) aqueous species, as calculated from the Davies Equation.

affected by ionic strength. Thus, metal transport in sites with highly saline conditions may be different from that for sites with low-salinity (*i.e..*, lower ionic strength) water.

7.2.3 Ion Hydration, Ion Hydrolysis, and Acid-Base Reactions

The activities of solutes impact the rate and extent of various phase-transfer processes and biotic uptake (by plants and microbes). Inorganic solutes are present as cations, anions, or neutral molecules. Major species present in natural waters affect the fate of contaminants via their mutual participation in chemical reactions. Major species include: (1) *nonhydrolyzing cations* (Na^+, Ca^{2+}, Mg^{2+}, K^+); (2) *hydrolyzing cations* (Al^{3+}, H^+); (3) *strong acid anions* (Cl^-, NO_3^-, SO_4^{2-}); and (4) *weak acid anions* (CO_3^{2-}, HCO_3^-, organic acids). Inorganic pollutants also fall into these four categories (See Table 7.1), but are often present at much lower concentrations than the major species.

TABLE 7.1 Classification of several important inorganic contaminants on the basis of their hydrolysis and acid-base behavior in aqueous solution.*

Nonhydrolyzing Cations:	$^{137}Cs^+$, $^{90}Sr^{2+}$
Hydrolyzing Cations:	Al^{3+}, Cr^{3+}, Co^{2+}, Ni^{2+}, Cu^{2+}, Zn^{2+}, Cd^{2+}, Hg^{2+}, Pb^{2+}, $U^{VI}O_2^{2+}$
Strong Acid Anions:	NO_3^-, $Se^{VI}O_4^{2-}$, $^{129}I^-$
Weak Acid Anions:	PO_4^{3-}, $Cr^{VI}O_4^{2-}$, $As^VO_4^{3-}$, $As^{III}(OH)_4^-$, $Se^{IV}O_3^{2-}$, MoO_4^{2-}

*Radioactive contaminants are indicated by the atomic weight of the relevant isotope. The oxidation state of redox active elements is shown in Roman numerals in the superscript.

The distinction between hydrolyzing and nonhydrolyzing cations relates to the strength of their bonding to water molecules. When ions are dissolved in solution, they become **hydrated** by forming **ion-dipole bonds** with the dipolar solvent water. Cations coordinate with the oxygen atoms in water, each of which has a partial negative charge, whereas anions coordinate with the hydrogen atoms (H), each of which has a partial positive charge (Figure 7.3). The number of water molecules that surround a given ion depends on its ionic radius (r), with larger ions coordinating with a larger number of water molecules. A useful parameter for prediction of ion behavior in water is termed the **ionic potential:** the ratio of charge (Z) to radius (r) of a given ion. The force of attraction between an ion and its hydration waters increases as the charge of the ion becomes more concentrated or, in other words, as ionic potential (Z/r) increases. The chemical behavior of main-group cations in aqueous solution depends strongly on the ionic potential of the cation. Small values of ionic potential give rise to hydrated cations, whereas higher values give rise to **hydrolysis products** of these hydrated cations, which can result in the formation of both **insoluble oxides or hydroxides** and **soluble oxyanions**.

The force of ion-dipole attraction influences the **hydrolysis** of ions, which strongly impacts their transport and fate in the environment. Ion hydrolysis—defined as the breaking of O-H bonds in water molecules that are attached to the ions—occurs in response to changes in the pH of the aqueous solution. These reactions are important for cations of high ionic potential (all the hydrolyzing cations listed in Table 7.1). The strong ion-dipole interaction (and also covalent bonding for transition metals) makes one or more of the protons (H$^+$ ions) in the coordinating water molecules susceptible to release to solution. This process is termed **proton dissociation**. Hydrolysis of the hydrated Pb^{2+} ion, for example, is given by:

$$Pb(H_2O)_6^{2+} (aq) \leftrightarrow PbOH(H_2O)_5^+ (aq) + H^+ (aq) \quad \text{(Eq. 7.5a)}$$

where (*aq*) indicates the species is dissolved in aqueous solution. This reaction clearly shows that the H$^+$ released to solution comes from one of the water molecules attached to the Pb^{2+} ion. More typically, the reaction depicted in Equation 7.5a is written in a simpler notation, leaving out the hydration waters:

$$Pb^{2+} (aq) + H_2O (l) \leftrightarrow PbOH^+ (aq) + H^+ (aq) \quad \text{(Eq. 7.5b)}$$

where (*l*) represents liquid water. Reactions 7.5 indicate that an increase in pH (*i.e., a decrease in H$^+$ activity*) will favor the hydrolysis reaction. [Recall: pH = −log(H$^+$)]. An **equilibrium constant** for this reaction may be written in terms of activities of reactants and products, and its magnitude depends on the temperature (T) and pressure (P) of the system. At standard temperature (25°C) and pressure (0.1 MPa) (as will be the case throughout this chapter unless otherwise noted):

$$K_{h1} = \left(\frac{(PbOH^+)(H^+)}{(Pb^{2+})(H_2O)} \right) = 10^{-7.7} \quad \text{(Eq. 7.6)}$$

where K$_{h1}$ is the first hydrolysis constant for aqueous Pb^{2+}. The value of this constant indicates that (assuming the activity of solvent H$_2$O = 1) the ratio of (PbOH$^+$)/(Pb^{2+}) ~ 1 at pH = 7.7, and it increases with pH. Further increases in pH likewise result in further hydrolysis:

$$PbOH^+ (aq) + H_2O (l) \leftrightarrow Pb(OH)_2^0 (aq) + H^+ (aq)$$
$$K_{h2} = 10^{-9.4} \quad \text{(Eq. 7.7)}$$

The uncharged Pb(OH)$_2^0$ aqueous species (favored at pH > 9.4) is relatively insoluble, which results in the formation of an insoluble hydroxide [Pb(OH)$_2$] solid. Similar pH-dependent hydrolysis equilibria occur for all of the hydrolyzing cations in Table 7.1.

These effects of pH on the speciation of inorganics are critical to their transport and fate behavior in aqueous systems (surface water, soil, groundwater). For example, as shown in Equation 7.7, an uncharged species of Pb is formed at higher pH values. This species has a low solubility in water, so it will tend to precipitate into a solid form. Lead in solid form is unlikely to be transported, which means a low probability of lead pollution for the surrounding area. In addition, because of their low solubilities at these higher pH's, lead and other metals are not as available to organisms (termed bioavailability). Thus, they do not pose as great a risk as they would in more soluble forms that occur at lower pH values (see Chapter 13).

Contaminant elements that occur in high oxidation states tend to form stronger covalent bonds with oxygen, resulting in the formation of **oxyanions**—compounds composed of another element combined with oxygen (see Table

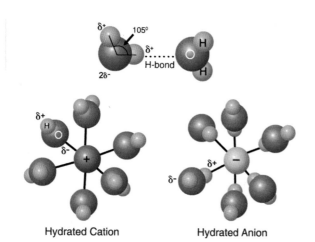

Figure 7.3 The dipole structure of water, which arises from the H-O-H bonding geometry of the water molecule, influences H-bonding between water molecules in solution (top), and the coordination of water molecules to form the primary hydration sphere around cations and anions in solution (bottom). The bond to an ion from the negative (for cations) and positive (for anions) poles of the H$_2$O molecule is called an ion-dipole bond.

7.1). These oxyanions can be considered as hydrolysis products that have undergone significant H^+ dissociation. Since this dissociation is pH dependent, some of these species (the weak acid oxyanions) form complexes with aqueous phase H^+ ions in the pH range of natural waters, thereby altering their charge and behavior.

The tools of acid-base chemistry permit us to evaluate the extent of protonation or proton dissociation of any acid as a function of pH. For example, the acid-base chemistry of arsenate (which is very similar to that of phosphate) is given by:

$$H_3AsO_4^0 \, (aq) \leftrightarrow H_2AsO_4^- \, (aq) + H^+ \, (aq)$$
$$K_{a1} = 10^{-2} \quad \text{(Eq. 7.8a)}$$

$$H_2AsO_4^- \, (aq) \leftrightarrow HAsO_4^{2-} \, (aq) + H^+ \, (aq)$$
$$K_{a2} = 10^{-7} \quad \text{(Eq. 7.8b)}$$

$$HAsO_4^{2-} \, (aq) \leftrightarrow AsO_4^{3-} \, (aq) + H^+ \, (aq)$$
$$K_{a3} = 10^{-12} \quad \text{(Eq. 7.8c)}$$

where K_{a1}, K_{a2} and K_{a3} are the first, second and third acid dissociation constants. The equilibrium constant for Eq. 7.8a is given by:

$$K_{a1} = \frac{(H_2AsO_4^-)(H^+)}{(H_3AsO_4^0)} \quad \text{(Eq. 7.9a)}$$

Taking $-\log_{10}$ of both sides and rearranging gives:

$$-\log(H^+) = -\log K_{a1} - \log\left[\frac{(H_2AsO_4^-)}{(H_3AsO_4^0)}\right] \quad \text{(Eq. 7.9b)}$$

Equation 7.9b is termed the *Henderson-Hasselbalch Equation*, and it provides a useful index for the speciation of weak acids. When $(H_3AsO_4^0) = (H_2AsO_4^-)$ (*i.e.*, activity of acid equals that of the conjugate base), the second term on the right side vanishes ($\log 1 = 0$), and pH = pK_a. Thus, given a knowledge of solution pH and the pK_a values of oxyanions, we can determine the relative predominance of the various weak acid species. In this case, the predominant species are: $H_3AsO_4^0$ below pH 2; $H_2AsO_4^-$ between pH 2 and 7; $HAsO_4^{2-}$ between pH 7 and 12; and AsO_4^{3-} at pH above 12. The transport behavior and bioavailability of As and other similar elements will be greatly affected by the effect of pH on speciation.

7.2.4 Aqueous-Phase Complexation Reactions

Dissolved ions can form stable bonds to other ions and molecules in solution giving rise to **aqueous-phase complexes**. Bonding between cations and anions may occur as *inner-sphere* or *outer-sphere complexes*. The former case involves direct ion-ion contact with no hydration waters interposed and may involve some degree of covalent bonding, whereas in the latter case, hydration waters are retained around each ion and the two are attracted electrostatically (Figure 7.4). In either case, the stable unit is termed a **metal-ligand complex**, where the cation is the *central metal group* and the coordinating anion is termed a *ligand*. The formation of aqueous phase complexes influences the solubility, fate, and transport of the constituent molecules.

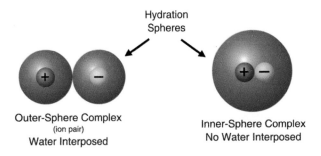

Figure 7.4 Aqueous phase complexation of inorganic ions can result in the formation of outer-sphere or inner-sphere complexes.

The tendency for metal-ligand complexes to form is characterized by a **thermodynamic stability constant** (K_{stab}). The K_{stab} is the equilibrium constant for a complex formation reaction written in terms of aqueous phase activities. For example, the mercuric (Hg^{2+}) cation forms a stable complex with the chloride (Cl^-) anion. The aqueous phase complexation of these two ions is given by:

$$Hg^{2+} \, (aq) + Cl^- \, (aq) \leftrightarrow HgCl^+ \, (aq)$$
$$K_{stab} = 10^{6.7} \quad \text{(Eq. 7.10)}$$

According to Eq. 7.10, if $Cl^- \, (aq)$ is present in solution at 1 mmol dm^{-3}, then the activity of $HgCl^+ \, (aq)$ will exceed that of the free $Hg^{2+} \, (aq)$ cation by more than 5000 times (assuming activity coefficients = 1). Since $HgCl^+ \, (aq)$ is a monovalent cationic species, it exhibits a lower affinity for negatively charged sites on soil particle surfaces (see Section 7.2.5) and, therefore, a greater mobility in soils and sediments than the Hg^{2+} cation. Hence, aqueous-phase speciation is an important determinant of environmental fate. Aqueous-phase complexation reactions are quite rapid, occurring on time scales ranging from μs (10^{-6} seconds) to minutes, and therefore can be considered to achieve equilibrium during the time scales of water movement through porous media.

The formation of dissolved complexes increases the total aqueous-phase concentration of a given element in equilibrium with solid-phase precipitates (see Section 7.2.6). In addition to inorganic anions (*e.g.*, Cl^- and SO_4^{2-}), organic molecules, including the low molecular weight organic acids (*e.g.*, oxalate, citrate, and salicylate) and fulvic acids, are important ligands in natural waters. These organic anions are particularly important to forming metal-ligand complexes in zones of high biological activity such as surficial soils and wetland sediments. Prediction of transport and fate of inorganic contaminants must include consideration of their aqueous-phase speciation, including complexation with organic and inorganic molecules.

7.2.5 Precipitation-Dissolution Reactions

The solubility of minerals can regulate the partitioning of inorganic contaminants between the aqueous and solid phases (see Figure 7.1). Soils and sediments contain a mixture of mineral solids, all of which may be subjected to dissolution or precipitation reactions, depending on aqueous-phase conditions at a given point in time. Although these processes typically proceed more slowly than adsorption-desorption reactions, they are very important over long-time frames (weeks to decades). Precipitation can result in contaminant immobilization into sparingly soluble solids, whereas mineral dissolution can result in replenishment of the bio-available pool of dissolved and exchangeable elements. Contaminants are incorporated into precipitated minerals both as **major elements** (> 100 mg contaminant per kg of solid) and also as minor or **trace elements** (< 100 mg kg^{-1}).

The solubility of chromium (III) hydroxide, for example, is governed by the following reaction:

$$Cr(OH)_3 \ (s) + 3 \ H^+ \ (aq) \leftrightarrow Cr^{3+} \ (aq)$$
$$+ 3 \ H_2O(l) \quad \text{(Eq. 7.11a)}$$

where dissolution of the $Cr(OH)_3$ solid proceeds from left to right, releasing $Cr^{3+}(aq)$ into solution. Precipitation proceeds from right to left, as the solid phase removes $Cr^{3+}(aq)$ from solution. This reaction can be characterized by the thermodynamic **dissolution equilibrium constant:**

$$K_{dis} = \frac{(Cr^{3+})(H_2O)^3}{(Cr(OH)_3)(H^+)^3} = 10^{12} \quad \text{(Eq. 7.11b)}$$

In considering the dissolution-precipitation of pure mineral solids in water-saturated systems, the activities of solvent H_2O and mineral solid ($Cr(OH)_3$ in this case) are assumed to be unity, and the dissolution equilibrium constant is readily transformed into a **solubility product constant** (K_{SO}), written in terms of activities of only aqueous phase species:

$$K_{so} = \frac{(Cr^{3+})}{(H^+)^3} = 10^{12} \quad \text{(Eq. 7.11c)}$$

The solubility product gives the activities of $Cr^{3+}(aq)$ and $H^+(aq)$ in equilibrium with the solid phase $Cr(OH)_3$.

One can readily determine whether precipitation or dissolution will occur in a given aqueous environmental system by comparing the *actual values* of species activities in that system with those that occur at equilibrium. To accomplish this, actual values for a given natural system (which may or may not be at equilibrium) are incorporated into a term, comparable to that of Equation 7.11c, called the **ion activity product (IAP)**:

$$IAP = \frac{(Cr^{3+})}{(H^+)^3} = \frac{\gamma_{Cr^{3+}}[Cr^{3+}]}{\gamma_{H^+}^3[H^+]^3} \quad \text{(Eq. 7.12)}$$

where activities in this case are determined from *measured* species concentrations, and activity coefficient terms are calculated from a model such as the Davies equation (Equation 7.4). The tendency for precipitation or dissolution to occur is then assessed on the basis of the **relative saturation** (Ω) of the system:

$$\Omega = \frac{IAP}{K_{so}} \quad \text{(Eq. 7.13)}$$

If $\Omega > 1$ (IAP $> K_{SO}$), the solution is termed *supersaturated* with respect to the selected solid phase (*e.g.*, $Cr(OH)_3$) and precipitation will occur. If $\Omega < 1$ (IAP $< K_{SO}$), the solution is termed *undersaturated* with respect to the selected solid phase and (if the solid phase is present) dissolution will occur. If $\Omega = 1$ (IAP $= K_{SO}$), the solution is termed *saturated* with respect to the selected solid phase; the solution and solid are at equilibrium and neither precipitation nor dissolution of the solid will occur.

The preceding example pertains to a single Cr(III) solid phase in which Cr^{3+} is the major cationic constituent. However, the natural environment contains numerous potential contaminant-bearing minerals, including those in which the contaminant exists as a minor species. For example, Cr^{3+} can be incorporated into iron oxides, such as the mineral goethite (α-FeOOH), where it substitutes for a very small fraction of the Fe^{3+} cations:

$$Fe_{(1-x)}Cr_xOOH \ (s) + 3 \ H^+ \ (aq) \leftrightarrow (1-x) \ Fe^{3+} \ (aq)$$
$$+ x \ Cr^{3+} \ (aq) + 2 \ H_2O(l) \quad \text{(Eq. 7.14)}$$

Since goethite is very insoluble (*i.e.*, low K_{SO}), natural waters may become supersaturated with the Cr-containing goethite phase (Eq. 7.14 proceeds from right to left) at Cr concentrations much lower than those that would result in precipitation of $Cr(OH)_3$ (s). This type of reaction, called **co-precipitation**, is very important for controlling the fate of inorganic contaminant species in natural waters, since contaminant concentrations are often much lower those of major mineral-building elements such as Si, Al and Fe. Prediction of geochemical fate of inorganic contaminants, therefore, requires the assessment of Ω for each of the potential contaminant-bearing solids. As discussed previously, the transport potential and bioavailability of metals and other inorganics is much lower for solid species than for aqueous (dissolved) species.

It is important to note that K_{SO}, IAP and Ω values are *written in terms of activities of particular aqueous species, normally the free ion of interest* (e.g., Cr^{3+} (aq) in this case). Therefore, speciation of solution for the mineral-building constituent is a pre-requisite for calculating the relative saturation with respect to any mineral phases. In cases where numerous potential solid phases exist for a given inorganic contaminant, speciation of the solution phase and calculation of Ω values for each of the solids is normally accomplished by using a computer equipped with a geochemical speciation program.

7.3 BASIC PROPERTIES OF ORGANIC CONTAMINANTS

7.3.1 Phases—Solids, Liquids, Gases

As with inorganic contaminants, a critical property to consider when evaluating the transport and fate behavior of an organic contaminant is its phase state. Under "natural" conditions (temperature $T = 25°C$, pressure $P = 0.1$ MPa), organic chemicals in their pure form exist as solids, liquids, or gases (See Table 7.2). The phase state of a contaminant

TABLE 7.2 Aqueous solubilities, vapor pressures, and Henry's coefficients for selected organic compounds.

	AQUEOUS SOLUBILITY (mg/l)	VAPOR PRESSURE (atm)	HENRY'S CONSTANT (−)	PHASE STATE (STANDARD CONDITIONS: T = 20°C, P = ATMOSPHERIC)
Benzene	1780	0.1	0.18	nonpolar liquid
Toluene	515	0.03	0.23	nonpolar liquid
Naphthalene	30	~ 0.00066	0.02	nonpolar solid
Phenol	82,000	0.00026	0.00005	polar liquid
Methane	24	275	27	nonpolar gas

Data from Verschueren, K. *Handbook of Environmental Data on Organic Chemicals*, 1983, and Schwarzenbach, R.P., Gschwend, P.M., and Imboden, D.M. *Environmental Organic Chemistry*, 2003.

has a significant impact on mass-transfer processes such as dissolution and evaporation (Figure 7.5).

7.3.2 Dissolution and Aqueous Solubilities of Organic Contaminants

The transfer of molecules from their pure state to water is called dissolution. The extent to which molecules of a compound will transfer from its pure phase into water is the aqueous solubility. The solubility of organic compounds depends strongly on the degree to which water and contaminant molecules interact. The well-known rule of thumb that "likes dissolve likes" generally holds for solubility. Water is a very polar solvent (see Figure 7.3) and, therefore, the solubility of organic compounds depends strongly on the degree of polarity of the molecules. Water can interact easily with other polar compounds. Therefore, the aqueous solubilities of ionic or polar organic compounds are relatively large. Conversely, it requires much more energy for water to interact with or solvate nonpolar organic compounds. Therefore, the solubilities of nonpolar compounds are generally much smaller than those of the polar and ionic compounds. This is illustrated in Table 7.2, where solubilities for several representative organic compounds are presented. Compare the solubilities of phenol, a polar liquid, with that of toluene, a nonpolar liquid. The solubility of phenol is 159 times greater than that of toluene's.

Organic compounds in their pure form exist as either a solid (*e.g.*, naphthalene), liquid (benzene, toluene, phenol), or gas (methane). Because dissolution (solubilization) requires breaking bonds between contaminant molecules, the solubility of organic compounds also depends on the form of the compound. For example, more energy is required to break bonds in solids than in liquids; the solubilization of solids can be thought of as a two-step process, where the solid must first "melt" (convert to a liquid), and then dissolve. Therefore, the solubilities of solid organic compounds are usually smaller than those of liquid compounds (See Table 7.2, compare naphthalene to toluene or benzene).

The concept of activity coefficients was introduced in the prior section. For organic compounds, the aqueous solubility is inversely proportional to the activity coefficient. In other words, compounds with larger activity coefficients will have lower solubilities, and vice versa. In contrast to inorganics, for which activity coefficients are generally less than one, the activity coefficients of organic compounds are greater than one. This reflects the fact that it is more difficult (requires more energy) for water to interact with most organics than most inorganics.

An important property of liquid organic compounds is their miscibility, or lack thereof, with water. A miscible organic liquid is one that can be mixed with water such that a single liquid phase results. Alcohols such as methanol and ethanol are examples of miscible liquids. They can be considered as having an infinite solubility in water. Conversely, an immiscible liquid is one that cannot be mixed with water. Benzene, an aromatic hydrocarbon and a major component of gasoline, is an example of an immiscible liquid. If a volume of pure liquid benzene is mixed with water, the two liquids will separate quickly after cessation of mixing because they are immiscible. However, a relatively small fraction of the benzene molecules will transfer into the water phase, thus becoming dissolved. The maximum amount of benzene that can be dissolved in water is the aqueous solubility of benzene. Even though the amount of benzene that can dissolve in water is very small (< 2 g L^{-1}), it can be very significant because the federal maximum contaminant level for benzene is 5 $\mu g \, L^{-1}$.

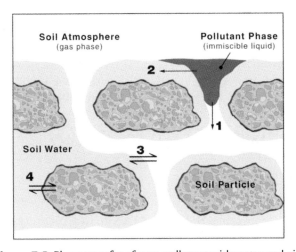

Figure 7.5 Phase transfer of pure pollutant with water and air phases: (1) evaporation, (2) solubilization, (3) volatilization, (4) sorption. From *Pollution Science* © 1996, Academic Press, San Diego, CA.

The solubility of organic compounds in water is a function of water-contaminant and contaminant-contaminant molecular interactions, and depends primarily on chemical properties of the compounds. However, solubility is also affected by environmental factors such as temperature. Many organic compounds become more soluble as the temperature increases, but a few behave in the opposite way. Generally, solubility changes by less than a factor of two in the temperature range of most natural systems (*e.g.*, 0–35 °C). Salinity or ionic strength usually causes a small decrease in the solubility of nonpolar organic compounds ("salting-out effect"). This effect is also small for typical environmental conditions (less than a factor of two change in solubility).

The presence of other contaminants can influence the solubilities of organic contaminants. For example, the presence of alcohols or detergents (surfactants) can increase the amount of contaminant in solution. Alcohols such as ethanol are added to gasoline, often at quantities of 10% or more, to boost their oxygen content. If this gasoline mixture leaks from a storage tank into the subsurface, the presence of the alcohol can result in greater transport of the gasoline components. Indeed, the ability of alcohols and surfactants to increase the solubilities of contaminants is also the basis for the enhanced-solubilization flushing method for subsurface remediation (See Chapter 19).

7.3.3 Evaporation of Organic Contaminants

Evaporation of a compound involves transfer from the pure liquid or solid contaminant-phase to the gas phase. The vapor pressure of a compound is the pressure of contaminant gas in equilibrium with the solid or liquid contaminant phase, and is an index of the degree to which a compound will evaporate. The vapor pressure can be thought of as the "solubility" of the compound in air. Evaporation can be an important transfer process when pure-phase contaminant exists in the vadose zone, such that contaminant molecules can evaporate into the soil atmosphere.

In contrast to solubility, which is governed by contaminant-contaminant and water-contaminant molecular interactions, evaporation is controlled primarily by contaminant-contaminant interactions (*i.e.*, the energy of bonding) in the solid or liquid phase. This is because inter-molecule interactions are very minor for most gases, where the space between molecules is relatively large compared to that for liquids or solids. Simply put, the greater the bonding energy between contaminant molecules, the lower will be the vapor pressure. The vapor pressures of liquids are therefore typically larger than those of solids (See Table 7.2). The vapor pressures for benzene and toluene are approximately 100 times larger than naphthalene's vapor pressure. However, the vapor pressure for phenol is smaller than naphthalene's; this is because phenol is a polar compound. The vapor pressures for gases are very large (See methane in Table 7.2). The vapor pressure is a strong function of temperature because of the strong influence of temperature on gas-phase interactions.

7.3.4 Volatilization of Organic Contaminants

Transfer of contaminant molecules between water and gas phases is an important component of the transport of many organic compounds in the vadose zone and the atmosphere. Volatilization is different from evaporation; we use the latter term to specify a transfer of contaminant molecules from their pure phase to the gas phase. For example, the transfer of benzene molecules from a pool of gasoline to the atmosphere is evaporation, whereas the transfer of benzene molecules from water (where they are dissolved) to the atmosphere is volatilization. The vapor pressure provides a rough idea of the extent to which a compound will volatilize. However, volatilization depends also on the aqueous-phase solubility of the compound and on environmental factors.

At equilibrium, the distribution of a contaminant between gas and aqueous phases is described by **Henry's law:**

$$C_g = HC_w \qquad \text{(Eq. 7.15)}$$

where:

C_g is concentration of pollutant in the gas phase $(M \ L^{-3})$

C_w is concentration of pollutant in the water phase $(M \ L^{-3})$

***H* is Henry's Constant** (dimensionless). Henry's law can be used to evaluate the preference of a contaminant for aqueous and gas phases (Information Box 7.1)

M and L^3 represent any consistent set of mass and volume units.

7.3.5 Multiple-Component Organic Phase

The preceding discussion dealt with the behavior of single organic contaminants. However, many important contaminants contain multiple components. The primary examples of this type of contamination are multi-component immiscible liquids such as gasoline, diesel fuel, and coal tar. Knowledge of the partitioning behavior of multi-component contaminants is essential to the prediction of their impact on environmental quality.

The transfer of individual components of a multiple-component contaminant into water is controlled by the aqueous solubility of the component and the composition of the liquid. A simple approach to estimating the solubility of multiple-component liquids involves an assumption of ideal behavior in both aqueous and organic phases and the application of **Raoult's Law:**

$$C_w^i = X_o^i S_w^i \qquad \text{(Eq. 7.16)}$$

where:

C_w^i is aqueous concentration $(mol \ L^{-1})$ of component i

S_w^i is aqueous solubility $(mol \ L^{-1})$ of component i

X_o^i is mole fraction of component i in the organic liquid

INFORMATION BOX 7.1

Henry's Constant

Consider the preference of three contaminants in three separate closed containers in which reside equal volumes of water and air.

The first contaminant has a Henry's Constant of 1. A value of one means that the contaminant concentration in the air is equal to the concentration in the water. Thus, this contaminant "likes" water and air equally.

The second contaminant has a Henry's Constant of 0.1. This means that the concentration of the contaminant in the air is ten times less than the concentration in the water; this contaminant prefers the water.

The third contaminant has a Henry's Constant of 10, which means that its concentration in air is ten times greater than it is in water.

Henry's law describes the distribution of contaminant mass at equilibrium. Instantaneous transfer and the achievement of equilibrium is not guaranteed in natural systems. In soils, however, the rate of contaminant transfer between water and gas phases is relatively rapid in comparison to other transport processes. Thus, assuming instantaneous transfer is often not a major problem. The magnitude of the Henry's Constant is influenced by temperature, with larger values obtained at higher temperatures.

7.4 SORPTION PROCESSES

Sorption is a major process influencing the transport and fate of many contaminants in the environment. The broadest definition of sorption (or retention) is the association of contaminant molecules with the solid phase (soil or sediment particles). The solid phase to which the contaminants sorb is often referred to as the sorbent; we will use this term for the following discussion.

A critical impact of sorption is that it generally slows or retards the rate of movement of contaminants (see Chapter 6). In addition, sorption can influence the magnitude and rate of other processes. For example, sorption can influence biodegradation rates by affecting the bioavailability of contaminants to microorganisms. Sorption processes are also a basic component of traditional water-treatment technologies, such as the use of granular activated carbon beds to remove organic contaminants from water or the use of ion-exchange beds to remove inorganics from water (see Chapter 24). Sorption can occur by numerous mechanisms, depending on the properties of the contaminant and of the sorbent; this will be discussed in the following sections.

7.4.1 Inorganics

As noted in Chapter 2, the surfaces of soil and sediment particles are important for the uptake and release of contaminants. Inorganic pollutants are attracted to particle surfaces largely because of their charge properties. Soil particles contain positively charged sites and negatively charged sites, which results in the uptake of anions and cations, respectively. **Isomorphic substitutions** in clay minerals give rise to *permanent charge*, which is dominantly negative. **pH-dependent ionization reactions** at surface hydroxyl groups of minerals and organic matter give rise to *variable charge*, which can be positive or negative, depending on acid-base reactions at surface sites. For example, many Fe and Al oxides are positively charged at pH < 8 and negatively charged at pH > 8. At any given point in time, a volume of soil contains some number of both positive-charged and negative-charged sites that contribute to the uptake of ionic contaminant species. As shown in Figure 7.1, the accumulation of ions or molecules at these particle surfaces during their removal from aqueous solution is termed **adsorption**. The release of adsorbed molecules from surfaces to aqueous solution is termed **desorption**.

A **surface complex** is formed when an **adsorbate** (the ion or molecule adsorbed at a surface) forms a stable bond with an **adsorbent** (the solid phase whose surface provides the site for contaminant adsorption). Mechanisms of ion adsorption include the formation of inner-sphere and outer-sphere **surface complexes** that are analogous to those that form in aqueous solution. Outer-sphere surface complexes are dominantly the result of electrostatic interaction. Inner-sphere complex formation involves some degree of covalent bonding and, therefore, chemical **specificity** between the adsorbate and adsorbent. For this reason, formation of outer-sphere com-

The mole fraction represents the concentration of component *i* in the immiscible liquid. In essence, Raoult's Law states that the aqueous concentration obtained for any given component is proportional to the amount of that component in the immiscible liquid.

For example, assume we have a two-component immiscible liquid, with the mole fraction of each component equal to 0.5. This means that there is an equal amount of each component in the liquid contaminant. Let us further assume that the aqueous solubility of component A is 100 mg L^{-1} and that of component B is 10 mg L^{-1}. We now wish to calculate the concentrations of A and B in a volume of water that is in contact with the immiscible liquid. Using Raoult's Law we find that the aqueous concentration for component A is 50 mg L^{-1} and for component B is 5 mg L^{-1}. Thus, the aqueous concentrations for the two components are half of their aqueous solubilities because they are not dissolving from their pure state, but rather from a mixture. One can think of this as the two components "competing" with each other to dissolve into water. Inspection of Equation 7.16 shows that when the mole fraction is equal to one (*i.e.*, a single-component liquid), the aqueous concentration is equal to the aqueous solubility. Raoult's Law can also be used to determine the concentrations of components in air that are in equilibrium with a multiple-component contaminant liquid.

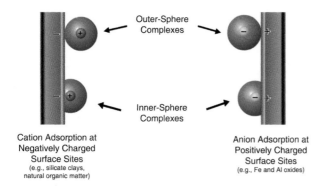

Figure 7.6 Mechanisms of inorganic contaminant adsorption: Outer-sphere and inner-sphere surface complexation.

plexes is often referred to as **physisorption**, whereas inner-sphere complex formation is termed **chemisorption**. Ions adsorbed in inner-sphere coordination are bound more strongly, and are less likely to desorb from the surface because of displacement by other ions in solution.

Adsorption mechanisms are dependent on the adsorbate and adsorbent composition. For example, the weak acid species of arsenate ($H_2AsO_4^-$, $HAsO_4^{2-}$) and phosphate ($H_2PO_4^-$, HPO_4^{2-}), are known to form strong, inner-sphere complexes with surface sites on oxides and hydroxides, whereas the oxyanions selenate (SeO_4^{2-}), sulfate (SO_4^{2-}) and nitrate (NO_3^-) form weaker, outer-sphere complexes. Likewise, the transition metal cations Cu^{2+} and Pb^{2+} form strong inner-sphere complexes at variable-charge surface hydroxyl groups, whereas Ni^{2+} and Mn^{2+} tend to form outer-sphere complexes (Figure 7.6).

Ions adsorbed as outer-sphere complexes are termed **exchangeable** because their electrostatic attraction to the surface is readily disrupted, leading to subsequent desorption, by introduction of an ion of similar or greater charge and surface affinity. Exchangeable ions are also considered to be "bioavailable," since they are a principal source for replenishment of ions that are removed from the solution phase by microbial or plant uptake. The number of moles of exchangeable cation charge that can be adsorbed per unit mass of soil or sediment is termed the **cation exchange capacity** (CEC). The corresponding value for anions is termed the **anion exchange capacity** (AEC).

In temperate zone soils dominated by permanent-charged silicate clays and organic matter, the CEC is typically much higher than the AEC, and adsorption of exchangeable cations exceeds that of exchangeable anions. The mobility of cations relative to anions is diminished as a result of adsorption processes. Cation exchange reactions are assumed to be reversible, and they conserve the number of moles of charge adsorbed to the negatively charged exchange sites, even when the exchanging cations have different valence.

For example, consider the exchange of adsorbed *bivalent* Cd^{2+} for adsorbed *monovalent* Na^+:

$$X_2Cd\ (s) + 2\ Na^+\ (aq) \leftrightarrow 2X Na\ (s) + Cd^{2+}\ (aq) \quad \text{(Eq. 7.17)}$$

where X represents one mole of negative charge on the exchanger, and (*s*) refers to the solid phase. Thus one mole of Cd^{2+} occupies two moles of exchanger charge, and $Cd^{2+} \rightarrow Na^+$ exchange involves two moles of Na^+ to maintain charge balance. The reaction 7.17 can be considered from the perspective of **Le Chatellier's Principal of Equilibrium Chemistry**. This principal states that if a system is perturbed from its equilibrium state, chemical reactions will proceed to return the system to equilibrium. Thus, an increase in aqueous-phase Na^+ concentration will result in the release of Cd^{2+} from exchange sites, enhancing their mobility in the dissolved state. The charge-based stoichiometry of cation exchange is further illustrated in Figure 7.7.

Highly weathered, silicate-clay-depleted soils, such as Oxisols and Ultisols of the humid tropics, contain large amounts of variable-charge iron and aluminum oxides. These soils can have AEC > CEC, particularly under acidic conditions when pH is low and variable-charge sites become positive-charged because of proton adsorption. In such systems, the mobility of anions is diminished relative to that of cations. Whereas ion exchange reactions involving outer-sphere complex formation can be quite rapid (time scales ranging from microseconds to minutes), longer term diffusion into small pores of adsorbent solids and the formation of inner-sphere complexes can extend equilibration timeframes into weeks or even months.

7.4.2 Organics

The mechanisms by which many organic contaminants are sorbed or retained by porous- media particles are usually quite different from those involved for inorganic contaminants. We can use the "likes dissolve likes" rule to help explain the sorption of nonpolar organic contaminants by sorbents. It is now generally accepted that for many natural sorbents (soil, sediment), organic contaminants interact primarily with organic material associated with the sorbent. This organic material is generally less polar than water and provides a more favorable environment for nonpolar organic contaminants. Thus, the sorption of nonpolar organics is driven primarily by the incompatibility between water and the organic compound, a phenomenon known as the "hydrophobic effect." In essence, this effect involves the expulsion of a nonpolar organic compound from the aqueous phase. This happens because its presence there requires the breaking up of the hydrogen-bonded structure of liquid water (see Figure 7.3) and there are no favorable bonds formed between water molecules and nonpolar organic molecules in return. Conversely, association with hydrophobic portions of organic matter does not require such disruption, and there are also weak bonding interactions (*e.g.*, van der Waals associations) that can contribute to the favorable overall energetics of sorption.

The mechanisms governing the sorption of polar or ionic organics are similar to some of those governing retention of inorganics (*e.g.*, electrostatic attraction, surface complex formation). Many of the important ionizable organic contaminants are negatively charged under environmental

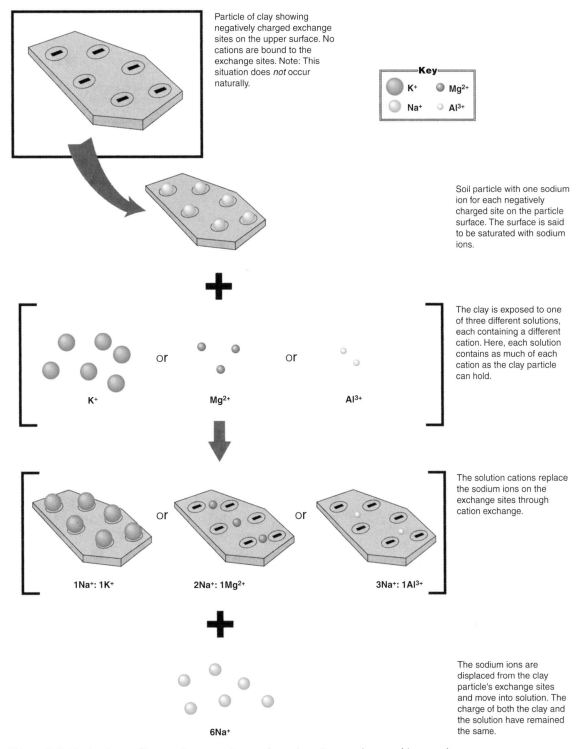

Particle of clay showing negatively charged exchange sites on the upper surface. No cations are bound to the exchange sites. Note: This situation does *not* occur naturally.

Key
K+   Mg²⁺
Na+   Al³⁺

Soil particle with one sodium ion for each negatively charged site on the particle surface. The surface is said to be saturated with sodium ions.

The clay is exposed to one of three different solutions, each containing a different cation. Here, each solution contains as much of each cation as the clay particle can hold.

K⁺   Mg²⁺   Al³⁺

The solution cations replace the sodium ions on the exchange sites through cation exchange.

1Na⁺: 1K⁺   or   2Na⁺: 1Mg²⁺   or   3Na⁺: 1Al³⁺

The sodium ions are displaced from the clay particle's exchange sites and move into solution. The charge of both the clay and the solution have remained the same.

6Na⁺

Figure 7.7 Mechanisms of inorganic contaminant adsorption: Outer-sphere and inner-sphere surface complexation. From *Pollution Science* © 1996, Academic Press, San Diego, CA.

conditions (*e.g.*, phenols, chlorophenols, carboxylic acids). Because most soil and sediment particles have a net negative charge, there is often repulsion between negatively charged organic contaminants and the sorbent, which results in little sorption to such particles. However, sorption to positive charged particles (*e.g.*, oxides of Fe and Al) can be significant, so the amount of surface area composed of such particles can be an important determinant of contaminant sorption.

Sorption processes are usually considered in terms of aqueous systems, such as sorption by soil or aquifer material in subsurface systems, sorption by sediments in surface-water bodies, and sorption by porous media in packed-bed waste-water treatment systems. In all these cases, the contaminant is dissolved in water that is flowing though or over the porous media. However, sorption processes may also occur in gas-phase systems, wherein volatile organic contaminants are associated with the atmosphere or soil atmosphere that is in contact with porous media. This process is often referred to as **vapor adsorption**. This issue is of particular interest for gas-phase contaminant transport in the vadose zone, and for potential transport of contaminants sorbed to particles suspended in air, which can be transported great distances by atmospheric processes. Vapor adsorption is strongly influenced by the amount of water present at the surfaces of the porous-medium grains. For example, vapor adsorption by oven-dry soil has been observed to be orders-of-magnitude larger than adsorption by water-saturated soil. However, once water starts coating the soil surfaces, the magnitude of observed adsorption decreases greatly. This effect is related to the ability of water to "out compete" organic contaminants for adsorption at the soil particle surfaces.

The properties of the contaminant (*e.g.*, nonpolar or polar organic, inorganic charge sign and magnitude) are key determinants of the degree to which the contaminant will be sorbed or retained by a sorbent. However, the physical/chemical properties of the sorbent are also important. For example, sorption of nonpolar organic contaminants is often controlled by sorbent organic matter. Therefore, the amount of organic matter associated with the sorbent is very important. The cation exchange capacity, clay content, and metal-oxide content are important properties for sorption of ionizable and ionic organic contaminants, as they are for inorganics.

7.4.3 Magnitude and Rate of Sorption

Sorption is quantified by measuring a sorption isotherm, which is simply a description of the relationship between the concentration of contaminant in the sorbed state and the concentration in the aqueous phase (or air for vapor-phase sorption). Many different forms of isotherms have been proposed and used to describe sorption. The simplest is the **Linear Isotherm** (the same idea as Henry's Law), which is given by:

$$S = K_d C_w \qquad \text{(Eq. 7.18)}$$

where:

S is the concentration of contaminant sorbed by the soil (M L^{-3})

K_d is the sorption coefficient (L^3 M)

The larger the value of K_d, the greater the degree to which a contaminant is sorbed by the sorbent.

A linear isotherm signifies that for all concentrations of contaminant in water, there will always be proportionally the same sorbed concentration. The sorption of many nonpolar organic contaminants is linear or close to linear. An example of a linear isotherm is shown in Figure 7.8. Some organic and many inorganic contaminants exhibit nonlinear sorption. An example of a widely used nonlinear isotherm is the **Freundlich Isotherm** given by:

$$S = K_f C_w^n \qquad \text{(Eq. 7.19)}$$

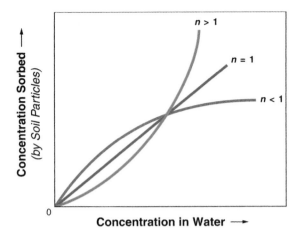

Figure 7.8 Sorption isotherms used to describe the relationship between sorbed concentrations and aqueous concentrations of a contaminant; linear (n=1) and nonlinear isotherms are shown. From *Pollution Science* © 1996, Academic Press, San Diego, CA.

where K_f is the Freundlich sorption coefficient and n is, in essence, a power function related to the sorption mechanism(s). Examples of nonlinear isotherms for the cases of n > 1 and n < 1 are presented in Figure 7.8. For a nonlinear isotherm, the distribution of contaminant between the sorbed and aqueous phases is proportionally different at different concentrations. For example, at higher concentrations, the proportional distribution is less than it is at lower concentrations for the n < 1 isotherm. This is noted by observing the slope of the isotherm line; the magnitude of sorption for a given aqueous concentration correlates to the slope at that concentration.

As the name implies, an isotherm is measured at one temperature. Temperature has a small but measurable effect on the sorption of contaminants. The effect of temperature will depend on the type of sorption mechanism involved. For low-polarity organic compounds, the sorption is governed by aqueous-solubility interactions as discussed above. So, if a change in temperature causes a change in solubility, it might be expected that a change in temperature may also cause a change in sorption. For relatively large organic compounds like anthracene, an increase in temperature causes an increase in solubility. Thus, the sorption of anthracene may decrease with increasing temperature. The sorption mechanisms for inorganic contaminants involves strong contaminant-sorbent interactions. An increase in temperature could increase the energetics of this interaction and therefore produce an increase in sorption. The effect of salinity or ionic strength on sorption is also dependent on the type of sorption mechanism involved. For nonpolar organic contaminants, the effect is usually relatively small, whereas it can be significant for inorganic contaminants.

The use of sorption isotherms presumes the existence of equilibrium between the sorbent and aqueous or gas phases. Research has shown that the rate of sorption of many nonpolar organic contaminants is very slow, taking anywhere from several hours to several months to reach equilibrium. This slow rate of sorption is often due to the slow diffusion of contaminant molecules into spaces in the soil that have very small openings or pores. We know that the sorption of many organic contaminants is dominated by sorbent organic matter. This organic matter has a polymeric type of structure. It can take a long time for contaminant molecules to move from the surface to the inside of the organic material. In addition, some sorbents have solid mineral particles that have small pore spaces. It can take considerable time for contaminant molecules to diffuse into all of these pores.

7.5 ABIOTIC TRANSFORMATION REACTIONS

The transformation of contaminants by microorganisms is discussed in Chapter 8. Some contaminants can also be transformed by abiotic (physical/chemical) processes. We will briefly discuss some major abiotic transformation processes in this section.

7.5.1 Hydrolysis

Water is an ubiquitous component of the environment. As a result, most contaminants will come into contact with water to some extent. It is important, therefore to understand if and when a contaminant will react with water when they are in contact. Reaction of a contaminant with water is termed hydrolysis. A generalized example of this reaction for organic compounds is given by:

$$R\text{-}X + H_2O \rightarrow R\text{-}OH + X^- + H^+ \qquad \text{(Eq. 7.20)}$$

where R-X is an organic compound with X representing a functional group such as a halide (*e.g.*, Cl). By reacting with water, the original compound (R-X) has been transformed to another compound (R-OH). The hydrolysis of inorganics was covered in Section 7.2.3.

The two key factors in hydrolysis reactions are the charge properties of the contaminant molecules and the pH. Hydrolysis is essentially an interaction between nucleophiles (substance with excess electrons, such as OH^-) and electrophiles (substance deficient in electrons, such as H^+). Thus, the charge properties of the molecules will govern its reactivity with water. For many compounds, hydrolysis may be catalyzed or enhanced under acidic or basic conditions. This means that the occurrence and rate of hydrolysis is often pH dependent. For example, a hydrolysis reaction catalyzed by OH^- would occur more rapidly at higher pH values because of larger OH^- concentrations. Hydrolysis can also be influenced by sorption interactions. For example, the pH at the surface of many soils is lower than the pH of the water surrounding the soil particles. Thus, an acid-catalyzed hydrolysis reaction could be enhanced when the contaminant is associated with the soil.

7.5.2 Oxidation-Reduction Reactions

Oxidation-reduction (**redox**) reactions are chemical reactions that involve the transfer of electrons between two molecular species. The two species involved can be organic or inorganic, and they may be present in any environmental phase (gas, liquid, or solid). In a full redox reaction, one species begins the reaction in its more reduced form and this species is **oxidized** (*i.e.*, loses one or more electrons) during the reaction while the other enters the reaction in its more oxidized form and is **reduced** (accepts one or more electrons). Figure 7.9 depicts this process schematically. Many of the environmentally important redox reactions are *catalyzed* (*i.e.*, made to proceed faster) by microorganisms, but they only proceed when favorable thermodynamically.

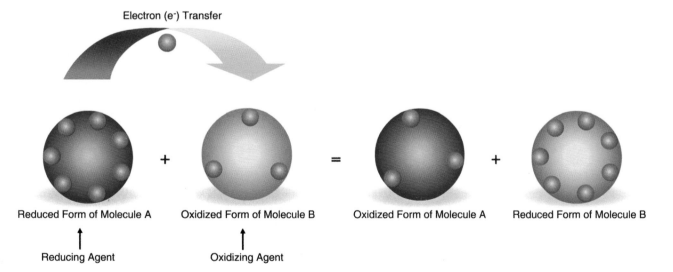

Figure 7.9 A full oxidation-reduction reaction involves the transfer of electrons from one species (the *reducing agent*) to another (the *oxidizing agent*).

Perhaps the best known example of a redox reaction is aerobic, heterotrophic **respiration**, with molecular oxygen (O_2) acting as an electron acceptor during the oxidation of carbohydrate (See also Chapter 5):

$$CH_2O\ (aq) + O_2\ (g) \leftrightarrow CO_2\ (g) + H_2O\ (l) \quad \text{(Eq. 7.21)}$$

In this reaction, one mole of carbon (C) is reduced from the 0 oxidation state in CH_2O, to the +4 oxidation state in CO_2 while two moles of oxygen (O), are reduced from the 0 oxidation state in O_2, to the −2 oxidation state (one mole ends up in H_2O and the other in CO_2; the third mole of O was already in the −2 oxidation state in CH_2O). Thus, a total of four moles of electrons are transferred per mole of CH_2O oxidized. When microorganisms catalyze the respiration of carbohydrates, they capture some of the energy released in the reaction. In a similar way, microbes can catalyze the oxidation of other organic compounds that contain reduced C, including many organic contaminants. Although many organic contaminants are oxidized much more slowly than "labile" forms of C, such as carbohydrates, they are eventually subjected to oxidation, and the process is most favorable energetically when oxygen is available to act as the electron acceptor (*i.e.*, in *oxic* environments).

Many subsurface environmental systems, including biologically active soils or sediments, are depleted of gaseous or dissolved O_2. This occurs when respiration consumes O_2 faster than it can be replenished by diffusion from the atmosphere. In these *anoxic* systems, alternative oxidizing agents must be used as electron acceptors in respiration. The major *alternative electron acceptors* in aqueous environments include reducible solutes and mineral solids. These include (in order of decreasing energy yield): nitrate (NO_3^-), manganese (IV) oxides, iron (III) oxides, sulfate (SO_4^{2-}). Oxidation of both natural and xenobiotic reduced C compounds can be coupled effectively to the reduction of these redox-active constituents.

In addition to these major alternative electron acceptors, inorganic contaminants can also be reduced in the absence of O_2. As discussed earlier, many inorganic contaminants (*e.g.*, As, Se, Cr, Hg and Pb) can occur in more than one oxidation state, depending on environmental conditions. Anoxic conditions favor the reduced forms of these elements. For example, the more toxic and mobile aqueous species of selenium is selenate (SeO_4^{2-}), with Se in the +6 oxidation state. Selenate can be reduced to the less toxic and less mobile species, *selenite* (SeO_3^{2-}):

$$2SeO_4^{2-}(aq) + 4\,H^+(aq) + 4\,e^-\ (aq) \leftrightarrow$$
$$2SeO_3^{2-}\ (aq) + 2\,H_2O\ (l);\ \log K = 60 \quad \text{(Eq. 7.22a)}$$

This reduction of selenate must be coupled to an oxidation reaction, such as the oxidation of carbohydrate, providing the necessary electrons:

$$CH_2O\ (aq) + H_2O\ (l) \leftrightarrow CO_2\ (g) + 4\,H^+\ (aq)$$
$$+ 4\,e^-(aq);\ \log K = 0.8 \quad \text{(Eq. 7.22b)}$$

Reaction 7.23a is considered a *reduction half-reaction*, whereas 7.23b is an *oxidation half-reaction*. Their sum provides the full, balanced redox reaction:

$$CH_2O\ (aq) + 2\,SeO_4^{2-}\ (aq) \leftrightarrow CO_2\ (g)$$
$$+ 2\,SeO_3^{2-}\ (aq) + H_2O\ (l);\ \log K = 60.8 \quad \text{(Eq. 7.22c)}$$

Here, the pairs CH_2O/CO_2 and SeO_4^{2-}/SeO_3^{2-} are real-world examples of molecules A and B in Figure 7.7, respectively. Note that the oxidation half reaction (Eq. 7.23c) could be replaced by one for an organic contaminant, in which case the oxidative transformation of an organic contaminant would be coupled directly to the reductive transformation of an inorganic contaminant.

7.5.3 Photochemical Reactions

Some chemical compounds that are present in the atmosphere, in the top several centimeters of surface waters, or at the land surface can be transformed under the influence of sunlight. Chemical transformations that are induced by light energy are termed **photochemical processes**. Photochemical processes most often result in oxidation or reduction as a result of the absorption of light energy by one or more reactant species. Thus, a prerequisite for photochemical transformation is the capability of a molecule to absorb discrete quantities of light energy called **photons**. The energy (in Joules, J) gained by absorption of a photon is given by:

$$E = h\nu = hc/\lambda \quad \text{(Eq. 7.23)}$$

where h is Planck's constant (6.626×10^{-34} J s^{-1}), ν is the frequency of light, c is the velocity of light (3×10^8 m s^{-1}) and λ is its wavelength (m). Equation 7.24 indicates that the shorter the wavelength of the light, the greater the energy it transfers to matter when absorbed.

Photochemical transformation of both organic and inorganic pollutants can result from either *direct* or *indirect* photolysis. In **direct photolysis**, the light absorbing substance itself is transformed. An example of direct photolysis is the conversion of a chlorinated, refractory organic contaminant into an hydroxylated organic compound, which is normally less toxic and less refractory. This is very similar to the hydrolysis reaction given in Equation 7.21, but here, the reaction is being promoted by the presence of light:

$$R\text{-}Cl(aq) \overset{h\nu}{\to} R\text{-}Cl^*(aq) \overset{H_2O}{\to} R\text{-}OH$$
$$+ H^+(aq) + Cl^-(aq) \quad \text{(Eq. 7.24)}$$

The asterisk indicates the **excited state** of the chlorinated pollutant *R-Cl* that results from photon absorption. This *photoactivation* is characterized by the transition of one electron from its ground state to a higher energy state, which makes the contaminant molecule more susceptible to the subsequent transformation reaction.

INFORMATION BOX 7.2

There are three main types of ionizing radiation found in both natural and anthropogenic sources: alpha, beta, and gamma/x-ray radiation.

Alpha particles are subatomic fragments consisting of two neutrons and two protons. Alpha radiation occurs when the nucleus of an atom becomes unstable (the ratio of neutrons to protons is too low) and alpha particles are emitted to restore balance. Alpha decay occurs in elements with high atomic numbers, such as uranium, radium, and thorium. The nuclei of these elements are rich in neutrons, which makes alpha particle emission possible. Alpha particles are relatively heavy and slow, and therefore have low penetrating power and can be blocked with a sheet of paper.

Beta radiation occurs when an electron is emitted from the nucleus of a radioactive atom. Beta decay also occurs in elements that are rich in neutrons. Just like electrons found in the orbital of an atom, beta particles have a negative charge and weigh significantly less than a neutron or proton. Beta particles can be blocked by a sheet of metal or plastic and are typically produced in nuclear reactors.

Gamma or x-ray radiation is produced during a nucleus' excited state following a decay reaction. Instead of releasing another alpha or beta particle, it purges the excess energy by emitting a pulse of electromagnetic radiation called a gamma ray. Gamma rays are similar in nature to light and radio waves except that it has very high energy. Gamma rays have no mass or charge. They can travel for long distances and thus pose external and internal hazards for people. An example of a gamma emitter is cesium-137, which is used to calibrate nuclear instruments.

Indirect photolysis occurs after sunlight produces highly-reactive, transient (short-lived) oxygen species in oxygenated waters. These products, which form from photolysis of dissolved species such as nitrate, nitrite, aqueous iron complexes, and dissolved organic matter, include singlet oxygen ($^1O_2$), superoxide anion (O_2^-), hydroperoxyl ($HO_2^{\bullet}$), hydrogen peroxide (H_2O), ozone (O_3), hydroxyl radical ($OH^{\bullet}$) and organic peroxy radicals ($ROO^{\bullet}$). These reactive photolysis products can then accelerate the oxidation of other compounds that are normally quite refractory to oxidation by the more common oxidizing agents. The reaction of contaminants with these photolysis products is termed *indirect* photolysis because the transformed pollutants are not themselves absorbing the light that induces their transformation.

7.5.4 Radioactive Decay

Radioactive decay is an important transformation reaction for a special class of contaminants—radioactive elements. Radioactive decay is caused by an instability of the nucleus in the atom, whereby either protons and neutrons or electrons are emitted in the form of radiation (see Information Box 7.2). This transformation is spontaneous, but the rate at which it occurs varies widely depending on the element concerned.

7.5.5 Quantifying Transformation Rates

Many transformation reactions can be described with a first-order equation:

$$\frac{\partial C}{\partial t} = -kC \qquad \text{(Eq. 7.25)}$$

where k is the transformation rate constant (1/T). This equation states that the rate of transformation depends on the amount of contaminant present. The minus sign on the right-hand side of the equation denotes that the concentration change is negative (*i.e.*, concentrations of reactants decrease with time). The first-order equation can be used for example to represent fixed-pH hydrolysis, radioactive decay, and biodegradation when there is minimal net change in microbial cell numbers.

For transformation reactions, it is useful to define a half-life, which is the time required for half of the original contaminant mass to be transformed. For reactions that follow first-order kinetics, the half-life ($T_{1/2}$) is defined as:

$$T_{1/2} = \frac{0.693}{k} \qquad \text{(Eq. 7.26)}$$

This equation can be used to estimate the time required for a contaminant to be transformed, which is related to its persistence in the environment. For example, less than 1% of the original contaminant mass remains after a time period equal to 7 half-lives. After 10 half-lives, less than 0.1% remains. For a contaminant that has a half life of 1 day, this would mean that only 0.1% would remain untransformed after 10 days. A comparison of approximate half-lives for five compounds is presented in Table 7.3. Glucose, a labile (readily biodegradable) compound would be expected to be fully degraded in approximately three weeks (7 half-lives $\times$ 3 days). Conversely, anthracene is expected to persist for many years.

TABLE 7.3 Half-lives for selected organic compounds.*

COMPOUND	APPROXIMATE HALF-LIFE IN SOIL
Glucose	Three days
Benzene	A few days
2,4-D	Several days
Polychlorinated Biphenyl Mixture	Hundreds of days
Anthracene	Hundreds of days

*These values are very approximate and are meant for purposes of comparison. Note that all processes contributing to "loss" are incorporated (*e.g.*, biodegradation, abiotic transformation, and volatilization). Sources: Mackay et al., 1992; Montgomery, 1996.

QUESTIONS AND PROBLEMS

1. List and briefly discuss some of the major differences in the properties and behavior of inorganic and organic compounds.

2. Calculate the time required for 99.9% of the mass to transform for compounds with half-lives of 1 day, 10 days, 100 days, and 1000 days.

3. Which type of radiation is generally the most potentially hazardous to humans, and why?

4. What is hydrolysis?

5. Discuss some examples of LeChatellier's Principal that occur in everyday life.

6. Why are the aqueous solubilities of many organic compounds very low?

7. How does a change in ionic strength of an aqueous solution affect the solubility of (a) inorganic elements? (b) organic compounds? (c) which is generally affected to a greater degree and why?

REFERENCES AND ADDITIONAL READING

Langmuir D. (1996) *Aqueous Environmental Geochemistry*. Prentice Hall, New York, New York.

Mackay D., Shiu W.Y., and Ma K.C. (1992) *Illustrated Handbook of Physical-Chemical Properties and Environmental Fate of Organic Chemicals*. Lewis Publ., Boca Raton, Florida.

Montgomery J.H. (1996) *Groundwater Chemicals*, 2nd ed. CRC/Lewis Publ., Boca Raton, Florida.

Schwarzenbach R.P., Gschwend P.M., and Imboden D.M. (2003). *Environmental Organic Chemistry*, 2nd ed. John Wiley & Sons, New York, New York.

Verschueren K. (1983) *Handbook of Environmental Data on Organic Chemicals*. John Wiley & Sons, New York, New York.

CHAPTER 8

BIOLOGICAL PROCESSES AFFECTING CONTAMINANT TRANSPORT AND FATE

R.M. Maier

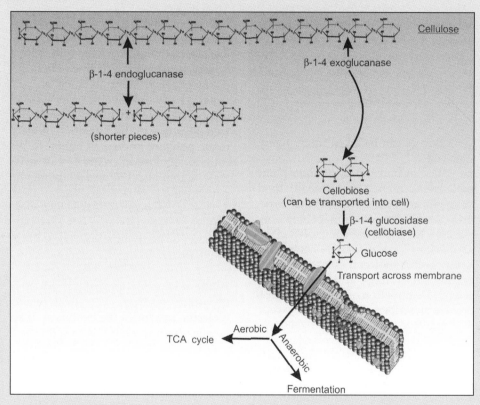

The degradation of cellulose begins outside the cell with a series of extracellular enzymes called cellulases. The resulting smaller glucose subunit structures can be taken up by the cell and metabolized. *From: Environmental Microbiology © 2000 Academic Press, San Diego, CA.*

8.1 BIOLOGICAL EFFECTS ON POLLUTANTS

Although physical and chemical factors affect the fate of pollutants in soil and water, it is apparent that these are not the only factors governing the fate of pollutants. If they were, the large-scale accumulation of contaminants, including environmental and natural organic substances would dominate the earth, rendering our current pollution problems minuscule by comparison. Thus we must look at a third basic factor—the biological component of soil and water (see Chapter 5). This component, which is responsible for degradation of naturally occurring organic matter, also mitigates the impact of pollutants on the environment. Biological interactions with pollutants are of great interest to scientists and engineers because of the growing use of biological approaches for the remediation of contaminated sites.

The presence of microorganisms in soil and water can affect the distribution, movement, and concentration of pollutants through a process called biodegradation. Indeed, some pollutants have very short lifetimes under normal environmental conditions because they readily serve as sources of food for actively growing microorganisms. For other pollutants, the effect of microorganisms may be limited for a variety of reasons. Low numbers of degrading microorganisms, microbe-resistant pollutant structures, or adverse environmental conditions can all cause extremely low rates of biodegradation.

In this chapter we will focus on understanding the interaction between microorganisms and pollutants in the environment. We will examine microbial interactions with both organic and inorganic pollutants, as well as the effects that environmental parameters and pollutant structure have on the extent and rate of these biological reactions.

8.2 THE OVERALL PROCESS OF BIODEGRADATION

Biodegradation is the breakdown of organic compounds ("organics") through microbial activity. Biodegradable organic compounds serve as the food source, or **substrate**, for microbes, and the availability of an organic to such microbes is the **bioavailability** of that organic. Bioavailability, which is one important aspect of the biodegradation of any substrate, depends largely on the water phase concentration of the organic. Microbial cells are 70–90% water, and the food they obtain comes from the water surrounding the cell. Thus, the bioavailability of a substrate refers to the amount of substrate in the water solution around the cell. Two important factors that reduce bioavailability are: (1) low water solubility (e.g., gasoline); and (2) sorption of substrate by soil (see Chapters 6 and 7).

Biodegradation of organic compounds is really "a series of biological degradation steps or a pathway which ultimately results in the oxidation of the parent compound." Often, the oxidation process generates energy (as described in Chapter 5). Complete biodegradation, or **mineralization**, involves oxidation of the parent compound to form carbon dioxide and water, a process that provides both carbon and energy for growth and reproduction of cells. Figure 8.1 illustrates the mineralization of any organic compound under aerobic conditions. Mineralization is composed of a series of degradation steps that have much in common, whether the carbon source is a simple sugar such as glucose, a plant polymer such as cellulose, or a pollutant molecule. Each degradation step in the pathway is facilitated by a specific catalyst, or **enzyme**, made by the degrading cell. Enzymes are most often found within a cell, but they are also made and released from the cell to help initiate degradation reactions. Enzymes found external to the cell are known as **exoenzymes**. Exoenzymes are important in the degradation of macromolecules such as the plant polymer cellulose because macromolecules must be broken down into smaller subunits to allow transport into the microbial cell. Both internal enzymes and exoenzymes are essential to the degradation process: degradation will stop at any step if the appropriate enzyme is not present (Figure 8.2). Lack of appropriate biodegrading enzymes is one common reason for persistence of some pollutants, particularly those with unusual chemical structures that existing enzymes do not recognize. Thus, we can see that degradation depends on chemical structure. Pollutants that are structurally similar to natural substrates usually degrade easily, while pollutants that are dissimilar to natural substrates often degrade slowly or not at all.

Mineralization can also be described as a **mass balance equation** that can be solved to determine the relationship between the amount of substrate consumed, oxygen and nitrogen utilized, and cell mass, carbon dioxide and water produced. For example, the mass balance equation for glucose (illustrated in Figure 8.1) can be written:

$$C_6H_{12}O_6 + NH_3 + O_2 \rightarrow C_5H_7NO_2 + CO_2 + H_2O \quad \text{(Eq. 8.1)}$$

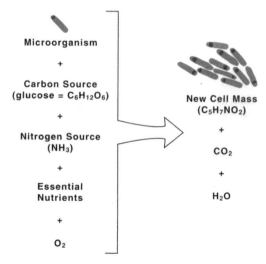

Figure 8.1 Aerobic mineralization of an organic compound. From *Pollution Science* © 1996 Academic Press, San Diego, CA.

Figure 8.2 Stepwise degradation of organic compounds. A different enzyme catalyzes each step of the biodegradation pathway. If any one enzyme is missing, the product of the reaction it catalyzes is not formed (denoted with a red cross). The reaction stops at that point and no further product is made (shown in gray). From *Pollution Science* © 1996, Academic Press, San Diego, CA.

As shown in Eq. 8.1 for glucose, many pollutant molecules—such as most gasoline components and many of the herbicides and pesticides used in agriculture—can be mineralized under the correct conditions (see Example Calculation 7.1).

Some organic compounds are only partially degraded. Incomplete degradation can result from the absence of the appropriate degrading enzyme or it may result from **cometabolism**. In cometabolism, a partial oxidation of the substrate occurs, but the energy derived from the oxidation is not used to support growth of new cells. This phenomenon arises when organisms possess enzymes that coincidentally degrade a particular pollutant; that is, their enzymes are nonspecific. Cometabolism can occur not only during periods of active growth, but also during periods in which resting (nongrowing) cells interact with an organic compound. Although difficult to measure in the environment, cometabolism has been demonstrated for some environmental pollutants. For example, the industrial solvent trichloroethene (TCE) can be oxidized cometabolically by **methanotrophic bacteria** while growing on their normal carbon source, methane. Trichloroethene is currently of great interest for several reasons. It is one of the most frequently reported contaminants at hazardous waste sites, it is a suspected carcinogen, and it is generally resistant to biodegradation. As shown in Figure 8.3, the first step in the methanotrophic oxidation of methane is catalyzed by the enzyme **methane monooxygenase**. This enzyme is so nonspecific that it can also catalyze the first step in the oxidation of TCE when both methane and TCE

are present. However, the methanotrophic bacteria receive no energy benefit from the oxidation of TCE, and so it is considered to be a cometabolic reaction. The subsequent degradation steps shown in Figure 8.3 for TCE may be catalyzed spontaneously, by other bacteria, or in some cases by the methanotrophs themselves. This type of cometabolic reaction has great significance in remediation and is not limited

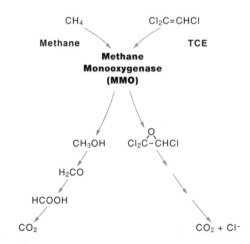

Figure 8.3 The oxidation of methane by methanotrophic bacteria. This is catalyzed by the enzyme methane monooxygenase. The same enzyme can act nonspecifically on TCE. Subsequent TCE degradation steps may be catalyzed spontaneously, by other bacteria, or in some cases by the methanotroph. From *Pollution Science* © 1996, Academic Press, San Diego, CA.

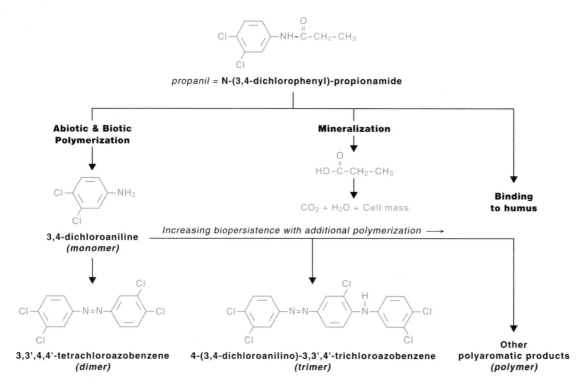

Figure 8.4 Polymerization reactions that occur with the herbicide propanil during biodegradation. Propanil is a selective post-emergence herbicide used in growing rice. It is toxic to many annual and perennial weeds. The environmental fate of propanil is of concern because it, like many other pesticides, is toxic to most noncereal crops. It is also toxic to fish. Care is used in propanil application to avoid contamination of nearby lakes and streams. From *Pollution Science* © 1996, Academic Press, San Diego, CA.

to methanotrophs. Other cometabolizing microorganisms that grow on toluene, propane, and even ammonia have also been identified. Currently, scientists are using this technique, the stimulation of cometabolic reactions, for remediation of some TCE-contaminated sites.

Partial or incomplete degradation can also result in **polymerization**, that is, the synthesis of compounds more complex and stable than the parent compound. This occurs when initial degradation steps, often catalyzed by exoenzymes, create highly reactive intermediate compounds, which can then combine either with each other or with other organic matter present in the environment. As illustrated in Figure 8.4, which shows some possible polymerization reactions that occur with the herbicide propanil during biodegradation, these include formation of stable dimers or larger polymers, both of which are quite stable in the environment. Such stability may be the result of low bioavailability (low water solubility, high sorption) or the absence of degrading enzymes.

8.3 MICROBIAL ACTIVITY AND BIODEGRADATION

It is often difficult to predict the fate of a pollutant in the environment because the interactions between the microbial, chemical, and physical components of the environment are

still not well understood. Total microbial activity depends on a variety of factors, such as microbial numbers, available nutrients, environmental conditions (including soil), and pollutant structure. In this section, we will discuss the impact of some of the most important factors affecting microbial activity, with the implicit understanding that microbial activity can be inhibited by any one of these factors even if all other factors are optimal.

8.3.1 Environmental Effects on Biodegradation

The environment around a microbial community—that is, the sum of the physical, chemical, and biological parameters that affect a microorganism—determines whether a particular microorganism will survive and/or metabolize. The occurrence and abundance of microorganisms in an environment are determined by nutrient availability, as well as by various physicochemical factors such as pH, redox potential, temperature, and soil texture and moisture. (These factors are described in detail in Chapters 2 and 5.) Because a limitation imposed by any one of these factors can inhibit biodegradation, the cause of the persistence of a pollutant is sometimes difficult to pinpoint.

Oxygen availability, organic matter content, nitrogen and phosphorus availability, and bioavailability are particularly significant in controlling pollutant biodegradation in

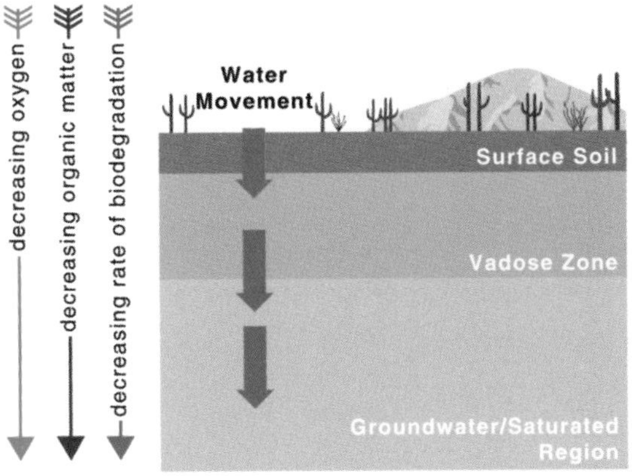

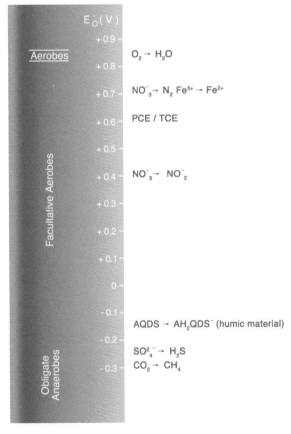

Figure 8.5 Contamination in different ecosystems. There are three major locations where contamination can occur in terrestrial ecosystems: surface soils, the vadose zone, and the saturated zone. The availability of both oxygen and organic matter varies considerably in these zones. As indicated, oxygen and organic matter both decrease with depth, resulting in a decrease in biodegradation activity with depth. From *Pollution Science* © 1996, Academic Press, San Diego, CA.

Some pollutants that degrade aerobically are not degradable anaerobically and vice versa. For example, the highly reduced hydrocarbons found in petroleum, some of which are shown in Figure 8.7, are readily degraded aerobically, but unless an oxygen atom is present in the structure initially, these compounds are quite stable under anaerobic conditions (Figure 8.8). This anaerobic stability explains why underground petroleum reservoirs, which contain no oxygen, have remained intact for thousands of years, even though microorganisms are present. In contrast, highly chlorinated organic compounds are more stable under aerobic conditions. That is, increasing chlorine content favors anaerobic dehalogenation (removal of chlorines) over aerobic dehalogenation.

In terms of oxygen availability, surface soils and the vadose zone are similar. They both contain significant amounts of air-filled pore spaces and thus tend to favor aerobic degradation of pollutants. However, these regions may contain pockets of anaerobic activity generated by high biodegradative activity or confined zones of water saturation that reduce oxygen levels. In contrast, the oxygen

the environment. Interestingly, the first three of these factors can change considerably, depending on the location of the pollutant. As Figure 8.5 shows, contamination can occur in terrestrial ecosystems in three major locations: surface soils, the vadose zone, and the saturated zone. The availability of both oxygen and organic matter varies considerably in these zones. In general, oxygen and organic matter both decrease with depth, so biodegradation activity also decreases with depth. There are exceptions to this rule in shallow-groundwater regions, which can have relatively high organic matter contents because the rates of groundwater recharge are high.

8.3.1.1 Oxygen and other terminal electron acceptors

Oxygen is very important in determining the extent and rate of biodegradation of pollutants—it serves as the **terminal electron acceptor (TEA)** in **aerobic** (oxygen is present) biodegradation reactions. As the TEA, oxygen is reduced to water as it accepts electrons from the pollutant substrate or **electron donor** that is being oxidized to carbon dioxide. At lower redox potentials that result in **anaerobic** conditions (oxygen is not available), an alternate TEA must be used. The electron tower shown in Figure 8.6 illustrates several different TEAs that can be used if available in the environment. Keeping in mind that the higher the place of the TEA on the electron tower, the more energy provided by the reaction, examination of this tower reveals that oxygen provides the most energy for biodegradation. Oxygen is followed closely by iron and nitrate and—at the bottom of the tower—sulfate and carbon dioxide. Because of the energy advantage associated with oxygen, both growth and biodegradation rates are much faster under aerobic conditions than under anaerobic conditions.

Figure 8.6 Terminal electron acceptors over a range of redox potentials. The electron tower shows the various terminal electron acceptors (TEAs) used under aerobic (oxygen) and anaerobic (all others) conditions. Note that the higher the place on the electron tower, the more energy that will be produced for metabolism and growth. From *Pollution Science* © 1996, Academic Press, San Diego, CA.

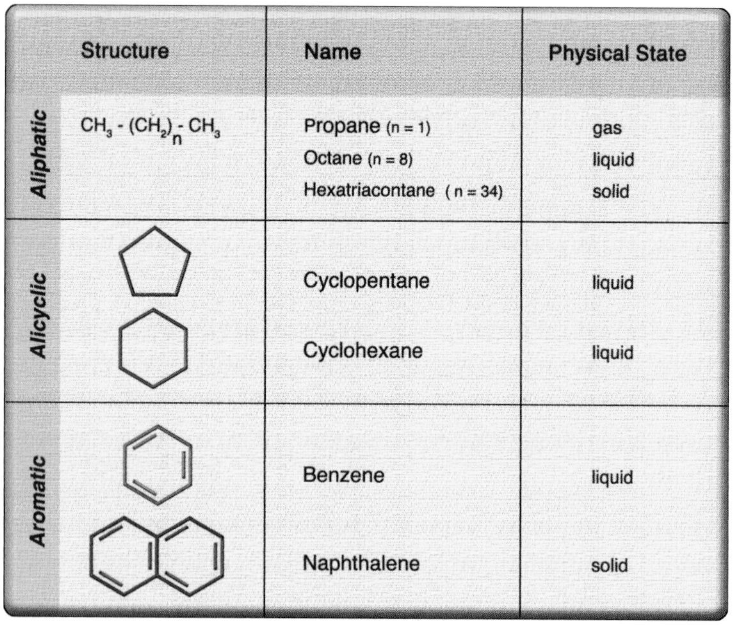

	Structure	Name	Physical State
Aliphatic	$CH_3 - (CH_2)_n - CH_3$	Propane (n = 1) Octane (n = 8) Hexatriacontane (n = 34)	gas liquid solid
Alicyclic	(pentagon) (hexagon)	Cyclopentane Cyclohexane	liquid liquid
Aromatic	(benzene ring) (naphthalene)	Benzene Naphthalene	liquid solid

Figure 8.7 Aliphatic, alicyclic, and aromatic hydrocarbons. Petroleum is usually composed of a mix of these hydrocarbon types. For example, familiar constituents of gasoline are octane and benzene. From *Pollution Science* © 1996, Academic Press, San Diego, CA.

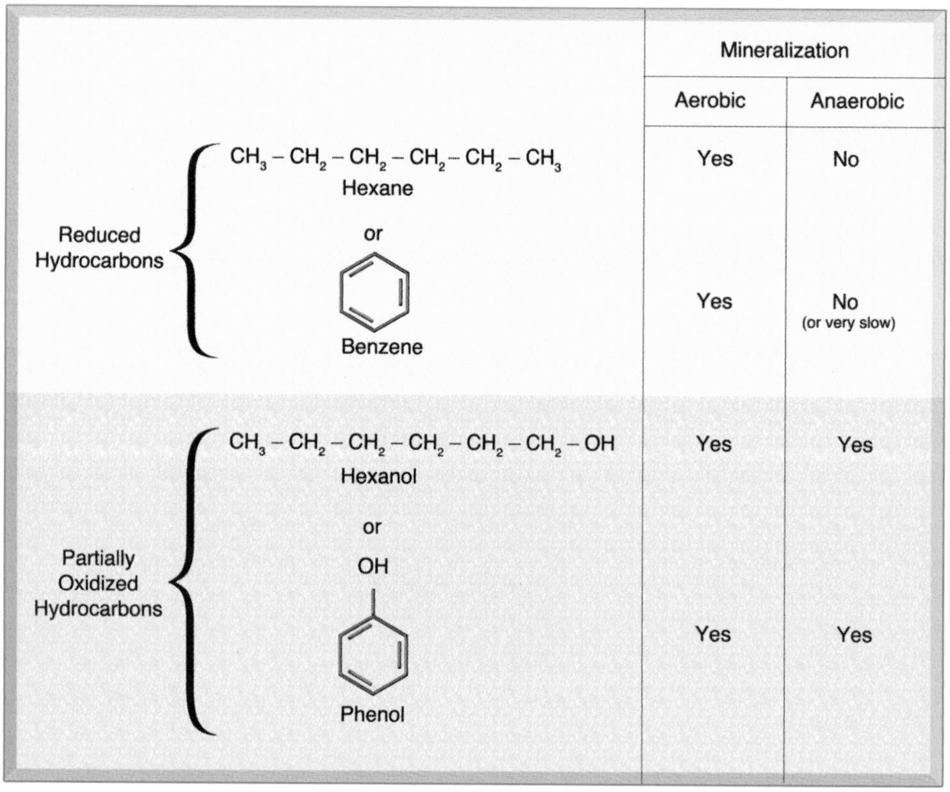

		Mineralization	
		Aerobic	Anaerobic
Reduced Hydrocarbons	$CH_3 - CH_2 - CH_2 - CH_2 - CH_2 - CH_3$ Hexane	Yes	No
	or (benzene ring) Benzene	Yes	No (or very slow)
Partially Oxidized Hydrocarbons	$CH_3 - CH_2 - CH_2 - CH_2 - CH_2 - CH_2 - OH$ Hexanol	Yes	Yes
	or OH (phenol ring) Phenol	Yes	Yes

Figure 8.8 The effects of oxidation on the biodegradability of aliphatic compounds. Hydrocarbons with no oxygen, such as hexane, are only degraded aerobically, while the addition of a single oxygen atom (hexanol) enables both aerobic and anaerobic degradation to occur.

concentrations in the groundwater or water-saturated regions are low. The only oxygen that exists in these regions is that which is dissolved in water, and the oxygen solubility in water is quite low (~9 mg/L). Therefore, if significant microbial activity occurs, the limited supply of dissolved oxygen is rapidly used up, causing anaerobic conditions to develop (see also Chapter 19). Addition of air or oxygen can often improve biodegradation rates, particularly in subsurface areas that are water saturated.

8.3.1.2 Microbial populations and organic matter content

Surface soils have large numbers of microorganisms. Culturable bacterial numbers generally range from 10^6 to 10^9 organisms per gram of soil. Fungal numbers are somewhat lower, 10^4 to 10^6 per gram of soil. In contrast, microbial populations in deeper regions, such as the vadose zone and groundwater region, are often lower by two orders of magnitude or more. This large decrease in microbial numbers with depth is primarily due to differences in organic matter content. Whereas the soil surface may be rich in organic matter, both the vadose zone and the groundwater region often have low amounts of organic matter. One consequence of low total numbers of microorganisms is that the population of pollutant degraders initially present is also low. Thus, biodegradation of a particular pollutant may be slow until a sufficient biodegrading population has been built up. The process in which degradation rates are initially low, but then increase over time is sometimes referred to as **adaptation** or **acclimation**. A second reason for slow biodegradation in the vadose zone and groundwater region is that the organisms in this region are often dormant because of the low amount of organic matter present. If microorganisms are dormant, their response to an added carbon source is slow, especially if the carbon source is a pollutant molecule to which they have not previously been exposed.

Given these two factors, oxygen availability and organic matter content, we can make several generalizations about surface soils, the vadose zone, and the groundwater region (see Figure 8.5):

1. Biodegradation in surface soils is primarily aerobic and rapid.
2. Biodegradation in the vadose zone is also primarily aerobic, but significant acclimation times may be necessary for large biodegrading populations to develop.
3. Biodegradation in the groundwater region is initially slow, owing to low numbers, and can rapidly become anaerobic due to lack of available oxygen.

8.3.1.3 Nitrogen

Nitrogen is another macronutrient that often limits microbial activity because it is an essential part of many key microbial metabolites and building blocks, including proteins and amino acids. As shown by the chemical formula for a cell (see Figure 8.1), nitrogen is a large component by mass of microorganisms. It is also subject to removal from the soil/water continuum by various processes such as leaching or denitrification (see Chapter 16). Many pollutants are carbon-rich and nitrogen-poor; thus nitrogen limitations can inhibit their biodegradation, whereas the simple addition of nitrogen-rich compounds can often improve it. For example, in the case of petroleum oil spills, where nitrogen shortages can be acute, biodegradation can be significantly accelerated by adding nitrogen fertilizers. In general, microbes have an average C:N ratio within their biomass of about 5:1 to 10:1, depending on the type of microorganism, so the C:N ratio of the material to be biodegraded must be 20:1 or less. The difference in the ratios is due to the fact that approximately 50% of the carbon metabolized is released as carbon dioxide, whereas almost all of the nitrogen metabolized is incorporated into the microbial biomass.

8.3.2 Pollutant Structure

The rate at which a pollutant molecule is degraded in the environment depends largely on its structure. If the molecule is not normally found in the environment—or if its structure does not resemble that of a molecule usually found in the environment—a biodegrading organism may not be present. In this case, chances for biodegradation to occur are low. The bioavailability of the pollutant is also extremely important in determining the rate of biodegradation. If the water solubility of the pollutant is extremely low, it will have low bioavailability. Many pollutant molecules that are persistent in the environment share the property of low water solubility. Examples include **d**ichloro**d**iphenyl**t**richloroethane (DDT), a pesticide that is now banned in the United States; **p**oly**c**hlorinated **b**iphenyls (PCBs), similarly banned and have not been manufactured in the U.S. since 1977, which are used as heat-exchange fluids; and petroleum hydrocarbons. Both PCBs and petroleum hydrocarbons are liquids at room temperature and actually form a phase that separates from water. Although microorganisms are not excluded from this phase, active metabolism seems to occur only in the aqueous phase or at the oil–water interface. The second factor that reduces bioavailability is sorption of the pollutant by soil. Compounds that have low water solubility, such as DDT, PCBs, and petroleum constituents, are also prone to sorption by soil surfaces (see Chapters 6 and 7).

Many pollutants have extensive branching or functional groups that block or sterically hinder the pollutant carbon skeleton at the reactive site, that is, the site at which the substrate and enzyme come into contact during a biodegradation step. For example, we now use biodegradable detergents, namely, the linear alkylbenzylsulfonates (ABSs). The only difference between these readily biodegradable detergents and the slowly biodegradable nonlinear ABSs is the absence of branching (Figure 8.9).

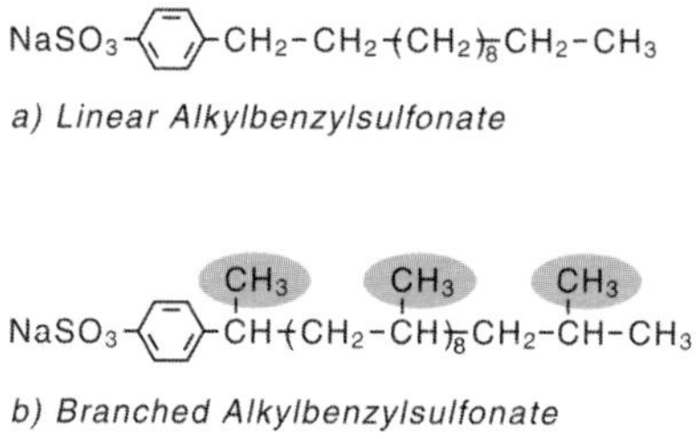

$$NaSO_3-\langle\rangle-CH_2-CH_2+(CH_2)_8 CH_2-CH_3$$

a) Linear Alkylbenzylsulfonate

$$NaSO_3-\langle\rangle-CH+(CH_2-CH)_8 CH_2-CH-CH_3$$

b) Branched Alkylbenzylsulfonate

Figure 8.9 Linear and branched alkylbenzylsulfonates (ABS) are commonly used surfactants. Since the linear variant (a) is readily biodegradable, and the branched form (b) is not, and both work equally well as detergents, the linear ABS has entirely supplanted the branched ABS in environmentally conscious markets. From *Pollution Science* © 1996, Academic Press, San Diego, CA.

As a result of our increasing knowledge of the effect of pollutant structure on biodegradation in the environment, efforts are focusing on developing and utilizing "environmentally friendly" compounds. For example, slowly biodegradable pesticides are being replaced by rapidly biodegradable ones, which are used in conjunction with integrated pest-management approaches (see Chapter 16). This approach means that pesticides are not used on a yearly basis, but instead are rotated. Thus, on the one hand target insects do not become fully acclimated to these easily degraded pesticides, but on the other hand, soil microorganisms degrade them rapidly so that they are active only during the intended time frame.

8.4 BIODEGRADATION PATHWAYS

The vast majority of the organic carbon available to microorganisms in the environment is material that has been photosynthetically fixed (plant material). Anthropogenic activity has resulted in the addition of many industrial and agricultural chemicals, including petroleum products, chlorinated solvents, and pesticides (see Information Box 8.1). Many of these chemicals are readily degraded in the environment because of their similarity to photosynthetically produced organic material. This allows degrading organisms to utilize preexisting biodegradation pathways. However, some chemical structures are unique or have unique components, which result in slow or little biodegradation. To help understand and predict biodegradation of organic contaminants in the environment, one can classify organic contaminants into one of three basic structural groups: the aliphatics, the alicyclics, and the aromatics (see Figure 8.7). Constituents of each of these groups can be found in all three physical states—gaseous, solid, and liquid. The general degradation pathways for each of these structural classes have been delineated. These pathways differ for aerobic and anaerobic conditions and can be affected by contaminant structural modifications.

INFORMATION BOX 8.1

The 2003 Agency for Toxic Substances and Disease Registry (ATSDR) Top Twenty Pollutants

By law, the ATSDR and Environmental Protection Agency (EPA) are required to prepare a list, in order of priority, of substances that are most commonly found at facilities on the National Priorities List (NPL) and which are determined to pose the most significant potential threat to human health due to their known or suspected toxicity and potential for human exposure at NPL sites (see Chapter 19). This list is revised every two years as additional information becomes available.

1 Arsenic	11 Chloroform
2 Lead	12 DDT
3 Mercury	13 Aroclor 1254
4 Vinyl Chloride	14 Aroclor 1260
5 Polychlorinated Biphenyls	15 Dibenzo(a,h)anthracene
6 Benzene	16 Trichloroethene
7 Cadmium	17 Chromium, Hexavalent
8 Polycyclic Aromatic Hydrocarbons	18 Dieldrin
9 Benzo(a)pyrene	19 Phosphorus, White
10 Benzo(b)fluoranthene	20 Chlordane

From: http://www.atsdr.cdc.gov/clist.html.

8.4.1 Biodegradation Under Aerobic Conditions

In the presence of oxygen, many heterotrophic microorganisms rapidly mineralize organic compounds (see Chapter 5). During degradation, some of the carbon is completely oxidized to carbon dioxide to provide energy for growth, and some carbon is used as structural material in the formation of new cells (see Figure 8.1). Energy used for growth is produced through a series of oxidation–reduction (redox) reactions in which oxygen is used as the TEA and reduced to water (see Figure 8.6).

8.4.1.1 Aliphatic hydrocarbons

Aliphatic hydrocarbons are straight-chain and branched-chain structures. Most aliphatic hydrocarbons introduced into the environment come from industrial solvent waste and the petroleum industry. Liquid aliphatics readily degrade under aerobic conditions, especially when the number of carbons is between 8 and 16. Longer-chain aliphatics are usually waxy substances. Biodegradation of these longer chains is slowed due to limited water solubility, while biodegradation of shorter chains may be impeded by the toxic effects of the short-chain aliphatic on microorganisms. In addition, several common structural modifications can result in severely reduced biodegradation. One of these modifications is extensive branching in the hydrocarbon chain (see Figure

8.9). Commonly found in petroleum, branched hydrocarbons constitute one of the slowest degraded fractions therein. Another common modification that slows biodegradation is halogen substitution, as seen in TCE and eight other compounds found on the ATSDR top twenty list (Information Box 8.1). The ATSDR list contains a variety of compounds (both aliphatic and nonaliphatic) that not only have limited biodegradability, but also pose serious toxicity problems. Thus, both the rate of biodegradation and toxicity must be considered in evaluating the potential hazard of pollutants in the environment.

Biodegradation of aliphatic compounds generally occurs by one of the three pathways shown in Figure 8.10. The most common is a direct enzymatic incorporation of molecular oxygen O_2 (Pathway I). All three of these pathways result in the formation of a common intermediate—a primary fatty acid. The fatty acid formed in the degradation of an alkane is subject to normal cellular fatty acid metabolism. This includes β-oxidation, which cleaves off consecutive two-carbon fragments. Each two-carbon fragment is removed by coenzyme A (CoA) as acetyl-CoA, and shunted to the tricarboxylic acid (TCA) cycle for complete degradation to CO_2 and H_2O. If the alkane has an even number of carbons, acetyl-CoA is the last residue. If the alkane has an odd number of carbons, propionyl-CoA is the last residue, which is also shunted to the TCA cycle after conversion to succinyl-CoA.

We know that both branching and halogenation slow biodegradation. In the former case, we can see that extensive branching causes interference between the degrading enzyme and the enzyme-binding site. In the latter case, however, we need to know something about the bonds and the reactions involved. For halogenated compounds, the relative strength of the carbon–halogen bond requires two things: (1) an enzyme that can act on the bond and (2) a large input of energy to break the bond. In general, monochlorinated alkanes are considered degradable; however, increasing halogen substitution results in increased inhibition of degradation. Halogenated aliphatics can be degraded by two types of re-

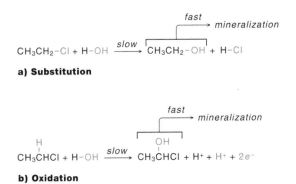

a) Substitution

b) Oxidation

Figure 8.11 Aerobic biodegradation of chlorinated aliphatic compounds. From *Pollution Science* © 1996, Academic Press, San Diego, CA.

actions that occur under aerobic conditions. The first is substitution, which is a nucleophilic reaction (the reacting species donates an electron pair) in which the halogen is substituted by a hydroxy group. The second is an oxidation reaction, which requires an external electron acceptor. These two reactions are compared in Figure 8.11. Although increasing halogenation generally slows degradation, aerobic oxidation of highly chlorinated aliphatics can occur cometabolically.

8.4.1.2 Aromatic hydrocarbons

The **aromatic hydrocarbons** contain at least one unsaturated ring system with the general structure C_6R_6, where R is any functional group (see Figure 8.7). The parent hydrocarbon of this class of compounds is benzene (C_6H_6), which exhibits the **resonance**, or delocalization of electrons, typical of unsaturated cyclic structures. Owing to its resonance energy, benzene is remarkably inert. (*Note*: As a group, the benzene-like "aromatics" tend to have characteristic aromas—hence the name.)

Aromatic compounds—including **polyaromatic hydrocarbons (PAHs)**, which contain two or more fused benzene rings—are synthesized naturally by plants. For example, they serve as a major component of lignin, a common plant polymer. Release of aromatic compounds into the environment occurs as a result of such natural processes as forest and grass fires. The major anthropogenic sources of aromatic compounds are fossil-fuel processing and utilization (burning). For example, benzene is one component of gasoline that is often released into the environment; it is of particular concern because it is, like many PAHs, a carcinogen.

Aromatic compounds, especially PAHs, are characterized by low water solubility and are therefore very hydrophobic. As is common with hydrophobic compounds, aromatics are often found sorbed to soil and sediment particles. The combination of low solubility and high sorption results in low substrate bioavailability and slow biodegradation rates. This is particularly true for PAHs having three or more rings because water solubility decreases as the number

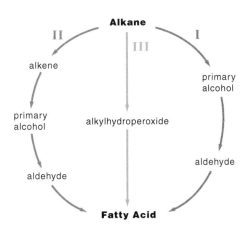

Figure 8.10 Aerobic biodegradation pathways for aliphatic compounds. From *Pollution Science* © 1996, Academic Press, San Diego, CA.

of rings increases. Thus, in general, PAHs having two or three condensed rings are biodegraded rapidly in the environment, often mineralizing completely, whereas PAHs with four or more condensed rings are transformed much more slowly, often as a result of cometabolic attack.

A wide variety of bacteria and fungi can degrade aromatic compounds. Under aerobic conditions, both groups of microbes incorporate oxygen as the first step in biodegradation; however, these two groups of microbes use different pathways, as shown in Figure 8.12. Bacteria use a **dioxygenase** enzyme that incorporates both atoms of molecular oxygen into the PAH to form a stereo-specific *cis*-dihydrodiol, and then the common intermediate catechol. The ring is then cleaved by a second dioxygenase, as shown in Figure 8.12, using either an ortho or a meta pathway. Which of these pathways is used is organism specific. Close examination of the ring cleavage products shows molecules that have a fatty acid character and that can now be degraded using normal cellular fatty acid metabolism, as explained for aliphatics.

In contrast to bacteria, fungi degrade aromatic compounds using a **monooxygenase** enzyme that incorporates only one atom of molecular oxygen into the PAH and reduces the second oxygen to water. The result is the formation of an arene oxide, followed by the enzymatic addition of water to yield a stereo-specific *trans*-dihydrodiol and then the common intermediate, catechol (see Figure 8.12). This catechol is then mineralized completely as for bacteria. Alternatively, the arene oxide can be isomerized to form a phenol, which can be conjugated with sulfate, glucuronic acid, glucose, or glutathione. These conjugates

are similar to those formed in higher organisms, such as humans, and seem to aid in detoxification and elimination of PAH.

8.4.1.3 Alicyclic hydrocarbons

Alicyclic hydrocarbons are saturated carbon chains that form ring structures (see Figure 8.7). Naturally occurring alicyclic hydrocarbons are common. For example, alicyclic hydrocarbons are a major component of crude oil, comprising 20 to 67% by volume. Other examples of complex, naturally occurring alicyclic hydrocarbons include camphor, which is a plant oil; cyclohexyl fatty acids, which are components of microbial lipids; and the paraffins from leaf waxes. Anthropogenic sources of alicyclic hydrocarbons to the environment include fossil-fuel processing and oil spills, as well as the use of such agrochemicals as the pyrethrin insecticides.

It is very difficult to isolate pure cultures of bacteria that can degrade alicyclic hydrocarbons. For this reason, biodegradation of alicyclic hydrocarbon degradation is thought to take place as a result of teamwork among mixed populations of microorganisms. Such a team is commonly referred to as a **microbial consortium**. Another unique aspect of alicyclic degradation is the formation of a lactone ring during one of the biodegradation steps that is one larger than the original ring. For example, in the degradation of cyclohexane, a six-membered ring, we get the formation of ε-caprolactone, a seven-membered ring. Thus, in the degradation of cyclohexane, one population in the microbial consortium performs the first two degradation steps, cyclohexane to cyclo-

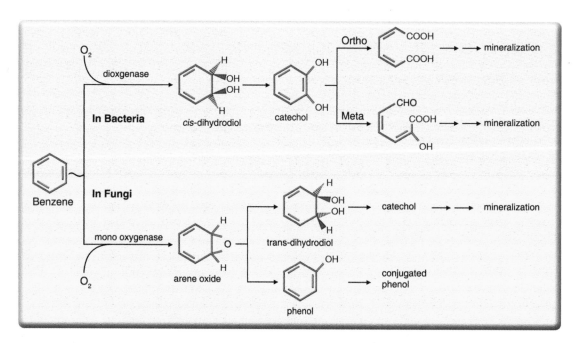

Figure 8.12 Aerobic biodegradation pathways of aromatic compounds in bacteria and fungi.
From *Pollution Science* © 1996, Academic Press, San Diego, CA.

hexanone via cyclohexanol, but is unable to lactonize and open the ring. Subsequently, a second population in the consortium, which cannot oxidize cyclohexane to cyclohexanone, performs the lactonization and ring-opening steps, and then degrades the compound completely (Figure 8.13).

Interestingly, cyclopentane and cyclohexane derivatives, which contain one or two hydroxyl, carbonyl, or carboxyl groups, degrade more readily in the environment than do their parent compounds. In fact, microorganisms capable of degrading cycloalkanols and cycloalkanones are ubiquitous in environmental samples.

8.4.2 Biodegradation Under Anaerobic Conditions

Anaerobic conditions are not uncommon in the environment. Most often, such conditions develop in water or saturated sediment environments. But even in well-aerated soils there are microenvironments with little or no oxygen. In all of these environments, **anaerobiosis** occurs when the rate of oxygen consumption by microorganisms is greater than the rate of oxygen diffusion through either air or water. In the absence of oxygen, organic compounds can be mineralized through **anaerobic respiration**, in which a TEA other than oxygen is used (see Figure 8.6). The series of alternative TEAs in the environment includes iron, nitrate, manganese, sulfate, and carbonate, which are listed in order from most oxidizing to most reducing conditions. This progression means they are usually utilized in this order because the amount of energy generated for growth depends on the oxidation potential of the TEA. Since none of these TEAs are as oxidizing as oxygen, growth under anaerobic conditions is never as efficient as growth under aerobic conditions although it can come close for TEAs such as iron or nitrate (Figure 8.6; see also Chapter 5).

Interestingly, many compounds that are easily degraded aerobically, such as saturated aliphatics, are far more difficult to degrade anaerobically. However, in at least one group of compounds—those that are highly chlorinated —the chlorine substituents are removed more rapidly under anaerobic conditions. But once dechlorination has occurred, the remaining molecule behaves more typically; that is, it is generally degraded more rapidly and extensively aerobically than anaerobically. As a consequence of this sequential process, technologies have been developed that utilize sequential anaerobic–aerobic treatments to optimize degradation of highly chlorinated compounds.

8.4.2.1 Aliphatic hydrocarbons

Saturated aliphatic hydrocarbons are degraded slowly, if at all, under anaerobic conditions. In general, the longer the hydrocarbon chain, the more likely biodegradation will occur although very slowly. Specifically, methane (CH_4) is not degraded, hexane (C_6H_{14}) may be very slowly or not degraded, and hexadecane ($C_{16}H_{34}$) is slowly degraded. We see evidence of this slow to nonexistent degradation in nature; for example, hydrocarbons in natural underground reservoirs of oil and methane (which are under anaerobic conditions) are not degraded, despite the presence of microorganisms. However, both unsaturated aliphatics and oxygen-containing aliphatics (aliphatic alcohols and ketones) are readily biodegraded anaerobically using a variety of TEAs. The suggested pathway of biodegradation for a saturated hydrocarbon is the addition of the 4-carbon molecule fumarate, which forms a fatty acid intermediate (Figure 8.14).

Chlorinated aliphatics can be partially or completely degraded under anaerobic conditions, but the mechanism is very dependent on the actual chlorinated compound in question. For example, C_1 molecules such as chloromethane (CH_3Cl) and dichloromethane (CH_2Cl_2) can support the growth of anaerobic microbes. These microbes first remove the chlorine using a **dehalogenase** enzyme, leaving a methyl group that is oxidized in a complex series of reactions to provide energy for growth. In contrast, C_2 molecules such as chloroethane (CH_3-CH_2Cl) do not support microbial growth under anaerobic conditions. However, for highly chlorinated

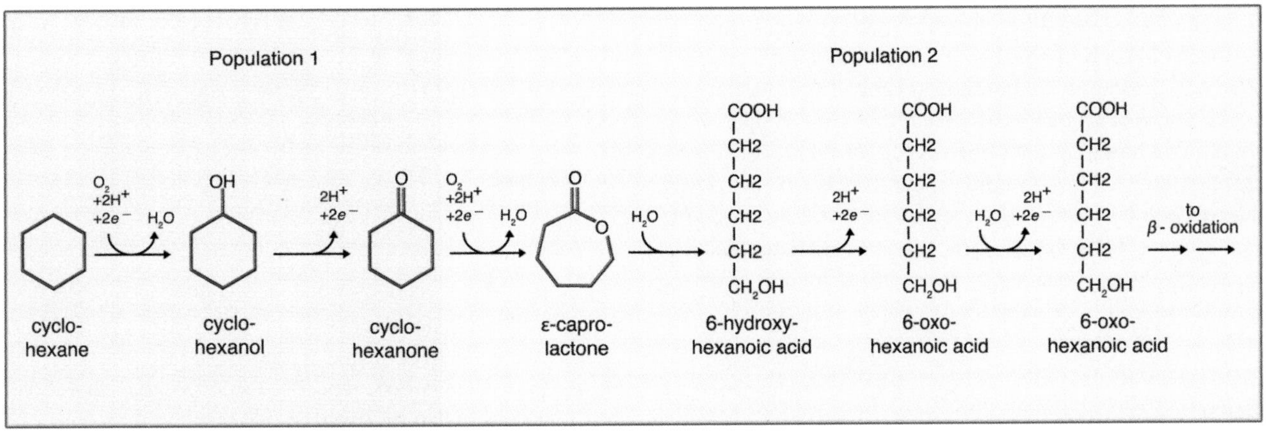

Figure 8.13 Aerobic biodegradation of cyclohexane by a microbial consortium. From *Pollution Science* © 1996, Academic Press, San Diego, CA.

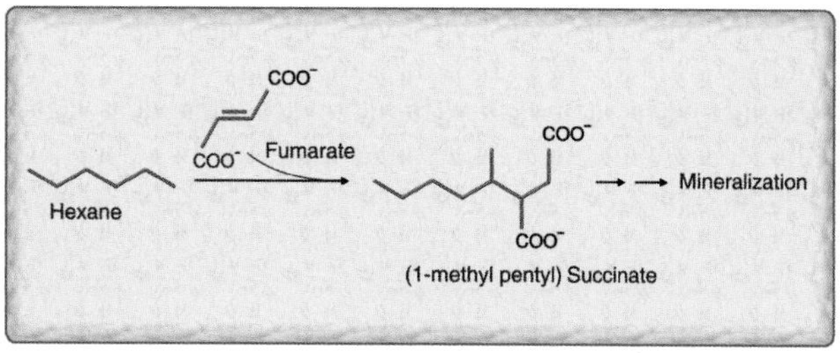

Figure 8.14 General anaerobic biodegradation pathway of an alkene.

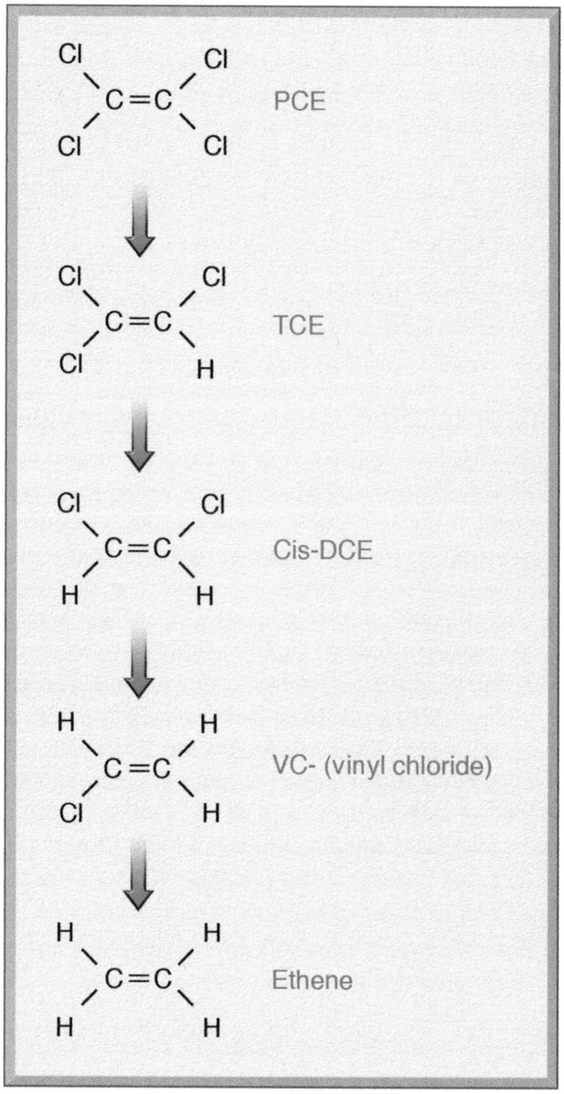

Figure 8.15 Halorespiration of perchloroethene (PCE). There are a number of anaerobic halorespiring microbes that can dehalogenate PCE to either TCE or DCE. Recently, there even has been one microbe described that can completely dechlorinate PCE to ethene!

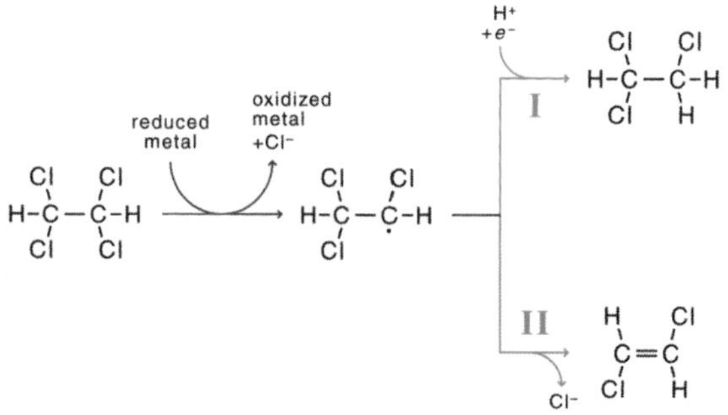

compounds, **reductive dehalogenation** is used to remove chlorines. In this case, the dechlorination reaction may be cometabolic or linked to respiration, a process called **halorespiration**, in which the chlorinated aliphatic acts as a TEA and the electron donor is either H_2 or a C_1 or C_2 carbon organic compound such as ethanol (CH_3CH_2OH). Halorespiration results in a compound with a reduced number of chlorine atoms that is now more amenable to aerobic biodegradation. A good example is the common groundwater contaminant perchloroethene (PCE). PCE is not known to degrade at all under aerobic conditions; however, as shown in Figure 8.15, PCE readily undergoes reductive dehalogenation. In cometabolic reductive dehalogenation, the process may be mediated by reduced transition-metal/metal complexes. The steps in this transformation are shown in Figure 8.16. In the first step, electrons are transferred from the reduced metal to the halogenated aliphatic, resulting in an alkyl radical and free halogen. Then, the alkyl radical can either scavenge a hydrogen atom (I) or lose a second halogen to form an alkene (II). In general, anaerobic conditions favor the degradation of highly halogenated compounds, while aerobic conditions favor the degradation of mono- and disubstituted halogenated compounds.

8.4.2.2 Aromatic hydrocarbons

Recent evidence indicates that nonsubstituted aromatics like benzene can be degraded only very slowly under anaerobic conditions. However, like aliphatic hydrocarbons, substituted aromatic compounds can be rapidly and completely degraded under anaerobic conditions (Figure 8.17). Anaerobic mineralization of aromatics often requires a mixed microbial community whose populations work together under different redox potentials. For example, mineralization of benzoate can be achieved by growing an anaerobic benzoate degrader in co-culture with an aerobic methanogen or sulfate reducer. In this consortium, benzoate is transformed by one or more anaerobes to yield aromatic acids, which in turn are transformed to methanogenic precursors such as acetate, carbon dioxide, or formate. These small molecules can then be utilized by methanogens (Figure 8.18). This process can be described as an anaerobic food chain because the organisms higher in the food chain cannot utilize acetate or other methanogenic precursors, while the methanogens cannot uti-

Figure 8.16 Cometabolic reductive dehalogenation of a chlorinated hydrocarbon in the presence of a metal to form an alkyl radical. (I) The alkyl radical scavenges a hydrogen atom. (II) The alkyl radical loses a second halogen to form an alkene. From *Pollution Science* © 1996, Academic Press, San Diego, CA.

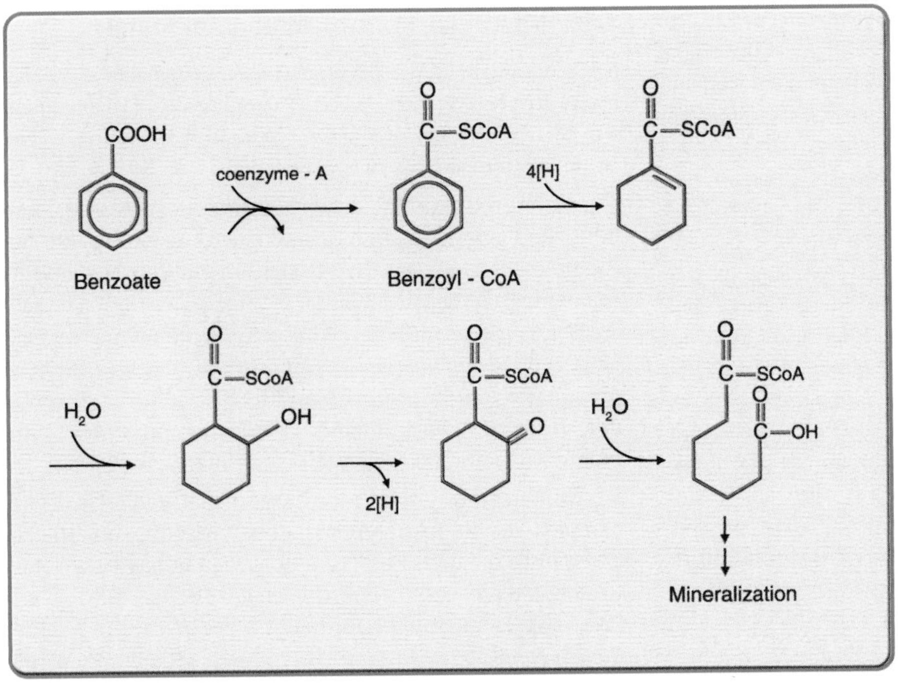

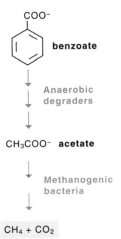

Figure 8.17 Anaerobic biodegradation of benzoate. Note that the intermediate benzoyl-CoA is a common intermediate found in the anaerobic degradation of aromatic compounds. Further note (in contrast to aerobic conditions) that anaerobic microbes completely saturate the ring during biodegradation before it is opened.

lize larger molecules such as benzoate. Methanogens utilize carbon dioxide as a terminal electron acceptor, thereby forming methane. (*Note*: Methanogens should not be confused with methanotrophic bacteria, which aerobically oxidize methane to carbon dioxide.)

Figure 8.18 An example of an aerobic food chain. Shown is the formation of simple compounds from benzoate by a population of anaerobic bacteria and the subsequent utilization of the newly available substrate by a second anaerobic population, the methanogenic bacteria. From *Pollution Science* © 1996, Academic Press, San Diego, CA.

8.5 TRANSFORMATION OF METAL POLLUTANTS

Metals compose a second important class of pollutants (see Information Box 8.1). However, metals are also essential components of microbial cells. For example, sodium and potassium regulate gradients across the cell membrane, while copper, iron, and manganese provide metalloenzymes for photosynthesis and electron transport. On the other hand, metals can also be extremely toxic to microorganisms. Although the most toxic metals are the nonessential metals such as arsenic, cadmium, lead, and mercury, even essential metals can become toxic in high concentrations.

What is the fate of metals in the environment? Metals and metal-containing contaminants are not degradable in the same sense that carbon-based molecules are for two reasons. First, unlike carbon, the metal atom is not the major building block for new cellular components. Secondly, while a significant amount of carbon is released to the atmosphere in gaseous form as carbon dioxide, the metals rarely enter the environment in a volatile, or gaseous, phase. (There are some exceptions to this—most notably, mercury and selenium, which can be transformed and volatilized by microorganisms under certain conditions.) In general, however, the nondegradability of metals means that it is difficult to eliminate metal atoms from the environment. Therefore, localized, elevated levels of metals are common, especially in industrially developed countries. Consequently, these metals can accumulate in biological systems, where their toxicity poses serious threats to human and environmental health.

Metals and metal-containing molecules can undergo transformation reactions, many of which are mediated by microorganisms. The nature of these reactions is important for consideration of metal toxicity in the environment, because toxic effects more often depend on the form and bioavailability of the metal than on the total metal concentration. In general, the most active form of added metals are free metal ions. The metals having the highest toxicity are the cations of mercury (Hg^{2+}) and lead (Pb^{2+}), although other metallic cations (arsenic, beryllium, boron, cadmium, chromium, copper, nickel, manganese, selenium, silver, tin, and zinc) also exhibit significant toxic effects. (The specific toxicities of selected metals to humans are discussed in Chapter 13.)

8.5.1 Effects of Metals on Microbial Metabolism

The nature of the interaction between heavy metals and microorganisms is complex. Metal toxicity requires uptake of the metal by a cell, which is dependent on many factors such as pH, soil type, and temperature. For example, accumulation of metals by cell-surface binding increases with increasing pH. Transport of metals into a cell is also pH-dependent, with maximal transport rates in the pH range 6.0–7.0. Once a metal is taken up by a cell, toxicity can result. After an initial period in which the toxic effects of the metals are evident, microorganisms often acquire tolerance mechanisms that enable them to repair metal toxicity damage, after which they can to start metabolizing and growing again at a nearly normal rate. The length of time required to develop tolerance mechanisms is influenced by both biotic and abiotic factors. The biotic factors of importance may involve the physiological state of the organism in question, such as the nutritional level, or genetic adaptations that result in metal resistance. Abiotic factors include the physicochemical characteristics of the environment such as pH, temperature, and redox potential, all of which affect the precipitation and complexation of metals.

The specific toxic effects of heavy metals on microorganisms are caused by the binding of the metal to cellular ligands such as proteins or nucleic acids. This metal–ligand binding leads to conformational changes and loss of normal ligand activity. For example, the particularly strong affinity of cationic metals for protein sulfhydryl groups can lead to alterations in protein folding. Both the ligand structure and the size of the metal affect the type binding. Large metal ions such as copper, silver, gold, mercury, and cadmium preferentially form covalent bindings with sulfhydryl groups. In contrast, small, highly electropositive metal ions such as aluminum, chromium, cobalt, iron, titanium, zinc, and tin preferentially complex with carboxyl, hydroxyl, phosphate, and amino groups.

8.5.2 Microbial Transformations of Metals

Microorganisms have developed various resistance mechanisms to prevent metal toxicity: among these mechanisms we will briefly discuss metal oxidation/reduction, metal complexation, and alkylation of metals.

Oxidation/reduction: Metal oxidation enhances metal mobility, stimulating metal movement away from the cell. For example, the Gram-positive bacterium *Bacillus megaterium* can oxidize elemental selenium to selenite, a reaction that increases selenium mobility. Alternatively, some microorganisms reduce metals such as chromium, causing them to precipitate and become immobilized. There is currently much interest in using microbial-reduction processes to immobilize metals and radionuclides within a site.

Complexation: Other microorganisms can effectively complex metals to polymeric materials either internal or external to the cell. For example, uranium has been found to accumulate extracellularly as needle-like fibrils in a layer approximately 2 μm thick on the surface of a yeast, *Saccharomyces cerevisiae*. In contrast, uranium accumulates as dense intracellular deposits in *Pseudomonas aeruginosa*.

Alkylation: Some microorganisms can transform metals by alkylation, which involves the transfer of one or more organic ligand groups (*e.g.*, methyl groups) to the metal, thus affording stable organometallic compounds. One such metal is mercury, as shown in Figure 8.19. The physical and chemical properties of organometals are different from those of pure metals. For example, volatility and solubility are increased, thus facilitating the movement of the organometal through the soil solution and, ultimately, into the atmosphere. In general, the extent of organometallic mobility depends on the nature and number of the organic ligands involved. But the presence of even a single methyl group can significantly increase both the volatility and **lipophilicity** (the affinity for lipids) of a metal. Transformations that increase volatility allow microorganisms to help remove a toxic metal from its environment; however, the concomitant increase in lipophilicity can cause biomagnification of the metal, resulting in toxic effects on other members of the ecosystem.

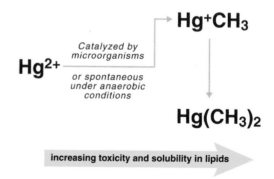

Figure 8.19 Microbial alkylation of mercury. From *Pollution Science* © 1996, Academic Press, San Diego, CA.

EXAMPLE CALCULATION 8.1
Leaking Underground Storage Tanks

The EPA estimates that more than 1 million underground storage tanks (USTs) have been in service in the United States alone. Over 436,000 of these have had confirmed releases into the environment. Although new regulations now require USTs to be upgraded, leaking USTs continue to be reported at a rate of 20,000/yr and a cleanup backlog of more than 139,000 USTs still exists. In this exercise we will calculate the amount of oxygen and nitrogen necessary for the biological remediation of a leaking UST site that has released 10,000 gallons of gasoline. To simplify this problem, we will assume that octane (C_8H_{18}) is a good representative of all petroleum constituents found in gasoline. We will use the following mass balance equation to calculate the biological oxygen demand (BOD) and the nitrogen demand:

$$a(C_8H_{18}) + b(NH_3) + c(O_2) \rightarrow d(C_5H_7NO_2)$$
$$\text{octane} \quad \text{ammonia} \quad \text{oxygen} \quad \text{cell mass}$$
$$+ e(CO_2) + f(H_2O)$$
$$\text{carbon dioxide} \quad \text{water}$$

In this equation, the coefficients a through f indicate the number of moles for each component. To solve the mass balance equation, we must be able to relate the amount of cell mass produced to the amount of substrate (octane) consumed. This is done using the cell yield (Y), where:

$$Y = \frac{\text{mass of cell mass produced}}{\text{mass of substrate consumed}}$$

Literature indicates that a reasonable cell yield value for octane is 1.2. Using the cell yield we can now calculate the coefficient d. We will start with 1 mole of substrate (a = 1) and use the following equation:

$$d \text{ (MW cell mass)} = a \text{ (MW octane) (Y)}$$
$$d \text{ (113 g/mol)} = 1 \text{ (114 g/mol) (1.2)} \quad \Rightarrow \quad d = 1.2$$

We can then solve for the other coefficients by balancing this equation.

We start with nitrogen. We know that there is one N on the right side of the equation in the biomass term. Examining the left side of the equation, we see that there is similarly one N as ammonia. We can set up a simple relationship for nitrogen and use this to solve for coefficient b:

For b: b(1 mol nitrogen) = d(1 mol nitrogen)

$$b(1) = 1.2(1)$$
$$b = 1.2$$

Next we balance carbon and solve for coefficient e:

For e: e(1 mol carbon) = a(8 mol carbon) − d(5 mol carbon)

$$e(1) = 1(8) - 1.2(5)$$
$$e = 2.0$$

Next we balance hydrogen and solve for coefficient f:

For f: f(2 mol hydrogen) = a(18 mol hydrogen) + b(3 mol hydrogen) − d(7 mol hydrogen)

$$f(2) = 1(18) + 1.2 (3) - 1.2(7)$$
$$f = 6.6$$

Finally we balance oxygen and solve for coefficient c:

For c: c(2 mol oxygen) = d(2 mol oxygen) + e(2 mol oxygen) + f(1 mol oxygen)

$$c(2) = 1.2(2) + 2(2.0) + 6.6(1)$$
$$c = 6.5$$

Thus, the solved mass balance equation is:

$$1(C_8H_{18}) + 1.2(NH_3) + 6.5(O_2) \rightarrow 1.2(C_5H_7NO_2) + 2.0(CO_2) + 6.6(H_2O)$$

Now we will use this mass balance equation to determine how much nitrogen and oxygen will be needed to remediate the site. First, we will convert gallons of gasoline into mol of octane using the assumption that octane is a good representative of gasoline.

Recall that we started with 10,000 gallons of gasoline:

convert to liters (L): 10,000 gallons (3.78 L/gallon) = 3.78 $\times 10^4$ L gasoline

convert to grams (g): 3.78 $\times 10^4$ L gasoline (690 g gasoline/L) = 2.6. $\times 10^7$ g gasoline in the site

convert to moles: $\dfrac{2.6 \times 10^7 \text{ g gasoline}}{114 \text{ g octane/mol}} = 2.3 \times 10^5$ mol octane in the site

Now we ask—how much nitrogen is needed to remediate this spill? From the mass balance equation we know that we need 1.2 mol NH_3/mol octane (see coefficient b).

$\Rightarrow \dfrac{1.2 \text{ mol } NH_3}{\text{mol octane}} \times 2.3 \times 10^5$ mol octane in site

$\Rightarrow 2.76 \times 10^5$ mol NH_3

$\Rightarrow 2.76 \times 10^5 \times \dfrac{17 \text{ g } NH_3}{\text{mol}}$

$\Rightarrow 4.7 \times 10^6$ g NH_3

$\Rightarrow 4.7 \times 10^6$ g (1 kg/1000g) (2.2046 lb/kg) = **This is 10,000 lb or 5 tons of NH_3!!!**

Finally we ask—how much oxygen is needed to remediate this spill? From the mass balance equation we know that we need 6.5 mol O_2/mol octane (see coefficient c).

$\Rightarrow 6.5$ mol O_2/mol octane (2.3 $\times 10^5$ mol octane in the site) = 1.5 $\times 10^6$ mol O_2

A gas takes up 22.4 L/mol, but remember that air is only 21% oxygen.

$\Rightarrow 1.5 \times 10^6$ mol O_2 (22.4 L/mol air) (1 mol air/0.21 mol O_2) = 1.6 $\times 10^8$ L air

1 cubic foot of air = 28.33 L

$\Rightarrow 1.6 \times 10^8$ L air (1 cubic foot/28.33 L gas) = **This is 5.5 $\times 10^6$ cubic feet of air or enough air to fill a football field to a height of 100 ft!!!!**

QUESTIONS AND PROBLEMS

1. Define biodegradation, making reference to the terms transformation, mineralization, and cometabolism.

2. a) Consider *n*-octane, an eight-carbon straight-chain aliphatic compound. Draw its structure.

   b) Is this compound biodegradable?

   c) Beginning with the structure you have drawn, show how you can alter it to make it less biodegradable.

   d) Compare the structure in part (a) with the equation (1) for biodegradation shown in Section 8.2, and consider the following situation: A site is contaminated by a leaking underground gasoline storage tank. The remediation firm that you work for would like to use bioremediation to clean the site. Given the structure of gasoline components (which typically include simple aliphatic, alicyclic, and aromatic compounds), what other nutrients may be required to complete bioremediation? Explain.

3. List and explain the factors that determine bioavailability of an organic compound.

4. Compare the following structures:

phenol	TCE	naphthalene
A	B	C

   a) Predict the order of bioavailability.

   b) Predict the order of biodegradability.

   c) What does a comparison of (a) and (b) tell you about the relationship between bioavailability and biodegradability?

   d) Which is the most likely type of biodegradation for each of the above compounds?

REFERENCES AND ADDITIONAL READING

Atlas R.M. and Bartha R. (1993) *Microbial Ecology*, 3rd ed. Benjamin Cummings, Menlo Park, California.

PART 2

MONITORING, ASSESSMENT, AND REGULATION OF ENVIRONMENTAL POLLUTION

CHAPTER **9**

PHYSICAL CONTAMINANTS

J. Walworth and I.L. Pepper

Aerosol production from tractor operations. *Photo courtesy J.P. Brooks.*

9.1 PARTICLE ORIGINS

Small particles, whether of natural or anthropogenic origin, can pollute air and water supplies. These particles pose a hazard to human health and to the environment in a variety of ways. We will explore the properties of particulate contaminants, the health threats they present, where they come from, and how they behave in the environment.

Particulate sources are divided into those arising from a single, well-defined emission source, which is called **point-source pollution**, and **nonpoint-source pollution**, which is generated from a wide area. Particulate emissions can have natural origins or be of human origin. Naturally occurring particulates may become a threat when mobilized by human activities, such as agriculture, logging, and construction, or they may arise from natural processes such as volcanoes or soil erosion. Human-made particulates include those created by industrial processes; combustion in power plants, wood stoves, fireplaces, and internal combustion engines; rubber particles from tire wear; and other sources.

9.2 PARTICLE SIZE

Particle behavior is, to a large extent, determined by size. The diameters of some particles and familiar substances are shown in Figure 9.1. Many particle properties, as well as their environmental and health impacts, are related to their

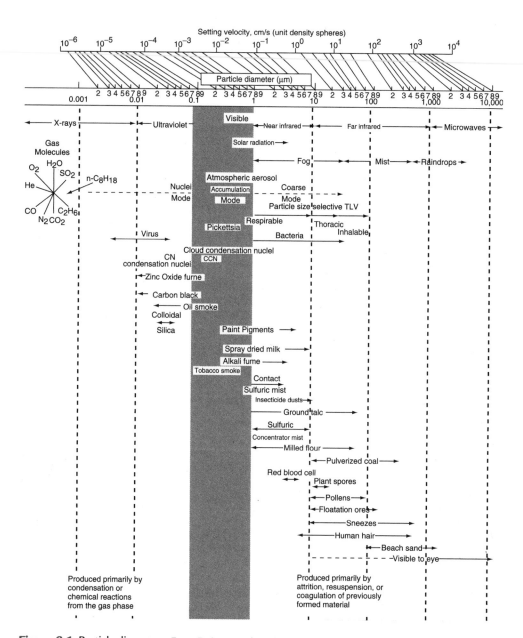

Figure 9.1 Particle diameters. From *Environmental Monitoring and Characterization* © 2004, Academic Press, San Diego, CA.

size. In general, smaller particles pose a greater health threat than do larger particles.

9.2.1 Nanoparticles

Nanotechnologies refer to "technology of the tiny," with dimensions in the range of nanometers (SI unit prefix for 10^{-9} or one-billionth, 0.000000001, of a meter). As an illustration of the scale of interest, a chain of 5 to 10 atoms is about 1 nm long, the helix of DNA has a diameter of about 2 nm, and the average human hair is about 80000 nm in diameter.

Nanotechnology and nanoscience research represent a key aspect of the development of innovative materials and new productive sectors. Nanomaterials (nanoparticles, nanospheres, nanotubes, and nanostructured surfaces) are used in ceramic, textile, cosmetic, optic, and chemical industries. In addition, they are applied in biomedicine as nanobiomaterials, nanospheres for drug release, and nanotubes for gene therapy. Most nanoparticles that are currently used today are made from transition metals including silicon, carbon, and metal oxides.

Human exposure to nanoparticles can occur as environmental (*e.g.*, elemental Pt^0 on larger Al_2O_3 carrier particles emitted from automotive catalytic converters, ultrafine TiO_2 in cosmetic ingredients such as sunscreen), occupational (*e.g.*, large-scale preparation of nanoparticles), and biomedical (*e.g.*, ultrafine TiO_2 for tumor tissue targeting and delivery of killing compounds for cancer cells by UV light). At the occupational level, there are four main groups of nanoparticles production processes: gas-phase, vapor deposition, colloidal, and attrition, all of which may potentially result in exposure by inhalation, dermal, or ingestion routes. All processes may give rise to exposure to agglomerated nanoparticles during recovery, powder handling, and production processing. In spite of the potential occupational and public exposure to nanomaterials that is dramatically increasing, information on the human health impact of nanoparticles is severely lacking.

9.3 PARTICLES IN AIR OR AEROSOLS

Particles suspended in air are called **aerosols**. These pose a threat to human health mainly through respiratory intake and deposition in nasal and bronchial airways. Smaller aerosols travel further into the respiratory system and generally cause more health problems than larger particles. For this reason the United States Environmental Protection Agency (U.S. EPA) has divided airborne particulates into two size categories: **PM$_{10}$**, which refers to particles with diameters less than or equal to 10 μm (10,000 nm), and **PM$_{2.5}$** which are particles less than or equal to 2.5 μm (2,500 nm) in diameter. For this classification, the diameter of aerosols is defined as the **aerodynamic diameter**:

$$d_{pa} = d_{ps} \left(\rho_p / \rho_w \right)^{1/2} \qquad \text{(Eq. 9.1)}$$

TABLE 9.1 U.S. PM$_{10}$ production from various sources.

	SOURCE	PM$_{10}$ (MILLIONS OF TONS)
Industrial processes	Chemical industries	0.070
	Metals processing	0.220
	Petroleum industries	0.041
	Other industries	0.530
	Solvent utilization	0.006
	Storage and transport	0.114
	Waste disposal and recycling	0.296
Fuel combustion	Electric utilities	0.290
	Industrial	0.314
	On-road vehicles	0.268
	Non-road sources	0.466
Other	Agriculture and forestry	4.707
	Fire and other combustion	1.015
	Unpaved roads	12.305
	Paved roads	2.515
	Construction	4.022
	Wind erosion	5.316

Council on Environmental Quality, 1997.

where:

d_{pa} = aerodynamic particle diameter (μm)
d_{ps} = Stokes' diameter (μm)
ρ_p = particle density (g cm^{-3})
ρ_w = density of water (g cm^{-3})

Atmospheric particulate concentration is expressed in micrograms of particles per cubic meter of air (μg/m^3). The U.S. EPA established a **National Ambient Air Quality Standard (NAAQS)** for PM$_{10}$ of 150 μg m^{-3} averaged over a 24-hour period, and 50 μg m^{-3} averaged annually. More recently, separate standards for PM$_{2.5}$ of 65 μg m^{-3} for 24 hours and 15 μg m^{-3} annually have been introduced.

Symptoms of particulate matter inhalation include decreased pulmonary function, chronic coughs, bronchitis, and asthmatic attacks. The specific causal mechanisms are poorly understood. One well-documented episode occurred in London in 1952, when levels of smoke and sulfur dioxide aerosols, largely associated with coal combustion, reached elevated levels due to local weather conditions. Over a 10-day period, approximately 4,000 deaths were attributable to cardiovascular and lung disorders brought on or aggravated by these aerosols. Sources of PM$_{10}$ in the US are shown in Table 9.1. In addition to the sources shown in Table 9.1, volcanoes and breaking waves also generate airborne particles, sometimes in very large quantities.

Airborne particles can travel great distances. Intense dust storms during 1998 and 2001 in the Gobi desert of Western China and Mongolia (Figure 9.2) elevated aerosol levels to concentrations near the health standard in Western North America several thousand miles away!

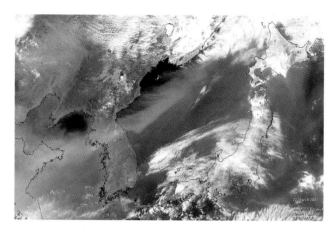

Figure 9.2 Mongolian dust over the Sea of Japan. Photo provided by NASA.

Smaller particles tend to travel greater distances than large particles. Stokes' Law (Equation 9.2) is used to describe the fall of particles through a dispersion medium, such as air or water.

$$V = [D^2 \times (\rho_p - \rho_l) \times g]/18\eta \qquad \text{(Eq. 9.2)}$$

where:

V = velocity of fall (cm s^{-1})
g = acceleration of gravity (980 cm s^{-2})
D = diameter of particle (cm)
ρ_p = density of particle (density of quartz particles is 2.65 g cm^{-3})
ρ_l = density of dispersion medium (air has a density of about 0.001213 g cm^{-3}; water has a density of about 1 g cm^{-3})
η = viscosity of the dispersion medium (about 1.83×10^{-4} poise or g cm^{-1} s^{-1} for air; 1.002×10^{-2} poise for water)

Using Stokes' Law, we can calculate the rate of fall in air (Information Box 9.1). Small particles are thus a greater concern than larger particles for several reasons. Small particles stay suspended longer. Therefore, they travel greater distances. Risk of exposure to small particles is amplified by their extended suspension times. Small particles also tend to

---

INFORMATION BOX 9.1

Influence of Particle Size on Velocity of Deposition of Particles in Air, Calculated Using Stokes' Law

Particle Diameter (mm)	Particle Type	Rate of Fall in Air (cm sec^{-1})
1	Sand	7880
0.1	Silt	79
0.001	Clay	7.9×10^{-5}

---

move further into the respiratory system, exacerbating their effects on health.

9.3.1 Aerosols of Concern

9.3.1.1 Asbestos

Asbestos particles are a special case of mineral aerosols that are known to lead to debilitation, disease, and death. They have been defined by the EPA as six naturally occurring minerals with well-defined compositions and their hazardous size mandated. When airborne, these exceedingly small particles can be identified using standard mineralogic analyses employing x-ray diffraction and optical and electron microscopy in transmission, diffraction, and scanning modes. The mineral classified asbestos, and many other silicate mineral species (*e.g.*, talc, erionite, and vermiculite), all may occur in fibrous form (Wilson and Spengler, 1996). The fibrous structure of these minerals permits the formation of sharp aerosols that can be embedded in the lungs. It is likely that no matter what the environment, at home or outdoors, people have been, and are, continually exposed to particulates and aggregates that are mixtures, and some will contain asbestos or other minerals.

Asbestosis is the disease that results when lungs generate scar tissue (fibrosis) as a result of high exposure. Lung scarring may continue post-asbestos exposure, and although affected portions of the lung can be rejected, the quality of life thereafter is abysmal. Lung cancer, although associated with asbestos mineral exposure, is more frequently directly associated with smoking. Many occupations, for example, construction workers or brake repair technicians, often include individuals who smoke, resulting in multiple opportunities for respiratory trauma and therefore multiplying the risks. Another deadly disease associated with asbestos particles is mesothelioma, a cancer of the pleura rather than of the lung tissues.

Though the public health issues related to asbestos have been aired, the specific mechanisms of fibrogenesis and carcinogenesis related to the exposures, especially at low doses (nonoccupational) are not fully elucidated, and remain under discussion and investigation. Asbestos in the built indoor environment is also a potential source of exposure (Skinner et al., 1988). Asbestos removal from buildings is closely regulated to prevent exposure to construction workers. The recent (2001) terrorist attack on the World Trade Center that generated widespread dust throughout lower New York City is a recent instance where asbestos issues were raised.

9.3.1.2 Silica

One of the most common natural materials and a major component of beach sand, quartz (SiO$_2$), may become an offending material causing a particulate-based disease. Silicosis is due to exposure to crystalline silica, and is exclusively occupational, with the size and morphologic characteristics of the

CASE STUDY 9.1

ASBESTOS EXPOSURE IN CALIFORNIA NEAR COALINGA FROM ASBESTOS-BEARING SERPENTINITE.

The deposit was one of the oldest (1885) and best known mercury mines in the country. Chrysotile, naturally occurring in serpentinite, has been mined at the site since the late 1950s, as it was available in pure mineral form and had many industrial applications. The grain size of the natural "short-fiber" chrysotile was milled, so the product probably approached the respirable size range and was certainly small enough to be airborne. The bagged powdery material was cheap and shipped to markets inside and outside the U.S. The site and the large mine dumps created are now the Clear Creek Management Area on the Superfund National Priorities List under the Bureau of Land Management. The climate, elevation, and topography of the site mean that there are increased amounts of dust in the local atmosphere blown by prevailing winds toward populated areas.

particle key to respiratory problems. Construction workers, especially those jack hammering or those blasting dirt off of building surfaces using a stream of silica, without nasal and mouth protection, are at great risk. Often biological as well as mineral materials become airborne in both cases.

There are several crystal forms of SiO_2, including diatoms, the source of diatomaceous earth. The use of these various silica materials is not monitored nor are those at risk necessarily aware of their exposure, but great efforts have been made by some industries and one can anticipate future actions, especially responding to OSHA regulations.

Silicosis is characterized by focal nodular lesions, which can be detected radiologically in the upper lung. This expression of silicosis is distinct from that of asbestosis, where fibrosis is usually diffuse and in the lower portions of the lung. Lung function may not be markedly affected initially although under continuing exposure, the nodules coalesce and the fibrosis becomes massive and pervasive for large parts of the lung. The formally pliable lung tissues become occluded by scarring and the deposition of the fibrous protein, collagen, often calcify or harden, further compromising respiration and the transmission of the essential gases in these portions of the respiratory system.

9.3.1.3 Human-made aerosols

Particulate matter in the atmosphere can be from direct emissions or pollution that enters the atmosphere as previously formed particles. These are called **primary particles**. Alternatively, **secondary particles** are formed in the atmosphere from precursor components, such as ammonia, volatile organics, or oxides of nitrogen (NO_x) and sulfur (SO_x). Primary particles may fall into the $PM_{2.5}$ or the PM_{10} size ranges (plumes), whereas secondary particles fall mainly into the $PM_{2.5}$ category.

Industrially generated primary particles arise largely from incomplete combustion processes and high-temperature metallurgical processes. Secondary particles, on the other hand, are produced from gases emitted from industrial processing and various combustion processes (including automobiles, power plants, wood burning, and incinerators) that undergo gas-to-particle conversion and then growth and coagulation. In the atmosphere, sulfur oxides are oxidized to form sulfuric acid and fine sulfate particles. Gases condense to form ultra-fine aerosols (less than 0.01 μm), either from supersaturated vapor produced in high temperature combustion processes or through photochemical reactions. These particles grow in size through condensation and coagulation to form larger particles (0.1–2.5 μm). The principal sources of SO_x in the U.S. include coal power plants, petroleum refineries, paper mills, and smelters. In contrast, NO_x is largely produced by industrial and automotive combustion processes.

Health threats from $PM_{2.5}$ and PM_{10} generated by human activities are much like those from naturally occurring particulates. Adverse health effects are most severe in senior citizens and those with pre-existing heart or lung problems. Recent studies estimate that with each 10 μg m^{-3} increase in PM_{10} above a base level of 20 μg m^{-3}, daily respiratory mortality is estimated to increase by 3.4%, cardiac mortality increases by 1.4%, hospitalizations increase by 0.8%, emergency room visits for respiratory illnesses increase by 1.0%, days of restricted activity due to respiratory symptoms increase by 9.5%, and school absenteeism increases by 4.1% (Vedal, 1995). Particles formed from incomplete combustion, such as those formed by wood-burning and diesel engines, contain organic substances that may have additional health effects. Diesel exhaust has been shown to increase lung tumors in rats and mice, and long-term human exposure to diesel exhaust may be responsible for a 20–50% increase in the risk of lung cancer (Koenig, 1999).

In addition to human health concerns, $PM_{2.5}$ associated with wood burning, automobile exhaust, and industrial activities is responsible for much of the atmospheric haze in the U.S. Aerosols (particularly $PM_{2.5}$) absorb and scatter light, producing haze and reducing visibility. When severe, this interferes with automobile and aviation navigation, posing safety threats. Atmospheric particulates can also be a nuisance by settling on and in cars and homes and other buildings.

TABLE 9.2 Aerosolized endotoxin concentrations detected downwind of biosolids operations, a wastewater treatment plant aeration basin, and a tractor operation.

Sample Type	# of samples collected	Distance from Site (m)	AEROSOLIZED ENDOTOXIN			
			Avg	Median	Minimum	Maximum
			EU m*			
Controls						
Background	12	NA	2.6	2.49	2.33	3.84
Biosolids Operations						
Loading	39	2–50	343.7	91	5.6	1807.6
Slinging	24	10–200	33.5	6.3	4.9	14.29
Biosolids Pile	6	2	103	85.4	48.9	207.1
Total Operation	33	10–200	133.9	55.6	5.6	623.6
Wastewater Treatment Plant						
Aeration Basin	6	2	627.3	639	294.4	891.1
Non Biosolids Field						
Tractor	6	2	469.8	490.9	284.4	659.1

* EU m^{-3}—Endotoxin units per m^3. From Brooks, 2004.

9.3.1.4 Bioaerosols

Biological contaminants include whole entities such as bacterial and viral human pathogens. They also include airborne toxins, which can be parts or components of whole cells. In either case, biological airborne contaminants are known as bioaerosols, which can be ingested or inhaled by humans (see also Chapter 26).

Coccidioidomycosis (also known as **Valley Fever**) is one disease caused by inhalation of spores of the fungus *Coccidioides immitis*, which is indigenous to hot, arid regions, including the Southwestern U.S. The fungus can travel from the respiratory tract to the skin, bones, and central nervous system and can result in systemic infection and death.

Endotoxin, also known as lipopolysaccharide, is ubiquitous throughout the environment and may be one of the most important allergens. Endotoxin is derived from the cell wall of Gram-negative bacteria and is continually released during both active cell growth and cell decay. Hence, endotoxin is found wherever Gram-negative bacteria are found. In soils, bacterial concentrations routinely exceed 10^8 per gram, with a majority of bacteria being Gram negative. Soil particles containing sorbed microbes can be aerosolized and hence act as a source of endotoxin. Farming operations such as driving a tractor across a field has been shown to result in endotoxin levels of 469 **endotoxin units (EU)** m^{-3} (Table 9.2). EU units are related to a turbidometric Limulus Amebocyte Assay. These values are comparable to those found during land application of biosolids operations. Daily exposures of as little as 10 EU m^{-3} from cotton dust can cause asthma and chronic bronchitis. However, dose response is dependent on the source of the material, the duration of exposure, and repeated exposures (Brooks, 2004).

When inhaled by humans, endotoxin has demonstrated the ability to cause a wide variety of health effects including fever, asthma, and shock.

Data in Table 9.2 illustrate that endotoxin aerosolization can occur during both wastewater treatment and land application of biosolids. However, the data also show that endotoxin of soil origin resulting from dust generated during tractor operations results in similar amounts of aerosolized endotoxin. Given that the major source of PM$_{10}$ in the U.S. are unpaved roads (Table 9.1), and that these particulates are of soil-borne origin, it is possible that endotoxin associated with wind blow soil particles is a major contributor to respiratory problems.

Mycotoxins are secondary metabolites produced by fungal molds. Fungi such as species of *Aspergillus*, *Alternaria*, *Fusarium*, and *Penicillium* are common soil-borne fungi capable of producing mycotoxins. Most notably, aflatoxin is produced by *Aspergillus flavus*. Aflatoxin is one of the most potent carcinogens known and is linked to a variety of health problems.

9.4 PARTICULATES IN WATER

In water, suspended particulates pose quite different risks than aerosols. Inorganic particles cause an increase in the turbidity of affected water, and the particles themselves can cause problems through sedimentation that can fill lakes, dams, reservoirs, and waterways. In the U.S., waterborne soil particles fill over 123 million cubic meters of reservoir capacity each year, reducing water storage capacity and necessitating expensive dredging operations. Suspended particles increase wear on pumps, hydroelectric generators, and related equipment. Also, soils from which suspended particles are derived suffer damage that can reduce agricultural productivity and land values. Severe soil erosion can threaten buildings, roads, and other structures. This issue is discussed in more detail in Chapter 16

Using Stokes' Law, settling rates can be calculated for various size particles suspended in water (see Information Box 9.2). Sedimentation is not an important factor for

INFORMATION BOX 9.2

Influence of Particle Size on Velocity of Deposition of Particles in Water, Calculated Using Stokes' Law

Particle Diameter (mm)	Particle Type	Rate of Fall in Water (cm sec^{-1})
1	Sand	90
0.1	Silt	0.9
0.001	Clay	9×10^{-5}

INFORMATION BOX 9.3

Soil Particle Diameter Sizes

Sand:	2–0.05mm
Silt:	0.05–0.002 mm
Clay:	< 0.002 mm (< 2 μm)

particles less than 0.0005 mm in diameter because they act as **colloids**, which are so small that once suspended in a dispersion medium, they can remain in suspension indefinitely and do not settle out through the force of gravity. Large particles remain suspended for only a short time, and their range of movement is limited. Turbulence, disorderly flow or mixing of the dispersion medium, can increase suspension times considerably, so the calculated settling rates are valid only in the absence of turbulence. Actual settling rates are generally higher.

Decomposition of organic particles such as plant and animal residues, composed of carbohydrates, proteins, and more complex compounds, occurs largely through microbially mediated oxidation processes, and can also impact waters. The oxidation of organic substrate via aerobic heterotrophic respiration is described in Chapter 5.

The oxidation of waterborne organic particulates consumes dissolved oxygen and produces CO_2. The use of oxygen in biological oxidation reactions is called **biological oxygen demand (BOD)**. Increasing BOD and resulting oxygen depletion, called **hypoxia**, can have a profound effect on aquatic animals, such as fish, that depend on the dissolved oxygen supply (see also Chapter 18).

Both organic and inorganic particles (such as soil) can act as carriers of other contaminants that are sorbed to particle surfaces. Nutrients, herbicides, insecticides, fuels, solvents, preservatives, and other industrial and agricultural chemicals can adsorb and desorb from waterborne particulates. The health threats associated with this very diverse group of chemicals are similarly broad. These contaminants are discussed in detail in Chapters 10, 16, and 17.

9.4.1 Soil Particles

Soil particles, which are natural contaminants, are not considered to be pollutants unless they move into the atmosphere or surface waters. They are classified on the basis of size into three categories in the U.S. Department of Agriculture (USDA) classification system (Information Box 9.3).

Sand and silt particles are dominated by primary minerals such as quartz, feldspar, and mica formed directly from molten magma. Clay particles, on the other hand, are composed largely of secondary layered silicate clay minerals (weathering products of primary minerals), hydrous oxides of aluminum and iron, and organic materials. Most of these particles have negative electric charge, although some can have a positive charge, and others no charge at all. This charge arises from either characteristics of their crystalline structure, in the case of layered silicate clay minerals, or due to surface chemical reactions (see also Chapter 7). Positively charged cations surrounding clay particles balance the surface charge of the particles. Negatively charged clay particles are almost always surrounded by a cloud of cations. Additionally, certain anions and organic chemicals can be sorbed to clay particles and transported by moving particles. Thus clay particles can act as carriers in the distribution and transport of associated cations.

9.4.1.1 Soil particle flocculation

As indicated above, the settling rates of particles suspended in water depends, among other things, on their effective diameter. This can be greatly increased when particles adhere to one another or **flocculate** to form aggregates made up of groups of soil particles. "Like" charged particles repel each other, so the forces of repulsion must be overcome for particles to aggregate. The negative charge on clay particles is balanced by a cloud of counterions (cations) surrounding the clay particle, called a **diffuse double layer**. In this double layer, the concentration of cations increases as the clay surface is approached, and, similarly, the concentration of anions decreases. The thickness of this diffuse double layer depends on several factors, the most important of which are counterion valence and concentration.

$$1/\chi = (K/z^2n)^{1/2} \qquad \text{(Eq. 9.3)}$$

where:

$1/\chi$ = the effective thickness of the double layer
z = valence of the counterions (cations)
K = a constant dependent on temperature and the dielectric constant of the solvent
n = electrolyte concentration in the equilibrium solution

The thickness of the double layer decreases as the valence of counterions increases, or as the electrolyte concentration in the solvent increases (Figure 9.3). As the double layer thickness decreases, attractive energy between particles becomes greater than repulsive forces, and the particles

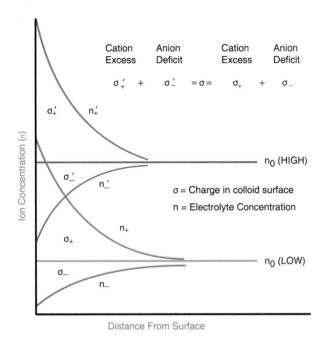

Figure 9.3 Thickness of the electrical double layer and ion distribution at two electrolyte concentrations. From Sumner M.E. and Stewart B.A., eds. (1992) *Soil Crusting and Physical Processes*. Lewis Publishers, Boca Raton, FL. Reprinted by CRC Press.

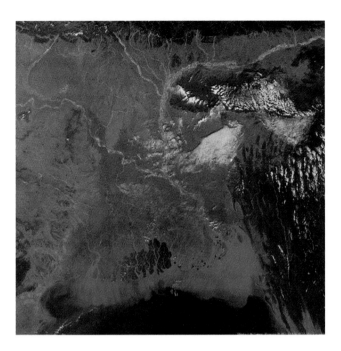

Figure 9.4 Sediment-laden water from the Ganges and Brahmaputra Rivers flows into the Bay of Bengal. Photo provided by NASA.

flocculate. The electrolyte concentration at which this occurs is called the **critical flocculation concentration (CFC)**, which is a function of $1/z^6$. Although this relationship does not completely explain the varying ability of different cations to flocculate soil particles, we can see that the effect of counterion valence is extremely important.

The relative flocculating power of common soil cations is shown in Table 9.3. In general, counterion valence is the most important factor determining CFC; therefore, higher valence cations are more effective flocculators than lower valence cations. The hydrated radius of cations is also important and explains much of the difference between cations of identical valence.

Soil particles suspended in water will tend to be dispersed if the electrolyte concentration is low and the dominant counterions are potassium or sodium, whereas flocculation will occur with a high electrolyte concentration or if the dominant counterions are of higher valence. Where rivers flow into the ocean, sediment laden river water mixes

with saline ocean water. Suspended particles flocculate and settle out to form river deltas (Figure 9.4). To flocculate and settle particles out of wastewater, high-valence flocculating agents such as aluminum, iron, and copper sulfates and chlorides are added as clarifiers. Organic polymers and synthetic polyelectrolytes that have anionic or cationic functional groups are also used as flocculants (see also Chapter 28).

Depending on the properties of the suspended particles and the impacted water, particles may flocculate into aggregates and quickly settle out of suspension, or they may remain dispersed. As indicated above, dispersed colloidal particles can remain in suspension indefinitely when in an unflocculated or dispersed state.

9.5 SUMMARY

Particulate contaminants can be of either natural or human-made origin. These particles can pollute both air and water. Naturally occurring particles include soil or other mineral particles, pollen, ocean spray, bacteria, viruses, and spores. Mineral particles can contaminate water and air via natural processes such as volcanic activity, wave action, and wind and water erosion. Wind and water erosion can be accelerated by human activities that leave soil in a susceptible condition, whereas many of the other processes are beyond human control. Human-made particles include nitrogen and sulfur oxides and a wide range of solid particles formed during combustion, and industrial processes that grind or abrade materials into fine particles.

TABLE 9.3 Relative flocculating power of common monovalent and divalent cations.

ION	RELATIVE FLOCCULATING POWER (RELATIVE TO NA^+)
Na^+	1.00
K^+	1.70
Mg^{2+}	27.00
Ca^{2+}	43.00

From Rengasamy P. and Sumner M.E. (1998) *Sodic Soils: Distribution, Properties, Management, and Environmental Consequences*. Oxford University Press, New York, NY.

Airborne particulates are a major health concern because they can penetrate the human respiratory system, causing a wide range of health problems. They can degrade visibility and become a health hazard by impeding air and ground vehicular traffic. Waterborne particulates can be a direct threat by clogging reservoirs and waterways, and by acting as carriers for other contaminants, such as phosphorus and pesticides that contaminate surface waters. Organic particles increase the BOD of contaminated waters, leading to oxygen depletion (hypoxia) that can harm aquatic biota.

There are many ways of reducing production of particulate contaminants or of removing them from industrial exhaust streams. However, all natural waters contain suspended particles, as does even the cleanest of air. Some of the processes that distribute particulate contaminants to air and water are natural and can be only partially controlled, if at all. On the other hand, human activities that contribute to particulate degradation of air and water often can be modified to minimize their impact on these natural resources (see Chapter 20).

QUESTIONS AND PROBLEMS

1. Define the terms PM_{10} and $PM_{2.5}$.
2. How do waterborne particles reduce dissolved O_2 in surface water? What are the resulting effects on aquatic biota?
3. Explain why sediments carried in freshwater rivers quickly settle out when these sediment-laden waters combine with ocean water.

4. Explain the differences between primary and secondary particles in terms of size, formation, and composition.
5. Calculate the rate of deposition of a particle with a diameter of 0.003 mm using Stokes' Law: (a) in air; and (b) in water.

REFERENCES AND ADDITIONAL READING

Brooks J.P. *Biological Aerosols Generated from the Land Application of Biosolids: Microbial Risk, Assessment.* 2004 Ph.D. Dissertation, The University of Arizona, Tucson, Arizona.

Council on Environmental Quality. (1997) *Environmental Quality: The 1997 Report of the Council on Environmental Quality.* U.S. Government Printing Office, Washington, DC.

Koenig J.Q. (1999) *Health Effects of Ambient Air Pollution.* Kluwer Academic Publishers, Boston, Massachusetts.

Liu D.H.F. and Liptak B.G. (2000) *Air Pollution.* Lewis Publishers, Boca Raton, Florida.

Rengasamy P. and Sumner M.E. (1998) Processes Involved in Sodic Behavior. In M.E. Sumner and R. Naidu (eds.) *Sodic Soils: Distribution, Properties, Management, and Environmental Consequences.* Oxford University Press, New York, New York.

Skinner H.C.W., Ross M. and Frondel C. (1988) *Asbestos and Other Fibrous Materials: Mineralogy, Crystal Chemistry and Health Effects.* Oxford University Press, New York, New York.

Sumner M.E. (1992) The Electrical Double Layer and Clay Dispersion. In M.E. Sumner and B.A. Stewart (eds.) *Soil Crusting: Chemical and Physical Processes.* Lewis Publishers, Boca Raton, Florida.

Wilson R. and Spengler J. (eds.) (1996) *Particles in Our Air: Concentrations and Health Effects.* Harvard University Press, Cambridge, Massachusetts.

Vedal S. (1995) *Health Effects of Inhalable Particles: Implications for British Columbia.* Air Resources Branch, Ministry of Environment, Lands and Parks.

CHAPTER 10

CHEMICAL CONTAMINANTS

M.L. Brusseau, C.M. McColl, G. Famisan, and J.F. Artiola

Chemical contamination is a major source of pollution. *Photo courtesy K.L. Josephson.*

10.1 INTRODUCTION

It can be argued that all matter in one form or another can become a contaminant when found out of its usual environment or at concentrations above normal. However, chemical contaminants become pollutants when accumulations are sufficient to adversely affect the environment, or to pose a risk to living organisms. Today, there are thousands of industrial chemicals that can be dangerous to humans and the environment. Fortunately, the vast majority of these chemicals are not produced in large enough quantities to be a human or environmental threat. However, there are more than 3000 natural and human-made chemicals that are toxic enough and are produced in sufficient quantities to be a potential environmental hazard. Thus, the production, storage, transport, and disposal of these chemicals are regulated by government agencies. There are numerous sources of chemical contaminants released to the environment, but these generally fall into a few general categories. This chapter will present an overview of the various types of chemical contaminants and their sources.

10.2 TYPES OF CONTAMINANTS

There are three basic categories of chemical contaminants: organic, inorganic, and radioactive. In turn, there are several classes of contaminants within each of these categories. Major classes of contaminants are listed in Table 10.1. Some of these contaminants are considered in greater detail in other chapters (16, 17, 18, and 31).

TABLE 10.1 Examples of organic, inorganic and radioactive chemical contaminants.

Organic Contaminants

Petroleum hydrocarbons (fuels)—Benzene, toluene, xylene, polycyclic aromatics

Chlorinated solvents—Trichloroethene, tetrachloroethene, trichloroethane, carbon tetrachloride

Pesticides—DDT (dichloro-diphenyl-trichloro-ethane), 2,4-D (2,4-Dichlorophenoxyacetic acid), atrazine

Polychlorinated biphenyls (PCBs)—insulating fluids, plasticizers, pigments

Coal tar/creosote—Polycyclic aromatics

Pharmaceuticals/food additives/cosmetics—Drugs, surfactants, dyes

Gaseous compounds—Chlorofluorocarbons (CFCs), hydrochlorofluorocarbons (HCFCs)

Inorganic Contaminants

Inorganic "salts"—Sodium, calcium, nitrate, sulphate
Heavy/trace metals—Lead, zinc, cadmium, mercury, arsenic

Radioactive Contaminants

Solid elements—Uranium, strontium, cobalt, plutonium

Gaseous elements—Radon

Thousands of chemicals are released into the environment every day. Thus, when conducting site characterization studies, it is important to prioritize the suite of chemicals under investigation. For most sites this is done by focusing on so-called priority pollutants, those that are regulated by federal, state, or local governments. The primary such list of priority pollutants is that governed by the National Primary Drinking Water Regulations, which provide legally enforceable standards that apply to all public water systems. These standards protect public health by limiting the levels of contaminants that are allowed to exist in drinking water. The organic and inorganic contaminants on this list are presented in Table 10.2. Note that the full list also includes microorganisms, radionuclides, and water disinfection byproducts.

The frequency of occurrence of the contaminants listed in Table 10.2, as well as other chemicals, differs greatly for each specific contaminated site. The contaminants that are most frequently encountered at U. S. Environmental Protection Agency (EPA) designated Superfund sites are presented in Table 10.3. It is quite likely that one or more of these contaminants will be present at most hazardous waste sites.

The U.S. EPA has developed special reporting rules for certain chemicals of concern under the Toxic Release Inventory program. These chemicals, listed in Table 10.4, are classified as **persistent, bioaccumulative, and toxic (PBT)** chemicals. These compounds pose increased risk to human health not only because they are toxic, but also because they remain in the environment for long periods of time, are not readily destroyed, and build up or accumulate in body tissue.

In a related development, an international treaty was recently enacted to control the future production of a class of chemicals termed **persistent organic pollutants (POPs)**. The Stockholm Convention is a global treaty to protect human health and the environment from POPs, which are chemicals that remain intact for long periods, become widely distributed geographically, accumulate in the fatty tissue of living organisms, and are toxic. There are 12 chemicals currently on the POP list: aldrin, chlordane, DDT, dieldrin, dioxins, endrin, furans, heptachlor, hexachlorobenzene, mirex, polychlorinated biphenyls, and toxaphene. Many of these chemicals are pesticides, and inspection of Table 10.4 shows some of them are also listed as PBTs.

10.3 SOURCES: AGRICULTURAL ACTIVITIES

Agricultural systems consist of highly controlled tracts of land that generally receive large inputs of chemical fertilizers and pesticides. The ultimate goal of these chemical additions is to generate optimum amounts of food and fiber. However, fertilizers are often applied in excess of the crop needs or are in chemical forms that make them very mobile in soil and water environments. For example, nitrate pollution of groundwater is often caused by excessive nitrogen fertilizer applications that result in leaching below the root zone. Agricultural activities can cause land, water, and air pollution, as discussed in Chapters 16, 17, 18, and 21.

TABLE 10.2 National primary drinking water standards.

INORGANIC CHEMICALS	MCL OR TT[1] (mg L^{-1})	POTENTIAL HEALTH EFFECTS FROM INGESTION OF WATER
Antimony	0.006	Increase in blood cholesterol; decrease in blood glucose
Arsenic	0.01	Skin damage; circulatory system problems; increased risk of cancer
Asbestos (fiber >10 micrometers)	7 MFL	Increased risk of developing benign intestinal polyps
Barium	2	Increase in blood pressure
Beryllium	0.004	Intestinal lesions
Cadmium	0.005	Kidney damage
Chromium (total)	0.1	Some people who use water containing chromium well in excess of the MCL over many years could experience allergic dermatitis.
Copper	TT[8]; Action Level = 1.3	Short-term exposure: Gastrointestinal distress. Long-term exposure: Liver or kidney damage. People with Wilson's Disease should consult their doctor if their water systems exceed the copper action level.
Cyanide (as free cyanide)	0.2	Nerve damage or thyroid problems
Fluoride	4	Bone disease (pain and tenderness of the bones); children may get mottled teeth.
Lead	TT[8]; Action Level = 0.015	Infants and children: Delays in physical or mental development. Adults: Kidney problems; high blood pressure.
Mercury (inorganic)	0.002	Kidney damage
Nitrate (measured as Nitrogen)	10	Methemoglobinemia or "blue baby syndrome" in infants under 6 months–life threatening without immediate medical attention. Symptoms: Infant looks blue and has shortness of breath.
Nitrite (measured as Nitrogen)	1	"Blue baby syndrome" in infants under 6 months–life threatening without immediate medical attention. Symptoms: Infant looks blue and has shortness of breath.
Selenium	0.05	Hair or fingernail loss; numbness in fingers or toes; circulatory problems
Thallium	0.002	Hair loss; changes in blood; kidney, intestine, or liver problems

ORGANIC CHEMICALS	MCL OR TT[1] (mg L^{-1})	POTENTIAL HEALTH EFFECTS FROM INGESTION OF WATER
Acrylamide	TT[9]	Nervous system or blood problems; increased risk of cancer
Alachlor	0.002	Eye, liver, kidney or spleen problems; anemia; increased risk of cancer
Atrazine	0.003	Cardiovascular system problems; reproductive difficulties
Benzene	0.005	Anemia; decrease in blood platelets; increased risk of cancer
Benzo(a)pyrene (PAHs)	0.0002	Reproductive difficulties; increased risk of cancer
Carbofuran	0.04	Problems with blood or nervous system; reproductive difficulties
Carbontetrachloride	0.005	Liver problems; increased risk of cancer
Chlordane	0.002	Liver or nervous system problems; increased risk of cancer
Chlorobenzene	0.1	Liver or kidney problems
2,4-dichlorophenoxyacetic acid	0.07	Kidney, liver, or adrenal gland problems
Dalapon	0.2	Minor kidney changes
1,2-Dibromo-3-chloropropane (DBCP)	0.0002	Reproductive difficulties; increased risk of cancer
o-Dichlorobenzene	0.6	Liver, kidney, or circulatory system problems
p-Dichlorobenzene	0.075	Anemia; liver, kidney or spleen damage; changes in blood
1,2-Dichloroethane	0.005	Increased risk of cancer
1,1-Dichloroethylene	0.007	Liver problems
cis-1,2-Dichloroethylene	0.07	Liver problems
trans-1,2-Dichloroethylene	0.1	Liver problems
Dichloromethane	0.005	Liver problems; increased risk of cancer
1,2-Dichloropropane	0.005	Increased risk of cancer
Di(2-ethylhexyl) adipate	0.4	General toxic effects or reproductive difficulties
Di(2-ethylhexyl) phthalate	0.006	Reproductive difficulties; liver problems; increased risk of cancer
Dinoseb	0.007	Reproductive difficulties
Dioxin (2,3,7,8-TCDD)	0	Reproductive difficulties; increased risk of cancer

(continued)

TABLE 10.2 National primary drinking water standards. *(continued)*

ORGANIC CHEMICALS	MCL OR TT[1] (mg L^{-1})	POTENTIAL HEALTH EFFECTS FROM INGESTION OF WATER
Diquat	0.02	Cataracts
Endothall	0.1	Stomach and intestinal problems
Endrin	0.002	Nervous system effects
Epichlorohydrin	TT[9]	Stomach problems; reproductive difficulties; increased risk of cancer
Ethylbenzene	0.7	Liver or kidney problems
Ethylene dibromide	0.00005	Stomach problems; reproductive difficulties; increased risk of cancer
Glyphosate	0.7	Kidney problems; reproductive difficulties
Heptachlor	0.0004	Liver damage; increased risk of cancer
Heptachlor epoxide	0.0002	Liver damage; increased risk of cancer
Hexachlorobenzene	0.001	Liver or kidney problems; reproductive difficulties; increased risk of cancer
Hexachlorocyclopentadiene	0.05	Kidney or stomach problems
Lindane	0.0002	Liver or kidney problems
Methoxychlor	0.04	Reproductive difficulties
Oxamyl (Vydate)	0.2	Slight nervous system effects
Polychlorinatedbiphenyls (PCBs)	0.0005	Skin changes; thymus gland problems; immune deficiencies; reproductive or nervous system difficulties; increased risk of cancer
Pentachlorophenol	0.001	Liver or kidney problems; increased risk of cancer
Picloram	0.5	Liver problems
Simazine	0.004	Problems with blood
Styrene	0.1	Liver, kidney, and circulatory problems
Tetrachloroethylene	0.005	Liver problems; increased risk of cancer
Toluene	1	Nervous system, kidney, or liver problems
Toxaphene	0.003	Kidney, liver, or thyroid problems; increased risk of cancer
2-(2,4,5-Trichlorophenoxy) propionic acid (silvex)	0.05	Liver problems
1,2,4-Trichlorobenzene	0.07	Changes in adrenal glands
1,1,1-Trichloroethane	0.2	Liver, nervous system, or circulatory problems
1,1,2-Trichloroethane	0.005	Liver, kidney, or immune system problems
Trichloroethylene	0.005	Liver problems; increased risk of cancer
Vinyl chloride	0.002	Increased risk of cancer
Xylenes (total)	10	Nervous system damage

[1]MCL = maximum contaminant level, the highest level of a contaminant that is allowed in drinking water; TT = treatment technique level.

From http://www.epa.gov/safewater/standards.html.

TABLE 10.3 Common pollutants found at Superfund sites.

Acetone	Mercury
Aldrin/Dieldrin	Methylene Chloride
Arsenic	Naphthalene
Barium	Nickel
Benzene	Pentachlorophenol
2-Butanone	Polychlorinated Biphenyls (PCBs)
Cadmium	
Carbon Tetrachloride	Polycyclic Aromatic Hydrocarbons (PAHs)
Chlordane	
Chloroform	Tetrachloroethylene
Chromium	Toluene
Cyanide	Trichloroethylene
DDT, DDE, DDD[1]	Vinyl Chloride
1,1-Dichloroethene	Xylene
1,2-Dichloroethane	Zinc
Lead	

[1]DDT = dichlorodiphenyltrichloroethane; DDE = dichlorodiphenyldichloroethene; DDD = dichlorodiphenyldichloroethane.

From www.epa.gov/superfund/resources/chemicals.htm.

Fertilizers, which are generally inorganic chemicals, are routinely applied at least once a year and include, in order of decreasing amounts, N, P, K, and metals. The annual applications of these chemicals range from 50 to 200 kg ha^{-1}, as N, P or K. Micronutrient (*e.g.*, Fe, Zn, Cu, B, and Mo) fertilizer additions are also applied regularly to agricultural fields but with less frequency because of lower crop requirements. These chemicals are applied to agricultural lands at average rates of 0.5–2 kg ha^{-1}, in their respective elemental forms, every 2–5 five years. A third group of inorganic chemicals applied to agricultural land consists of soil amendments. These materials are applied to agricultural fields with some frequency for two reasons: (1) to decrease or increase soil pH, decrease soil salinity, and improve soil structure, and (2) to replenish macronutrients like Ca^{++}, Mg^{++}, K$^+$, and SO$_4^=$. To control macronutrient deficiencies, the application rates of these chemicals range from 50 to 500 kg ha^{-1}. To control soil pH and salinity, applications typically range from 2,000 to 10,000 kg ha^{-1}. The common

TABLE 10.4 Persistent, bioaccumulative, and toxic chemicals.

CHEMICAL CATEGORIES	SOURCES
Dioxin and dioxin-like compounds	Chemical manufacturing and processing byproducts, waste combustion
Lead compounds	Mining, manufacturing, leaded fuels
Mercury compounds	Mining, manufacturing
Polycyclic aromatic compounds	Petroleum production, combustion, coal tar

CHEMICALS	SOURCES
Aldrin	Pesticide
Benzo(g,h,i)perylene	Petroleum refining, fuel combustion
Chlordane	Pesticide
Heptachlor	Pesticide
Hexachlorobenzene	Pesticide
Isodrin	Pesticide
Lead	See above
Mercury	See above
Methoxychlor	Pesticide
Octachlorostyrene	Chemical manufacturing byproducts, combustion
Pendimethalin	Pesticide
Pentachlorobenzene	Pesticide production, combustion
Polychlorinated biphenyl (PCBs)	Transformer fluids, lubricants, flame retardants, water-proofing agents
Tetrabromobisphenol A	Flame retardant
Toxaphene	Pesticide
Trifluralin	Pesticide

From http://www.epa.gov/tri/lawsandregs/pbt/pbtrule.htm

forms of these chemicals are listed in Table 10.5 in order of decreasing probable impact to the environment.

Table 10.5 illustrates that the inorganic chemicals listed above can act as a nutrient and as a pollutant, depending on the amounts applied, the location of application, and soil-plant-water dynamics. For example, Figure 10.1 shows the soil nitrogen cycle, which illustrates the transformations, sinks, and sources of this element. Plants and some soil minerals can act as sinks for the two major forms of N. Conversely, some plants, animals, the atmosphere, and humans (fertilizer additions) can contribute to excessive N (NO_3^-)

concentrations that lead to groundwater pollution. Groundwater polluted with high levels of nitrate has been shown to cause methemoglobinemia (blue baby syndrome) in infants and some adults. Methemoglobinemia occurs when nitrate is converted to nitrite by the digestive system. Nitrite reacts with oxyhemoglobin (oxygen carrying blood protein), forming methemoglobin. Methemoglobin cannot carry oxygen resulting in a decreased ability of the blood to carry oxygen. Consequently, oxygen deprivation in body tissues can occur. Infants suffering from methemoglobinemia develop a blue coloration of their mucous membranes and possible digestive and respiratory problems.

Most pesticides are organic compounds and are often applied in agricultural systems at least once a year, albeit in much smaller quantities than fertilizers. However, synthetic pesticides, designed to be very toxic to plants and pests, may have deleterious effects at very low concentrations. Most synthetic pesticides are broadly classified as insecticides, herbicides, and fungicides. While most pesticides are solids, they are usually dissolved in water or oil to facilitate their handling and application. Fumigants are gaseous pesticides typically used to control insects. A list of common organic pesticides is presented in Table 10.6. Less common forms of inorganic pesticides are used to control roaches and rats. These chemicals, which have all too often been used in close proximity to humans, have, as their primary acting agent, toxic forms of arsenic (AsO_4^{3-}), boron (H_3BO_3), and S (SO_2).

The chemical structure of organic pesticides controls their water solubility, mobility, environmental persistence, and toxicity. The first generation of organic pesticides had multiple chlorine groups inserted into their structures to give them a broad spectrum of biotoxic effects. However, the chlorine groups also made them very difficult to degrade, making them very persistent (see Chapter 8). The next step in pesticide development sought a compromise between persistence and toxicity, with chemical structures that were moderately soluble in water and with more targeted toxicity effects. The next generation of pesticides again sought to decrease the persistence of these chemicals in the environment by making them even more water soluble and continued to focus their toxic effects. This class of pesticides seldom bioaccumulate in humans or animals and have short life

TABLE 10.5 Common fertilizer and soil amendments materials and potential contaminant forms.

FERTILIZERS	NUTRIENT FORM	POLLUTANT PROPERTIES
NH_3(gas),$CO(NH_2)_2$ (urea), NH_4 — PO_4 solutions.	NH_4NO_3, $(NH_4)_2SO_4$, KNO_3 NO_3^-, NH_4^+ PO_4^{3-}	-very mobile, promotes microbial growth -toxic, volatile as NH_3 -promotes eutrophicationl gral l
Superphosphate, triple superphosphate, N-P solutions	PO_4^{3-},Ca^{++}	-variable mobility, promotes microbial growth -increases water hardness
Ammonium phosphate	NO_3^-, NH_4^+, PO_4^{3-}	-see above
Calcite ($CaCO_3$)	Ca^{++}, CO_3^-	-increases soil water alkalinity
Gypsum ($CaSO_4.2H_2O$)	Ca^{++}, $SO_4^=$	-mobile, may pollute water sources
Micronutrients, salt forms, chelates	Fe^{++}, Mn^{++}, Zn^{++}, Cu^{++}, $MoO_4^=$, H_3BO_3, Cl^-	-cations are mobile in acid soils -anions are mobile in alkaline soils

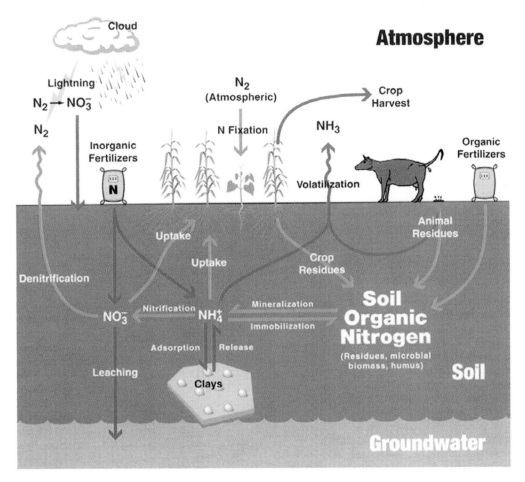

Figure 10.1 Soil-nitrogen transformations. From *Environmental Monitoring and Characterization* © 2004, Elsevier Academic Press, San Diego, CA.

spans (days) in the environment. However, when misused, these chemicals can be found in water sources. For example, today the members of the triazine family are the most commonly found pesticides in surface and groundwater resources. Conversely, chlorinated pesticides are seldom found in water but can still be found in soils and sediments (see Chapter 16).

Animals generate significant amounts of residues that are benign to the environment in open environments with low concentrations of animals. However, in the last 100 years, large-scale animal production systems have created concentrated sources of animal-derived contaminants.

Large-scale animal feeding operations include feedlots for beef, swine, and poultry production, dairies, and fish farms. These operations act as point sources for the common chemicals listed in Table 10.5 (see Figure 10.2). Nitrate-N, ammonium-N, and phosphate-P are the three most common contaminants derived from unregulated animal waste disposal practices. These three chemicals are usually found at concentrations ranging from 1,000 to 50,000 mg kg^{-1} (elemental form) in animal wastes. Nitrates are very mobile in the environment and can only be controlled by plant and microorganism uptake or by the process of denitrification. Large releases of ammonium can have several detrimental

TABLE 10.6 Major classes of organic pesticides and their potential pollutant properties.

CLASS/ELEMENTAL COMPOSITION	COMMON EXAMPLES	POLLUTANT PROPERTIES
Organochlorines	DDT	Resistant to degradation (persistent)
Organophosphates	Chlorpyrifos	Mobile in the soil environment
Carbamates	Carbaryl	Very mobile in the soil environment
Triazines	Atrazine	Very mobile in the soil environment
Plant Insecticides	Pyrethroids	Some toxic to fish
Fumigants	Dichloropropene	Toxic to animals, volatile

Note: All of these chemicals have some degree of toxicity (acute and/or chronic) toxicity to humans.

Figure 10.2 Runoff from feedlots may enter nearby surface water and degrade water quality. Photo courtesy USDA National Resources Conservation Service.

effects in the environment. First, the ammonium ion is unstable; it can volatilize in alkaline water or be oxidized to nitrate, increasing the pool of this anion. Second, the ammonium ion is very toxic to fish. Finally, the process of ammonium oxidation to nitrate (nitrification) releases acidity into the environment.

Phosphates are much less mobile in the environment. However, small quantities (>1 mg L^{-1}) can be extremely deleterious to stagnant water bodies because phosphates can trigger excessive microbial growth that leads to eutrophication. Information Box 10.1 shows a list of contaminants, in addition to N and P, that concentrated animal wastes can introduce in significant amounts into the environment.

10.4 SOURCES: INDUSTRIAL AND MANUFACTURING ACTIVITIES

There are numerous sources of industrial chemical contaminants, the result of controlled or uncontrolled waste disposal and releases into the environment. Industrial wastes may contain contaminants classified by the Federal government as hazardous and nonhazardous. However, this classification primarily separates wastes containing high concentrations of pollutants versus wastes that contain low concentrations. For example, metal-plating industrial wastes contain high concentrations of toxic metals such as Cr, Ni, and Cd and are usually classified as hazardous. However, municipal wastes, classified as nonhazardous, also contain these metals and many others, but at much lower concentrations.

Most industrial contaminants originate from a few general categories of industrial wastes. These are summarized in Information Box 10.2, with examples of industries and their common classes of contaminants. Industrial and manufacturing activities have produced many pollution problems for surface water and groundwater resources (see Chapters 17 and 18).

10.5 SOURCES: MUNICIPAL WASTE

Municipal wastewater treatment plants produce wastes that contain many potential contaminants (see Chapter 25). Reclaimed wastewater is usually clean enough to be used for irrigation, but routinely contains higher ($\sim$1.5 times) concentrations of dissolved solids than the source water. Also, chlorine-disinfected reclaimed water can contain

---

INFORMATION BOX 10.1

Pollutants Released from Animal Wastes

- **Total dissolved solids (TDS)** (Na, Cl, Ca, Mg, K, soluble N and P forms): Most animal wastes are high ($\gg$10,000 mg L^{-1}) in TDS.

- **Organic Carbon:** Excessive amounts of soluble carbon together with soluble P can quickly reduce O_2 availability in water by raising the biochemical oxygen demand.

- **Residual pesticides:** Used to control pests in animal facilities.

- **Residual Metals:** Cu, As, from animal diets and pesticides.

- **Medicines:** Antibiotics, growth regulators.

- **Gases:** From waste storage facilities and waste disposal activities, Greenhouse —(CO_2, N_2O), toxic (NH_3, H_2S), Odors—H_2S, mercaptans, indoles, org-sulfides.

INFORMATION BOX 10.2

Industrial Wastes and Sources of Contaminants

Solid, liquid and slurry wastes with high concentrations of metals, salts, and solvents

> *Industries*: Metal plating, painting.
> *Types of pollutants*: Metals, solvents, toxic aromatic and non-aromatic hydrocarbons.

Liquid wastes with high concentrations of hydrocarbons and solvents

> *Industries*: Chemical manufacturing, electronics manufacturing, plastics manufacturing.
> *Types of pollutants*: Chlorinated solvents, hydrocarbons, plastics, plasticizers, metals, catalysts, cyanides, sulfides.

Wastewaters containing organic chemicals

> *Industries*: Paper processing, tanneries, food processing, industrial wastewater treatment plants, pharmaceuticals.
> *Types of pollutants*: Various organic chemicals.

Figure 10.3 Stormwater runoff will flow over impervious surfaces, acquiring pollutants. Photo courtesy USGS.http://www. umesc.usgs.gov/flood_2001/surface.html

significant trace amounts of disinfection byproducts such as trihalomethanes and haloacetic acids. In addition, an emerging issue for municipal wastewater treatment is pharmaceutical waste. There is growing concern that pharmaceuticals (including hormones from birth control pills and antibiotics) that are excreted in urine and disposed of in wastes may end up in water-supply resources. Many of these compounds are not fully treated in current wastewater treatment systems. There is concern about the effects that these compounds may have on humans and wildlife. Antibiotic resistance and endocrine disruptors will be discussed in detail in Chapters 29 and 30, respectively.

The solid residues of wastewater treatment plants, called biosolids, typically contain common inorganic chemicals such as those listed in Table 10.5, and may also contain heavy metals, synthetic organic compounds found in household products, and microbial pathogens. Since biosolids usually contain macro- and micronutrients and organic carbon, they are routinely applied to agricultural lands as fertilizer and soil amendments (see Chapter 26). Regulations in many states allow for the annual application of up to 8 tons (dry weight) of biosolids on farmland, depending on the metal content of each biosolids source. Land disposal of biosolids completes the natural C and N cycle in the environment. However, repeated application of biosolids often increases the concentrations of metals, P, and some salts in the soil environment. In addition, excessive, concentrated, or uneven applications of biosolids can result in surface and groundwater pollution.

Stormwater is a source of nonpoint-source pollution for both urban and rural communities. Stormwater runoff picks up pollutants as it flows over the ground surface. In urban areas, stormwater runoff will flow over a variety of impervious surfaces, including driveways, sidewalks, and streets, acquiring pollutants such as dirt, debris, and hazardous wastes such as insecticides, pesticides, paint, solvents, used motor oil, and other auto fluids (Figure 10.3). In agricultural areas, stormwater runoff may include dirt, debris, excess nutrients, pesticides, bacteria, and other pathogens. Stormwater will either flow into a sewer system or directly into a lake, stream, river, wetland, or coastal water. In some cities, stormwater runoff flows into a storm sewer system and the collected water is discharged untreated into water bodies. In many areas, stormwater and municipal wastewater enter the same sewer system. During large storm events, wastewater treatment facilities often receive more municipal and storm water than the facility can handle. When facilities are unable to handle incoming waste, untreated municipal wastewater and stormwater are discharged without treatment.

Septic systems are another repository for municipal waste (Figure 10.4). One-fourth of all homes in the United

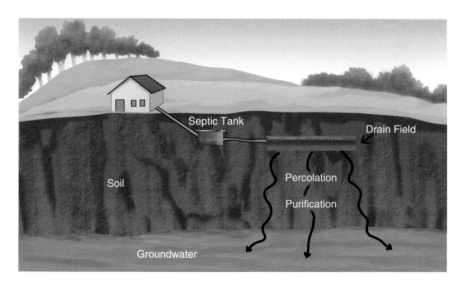

Figure 10.4 A septic system is composed of a septic tank and a drain field. Wastewater in the drain field will percolate through the subsurface, which acts as a purification system. If the system is working properly, wastewater is free of pollutants before reaching groundwater. Image taken from *A Homeowner's Guide to Septic Systems. EPA–832–B–02–005, December 2002. U.S. EPA, Washington, D.C.*

TABLE 10.7 Items that can clog or damage septic systems.

CLOGGERS	DAMAGE MICROBIAL COMMUNITIES
Diapers	Household chemicals
Cat litter	Gasoline
Cigarette filters	Oil
Coffee grounds	Pesticides
Grease	Antifreeze
Feminine hygiene products	Paint

Adapted from U.S. EPA, *A Homeowner's Guide to Septic Systems.*

States use a septic system for household wastewater disposal, with more than 4 billion gallons of wastewater disposed below the ground surface daily. Septic systems utilize microbial communities to decompose and digest waste. Most bacteria recover quickly after small amounts of cleaning products have entered the system. However, excess chemical use can cause a septic system to fail. Table 10.7 shows examples of items that can either clog a septic system or kill the microbial populations in the system. To prevent pollutants in household wastewater from entering the groundwater, it is extremely important to maintain household septic systems and to make sure they are functioning properly. Typical household wastewater pollutants include nitrogen, phosphorous, and disease-causing bacteria and viruses. To ensure that a septic system is working properly, it should be inspected every three years and pumped every three to five years.

Municipal solid waste, more commonly known as trash or garbage, is another potential source of pollution. Municipal solid waste consists of items such as paper, food scraps, grass clippings, product packaging, bottles, clothes, and furniture. Many households also improperly discard hazardous household waste into their municipal waste receptacles. Hazardous household waste products can be dangerous to human health and the environment, and should be sent to a proper disposal facility. Examples of hazardous household waste include paint, cleaners, oils, pesticides, and batteries. Municipal solid waste is collected and disposed of by landfill or combustion/incineration. Burning municipal solid waste will reduce its volume by up to 90% and its weight by up to 75%. However, air emissions pose an environmental concern. Landfilling municipal solid waste also causes an environmental concern. Landfills produce carbon dioxide and methane, both of which are greenhouse gases. Many landfills capture methane to use as an energy source. Another source of landfill pollution is landfill leachate, which is formed when water percolates through the landfill, dissolving compounds along the way. Landfill leachate may contain heavy metals, ammonia, toxic organic compounds, and pathogens, and is of concern as a groundwater pollutant (see Chapter 17).

10.6 SOURCES: SERVICE-RELATED ACTIVITIES

There are many service activities that produce waste materials that are potential sources of environmental pollution,

especially for groundwater (see Chapter 17). The service industries that produce substantial amounts of waste include dry cleaners and laundry plants, automotive service and repair shops, and fuel stations. These facilities are subject to regulation under the Resource Conservation and Recovery Act (RCRA) if they generate wastes that fall under RCRA's definition of a hazardous waste (see Chapter 15).

Dry cleaning, a service industry involved in the cleaning of textiles, uses solvents in the cleaning process that are considered as hazardous waste. These solvents include tetrachloroethene, petroleum solvents, and 1, 1, 1-trichloroethane. Along with spent solvents, other wastes produced are solvent containers, spent filter cartridges, residues from solvent distillation, and solvent-contaminated wastewater.

Underground storage tanks (USTs) are used to hold petroleum products and certain hazardous substances for several service-related activities. Until 1984, many USTs were not equipped with spill, overfill, and corrosion protection. As a result, these USTs have leaked and polluted soil and groundwater (Figure 10.5). Vapors and odors from leaking underground storage tanks (LUSTs) can collect in basements, utility vaults, and parking garages. Collected vapors can cause explosions, fires, asphyxiation, or other adverse health effects. Petroleum-based fuels, such as gasoline, diesel fuel, and aviation fuels, are ubiquitous sources of contamination at automotive, train, and aviation fuel stations. The lower molecular weight, more soluble constituents, such as benzene and toluene, are of special concern with respect to groundwater contamination potential. In addition, some fuel additives may also be of concern. For example, **methyl-tertiary-butyl ether (MTBE)** is a hydrocarbon derivative that has been added to gasoline for the past several years to boost the oxygen content of the fuel. This was done in accordance with federal regulations formulated to improve air quality. However, MTBE is a very soluble compound that is also resistant to biodegradation. It is a very mobile and persistent compound, and this nature has lead to widespread groundwater contamination. Low levels of MTBE can make water supplies undrinkable due to

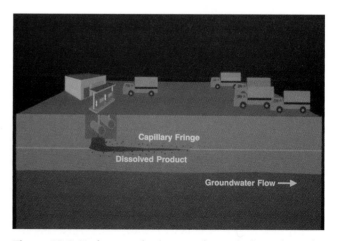

Figure 10.5 Underground storage tanks can leak causing pollution of soil and groundwater. Image courtesy U.S. EPA.: http://www.epa.gov/swerust1/graphics/cca017.jpg.

its offensive taste and odor. The use of MTBE in gasoline is now being phased out as a result of this situation.

Automotive service and repair shops can be a source of numerous contaminants. Various types of solvents are use to degrease and clean engine parts. Metal contaminants can originate from batteries, circuit boards, and other vehicle components. Fuel-based contaminants are also typically present.

10.7 SOURCES: RESOURCE EXTRACTION/PRODUCTION

Mineral extraction (mining) and petroleum and gas production are major resource-extraction activities that provide the raw materials to support our economic infrastructure. An enormous amount of pollution is generated from the extraction and use of natural resources. The Environmental Protection Agency's Toxic Releases Inventory report lists mining as the single largest source of toxic waste of all industries in the United States. Mineral extraction sites, which include strip mines, quarries, and underground mines, contribute to surface and groundwater pollution, erosion, and sedimentation (see Chapter 16). The mining process involves the excavation of large amounts of waste rock in order to remove the desired mineral ore (Figure 10.6). The ore is then crushed into finely ground tailings for chemical processing and separation to extract the target minerals. After the minerals are processed, the waste rock and mine tailings are stored in large aboveground piles and containment areas (see also Chapter 16). These waste piles, along with the bedrock walls exposed from mining, pose a huge environmental problem because of the metal pollution associated primarily with acid mine drainage. Acid mine drainage is caused when water draining through surface mines, deep mines, and waste piles comes in contact with exposed rocks containing pyrite, an iron sulfide, causing a chemical reaction. The resulting water is high in sulfuric acid and contains

elevated levels of dissolved iron. This acid runoff also dissolves heavy metals such as lead, copper, and mercury, resulting in surface and groundwater contamination. Wind erosion of mine tailings is also a significant problem.

Petroleum and natural gas extraction pose environmental threats such as leaks and spills that occur during drilling and extraction from wells, and air pollution as natural gas is burned off at oil wells. The petroleum and natural gas extraction process generates production wastes including drilling cuttings and muds, produced water, and drilling fluids. Drilling fluids, which contain many different components, can be oil based, consisting of crude oil or other mixtures of organic substances like diesel oil and paraffin oils, or water based, consisting of freshwater or seawater mixed with bentonite and barite. Each component of a drilling fluid has a different chemical function. For example, barite is used to regulate hydrostatic pressure in drilling wells. As a result of being exposed to these drilling fluids, drilling cuttings and muds contain hundreds of different substances. This waste is usually stored in waste pits, and if the pits are unlined, the toxic chemicals in the spent waste cuttings and muds, such as hydrocarbon based lubricating fluids, can pollute soil, surface, and groundwater systems. **Produced water** is the wastewater created when water is injected into oil and gas reservoirs to force the oil to the surface, mixing with formation water (the layer of water naturally residing under the hydrocarbons). At the surface, produced water is treated to remove as much oil as possible before it is reinjected, and eventually when the oil field is depleted, the well fills with the produced water. Even after treatment, produced water can still contain oil, low-molecular-weight hydrocarbons, inorganic salts, and chemicals used to increase hydrocarbon extraction.

Mined and extracted resources can also be potential pollutants once they are used for production. For example, fossil fuels are key resources for energy production. Coal-burning power plants produce nitrogen and sulfur oxides, which are known to be the primary causes of acid rain (see Chapter 21). In addition, fossil fuel combustion produces carbon dioxide, which is a primary culprit in global warming (see Chapter 23).

10.8 SOURCES: RADIOACTIVE CONTAMINANTS

Radioactive waste primarily originates from nuclear fuel production and reprocessing, nuclear power generation, military weapons development, and biomedical and industrial activities. The largest quantities of radioactive waste, in terms of both radioactivity and volume, are generated by commercial nuclear power and military nuclear weapons production industries, and by activities that support these industries, such as uranium mining and processing. However, radioactive material can also originate from natural sources. Groundwater contamination by radioactive waste

Figure 10.6 Acid mine drainage has collected at the bottom of this pit mine in Bisbee, Arizona. Photo courtesy Alex Merrill.

TABLE 10.8 Selected natural and anthropogenic radioisotopes.

ELEMENT	RADIOISOTOPE	ORIGIN	ACTIVITY
Uranium	^{238}U	Natural, enriched	Uranium mining
Radium	^{226}Ra	Natural, enriched	Uranium mining
Radon	^{222}Rn	Natural, enriched	Uranium mining, construction
Strontium	^{90}Sr	Fission product	Reactors, weapons
Cesium	^{137}Cs	Natural, fission product	Reactors, weapons

is a major problem at several Department of Energy facilities in the U.S. Selected examples of radioisotopes are presented in Table 10.8.

Naturally occurring sources of radioactive materials, including soil, rocks, and minerals that contain radionuclides, can be concentrated and exposed by human industrial activities such as uranium mining, oil and gas production, and phosphate fertilizer production. For example, when uranium is mined using in-situ leaching or surface methods, bulk waste material is generated from excavated topsoil, uranium waste rock, and subgrade ores, all of which can contain radionuclides of radium, thorium, and uranium. Other extraction and processing practices that can generate and accumulate radioactive wastes similar to that of uranium mining are aluminum and copper mining, titanium ore extraction, and petroleum production. According to EPA reports, the total amounts of naturally occurring radioactive waste that are enhanced by industrial practices number in excess of 1 billion tons annually. Sometimes, the levels of radiation are relatively low in comparison to the large volume of material that contains the radioactive waste. This causes a problem because of the high cost of disposing of radioactive waste in comparison with the relatively low value of the product from which the radioactive waste is separated. Additionally, relatively few landfills or other licensed disposal locations can accept radioactive waste.

Radioactive wastes are classified for disposal according to their physical and chemical properties, along with the source from which the waste originated. The half-life of the radionuclide and the chemical form in which it exists are the most influential of the physical properties that determine waste management. The United States divides its radioactive waste into the following categories: high-level waste, transuranic waste, and low-level waste. High-level waste consists of spent irradiated nuclear fuel from commercial reactors, and the liquid waste from solvent extraction cycles along with the solids that liquid wastes have been converted into from reprocessing. Transuranic wastes are alpha-emitting residues that contain elements with atomic numbers greater than 92, which is the atomic number of uranium. Wastes are considered transuranic when the elements have half-lives greater than 20 years and concentrations exceeding 100 nCi g^{-1}. Wastes in this category originate primarily from military manufacturing, with plutonium and americium being the principal elements of concern. Low-level waste encompasses the radioactive waste that is not classified under the above categories. Low-level wastes are separated

into subcategories: Classes A, B, C, and Greater-Than-Class-C (GTCC), with Class A being the least hazardous and GTCC being the most hazardous. Commercial low-level waste is generated by industry, medical facilities, research institutions and universities, and a few government facilities.

In some commercial and military activities, radioactive wastes are mixed with hazardous waste, creating a complex environmental problem. Mixed waste is dually regulated by the EPA and the United States Nuclear Regulatory Commission, and waste handlers must comply with both the Atomic Energy Act and the Resource Conservation and Recovery Act statutes and regulations once a waste is deemed a mixed waste. Military sources are regulated by the Department of Energy and comply with the Atomic Energy Act in regard to radiation safety.

Radon, a naturally occurring radioactive gas that is produced by the radioactive decay of uranium in rock, soil, and water, is of great concern because of the potential for the gas to become concentrated in buildings and homes (see also Chapter 22). The higher the uranium levels in the rocks, the greater the chances that a home or building may have radon gas contamination. Once the parent material decays into radon, it dissolves into the water contained in the pore spaces between soil grains. A fraction of the radon in the pore water volatilizes into the soil atmosphere gas, rendering it more mobile via gas-phase diffusion.

Exposure of humans to radon occurs in several ways. Decay products of radon are electrically charged when formed, so they tend to attach themselves to atmospheric dust particles that are normally present in the air. This dust can be inhaled, and while the inert gases are mostly exhaled immediately, a fraction of the dust particles deposit on the lungs, building up with every breath. Radon dissolved in groundwater is another source of human exposure, mainly because radon gas is released into the home atmosphere from water as it exits the tap. Another source of human exposure in home and building settings is the tendency for radon gas to enter structures via diffusion through their foundations and from certain construction materials. Radon gas availability in structures is mainly associated with the concentration of radon in the rock fractures and soil pores surrounding the structure and the permeability of the ground to gases. Slight pressure differentials between structure and soil foundations, which can be caused by barometric changes, winds, and temperature differentials, creates a gradient for radon gas to move from soil gas, through the foundations, and into the structural atmosphere.

CASE STUDY 10.1

ARSENIC POLLUTION IN BANGLADESH (HARUN-UR-RASHID, 1998).

Arsenic occurs naturally in aquifers of the country of Bangladesh. As a result, perhaps as many as 50% of the 125 million people of this country may be exposed to abnormally high (from 50 to $<1000\ \mu g\ L^{-1}$) arsenic (As) concentrations found in their drinking water. Long term chronic exposure to As promotes several skin diseases (from dermatitis to depigmentation). More advanced stages of As exposure produce gastroenteritis, gangrene, and cancer, among other diseases. More than 2 million people in Bangladesh suffer from one or more of these As-induced diseases. High As concentrations in the groundwater have been associated with As-rich sediments from the Holocene period. These sediments are primarily found in the flood and delta plains of Bangladesh. In these areas, >60% of the wells have elevated As concentrations.

Arsenic exists in two oxidation states—arsenate, As(III), and arsenite, As(V)—both of which are anions (see also Chapter 7). Although both forms are toxic, arsenite is much more toxic and is also very soluble and mobile in water environments. The exact mechanism of As enrichment in the groundwater of Bangladesh is not known but may be likely related to presence of arsenite-bearing minerals and the reductive dissolution of arsenate to the much more soluble form of arsenite. Iron reacts with As anions and can form insoluble and eventually very stable Fe-As complexes that remove As from water. In fact, amorphous Fe oxide is commonly used by water utilities to decontaminate drinking water. Another possible means of treating As-contaminated water include the use of natural soil material (as filtering devices) that contain high concentrations of iron minerals such as goethite and hematite, which can adsorb and chemically bond As.

No country is immune to the effect of this natural pollutant. In the U.S., the drinking water standard has recently been lowered to $10\ \mu g\ L^{-1}$ by the EPA. The annual added costs for drinking-water source treatment needed to comply with the new As standard are estimated to be in the billions of dollars. The states most likely to have groundwater sources with elevated As concentrations include Arizona, New Mexico, Nevada, Utah, and California.

10.9 NATURAL SOURCES OF CONTAMINANTS

The contaminant sources presented above are associated with human activities involving the production, use, and disposal of chemicals and products. It is important to realize that there are also natural sources of contaminants. A major source of such contaminants is drinking water pumped from aquifers composed of sediments and rocks containing naturally occurring elements that dissolve into the groundwater. One example, that of radioactive contaminants such as radon, was discussed in the previous section. Another major example is arsenic, which has become of great concern in recent years (see case study).

QUESTIONS AND PROBLEMS

1. What are "POPs," and why are they of such great environmental concern?
2. Describe the evolution of organic pesticide structure and persistence.
3. Describe three concerns associated with disposal of municipal solid waste.
4. What is MTBE, what was it used for, and why is it an environmental concern?
5. What are the processes that lead to acid mine drainage?
6. Why are mining processes responsible for the greatest amount of radioactive pollutants?

REFERENCES AND ADDITIONAL READING

Eisenbud M. and Gesell T.F. (1997) *Environmental Radioactivity: From Natural, Industrial, and Military Sources*. Academic Press, San Diego, California.

Harun-ur-Rashid Mridna A.K. (1998) Arsenic Contamination of Groundwater in Bangladesh. 24th WEDC conference, Islamabad, Pakistan, 1998.

Kathren R.L. (1984) *Radioactivity in the Environment: Sources Distribution, and Surveillance*. Harwood Academic Publishers, New York, New York.

Schwartzenbach R.P. Gschwend P.M. and Imboden D.M. (2003) *Environmental Organic Chemistry*. Wiley, Hoboken, New Jersey.

CHAPTER 11

MICROBIAL CONTAMINANTS

C.P. Gerba and I.L. Pepper

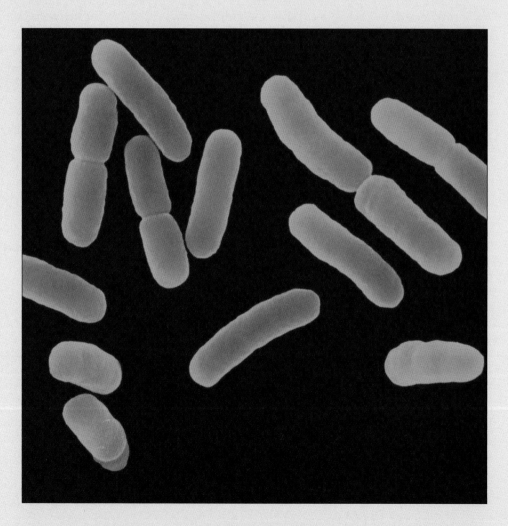

Rod-shaped *Escherichia coli* bacteria 0157:H7 (magnification 22, 245x), on nucleopore membrane filtration. Copyright Dennis Kunkel Microscopy, Inc.

11.1 WATER-RELATED MICROBIAL DISEASE

London's Dr. John Snow (1813–58) was one of the first to make a connection between certain infectious diseases and drinking water contaminated with sewage. In his famous study of London's Broad Street pump, published in 1854, he noted that people afflicted with cholera were clustered in a single area around the Broad Street pump, which he identified as the source of the infection. When, at his insistence, city officials removed the handle of the pump, Broad Street residents were forced to obtain their water elsewhere. Subsequently, the cholera epidemic in that area subsided. However successful the effect, Snow's explanation of the cause was not generally accepted because disease-causing germs had not been discovered at the time.

In the United States, the concept of **waterborne disease** was equally poorly understood. During the Civil War (1861–1865), encamped soldiers often disposed of their waste upriver, but drew drinking water from downriver. This practice resulted in widespread dysentery. In fact, dysentery, together with its sister disease typhoid, was the leading cause of death among the soldiers of all armies until the 20th century. It was not until the end of the 19th century that this state of affairs began to change. By that time, the germ theory was generally accepted, and steps were taken to properly treat wastes and protect drinking water supplies.

In 1890, more than 30 people out of every 100,000 in the United States died of typhoid. But by 1907 water filtration was becoming common in most U.S. cities, and in 1914 chlorination was introduced. Because of these new practices, the national typhoid death rate in the United States between 1900 and 1928 dropped from 36 to 5 cases per 100,000 people. The lower death toll was largely the result of a reduced number of outbreaks of waterborne diseases. In Cincinnati, for instance, the yearly typhoid rate of 379 per 100,000 people in the years 1905–1907 decreased to 60 per 100,000 people between 1908 and 1910 after the inception of sedimentation and filtration treatment. The introduction of chlorination after 1910 decreased this rate even further (Figure 11.1).

Poor water quality and sanitation account for 1.7 million deaths a year worldwide, mainly through infections and diarrhea. Nine out of 10 are children and virtually all are from developing countries.

Although many diseases have been eliminated or controlled in the developed countries, microorganisms continue to be the major cause of waterborne illness today. Most outbreaks of such diseases are attributable to the use of untreated water, inadequate or faulty treatment (*i.e.*, no filtration or disinfection), or contamination after treatment. In addition, some pathogens, such as *Cryptosporidium*, are very resistant to removal by conventional drinking water treatment and disinfection. Moreover, an increasing proportion of waterborne disease outbreaks are associated with nonbacterial microorganisms such as enteric viruses and protozoan parasites, because of their successful resistance to water treatment processes.

The true incidence of waterborne disease in the United States is not known because neither investigation nor reporting of waterborne disease outbreaks is required. Investigations are difficult because waterborne disease is not easily recognized in large communities and epidemiological studies are costly to conduct. Nevertheless, between 12 and 20 waterborne disease outbreaks per year are documented in the United States, and the true incidence may be 10 to 100 times greater (Figure 11.2).

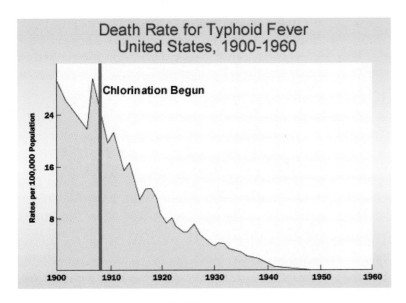

Figure 11.1 Death rate for typhoid fever. From U.S. Centers for Disease Control and Prevention, Summary of Notifiable Diseases, 1997.

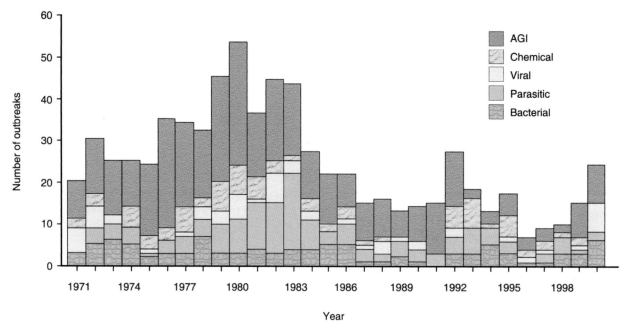

Figure 11.2 Number of waterborne-disease outbreaks associated with drinking water, by year and etiologic agent—United States, 1971–2000 (n = 730).*

*The total from previous reports has been corrected from n = 691 to n = 688.
†Acute gastrointestinal illness of unknown etiology.
From Centers for Disease Control 2004.

11.2 CLASSES OF DISEASES AND TYPES OF PATHOGENS

Disease-causing organisms, or pathogens, that are related to water can be classified into four groups (as shown in Information Box 11.1).

Waterborne diseases (Table 11.1) are those transmitted through the ingestion of contaminated water that serves as the passive carrier of the infectious or chemical agent. The classic waterborne diseases, cholera and typhoid fever, which frequently ravaged densely populated areas throughout human history, have been effectively controlled by the protection of water sources and by treatment of contaminated water supplies. In fact, the control of these classic diseases gave water supply treatment its reputation and played an important role in the reduction of infectious diseases. Other diseases caused by bacteria or by viruses, protozoa, and helminths may also be transmitted by contaminated drinking water. However, it is important to remember that waterborne diseases are transmitted through the fecal-oral route, from human to human or animal to human, so that drinking water is only one of several possible sources of infection.

Water-washed diseases are those closely related to poor hygiene and improper sanitation. In this case, the availability of a sufficient quantity of water is generally considered more important than the quality of the water. The lack of water for washing and bathing contributes to diseases that affect the eye and skin, including infectious conjunctivitis and

trachoma, as well as to diarrheal illnesses, which are a major cause of infant mortality and morbidity in the developing countries. The diarrheal diseases may be directly transmitted through person-to-person contact or indirectly transmitted through contact with contaminated foods and utensils used by persons whose hands are fecally contaminated. When enough water is available for hand washing, the incidence of diarrheal diseases has been shown to decrease, as has the prevalence of enteric pathogens such as *Shigella*.

Water-based diseases (Table 11.1 and 11.2) are caused by pathogens that either spend all (or essential parts) of their lives in water or depend upon aquatic organisms for the completion of their life cycles. Examples of such organisms are the parasitic helminth *Schistosoma* and the bacterium *Legionella*, which cause schistosomiasis and Legionnaires' disease, respectively.

The three major schistosome species that develop to maturity in humans are *Schistosoma japonicum*, *S. haematobium*, and *S. mansoni*. Each has a unique snail host and a different geographic distribution. It is estimated that more than 200 million people in Asia, Africa, South America, and the Caribbean are currently infected with one, or perhaps two, of these schistosome species. Although schistosomiasis is not indigenous to North America, schistosomiasis dermatitis has been documented in the United States, immigrants to the United States have been found to be infected with schistosomiasis, and some 300,000 persons in Puerto Rico are probably infected. The economic effects of schistosomiasis have been estimated at some $642 million annually—a figure that

---

INFORMATION BOX 11.1

Classification of Water-Related Illnesses Associated with Microorganisms

Class	Cause	Example
Waterborne	Pathogens that originate in fecal material and are transmitted by ingestion.	Cholera, typhoid
Water-washed	Organisms that originate in feces and are transmitted through contact because of inadequate sanitation or hygiene.	Trachoma
Water-based	Organisms that originate in the water or spend part of their life cycle in aquatic animals and come in direct contact with humans in water or by inhalation.	Schistosomiasis
Water-related	Microorganisms with life cycles associated with insects that live or breed in water.	Yellow fever

Modified from White and Brandley (1972).

---

includes only the resource loss attributable to reduced productivity, not the cost of public health programs, medical care, or compensation of illness.

Legionella pneumophila, the cause of *Legionnaires' disease*, was first described in 1976 in Philadelphia, Pennsylvania. This bacterium is ubiquitous in aquatic environments. Capable of growth at temperatures above 40°C, it can proliferate in cooling towers, hot water heaters, and water fountains. If growth occurs at high temperatures, these bacteria become capable of causing pneumonia in humans if they are inhaled as droplets or in an aerosol.

Water-related diseases, such as yellow fever, dengue, filariasis, malaria, onchocerciasis, and sleeping sickness, are transmitted by insects that breed in water (like the mosquitoes that carry malaria) or live near water (like the flies that transmit the filarial infection onchocerciasis). Such insects are known as **vectors**.

11.3 TYPES OF PATHOGENIC ORGANISMS

Pathogenic organisms identified as capable of causing illness when present in water include such microorganisms as viruses, bacteria, protozoan parasites, and blue-green algae,

as well as some macroorganisms—the helminths, or worms—which can grow to considerable size. Some of the characteristics of these organisms are listed in Table 11.2.

- **Viruses** are organisms that usually consist solely of nucleic acid (which contains the genetic information) surrounded by a protective protein coat or **capsid**. The nucleic acid may be either ribonucleic acid (RNA) or deoxyribonucleic acid (DNA). They are always obligate parasites; as such, they cannot grow outside of the host organism (*i.e.*, bacteria, plants, or animals), but they do not need food for survival. Thus, they are potentially capable of surviving for long periods of time in the environment. Viruses that infect bacteria are called **bacteriophages** and those bacteriophages that infect intestinal, or coliform, bacteria are known as **coliphages**.

- **Bacteria** are prokaryotic single-celled organisms surrounded by a membrane and cell wall. Bacteria that grow in the human intestinal or gastrointestinal (GI) tract are referred to as **enteric bacteria**. Enteric bacterial pathogens usually cannot survive for prolonged periods of time in the environment.

- **Protozoa** are single-celled animals. Protozoan parasites that live in the GI tract are capable of producing environmentally resistant cysts or oocysts. These oocysts have very thick walls, which make them very resistant to disinfection.

- **Helminths** (literally "worms") are multicellular animals that parasitize humans. They include roundworms, hookworms, tapeworms, and flukes. These organisms usually have both an intermediate and a final host. Once these parasites enter their final human host, they lay eggs that are excreted in the feces of infected persons and spread by wastewater, soil, or food. These eggs are very resistant to environmental stresses and to disinfection.

- **Blue-green algae**, or **cyanobacteria**, are prokaryotic organisms that do not contain an organized nucleus—unlike the green algae. Cyanobacteria, which may occur as unicellular, colonial, or filamentous organisms, are responsible for algal blooms in lakes and other aquatic environments. Some species produce toxins that may kill domestic animals or cause illness in humans.

11.3.1 Viruses

More than 140 different types of viruses are known to infect the human intestinal tract, from which they are subsequently excreted in feces. Viruses that infect and multiply in the intestines are referred to as **enteric viruses**. Some enteric viruses are capable of replication in other organs such as the liver and the heart, as well as in the eye, skin, and nerve tissue. For example, hepatitis A virus infects the liver, causing hepatitis. Enteric viruses are generally very host specific; therefore, human enteric viruses cause disease only in humans and sometimes in other primates. During infection, large numbers of virus particles, up to 10^8–10^{12} per gram, may be excreted in feces, whence they are borne to sewer systems.

TABLE 11.1 Waterborne and water-based human pathogens.

GROUP	PATHOGEN	DISEASE OR CONDITION
Viruses	Enteroviruses (polio, echo, coxsackie)	Meningitis, paralysis, rash, fever, mycocarditis, respiratory disease, diarrhea
	Hepatitis A and E	Hepatitis
	Norovirus	Diarrhea
	Rotavirus	Diarrhea
	Astrovirus	Diarrhea
	Adenovirus	Diarrhea, eye infections, respiratory disease
Bacteria	*Salmonella*	Typhoid dysentery, diarrhea
	Shigella	Diarrhea
	Campylobacter	Diarrhea
	Vibrio cholerae	Diarrhea, cholera
	Yersinia enterocolitica	Diarrhea
	Escherichia coli (certain strains)	Diarrhea
	Legionella	Pneumonia, other respiratory infection
Protozoa	*Naegleria*	Meningoencephalitis
	Entamoeba histolytica	Amoebic dysentery
	Giardia lamblia	Diarrhea
	Cryptosporidium	Diarrhea
	Toxoplasma	Mental retardation, loss of vision
Blue-green algae	*Microcystis*	Diarrhea, possible production of carcinogens
	Anabaena	
	Aphanitomenon	
Helminths	*Ascaris lumbricoides*	Ascariasis
	Trichuris trichiura	Trichuriasis-whipworm
	Necater americanus	Hookworm
	Taenia saginata	Beef tapeworm
	Schistosoma mansoni	Schistosomiasis (complications affecting the liver, bladder, and large intestines)

Enteroviruses, which were the first enteric viruses ever isolated from sewage and water, have been the most extensively studied viruses. The more common enteroviruses include the polioviruses (3 types), coxsackieviruses (30 types), and the echoviruses (34 types). Although these pathogens are capable of causing a wide range of serious illness, most infections are mild. Usually only 50% of the people infected actually develop clinical illness. However, coxsackieviruses can cause a number of life-threatening illnesses, including heart disease, meningitis, and paralysis; they may also play a role in insulin-dependent diabetes.

Infectious viral hepatitis is caused by **hepatitis A** virus (HAV) and **hepatitis E** virus (HEV). These types of viral hepatitis are spread by fecally contaminated water and food, whereas other types of viral hepatitis, such as hepatitis B virus (HBV), are spread by exposure to contaminated blood. Hepatitis A and E virus infections are very common in the developing world, where as much as 98% of the population may exhibit antibodies against HAV. HAV is not only associated with waterborne outbreaks, but is also commonly associated with foodborne outbreaks, especially shellfish. HEV has been associated with large waterborne outbreaks in Asia and Africa, but no outbreaks have been documented in developed countries. HAV is one of the enteric viruses that is very resistant to inactivation by heat.

Rotaviruses (5 types) have been identified as the major cause of infantile gastroenteritis, that is, acute gastroenteritis in children under 2 years of age. This condition is the leading

TABLE 11.2 Characteristics of waterborne and water-based pathogens.

ORGANISM	SIZE (μm)	SHAPE	ENVIRONMENTALLY RESISTANT STAGE
Viruses	0.01–0.1	variable	virion
Bacteria	0.1–10	rod, spherical, spiral, comma	spores or dormant cells
Protozoa	1–100	variable	cysts, oocysts
Helminths	$1-10^9$	variable	eggs
Blue-green algae	1–100	coccoid, filamentous	cysts

cause of mortality in children and is responsible for millions of childhood deaths per year in Africa, Asia, and Latin America. These viruses are also responsible for outbreaks of gastroenteritis among adult populations, particularly among the elderly, and can cause "traveler's diarrhea" as well. Several waterborne outbreaks have been associated with rotaviruses.

The **norovirus**, first discovered in 1968 after an outbreak of gastroenteritis in Norwalk, Ohio, causes an illness characterized by vomiting and diarrhea that lasts a few days. This virus is the agent most commonly identified during water and foodborne outbreaks of viral gastroenteritis in the United States (Figure 11.3). It has not yet been grown in the laboratory. Norovirus is a genus in the calicivirus family.

The ingestion of just a few viruses is enough to cause infection. But because enteric viruses usually occur in relatively low numbers in the environment, large volumes of environmental samples must usually be collected before the presence of these viruses can be detected. For example, from 10 to 1,000 L of water must be collected in order to assay these pathogens in surface and drinking water. This volume must first be reduced in order to concentrate the viruses. The water sample is thus passed through microporous filters to which the viruses adsorb; then the adsorbed viruses are eluted from the filter. This process is followed by further concentration, down to a few milliliters of sample, leaving a highly concentrated virus population. Next, the concentrate is assayed by using either cell culture or molecular techniques. Cell culture techniques involving animal cells are effective (Figure 11.4a), but they may require several weeks for results; thus, bacteriophages may sometimes be used as

timely and cost-effective surrogates (Figure 11.5). For example, coliphages are commonly used as models to study virus fate during water and wastewater treatment and in natural waters.

11.3.2 Bacteria

11.3.2.1 Enteric bacteria

The existence of some enteric bacterial pathogens has been known for more than a hundred years. At the beginning of the 20th century, modern conventional drinking water treatment involving filtration and disinfection was shown to be highly effective in the control of such enteric bacterial diseases as typhoid fever and cholera. Today, outbreaks of bacterial waterborne disease in the United States are relatively rare: they tend to occur only when the water treatment process breaks down, when water is contaminated after treatment, or when nondisinfected drinking water is consumed. The major bacteria of concern are members of the genus *Salmonella, Shigella, Campylobacter, Yersinia, Escherichia,* and *Vibrio.*

Salmonella is a very large group of bacteria comprising more than 2,000 known serotypes. All these serotypes are pathogenic to humans and can cause a range of symptoms from mild gastroenteritis to severe illness or even death. *Salmonella* are capable of infecting a large variety of both cold- and warm-blooded animals. Typhoid fever, caused by *S. typhi,* is an enteric fever that occurs only in humans and primates. In the United States, salmonellosis is primarily due

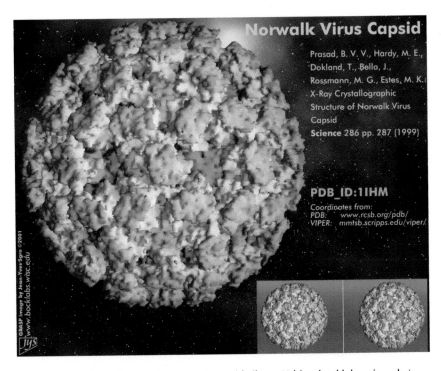

Figure 11.3 Norwalk virus (norovirus) capsid. (http://rhino.bocklabs.wisc.edu/cgi-bin/virusworld/virustable.pl) © 2005 Virusworld, Jean-Yves Sgro, Institute for Molecular Virology, University of Wisconsin-Madison.

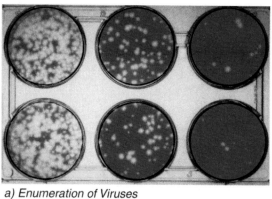

a) Enumeration of Viruses

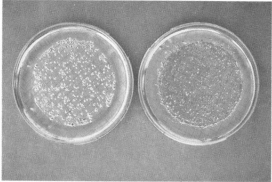

b) Enumeration of Bacteria

Figure 11.4 Quantitative assays for viruses and bacteria: (a) Viruses can be enumerated by infection of buffalo green monkey kidney cells (BGMK cells). The infections lead to lysis of the host animal cells resulting in a clearing (plaque) on the medium. This yields a measure of the infectious virus in terms of plaque forming units (PFU). (b) Fecal coliforms can be enumerated by growth on a selective medium, such as the above-pictured m-Endo agar. The resulting bacterial colonies are counted as colony forming units (CFU). Note that in both (a) and (b), more than one original virus or bacterium may have participated in the formation of macroscopic entities. Photographs courtesy of (a) C.P. Gerba and (b) I.L. Pepper. From *Environmental Microbiology* © 2000, Academic Press, San Diego, CA.

to foodborne transmission, because the bacteria infect beef and poultry and are capable of growing in foods. The pathogen produces a toxin that causes fever, nausea, and diarrhea, and may be fatal if not properly treated.

Shigella spp. infect only human beings, causing gastroenteritis and fever. They do not appear to survive long in the environment, but outbreaks from drinking and swimming in untreated water continue to occur in the United States. *Campylobacter* and *Yersinia* spp. occur in fecally contaminated water and food and are believed to originate primarily from animal feces. *Campylobacter*, which infects poultry is often implicated as a source of foodborne outbreaks; it is also associated with the consumption of untreated drinking water in the United States. *Escherichia coli* is found in the gastrointestinal tract of

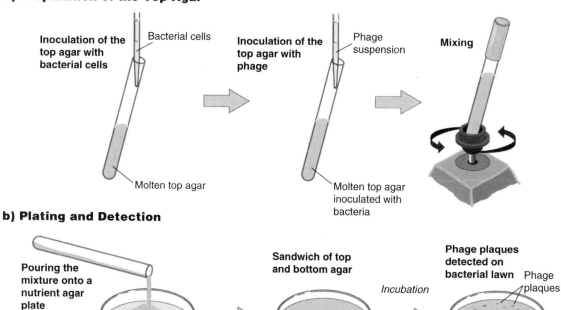

Figure 11.5 Technique for performing a bacteriophage assay. From Environmental Microbiology © 2000, Academic Press, San Diego, CA.

all warm-blooded animals and is usually considered a harmless organism. However, several strains are capable of causing gastroenteritis; these are referred to as **enterotoxigenic (ETEC)**, **enteropathogenic (EPEC)**, or **enterohemorrhagic (EHEC)** strains of *E. coli*. Enterotoxigenic *E. coli* causes a gastroenteritis with profuse watery diarrhea accompanied by nausea, abdominal cramps, and vomiting. This bacterium is another common cause of traveler's diarrhea. EPEC strains are similar to ETEC isolate but contain toxins similar to those found in the shigellae. Enterohemorrhagic *E. coli* almost always belong to the single serological type **0157:H7**. This strain generates a potent group of toxins that produce bloody diarrhea and damage the kidneys. It can be fatal in infants and the elderly. This organism can contaminate both food and water. Cattle are a major source of this organism in the environment (Figure 11.6).

The genus ***Vibrio*** comprises a large number of species, but only a few of these species infect human beings. One such is *V. cholerae*, which causes cholera exclusively in humans. Cholera can result in profuse diarrhea with rapid loss of fluid and electrolytes. Fatalities exceed 60% for untreated cases, but death can be averted by replacement of fluids. Cholera outbreaks were unknown in the Western Hemisphere in this century until 1990, when an outbreak that began in Peru spread through South and Central America. The only cases that occur in the United States are either imported or result from consumption of improperly cooked crabs or shrimp harvested from Gulf of Mexico coastal waters. *Vibrio cholera* is a native marine microorganism that occurs in low concentrations in warm coastal waters.

Usually, the survival rate of enteric bacterial pathogens in the environment is just a few days, which is less than the survival rates of enteric viruses and protozoan parasites. They are also easily inactivated by disinfectants commonly used in drinking water treatment. Analysis of environmental samples for enteric bacteria is not often performed because they are difficult to isolate. Instead, indicator bacteria are used to indicate their possible presence.

11.3.2.2 Legionella

The pathogen *Legionella pneumophila* was unknown until 1976, when 34 people died after an outbreak at the annual convention of the Pennsylvania Department of the American Legion in Philadelphia. **Legionellosis**, the acute infection resulting from *L. pneumophila*, is currently associated with two different diseases: Pontiac fever and Legionnaires' disease. Since 1976, numerous deaths from Legionnaires' disease have been reported. Pontiac fever is a milder type of legionellosis. Both these diseases are *noncommunicable*, that is, not transmitted person-to-person. The Centers for Disease

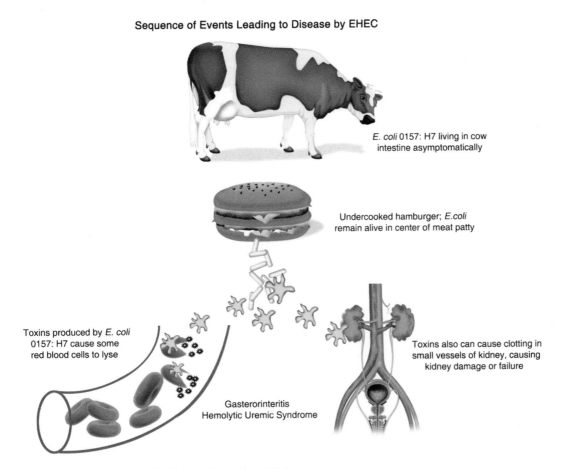

Sequence of Events Leading to Disease by EHEC

E. coli 0157: H7 living in cow intestine asymptomatically

Undercooked hamburger; *E.coli* remain alive in center of meat patty

Toxins produced by *E. coli* 0157: H7 cause some red blood cells to lyse

Toxins also can cause clotting in small vessels of kidney, causing kidney damage or failure

Gasterorinteritis
Hemolytic Uremic Syndrome

Figure 11.6 Sequence of events leading to disease by EHEC.

Control estimates that between 50,000 and 100,000 cases of legionellosis occur annually in the United States, an unknown number of which are due to contaminated drinking water.

Scientists, however, point out the error of referring to *L. pneumophila* as a classical contaminant. Although this organism occupies an ecological niche (just as do hundreds of other microorganisms in the water environment), no outbreak of legionellosis has yet been directly associated with a natural waterway such as a lake, stream, or pond. The only scientifically documented habitats for *Legionella pneumophila* are damp or moist environments. Evidently, it takes human activity—and certain systems like cooling towers, plumbing components, or even dentist water lines—to harbor or grow the organisms. Therefore, while *L. pneumophila* may be common to natural water, they can proliferate only when taken into distribution systems where water is allowed to stagnate and temperatures are favorable.

Legionella pneumophila can grow to a level that can cause disease in areas that restrict water flow and cause buildup of organic matter. Moreover, the optimum temperature for the growth of *L. pneumophila* is 37°C. Thus, *L. pneumophila* has been discovered in the hot water tanks of hospitals, hotels, factories, and homes (Figure 11.7). Ironically, some hospitals and hotels keep their water-heater temperatures low to save money and to avert lawsuits from people burned by hot water, thereby rendering themselves vulnerable to *Legionella* growth. Once established, *Legionella* tends to be persistent. One survey of a hospital water system showed that *L. pneumophila* can exist for long periods under such conditions, collecting in showerheads and faucets in the system. It

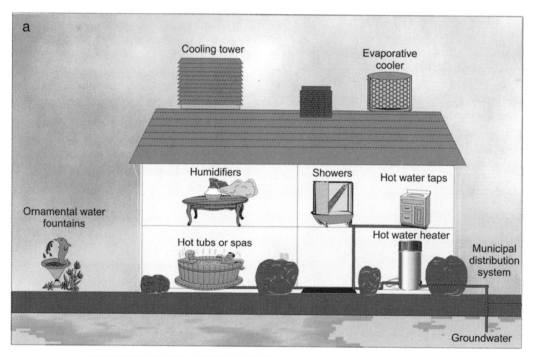

Figure 11.7 (a) Sources of *Legionella* in the environment. (b) Cooling towers. Outbreaks of Legionnaires' disease have been commonly traced to cooling towers. From *Environmental Microbiology* © 2000, Academic Press, San Diego, CA.

is believed that showerheads and faucets can emit aerosols composed of very small particles that harbor *L. pneumophila*. Such aerosols, owing simply to their smaller size, can reach the lower respiratory tract of humans.

A link between the presence of *L. pneumophila* in the water system and Legionnaires' disease in susceptible hospital patients has been established by the medical community. It is this abundance of susceptible people, together with the nature of the water system, that has resulted in outbreaks in hospitals. The great majority of people who have contracted Legionnaires' disease were immunosuppressed or compromised because of illness, old age, heavy alcohol consumption, or heavy smoking. Although some healthy people have come down with Legionnaires' disease, outbreaks that included healthy individuals have usually resulted in the milder Pontiac fever. But the fact that *L. pneumophila* exists in a water system does not necessarily mean disease is inevitable. *Legionella* bacteria have been detected in systems where no disease or only a few random cases were found. Therefore, the condition or susceptibility of the host or patient is considered to be the single most important factor in whether the infection develops.

Legionella has the ability to survive conventional water treatment. It appears to be considerably more resistant to chlorination than coliform bacteria is, and can survive for extended periods in water with low chlorine levels. In addition, it can gain access to municipal water systems through broken or corroded piping, water-main work, and cross connections.

11.3.2.3 Opportunistic bacterial pathogens

Some bacteria common in water and soil are, at times, capable of causing illness. This are referred to as opportunistic pathogens. Segments of the population particularly susceptible to opportunistic pathogens are the newborn, the elderly, and the sick. This group includes heterotrophic Gram-negative bacteria belonging to the following genera: *Pseudomonas, Aeromonas, Klebsiella, Flavobacterium, Enterobacter, Citrobacter, Serratia, Acinetobacter, Proteus,* and *Providencia*. These organisms have been reported in high numbers in hospital drinking water, where they may attach to water distribution pipes or grow in treated drinking water. However, their public health significance with regard to the population at large is not well understood. Other opportunistic pathogens are the nontubercular mycobacteria, which cause pulmonary and other diseases. The most frequently isolated nontubercular mycobacteria belong to the species *Mycobacterium avium intracellular*. Potable water, particularly that found in hospital water supplies, can support the growth of these bacteria, which may be linked to infections of hospital patients.

11.3.2.4 Indicator bacteria

The routine examination of water for the presence of intestinal pathogens is currently a tedious, difficult, and time-consuming task. Thus, scientists customarily tackle such examinations by looking first for certain indicator bacteria whose presence indicates the possibility that pathogenic bacteria may also be present. Developed at the turn of the 19th century, the indicator concept depends upon the fact that certain nonpathogenic bacteria occur in the feces of all warm-blooded animals. These bacteria can easily be isolated and quantified by simple bacteriological methods. Detecting these bacteria in water means that fecal contamination has occurred and suggests that enteric pathogens may also be present.

For example, **coliform bacteria**, which normally occur in the intestines of all warm-blooded animals, are excreted in great numbers in feces. In polluted water, coliform bacteria are found in densities roughly proportional to the degree of fecal pollution. Because coliform bacteria are generally hardier than disease-causing bacteria, their absence from water is an indication that the water is bacteriologically safe for human consumption. Conversely, the presence of the coliform group of bacteria is indicative that other kinds of microorganisms capable of causing disease also may be present, and that the water is unsafe to drink.

The coliform group, which includes *Escherichia, Citrobacter, Enterobacter,* and *Klebsiella* genus, is relatively easy to detect; specifically, this group includes all aerobic and facultatively anaerobic, Gram-negative, nonspore-forming, rod-shaped bacteria that produce gas upon lactose fermentation in prescribed culture media within 48 hours at 35°C. In short, they're hard to miss.

Scientists commonly use three methods to identify total coliforms in water. These are the **most probable number (MPN)**, the **membrane filter (MF)**, and the **presence–absence (P–A)** tests.

11.3.2.5 The most probable number (MPN) test

The MPN test allows scientists to detect the presence of coliforms in a sample and to estimate their numbers. This test consists of three steps: a presumptive test, a confirmation test, and a completed test. In the **presumptive test** (Figure 11.8), lauryl sulfate tryptose lactose broth is added to a set of test tubes containing different dilutions of the water to be tested. Usually, three to five test tubes are prepared per dilution. These tests tubes are incubated at 35°C for 24 to 48 hours, then examined for the presence of coliforms, which is indicated by gas and acid production. Once the positive tubes have been identified and recorded, it is possible to estimate the total number of coliforms in the original sample by using an MPN table that gives numbers of coliforms per 100 mL conformation (Table 11.3; Information Box 11.2).

In the **confirmation test**, the presence of coliforms is verified by inoculating such selective bacteriological agars as Levine's Eosin Methylene Blue (EMB) agar or Endo agar with a small amount of culture from the positive tubes. Lactose-fermenting bacteria are indicated on the media by the production of colonies with a green sheen or colonies with a dark center. In some cases, a **completed test** (not shown in Figure 11.8) is performed in which colonies from the agar

a) Presumptive Test

Transfer the specified volumes of sample to each tube.
Incubate 24 h at 35°C.

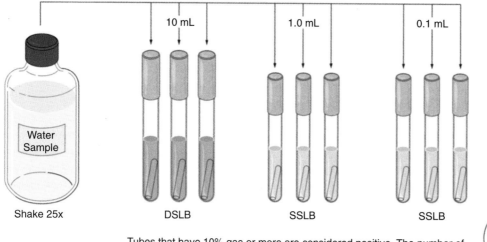

Shake 25x   DSLB   SSLB   SSLB

Tubes that have 10% gas or more are considered positive. The number of positive tubes in each dilution is used to calculate the MPN of bacteria.

b) Confirmed Test

One of the positive tubes is selected, as indicated by the presence of gas trapped in the inner tube, and used to inoculate a streak plate of Levine's EMB agar and Endo agar. The plates are incubated 24 h at 35°C and observed for typical coliform colonies.

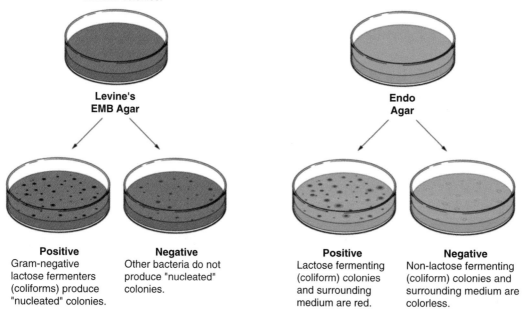

Levine's
EMB Agar

Endo
Agar

Positive
Gram-negative lactose fermenters (coliforms) produce "nucleated" colonies.

Negative
Other bacteria do not produce "nucleated" colonies.

Positive
Lactose fermenting (coliform) colonies and surrounding medium are red.

Negative
Non-lactose fermenting (coliform) colonies and surrounding medium are colorless.

Figure 11.8 Procedure for performing an MPN test for coliforms on water samples: (a) presumptive test and (b) confirmed test. From *Environmental Microbiology: A Laboratory Manual*, 2nd ed. © 2005, Elsevier Academic Press, San Diego, CA.

are inoculated back into lauryl sulfate tryptose lactose broth to demonstrate the production of acid and gas.

11.3.2.6 The membrane filter (MF) test

The membrane filter (MF) test also allows scientists to confirm the presence and estimate the number of coliforms in a sample, but it is easier to perform than the MPN test because

it requires fewer test tubes and less handling (Figure 11.9). In this technique, a measured amount of water (usually 100 mL for drinking water) is passed through a membrane filter (pore size 0.45 μm) that traps bacteria on its surface. This membrane is then placed on a thin absorbent pad that has been saturated with a specific medium designed to permit growth and differentiation of the organisms being sought. For example, if total coliform organisms are sought, a modified Endo

TABLE 11.3 Most Probable Number (MPN) table used for evaluation of the data in this experiment, using three tubes in each dilution.

NUMBER OF POSITIVE TUBES IN DILUTIONS				NUMBER OF POSITIVE TUBES IN DILUTIONS			
10 mL	1 mL	0.1 mL	MPN per 100 mL	10 mL	1 mL	0.1 mL	MPN per 100 mL
0	0	0	<3	2	0	0	9.1
0	1	0	3	2	0	1	14
0	0	2	6	2	0	2	20
0	0	3	9	2	0	3	96
0	1	0	3	2	1	0	15
0	1	1	6.1	2	1	1	20
0	1	2	9.2	2	1	2	27
0	1	3	12	2	1	3	34
0	2	0	6.2	2	2	0	21
0	2	1	9.3	2	2	1	28
0	2	2	12	2	2	2	35
0	2	3	16	2	2	3	42
0	3	0	9.4	2	3	0	29
0	3	1	13	2	3	1	36
0	3	2	16	2	3	2	44
0	3	3	19	2	3	3	53
1	0	0	3.6	3	0	0	23
1	0	1	7.2	3	0	1	39
1	0	2	11	3	0	2	64
1	0	3	15	3	0	3	95
1	1	0	7.3	3	1	0	43
1	1	1	11	3	1	1	75
1	1	2	15	3	1	2	120
1	1	3	19	3	1	3	160
1	2	0	11	3	2	0	93
1	2	1	15	3	2	1	150
1	2	2	20	3	2	2	210
1	2	3	24	3	2	3	290
1	3	0	16	3	3	0	240
1	3	1	20	3	3	1	460
1	3	2	24	3	3	2	1100
1	3	3	29	–	–	–	–

medium is used. For coliform bacteria, the filter is incubated at 35°C for 18 to 24 hours. The success of the method depends on using effective differential or selective media that can facilitate identification of the bacterial colonies growing on the membrane filter surface (Figure 11.9). To determine the number of coliform bacteria in a water sample, the colonies having a green sheen are enumerated.

11.3.2.7 The presence–absence (P–A) test

Presence–absence tests are not quantitative tests—rather they answer the simple question of whether the target organism is present in a sample or not. The use of a single tube of lauryl sulfate tryptose lactose broth as used in the MPN test, but without dilutions, would be used as a P–A test. Enzymatic assays have been developed that allow for the detection of both total coliform bacteria and *E. coli* in water and wastewater at the same time. These assays can be a simple P–A test or an MPN assay. One commercial P–A test commonly used is the Colilert® test, also called the ONPG-MUG (for *O*-nitrophenyl-β-D-galactopyranoside 4-methylumbelliferyl-β-D-glucuronide) test (Figure 11.10). The test is performed by adding the sample to a single bottle (P–A test) or MPN tubes that contain(s) powdered ingredients consisting of salts and specific enzyme substrates that serve as the only carbon

---

INFORMATION BOX 11.2

MPN Calculations

Consider the following: If you had gas in the first three tubes and gas only in one tube of the second series, but none in the last three tubes, your test would be read as 3-1-0. Table 11.3 indicates that the MPN for this reading would be 43. This means that this particular sample of water would have approximately 43 organisms per 100 mL. Keep in mind that the MPN of 43 is a statistical probability number.

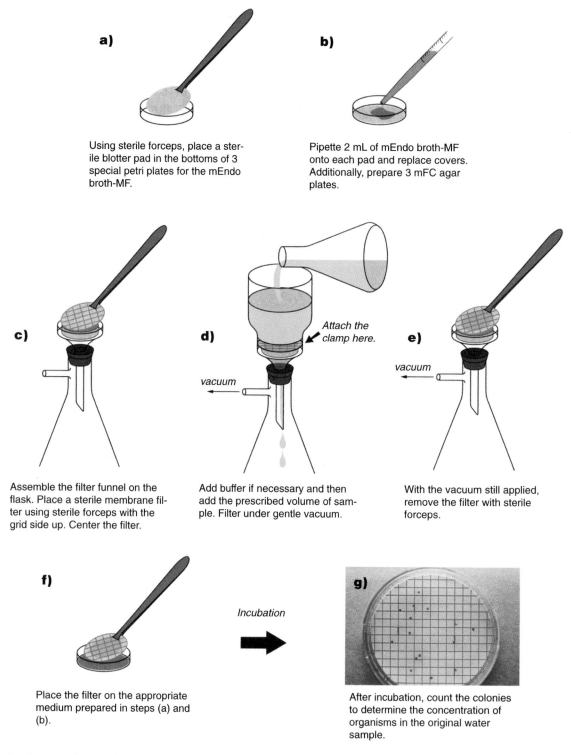

a) Using sterile forceps, place a sterile blotter pad in the bottoms of 3 special petri plates for the mEndo broth-MF.

b) Pipette 2 mL of mEndo broth-MF onto each pad and replace covers. Additionally, prepare 3 mFC agar plates.

c) Assemble the filter funnel on the flask. Place a sterile membrane filter using sterile forceps with the grid side up. Center the filter.

d) Attach the clamp here. vacuum. Add buffer if necessary and then add the prescribed volume of sample. Filter under gentle vacuum.

e) vacuum. With the vacuum still applied, remove the filter with sterile forceps.

f) Incubation. Place the filter on the appropriate medium prepared in steps (a) and (b).

g) After incubation, count the colonies to determine the concentration of organisms in the original water sample.

Figure 11.9 The membrane filtration method for determining the coliform count in a water sample. From *Environmental Microbiology* © 2000, Academic Press, San Diego, CA.

source for the organisms (Figure 11.11a). The enzyme substrate used for detecting total coliform is ONPG and that used for detecting of *E. coli* is MUG. After 24 hours of incubation, samples positive for total coliforms turn yellow (Figure 11.11b), whereas *E. coli*-positive samples fluoresce under long-wave UV illumination (Figure 11.11c).

Although the total coliform group has served as the main indicator of water pollution for many years, many of the organisms in this group are not limited to fecal sources. Thus methods have been developed to restrict the enumeration to those coliforms that are more clearly of fecal origin—that is, the **fecal coliforms**. These organisms, which include

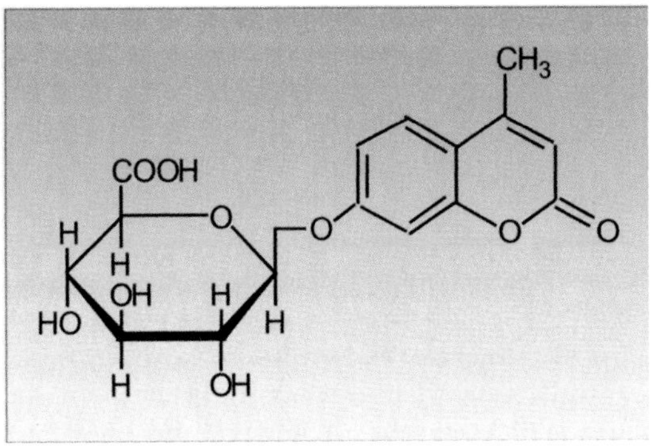

Figure 11.10 The structure of 4-methylumbelliferyl- β-d-glucuronide (MUG). From *Environmental Microbiology* © 2000, Academic Press, San Diego, CA.

INFORMATION BOX 11.3

Criteria for an Ideal Indicator Organism

· The organism should be useful for all types of water.

· The organism should be present whenever enteric pathogens are present.

· The organism should have a reasonably longer survival time than the hardiest enteric pathogen.

· The organism should not grow in water.

· The testing method should be easy to perform.

· The density of the indicator organism should have some direct relationship to the degree of fecal pollution.

the genera *Escherichia* and *Klebsiella*, are differentiated in the laboratory by their ability to ferment lactose with the production of acid and gas at 44.5°C within 24 hours. In general, then, this test indicates fecal coliforms; it does not, however, distinguish between human and animal contamination.

Although coliform and fecal coliform bacteria have been successfully used to assess the sanitary quality of drinking water, they have not been shown to be useful indicators of the presence of enteric viruses and protozoa. While outbreaks of enteric bacterial waterborne disease are rare in the United States, outbreaks of waterborne disease have occurred in which coliform bacteria were not found. Enteric viral and protozoan pathogens are more resistant to inactivation by disinfectants than bacteria. Viruses are also more difficult to remove by filtration, due to their smaller size. For this reason, other potential indicators have been investigated. These include bacteriophages (*i.e.*, bacterial viruses) of the coliform bacteria (known as coliphages), fecal streptococcus, enterococci, or *Clostridium perfringens*. The criteria for an ideal indicator organism are shown in In-

formation Box 3. None of these potential indicators has yet been proven ideal.

11.3.2.8 Fecal streptococci

The fecal streptococci are a group of Gram-positive Lancefield group D streptococci. The fecal streptococci belong to the genera *Enterococcus* and *Streptococcus*. The genus *Enterococcus* includes all streptococci that share certain biochemical properties and have a wide range of tolerance of adverse growth conditions. They are differentiated from other streptococci by their ability to grow in 6.5% sodium chloride, pH 9.6, and 45°C and include *Ent. avium, Ent. faecium, Ent. durans, Ent. faecalis,* and *Ent. gallinarum*. In the water industry the genus is often given as *Streptococcus* for

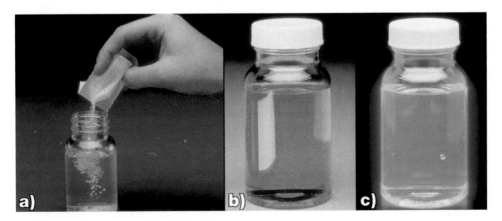

Figure 11.11 Detection of indicator bacteria with Colilert® reagent. (a) Addition of salts and enzyme substrates to water sample; (b) yellow color indicating the presence of coliform bacteria; (c) fluorescence under long-wave ultraviolet light indicating the presence of *E. coli*. Photographs used with permission of IDEXX Laboratories, Inc., Westbrook, Maine. From *Pollution Science* © 1996, Academic Press, San Diego, CA.

this group. Of the genus *Streptococcus*, only *S. bovis* and *S. equinus* are considered to be true fecal streptococci. These two species of *Streptococcus* are predominately found in animals; *Ent. faecalis* and *Ent. faecium* are more specific to the human gut. Fecal streptococci are considered to have certain advantages over the coliform and fecal coliform bacteria as indicators.

- They rarely multiply in water.
- They are more resistant to environmental stress and chlorination than coliforms.
- They generally persist longer in the environment

A relationship between the number of enterococci in water and gastroenteritis in bathers has been observed in several studies (National Research Council, 2004) and has been used in standards for recreational waters.

11.3.2.9 Heterotrophic plate count

An assessment of the numbers of aerobic and facultatively anaerobic bacteria in water that derive their carbon and energy from organic compounds is conducted via the **heterotrophic plate count (HPC)**. This group includes Gram-negative bacteria belonging to the following genera: *Pseudomonas, Aeromonas, Klebsiella, Flavobacterium, Enterobacter, Citrobacter, Serratia, Acinetobacter, Proteus, Alcaligenes, Enterobacter,* and *Moraxella*. In drinking water, the number of HPC bacteria may vary from less than 1 to more than 10^4 colony forming units (CFU)/mL, and their numbers are influenced mainly by temperature, presence of residual chlorine, and the level of organic matter.

In reality, these counts themselves have no or little health significance. However, there has been concern because the HPC can grow to large numbers in bottled water and charcoal filters on household taps. In response to this concern, studies have been performed to evaluate the impact of HPC on illness. These studies have not demonstrated a conclusive impact on illness in persons who consume water with high HPC. Although the HPC is not a direct indicator of fecal contamination, it does indicate variation in water quality and potential for pathogen survival and regrowth. These bacteria may also interfere with coliform and fecal coliform detection when present in high numbers. It has been recommended that the HPC should not exceed 500 per mL in tap water (LeChevallier et al., 1980).

Heterotrophic plate counts are normally done by the spread plate method using yeast extract agar incubated at 35°C for 48 hours. A low-nutrient medium, R_2A (Reasoner and Geldreich, 1985), has seen widespread use and is recommended for disinfectant-damaged bacteria. This medium is recommended for use with an incubation period of 5–7 days at 28°C. HPC numbers can vary greatly depending on the incubation temperature, growth medium, and length of incubation (Figure 11.12).

11.3.2.10 Bacteriophage

Because of their constant presence in sewage and polluted waters, the use of bacteriophage (or bacterial viruses) as appropriate indicators of fecal pollution has been proposed. These organisms have also been suggested as indicators of viral pollution. This is because the structure, morphology, and size, as well as behavior in the aquatic environment of many bacteriophage closely resemble those of enteric viruses. For these reasons, they have also been used extensively to evaluate virus resistance to disinfectants, to evaluate virus fate during water and wastewater treatment, and as surface and groundwater tracers. The use of bacteriophage as indicators of fecal pollution is based on the assumption that their presence in water samples denotes the presence of bacteria capable of supporting the replication of the phage. Two groups of phage in particular have been studied: the **somatic coliphage**, which infect *E. coli* host strains through cell wall receptors, and the **F-specific RNA coliphage**, which infect strains of *E. coli* and related bacteria through the F^+ or sex pili. A significant advantage of using coliphage is that they can be detected by simple and inexpensive techniques that yield results in 8–18 hours. Both a plating method (the agar overlay method) and the MPN method can be used to detect coliphage (Figure 11.5) in volumes ranging from 1 to 100 ml. The F-specific coliphage (male-specific phage) have received the greatest amount of attention because they are similar in size and shape to many of the pathogenic human enteric viruses. Because F-specific phage are infrequently detected in human fecal matter and show no direct relationship to the fecal pollution level, they cannot be considered indicators of fecal pollution. However, their presence in high numbers in wastewaters and their relatively high resistance to chlorination contribute to their consideration as an index of wastewater contamination as potential indicator of enteric viruses.

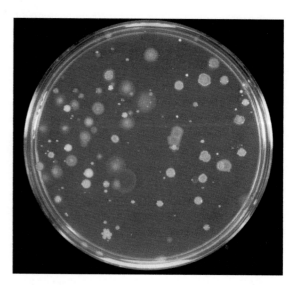

Figure 11.12 Heterotrophic plate count bacteria. Courtesy E.M. Jutras. From *Pollution Science*, First Edition © 1996, Academic Press, San Diego, CA.

11.3.3 Protozoa

11.3.3.1 Giardia

Anton van Leeuwenhoek (1632–1723), the inventor of the microscope, was the first person to identify the protozoan *Giardia* in 1681. However, *Giardia lamblia*, the specific microorganism responsible for giardiasis, was unknown in the United States until 1965, when the first case of giardiasis was reported in Aspen, Colorado. Giardiasis is a particularly nasty disease whose acute symptoms include gas, flatulence, explosive watery foul diarrhea, vomiting, and weight loss. In most people these symptoms last from one to four weeks, but have been known to last as little as three or four days or as long as several months. The incubation period ranges from one to three weeks.

Giardia lamblia occurs in the environment—usually water—as a cyst, which can survive in cold water for months and is fairly resistant to chlorine disinfection.

Humans become infected with *G. lamblia* by ingesting the environmentally resistant stage, the cyst (Figures 11.13). Once ingested, it passes through the stomach and into the upper intestine. The increase in acidity via passage through the stomach stimulates the cyst to excyst, which releases two trophozoites into the upper intestine. The trophozoites attach to the epithelial cells of the small intestines (Figure 11.14). It is believed that the trophozoites use their sucking disks to adhere to epithelial cells. It can cause both acute and chronic diarrhea within 1–4 weeks of ingestion of cysts, resulting in foul-smelling, loose, and greasy stools.

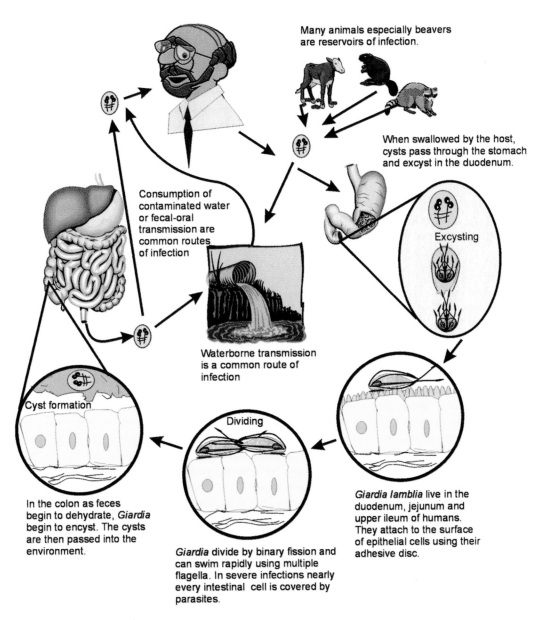

Many animals especially beavers are reservoirs of infection.

Consumption of contaminated water or fecal-oral transmission are common routes of infection

When swallowed by the host, cysts pass through the stomach and excyst in the duodenum.

Excysting

Waterborne transmission is a common route of infection

Cyst formation

In the colon as feces begin to dehydrate, *Giardia* begin to encyst. The cysts are then passed into the environment.

Dividing

Giardia divide by binary fission and can swim rapidly using multiple flagella. In severe infections nearly every intestinal cell is covered by parasites.

Giardia lamblia live in the duodenum, jejunum and upper ileum of humans. They attach to the surface of epithelial cells using their adhesive disc.

Figure 11.13 Life cycle of *Giardia lamblia*. From *Environmental Microbiology* © 2000, Academic Press, San Diego, CA.

Giardia Lamblia
Infectious dose: approx. 200   *Size: 8 to 10 microns in length*

Figure 11.14 Trophozoites of *Giardia lamblia* the reproductive stage of this waterborne protozoa. Photo courtesy E.A. Myer.

Outbreaks of giardiasis have occurred throughout the United States, but most commonly in mountainous areas in the New England, Rocky Mountain, and Pacific Northwestern states, where high-quality water is often expected. That is, most of these areas are mistakenly thought to be void of sewage or microbial contamination, and communities there tend to use smaller, less comprehensive treatment plants. The majority of these outbreaks are the result of consumption or contact with surface water that is either untreated or treated solely with chlorine. In fact, outbreaks in swimming pools used by small children have also been identified.

Researchers have identified three factors contributing to the greatest risk of *Giardia* infection: drinking untreated water; contact with contaminated surface water; and having children in a day-care center.

Giardia cysts may be constantly present at low concentrations, even in isolated and pristine watersheds. Moreover, wild animals have been implicated as the cause of giardiasis outbreaks. For example, beavers have been blamed as the source that originally transferred the disease to humans. Another study showed that although giardiasis outbreaks occur most often in surface water systems, *Giardia* cysts may also be present in groundwater supplies, such as springs.

Giardia outbreaks occur most often in the summer months, especially among visitors of recreational areas.

11.3.3.2 Cryptosporidium

Cryptosporidium, an enteric protozoan first described in 1907, has been recognized as a cause of waterborne enteric disease in humans since 1980. Within five years of its

recognition as a human pathogen, the first disease outbreak associated with *Cryptosporidium* was described in the United States. Since then, many outbreaks have been reported, and several studies have documented that *Cryptosporidium parvum* is widespread in U.S. surface waters. Moreover, this species is responsible for infection in both human beings and domesticated animals. For example, this species infects cattle, which, in turn, serve as a major source of the organism in surface waters.

The prevalence of *Cryptosporidium* infection is largely attributable to its life cycle (Figure 11.15). The organism produces an environmentally stable oocyst that is released into the environment in the feces of infected individuals. The spherical oocysts of *Cryptosporidium parvum* range in size from 3 to 6 μm in diameter. After ingestion, the oocysts undergoes excystation, releasing sporozoite, which then initiate the intracellular infection within the epithelial cells of the gastrointestinal tract (Figure 11.15). Once in the GI tract, *Cryptosporidium* in humans causes **cryptosporidiosis**, characterized by profuse watery diarrhea, which can result in fluid losses averaging 3 or more liters a day. Other symptoms of cryptosporidiosis may include abdominal pain, nausea, vomiting, and fever. These symptoms, which usually set in about three to six days after exposure, can be severe enough to cause death.

Studies have indicated that *Cryptosporidium* oocysts occur in 55 to 87% of the world's surface waters. Thus, *Cryptosporidium* oocysts are generally more common in surface water than are the cysts of *Giardia*. But like *Giardia* cysts, *Cryptosporidium* oocysts are very stable in the environment, especially at low temperatures, and may survive for many weeks. Levels are lowest in pristine waters and protected watersheds, where human activity is minimal and domestic animals are scarce.

Occurrence of *Cryptosporidium* outbreaks associated with conventionally treated drinking water suggests that the organism is unusually resistant to removal by this process. So far, the oocysts of *Cryptosporidium* have proved to be the most resistant of any known enteric pathogens to inactivation by common water disinfectants. Concentrations of chlorine commonly used in drinking water treatment (1–2 mg L^{-1}) are not enough to kill the organism. Water filtration is the primary technique used to protect water supplies from contamination by this organism. However, the oocysts are easily inactivated by ultraviolet light disinfection. Several large outbreaks of waterborne disease in the United States and Europe attest to the fact that conventional drinking water treatments involving filtration and disinfection may not be sufficient to prevent disease outbreaks when large concentrations of this organism occur in surface waters. For example, one U.S. outbreak in Carrollton, Georgia, involved illness in 13,000 people—fully one-fifth of the county's total population. In this case, oocysts were identified in the drinking water and in the stream from which the conventional drinking water plant drew its water. Another outbreak of *Cryptosporidium*,

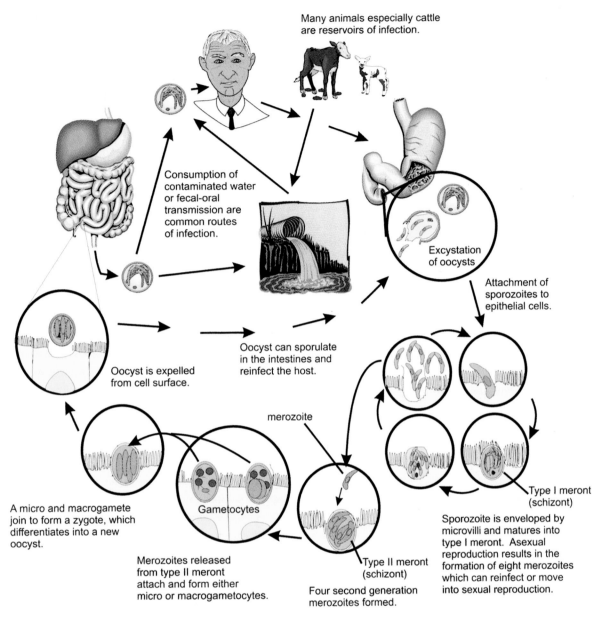

Figure 11.15 Life cycle of *Cryptosporidium parvum*. From *Environmental Microbiology* © 2000, Academic Press, San Diego, CA.

which occurred in Milwaukee, Wisconsin, was the largest outbreak of waterborne disease ever documented in the United States (see Case Study: *Cryptosporidiosis* in Milwaukee). In other cases, *Cryptosporidium* infection has been transmitted by contact with, or swimming in, contaminated water.

As in the case with other enteric organisms, only low numbers of *Giardia* cysts and *Cryptosporidium* oocysts need to be ingested to cause infection. Thus, large volumes of water are sampled (from 100 to 1,000 L) for analysis. The organisms are entrapped on filters with a pore size smaller than the diameter of the cysts and oocysts. After extraction from the filter, they can be further concentrated and detected by

observation with the use of a microscope. Antibodies tagged with a fluorescent compound are used to aid in the identification of the organisms.

11.3.4 Helminths

In addition to the unicellular protozoa, some multicellular animals—the helminths—are capable of parasitizing humans. These include the Nematoda (roundworms) and the Platyhelminthes, which are divided into two subgroups: the Cestoda, or tapeworms, and the Trematoda, or flukes. In these parasitic helminths, the microscopic ova, or eggs, constitute the infectious stage. Excreted in the feces of infected

CASE STUDY 11.1

CRYPTOSPORIDIOSIS IN MILWAUKEE

Early in the spring of 1993, heavy rains flooded the rich agricultural plans of Wisconsin. These rains produced an abnormal run-off into a river that drains into Lake Michigan, from which the city of Milwaukee obtains its drinking water. The city's water treatment plant seemed able to handle the extra load; it had never failed before, and all existing water quality standards for drinking water were properly met. Nevertheless, by April 1, thousands of Milwaukee residents came down with acute watery diarrhea, often accompanied by abdominal cramping, nausea, vomiting, and fever. In a short period of time, more than 400,000 people developed gastroenteritis, and more than 100 people—mostly elderly and infirm individuals—ultimately died, despite the best efforts of modern medial care. Finally, after much testing, it was discovered that *Cryptosporidium* oocysts were present in the finished drinking water after treatment. These findings pointed to the water supply as the likely source of infection; and on the evening of April 7, the city put out an urgent advisory for residents to boil their water. This measure effectively ended the outbreak. All told, direct costs and loss of life are believed to have exceeded $150 million dollars.

The Milwaukee episode was the largest waterborne outbreak of disease ever documented in the United States. But what happened? How could such a massive outbreak occur in a modern U.S. city in the 1990s? And how could so many people die? Apparently, high concentrations of suspended matter and oocysts in the raw water resulted in failure of the water treatment process—a failure in which *Cryptosporidium* oocysts passed right through the filtration system in one of the city's water treatment plants, thereby affecting a large segment of the population. Among this general population were many whose systems could not withstand the resulting illness. In immunocompetent people, *Cryptosporidiosis* is a self-limiting illness; it's very uncomfortable, but it goes away of its own accord. However, in the immunocompromised, *Cryptosporidiosis* can be unrelenting and fatal.

---

persons and spread by wastewater, soil, or food, these ova are very resistant to environmental stresses and to disinfection. The most important parasitic helminths are listed in Table 11.4.

Ascaris, a large intestinal roundworm, is a major cause of nematode infections in humans. This disease can be acquired through ingestion of just a few infective eggs (Figure 11.16). Since one female *Ascaris* can produce approximately 200,000 eggs per day and each infected person can excrete a large quantity of eggs, this nematode is very common. Worldwide estimates indicate that between 800 million and 1 billion people are infected, with most infections being in the tropics or subtropics. In the United States, infections tend to occur in the Gulf Coast. The life cycle of this parasite includes a phase in which the larvae migrate through the lungs and cause pneumonitis (known as Loeffler's syndrome). Although the eggs are dense, and hence readily removed by

TABLE 11.4 Pathogens in the environment.

ORGANISM	DISEASE (MAIN SITE AFFECTED)
Nematodes (roundworms)	
Ascaris lumbricoides	Ascariasis: intestinal obstruction in children (small intestine)
Trichuris trichiura	Whipworm (trichuriasis): (intestine)
Hookworms	
Necator americanus	Hookworm disease (intestine)
Ancylostoma duodenale	Hookworm disease (intestine)
Cestodes (tapeworms)	
Taenia saginata	Beef tapeworm: results in abdominal discomfort, hunger pains, chronic indigestion
Taenia solium	Pork tapeworm (intestine)
Trematodes (flukes)	
Schistosoma mansoni	Schistosomiasis (liver [cirrhosis], bladder, and large intestine)

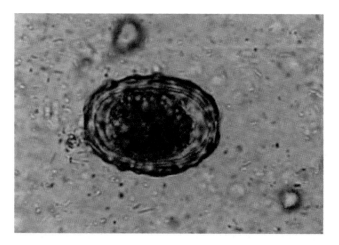

Figure 11.16 Egg of *Ascaris lumbriodes*. From U.S. Environmental Protection Agency.

sedimentation in wastewater treatment plants, they are quite resistant to the action of chlorine. Moreover, they can survive for long periods of time in sewage sludge after land application, unless they were previously removed by sludge treatment.

Although ***Taenia saginata*** (beef tapeworm) and ***Taenia solium*** (pig tapeworm) are now relatively rare in the United States, they can still be found in developing countries around the world. These parasites develop in an intermediate animal host, where they reach a larval stage called **cysticercus**. Then these larvae are passed, via meat products, to humans, who serve as final hosts. For instance, cattle that ingest the infective ova while grazing serve as intermediate hosts for *Taenia saginata*, while pigs are the intermediate hosts for *Taenia solium*. The cysticerci invade the muscle, eye, and brain tissue of the intermediate host and can cause severe enteric disturbances, such as abdominal pains and weight loss in their final hosts.

11.3.5 Blue-Green Algae

Blue-green algae, or **cyanobacteria** (Figure 11.17), occur commonly in all natural waters, where they play an important role in the natural cycling of nutrients in the environment and the food chain. However, a few species of blue-green algae, such as *Microcystis, Aphanizomenon,* and *Anabaena*, produce toxins capable of causing illness in humans and animals. These toxins can cause gastroenteritis, neurological disorders, and possibly cancer. In this case, illness is caused by the ingestion of the toxin produced by the organisms, rather than ingestion of the organism itself, as is the case with helminths. Numerous cases of livestock, pet, and wildlife poisonings by the ingestion of water blooms of cyanobacteria have been reported, and evidence has been mounting that humans are also affected. Heavy blooms of cyanobacteria can occur in surface waters when sufficient nutrients are available, resulting in sewage contaminated water supplies.

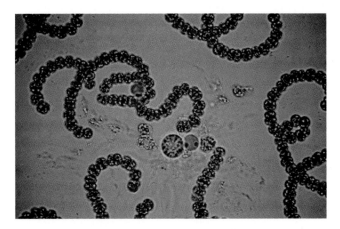

Figure 11.17 Blue-green algae. Copyright Trustees, University of Pennsylvania. cal.vet.upenn.edu/poison/plants/ppblueg.htm.

11.4 SOURCES OF PATHOGENS IN THE ENVIRONMENT

Waterborne enteric pathogens are excreted, often in large numbers, in the feces of infected animals and humans, whether or not the infected individual exhibits the symptoms of clinical illness. In some cases, infected individuals may excrete pathogens without ever developing symptoms; in other cases, infected individuals may excrete pathogens for many months—long after clinical signs of the illness have passed. Such **asymptomatic** infected individuals are known as carriers, and they may constitute a potential source of infection for the community. Owing to such sources, pathogens are almost always present in the sewage of any community. However, the actual concentration of pathogens in community sewage depends on many factors: the incidence of enteric disease (*i.e.*, the number of individuals with the disease in a population), the number of carriers in the community, the time of year, sanitary conditions, and per capita water consumption.

The peak incidence of many enteric infections is seasonal in temperate climates. Thus, the highest incidence of enterovirus infection is during the late summer and early fall, while rotavirus infections tend to peak in the early winter, and *Cryptosporidium* infections peak in the early spring and fall. The reason for the seasonality of enteric infections is not completely understood, but several factors may play a role. It may be associated with the survival of different agents in the environment during the different seasons: *Giardia*, for example, can survive winter temperatures very well. Alternatively, excretion differences among animal reservoirs may be involved, as is the case with *Cryptosporidium*. It may well be that greater exposure to contaminated water, as in swimming, is the explanation for increased incidence in the summer months.

Certain populations and subpopulations are also more susceptible to infection. For example, enteric infection is more common in children, because they usually lack previous protective immunity. Thus, the incidence of enteric virus and protozoa infections in day-care centers, where young children are in close proximity, is usually much higher than that in the general community (Table 11.5). A greater

TABLE 11.5 Incidence and concentration of enteric viruses and protozoa in feces in the United States.

PATHOGEN	INCIDENCE (%)	CONCENTRATION IN STOOL (per gram)
Enterovirus	10–40	10^3–10^8
Hepatitis A	0.1	10^8
Rotavirus	10–29	10^{10}–10^{12}
Giardia	3.8	10^6
	18–54[a]	10^6
Cryptosporidium	0.6–20	10^6–10^7
	27–50[a]	10^6–10^7

[a]Children in day-care centers.

TABLE 11.6 Estimated levels of enteric organisms in sewage and polluted surface water in the United States.

Organism	CONCENTRATION (per 100 mL)	
	Raw Sewage	Polluted Stream Water
Coliforms	10^9	10^5
Enteric viruses	10^2	$1–10$
Giardia	10	$0.1–1$
Cryptosporidium	$0.1–1$	$0.1–10^{2*}$

From U.S. EPA, 1988.

*Greatest numbers in surface waters result from animal (cattle) sources.

incidence of enteric infections is also evident in lower socioeconomic groups, particularly where lower standards of sanitary conditions prevail. Concentrations of enteric pathogens are much greater in sewage in the developing world than in the industrialized world. For example, the average concentration of enteric viruses in sewage in the United States has been estimated at 10^3 L^{-1} (Table 11.6), while concentrations as high as 10^5 L^{-1} have been observed in Africa and Asia.

11.4.1 Sludge

During municipal sewage treatment, **biosolids**—(or sludges)—are produced (Chapter 26). Biosolids are a byproduct of physical (primary treatment), biological (activated sludge), and physicochemical precipitation of suspended solids (by chemicals) treatment processes. Although treatment by anaerobic or aerobic digestion and/or dewatering reduces the numerical population of disease agents in these biosolids, significant numbers of the pathogens present in raw sewage often remain in biosolids. On a volume basis, the concentration of pathogens in biosolids can be fairly high because of settling (of the large organisms, especially helminths) and adsorption (especially viruses). Moreover, most microbial species found in raw sewage are concentrated in sludge during primary sedimentation. And although enteric viruses are too low in mass to settle alone, they are also concentrated in sludge because of their strong binding affinity to particulates.

The densities of pathogenic and indicator organisms in **primary sludge** shown in Table 11.7 represent typical, average values detected by various investigators. Note that the indicator organisms are normally present in fairly constant amounts. But bear in mind that different sludges may contain significantly greater or fewer numbers of any organism, depending upon the kind of sewage from which the sludge was derived. Similarly, the quantities of pathogenic species are especially variable because these figures depend on which kind are present in a specific community at a particular time. Finally, note that concentrations determined in any study are dependent on assays for each microbial species; thus, these concentrations are only as accurate as the assays themselves, which may be compromised by such factors as inefficient recovery of pathogens from environmental samples.

TABLE 11.7 Densities of microbial pathogens and indicators in primary sludges.

TYPE	ORGANISM	DENSITY (NUMBER PER GRAM DRY WEIGHT)
Virus	Various enteric viruses	$10^2–10^4$
	Bacteriophages	10^5
Bacteria	Total coliforms	$10^8–10^9$
	Fecal coliforms	$10^7–10^8$
	Fecal streptococci	$10^6–10^7$
	Salmonella spp.	$10^2–10^3$
	Clostridium spp.	10^6
	Mycobacterium tuberculosis	10^6
Protozoa	*Giardia* spp.	$10^2–10^3$
Helminths	*Ascaris* spp.	$10^2–10^3$
	Trichuris trichivra	10^2
	Toxacara spp.	$10^1–10^2$

Modified from Straub et al., 1993.

Secondary sludges are produced following the biological treatment of wastewater. Microbial populations in sludges following these treatments depend on the initial concentrations in the wastewater, die-off or growth during treatments, and the association of these organisms with sludge. Some treatment processes, such as the activated sludge process, may limit or destroy certain enteric microbial species. Viral and bacterial pathogens, for example, are reduced in concentration by activated sludge treatment. Even so, the ranges of pathogen concentration in secondary sludges obtained from this and most other secondary treatments are usually not significantly different from those of primary sludges, as shown in Table 11.8.

11.4.2 Solid Waste

Municipal solid waste may contain a variety of pathogens, a source of which is often disposable diapers—it has been found that as many as 10% of the fecally soiled disposable diapers entering landfills contain enteroviruses. Another primary source of pathogens is sewage biosolids, where co-disposal is prac-

TABLE 11.8 Densities of pathogenic and indicator microbial species in secondary sludge biosolids.

TYPE	ORGANISM	DENSITY (NUMBER PER GRAM DRY WEIGHT)
Virus	Various enteric viruses	3×10^2
Bacteria	Total coliforms	7×10^8
	Fecal coliforms	8×10^6
	Fecal streptococci	2×10^2
	Salmonella spp.	9×10^2
Protozoa	*Giardia* spp.	$10^2–10^3$
Helminths	*Ascaris* spp.	1×10^3
	Trichuris trichivra	$< 10^2$
	Toxacara spp.	3×10^2

Modified from Straub et al., 1993.

ticed. Pathogens may also be present in domestic pet waste (*e.g.*, cat litter) and food wastes. Municipal solid wastes from households have been found to average 7.7×10^8 coliforms and 4.7×10^8 fecal coliforms per gram. *Salmonella* have also been detected in domestic solid waste. In unlined landfills, such pathogens may be present in the leachate beneath landfills.

11.5 FATE AND TRANSPORT OF PATHOGENS IN THE ENVIRONMENT

There are many potential routes for the transmission of excreted enteric pathogens. The ability of an enteric pathogen to be transmitted by any of these routes depends largely on its resistance to environmental factors, which control its survival, and its capacity to be carried by water as it moves through the environment. Some routes can be considered "natural" routes for the transmission of waterborne disease, but others—such as the use of domestic wastewater for groundwater recharge, large-scale aquaculture projects, or land disposal of disposable diapers—are actually new routes created by modern human activities.

Human and animal excreta are sources of pathogens. Humans become infected by pathogens through consumption of contaminated foods, such as shellfish from contaminated waters or crops irrigated with wastewater; from drinking contaminated water; and through exposure to contaminated surface waters as may occur during bathing or at recreational sites. Furthermore, those individuals infected by the above processes become sources of infection through their excrement, thereby completing the cycle.

In general, viral and protozoan pathogens survive longer in the environment than enteric bacterial pathogens (Figure

---

INFORMATION BOX 11.4

Environmental Factors Affecting Enteric Pathogen Survival in Natural Waters

Factor	Remarks
Temperature	Probably the most important factor; longer survival at lower temperatures; freezing kills bacteria and protozoan parasites, but prolongs virus survival.
Moisture	Low moisture content in soil can reduce bacterial populations.
Light	UV in sunlight is harmful.
pH	Most are stable at pH values of natural waters. Enteric bacteria are less stable at pH > 9 and < 6.
Salts	Some viruses are protected against heat inactivation by the presence of certain cations.
Organic matter	The presence of sewage usually results in longer survival.
Suspended solids or sediments	Association with solids prolongs survival of enteric bacteria and virus.
Biological factors	Naive microflora is usually antagonistic.

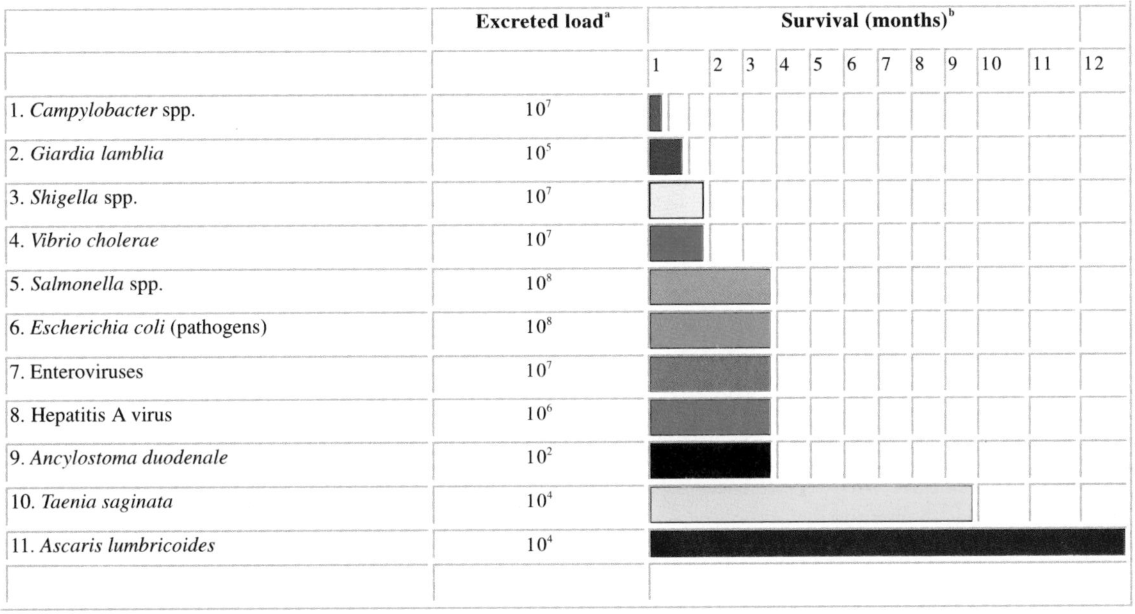

11.18). How long a pathogen survives in a particular environment depends on a number of complex factors, which are listed in Information Box 11.4. Of all the factors, temperature is probably the most important. Temperature is a well-defined factor with a consistently predictable effect on enteric

	Excreted load[a]	Survival (months)[b]											
		1	2	3	4	5	6	7	8	9	10	11	12
1. *Campylobacter* spp.	10^7												
2. *Giardia lamblia*	10^5												
3. *Shigella* spp.	10^7												
4. *Vibrio cholerae*	10^7												
5. *Salmonella* spp.	10^8												
6. *Escherichia coli* (pathogens)	10^8												
7. Enteroviruses	10^7												
8. Hepatitis A virus	10^6												
9. *Ancylostoma duodenale*	10^2												
10. *Taenia saginata*	10^4												
11. *Ascaris lumbricoides*	10^4												

[a]Typical average number of organisms/g feces
[b]Estimated average life of infective stage at 20-30°C.
(Modified from Feachem et al., 1983).

Figure 11.18 Survival times of enteric pathogens in water, wastewater, soil and on crops.

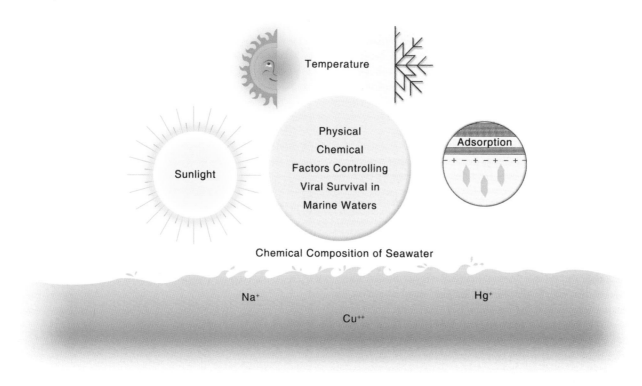

Figure 11.19 Factors affecting the survival of enteric bacteria and viral pathogens in seawater.

pathogen survival in the environment. Usually, the lower the temperature, the longer the survival time. But freezing temperatures generally result in the death of enteric bacteria and protozoan parasites. Viruses, however, can remain infectious for months or years at freezing temperatures. Moisture—or lack thereof—can cause decreased survival of bacteria. The UV light from the sun is a major factor in the inactivation of indicator bacteria in surface waters; thus, die-off in marine waters can be predicted by amount of exposure to daylight. Viruses are much more resistant to inactivation by UV light.

Many laboratory studies have shown that the microflorae of natural waters and sewage are antagonistic to the survival of enteric pathogens. It has been shown, for example, that enteric pathogens survive longer in sterile water than in water from lakes, rivers, and oceans. Bacteria in natural waters can feed upon indicator bacteria. Suspended matter (clays, organic debris, and the like) and fresh or marine sediments has been shown to prolong their survival time (Figure 11.19).

11.6 STANDARDS AND CRITERIA FOR INDICATORS

Bacterial indicators such as coliforms have been used for the development of **water quality standards**. For example, the U.S. Environmental Protection Agency (U.S. EPA) has set a standard of no detectable coliforms per 100 ml of drinking water. A drinking water standard is legally enforceable in the United States (see Chapter 15). If these standards are violated by water suppliers, they are required to take corrective action or they may be fined by the state or federal government. Authority for setting drinking water standards was given to the U.S. EPA in 1974, when Congress passed the Safe Drinking Water Act (see Chapter 15). Similarly, authority for setting standards for domestic wastewater discharges is given under the Clean Water Act. In contrast, standards for recreational waters and wastewater use are determined by the individual states. Microbial standards set by various government bodies in the United States are shown in Table 11.9. Standards used by the European Union are given in Table 11.10.

TABLE 11.9 U.S. federal and state standards for microbial quality of water.

AUTHORITY	STANDARDS
U.S. EPA	
Safe Drinking Water Act	0 coliforms/100 ml
Clean Water Act	
Wastewater discharges	200 fecal coliforms/100 ml
Sewage sludge	<1000 fecal coliforms/4 g
	<3 *Salmonella*/4 g
	<1 enteric virus/4g
	<1 helminth ova/4g
California	
Wastewater reclamation for irrigation	≤2.2 MPN coliforms
Arizona	
Wastewater reclamation for irrigation of golf courses	25 fecal coliforms/100 ml
	125 enteric virus/40 liters
	No detectable *Giardia*/40 liters

TABLE 11.10 Drinking water criteria of the European Union.

Tap Water

Escherichia coli	0/100 ml
Fecal Streptococci	0/100 ml
Sulfite-reducing clostrida	0/20 ml

Bottled Water

Escherichia coli	0/250 ml
Fecal streptococci	0/250 ml
Sulfite-reducing clostrida	0/50 ml
Pseudomonas aeruginosa	0/250 ml

Criteria and **guidelines** are terms used to describe recommendations for acceptable levels of indicator microorganisms. They are not legally enforceable but serve as guidance indicating that a potential water quality problem exists. Ideally, all standards would indicate that an unacceptable public health threat exists or that some relationship exists between the amount of illness and the level of indicator organisms. Such information is difficult to acquire because of the involvement of costly epidemiological studies that are often difficult to interpret because of confounding factors. An area where epidemiology has been used to develop criteria is that of recreational swimming. Epidemiological studies in the United States have demonstrated a relationship between swimming-associated gastroenteritis and the densities of enterococci (Figure 11.20) and fecal coliforms. No relationship was found for coliform bacteria (Cabelli, 1989). It was suggested that a standard geometric average of 35 enterococci per 100 ml be used for marine bathing waters. This would mean accepting a risk of 1.9% of the bathers developing gastroenteritis. Numerous other epidemiological studies of bathing-acquired illness have been conducted. These studies have shown slightly different relationships to illness and that other bacterial indicators were

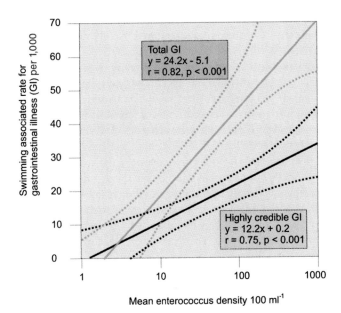

Figure 11.20 Dose-response relationships. Produced by the work of Cabelli et al. (1982). From *Environmental Microbiology*, © 2000, Academic Press, San Diego, CA.

more predictive of illness rates. These differences probably arise because of the different sources of contamination (raw versus disinfected wastewater), types of recreational water (marine versus fresh), types of illness (gastroenteritis, eye infections, and skin complaints), immune status of the population, length of observation, and so on. Various guidelines for acceptable numbers of indicator organisms have been in use (Table 11.11), but there is no general agreement on standards.

The use of microbial standards also requires the development of standard methods and quality assurance or quality control plans for the laboratories that will do the monitoring.

TABLE 11.11 Guidelines for recreational water quality standards.

COUNTRY OR AGENCY	REGIME (SAMPLES/TIME)	CRITERIA OR STANDARD[a]
U.S. EPA[b]	5/30 days	200 fecal coliforms/100 ml
		<10% to exceed 400/ml
		Fresh water[b]
		33 enterococci/100 ml
		126 E. coli/100 ml
		Marine waters[b]
		35 enterococci/100 ml
European Economic Community	2/30 days[c]	500 coliforms/100 ml
		100 fecal coliforms/100 ml
		100 fecal streptococci/100 ml
		0 *Salmonella*/liter
		0 Enteroviruses/10 liters
Ontario, Canada	10/30 days	≤1000 coliforms/100 ml
		≤100 fecal coliforms/100 ml

From U.S. EPA, 1986.

[a]All bacterial numbers in geometric means.

[b]Acceptance is by individual state.

[c]Coliforms and fecal coliforms only.

Knowledge of how to sample and how often to sample is also important. All of this information is usually defined in the regulations when a standard is set. For example, frequency of sampling may be determined by the size (number of customers) of the utility providing the water. Sampling must proceed in some random fashion so that the entire system is characterized. For drinking water, no detectable coliforms are allowed in the United States (Table 11.9). However, in other countries some level of coliform bacteria is allowed. Because of the wide variability in numbers of indicators in water, some positive samples may be allowed or tolerance levels or averages may be allowed. Usually, **geometric averages** are used in standard setting because of the often skewed distribution of bacterial numbers. This prevents one or two high values from giving overestimates of high levels of contamination, which would appear to be the case with **arithmetic averages** (Table 11.12).

Geometric averages are determined as follows:

$$\log \bar{x} = \frac{\Sigma(\log x)}{N} \qquad \text{(Eq. 11.1)}$$

$$\bar{x} = \text{antilog} \ (\log \bar{x}) \qquad \text{(Eq. 11.2)}$$

TABLE 11.12 Arithmetic and geometric averages of bacterial numbers in water.

MPN[a]	LOG
2	0.30
110	2.04
4	0.60
150	2.18
1100	3.04
10	1.00
<u>12</u>	<u>1.08</u>
198 = arithmetic average	1.46 = log
	antilog = 29
	29 = geometric average

[a]MPN, most probable number.

where N is the number of samples $\bar{x}$ is and the geometric average, and x is the number of organisms per sample volume.

As can be seen, standard setting and the development of criteria is a difficult process and there is no ideal standard. A great deal of judgment by scientists, public health officials, and the regulating agency is required.

QUESTIONS AND PROBLEMS

1. What are pathogens? What is an enteric pathogen?
2. What is the difference between a waterborne and a water-based pathogen?
3. Which group of enteric pathogens survives the longest in the environment and why?
4. Describe some of the methods that can be used to detect indicator bacteria in water.
5. What are some of the criteria for indicator bacteria?
6. Protozoan parasites are the leading cause of waterborne disease outbreaks in the United States when an agent can be identified. Why?
7. What are some of the factors that control the survival of enteric pathogens in the environment?
8. Why are geometric means used to report average concentrations of indicator organisms?
9. Calculate the arithmetic and geometric averages for the following data set: Fecal coliforms/100 ml on different days at a bathing beach were reported as 3, 7, 1000, 125, 150, and 3000.

Did the bathing beach make the standard of 200 fecal coliforms per 100 ml for bathing?

10. Calculate the most probable number (MPN) for the following data set.

SAMPLE	
1 ml	3 positive tubes
0.1 ml	2 positive one negative tube
10^{-2}	3 negative tubes

11. What are enterococci?
12. Why have bacteriophage been suggested as standards for water quality?
13. Why is *Cryptosporidium* so difficult to control in drinking water treatment? What animal is a source of *Cryptosporidium* in water?
14. Why do enteric viruses and protozoa survive longer in the environment than enteric bacteria?
15. What are some of the niches that *Legionella* bacteria can grow to high numbers?

REFERENCES AND ADDITIONAL READING

Bitton G. (2005) *Wastewater Microbiology*, 3rd ed. Wiley-Liss, New York, New York.

Cabelli V.J., Dufour A.P., McCabe L.J. and Levin, M.A. (1982) Swimming-Associated Gastroenteritis and Water Quality. *Am. Epidemiol.* **115**,606–615.

Cortruvo J.A., Dufour A., Rees G., Bartram J., Carr R., Cliver D.O., Craun G.F., Fayer R. and Gannon V.P.J. (2004) *Waterborne Zoonoses*. IWA Publishing, London.

European Union (EU). (1995) Proposed for a Council Directive Concerning the Quality of Water Intended for Human Consumption. Com (94) 612 Final. *Offic. J. Eur. Union* 131:5–24.

Feachem R.G., Bradley D.J., Garelick H. and Mara D.D. (1983) *Sanitation and Disease Health Aspects of Excreta and Wastewater Management*. World Bank, Washington, D.C.

LeChevallier M.W., Seidler R.J. and Evans T.M. (1980) Enumeration and Characterization of Standard Plate Count Bacteria in Chlorinated and Raw Water Supplies. *Appl. Environ. Microbiol.* **40**,922–930.

Maier R.M., Pepper I.L. and Gerba C.P. (2000) *Environmental Microbiology*. Academic Press, San Diego, California.

National Research Council. (2004) *Indicators for Waterborne Pathogens*. The National Academies Press, Washington, D.C.

Pepper I.L., Gerba C.P. and Brendecke J.W. (2000) *Environmental Microbiology: A Laboratory Manual*, 2nd ed. Academic Press, San Diego, California.

Prasad B.V.V., Hardy M.E., Dokland T., Bella J., Rossmann M.G. and Estes M.K. (1999) X-ray Crystallographic Structure of Norwalk Virus Capsid. *Science* **286**,287–290.

Reasoner D.J. and Geldreich E.E. (1985) A New Medium for Enumeration and Subculture of Bacteria from Potable Water. *Appl. Environ. Microbiol.* **49**,1–7.

Straub T.M., Pepper I.L. and Gerba C.P. (1993) Hazards of Pathogenic Microorganisms in Land-disposed Sewage Sludge. *Reviews of Environmental Contamination and Toxicology.* **132**,55–91.

United States Environmental Protection Agency. (1986). Ambient Water Quality. Criteria—1986. EPA440/5–84–002, Washington, D.C.

United States Environmental Protection Agency. (1988) Comparative Health Effects Assessment of Drinking Water. United States Environmental Protection Agency, Washington, DC.

White G.F. and Bradley D.J. (1972) *Drawers of Water*. University of Chicago Press, Chicago, Illinois.

CHAPTER **12**

THE ROLE OF ENVIRONMENTAL MONITORING IN POLLUTION SCIENCE

J.F. Artiola and M.L. Brusseau

Yellowstone National Park. *Photo courtesy J.F. Artiola.*

12.1 INTRODUCTION

Environmental monitoring is based on scientific observations of changes that occur in our environment. Scientists need to observe changes to study the dynamics of not only natural cycles, but also anthropogenic-based impacts. The effects of pollution in the environment, as in humans, can be slow (chronic) and may require multiple observations over time, or they can be fast-acting (acute) and be assessed with simultaneous observations. The effects of pollution occur at all spatial scales; therefore, observations are also made at multiple scales of space. However, since the environment is a continuum, observations must be made using physical, chemical, and biological methods. Only science-based observations, standards-based data processing, and objective interpretations can produce the knowledge and level of understanding required to accurately evaluate and solve environmental problems. There are numerous examples of knowledge-based regulations that benefit modern society and protect the environment from pollution. These include waste management regulations involving disposal, treatment, or reuse; regulations governing the protection of water resources including natural and public water supplies; and regulations protecting endangered species.

There are numerous agencies and world institutions involved in environmental monitoring. In the U.S., pollution monitoring and prevention is the primary focus of the Environmental Protection Agency (EPA). This agency is mandated to develop and enforce laws and regulations that are protective of our health and the environment.

12.2 SAMPLING AND MONITORING BASICS

Developing a program to monitor the extent or effects of pollution in a particular environment requires careful consideration of the following: (1) The *purpose* of the monitoring program—this includes determining, for example, pollution level changes in space and time; (2) the *objectives*, which may include specific chemical, physical, *or* biological analyses; and (3) the *approach*, which will assist in defining the number and types of measurements. To achieve our objectives, we must also consider the environmental characteristics or uniqueness of each environment. Finally, we must consider sampling methods, including locations, timing, and type. There are several types of sampling methods. **Random** sampling, for example, assumes that all units from an environment have an equal chance of being selected. In contrast, **systematic** sampling selects sampling locations at predefined intervals in space or time.

Specific sampling plans are needed to insure that all aspects of monitoring are detailed and described. The U.S. EPA defines seven critical elements of a sampling plan to insure that all the data quality objectives are met; these are described in Information Box 12.1. In the development of sampling plans, environmental scientists must be thoroughly

---

INFORMATION BOX 12.1

Data Quality Objectives of a Sampling Plan

Data quality control and objectives are needed in a sampling plan. Although the following requirements are borrowed from U.S. EPA pollution monitoring guidelines, these are generic enough that they should be included in any type of environmental sampling plan.

Quality: Discusses statistical measures of:

Accuracy (bias): Determines how data will be compared to reference values when known. Estimates overall bias of the project based on criteria and assumptions made.

Precision: Discusses the specific (sampling methods, instruments, measurements) variances and overall variances of the data or data sets when possible using relative standard deviations (%CV).

Defensible: Insures that sufficient documentation is available after the project is complete to trace the origins of all data.

Reproducible: Insures that the data can be duplicated by following accepted sampling protocols, methods of analyses, sound statistical evaluations, and so on.

Representative: Discusses the statistical principles used to insure that the data collected represents the environment targeted in the study.

Useful: Insures that the data generated meets regulatory criteria and sound scientific principles.

Comparable: Shows similarities or differences between this and other data sets, if any.

Complete: Addresses any incomplete data and how this might affect decisions derived from these data.

Adopted from Artiola et al., 2004.

---

familiar with types of sampling such as destructive and nondestructive methods. They should also know the accepted methods of analyses by consulting the latest scientific and regulatory reference manuals and books. Environmental scientists should also be familiar with the basic concepts and applications of analysis and measurement principles.

All data must be reported with appropriate degree of precision by considering the sampling methods, instrumental methods, and number of samples analyzed. Typically, the more uncertainty associated with a data value, the lower the number of digits it has. Thus, a number with 2 significant figures is less precise than a number with 4 significant figures. Often, data values are derived from various data manipulations, including statistical (such as the means of two or more measurements), method-dependent (such as dilutions), and field sampling (means of two or more samples). As a rule, precision of the final value is determined by the least precise step. For example, if an instrument measurement detects 235 $\mu g \, L^{-1}$ of total Pb in a water sample, and the sample had been diluted 5 times using a pipette with only two digits of precision, the final result should be reported as 1.2 mg L^{-1} total Pb ($0.235 \times 5.0 = 1.2$).

TABLE 12.1 Statistical symbols and their definitions.

TERM	SYMBOL	DEFINITION
Sample mean or estimated mean	$\bar{x} = \dfrac{x_1 + x_2 + \ldots + x_n}{n}$	Mean estimated from the n values $x_1, x_2, \ldots, x_n$. (Specific cases are identified by xbar, ybar, ybar.)
Mean	μ	Population or true mean. Limiting value of xbar as n becomes large or if all possible values of x_i are included.
Sample variance	$s^2 = \dfrac{\sum\limits_{i=1}^{n}\left(x_i - \bar{x}\right)^2}{n-1}$	Estimate of variance based on n values given by $x_1, x_2, \ldots, x_n$. (If all possible values of n are included, then use n in place of $n-1$ for denominator).
Variance	σ^2	True variance for the population.
Standard deviation	s, σ	Defined for s^2 and σ^2 above. (Specific cases are identified by subscripts, such as s_x and s_y).
Coefficient of variation	$CV = \dfrac{\sigma}{\mu}$ or $\dfrac{s}{\bar{x}}$	A relative standard deviation. Can also be expressed as a percentage.
Correlation coefficient	$r = \dfrac{\sum\limits_{i=1}^{n}(X_i - \bar{X})(Y_i - \bar{Y})}{(n-1)s_X s_Y} = \dfrac{b\, s_X}{s_Y}$	Linear relationship between two variables measured n times. The data pairs are $(X_1, Y_1), (X_2, Y_2), \ldots (X_n, Y_n)$. Range of r is -1 to $+1$.
Coefficient of determination	r^2	Square of r (above) for linear correlation. Range is 0 to 1.
Slope	$b = \dfrac{\sum\limits_{i=1}^{n}(X_i - \bar{X})(Y_i - \bar{Y})}{(n-1)s_X{}^2}$	Slope for linear regression for n data pairs.
Intercept	$a = \bar{Y} - b\bar{X}$	Y-intercept for linear regression for n data pairs.
Predicted value	$Y = a + bX$	Predicted value of dependent variable Y.

Modified from *Pollution Science*, 1996 Academic Press, San Diego, CA.

It is also important to report environmental data using commonly accepted units, preferably conforming to the SI (International System of Units), used worldwide. Unfortunately, in the U.S., environmental scientists and engineers often must convert units across two or more systems, a process that often leads to errors in data reporting.

12.3 STATISTICS AND GEOSTATISTICS

Statistical methods are necessary in pollution science because it is impossible to characterize all properties of an environment everywhere all of the time. Statistics are used to select **samples** from a **population** in an unbiased manner. They also help to interpret the data with the appropriate degree of confidence. **Descriptive statistics** are very useful in environmental science, because they provide a summary of the properties or characteristics of an environment. Descriptive statistics include sample means, standard deviations, and coefficients of variation (Table 12.1).

Data samples should be collected randomly or systematically from an environment. Biased sampling (usually based on convenience) will produce biased data. The range of values that can be expected from sampling an environment varies randomly, but values have a likelihood of occurrence that is defined by a probability distribution (Figure 12.1).

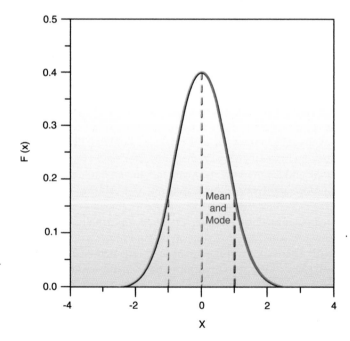

Figure 12.1 Frequency distribution *f(x)* for a normal distribution. The dashed lines show the center and $\pm$ standard deviation (σ) from the mean. (From Warrick et al., 1996). Adapted from *Environmental Monitoring and Characterization* © 2004, Elsevier Academic Press, San Diego, CA.

A reoccurring question in pollution science is how many samples (n) are needed to define the extent of pollution with some statistical certainty. This is a near impossible task, since pollution distributions are not usually known beforehand. However, we can estimate the number of samples needed, assuming that the sample standard deviation (s) from previous studies is equivalent to the population standard deviation (σ), and that the sample mean (x) is equivalent to the population mean (μ). Also we must assume a **normal distribution** (see Figure 12.1) of values and be willing to accept that our sample mean value will be within a specific confidence interval.

Thus, the number of samples (sample size n) needed may be defined as:

$$n = [(z_{\alpha/x}\sigma)/d]^2$$

where:

$z_{\alpha/x}$ is the probability function value for a given confidence interval (example: $z_{\alpha/x} \simeq 2$ for a 95% confidence interval and n > 30), and

d is the maximum tolerance or error acceptable of the mean value (x).

See Chapter 3 of Artiola et al., 2004 for a complete description of sample size estimation.

Inferential statistics are also widely used in pollution science to help infer or interpolate information from a partial or incomplete set of data. For example, regression methods can be used to correlate one or more sample attributes with a measurable quantity. This is commonly the case in analytical chemistry, where the response of an instrument (measure as light emitted or absorbed) is correlated with a particular parameter (such as concentration). For example, to calibrate a spectrophotometer for aluminum measurement, we would first dilute a certified standard several times to make a series of known solutions, as listed in Table 12.2. When we analyze these solutions on the spectrophotometer, we obtain a reading that corresponds to the concentration of aluminum in each solution. We can then plot the solution concentration (*x*-axis) against the instrument response (*y*-axis) and perform a regression (see Table 12.2). The resulting equation defines the best fit among the (*x,y*) points. Thus, we have a curve of the true values for each concentration.

We may want to estimate the concentration of a pollutant at a field location that was not sampled. In this case, spatial statistics can be used to estimate the concentration of this pollutant. For this application, the interpolation method is used to relate the location (coordinates) of the unknown sample to locations for which concentrations are known. The inverse distance weighting method considers the distances to locations (with unknown values) to be inversely related to the values at those locations. This and other more advanced methods like kriging are often used in the development of maps that show contour lines of pollutant distributions, see chapter 3 of Artiola et al., (2004) for a brief description of these methods.

TABLE 12.2 Calibration of a spectrophotometer to measure aluminum in an unknown sample.

(AL) IN STANDARD SOLUTION (mg L^{-1})	SPECTROPHOTOMETER RESPONSE (INTENSITY UNITS)
0.100	4,142
0.500	17,315
2.00	63,305
5.00	161,486
10.0	320,087
Unknown sample	250,090

Statistical Evaluation	
Linear regression equation	$y = a + bx$
y	Instrument response
x	[Al]
a	803.4
b	3.195×10^4
r^2	0.9999
[Al] in the unknown sample	7.80 mg L^{-1}

From *Pollution Science* © 1996, Academic Press, San Diego, CA.

12.4 SAMPLING AND MONITORING TOOLS

Modern data collection uses automated data acquisition methods to monitor environmental variables related to pollution. For example, urban air quality is monitored using a network of automated stations that measure pollutants such as carbon monoxide (CO) and ozone (O_3) at close intervals. Coupled with weather stations, these gas monitors produce near-realtime environmental information, which is in turn used to predict future pollution-related events.

The basic components of automated data acquisition systems are shown in Figure 12.2.

A critical component of the system is the **sensor**, which responds to an environmental stimulus such as temperature. Thermistors, for example, are able to respond to temperature changes by changing their internal resistance. The **data acquisition system (DAS)** usually includes an analog-to-digital converter used to convert analog (continuous) signals from the sensor to a digital (discrete) value that can be stored and manipulated by computer processors. Data storage and transmission systems are also needed in modern DAS to collect raw and processed data, and send it to the user.

Environmental scientists that use modern data collection systems must have a rudimentary working knowledge of electricity, computer processing, and computer programming, and be familiar with basic laws of environmental physics and chemistry.

12.4.1 Maps

Until recently, drawn maps and aerial photographs were the only means at the disposal of environmental scientist to locate places and land features and to navigate (with a magnetic compass). Today, the use of the **geographic positioning**

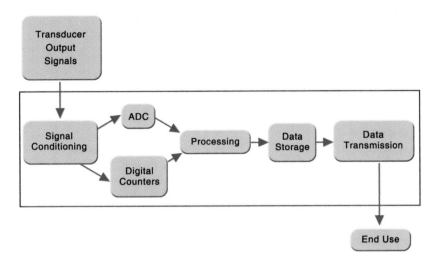

Figure 12.2 Block diagram showing the basic components of a data acquisition system (DAS). ADC = Analog to digital converter. From *Environmental Monitoring and Characterization* © 2004, Elsevier Academic Press, San Diego, CA.

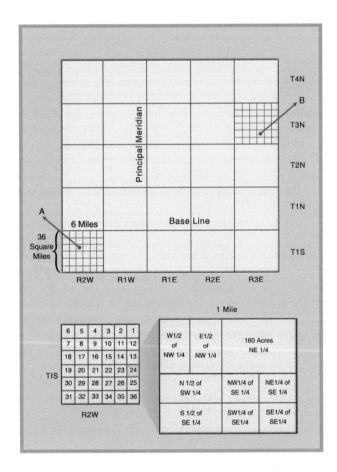

Figure 12.3 Townships, ranges, and further subdivisions of the U.S. Public Land Survey. Adapted from *Environmental Monitoring and Characterization* © 2004, Elsevier Academic Press, San Diego, CA.

system (GPS) of satellites makes it much easier to perform these tasks. However, since maps are small portable representations of portions of our environment, we must be familiar with their principles and types. There are three major types of maps: **planimetric** maps, which present scaled information in two dimensions; **topographic** maps, which add elevation information; and **thematic** maps (see GIS maps). All maps must have a scale, which is the ratio between the map's distances and the real (earth) distances. Thus, most scales are a unitless ratio or fraction of two distances. For example, if 1 cm on the map represents 1 km in reality, then the map scale is 1/100,000.

The major types of locational maps include the **latitude** and **longitude** system developed in England, which divides the earth into parallel (north-south) and meridian (east-west) sections in degrees.

In the U.S., the **Public Land Survey System** is used exclusively to locate property lines in legal documents. This system is based on the 34 points that defined the origins of principal meridians and baselines coordinates from which townships (6 square miles) and sections (1 square mile) originate (Figure 12.3).

For example, the filled square (point A) shown on Figure 12.3 represents the NW1/4 corner of section 17 of Township 1S, Range 2W (abbreviated: NW1/4,S17,T1S, R2W).

Topographic maps add relief to land maps with the use of contour lines that define equal elevations drawn at fixed elevation intervals. The U.S. Geologic Survey generates topographic maps that are used in many disciplines, including environmental science. Contour lines are very useful in identifying unique land features such as watersheds, rivers, depressions, and mountains.

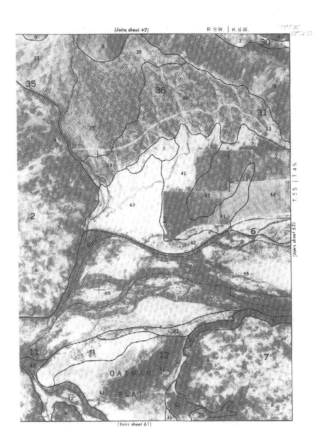

Figure 12.4 A portion of a detailed soil map showing an area along the Gila River northwest of Gila Bend, Arizona. From U.S. Department of Agriculture and Natural Resources Conservation Service, 1997. *Soil Survey of Gila Bend-Ajo Area, Arizona, Parts of Maricopa and Pima Counties.* Issued May 1997.

Soil survey maps combine land features (aerial photographs) with thematic soil data. These maps are very useful in determining the suitability (use) of land for certain activities, such as agriculture, land development, and septic

field location. These maps provide a wealth of data on soil characteristics related to pollution (such as infiltration, soil texture, and general composition), which makes them very valuable to soil and environmental scientists. An example of a soil survey map is given in Figure 12.4.

Geographic information system (GIS) maps are very useful in environmental applications, because they provide a way to arrange and present layers of data or themes as maps. Thus, GIS maps are decision-making tools that provide a visual representation of large amounts and types of data (Figure 12.5), which can be easily sorted because all the data and coordinates are in a digital form. For example, a city manager with a GIS map with overlaid soil properties could ask the question, "Show which areas in district X are suitable for septic field systems."

Because GIS uses digital data that can be manipulated using Boolean operations (binary 0 and 1 values), data can be reclassified, sorted, overlaid, and used to measure distances. GIS data is fast becoming an accessible community resource, often available on the Internet, for example, to search and locate parcels of land. The future of GIS will also include actual site visualization using digital photography and data, and virtual reality 3-D modeling.

12.4.2 Remote Sensing

Remote sensing is the use of space-based sensors, usually satellite mounted, to observe biophysical and geochemical phenomena in earth's atmosphere and land/ocean surfaces. Most remote sensors measure light from different parts of the electromagnetic spectrum that is reflected from earth's atmosphere and surface. Regions of the electromagnetic spectrum that are commonly used are listed in Table 12.3.

The UV region is commonly used to measure air pollutants (see Chapter 23). But often these gases and others like water vapor, CO, and CH_4 can interfere with land surface radiation measurements. Remote sensing imagery is captured

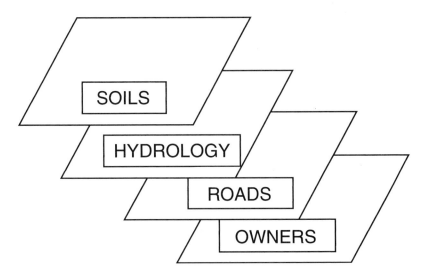

Figure 12.5 Combining data layers in a geographic information system. From *Environmental Monitoring and Characterization* © 2004, Elsevier Academic Press, San Diego, CA.

TABLE 12.3 Regions of the electromagnetic spectrum used in environmental monitoring.

SPECTRAL REGION	WAVELENGTHS	APPLICATION
Ultraviolet (UV)	0.003 to 0.4 μm	Air pollutants
Visible (VIS)	0.4 to 0.7 μm	Pigments, chlorophyll, iron
Near infrared (NIR)	0.7 to 1.3 μm	Canopy structure, biomass
Middle infrared (MIR)	1.3 to 3.0 μm	Leaf moisture, wood, litter
Thermal infrared (TIR)	3 to 14 μm	Drought, plant stress
Microwave	0.3 to 300 cm	Soil moisture, roughness

From *Environmental Monitoring and Characterization* © 2004, Elsevier Academic Press, San Diego, CA.

Figure 12.6 MODIS image showing sediment transport at the mouth of the Yellow River on February 28, 2000. Soil erosion from the Loess Plateau is proceeding at a very high rate. (Courtesy Jacques Descloitres, MODIS Land Rapid Response Team, NASA/GSFC.) From *Environmental Monitoring and Characterization* © 2004, Elsevier Academic Press, San Diego, CA.

using digital cameras that have finite optical pixels and therefore finite spatial resolution in terms of surface area per pixel. Sensors are rated in terms of coarse, medium or fine resolution. For example, high-resolution sensors are able to collect data ranging in resolution ~100–<1 m in resolution. These sensors are commonly used to study land cover patterns and changes that result from human activities, including agriculture, plant disease, urbanization, and deforestation. Coarse resolution sensors provide data ranging from 100 to <1000 m in resolution and are useful to study cloud cover, dust, and sediment transport (Figure 12.6).

Other remote-sensing applications include observing and measuring vegetation, snow cover changes, weather changes and forecasting, and physical land degradation, including erosion. Satellite imagery can also be used to study pollution migration from waste disposal sites. For example, multi-spectral data were collected from abandoned mines and other open waste disposal sites using remote sensing. Since minerals have unique spectral signatures such that they emit light in different areas of the spectrum, areas with minerals, such as metal sulfides, that are prone to fast weathering and thus have higher potential to release metals into the environment can be identified and mapped.

12.5 SOIL AND VADOSE ZONE SAMPLING AND MONITORING

The soil and vadose zones are defined as regions of porous materials below land surface that are usually not fully saturated with water. Soils, being exposed to the atmosphere, are composed of weathered, heterogeneous, unconsolidated minerals, and biological organic matter including plants, animals, and microorganisms. The heterogeneity of soils, which exists at all scales from landscape to soil particle size, arises from five soil-forming processes: parent material, climate, topography, biological activity, and time (see Chapter 2). Under the influence of all these factors, soils develop unique physical, chemical, and biological characteristics that have been used to classify soils. Classification consists of 12 orders, according to their degree of weathering and the major diagnostic horizons. (For a detailed discussion on soil classification, see Brady & Weil 1996.) Horizons in soils appear as the result of two factors, deposition and weathering. For example, clay mineral-rich (argillic) horizons develop as the results of intense chemical weathering conditions (high rainfall and heat) and the translocation of clay particles to subsurface layers.

Therefore, soil and vadose sampling strategies must consider changes in space at the meso and micro scales, and in time at geologic as well as daily-hourly time intervals. Changes such as physical weathering, which occur slowly, can be monitored at long time intervals such as years or decades. But changes such as those induced by improper waste disposal (soil and water contamination) usually require closer sampling intervals such as days to months. Soil sampling is best accomplished by considering (a priori) known soil properties. Since most counties in the U.S. have been surveyed, there is information available in Soil Survey Reports on the major soil properties and distributions of soil types by county.

Polluted soils should be sampled in an unbiased matter if the source of the pollution is not known. Systematic and

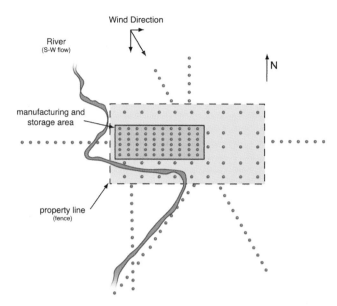

Figure 12.7 Grid and directional (exploratory) soil sampling patterns assume a pollution point of origin. Note that sampling can be done in the direction of potential pathways of pollution (transport via air and water). Dots represent sampling points. From *Environmental Monitoring and Characterization* © 2004, Elsevier Academic Press, San Diego, CA.

directional soil sampling may be more appropriate in cases where the source of pollution is known and may be correlated to spatial location (Figure 12.7).

There are numerous types of soil and vadose zone samplers, generally classified as handheld (manually operated) and mechanical (power assisted). Augers and sampling tubes are commonly used to collect soil root zone (~1.5 m depth) samples. Sampling tubes are more appropriate to collect soil

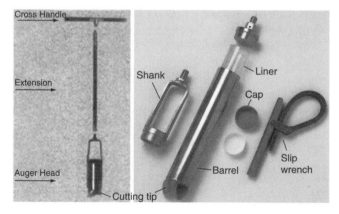

Figure 12.8 *Left*, basic hand-operated soil auger showing the auger head, an extension rod, and a handle. *Right*, details on the auger head showing cutting tip, barrel, slip wrench, and shank for attachment to extension rods. (From Ben Meadows company, a division of Lab Safety Supply, Inc.) From *Environmental Monitoring and Characterization* © 2004, Elsevier Academic Press, San Diego, CA.

samples for pollution-related measurements as they include an inert plastic or metal liner that is used to protect and store the sample (Figure 12.8).

Mechanical hollow-stem augers are commonly used to collect samples from the vadose zone. The hollow stem auger is used to drill to the sampling depth by engine power. Then a center rod that plugs the hollow stem during drilling is removed and a clean sampling tube is inserted to collect a soil sample.

Soil pore water is often sampled to monitor movement of water-soluble pollutants in the vadose zone. The use of suction lysimeters to collect soil and pore water extends to industrial municipal and agricultural waste treatment and disposal sites (landfills, ponds, septic systems, wetlands, irrigation). Porous cups made of ceramic or stainless steel may be inserted into the soil/vadose zone by drilling a hole, as shown in Figure 12.9.

The types of pollutants that can be monitored using these devices include salts, nitrates, soluble metals (like CrVI), and soluble organic carbon compounds.

12.6 GROUNDWATER SAMPLING AND MONITORING

Groundwater sampling is often conducted under the requirements of state or federal regulatory programs. Federal regulatory programs that include requirements or guidelines for groundwater monitoring include (1) **Comprehensive Environmental Response, Compensation, and Liability Act (CERCLA)**, commonly known as the "Superfund" program, which regulates characterization and cleanup at uncontrolled hazardous waste sites; (2) **Resource Conservation and Recovery Act (RCRA)**, which regulates management and remediation of waste storage and disposal sites; (3) **Safe Drinking Water Act (SDWA)**, which regulates drinking water quality for public water supplies and also includes the Underground Injection Control (UIC) Program and the Well Head Protection Program; and (4) **Surface Mining Control and Reclamation Act (SMCRA)**, which regulates permitting for open pit mining operations. In addition to the federal regulatory programs, many states have regulatory programs that are concerned with groundwater monitoring (see also Chapter 15).

Much of the groundwater sampling conducted during the past two decades has been focused on characterization and cleanup of sites where groundwater has become contaminated through spills, leaks, or land disposal of wastes (see Chapters 17 and 19). In addition to satisfying regulatory requirements, groundwater sampling programs at contaminated sites are usually conducted to obtain data necessary for making decisions on site management or cleanup. The specific objectives of the groundwater sampling program may change as the site becomes better characterized and as site remediation progresses. The frequency of sampling may also be adjusted depending on how rapidly groundwater conditions are changing at the site.

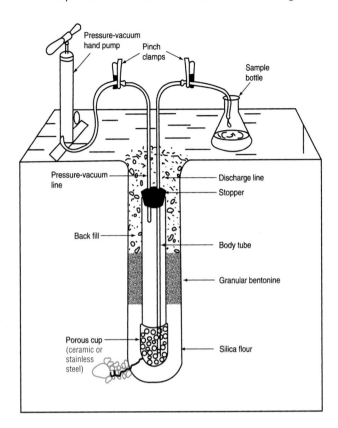

Figure 12.9 Design and operation of a pressure-vacuum lysimeter used to collect soil pore water samples at shallow depths. (After Fetter, 1999.) Modified from *Environmental Monitoring and Characterization* © 2004, Elsevier Academic Press, San Diego, CA.

Groundwater sampling is sometimes conducted to establish baseline conditions prior to development of a new facility or to characterize ambient conditions across a large area such as in basin-wide studies. Baseline sampling is usually meant to provide a "snapshot" of groundwater conditions at a particular time. The analytical suite for baseline sampling usually includes common anions and cations, and, depending on the objectives of the study, may include trace metals, pesticides, and a range of organic compounds (often those on the priority pollutant list presented in Chapter 10). If baseline sampling is being conducted prior to industrial development of a site, the analytical suite will typically include potential contaminants that may be present on the site once the facility is in operation.

Leak detection monitoring, sometimes referred to as compliance monitoring, is commonly conducted at landfills, hazardous waste storage facilities, and chemical storage and manufacturing facilities. This type of monitoring is designed to provide early warning of contamination associated with releases from these facilities. To be effective, leak detection monitoring must be conducted at wells that are properly located, generally in areas downgradient from the facility of interest. The sampling frequency and list of analytes must be adequate to detect a release before the contaminants have migrated to any sensitive downgradient receptors. Typically, upgradient monitor wells are also included in leak detection monitoring to provide baseline water quality data for groundwater moving to the site, to allow for comparative evaluations.

Groundwater monitoring is typically conducted by collecting samples from a network of monitor wells. There are many site-specific variables that must be considered in the design of a monitor well network. If the monitor wells are not located or constructed appropriately, the data collected may be misleading. For example, if monitor wells are too widely spaced or improperly screened, the zones of contaminated groundwater may be poorly defined or even completely undetected. The task of designing a monitor well network is generally more difficult for sites with more complex or heterogeneous subsurface conditions compared to sites with more homogeneous subsurface conditions. More details on this topic may be found in Chapter 8 of Artiola et al. (2004).

Although many site-specific variables must be considered in their design, most monitor wells share some common design elements (Figure 12.10). Typically, the upper section of the well is protected with a short section of steel casing called surface casing. The surface casing protects the well casing from physical damage by vehicles or from frost heaving in cold climates. The surface casing may also be grouted to provide a seal against surface water infiltration. The well casing, usually composed of steel or PVC (polyvinyl chloride), extends downward through the surface casing. Grout materials are installed around the outside of the well casing to prevent movement of surface water down the borehole. A well screen is installed in the interval where the groundwater is to be monitored. In monitor wells, the well screen is typically either wire wrapped (also known as continuous slot) or slotted. Surrounding the well screen is a gravel pack, which serves to filter out fine sand and silt particles that would otherwise enter the well through the well screen.

Many groundwater-sampling devices are currently commercially available. The use of appropriate sampling equipment is critical for implementation of a successful sampling program. There is no one sampling device that is ideal for all circumstances. The choice of sampling equipment is dependent on several site-specific factors, including the monitor well design and diameter, the depth to groundwater, the constituents being monitored, the monitoring frequency, and the anticipated duration of the sampling program.

The actual process of collecting a groundwater sample, and what is done with it after collection, is an important component of the groundwater-sampling program. The methods used for sample collection and processing depend on the type of contaminants present or suspected of being present. The primary concern of sample collection and processing is to maintain the integrity of the samples so that the concentrations of analytes are the same as when the samples were collected.

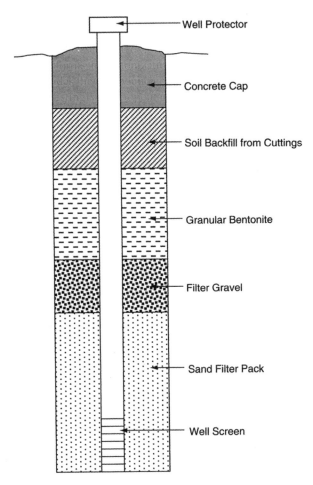

Figure 12.10 Schematic of generic monitor well design elements. From *Environmental Monitoring and Characterization* © 2004, Elsevier Academic Press, San Diego, CA.

12.7 SURFACE WATER SAMPLING AND MONITORING

Monitoring pollutants in water has become a global concern due to the limited and diminishing water resources and the negative impact of pollutants on ocean and fresh water sources. We have water standards that are protective of our health (National Drinking Water Standards) (see also Chapter 28), and that are protective of the environment by setting pollutant discharge limits from wastewaters into water sources (National Water Quality Criteria). Excessive nutrients in runoff waters, being addressed with new Ecoregional Nutrient Criteria, are also of concern, because they can produce a biological and chemical imbalance in natural surface water systems. The quality of water used in agriculture (which accounts for most of the fresh water use in the world) is also important as increased salinity and concentrations of toxic elements reduce plant yields and degrade agricultural soils. Figure 12.11 presents a list of water quality properties, including the major types of contaminants found in surface water.

Sampling the surface water environment may seem trivial at first glance, since often we assume that mixing of pollutants is complete. However, surface water bodies often exhibit significant spatial and temporal variability. For example, enclosed bodies of water such as lakes and reservoirs often develop water quality stratification due to temperature changes. These in turn affect oxygen saturation, pH, and total dissolved solids (TDS) by depth.

Field portable devices are commonly used to measure (in situ) major water quality parameters such as pH, TDS, oxygen saturation, temperature, and turbidity (see Figure 12.12).

Water sampling followed by analysis in the laboratory is common to monitor specific pollutants such as trace organic and inorganic chemicals (pesticides, hydrocarbons, toxic metals), and pathogens like *E. coli* bacteria. Field portable water quality kits, to measure basic water quality parameters and some types of specific pollutants, are becoming popular. They are convenient and inexpensive, but lack the automation and accuracy of conventional laboratory testing.

12.8 ATMOSPHERE SAMPLING AND MONITORING

Legislative efforts to control air pollution have evolved over a long period of time. Major efforts in the U. S. to control and prevent air pollution did not begin until the mid-20th century. Monitoring of air pollution is the result of legislation requiring polluters to demonstrate compliance with air quality standards and emissions limits. The legislative framework behind these requirements was, and still is, motivated largely by concerns for human health, as well as concerns about the impacts of air pollution on natural and agricultural ecosystems and global climate. Efforts by the federal government to control air pollution nationwide began in 1955 with enactment and promulgation of the Air Pollution Control Act. This act permitted federal agencies to aid state and local governments who requested assistance to carry out research on air pollution problems within their jurisdictions. This act was important in that it established the ongoing principle that state and local governments are ultimately responsible for air quality within their jurisdictions.

The overall objectives of the EPA air quality monitoring program are:

- To judge compliance with and/or progress made towards meeting ambient air quality standards;
- To activate emergency control procedures that prevent or alleviate air pollution episodes;
- To observe pollution trends;
- To provide a database for research evaluation of effects (urban, land use, transportation planning);
- The development and evaluation of abatement strategies; and
- The development and validation of diffusion models.

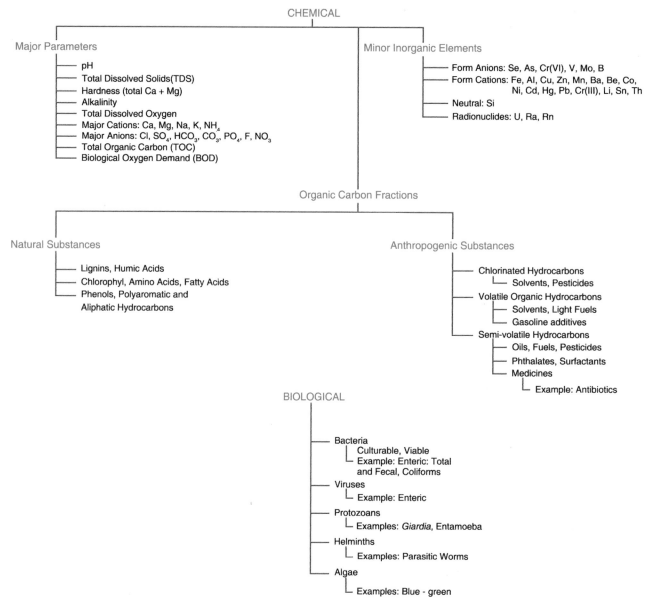

Figure 12.11 Major water properties. Adapted from *Environmental Monitoring and Characterization* © 2004, Elsevier Academic Press, San Diego, CA.

The monitoring is accomplished through the use of networks of air monitoring stations (Figure 12.13). How are networks established to meet monitoring requirements? First, the objectives of the network need to be identified. These generally include determining the highest concentrations expected to occur in the area covered by the network; determining representative concentrations in areas of high population density; determining the impact of significant sources or source categories on ambient pollution levels; and determining general background concentration levels. Site selection is obviously an important consideration when attempting to meet these objectives.

General EPA guidelines for selecting monitoring sites include locating sites in areas where pollutant concentrations are expected to be highest; locating sites in areas representative of high population density; locating sites near significant sources; locating sites upwind of the target region to establish background conditions; considering topographic features and meteorological conditions (*e.g.*, valleys, inversion frequency); and considering the effects of nearby obstacles,

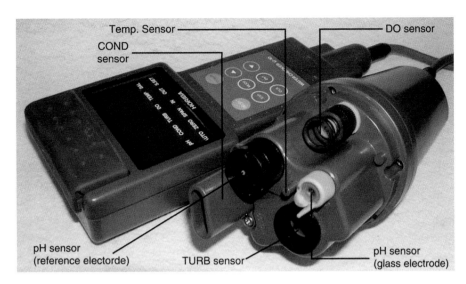

Figure 12.12 Field portable multiprobe water quality meter (manufacturer: Horiba Ltd.). The delicate multi-sensor array is covered with a protective sleeve. Each sensor requires a different level of care and periodic calibration. From *Environmental Monitoring and Characterization* © 2004, Elsevier Academic Press, San Diego, CA.

such as buildings and trees, on air flow patterns. Once the sites are selected, consideration must be given to the control of the sampling environment. An air-conditioned trailer is often necessary at each site to properly maintain analytic instruments, data recorders, and calibration equipment. This requires access to electrical power, surge protectors. and backup power.

Most monitoring stations include a standard set of meteorological instrumentation, such as a radiation-shielded thermometer to measure air temperature, an anemometer and wind vane to record wind speed and direction, and a pyranometer to measure solar radiation (needed for ozone monitoring). In addition, other specialized instrumentation is used for specific purposes (see Artiola et al., 2004).

12.9 CONCLUSIONS

Environmental monitoring is critical to the protection of human health and the environment. As the human population continues to increase, and as industrial development and energy use continues to increase, the continued production of pollution remains inevitable. Thus, the need for environmental monitoring is as great as ever. Continued advances in the development and application of monitoring devices are needed to enhance the accuracy and cost-effectiveness of monitoring programs. Equally as important is the need to produce more scientists and engineers who have the knowledge and training required to successfully develop and operate monitoring devices.

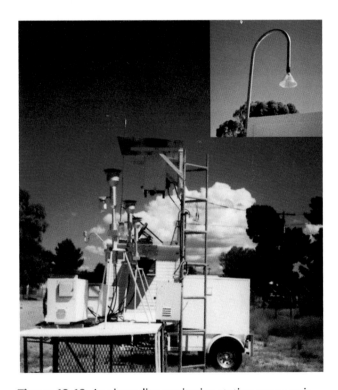

Figure 12.13 An air quality monitoring station near a residential area in Tucson, Arizona. Meteorological sensors and particulate matter samplers are visible in photograph. Note the air-sampling probe (*insert*) through which air is drawn to analytical instruments inside the trailer. (Photo courtesy A. Matthias.) From *Environmental Monitoring and Characterization* © 2004, Elsevier Academic Press, San Diego, CA.

QUESTIONS AND PROBLEMS

1. The figure below shows an area divided into 20 squares with X and O points marked. Which sampling locations appear to be random and which are systematic? Use the 20 data points lead values (mg kg^{-1}) to compute the sample mean, variance, standard deviation, and coefficient of variation values of the data with the appropriate number of significant digits (see Table 12.1). Estimate the 95% confidence interval of the data (assume that this is ~ sample mean $\bar{x} \pm 2s/\sqrt{m}$);

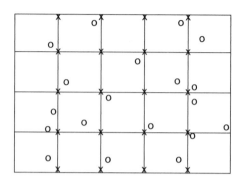

ID	LEAD (mg kg^{-1})	ID	LEAD (mg kg^{-1})
01	18.25	11	21.25
02	30.25	12	16.75
03	20.00	13	55.00
04	19.25	14	122.2
05	151.5	15	127.7
06	37.50	16	25.75
07	80.00	17	21.50
08	46.00	18	4.00
09	10.00	19	4.25
10	13.00	20	9.50

2. Using the calibration data shown in Table 12.2, plot the data points and fit a linear regression using Excel spreadsheet functions. Calculate the acid extractable aluminum concentration (in mg kg^{-1}) in a soil sample with the following information: (a) the soil acid extract had an instrument response of 110,021; and (b) the soil:acid extract ratio was 1:5 on a mass basis, that is, 10 g of soil were extracted with 50 ml of acid.

3. Using U.S. Public Land Survey notation, define the bottom half location of filled square in location B of Figure 12.3 to the nearest ¼ of a section.

4. Which regions of the electromagnetic spectrum are most useful in monitoring vegetation cover? Explain your answer.

5. Explain why in Figure 12.7 there are more sampling locations marked below than above the manufacturing plant.

6. Why has groundwater sampling and monitoring has become such an important issue in the past two decades? Explain your answer.

7. Why has surface water sampling and monitoring become such an important issue in the past two decades? Explain your answer.

8. What water quality parameters can be routinely monitored directly in the field using portable monitoring devices?

9. What is the Air Pollution Control Act and how does it impact our local air quality?

REFERENCES AND ADDITIONAL READING

Artiola J.F., Pepper I.L., and Brusseau M.L. (eds). (2004) *Environmental Monitoring and Characterization.* Elsevier Academic Press, San Diego, California.

Brady N.C. and Weil R.R. (1996) *The Nature and Properties of Soils,* 11th ed. Prentice Hall, Upper Saddle River, New Jersey.

Fetter C.W. (1999) *Contaminant Hydrogeology,* 2nd ed. Prentice Hall PTR, Upper Saddle River, New Jersey.

Pepper I.L., Gerba C.P. and Brusseau M.L. (1996) *Pollution Science.* Academic Press, San Diego, California.

CHAPTER 13

ENVIRONMENTAL TOXICOLOGY

C.P. Gerba

Flow through system for fish toxicity testing.
Source: Broxham Environmental Laboratory (www.brixham-lab.com)

13.1 HISTORY OF MODERN TOXICITY IN THE UNITED STATES

In **toxicology,** we study both the adverse effects of chemicals on health and the conditions under which those effects occur. A natural outgrowth of biology and chemistry, toxicology began to assume a well-defined shape in just the past four to five decades. Newer still, environmental toxicology is concerned with the effects of chemical contaminants on various ecological systems, both large and small. Here, we focus on some of the basic principles of toxicology as related to environmental contaminants.

The first federal law for regulating potentially toxic substances was the Pure Food and Drug Act, passed by Congress in 1906. Much of the impetus for this law came from the work of Harvey Wiley, who was the chief chemist of the Department of Agriculture under Theodore Roosevelt. Wiley and his "Poison Squad" had a very personal interest in their work—he and his team of chemists not infrequently dosed themselves with suspect chemicals to test for their deleterious effects (Rodricks, 1992).

The systematic study of toxic effects in laboratory animals (other than chemists) began in the 1920s, spurred by concerns about the unwanted side effects of food additives, drugs, and pesticides. (In this era DDT and related pesticides first became available.) The 1930s saw issues raised about occupational cancers and other chronic diseases resulting from chemical exposures. These concerns and issues prompted increased legislative activity culminating in the modern version of the Food, Drug, and Cosmetic Act. This law, enacted by Congress in 1938, was passed in response to a tragic episode in which more than 100 people died from acute kidney failure after ingesting contaminated sulfanilamide—the antibiotic had been improperly prepared in a diethylene glycol solution (Rodricks, 1992).

The real growth of toxicology largely paralleled that of the chemical industry, especially after World War II. The thousands of new compounds produced by chemical manufacturers created a need for information about their possible harmful effects. This growth received a significant stimulus from public opinion. Sporadically during the 1940s and 1950s, the public was presented with a series of seemingly unconnected announcements about poisonous pesticides in their foods, food additives of dubious safety, chemical disasters in the workplace, and air pollution episodes that claimed thousands of victims in urban centers throughout the world. Then, in 1962, marine biologist Rachel Carson (1907–1964) drew together these various environmental horror stories in her book, *Silent Spring.* Menaced by the presence of synthetic chemical killers in the environment, the public responded with a predictable howl of outrage, which (among other things) fostered renewed interest in the science of toxicology. It also helped pave the way for the introduction of several major federal environmental laws in the late 1960s and early 1970s, and for the creation of the EPA in 1970.

The relative newness of the science of toxicology is reflected in the fact that, even today, we have little solid information about the toxicity of a large number of chemicals. Of the 6,000,000 known chemicals, about 50,000 are in common use, and detailed chronic toxicity tests have been performed on only a few hundred of these. Even for those that have been tested, many questions remain about the interpretation of the results obtained, including serious reservations about the applicability of laboratory test results to human populations in everyday situations. In many cases, then, we lack a basic understanding of how toxicants act.

13.2 TOXIC VERSUS NONTOXIC

The term *safe* commonly means "without risk." But this common definition has no meaning in scientific study. Scientists cannot ascertain conditions under which a given chemical exposure is absolutely without risk of any type. Conversely, they can describe conditions under which risks are so low that they have no practical consequence to a specific population. In technical terms, the safety of chemical substances—whether in food, drinking water, air, or the workplace—has typically been defined as a condition of exposure under which there is a "practical certainty" that no harm will result to exposed individuals. In terms of mortality, this is usually accepted as a risk of 1:1,000,000 chance of dying during a lifetime (see Chapter 14).

Another fundamental concept is the classification of chemical substances as either *safe* or *unsafe* (or as *toxic* and *nontoxic*). This type of classification can be highly problematic. All substances, even those that we consume in high amounts every day, can be made to produce a toxic response under some conditions of exposure. In this sense, all substances can be "toxic." Thus, safety involves not simply the degree of toxicity of a substance, but rather the degree of risk under given conditions. In other words, we ask, "What is the probability that the toxic properties of a chemical will be expressed under actual or anticipated conditions of human or animal exposure?" The science of risk assessment attempts to link toxicological information on adverse effects to the probability of toxic effects during likely exposure scenarios (see Chapter 14).

13.3 EXPOSURE AND DOSE

Humans and other organisms can be exposed to substances in different environmental media—air, water, soil, or food—or they may have direct contact with a sample of the substance. The **exposure concentration** is the amount of a substance present in the medium with which an organism has contact. The **dose** is the amount of the chemical that is received by the target (organ). The exposure concentration may differ from the dose owing to biochemical transformations in living organisms.

Suppose, for example, a substance is present in drinking water. The amount of this substance in the water is the exposure concentration. For many environmental substances, this amount ranges from less than 1 microgram (μg) to greater than 1 milligram (mg), and is usually reported as milligrams or micrograms of the substance present in 1 liter of water (*i.e.*, in mg L^{-1} or μg L^{-1}).[1]

An individual's intake—or *dose*—of this substance depends on the amount present in a given volume of water and on the amount consumed in a given period of time. Given the concentration of the substance in water (say, in ppm) and the human consumption of water per unit of time, it is possible to estimate the total amount of the substance an individual will consume through use of contaminated water. For instance, adults are assumed to consume 2 L of water each day through all uses (see Table 14.4). Thus, if a substance is present at 10 mg L^{-1} (= 10 ppm) in water, the average daily individual intake of the substance is

$$10 \text{ mg } L^{-1} \times 2 \text{ L day}^{-1} = 20 \text{ mg day}^{-1}$$

Toxicity measures must also take body size differences into account, usually by dividing daily intake by the weight of the individual. That is, the toxicity of a substance is usually dependent upon concentration per unit of body weight. Thus, for a man of average weight (usually assumed to be 70 kg), the daily dose of this substance is

$$20 \text{ mg day}^{-1}/70 \text{ kg} = 0.29 \text{ mg kg}^{-1} \text{ day}^{-1}$$

For a person of lower weight, such as a female or child, the **daily dose** at the same intake rate would be larger. For example, a 50-kg woman ingesting this substance would receive a dose of

$$20 \text{ mg day}^{-1}/50 \text{ kg} = 0.40 \text{ mg kg}^{-1} \text{ day}^{-1}$$

Using the same equation, a child of 10 kg would receive a dose of 2.0 mg kg^{-1} day^{-1}. However, children drink less water each day than do adults (say, 1 L), so a child's dose would be

$$10 \text{ mg } L - 1 \times 1 \text{ L day}^{-1}/10 \text{ kg} = 1.0 \text{ mg kg}^{-1} \text{ day}^{-1}$$

In general, the smaller the body size, the greater the dose (in mg kg^{-1} day^{-1}) received from drinking water. This is also true of experimental animals. Usually rats or mice will receive a much higher dose of drinking water *contaminants* than humans because of their much smaller body size.

Because each medium (air, soil, water) of exposure must be treated separately, some calculations are more complex than those of dose per liter of water. Many calculations may simply be additive. For example, a human may be simultaneously exposed to the same substance through several media (*e.g.*, through inhalation, ingestion, and dermal contact). Thus, if an individual can both ingest and inhale (say, in the shower) some volatile compound in tap water, the

total dose received by that individual is the sum of doses received through each individual route. In some cases, however, it is inappropriate to add doses in this fashion because the toxic effects of a substance may depend on the route of exposure. For example, inhaled chromium is carcinogenic to the lung, but it appears that ingested chromium is not. In general, though, as long as a substance acts at an internal body site (*i.e.*, acts systematically rather than at a particular point of initial contact), it is usually acceptable procedure to add doses received from several routes.

Absorption, or **absorbed dose**, is another factor that requires special attention when considering dose and exposure. When a substance is ingested in food or drinking water, it enters the gastrointestinal tract. When it is present in air (*e.g.*, as a gas, aerosol, particle, dust, or fume), it enters the upper airways and lungs. A substance may also come into contact with the skin and other body surfaces as a gas, liquid, or solid. Some substances may cause toxic injury at the point of initial contact (the skin, gastrointestinal tract, upper airways, lungs, or eyes). Indeed, at high concentrations, most substances do cause at least irritation at these points of contact. However, for many substances, toxicity occurs after they have been absorbed, that is, after they pass through certain barriers (*e.g.*, the wall of the gastrointestinal tract or the skin itself), enter blood or lymph, and gain access to the various organs or systems of the body. Some chemicals may be distributed in the body in various ways and then excreted. However, some chemical types—usually lipid-soluble substances such as the pesticide dichlorodiphenyl-trichloroethane (DDT)—can be stored for long periods of time, usually in body fat.

Substances vary widely in extent of absorption. The fraction of a dose that passes through the wall of the gastrointestinal tract may be very small (1 to 10% for some metals) or it may be substantial (close to 100% for certain types of organic molecules). Absorption rates also depend on the medium in which a chemical is present; a substance present in water might be absorbed differently than the same substance present in, say, a fatty diet. Absorption rates also vary among animal species and among individuals within a species. Ideally, an estimation of a **systemic dose** should consider absorption rates. Unfortunately, data on absorption are limited for most substances, especially in humans, so absorption is not always included in dose estimation. In some cases, dose estimates may be crudely adjusted on the basis of the molecular characteristics of a particular substance and/or general principles of absorption. In many cases, however, absorption is simply considered to be complete by default.

The technique of **extrapolation**, or drawing inferences, from experimentally observed results can also be a major factor in predicting the likelihood of toxicity, say, from one route of exposure to other routes, or from one organism to another. Experiments for studying toxicity typically involve intentional administration of substances to subjects (usually mice or rats) through ingested food or inhaled air, or through direct application to skin. In other cases, they may include other routes of administration, such as injection under the

---

[1] These two units are sometimes expressed as the more ambiguous units of parts per million (ppm) or parts per billion (ppb), respectively.

skin (subcutaneous), into the blood (usually intravenous), or into body cavities (intraperitoneal). Such toxicity studies in experimental animals are of greatest value when experimental exposures mimic the mode of human exposure. Thus, if both animals and humans are exposed to a contaminant via drinking water, it is generally assumed that the data in animals can be applied directly to humans. But when experimental routes differ from human routes (*e.g.*, animal exposure via injection; human exposure via drinking water), a correction or safety factor must be used to apply such data to human exposures (see Chapter 14.2.3).

13.4 EVALUATION OF TOXICITY

Information on the toxic properties of chemical substances is obtained through plant, bacterial, and animal studies; controlled epidemiological investigations of exposed human populations or microcosms; and clinical studies or case reports of exposed humans or ecosystem studies (*e.g.*, oil spills). Other information bearing on toxicity derives from experimental studies in systems other than whole animals (*e.g.*, isolated organics in cells or subcellular components) and from analysis of the molecular structures of the substances of interest. These last two sources of information are generally less certain as indicators of toxic potential.

Many types of toxicity studies can be conducted to identify the nature of health damage produced by a substance and the range of doses over which such damage is produced. Each of the many different types of toxicological studies has a different purpose. The usual starting point for such investigations is a study of the **acute (single-dose) toxicity** of a chemical in experimental animals, plants, or bacteria. Acute toxicity studies are used to calculate doses that will not be lethal to any organism and can be used in toxicity studies of longer duration. Moreover, such studies provide an estimate of the compound's comparative toxicity and may indicate the target organ system (*e.g.*, kidney, lung, or heart) affected in an animal. Once the acute toxicity is known, organisms may be exposed repeatedly or continuously for several weeks or months in **subchronic toxicity** (Table 13.1) studies, or for close to their full lifetimes in **chronic toxicity** (Table 13.2) studies.

When toxicologists examine the lethal properties of a substance, they estimate its **LD$_{50}$**, which is the lethal dose for 50% of an exposed population (Figure 13.1). A group of well-known substances and their LD$_{50}$ values are listed in Table 13.3. LD$_{50}$ studies reveal one of the basic principles of toxicology:

• *Not all individuals exposed to the same dose of a substance will respond in the same way.*

Thus, the same dose of a substance that leads to the death of some experimental individuals will impair other organisms and not affect other organisms at all.

The premise underlying animal toxicity studies is the long-standing assumption that effects in humans can be inferred from effects observed in animals. This principle of extrapolating animal data to humans has been widely accepted in scientific and regulatory communities. For instance, all of the chemicals that have been demonstrated to be carcinogenic in humans are carcinogenic in some, although not all, animal species typically used for toxicological studies. In addition, the acutely toxic doses of many chemicals are assumed to be similar in humans and a variety of experimental animals. This inference is based on the evolutionary relationships between animal species. That is, at least among mammals, the basic anatomical, physiological, and biochemical parameters are expected to be much the same across species.

On the whole, the general principle of making such **interspecies inferences** is well founded. But some exceptions have been noted; for example, guinea pigs are much

---

TABLE 13.1 Subchronic toxicity tests are employed to determine toxicity likely to arise from repeated exposures of several weeks to months.

Species

Rodents (*usually rats*) preferred for oral and inhalation studies; rabbits for dermal studies; non-rodents (*usually dogs*) recommended as a second species for oral tests

Age

Young adults

Number of animals

10 of each sex for rodents, 4 of each sex for non-rodents per dose level

Dosage

Three dose levels plus a control group; include a toxic dose level plus NOAEL*; exposures are 90 days

Observation period

90 days (*same as treatment period*)

*See Section 13.4.6.

TABLE 13.2 Chronic toxicity tests determine toxicity from exposure for a substantial portion of a subject's life.

Species

Two species recommended; rodent and non-rodent (*rat and dog*)

Age

Young adults

Number of animals

20 of each sex for rodents, 4 of each sex for non-rodents per dose level

Dosage

Three dose levels recommended; includes a toxic dose level and NOAEL*; exposures generally for 12 months; FDA requests 24 months for food chemicals

Observation period

12–24 months

*See Section 13.4.6.

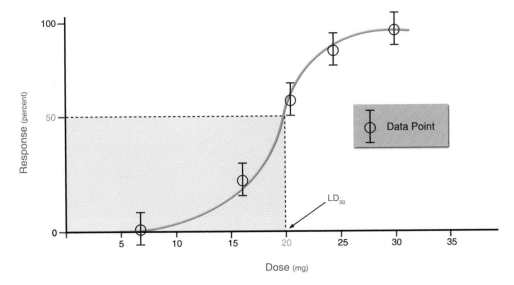

Figure 13.1 LD$_{50}$ is the dose (in mg) lethal to 50% of the animals administered the dose.

more sensitive to dioxin (2,3,7,8-tetrachlorodibenzo-*p*-dioxin) than are other laboratory animals. Many of these exceptions arise from differences in the ways various species handle exposure to a chemical and to differences in **pharmacokinetics**, which includes the rates at which a specific chemical is distributed among tissues, the manner in which it is excreted, and the types of metabolic changes it causes. Because of these potential differences, it is essential to evaluate all interspecies differences carefully when inferring human toxicity from animal toxicological studies.

In the particular case of long-term animal studies conducted to assess the carcinogenic potential of a compound, certain general observations increase the overall strength of the evidence that the compound is carcinogenic. Thus, for example, an increase in the number of tissue sites affected by

the agent is a strong indicator of carcinogenicity, as is an increase in the number of animal species, strains, and sexes showing a carcinogenic response. Other observations that affect the strength of the evidence may involve a high level of statistical significance of the increase of tumor incidence in treated versus control animals, as well as clear-cut dose–response relationships in the data evaluated, such as dose-related shortening of the time-to-tumor occurrence or time-to-death with tumor and a dose-related increase in the proportion of tumors that are malignant.

13.4.1 Manifestations of Toxicity

Toxic effects can take various forms. A toxic effect can be immediate, as in strychnine poisoning, or delayed, as in lung cancer. Indeed, cancer typically affects an individual many years after continuous or intermittent exposure to a carcinogen. An effect can be local (*i.e.*, at the site of application) or systemic (*i.e.*, carried by the blood or lymph to different parts of the body). When examining toxic effects, there are several important factors to consider, some of which are dosage-related.

1. The *severity* of injury can increase as the dose increases and vice versa. Some organic chemicals, for example, are known to affect the liver. High doses of such a chemical (*e.g.*, carbon tetrachloride) will kill liver cells—perhaps killing so many cells that the whole liver is destroyed, so that most or all of the experimental animals die. As the dose is lowered, fewer cells are killed, but the liver exhibits other forms of injury that indicate impairment in cell function and/or structure. At still lower doses, no cell deaths may occur, and only slight changes are observed in cell function or structure. Finally, a dose may be so low that no effect is observed or the biochemical alterations that are present have no known adverse effects on the health of the animal. One

TABLE 13.3 Approximate oral LD$_{50}$ in a species of rat for a group of well-known chemicals.

CHEMICAL	LD$_{50}$ (mg kg^{-1})
Sucrose (table sugar)	29,700
Ethyl alcohol	14,000
Sodium chloride (common salt)	3,000
Vitamin A	2,000
Vanillin	1,580
Aspirin	1,000
Chloroform	800
Copper sulfate	300
Caffeine	192
Phenobarbital, sodium salt	162
DDT	113
Sodium nitrite	85
Nicotine	53
Aflatoxin B1	7
Sodium cyanide	6.4
Strychnine	2.5

From U.S. EPA, 1989.

of the goals of toxicity studies is to determine this dose level, known as the **no-observed-adverse-effect level (NOAEL)** (see Chapter 14.2.3).

2. The *incidence* of an effect, but not its severity, may increase with increasing dosage. In such cases, as the dose increases, the fraction of experimental organisms experiencing diverse effects (*i.e.*, disease or injury) increases. At sufficiently high doses, all experimental subjects will experience the effect. Thus, increasing the dose increases the probability (*i.e.*, the risk) that an abnormality will develop in an exposed population.

3. Both the severity and the incidence of a toxic effect may increase as the level of exposure increases. The increase in severity is a result of increased damage at higher doses, while the increase in incidence is a result of differences in individual sensitivity. In addition, the site at which a substance acts (*e.g.*, liver, kidney) may change as the dosage changes. Many toxic effects, including cancer, fall in this category. Generally, as the duration of exposure increases, the critical NOAEL dose decreases; in some cases, new effects not seen with exposures of short duration appear after long-term exposure.

4. The *seriousness* of a toxic effect must also be considered. Certain types of toxic damage, such as asbestosis caused by inhalation of asbestos fibers, are clearly adverse and are a definite threat to health. However, the health significance of other types of effects observed during toxicity studies may be ambiguous. For example, at a given dose, a chemical may produce a slight increase in body temperature. If no other effects are observed at this dose, researchers cannot be sure that a true adverse response has occurred. Determining whether such slight changes are significant to health is one of the critical issues in assessing safety.

5. Toxic effects also vary in degree of reversibility. In some cases, an adverse health effect will disappear almost immediately following cessation of exposure. At the other extreme, some exposures will result in a permanent injury—for example, a severe birth defect arising from exposure to a substance that irreversibly damaged the fetus at a critical moment of its development. Furthermore, some tissues, such as the liver, can repair themselves relatively quickly, while others, such as nerve cells, have no ability to repair themselves. Most toxic responses fall somewhere between these extremes.

13.4.2 Toxicity Testing

Any organism can be used to assess the toxicity of a substance. The choice of test organism depends on several factors, including budget, time, and the organism's occurrence in a given environment. The simplest and least costly tests are performed with unicellular animals or plants and may last only a few hours or days. In a water environment, for example, *Daphnia* or algae may be used in testing for potential aquatic pollutants. Short-term tests may look at the death or

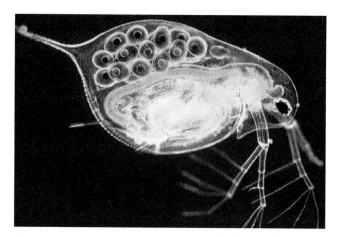

Figure 13.2 *Ceriodaphnia dubia,* an invertebrate used for toxicity testing. From South Dakota Department of Natural Resources (www.state.5d.us).

immobilization of swimming *Daphnia* (Figure 13.2), while longer term tests may look at the growth of the organisms (increase in biomass) or numbers of offspring. Animals higher in the food chain are also important in aquatic toxicity tests, and experiments involving fish, amphibians, and other macroinvertebrates are familiar standbys (Table 13.4). For terrestrial toxicity tests, higher plant, rodent, or bird toxicity tests can be used. Strains of genetically characterized rats and mice, for example, are often used in such studies. Avian toxicity tests have also been developed; for instance, birds are frequently used in evaluating the effects of pesticides on nontarget species.

In environmental toxicology, researchers often perform toxicity tests in artificially contained communities to assess the environmental impacts of toxic substances after release into the environment. These artificial communities, which serve as laboratory models of natural ecosystems, are

TABLE 13.4 Organisms commonly used in toxicity testing.

TYPES OF ORGANISMS	ORGANISM
Invertebrates	*Daphnia magna*
	Crayfish
	Mayflies
	Midges
	Plandria
Aquatic Vertebrates	Rainbow Trout
	Goldfish
	Fathead minnow
	Catfish
Algae	*Chlarnydomonas reinhardi* (green algae)
	Microcystis aeruginosa (blue-green algae)
Mammals	Rats
	Mice
Avian Species	Bobwhite
	Ring-necked pheasant

From *Pollution Science* © 1996, Academic Press, San Diego, CA.

Figure 13.3 Laboratory microcosms. (www.tamut.edu/~allard/ Biology/labs/general_ecology/fall2001.htm).

referred to as **microcosms**. While many microcosms are elaborate systems that effectively mimic whole ecosystems, some microcosms may be nothing more than a set of glass jars containing soil or water with sediment at the bottom (Figure 13.3). But even simple glass jars allow researchers to examine the effect of substances on multi species, such as algae, bacteria, and microinvertebrates.

Toxicity experiments vary widely in design and protocols. Some tests and research-oriented investigations are conducted using prescribed study designs, as is the case with carcinogenicity assays in fish. In connection with premarket testing requirements for certain classes of chemicals, however, regulatory and public agencies have developed relatively few standardized tests for various types of toxicity.

Rats and mice are the most commonly used laboratory animals for toxicity testing. These rodents are inexpensive and can be handled relatively easily; moreover, the genetic background and disease susceptibility of these species are well established. In addition, the full life span of these small rodents is complete in 2 to 3 years; thus, the effects of lifetime exposure to a substance can be measured relatively quickly. Other rodents such as hamsters and guinea pigs are also common laboratory subjects, as are rabbits, dogs, and primates. Usually, the choice of experimental animal depends on the system being studied. Reproductive studies, for example, often use primates such as monkeys or baboons because their reproductive systems are similar to that of humans. Similarly, rabbits are often used for testing dermal toxicity because their shaved skin is more sensitive than that of other animals.

Animals are usually exposed by a route that is as close as possible to the route by which humans will be exposed. In some cases, however, it may be necessary to use other routes or conditions of dosing to achieve the desired experimental dose. For example, some substances are administered by stomach tube (gavage) because they are too volatile or unpalatable to be placed in the animals' feed at the high levels needed for toxicity studies.

A toxicity experiment is of limited value unless researchers find a dose of sufficient magnitude to cause some type of adverse effect within the duration of the experiment. If no effects are seen at any dose administered, the toxic properties of the substance cannot be characterized; thus, experiments may be repeated at higher doses or for longer times until distinct adverse effects are observed. The most distinctive adverse effect is, of course, death. Therefore, researchers frequently begin their experiments by determining the LD_{50}, since the endpoint of this experiment (death) is easily measured. Next, researchers usually look at the effects of lower doses administered over longer periods to find the range of doses over which adverse effects occur and to identify the NOAEL for these effects.

Studies may be characterized according to the **duration of exposure**. Acute toxicity studies involve a single dose or exposures of very short duration (*e.g.*, 8 hours of inhalation). Chronic studies involve exposures for nearly the full lifetime of the experimental animals, while subchronic studies vary in duration between these two extremes. Although many different dose levels are needed to develop a well-characterized dose–response relationship, practical considerations usually limit the number to two or three, especially in chronic studies. Experiments involving a single dose are frequently reported, but these leave great uncertainty about the full range of doses over which effects are expected.

13.4.3 Toxicity Tests for Carcinogenicity

One of the most complex and important of the specialized tests is the **carcinogenesis bioassay**. This type of experiment is used to test the hypothesis of carcinogenicity—that is, the capacity of a substance to produce malignant tumors.

Usually, a test substance is administered over most of the adult life of a laboratory animal, then the animal is observed for formation of tumors. In this kind of testing, researchers generally administer high doses of the chemical to be tested—specifically, the **maximum tolerated dose (MTD)**, which is the maximum dose that an animal can tolerate for a major portion of its lifetime without significant impairment of growth or observable toxic effect other than carcinogenicity. The MTD and one-half of that, or MTD_{50}, are the usual doses used in a National Cancer Institute (NCI) carcinogenicity bioassay, so that the animals survive in relatively good health over their normal lifetime. The main reason for using the MTD as the highest dose in a bioassay is that these very high doses help to overcome the statistical insensitivity inherent in small-scale experimental studies.

Owing largely to cost considerations, experiments are carried out with relatively small groups of animals—typically, 50 or 60 animals of each species and sex at each dose level (Table 13.5), including the control group. At the end of such an experiment, the incidence of cancer (including tumor incidence in control animals) is tabulated and plotted as a function of dose. Then the data are analyzed to determine whether any observed differences in tumor incidence (the fraction of animals having a tumor of a certain

TABLE 13.5 Carcinogenicity tests are similar to chronic toxicity tests. However, they extend over a longer period of time and require larger groups of animals.

Species

Testing in two rodent species, the rat and mouse, preferred due to relatively short life spans

Age

Young adults

Number of animals

50 of each sex per dose level

Dosage

Three dose levels recommended; highest should produce minimal toxicity; exposure periods are at least 18 months for mice and 24 months for rats

Observation period

12–24 months for mice and 24–30 months for rats

type) are due to exposure to the substance under study or to random variations. In an experiment of this size, assuming none of the control animals develop tumors, the lowest incidence of cancer that is detectable with statistical reliability is in the range of 5%, which is equivalent to 3 out of 60 animals developing tumors. If control animals develop tumors (as they frequently do), the lowest range of statistical sensitivity is even higher. A cancer incidence of 5% is very high; but ordinary experimental studies are not capable of detecting lower rates, and most are even less sensitive.

13.4.4 Epidemiological Studies

Information on adverse health effects in human populations is obtained from four major sources: (1) summaries of self-reported symptoms in exposed persons; (2) case reports prepared by medical personnel; (3) correlation studies, in which differences in disease rates in human populations are associated with differences in environmental conditions; and (4) epidemiological studies. The first three of these sources are characterized as descriptive epidemiology, while the fourth category—**epidemiological studies**—is generally reserved for studies that compare the health status of a group of persons who have been exposed to a suspected agent with that of a nonexposed control group. (*Note:* Such studies cannot identify cause-and-effect relationships between exposure to a substance and particular diseases or conditions; however, they can draw attention to previously unsuspected problems and generate hypotheses that can be tested further.)

Most epidemiological studies are either case-control studies or cohort studies. **Case-control studies** (Figure 13.4)

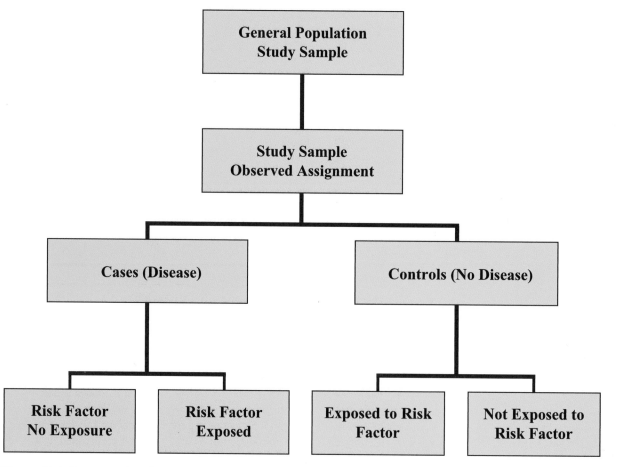

Figure 13.4 Case control study design.

first identify a group of individuals who have a specific disease, then attempt to ascertain commonalities in exposures that the group may have experienced. For example, the carcinogenic properties of diethylstilbestrol (DES), a drug once used to prevent miscarriages, were brought to light through studies of women afflicted with certain types of vaginal and cervical cancer. **Cohort studies** (Figure 13.5), on the other hand, begin by examining the health status of individuals known to have had a common exposure. These studies then attempt to determine whether any specific condition is associated with that exposure by comparing the exposed group's health with that of an appropriately matched control population. For example, a cohort study of lab workers exposed to benzene revealed an excessively high incidence of leukemia, thereby providing strong evidence in support of a benzene leukemogenesis hypothesis. Generally, epidemiologists have used individuals who belong to an identifiable group, such as those in certain occupational settings or patients treated with certain drugs, to conduct such studies—hence the name "cohort."

Convincing results from epidemiological investigations can be enormously beneficial, because the data provide information about humans under actual conditions of exposure to a specific agent. Therefore, results from well-designed, properly controlled studies are usually given more weight than results from animal studies. Although no study can provide complete assurance that a chemical is harmless, negative data from epidemiological studies of sufficient size can assist in establishing the maximum level of risk due to exposure to the agent.

Obtaining and interpreting epidemiological results, however, can be quite difficult. Appropriately matched control groups are difficult to identify, because the factors that lead to the exposure of the study group (*e.g.*, occupation or residence) are often inextricably linked to the factors that affect health status (*e.g.*, lifestyle and socioeconomic status). Thus, controlling for related risk factors (*i.e.*, cigarette smoking) that have strong effects on health is difficult. Moreover, the statistical detection power of epidemiological studies depends on the use of very large populations; thus, data may be hard to come by or incomplete. Few types of health effects—other than death—are recorded systematically in human populations, and even the information on cause of death is limited in reliability. For example, infertility, miscarriages, and mental illness are not, as a rule, systematically recorded by public health agencies, while death is often attributed to "heart failure," whatever its proximate cause.

In addition, accurate data on the degree of exposure to potentially hazardous substances are only rarely available, especially when exposures have taken place years or decades

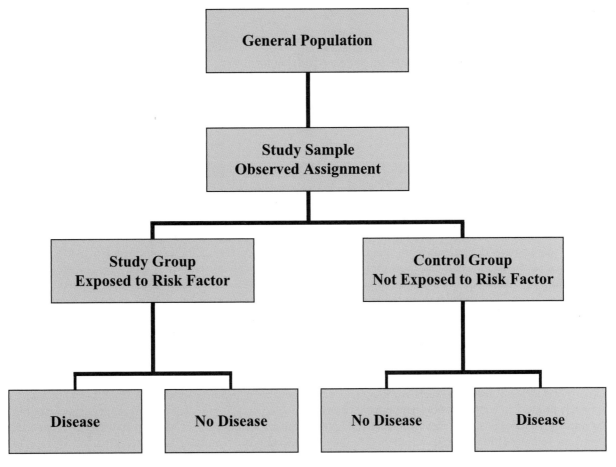

Figure 13.5 Cohort study design.

earlier. Establishing dose–response relationships is therefore frequently impossible. Nor can current data, however carefully obtained, immediately help researchers who are investigating slowly developing diseases such as cancer; rather, epidemiologists must wait many years to determine the absence of an effect. Meanwhile, exposure to suspect agents could continue during these extended periods of time, thereby increasing risk further.

For these and other reasons, interpretations of epidemiological studies are sometimes subject to extreme uncertainties. Independent confirmatory evidence is usually necessary, including supporting results from a second epidemiological study or supporting data from experimental studies in animals. Such confirmatory evidence is particularly necessary in the case of negative findings, which must be interpreted with great caution (EPA, 1989).

For example, suppose we have a drinking-water contaminant that is known to cause cancer in 1 out of every 100 people exposed to 10 mg L^{-1}. Further suppose that the average time required for cancer to develop from 10 mg L^{-1} of exposure is 30 years (which is not uncommon for a carcinogen). After our townspeople have been exposed to the drinking-water contaminant for 15 years, we conduct a study. For this study, we collect death certificates of 20 people exposed to the contaminant, but we have little information on actual exposure. We know that some of the deceased were exposed when the contaminant was first introduced into the water supply and that others were exposed several years later. When we turn to the health records, we find that they are incomplete. Finally, the results of our study reveal that 20 cancer deaths is not an excessive number when compared to an appropriate control group. Is it then correct for us to conclude that our known carcinogen is not carcinogenic?

13.4.5 Short-Term Tests for Toxicity

The lifetime animal study is the primary method used for detecting the carcinogenic properties and general toxicity of a substance. Short-term tests for toxicity, however, are used to measure effects that appear to be correlated with specific toxic effects. For example, those for carcinogenicity include assays for gene mutations in bacteria, yeast, fungi, insects, and mammalian cells; mammalian cell transformation assays; assays for DNA damage and repair, and *in vitro* (outside the animal) or *in vivo* assays (within the animal) for chromosomal mutations in animal cells. There are also a number of short-term toxicity assays that are based on inhibiting the functions of necessary enzymes in organisms, such as ATPases, phosphatases, and dehydrogenase. Phosphatase measurements, for example, can be used to assess the activity of specific substances, such as the toxicity of heavy metals in soils. In addition, short-term bioassays can use enzymes or microorganisms to assess general toxicity of environmental samples (Bitton, 1999).

Several tests involving whole animals are also available. These tests, which are usually of intermediate duration,

include the induction of skin and lung tumors in female mice, breast cancer in certain species of female rats, and anatomical changes in the livers of rodents.

Many carcinogenic (cancer-causing), mutagenic (mutation-causing), and teratogenic (defect-causing) agents act in the same way: they cause changes in DNA that eventually affect cell development. Because of this relationship, initial screening for such substances can often be accomplished quickly by testing their capacity to cause mutations in a particular strain of bacteria—*Salmonella typhimurium*. This unique strain of bacteria requires the essential amino acid histidine to grow, so it can only grow on histidine-free media if it has first mutated. Thus, if we observe these bacteria growing on histidine-free media after exposure to a test chemical, we can safely assume that the chemical caused the bacteria to mutate. The test chemical is therefore likely to be a carcinogen, mutagen, or teratogen. This short-term test is called the **Ames test** after its developer, Dr. Bruce Ames of the University of California at Berkeley (Figure 13.6).

Another short-term test is a bioassay based on the light output of the bioluminescent bacterium, *Photobacterium phosphoreum*. This bioassay has been used to assess the general toxicity of wastewater effluents, industrial wastes, sediment extracts, and hazardous waste leachates. As toxic substances diminish the viability of the bacterium, bioluminescent activity decreases, and the light output can be quantitatively measured by an instrument (Figure 13.7).

13.4.6 Threshold Effects

Commonly accepted theory suggests that most biological effects of noncarcinogenic chemical substances occur only after a certain concentration or level is achieved. This level, known as the threshold dose, is approximated by the NOAEL (see Chapter 14.2.3). Another widely accepted premise, at least in the setting of public health standards, is that the human population is likely to have a much more variable response to toxic agents than do the small groups of well-controlled, genetically homogeneous animals that are routinely used in experiments. The NOAEL is itself subject to some uncertainty owing to variabilities in the data from which it was obtained. For these reasons, public health agencies divide experimental NOAELs by large uncertainty factors, known as safety factors, when examining substances that display threshold effects (see Chapter 14.2.3). The magnitude of these safety factors varies according to the following: the nature and quality of the data from which the NOAEL is derived; the seriousness of the toxic effects; the type of protection being sought (*e.g.*, protection against acute, subchronic, or chronic exposures); and the nature of the population to be protected (*i.e.*, the general population versus identifiable subpopulations expected to exhibit a narrower range of susceptibilities). Safety factors of 10, 100, 1000, and 10,000 have been used in various circumstances.

At present, only agents displaying carcinogenic properties are treated as if they display no thresholds (see

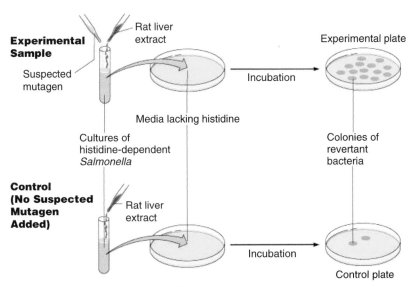

Experimental Sample

Suspected mutagen

Rat liver extract

Media lacking histidine

Incubation

Experimental plate

Colonies of revertant bacteria

Cultures of histidine-dependent *Salmonella*

Control (No Suspected Mutagen Added)

Rat liver extract

Incubation

Control plate

1 Two cultures are prepared of *Salmonella* bacteria that have lost the ability to synthesize histidine (histidine-dependent).

2 The suspected mutagen is added to the experimental sample only; rat liver extract (an activator) is added to both samples.

3 Each sample is poured onto a plate of medium lacking histidine. The plates are then incubated at 37°C for two days. Only bacteria whose histidine-dependent phenotype has mutated back (reverted) to histidine-synthesizing will grow into colonies.

4 The numbers of colonies on the experimental and control plates are compared. The control plate may show a few spontaneous histidine-synthesizing revertants. The test plates will show an increase in the number of histidine-synthesizing revertants if the test chemical is indeed a mutagen and potential carcinogen. The higher the concentration of mutagen used, the more revertant colonies will result.

Copyright ©2004 Pearson Education, Inc., publishing as Benjamin Cummings.

Figure 13.6 The Ames Test. Copyright © 2004 Pearson Education, Inc., published by Benjamin Cummings.

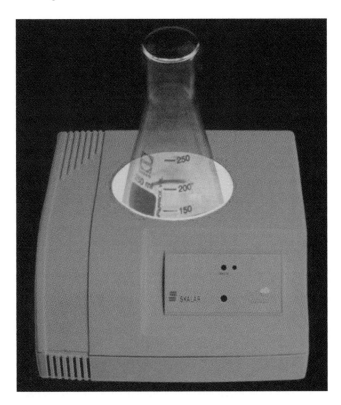

Figure 13.7 Tox Tracer System is a bioassay based on the principal of inhibition of the natural bioluminescence of the marine bacterium *Vibrio fischeri*. From www.skalar.com.

Chapter 14.2.3). Thus, the dose–response curve for carcinogens in the human populations achieves zero risk only at zero dose; as the dose increases above zero, the risk immediately becomes finite and thereafter increases as a function of dose. Risk in this case is the probability of producing cancer, and at very low doses the risk can be extremely small.

13.5 RESPONSES TO TOXIC SUBSTANCES

In general, an organism's response to a toxic chemical depends on the dose administered. However, once a toxicant enters the body, the interplay of four processes—absorption, distribution, excretion, and metabolism—determines the actual effect of a toxic chemical on the target organ, which is the part of the body that can be damaged by that particular chemical. Carbon tetrachloride, for example, affects the liver and kidneys, while benzene affects the blood-cell forming system of the body. Figure 13.8 summarizes routes of absorption, distribution, and excretion.

13.5.1 Absorption

Absorption of toxicants across body membranes and into the bloodstream can occur in the gastrointestinal (GI) tract, in the lungs, and through the skin. Contaminants present in

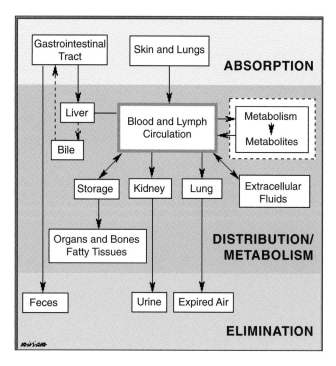

Figure 13.8 Key routes of chemical absorption, distribution and excretion.

drinking water, for example, enter the body primarily through the GI tract. Once they enter the GI tract, most chemicals must be absorbed to exert their toxic effect. Owing largely to differences in solubility, some compounds are absorbed more readily than others. Lipid-soluble, nonionized organic compounds, such as DDT and polychlorinated biphenyls (PCBs), are more readily absorbed by diffusion in the GI tract than are lipid-insoluble, ionized compounds, such as lead and cadmium salts. The GI tract also employs specialized active transport systems for compounds, such as sugars, amino acids, pyrimidines, calcium, and sodium. Although these active transport systems do not generally play a major role in absorption of toxicants, they can contribute to their absorption in some cases; lead, for example, can be absorbed via the calcium transport system.

The behavior of toxicants in the GI tract also depends on the action of digestive fluids. These digestive fluids can be beneficial or harmful. For example, snake venom, a protein that is quite toxic when injected, is nontoxic when administered orally because stomach enzymes attack the protein structure, breaking it down into amino acids. However, in the GI tract, these enzymes can also contribute to the conversion of nitrates to carcinogenic compounds known as nitrosamines.

Age is also an important factor affecting the intestine's ability to act as a barrier to certain toxicants. The GI tract of newborns, for instance, has a higher pH and a higher number of *E. coli* bacteria than that in adults. These conditions promote the conversion of nitrate, a common drinking water pollutant from agricultural runoff, into the more toxic chemical nitrite. The resulting nitrite then interferes with the blood's ability to carry oxygen, causing methemoglobinemia or "blue-baby syndrome." Lead is also absorbed more readily in newborns than in adults. (*Note*: Even though a chemical has been absorbed through the GI tract, it can still be excreted or metabolized by the intestine or liver before it reaches systemic circulation.)

The lungs are anatomically designed to absorb and excrete chemicals, as is shown by their continuous absorption of oxygen and excretion of carbon dioxide. The alveoli have a large surface area (50–100 m^2) and are well supplied with blood, and the blood is very close (10 μm) to the air space within the alveoli. These characteristics make the lungs particularly good vehicles for the absorption of toxicants. Toxicants may have to pass through as few as two cells to travel from the air into the bloodstream.

In contrast, the skin is relatively impermeable to toxicants. However, some toxicants, such as carbon tetrachloride, can be absorbed through the skin in sufficient quantities to cause live injury. In addition, a few chemicals, such as dimethyl sulfoxide (DMSO), have been shown to penetrate the skin fairly readily. Absorption through the skin is possible through the hair follicles, through the cells of the sweat glands and sebaceous glands, and through cuts or abrasions, which increase the rate and degree of absorption. The sole means of absorption through the skin appears to be passive diffusion.

13.5.2 Distribution

Distribution of a toxicant to various organs depends on the ease with which it crosses cell membranes, its affinity for various tissues, and the blood flow through the particular organ. A toxicant's site of concentration is not necessarily the target organ of toxicity. For example, lead can be stored harmlessly in bone, and many lipid-soluble toxicants (such as the chlorinated hydrocarbon insecticides) are stored in fat, where they cause relatively little harm. However, a stored contaminant can be released back into the bloodstream under various conditions. Thus, fat-stored chlorinated pesticides can be released during starvation, dieting, or illness, when fat is consumed.

A number of anatomical barriers in the body are thought to prevent or hinder the entrance of certain toxicants into organs. However, these barriers are not impenetrable walls. The so-called blood-brain barrier, for example, does not prevent toxicants from entering the **central nervous system (CNS)**; rather, the physiological conditions at the blood-brain interface make it more difficult for some toxicants to leave the blood and enter the CNS. In general, lipid-soluble toxicants can cross the blood-brain barrier, but some water-soluble toxicants cannot. There is also the "placental barrier," which is even less of a barrier; the fact is that any chemical absorbed into the mother's bloodstream can and will cross her placenta and enter the bloodstream of the fetus to some degree.

13.5.3 Excretion

Chemicals can be excreted from the body in several ways. The kidney removes toxicants from the blood in the same way that the end products of metabolism are eliminated, that is, by glomerular filtration, passive tubular diffusion, and active secretion. Glomerular filtration is simply a filtration process in which compounds below a certain molecular weight (and hence bulk) pass through pores in a part of the kidney known as the glomeruli. All compounds whose molecular weight is less than 60,000 can filter through the glomeruli unless they are bound to plasma proteins. The molecular weight of most toxicants in drinking water is between 100 and 500; thus, these compounds easily pass through the glomeruli. The toxicants then pass through collecting ducts and tubules, through which water-soluble toxicants may be excreted with urine. However, lipid-soluble toxicants can defeat the excretion process at this point by moving (via passive diffusion) through the tubule wall and back into the bloodstream.

The liver eliminates toxicants through the bile, which passes into the intestine through the gall bladder and bile duct, and finally exits from the body via the feces. As in the kidney, the transport mechanisms used are passive diffusion and carrier-mediated transport. Toxicants that have been excreted into the intestine through the bile can be reabsorbed (especially if they are lipid soluble) into the bloodstream while in the intestine.

Toxicants are also excreted through several other routes, including the lungs, GI tract, cerebrospinal fluid, milk, sweat, and saliva. Milk, for example, has a relatively high concentration of fat (3.5%); thus, lipid-soluble compounds such as DDT and PCBs can concentrate in milk. In addition, because milk is slightly acidic, with a pH of 6.5, basic compounds can also concentrate in it. In this way, toxicants may be passed from mother to child, or from cows to humans.

An important concept in excretion is a toxicant's half-life $T_{1/2}$, which is the time it takes for one-half of the chemical to be eliminated from the body. Thus, if a chemical has a half-life of 1 day, 50% of it will remain with the body one day after absorption, 25% will remain after two days, 12.5% after three days, and so on. The concept of the half-life is important, because it indicates how long a compound will remain within the body. Generally a compound is considered eliminated after a period of seven half-lifes.

13.5.4 Metabolism

Because lipid-soluble compounds can cross cell membranes to be reabsorbed in the kidney and intestine, they are subject to metabolic processes, which are the biochemical reactions by which cells transform food into energy and living tissue. Metabolism is the sum of biochemical changes occurring to a molecule within the body. In many cases, the body metabolizes these toxicants into water-soluble compounds, which can be excreted easily. However, in some instances, metabolism of a chemical creates a more toxic chemical or does not change the chemical's toxicity.

Two types of reactions occur in metabolism: (1) relatively simple reactions involving oxidation, reduction, and hydrolysis; and (2) more complex reactions involving conjugation and synthesis. All of these reactions occur primarily in the liver. Oxidation is the mechanism of metabolism for many compounds. When considering metabolism, researchers also look at species, strain, and gender differences. Age is also an important factor in both humans and laboratory animals; both the very young and very old are more susceptible to certain chemicals.

13.5.5 Biotransformation of Toxicants

Biotransformation is the process by which substances that enter the body are changed from hydrophobic to hydrophilic molecules to facilitate elimination from the body. This process usually generates products with few or no toxicological effects. Biotransformations sometimes yield toxic metabolites, through a process known as **bioactivation**. The chemical reactions responsible for changing a lipophilic toxicant into a chemical form are known as **Phase I** and **Phase II biotransformations**. The two groups are defined based on the reactions that are catalyzed. Phase I reactions transform hydrophobic chemicals to more polar products via oxidation, hydrolysis, or similar reactions (Figure 13.9). Phase II processes involve conjugation reactions that add polar functional groups, such as glucose or sulfate, to the Phase I products, to produce what are often even more polar metabolites. Thus, these become even more water soluble and can be readily excreted. Many biotransformation enzymes exhibit broad substrate specificity, providing a mechanism for enhancing the excretion of a wide range of hazardous compounds (Watts, 1998).

At physiologic pH, a toxicant or its metabolites that are water soluble will undergo dissociation into ions or become ionized. Ionized molecules are the molecules that *react* in living systems. These ionized molecules (*e.g.*, toxic metabolites), with their positively or negatively charged regions, are the molecules that are more readily transported across cell membranes. On occasion, biotransformation produces intermediate or final metabolites possessing toxic properties not found in the original parent chemical. The liver is the most important organ of bioactivation because of the high concentration of enzymes that catalyze biotransformation reactions.

13.5.6 Phase I Transformations

During Phase I biotransformation reactions, a small polar group is either exposed ("unmasked") in the toxicant or added to the toxicant (Table 13.6). The polar group enhances the solubility of the toxicant in water, which favors elimination. The reactions are catalyzed by nonspecific enzyme systems.

Toxicants undergoing Phase I biotransformation will result in metabolites that are sufficiently ionized, or

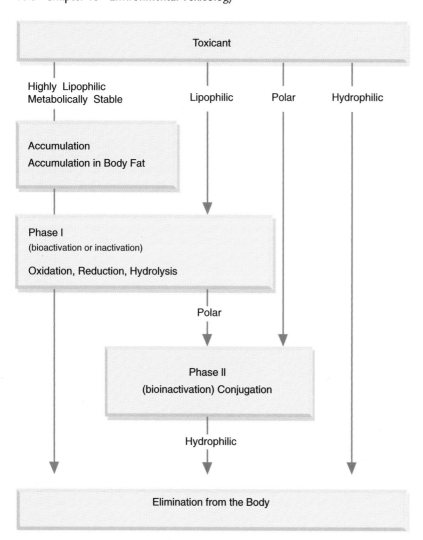

Figure 13.9 The various steps of biotransformation. This figure gives a general overview of biotransformation, showing Phase I and Phase II reactions. The final block represents the various excretory processes. Some lipophilic substances cannot be processed by the biotransformation system, for example, because they are not suitable substrates for the enzymes of the system. Examples of such substances are polychlorinated biphenyls (PCBs) and DDT. Such substances will therefore accumulate in the body, in particular in body fat.

hydrophilic, to be either readily eliminated from the body without further biotransformation reactions or rendered as an intermediate metabolite ready for Phase II biotransformation. Some intermediate or final metabolites may be more toxic than the parent chemical.

13.5.7 Phase II Transformations

On completion of a Phase I reaction, the new intermediate metabolite produced contains a reactive chemical group (*e.g.*, hydroxyl, −OH; amino, −NH₂; or carboxyl, −COOH). For many intermediate metabolites, the reactive sites, which were either exposed or added during Phase I biotransformation, do not confer sufficient hydrophilic properties to permit

elimination from the body. These metabolites must undergo additional biotransformation, called a Phase II reaction (Hughes, 1996).

During Phase II reactions, a molecule provided by the body must be added to the reactive site produced during Phase I. Phase II reactions are referred to as conjugation reactions. These reactions produce a conjugate metabolite that is more water soluble than the original toxicant or Phase I metabolite. In most instances, the hydrophilic Phase II metabolite can be readily eliminated from the body (Hughes, 1996).

One of the most common molecules added directly to the toxicant or its Phase I metabolite is glucuronic acid, a molecule derived from glucose, a common carbohydrate that is the primary source of energy for cells.

TABLE 13.6 Representative Phase I biotransformation reactions.

REACTION	EXAMPLE
Nitrogen oxidation	$RNH_2 \rightarrow RNHOH$
Sulfur oxidation	$\begin{matrix} R_1 \\ \backslash \\ S \\ / \\ R_2 \end{matrix} \rightarrow \begin{matrix} R_1 \\ \backslash \\ S=O \\ / \\ R_2 \end{matrix}$
Carbonyl reduction	$\begin{matrix} RCR' \\ \| \\ O \end{matrix} \rightarrow \begin{matrix} RCHR' \\ \| \\ OH \end{matrix}$
Hydrolysis (Esters)	$R_1COOR_2 \rightarrow R_1COOH + R_2OH$
Desulfuration	$\begin{matrix} R_1 \\ \backslash \\ C=S \\ / \\ R_2 \end{matrix} \rightarrow \begin{matrix} R_1 \\ \backslash \\ C=O \\ / \\ R_2 \end{matrix}$
Dehydrogenation	$RCH_2 \rightarrow RCHO$

From Hughes, 1996.

13.6 CARCINOGENS

Carcinogens are agents that cause cancer, which is the uncontrolled growth of cells. Every human being is made up of approximately 100 trillion cells, and any one of these cells can be transformed to a malignant or cancerous cell by a variety of agents, which may be chemical (*e.g.*, disinfection byproducts), biological (*e.g.*, cancer-causing viruses), or physical (*e.g.*, ultraviolet light, gamma irradiation) in origin. Approximately 100 different types of cancer, which can be found in every organ and system of the body, have been identified.

As of the mid-1990s in the United States, cancer is second only to heart disease as a cause of death. Figures 13.10 and 13.11 show an estimate of yearly cancer deaths in the United States, broken down by site and sex. More than 500,000 people die from cancer each year, with lung cancer by far the leading killer. In fact, lung cancer has increased more than 200% during the last 35 years. Cancer ultimately kills one out of about every four Americans. However, aside from the increased incidence of lung cancer, the incidence of all other forms of cancer has collectively declined by about 13% over the past 30 years.

Carcinogens trigger uncontrolled cell growth in many different ways. In general, we can think of carcinogens as initiators or promoters, depending upon the stage of carcinogenesis in which they are active.

Initiation—the first stage of carcinogenesis, or conversion of a normal cell to a cancer cell—is a rapid, essentially irreversible change caused by the interaction of a carcinogen with cell DNA (Figure 13.12). This step in the development of cancer involves an **initiator**, a type of carcinogen that structurally modifies a gene that normally controls cell growth. A gene is a specific segment of a DNA molecule, so these altered growth-regulating genes are called **oncogenes** (from *onco*, meaning tumor, and gene). For a cell to begin to

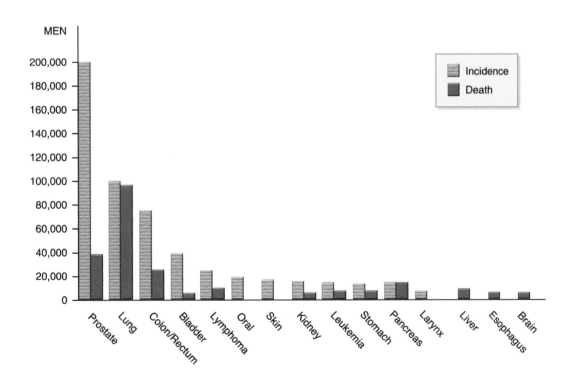

Figure 13.10 Cancer incidence rates in 1994 among men. From www.biorap.org/tg/tgcancerdirt.html.

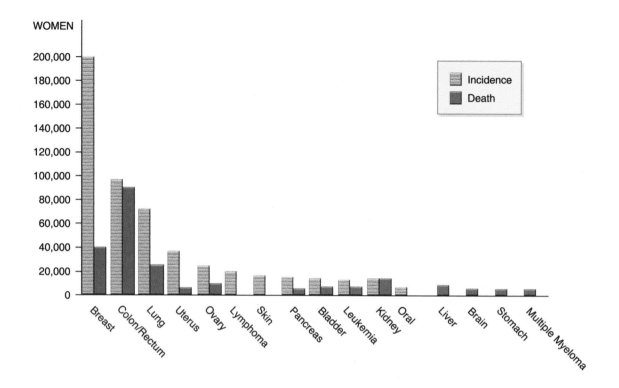

Figure 13.11 Cancer incidence rates in 1994 among women. From www.biorap.org/tg/tgcancer-dirt.html.

grow uncontrollably, at least two different growth-regulating genes must be altered. Such changes prime the cell for subsequent neoplastic development (from *neo*, meaning new, and *plastic*, referring to something formed, literally, new growth).

Some chemicals are initiators in their own right. For instance, formaldehyde, a widely used chemical in industrial glues, is thought to be an initiator. We know that vapors of formaldehyde trigger the development of malignant tumors in the respiratory tracts of rats, even though this compound

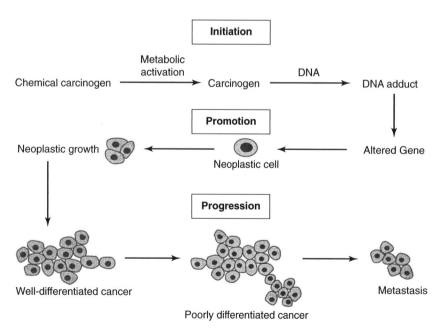

Figure 13.12 The three steps in cancer development. From Sullivan J.B. and Krieger G.R, (1992) *Hazardous Materials Toxicology: Clinical Principles of Environmental Health.* Williams and Wilkins, Baltimore, MD.

has not been shown to cause cancer in humans. In other cases, however, an initiator can be a metabolic byproduct. Thus, benzo(a)pyrene, a natural product of the incomplete combustion of organic materials (including tobacco), is not itself an initiator, but the body metabolizes it to a related chemical, benzopyrene-7,8-diol 9,10-epoxide, which is an initiator **bioactivation** (Figure 13.13).

The **promotion** process triggers the progressive multiplication of abnormal cells known as neoplastic development. **Promoters**, then, are carcinogens that activate the oncogenes, which would otherwise remain dormant. These carcinogens may act in several ways. For example, normal cells appear to prevent the activation of an oncogene in an adjacent initiated cell. A substance may act as a promoter by killing the normal cells that surround an initiated cell. Alternatively, promoters may activate oncogenes by inhibiting the action of **suppressor genes**, which prevent oncogenes from initiating uncontrolled cell growth. If suppressor genes are inactivated, oncogenes can then spur tumor formation. The final stage in carcinogenicity is the growth of neoplastic cells followed by *progression*, or spreading throughout the body. This final step may be subdivided into two stages: invasion and metastasis. *Invasion* is the localized movement of neoplastic cells into adjoining tissues, and *metastasis* is the more distant movement throughout the body. Invasion and metastasis are consistent with a physiological system that is out of control.

Many compounds, some of them seemingly innocuous, can act as promoters. For instance, dietary factors such as salts and fats apparently act as promoters by killing normal cells. Ingestion of such promoters is not immediately harmful, but lifelong exposure significantly increases the risk of cancer. Thus, people whose diets are high in salts or fats are more likely to develop stomach cancer and colon cancer, respectively. Fortunately, removing promoters from the area of an initiated cell that has not yet completed the promotion stage prevents the formation of a cancer cell. Therefore, if we reduce the salt and fat we consume, we can greatly reduce the risk of developing cancer.

Carcinogens and toxins differ in one very important respect: the incidence of cancer (number of cases per million population) is dose-dependent, but the severity of the response (cancer) is independent of dose. This means that we would expect more cases of cancer to develop as a population is exposed to higher levels of a carcinogen, just as we would expect more cases of poisoning in a population exposed to higher levels of a toxin. But while the severity of toxic response is also dose-dependent, the dose of the carcinogenic agent has little to do with the severity of the disease once an individual has contracted cancer. This distinction between toxins and carcinogens explains why exposure regulations for substances classified as carcinogens are much more stringent than for toxins.

13.7 MUTAGENS

Like carcinogens, mutagens and teratogens affect DNA. **Mutagens** cause **mutations**, which are inheritable changes in the DNA, sequences of chromosomes. Mutations involve a random change in the natural functioning of chromosomes or their component genes; such changes rarely benefit the organism's offspring.

A mutation is a change in the genetic code that may or may not have an effect on the organism. Harmful effects from mutations depend on the type of cell that is affected and whether the mutation leads to metabolic malfunctions. Mutations that occur in somatic cells (nonreproductive cells of the body) may or may not prove to be a threat to the organism. Mutations occur naturally, most commonly from ionizing radiation. Humans and other organisms have enzymes that can repair damaged DNA. However, not all mutations are

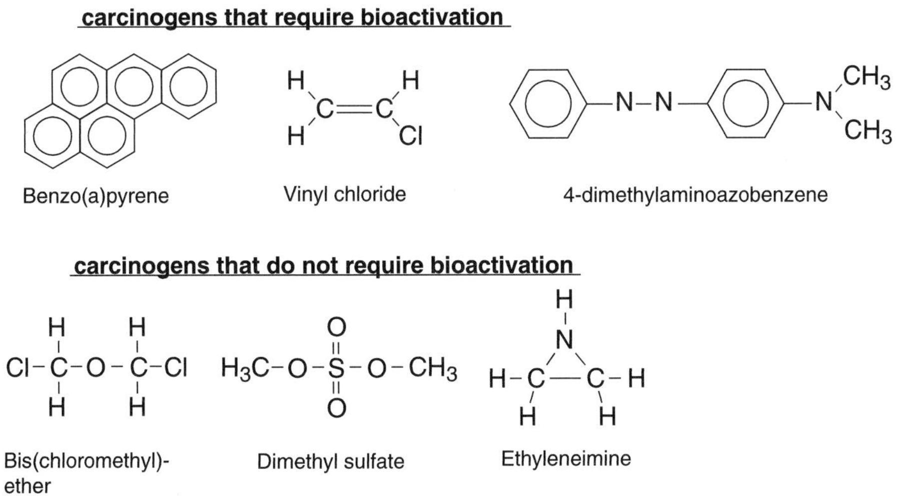

carcinogens that require bioactivation

Benzo(a)pyrene   Vinyl chloride   4-dimethylaminoazobenzene

carcinogens that do not require bioactivation

Bis(chloromethyl)-ether   Dimethyl sulfate   Ethyleneimine

Figure 13.13 Examples of the major classes of naturally occurring and synthetic carcinogens, some of which require bioactivation and others of which act directly.

repaired, and some mutations will cause the cell's metabolism to become out of control, which may result in cancer.

Numerous types of mutations have been documented. The simplest type of genetic damage results when there is a mutation of the DNA sequence. Referred to as a **point-mutation**, this mutation represents a change in the chromosome involving a single nucleotide (base) within the gene. These changes may result in the substitution, deletion, or insertion of a base (Figure 13.14).

A **base substitution** occurs when a nucleotide is substituted for a normally occurring base. If the substituted

TABLE 13.7 The effects of selected physical and chemical mutagens.

MUTAGEN	EFFECT ON DNA/RNA	TYPE OF MUTATION
Ultraviolet radiation	C, T, and U dimers that cause base substitutions, deletions, and insertions	No effect, missense, or nonsense
X rays	Breaks in DNA	Chromosomal rearrangements and deletions
Acridines (tricyclic ring present in dyes)	Adds or deletes a nucleotide	Missense or nonsense
Alkylating agents	Interferes with specificity of base pairing (e.g., C with T or A, instead of G)	No effect, missense, or nonsense
5-Bromouracil	Paris with A and G, replacing AT with GC, or GC with AT	No effect, missense, or nonsense

C = cytosine

T = thymine

U = uracil

base does not alter the amino acid coded for in that position, then it will have no effect on the protein being made from the DNA. This outcome is possible, since each amino acid is coded for by more than one codon (see Figure 13.14). Two additional outcomes, *missense* and *nonsense*, result when the mutated triplet codon codes for a different amino acid or stop signal, respectively. Base substitutions usually do not result in significant mutagenic activity. Because of the redundancy of the genetic code, one base substitution will not likely result in a major change in the translation of the genetic information. Moreover, if a mutation results in an inappropriate amino acid in a protein that functions as an enzyme, it may not even change the enzymatic activity if it is not at or near the active site on the enzyme.

Frameshift mutations are the result of the addition between base pairs, which shifts the triplet code down the DNA strand. Such shifts essentially change the entire coding of proteins, with consequent high potential for malfunction of the protein. The effects of some physical and chemical mutagens are shown in Table 13.7.

13.8 TERATOGENS

Teratogens affect the DNA in a developing fetus, often causing gross abnormalities or severe deformities such as the shortening or absence of arms or legs.

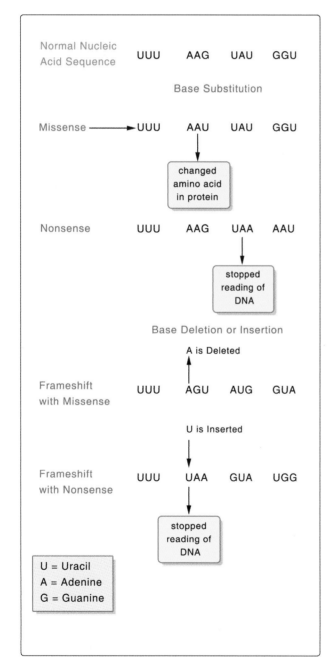

Figure 13.14 Types of base mutations.

Perhaps the most famous (or infamous) teratogen is thalidomide, a sedative that was taken by thousands of pregnant women during the early 1960s. Sometimes, however, the deleterious effect of a teratogen does not appear until many years after the mother has been exposed. This was the case for diethylstilbestrol (DES), a drug that was prescribed for pregnant women in the United States for more than 30 years. Developed to prevent miscarriages, DES has been implicated in cervical and vaginal abnormalities in the daughters of women who had used DES during pregnancy. Nor are drugs the only teratogens. For example, the rubella virus, which causes a mild viral infection (German measles), is a teratogen during the first trimester of pregnancy. This virus can cross the transplacental barrier to produce cardiac defects and deafness in the offspring.

13.9 CHEMICAL TOXICITY: GENERAL CONSIDERATIONS

The toxic effects of a substance on a particular individual depend on both the chemical and the individual. However, the variability in the toxic potential of different compounds greatly exceeds the variability in toxic response from individual to individual. That is, if we expose a particular individual to a whole series of different chemicals, we would see that some substances cause toxic effects in minute amounts, whereas others must be present in huge quantities. The range for toxic effects is enormous: the toxicity of one chemical can be millions or billions of times greater than that of another chemical. Thus, it can take millions or billions of times as much of one chemical to cause the same effect as another. The range of human variability is not nearly so great. If a particular chemical causes an effect in one individual when a particular amount is administered, it is not likely that an amount a billion times less will cause a toxic effect in another individual. The exact range of human variability is not well established, but it is probable that it is closer to a tenfold than a billionfold.

In considering toxicity, we cannot make a distinction between human-made (synthetic) and naturally occurring chemicals—that is, everything is chemical in composition. It is not the source of the chemical that is important, but its characteristics. In Figure 13.15, for example, we see the structures of some natural and synthetic compounds, each of which is considered a toxin at certain dosages. However, most chemicals, synthetic or natural, are not very toxic.

The molecular shape or structure of a chemical is one of the most important characteristics to consider in determining its toxicity. Current theory suggests that a living organism "recognizes," and hence reacts to, most chemicals that enter the body by their shape. These body-recognition responses can be very sensitive to subtle differences in shape or conformation. Two molecules, for example, might be very similar in structure, but exhibit slightly different configurations or three-dimensional isomers, one of which

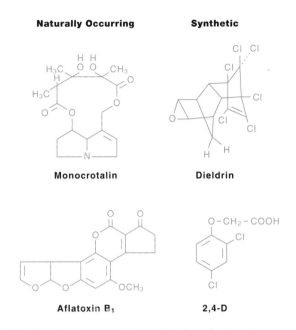

Figure 13.15 Structures of natural and synthetic toxic chemicals found in the environment. From *Pollution Science* © 1996, Academic Press, San Diego, CA.

one may induce a toxic response in the body, while the other will not (Figure 13.16). In fact, toxicity is frequently the result of recognition gone wrong, in which, at a cellular or chemical level, the body "prefers" the toxin over its structurally similar counterpart.

A second important characteristic in determining the degree of toxicity of a chemical is its solubility in different solvents. In particular, compounds are divided into those that are polar and hence soluble in water or water-like (polar) solvents, and those that are nonpolar and are soluble in fat (oil) or fat-like (nonpolar) solvents. This difference is very important in determining how easily a chemical can enter the body, how it is distributed inside the body, and how easily it can be excreted. Animals are most efficient at excreting

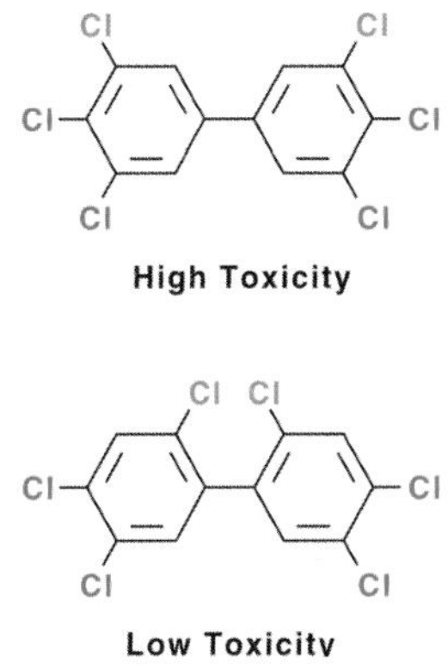

Figure 13.16 Example of structurally similar chemicals with different toxic potency. From *Pollution Science* © 1996, Academic Press, San Diego, CA.

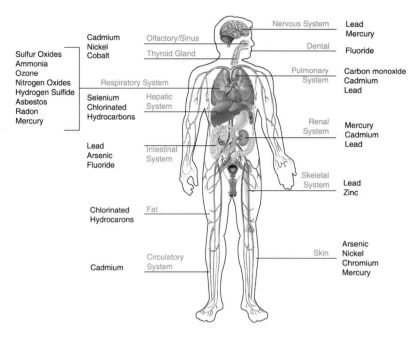

Figure 13.17 Target organ toxicity.

polar compounds, so an ionic compound like sodium chloride, which is very soluble in water, is easily excreted. But a nonpolar compound like the pesticide DDT, which dissolves in fat, is not so easily eliminated. In some cases, the body is capable of converting nonpolar compounds into polar variants, thus facilitating their removal from the body. Unfortunately, this is not true of DDT, which can remain in the body for longer periods of time.

Target organ toxicity (Figure 13.17) is defined as adverse effects manifested in specific organs in the body. Toxicity is unique for each organ, since each organ is a unique assemblage of tissues, and each tissue is a unique assemblage of cells. Toxicity may be enhanced by distribution features that deliver a high concentration of the toxicant to a specific organ

or by inherent features of the cells and tissues of the organ that render it highly susceptible to the toxicant (Table 13.8).

13.10 CHEMICAL TOXICITY: SELECTED SUBSTANCES

13.10.1 Heavy Metals

Heavy metals, including lead, mercury, cadmium, chromium, arsenic, and selenium, constitute a major category of inorganic pollutants. Heavy-metal contamination often originates from manufacturing and the use of various synthetic products (*e.g.*, pesticides, paints, and batteries), but these chemicals

TABLE 13.8 Additional examples of target organ toxicity.

ORGAN	TOXICANT	MECHANISM	TOXICITY
Heart	Fluorocarbons (Freon)	? Sensitizes heart to epinephrine	Decreased heart rate, contractility, and conduction
	Carbon monoxide (CO)	Interferes with energy metabolism	Myocardial infarction, increase or decrease in heart rate
	Cobalt (Co)	Competes with Ca^{2+}	Heart failure
Testis	Lead (Pb)	? Mutations in sperm	Decreased male fertility, increased spontaneous abortions in females
	Carbon disulfide (CS_2)	CNS effect on ejaculation	Reduced sperm counts
Ovary/uterus	Solder fumes	? Unknown	Increased spontaneous abortions
	Polycyclic aromatic hydrocarbons (PAH)	? Unknown	Damaged oocytes
Eye	Busulfan (chemotherapeutic agent)	? Alters mitosis in lens cells	Formation of cataracts
	Methanol (CH_3OH)	Produces optic atrophy	Permanent visual impairment, blindness

From Hughes, 1996.

may also occur naturally (see Chapter 10). Heavy metals, many of which are toxic to plants as well as animals, tend to be mobile in the food chain, which means they can be bioconcentrated in animals, including humans, who are at the top of the food chain. The toxicity of some heavy metals such as lead and mercury has been known for centuries, but was not fully understood until recent times. Because of their widespread use and/or occurrence in nature, some heavy metals are of particular concern.

Lead borders on ubiquitous (Figure 13.18). It can be found in drinking water, where it comes from several sources. The most significant of these sources is lead solder and piping in water distribution systems, particularly when in contact with corrosive water. While the use of lead solder and piping in repairs and construction of water systems has been banned, many existing water distribution systems still contain lead materials. Lead is also found in food, tetraethyl lead in gasoline (which ends up in the air, soil, and water), lead-based paint, and improperly glazed earthenware. In addition, it can come from industrial sources, such as smelters and lead-acid battery manufacturing. Since the toxic effects of lead depend on total exposure, environmental assessment must take into account all of these sources.

Compared with fat-soluble substances, lead is relatively poorly absorbed, as lead compounds are water soluble. In adults, only about 10% of the lead ingested through the GI tract is absorbed into the bloodstream. Lead is initially distributed to the kidney and liver and then redistributed, mostly to bone (about 95%). Moreover, lead does not readily enter the central nervous system in adults because the blood-brain barrier can keep it out. However, in children, the blood-brain barrier is not yet fully developed, so lead exposure in children can affect their mental development.

Acute lead poisoning is rare; however, chronic lead poisoning is not uncommon. In adults, chronic lead poisoning sometimes results in the painful gastrointestinal symptoms known as lead colic. Because of the pain, lead colic often compels exposed persons to seek medical help, whereupon an accurate diagnosis of lead exposure can prevent the development of more serious problems. Lead can also affect the neuromuscular system, decreasing muscle tone in the wrists and feet. Exposure to lead also affects the body's blood-forming system. Lead can interfere with the synthesis of heme (part of the oxygen-carrying compound hemoglobin), thereby causing anemia; it can also damage red blood cells in a condition known as basophilic stippling. The most serious effect of lead is the brain-degenerative condition called lead encephalopathy. This condition can occur in adults (Table 13.9). For example, historians now speculate that Caligula, the most insane of the Roman emperors, suffered from lead encephalopathy caused by eating food from lead-containing pewter dishes. Today, however, this condition is more common in children and can be quite serious; approximately 25% of children with lead encephalopathies die, and about 40% of the survivors experience neurologic after-effects. Finally, lead has been shown to affect the kidneys of laboratory animals, causing impairment of function and cancer.

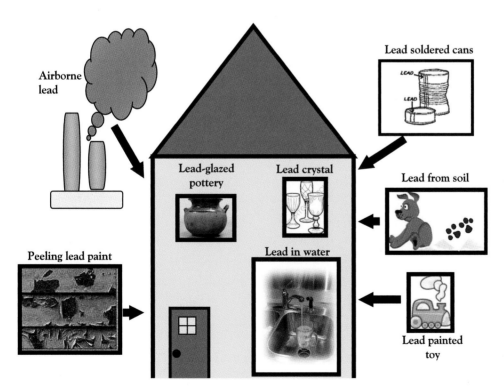

Figure 13.18 Sources of lead in the environment.

TABLE 13.9 Toxic responses to increased lead concentrations in the blood.

| | | HEALTH EFFECTS | |
| | | BLOOD LEAD CONCENTRATION (μg/l) | |
		Adults	Children
100	↓	Hypertension may begin to occur	Crosses placenta, developmental toxicity
			Impairment of IQ
			Increased erythrocyte protoporphyrin
200	↓		Beginning impairment of nerve conduction velocity
300	↓	Systolic hypertension, decreased hearing	Impaired vitamin D metabolism
400	↓	Infertility in males, renal effects, neuropathy	Hemoglobin synthesis impaired
		Fatigue, headache, abdominal pain	
500	↓	Decreased heme synthesis, decreased hemoglobin,	Abdominal pain, neuropathy
		anemia, intestinal symptoms, headache, tremor	Encephalopathy, anemia, nephropathy, seizures
		Lethargy, seizures	
1000		Encephalopathy	

Modified from Sullivan and Krieger, 1992.

Cadmium, like lead, has many sources. It is a byproduct of lead and zinc mining, it is used as a pigment, and it is found in corrosion-resistant coatings and nickel-cadmium batteries. It is also released when fossil fuels are burned. Cadmium can enter drinking water when corrosive water contacts certain types of water piping. Historically it entered the food chain through application of sewage sludge to the land, where it was taken up by plants and stored in leaves and seeds. However, point source controls have reduced this problem (see also Chapter 27). Cadmium's adverse effect on health was originally made public by an incident in Japan, where rice paddies were contaminated with cadmium-rich drainage from zinc mines. Rice grown on the paddies concentrated the cadmium, and those eating it suffered such characteristic symptoms as easily broken bones and extreme joint pain; thus, cadmium poisoning is known as *itai-itai* ("ouch-ouch") disease.

Since cadmium is water soluble, only 1–5% of a given dose is absorbed in the GI tract (although 10–40% can be absorbed through the lungs), and cadmium distributes to the kidney and liver (Figure 13.19). Acute cadmium poisoning causes GI disturbances. Chronic cadmium poisoning most severely affects the kidney. Animal studies have shown cadmium to be carcinogenic, and some researchers have suggested that it may increase the incidence of prostate cancer in elderly men.

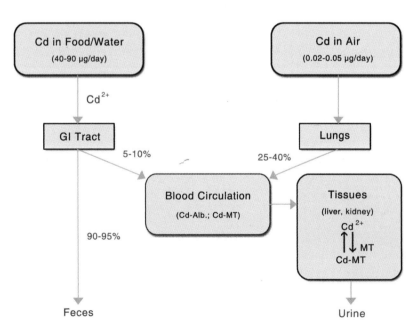

Figure 13.19 Metabolism of Cd in humans. Cd-Alb: Cd attached to albumin; Cd-MT: Cd attached to metallothionein. From Yu, 2001.

Arsenic occurs naturally in bedrock and soil (Figure 13.20) and is a waste product from smelting operations as well as the manufacture of products such as pesticides and herbicides. Airborne particles of arsenic may travel considerable distances and penetrate deeply into the lungs. Arsenic is also readily taken up by food plants, with the degree of uptake being dependent on soil pH. Arsenic is considered an essential dietary element, although in very small amounts. There are four main types of arsenic: organoarsenics, arsenate, arsenite, and arsine gas.

Arsenic can be excreted relatively quickly, and it has a half-life of 2 days. Although the effects of acute arsenic poisoning are seen primarily in mystery fiction, chronic arsenic poisoning manifests in a wide variety of chronic toxic effects. Many of these effects stem from the capacity of arsenic to increase the permeability of capillaries in various locations in the body, and the structural similarity of arsenate to phosphate, for which it can substitute. Increased permeability is, however, harmful; it allows plasma to leak into the tissues, sometimes leading to severe diarrhea and kidney injury. Arsenic also has the potential to damage the central nervous system, inflaming peripheral nerves and causing brain injuries, and damage the liver, infiltrating fatty deposits and causing tissue necrosis. Moreover, the EPA has classified arsenic as a Type A human carcinogen (*i.e.*, a human carcinogen based on epidemiological studies), with skin and lung cancer as the two principal types of cancer arising from arsenic exposure.

Mercury is a naturally occurring metal dispersed throughout the ecosystem. Mercury contamination of the environment is caused by both natural and anthropogenic sources. Natural sources include volcanic action, erosion of Hg-containing sediments, and gaseous emissions from the earth's crust. The majority of Hg comes from anthropogenic sources or activities including mining, combustion of fossil fuels (Hg content of coal is about 1 ppm), transporting Hg ores, processing pulp and paper, incineration, use of Hg compounds as seed dressings in agriculture, and exhaust from metal smelters (Yu, 2001). In addition, Hg waste is found as a byproduct of chlorine manufacturing plants, used batteries, light bulbs, and gold recovery processes.

Mercury compounds are added to paints as preservatives. In addition, Hg is used in jewelry making, pesticides, and other manufacturing processes. The light emitted by electrical discharge through Hg vapor is rich in ultraviolet rays, and lamps of this kind in fused quartz envelopes are widely used as sources of UV light. High-pressure Hg-vapor lamps are now widely used for lighting streets and highways.

In the U.S., the largest user of Hg is the chlor-alkali industry, in which chlorine and caustic soda are produced by electrolysis of salt (NaCl) solution (Yu, 2001). In some methods of producing chlorine, an Hg cathode is used. The Na^+ ions discharge at the Hg surface, forming sodium amalgam. The resultant amalgam is continuously drained away and treated with water to produce NaOH solution and Hg:

$$Hg - Na \xrightarrow{H_2O} NaOH(solution) + Hg \quad \text{(Eq. 13.1)}$$

Various forms of Hg are present in the environment. Conversion of one form to another occurs in sediment, water, and air, and is catalyzed by various biological systems. In addition, Hg frequently finds it way to lakes and seas. Microorganisms then convert the elemental Hg into methylmercury (MeHg) through a process called methylation, allowing it to enter the food chain and be biomagnified (Figure 13.21).

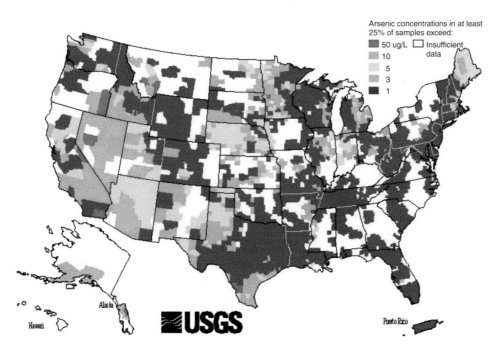

Arsenic concentrations in at least 25% of samples exceed:
50 ug/L
10
5
3
1
Insufficient data

Figure 13.20 Arsenic concentration in wells and springs (1973–2001). From http://co.water.usgs.gov/trace/pubs/geo_v46n11/fig2.jpeg.

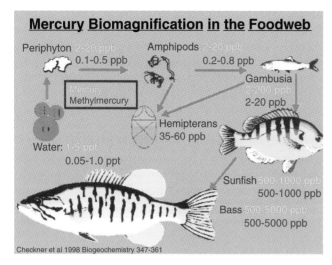

Figure 13.21 Mercury biomagnification in the food web. From Cleckner et al, (1998) Trophic transfer of methylmercury in the northern Everglades. *Biogeochemistry*, Vol 40, pp.347-361. Reprinted with kind permission of Springer Science and Business Media.

Mercury is a nerve toxin, and the main health concern is its effect on the brain, particularly in the growing fetus and the young. The phrase "mad as a hatter" stems from the mercury poisoning of hat makers, who used the metal for curing felt. Mercury can damage reproduction in mammals by interfering with the formation of sperm. Neurological and reproductive effects have also been seen in birds. In fish, its effects include a decreased sense of smell, damage to the gills, blindness, and changes in the ability to absorb nutrients in the intestines. Plants can also be sensitive to mercury, and high concentrations can lead to reduced growth.

13.10.2 Inorganic Radionuclides

Certain unstable elements spontaneously decay into different atomic configurations, in the process releasing radiation consisting of alpha particles, beta particles, or gamma rays (see also Chapter 21). These particles and rays can damage living tissue and/or cause cancer to develop, with the degree of damage depending on the type of radiation and means of exposure (*i.e.*, inhalation, ingestion, or external radiation). As an element undergoes radioactive decay, it progresses through a series of atomic configurations, known as isotopes. Isotopes are simply atoms identified by atomic weight, a number indicating the number of neutral and charged particles in the nucleus (Figure 13.22). The uranium in pitchblende, for example, is a mixture of three isotopes that have atomic weights of 234, 235, and 238.

Radionuclides are atoms (nuclei) that are undergoing spontaneous decay, emitting radiation as they disintegrate to form isotopes of lower atomic weight. The radioactivity of an isotope is expressed in units called picocuries (pCi),[2] which represents the isotope's number of disintegrations per

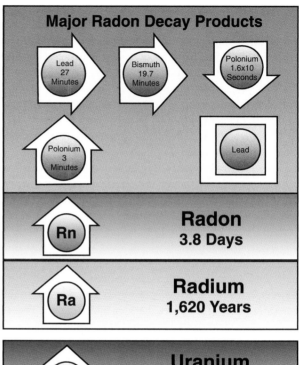

Figure 13.22 Natural decay of uranium. From http://energy.cr. usgs.gov/radon/georadon/page3.gif.

second; 1 pCi is equal to 3.7×10^{-2} disintegrations per second. (*Note*: Radionuclides continue to decay until they achieve a stable configuration, which is often the configuration of another element altogether.)

Radioactive decay is a natural process. In fact, everyone is exposed to some background radiation both from cosmic rays and from radioactive soil and rock (Figure 13.23). Three naturally occurring series of isotopes arise from the decay of the isotopes uranium-238, uranium-235, and thorium-232. These naturally occurring radionuclides can be found in drinking water, with the uranium-238 series (whose decay isotopes include uranium-234, radium-226, and radon-222) and the thorium series (whose decay products include radium-228) — being of greatest concern. Synthetic radioactive isotopes, such as strontium-90, also pose health risks, but such isotopes generally occur in lower concentrations in the environment than the naturally occurring radionuclides. However, site-specific contamination, such as nuclear waste disposal sites and nuclear power plant accidents (*e.g.*, Chernobyl), may release concentrations of anthropogenic radionuclides that threaten human and environmental health.

The concern over radionuclides focuses largely on their potential to cause cancer. Radium-226, which has a half-life of 1622 years, is perhaps the single most important radioactive isotope found in drinking water. It is deposited in bone and can

---

[2] Another unit commonly used to express radioactivity is the becquerel (Bq), which is defined as disintegrations per second (s^{-1}). 1 Bq = 3.7×10^2 pCi.

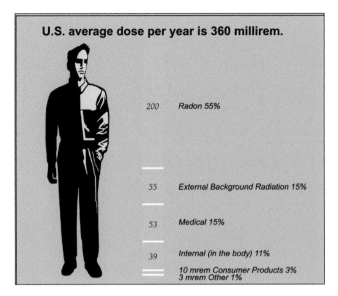

Figure 13.23 U.S. average dose per year is 360 millirem. From www.ocrwm.doe.gov.

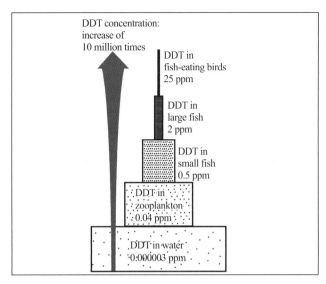

Figure 13.24 Biomagnification of DDT in the food change. Copyright © 2004 Pearson Education, Inc., published by Benjamin Cummings.

cause bone cancer. One of the decay products of radium-226 is radon gas, that is, the isotope radon-222, which has a half-life of 3.85 days. Inhalation of the short-lived decay products of radon-222 can cause lung cancer; however, less is known about the risks of ingested radon. This isotope has recently become the subject of great public concern because it has been discovered in homes and other buildings. Most of the total amount of radon-222 that enters homes comes through the soil, but it can also enter by degassing from a dissolved state, in drinking and washing water. This degassing occurs when water is heated and/or aerated, as for example, during showering, bathing, and clothes and dish washing.

13.10.3 Insecticides

Insecticides can be divided into organochlorine, organophosphorus, and carbamate compounds, as well as botanical insecticides (see also Chapters 10 and 16). Within each group, the pesticides have similar characteristics.

Organochlorine Insecticides: This category, which was developed in the 1930s and 1940s, includes the chlorinated ethanes, chlorinated cyclodienes, and other chlorinated compounds. Dichlorodiphenyltrichlorethane (DDT) is the most famous of the chlorinated insecticides. First synthesized in 1943, it was used extensively (worldwide) in agriculture from the end of World War II until 1972, when it was banned in the United States. It was first used to control disease-carrying insects such as body lice and mosquitoes that spread malaria. DDT also provided effective against a variety of agricultural pests and was extensively used on crops. This highly lipid-soluble compound is stored in fat—in fact, the fat of most U.S. residents contains DDT concentrations of 5–7 mg $kg^{-1,3}$. DDT is very persistent in the environment and is **biomagnified** (Figure 13.24) in the food chain. That is, smaller organisms absorb the compound, then they are eaten by larger

organisms, and the progression continues until DDT attains a relatively high concentration in macrovertebrates such as fish, which are then eaten by humans and other large animals.

In general, DDT is toxic to humans and most other higher animal life only in extremely high doses. However, because of its low toxicity, it was applied in much greater quantities than were necessary. Then, in the 1960s, the effects of these massive applications became noticeable. For example, certain birds, such as the peregrine falcon, began to produce overly fragile egg shells that broke before hatching, thereby threatening their survival as a species. Fish, too, are extremely vulnerable to DDT, and die-offs occurred following heavy rains, when the pesticide was washed into streams and rivers.

DDT and other chlorinated hydrocarbons are very resistant to metabolic breakdown. In animals and humans, DDT is degraded to DDE (1,1-dichloro-2,2-bis(p-chlorophenyl ethylene)) or dichlorodiphenyl dichloroethylene) or DDD (1,1-dichloro-2, 2-bis(p-chlorophenyl ethane)) (Figure 13.25).

The other chlorinated hydrocarbons, such as lindane, toxaphene, mirex, and kepone, are similar to DDT. In general, then, we can say that all organochlorine insecticides cause some central nervous system (CNS) stimulation, increase cancer incidence in laboratory animals, and persist in the environment to some degree.

Organophosphorus Insecticides: Organophosphorus insecticides are the most toxic of the insecticides; they are dangerous not only to insects but also to mammals. Many of the compounds, such as parathion, paraoxon and tetram are in the "supertoxic" category for humans. Human fatal doses for those toxicants are <5 mg/kg. As little as 2 mg of parathion has been known to kill children (Yu, 2001).

Organophosphorus compounds do not persist in the environment and have an extremely low potential to produce

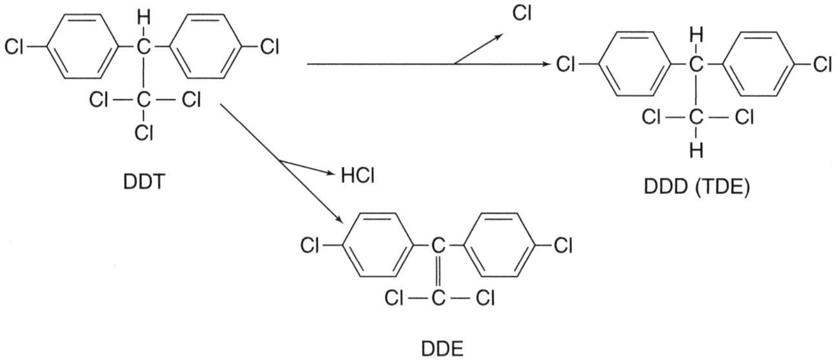

Figure 13.25 Metabolism of DDT.

cancer; thus, insecticides based on these compounds have largely replaced the chlorinated hydrocarbon insecticides. However, these phosphorous-containing compounds have a much higher acute toxicity in humans than the organochlorines, and are the most frequent cause of human insecticide poisoning. Fortunately, both laboratory tests and antidotes are available for acute organophosphorus poisoning.

A typical organophosphorus insecticide is parathion, which must be metabolized to the compound paraoxon to exert its toxic effect. This toxic effect stems from the compound's ability to inhibit the enzyme cholinesterase, a crucial chemical for the regulation of the nerve transmitter acetylcholine. Thus, acute effects of poisoning with organophosphorus insecticides include fibrillation of muscles, low heart rate, paralysis of respiratory muscles, confusion, convulsions, and eventually death. Other organophosphorus pesticides, such as malathion, are less toxic in acute doses than parathion.

Carbamate Insecticides: The carbamate insecticides, which include carbaryl and aldicarb (Figure 13.26), have toxicities very similar to those of the organophosphorus insecticides. Like the organophosphorus pesticides, these widely used chemicals also act by inhibiting cholinesterase, but the toxic effects of carbamates may be more easily reversed than those of the organophosphorus compounds. In addition, current evidence does not seem to suggest carcinogenicity as a toxic effect of the carbamates. Although most of these chemicals are not persistent in the environment, aldicarb may be the excep-

tion. Used on potato crops in Long Island, New York, aldicarb has contaminated groundwater there. It has been estimated that levels of 6 μg L^{-1} may persist up to 20 years.

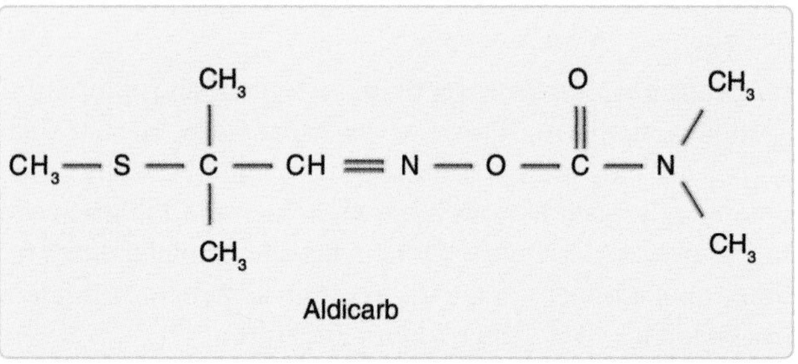

13.10.4 Herbicides

In the United States, herbicides—chemicals that kill plants—are used in greater quantities than insecticides. Herbicides are added directly to the soil; thus they can, in some cases, readily enter the groundwater. Many herbicides originally on the U.S. market, have subsequently been banned. Chlorophenoxy compounds include 2,4-dichlorophenoxyacetic acid (2,4-D); 2,4,5-trichlorophenoxyacetic acid (2,4,5-T); and 2,4,5-trichlorophenoxypropionic acid (2,4,5-TP or silvex). These herbicides act as growth hormones, forcing plant growth to outstrip the ability to provide nutrients. Although these compounds can have toxic effects on the liver, kidney, and central nervous system, clinical reports of poisoning are rare. The compounds, which have a half-life of about 24 hours, are rapidly excreted in humans via urine.

One of the best known, and most controversial, of the chlorophenoxy compounds is 2,4,5-T, which was combined with 2,4-D to create the defoliant Agent Orange, used in the Vietnam War. However, during the industrial synthesis of 2,4,5-T and 2,4-D, a hazardous byproduct can also be inadvertently formed : tetrachlorodioxin (TCDD) (Figure 13.27) or **dioxin** for short. Dioxin, which can also be produced during the combustion of certain substances, is the most toxic manufactured chemical known. In sufficient doses, dioxin is a potent teratogen and carcinogen in laboratory animals, and causes liver injury and general tissue wasting. At lower doses, it causes a form of acne called chloracne, which concentrates between the eyes and hairline.

Clinical reports of acute dioxin poisoning are rare. Dramatic interspecies differences exist for the effects of dioxin; the LD$_{50}$ for guinea pigs is about 1/10,000 of the LD$_{50}$ for hamsters. Fortunately, current evidence indicates that the human reaction to dioxin tends to resemble that of the hamster rather than that of the guinea pig. However, dioxin contamination must be considered a serious environmental problem. In Times Beach, Missouri, for exam-

Figure 13.26 Chemical structure of aldicarb.

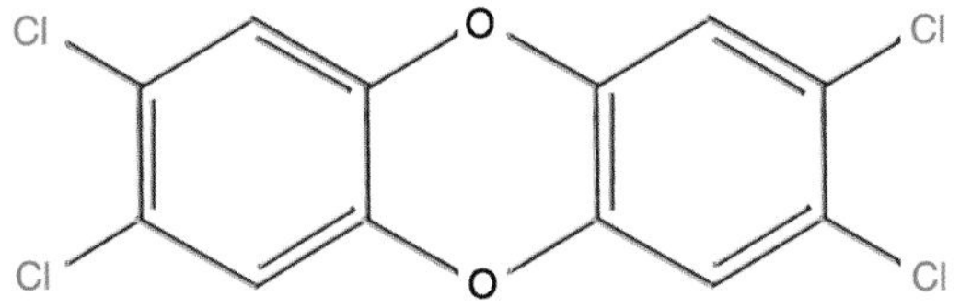

Figure 13.27 Chemical structure of dioxin (TCDD).

ple, dioxin contamination has forced the abandonment of 800 homes since 1984.

13.10.5 Halogenated Hydrocarbons

Halogenated hydrocarbons are common because they are widely used as effective, yet relatively nonflammable solvents, unlike kerosene or gasoline. Halogenated hydrocarbons are also formed during the chlorination of drinking water when chlorine combines with organic material in the water. One of the oldest and simplest of these compounds is carbon tetrachloride (CCl_4), which was extensively used as a solvent and dry-cleaning agent; it is also so nonflammable that it was used in fire extinguishers. (The use of carbon tetrachloride as a dry-cleaning agent was banned after it was shown to cause liver damage.)

The halogenated hydrocarbons (including trichloroethene, halothane, and chloroform) tend to be similar to carbon tetrachloride in their health effects. In high doses, carbon tetrachloride causes CNS depression—so much so that it

was once used as an anesthetic. It can also sensitize the heart muscle to catecholamines (hormones such as epinephrine) and thus can cause heart attacks. It can also cause kidney injury, liver injury, and cancer in laboratory animals. High blood alcohol levels can act as a potentiator for carbon tetrachloride's damaging effects on internal organics. It owes its toxicity to the fact that it is converted in the liver to carbon trichloride ($\cdot CCl_3$), which is a free radical capable of inducing peroxidation of lipid double bonds and poisoning protein-synthesizing enzymes.

Table 13.10 summarizes the health effects of various halogenated hydrocarbons. Plus signs ($+$) indicate a harmful effect, minus signs ($-$) indicate a lack of effect, and both a plus and a minus indicates a less significant effect. Among the methanes is chloroform, a trihalomethane formed during the chlorination of drinking water. Chloroethylene (vinyl chloride) receives three plus signs under cancer because it has been established as a human carcinogen, while the others have been established as probable human carcinogens based on results of animal studies. Also noteworthy are trichloroethylene and tetrachloroethylene (perchlorethylene). These chemicals are very common contaminants of groundwater (see Chapter 17). Although they are listed as probable carcinogens, they cause cancer in laboratory animals only in very high doses. Also, it has been observed that when perchlorethylene and tetrachloroethylene degrade naturally in groundwater, vinyl chloride is formed as a degradation product. The toxic effects of other common hazardous compounds are shown in Table 13.11.

TABLE 13.10 Health effects of some chlorinated halogenated hydrocarbons, ($+$) = harmful effect, ($-$) = no effect, ($+-$) = a less significant effect.

	CNS Depression	Sensitization of Heart	Liver Injury	Kidney Injury	Cancer
Methanes					
Carbon tetrachloride	+	+	++++	++	+
Chloroform	+	+	+++	+++	+
Dichloromethane (methylene chloride)	+	−	+−	−	+
Ethanes					
1,1-Dichloroethane	+	+	+	NDA	NDA
2,2-Dichloroethane	+	NDA[a]	+	−	+
1,1,1-Trichloroethane	+	+	+−	−	−
1,1,2-Trichloroethane	+	NDA	++	−	+
1,1,2,2-Tetrachlorethane	+	NDA	++	++	NDA
Hexachloroethane	+	NDA	NDA	+	+
Ethylenes					
Chloroethylene (vinyl chloride)	+	NDA	++	−	+++
1,1-Dichloroethylene (vinylidine chloride)	+	NDA	+++	−	+
1,2-*trans*-Dichloroethylene	+	NDA	++	NDA	NDA
Trichloroethylene	+	+	+	+−	+
Tetrachloroethylene (perchloroethylene)	0	−	+−	+−	0

[a]NDA = No data available.

Adapted from U.S. EPA, 1989.

TABLE 13.11 Toxic effects of common hazardous compounds.

CHEMICAL	ACUTE EFFECTS	CHRONIC EFFECTS
Aliphatic Hydrocarbons		
Alkanes	Central nervous system impairment	No known carcinogenicity or other chronic effects
Monocyclic Aromatic Hydrocarbons		
Toluene	Central nervous system depression, including agitation, delirium, coma	Central nervous system impairment; no proven carcinogenic activity
Xylenes	Liver toxicity, including steatosis, hepatic cell necrosis, and partial tract enlargement	
Nonhalogenated Solvents		
Acetone	Low toxicity, some CNS effects	Minimal chronic effects
Pesticides		
Aldrin and Dieldrin	Tremors, seizures, and coma	Carcinogenicity
DDT	Affects sodium-potassium pump of neural membrane pesticides produce headaches, nausea, vomiting, and tremors	Minimal chronic effects in humans; no proven human carcinogenicity
Pentachlorophenol	Uncoupling of oxidative phosphorylation	Liver toxicity, including fatty tissue infiltration and elevated enzymes
Industrial Intermediates		
Phenol	A range of health effects including cardiac dysrhythmia, dermal necrosis, and elevated liver enzymes	Not a human carcinogen; some evidence that phenol is a complete carcinogen in mice
Chlorinated benzenes	Dizziness, headaches	Porphyria cutanea, aplastic anemia, leukemia
Polychlorinated Biphenyls		
PCBs	Minimal acute toxicity (0.5 g/kg to 11.3 g/kg)	Chloracne; increased liver enzymes; possible reproduction effects; act as cancer promoters
Dioxins and Furans		
PCDDs/PCDFs	Chloracne, headaches, peripheral neuropathy	Induction of microsomal enzymes; altered liver metabolism; altered T-cell subsets; immunotoxicity; strongly implicated in carcinogenicity (may be a promoter)
Inorganic Compounds		
Arsenic	Loss of blood, intestinal injuries, acute respiratory failure	Myelogeneous leukemia, cancer of skin, lungs, lymph glands, bladder, kidney, prostate, and liver
Cadmium	Vomiting, cramping, weakness, and diarrhea	Oral ingestion results in renal necrosis and dysfunction; induces lung, prostate, kidney, and stomach cancer in animals; no documented human cancer
Hexavalent chromium	Readily absorbed by the skin, where it acts as an irritant and immune-system sensitizer; oral absorption results in acute renal failure	
Mercury	Central nervous system impairment including injury to motor neurons; renal dysfunction	Central nervous system dysfunction, memory deficits, decrease in psychomotor skill, tremors. Immune system effects resulting in allergic contact dermatitis
Nickel	Not highly toxic; headache, shortness of breath	

Modified from Watts, 1997. From U.S. EPA, 1989.

QUESTIONS AND PROBLEMS

1. What event in the year 1962 had an impact in creating an interest in the environment? Why?

2. Why are small animals used in laboratory tests involving toxic materials?

3. How is safety of chemical substances defined with regard to exposure?

4. What is exposure concentration? Exposure dose? How are the two related?

5. Adults are assumed to consume 2 L of water daily. If a substance is present at 10 mg L^{-1}, give the average individual intake of the substance.

6. Explain the similarities and differences between carcinogens, mutagens, and teratogens.

7. What is LD$_{50}$? How is it used?

8. What are the advantages of short-term toxicity testing?

9. What role do initiators, promoters, and suppressor genes play in cancer formation?

10. Can one predict if a new chemical is a carcinogen by its chemical structure? How would you do this?

11. Give an example of a chemical that is made more toxic through biotransformation in the body.

12. What are some of the eliminations in using epidemiological studies to assess the toxicity or carcinogenicity of a chemical? Do negative findings indicate that the substance is not a carcinogen? Why or why not?

13. What are some bioassays that can be used to assess the toxicity of effluents from wastewater treatment plants?

14. A section of DNA has the following sequence of nitrogenous bases.

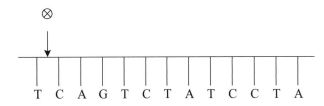

T C A G T C T A T C C T A

If a frameshift mutation occurs at ⊗, list the triplet codes before and after the mutation.

A person is exposed to a highly water-soluble chemical with a constant ingestion rate of 0.12 mg/h. Its transformation in the liver is described by a first-order rate constant of 0.1 h^{-1}. The compound and its metabolites are cleared through the kidneys at a rate of 0.03 mg/h. If the volume of body fluids is 15 L, determine the steady-state concentration of the chemical in the person's vascular system.

15. List two lipophilic substances that cannot be biotransformed by Phase I and II reactions.

16. List the following contaminants in increasing order of storage in the body:

Acetone

DDT

Phenol

Lead

17. How does chemical structure affect the toxicity of a chemical?

18. How do halogenated compounds get into drinking water?

REFERENCES AND ADDITIONAL READING

Bitton G. (1999) *Wastewater Microbiology*. Wiley-Liss, New York.

Chiras D.D. (1985) *Environmental Science*. Benjamin/ Cummings, Menlo Park, California.

Cleckner L.B., Garrison P.J., Hurley J.P., Olson M.L. and Krabbenhoft D.P. (1998) Trophic transfer of methylmercury in the northern Everglades. *Biogeochemistry*. Vol. 40, pp. 347–361.

Cockerham L.G. and Shane B.S., eds. (1994) *Basic Environmental Toxicology*. CRC Press, Boca Raton, Florida.

Hodgson E. and Levi P.E. (1987) *A Textbook of Modern Toxicology*. Appleton and Lange, Norwalk, Connecticut.

Hughes W.W. (1996) *Essentials of Environmental Toxicology*. Taylor and Francis, Washington, D.C.

Kolluru R.V. and Kamrin M.A. (1993) *Environmental Strategies Handbook*. McGraw-Hill, New York.

Landis W.G. and Yu M.H. (1995) *Introduction to Environmental Toxicology*. Lewis Publishers, Boca Raton, Florida.

Maraham S.E. (2003) *Toxicological Chemistry and Biochemistry*, 3rd ed. CRC Press, Boca Raton, Florida.

Philip R.B. (1995) *Environmental Hazards and Human Health*. Lewis Publishers, Boca Raton, Florida.

Rodricks J.V. (1992) *Calculated Risks: The Toxicity and Human Health Risks of Chemicals in Our Environment*. Cambridge University Press, Cambridge, England.

Stine K.E. and Brown T.M. (1996) *Principles of Toxicology*. CRC Press, Boca Raton, Florida.

Sullivan J.B. and Krieger G.R. (1992) *Hazardous Materials Toxicology: Clinical Principles of Environmental Health*. Williams and Wilkins, Baltimore, Maryland.

Tortora G.J., Funke Berdell R., Case, C.L. (2004) *Microbiology, An Introduction, 8th ed.* Pearson Education Inc., Benjamin Cummings, San Francisco, California.

U.S. EPA. (1989) *Risk Assessment, Management and Communication of Drinking Water Contamination*. EPA/625/4–89/024. U.S. Environmental Protection Agency, Washington, D.C.

Watts R.J. (1998) *Hazardous Wastes: Sources, Pathways, Receptors*. Wiley, New York.

Yu M.H. (2001) *Environmental Toxicology*. CRC Press, Boca Raton, Florida.

CHAPTER 14

RISK ASSESSMENT

C.P. Gerba

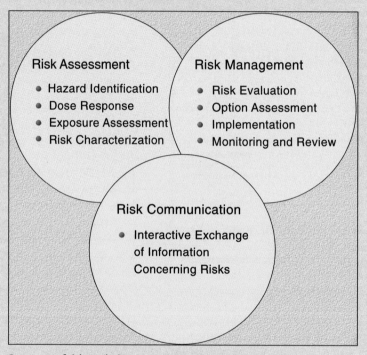

Structure of risk analysis.

14.1 THE CONCEPT OF RISK ASSESSMENT

Risk, which is common to all life, is an inherent property of everyday human existence. It is therefore a key factor in all decision making. Risk assessment or analysis, however, means different things to different people: Wall Street analysts assess financial risks and insurance companies calculate actuarial risks, while regulatory agencies estimate the risks of fatalities from nuclear plant accidents, the incidence of cancer from industrial emissions, and habitat loss associated with increases in human populations. What all these seemingly disparate activities have in common is the concept of a measurable phenomenon called risk that can be expressed in terms of probability. Thus, we can define **risk assessment** as the process of estimating both the probability that an event will occur, and the probable magnitude of its adverse effects—economic, health/safety-related, or ecological—over a specified time period. For example, we might determine the probability that a chemical reactor will fail and the probable effect of its sudden release of contents on the immediate area in terms of injuries and property loss over a period of days. In addition, we might estimate the probable incidence of cancer in the community where the chemical was spilled. Or, in yet another type of risk assessment, we might calculate the health risks associated with the presence of pathogens in drinking water or pesticide in food.

There are, of course, several varieties of risk assessment. Risk assessment as a formal discipline emerged in the 1940s and 1950s, paralleling the rise of the nuclear industry. Safety-hazard analyses have been used since at least the 1950s in the nuclear, petroleum-refining, and chemical-processing industries, as well as in aerospace. Health-risk assessments, however, had their beginnings in 1976 with the EPA's publication of the *Carcinogenic Risk Assessment Guidelines*.

In this chapter, we are concerned with two types of risk assessment:

- **Health-based risks**. For these risks, the focus is on general human health, mainly outside the workplace. Health-based risks typically involve high-probability, low-consequence, chronic exposures whose long latency periods and delayed effects make cause-and-effect relationships difficult to establish. This category also includes microbial risks, which usually have acute short-term effects. However, the consequences of microbial infection can persist throughout an individual's lifetime.

- **Ecological risks**. For these risks, the focus is on the myriad interactions among populations, communities, and ecosystems (including food chains) at both the micro and the macro level. Ecological risks typically involve both short-term catastrophes, such as oil spills, and long-term exposures to hazardous substances.

Whatever its focus, the **risk assessment process** consists of four basic steps:

- **Hazard identification**—Defining the hazard and nature of the harm; for example, identifying a chemical contaminant, say, lead or carbon tetrachloride, and documenting its toxic effects on humans.

- **Exposure assessment**—Determining the concentration of a contaminating agent in the environment and estimating its rate of intake in target organisms. An example would be finding the concentration of aflatoxin in peanut butter and determining the dose an "average" person would receive.

- **Dose–response assessment**—Quantifying the adverse effects arising from exposure to a hazardous agent based on the degree of exposure. This assessment is usually expressed mathematically as a plot showing the response in living organisms to increasing doses of the agent.

- **Risk characterization**—Estimating the potential impact of a hazard based on the severity of its effects and the amount of exposure.

Once the risks are characterized, various regulatory options are evaluated in a process called **risk management**, which includes consideration of social, political, and economic issues, as well as the engineering problems inherent in a proposed solution. One important component of risk management is **risk communication**, which is the interactive process of information and opinion exchange among individuals, groups, and institutions. Risk communication includes the transfer of risk information from expert to nonexpert audiences. In order to be effective, risk communication must provide a forum for balanced discussions of the nature of the risk, lending a perspective that allows the benefits of reducing the risk to be weighed against the costs.

In the United States, the passage of federal and state laws to protect public health and the environment has expanded the application of risk assessment. Major federal agencies that routinely use risk analysis include the Food and Drug Administration (FDA), the Environmental Protection Agency (EPA), and the Occupational Safety and Health Administration (OSHA). Together with state agencies, these regulatory agencies use risk assessment in a variety of situations (Information Box 14.1).

Risk assessment provides an effective framework for determining the relative urgency of problems and the allocation of resources to reduce risks. Using the results of risk analyses, we can target prevention, remediation, and control efforts toward areas, sources, or situations in which the greatest risk reductions can be achieved with the resources available. However, risk assessment is not an absolute procedure carried out in a vacuum; rather, it is an evaluative, multifaceted, comparative process. Thus, to evaluate risk, we must inevitably compare one risk to a host of others. In fact, the comparison of potential risks associated with several problems or issues has developed into a subset of risk assessment called **comparative risk assessment**. Some commonplace risks are shown in Table 14.1. Here we see, for example, that risks from chemical exposure are fairly small relative to those associated with driving a car or smoking cigarettes.

Comparing different risks allows us to comprehend the uncommon magnitudes involved and to understand the

INFORMATION BOX 14.1

Applications of Risk Assessment

· Setting standards for concentrations of toxic chemicals or pathogenic microorganisms in water or food.
· Conducting baseline analyses of contaminated sites or facilities to determine the need for remedial action and the extent of cleanup required.
· Performing cost/benefit analyses of contaminated-site cleanup or treatment options (including treatment processes to reduce exposure to pathogens).
· Developing cleanup goals for contaminants for which no federal or state authorities have promulgated numerical standards; evaluating acceptable variance from promulgated standards and guidelines (e.g., approving alternative concentration limits).
· Constructing "what-if" scenarios to compare the potential impact of remedial or treatment alternatives and to set priorities for corrective action.
· Evaluating existing and new technologies for effective prevention, control, or mitigation of hazards and risks.
· Articulating community public health concerns and developing consistent public health expectations among different localities.

TABLE 14.1 Examples of some commonplace risks in the United States.*

RISK	LIFETIME RISK OF MORTALITY
Cancer from cigarette smoking (one pack per day)	1:4
Death in a motor vehicle accident	2:100
Homicide	1:100
Home accident deaths	1:100
Cancer from exposure to radon in homes	3:1,000
Exposure to the pesticide aflatoxin in peanut butter	6:10,000
Diarrhea from rotavirus	1:10,000
Exposure to typical EPA maximum chemical contaminant levels	1:10,000–1:10,000,000

*Based on data in Wilson and Crouch (1987) and Gerba and Rose (1992). From *Pollution Science* © 1996, Academic Press, San Diego, CA.

level, or magnitude, of risk associated with a particular hazard. But comparison with other risks cannot itself establish the *acceptability* of a risk. Thus, the fact that the chance of death from a previously unknown risk is about the same as that from a known risk does not necessarily imply that the two risks are equally acceptable. Generally, comparing risks along a single dimension is not helpful when the risks

are widely perceived as qualitatively different. Rather, we must take into account certain qualitative factors that affect risk perception and evaluation when selecting risks to be compared. Some of these qualifying factors are listed in Table 14.2. We must also understand the underlying premise that *voluntary risk is always more acceptable than involuntary risk*. For example, the same people who cheerfully drive their cars every day—thus incurring a 2:100 lifetime risk of death by automobile—are quite capable of refusing to accept the 6:10,000 involuntary risk of eating peanut butter contaminated with aflatoxin.

In considering risk, then, we must also understand another principle—the *de minimis* principle, which means that there are some levels of risk so trivial that they are not worth

TABLE 14.2 Factors affecting risk perception and risk analysis.

FACTOR	CONDITIONS ASSOCIATED WITH INCREASED PUBLIC CONCERN	CONDITIONS ASSOCIATED WITH DECREASED PUBLIC CONCERN
Catastrophic potential	Fatalities and injuries grouped in time and space	Fatalities and injuries scattered and random
Familiarity	Unfamiliar	Familiar
Understanding	Mechanisms or process not understood	Mechanisms or process understood
Controllability (personal)	Uncontrollable	Controllable
Voluntariness of exposure	Involuntary	Voluntary
Effects on children	Children specifically at risk	Children not specifically at risk
Effects manifestation	Delayed effects	Immediate effects
Effects on future generations	Risk to future generations	No risk to future generations
Victim identity	Identifiable victims	Statistical victims
Dread	Effects dreaded	Effects not dreaded
Trust in institutions	Lack of trust in responsible institutions	Trust in responsible institutions
Media attention	Much media attention	Little media attention
Accident history	Major and sometimes minor accidents	No major or minor accidents
Equity	Inequitable distribution of risks and benefits	Equitable distribution of risks and benefits
Benefits	Unclear benefits	Clear benefits
Reversibility	Effects irreversible	Effects reversible
Origin	Caused by human actions or failures	Caused by acts of nature

Source: Covello et al. (1988). From *Pollution Science* © 1996, Academic Press, San Diego, CA.

bothering about. However attractive, this concept is difficult to define, especially if we are trying to find a *de minimis* level acceptable to an entire society. Understandably, regulatory authorities are reluctant to be explicit about an "acceptable" risk. (How much aflatoxin would you consider acceptable in *your* peanut butter and jelly sandwich? How many dead insect parts?) But it is generally agreed that a lifetime risk on the order of one in a million (or in the range of 10^{-6} to 10^{-4}) is trivial enough to be acceptable for the general public. Although the origins and precise meaning of a one-in-a-million acceptable risk remain obscure, its impact on product choices, operations, and costs is very real— running, for example, into hundreds of billions of dollars in hazardous waste site cleanup decisions alone. The levels of acceptable risk can vary within this range. Levels of risk at the higher end of the range (10^{-4} rather than 10^{-6}) may be acceptable if just a few people are exposed rather than the entire populace. For example, workers dealing with food additives can often tolerate higher levels of risk than can the public at large. These higher levels are justified because workers tend to be a relatively homogeneous, healthy group and because employment is voluntary; however, the sum level of risks would not be acceptable for those same food additives in general.

14.2 THE PROCESS OF RISK ASSESSMENT

14.2.1 Hazard Identification

The first step in risk assessment is to determine the nature of the hazard. For pollution-related problems, the hazard in question is usually a specific chemical, a physical agent (such as irradiation), or a microorganism identified with a specific illness or disease. Thus the hazard identification component of a pollution risk assessment consists of a review of all relevant biological and chemical information bearing on whether or not an agent poses a specific threat. For example, in the *Guidelines for Carcinogen Risk Assessment* (U.S. EPA, 1986), the following information is evaluated for a potential carcinogen:

- Physical/chemical properties, routes, and patterns of exposure
- Structure/activity relationships of the substance
- Absorption, distribution, metabolism, and excretion characteristics of the substance in the body
- The influence of other toxicological effects
- Data from short-term tests in living organisms
- Data from long-term animal studies
- Data from human studies

Once these data are reviewed, the animal and human data are both separated into groups characterized by degree of evidence:

- Sufficient evidence of carcinogenicity
- Limited evidence of carcinogenicity

- Inadequate evidence
- No data available
- No evidence of carcinogenicity

The available information on animal and human studies is then combined into a weight-of-evidence classification scheme to assess the likelihood of carcinogenicity. This scheme—which is like that developed by the EPA—gives more weight to human than to animal evidence (when it is available) and includes several groupings (Table 14.3).

Clinical studies of disease can be used to identify very large risks (between 1/10 and 1/100), most epidemiological studies can detect risks down to 1/1,000, and very large epidemiological studies can examine risks in the 1/10,000 range. However, risks lower than 1/10,000 cannot be studied with much certainty using epidemiological approaches. Since regulatory policy objectives generally strive to limit risks below 1/100,000 for life-threatening diseases like cancer, these lower risks are often estimated by extrapolating from the effects of high doses given to animals.

14.2.2 Exposure Assessment

Exposure assessment is the process of measuring or estimating the intensity, frequency, and duration of human exposure to an environmental agent. Exposure to contaminants can occur via inhalation, ingestion of water or food, or absorption through the skin upon dermal contact. Contaminant sources, release mechanisms, transport, and transformation characteristics are all important aspects of exposure assessment, as are the nature, location, and activity patterns of the exposed population. This explains why it is critical to understand the factors and processes influencing the transport and fate of a contaminant (see Chapters 6, 7, 8, and 17).

An **exposure pathway** is the course that a hazardous agent takes from a source to a receptor (*e.g.*, human or animal) via environmental carriers or media—generally, air (volatile compounds, particulates) or water (soluble compounds) (Figure 14.1). An exception is electromagnetic radiation, which needs no medium (see Chapter 21). The **exposure route**, or intake pathway, is the mechanism by which the transfer occurs—usually by inhalation, ingestion, and/or dermal contact. Direct contact can result in a local effect at the point of entry and/or in a systemic effect.

TABLE 14.3 EPA categories for carcinogenic groups.

CLASS	DESCRIPTION
A	Human carcinogen
B	Probable carcinogen
B_1	Linked human data
B_2	No evidence in humans
C	Possible carcinogen
D	No classification
E	No evidence

From U.S. EPA, 1986.

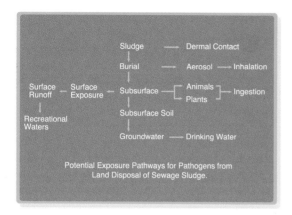

Figure 14.1 Exposure pathways for potential contaminants. Modified from Straub et al., 1993.

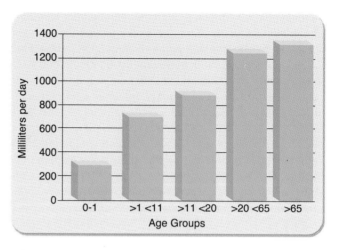

Figure 14.2 Average tap water ingestion rates in the United States by age. From Roseberry and Burmaster, 1992.

The quantification of exposure, intake, or potential dose can involve equations with three sets of variables:

- Concentrations of chemicals or microbes in the media
- Exposure rates (magnitude, frequency, duration)
- Quantified biological characteristics of receptors (*e.g.*, body weight, absorption capacity for chemicals; level of immunity to microbial pathogens)

Exposure concentrations are derived from measured and/or modeled data. Ideally, exposure concentrations should be measured at the points of contact between the environmental media and current or potential receptors. It is usually possible to identify potential receptors and exposure points from field observations and other information. However, it is seldom possible to anticipate all potential exposure points and measure all environmental concentrations under all conditions. In practice, a combination of monitoring and modeling data, together with a great deal of professional judgment, is required to estimate exposure concentrations.

In order to assess exposure rates via different exposure pathways, we have to consider and weigh many factors. For example, in estimating exposure to a substance via drinking

water, we first have to determine the average daily consumption of that water. But this isn't as easy as it sounds. Studies have shown that daily fluid intake varies greatly from individual to individual. Moreover, tap water intake depends on how much fluid is consumed as tap water, and how much is ingested in the form of soft drinks and other non-tap-water sources. Tap water intake also changes significantly with age (Figure 14.2), body weight, diet, and climate. Because these factors are so variable, the EPA has suggested a number of very conservative "default" exposure values that can be used when assessing contaminants in tap water, vegetables, soil, and the like (Table 14.4).

One important route of exposure is the food supply. Toxic substances are often bioaccumulated, or concentrated, in plant and animal tissues, thereby exposing humans who ingest those tissues as food. Moreover, many toxic substances tend to be biomagnified in the food chain, so that animal tissues contain relatively high concentrations of toxins. Take fish, for example. It is relatively straightforward to estimate concentrations of contaminants in water. Thus, we can use a **bioconcentration factor (BCF)** to estimate the tendency for a substance in water

TABLE 14.4 EPA standard default exposure factors.

LAND USE	EXPOSURE PATHWAY	DAILY INTAKE	EXPOSURE FREQUENCY (DAYS/YEAR)	EXPOSURE DURATION (YEARS)
Residential	Ingestion of potable water	2 L day^{-1}	350	30
	Ingestion of soil and dust	200 mg (child)	350	6
		100 mg (adult)		24
	Inhalation of contaminants	20 m^3 (total)	350	30
		15 m^3 (indoor)		
Industrial and commercial	Ingestion of potable water	1 liter	250	25
	Ingestion of soil and dust	50 mg	250	25
	Inhalation of contaminants	20 m^3 (workday)	250	25
Agricultural	Consumption of homegrown produce	42 g (fruit)	350	30
		80 g (vegetable)		
Recreational	Consumption of locally caught fish	54 g	350	30

Modified from Kolluru (1993). From *Pollution Science* ©1996, Academic Press, San Diego, CA.

to accumulate in fish tissue. The concentration of a chemical in fish can be estimated by multiplying its concentration in water by the BCF. The greater the value of the BCF, the more the chemical accumulates in the fish and the higher the risk of exposure to humans.

The units of BCF—liters per kilogram (L kg^{-1}) —are chosen to allow the concentration of a chemical to be expressed as milligrams per liter (mg L^{-1}) of water and the concentration in fish to be in milligrams per kilogram (mg kg^{-1}) of fish body weight. In Table 14.5, we see the BCFs of several common organic and inorganic chemicals. Note the high values of BCF for the chlorinated hydrocarbon pesticides like dichlorodiphenyltrichloroethane (DDT) and polychlorinated biphenyls (PCBs). This exemplifies the concern we have with such compounds, as discussed in Chapter 10.

14.2.3 Dose–Response Assessment

Chemicals and other contaminants are not equal in their capacity to cause adverse effects. To determine the capacity of agents to cause harm, we need quantitative toxicity data. Some toxicity data are derived from occupational, clinical, and epidemiological studies. Most toxicity data, however, come from animal experiments in which researchers expose laboratory animals, mostly mice and rats, to increasingly higher concentrations or doses and observe their corresponding effects. The result of these experiments is the *dose–response relationship*—a quantitative relationship that indicates the agent's degree of toxicity to exposed species. Dose is normalized as milligrams of substance or pathogen ingested, inhaled, or absorbed (in the case of chemicals) through the skin per kilogram of body weight per day (mg kg^{-1} day^{-1}). Responses or effects can vary widely—from no observable effect, to temporary and reversible effects (*e.g.*, enzyme depression caused by some pesticides or diarrhea caused by viruses), to permanent organ injury (*e.g.*, liver and kidney damage caused by chlorinated solvents, heavy metals, or viruses), to chronic functional impairment (*e.g.*, bronchitis or emphysema arising from smoke damage), to death.

TABLE 14.5 Bioconcentration factors (BCFs) for various organic and inorganic compounds.

CHEMICAL	BCF (L kg^{-1})
Aldrin	28
Benzene	44
Cadmium	81
Chlordane	14,000
Chloroform	3.75
Copper	200
DDT	54,000
Formaldehyde	0
Nickel	47
PCBs	100,000
Trichloroethylene	10.6
Vinyl chloride	1.17

From U.S. EPA, 1990.

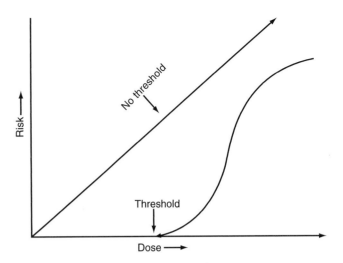

Figure 14.3 Relationship between a threshold and nonthreshold response.

The goal of a dose–response assessment is to obtain a mathematical relationship between the amount (concentration) of a toxicant or microorganism to which a human is exposed and the risk of an adverse outcome from that dose. The data resulting from experimental studies is presented as a dose–response curve, as shown in Figure 14.3. The abscissa describes the dose, while the ordinate measures the risk that some adverse health effect will occur. In the case of a pathogen, for instance, the ordinate may represent the risk of infection, and not necessarily illness.

Dose–response curves derived from animal studies must be interpreted with care. The data for these curves are necessarily obtained by examining the effects of large doses on test animals. Because of the costs involved, researchers are limited in the numbers of test animals they can use—it is both impractical and cost-prohibitive to use thousands (even millions) of animals to observe just a few individuals that show adverse effects at low doses (e.g., risks of 1:1,000 or 1:10,000). Researchers must therefore extrapolate low-dose responses from their high-dose data. And therein lies the rub: Dose–response curves are subject to controversy because their results change depending on the method chosen to extrapolate from the high doses actually administered to laboratory test subjects to the low doses humans are likely to receive in the course of everyday living.

This controversy revolves around the choice of several mathematical models that have been proposed for extrapolation to low doses. Unfortunately, no model can be proved or disproved from the data, so there is no way to know which model is the most accurate. The choice of models is therefore strictly a policy decision, which is usually based on understandably conservative assumptions. Thus, for noncarcinogenic chemical responses, the assumption is that some *threshold* exists below which there is no toxic response; that is, no adverse effects will occur below some very low dose (say, one in a million) (Figure 14.3). Carcinogens, however, are considered *nonthreshold*—that is, the conservative assumption is

TABLE 14.6 Primary models used for assessment of nonthreshold effects.

MODEL[a]	COMMENTS
One-hit	Assumes (1) a single stage for cancer and (2) malignant change induced by one molecular or radiation interaction *Very conservative*
Linear multistage	Assumes multiple stages for cancer *Fits curve to the experimental data*
Multihit	Assumes several interactions needed before cell becomes transformed *Least conservative model*
Probit	Assumes probit (lognormal) distribution for tolerances of exposed population *Appropriate for acute toxicity; questionable for cancer*

[a]All these models assume that exposure to the pollutant will always produce an effect, regardless of dose.

Modified from Cockerham and Shane, 1994. From *Pollution Science* © 1996, Academic Press, San Diego, CA.

TABLE 14.7 Lifetime risks of cancer derived from different extrapolation models.

MODEL APPLIED	LIFETIME RISK (1.0 mg kg^{-1} day^{-1}) OF TOXIC CHEMICAL[a]	
One-hit	6.0×10^{-5}	(1 in 17,000)
Multistage	6.0×10^{-6}	(1 in 167,000)
Multihit	4.4×10^{-7}	(1 in 2.3 million)
Probit	1.9×10^{-10}	(1 in 5.3 billion)

[a]All risks for a full lifetime of daily exposure. The lifetime is used as the unit of risk measurement, because the experimental data reflect the risk experienced by animals over their full lifetimes. The values shown are upper confidence limits on risk.

Source: U.S. EPA, 1990. From *Pollution Science* © 1996, Academic Press, San Diego, CA.

that exposure to any amount of carcinogen creates some likelihood of cancer. This means that the only "safe" amount of carcinogen is zero, so the dose–response plot is required to go through the origin (0), as shown in Figure 14.3.

There are many mathematical models to choose from, including the one-hit model, the multistage model, the multihit model, and the probit model. The characteristics of these models for nonthreshold effects are listed in Table 14.6.

The **one-hit model** is the simplest model of carcinogenesis in which it is assumed:

1. That a single chemical "hit," or exposure, is capable of inducing malignant change (*i.e.*, a single hit causes irreversible damage of DNA, leading to tumor development). Once the biological target is hit, the process leading to tumor formation continues independently of dose.

2. That this change occurs in a single stage.

The **multistage model** assumes that tumors are the result of a sequence of biological events, or stages. In simplistic terms, the biological rationale for the multistage model is that there are a series of biological stages that a chemical must pass through (*e.g.*, metabolism, covalent bonding, DNA repair, and so on) without being deactivated, before the expression of a tumor is possible.

The rate at which the cell passes through one or more of these stages is a function of the dose rate. The multistage model also has the desirable feature of producing a linear relationship between risk and dose.

The **multihit model** assumes that a number of dose-related hits are needed before a cell becomes malignant. The most important difference between the multistage and multihit model is that in the multihit model, all hits must result from the dose, whereas in the multistage model, passage through some of the stages can occur spontaneously. The practical implication of this is that the multihit models are

generally much flatter at low doses and consequently predict a lower risk than the multistage model.

The **probit model** is not derived from mechanistic assumptions about the cancer process. It may be thought of as representing distributions of tolerances to carcinogens in a large population. The model assumes that the probability of the response (cancer) is a linear function of the log of the dose (log normal). While these models may be appropriate for acute toxicity they are considered questionable for carcinogens. These models would predict the lowest level of risk of all the models.

The effect of models on estimating risk for a given chemical is shown in Table 14.7 and Figure 14.4. As we can see, the choice of models results in order-of-magnitude differences in estimating the risk at low levels of exposure.

The **linear multistage model,** a modified version of the multistage model, is the EPA's model of choice, because this agency chooses to err on the side of safety and overemphasize risk. This model assumes that there are multiple stages for cancer (*i.e.*, a series of mutations or biotransformations) involving many carcinogens, co-carcinogens, and promoters (see Chapter 13) that can best be modeled by a series of mathematical functions. At low doses, the slope of the

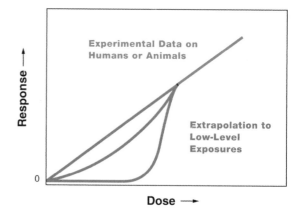

Figure 14.4 Extrapolation of dose–response curves. Adapted from U.S. EPA, 1990. From *Pollution Science* © 1996, Academic Press, San Diego, CA.

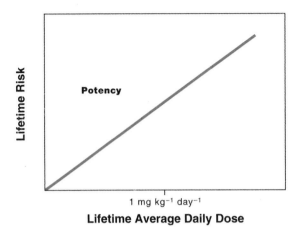

Figure 14.5 Potency factor is the slope of the dose–response curve at low doses. At low doses, the slope of the dose–response curve produced by the multistage model is called the potency factor. It is the risk produced by a lifetime average dose of 1 mg kg^{-1} day^{-1}. Adapted from U.S. EPA, 1990. From *Pollution Science* © 1996, Academic Press, San Diego, CA.

TABLE 14.8 Chemical RfDs for chronic noncarcinogenic effects of selected chemicals.

CHEMICAL	RfD (mg kg^{-1} day^{-1})
Acetone	0.1
Cadmium	0.0005
Chloroform	0.01
Methylene chloride	0.06
Phenol	0.04
Polychlorinated biphenyl	0.0001
Toluene	0.3
Xylene	2.0

From U.S. EPA, 1990.

dose–response curve produced by the linear multistage model is called the **potency factor (PF)** or **slope factor (SF)** (Figure 14.5), which is the reciprocal of the concentration of chemical measured in milligrams per kilogram of animal body weight per day, that is, 1/(mg kg^{-1} day^{-1}), or the risk produced by a lifetime **average dose (AD)** of 1 mg kg^{-1} day^{-1}. Thus the dose–response equation for a carcinogen is

$$\text{Lifetime Risk} = \text{AD} \times \text{PF} \qquad \text{(Eq. 14.1)}$$

The probability of *getting* cancer (not the probability of *dying* of cancer) and the associated dose, consist of an average taken over an assumed 70-year human lifetime. This dose is called the lifetime average daily dose or **chronic daily intake**.

The dose–response effects for noncarcinogens allow for the existence of thresholds, that is, a certain quantity of a substance or dose below which there is **no observable toxic effect (NOAEL**; see Chapter 13) by virtue of the body's natural repair and detoxifying capacity. If a NOAEL is not available, a LOAEL (lowest observed adverse effect level) may be used, which is the lowest observed dose or concentration of a substance at which there is a detectable adverse health effect. When a LOAEL is used instead of a NOAEL, an additional uncertainty factor is normally applied. Examples of toxic substances that have thresholds are heavy metals and polychlorinated biphenyls (PCBs). These thresholds are represented by the **reference dose**, or **RfD**, of a substance, which is the intake or dose of the substance per unit body weight per day (mg kg^{-1} day^{-1}) that is likely to pose no appreciable risk to human populations, including such sensitive groups as children (Table 14.8). A dose–response plot for carcinogens therefore goes through this reference point (Figure 14.6).

In general, substances with relatively high slope factors and low reference doses tend to be associated with higher toxicities. The RfD is obtained by dividing the NOAEL (see Chapter 13) by an appropriate uncertainty factor, sometimes called a **safety factor** or **uncertainty factor**. A 10-fold uncertainty factor is used to account for differences in sensitivity between the most sensitive individuals in an exposed human population. These include pregnant women, young children, and the elderly, who are more sensitive than "average" people. Another factor of 10 is added when the NOAEL is based on animal data that are extrapolated to humans. In addition, another factor of 10 is sometimes applied when questionable or limited human and animal data are available. The general formula for deriving an RfD is

$$\text{RfD} = \frac{\text{NOAEL}}{\text{VF}_1 \times \text{VF}_2 \dots \times \text{VF}_n} \qquad \text{(Eq. 14.2)}$$

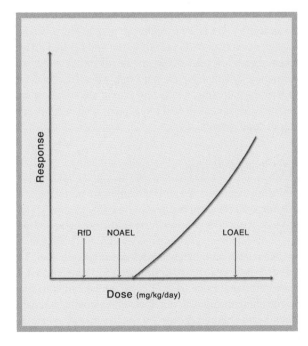

Figure 14.6 Relationships between RfD, NOAEL, and LOAEL for noncarcinogens.

where VF_i are the uncertainty factors. As the data become more uncertain, higher safety factors are applied. For example, if data are available from a high-quality epidemiological study, a simple uncertainty factor of 10 may be used by simply dividing the original value for RfD by 10 to arrive at a new value of RfD, which reflects the concern for safety. The RfDs of several noncarcinogenic chemicals are shown in Table 14.8.

The RfD (Figure 14.6) can be used in quantitative risk assessments by using the following relationship:

$$\text{Risk} = \text{PF (CDI} - \text{RfD)} \qquad \text{(Eq. 14.3)}$$

where CDI is the chronic daily intake, and the *potency factor (PF)* is the slope of the dose–response curve. Table 14.9 contains potency factors for some potential carcinogens:

$$\text{CDI (mg kg}^{-1} \text{ day}^{-1}) = $$
$$\frac{\text{Average daily dose (mg day}^{-1})}{\text{Body weight (kg)}} \qquad \text{(Eq. 14.4)}$$

This type of risk calculation is rarely performed. In most cases, the RfD is used as a simple indicator of potential risk in practice. That is, the chronic daily intake is simply compared with the RfD, then, if the CDI is below the RfD, it is assumed that the risk is negligible for almost all members of an exposed population.

14.2.4 Risk Characterization

The final phase of risk assessment process is risk characterization. In this phase, exposure and dose–response assessments are integrated to yield probabilities of effects occurring in humans under specific exposure conditions. Quantitative risks are calculated for appropriate media and pathways. For example, the risks of lead in water are estimated over a lifetime assuming: (1) that the exposure of 2 liters of water per day is ingested over a 70-year lifetime; and (2) that different concentrations of lead occur in the drinking water. This information can then be used by risk managers to develop standards or guidelines for specific toxic chemicals or infectious microorganisms in different media, such as the drinking water or food supply.

14.2.4.1 Cancer risks

If the dose–response curve is assumed to be linear at low doses for a carcinogen, then:

Incremental lifetime risk of cancer = (CDI) (PF)   (Eq. 14.5)

The linearized multistage model assumptions (see Table 14.6) estimates the risk of getting cancer, which is not necessarily the same as the risk of dying of cancer, so it should be even more conservative as an upper-bound estimate of cancer deaths. Potency factors can be found in the EPA database on toxic substances called the Integrated Risk Information System (IRIS) (see Information Box 14.2). Table 14.9 contains the potency factor for some of these chemicals.

TABLE 14.9 Toxicity data for selected potential carcinogens.

CHEMICAL	POTENCY FACTOR ORAL ROUTE (mg kg day^{-1})
Arsenic	1.75
Benzene	2.9×10^{-2}
Carbon tetrachloride	0.13
Chloroform	6.1×10^{-3}
DDT	0.34
Dieldrin	30
Heptachlor	3.4
Methylene chloride	7.5×10^{-3}
Polychlorinated biphenyls (PCBs)	7.7
2,3,7,8-TCDD (dioxin)	1.56×10^5
Tetrachloroethylene	5.1×10^{-2}
Trichloroethylene (TCE)	1.1×10^{-2}
Vinyl chloride	2.3

From U.S. EPA, www.epa.gov/iris.

The mean exposure concentration of contaminants is used with exposed population variables and the assessment determined variables to estimate contaminant intake. The general equation for chemical intake is

$$\text{CDI} = \frac{\text{C} \times \text{CR} \times \text{EFD}}{\text{BW}} \times \frac{1}{\text{AT}} \qquad \text{(Eq. 14.6)}$$

where:

CDI = chronic daily intake; the amount of chemical at the exchange boundary (mg/kg-day)

C = average exposure concentration over the period (*e.g.*, mg/L for water or mg/m^3 for air)

CR = contact rate, the amount of contaminated medium contacted per unit time (L/day or m^3/day)

EFD = exposure frequency and duration, a variable that describes how long and how often exposure occurs. The EFD is usually divided into two terms:
EF = exposure frequency (days/year) and
ED = exposure duration (years)

BW = average body mass over the exposure period (kg)

AT = averaging time; the period over which the exposure is averaged (days)

Determination of accurate intake data is sometimes difficult; for example, exposure frequency and duration vary among individuals and must often be estimated; site-specific information may be available; and professional judgment may be necessary. Equations for estimating daily contamination intake rates from drinking water, the air, and contaminated food, and for dermal exposure while swimming, have been reported by the EPA. Two of the most common routes of exposure are

---

INFORMATION BOX 14.2

Integrated Risk Information System (IRIS)

The Integrated Risk Information System (IRIS), prepared and maintained by the U.S. Environmental Protection Agency (U.S. EPA), is an electronic database containing information on human health effects that may result from exposure to various chemicals in the environment (www.epa.gov/iris). IRIS was initially developed for EPA staff in response to a growing demand for consistent information on chemical substances for use in risk assessments, decision-making, and regulatory activities. The information in IRIS is intended for those without extensive training in toxicology, but with some knowledge of health sciences. The heart of the IRIS system is its collection of computer files covering individual chemicals. These chemical files contain descriptive and quantitative information in the following categories:

· Oral reference doses and inhalation reference concentrations (RfDs) for chronic noncarcinogenic health effects.
· Hazard identification, oral slope factors, and oral and inhalation unit risks for carcinogenic effects.

Oral RfD Summary for Arsenic.

CRITICAL EFFECT	EXPERIMENTAL DOSES*	UF	RFD
Hyperpigmentation, keratosis and possible vascular complications Human chronic oral exposure Tseng, 1977; Tseng et al., 1968	NOAEL: 0.009 mg/L converted to 0.0008 mg/kg-day LOAEL: 0.17 mg/L converted to 0.014 mg/kg-day	3	3E-4 mg/kg-day

*Conversion Factors—NOAEL was based on an arithmetic mean of 0.009 mg/L in a range of arsenic concentration of 0.001 to 0.017 mg/L. This NOAEL also included estimation of arsenic from food. Since experimental data were missing, arsenic concentrations in sweet potatoes and rice were estimated as 0.002 mg/day. Other assumptions included consumption of 4.5 L water/day and 55 kg body weight (Abernathy et al., 1989). NOAEL = [(0.009 mg/L × 4.5 L/day) + 0.002 mg/day]/55 kg = 0.0008 mg/kg-day. The LOAEL dose was estimated using the same assumptions as the NOAEL starting with an arithmetic mean water concentration from Tseng (1977) of 0.17 mg/L. LOAEL = [(0.17 mg/L × 4.5 L/day) + 0.002 mg/day]/55 kg = 0.014 mg/kg-day.
UF = Uncertainty Factor or Safety Factor.

---

through drinking contaminated water and breathing contaminated air. The intake for ingestion of waterborne chemicals is

$$CDI = \frac{CW \times IR \times EF \times ED}{BW \times AT}$$ (Eq. 14.7)

where

CDI = chronic daily intake by ingestion (mg/kg-day)

CW = chemical concentration in water (mg/L)

IR = ingestion rate (L/day)

EF = exposure frequency (days/year)

ED = exposure duration (years)

BW = body weight (kg)

AT = averaging time (period over which the exposure is averaged—days)

Some of the values used in Equation 14.5 are

CW: site-specific measured or modeled value

IR: 2 L/day (adult, 90th percentile); 1.4 L/day (adult, average)

EF: pathway-specific value (dependent on the frequency of exposure-related activities)

ED: 70 years (lifetime; by convention); 30 years [national upper-bound time (90th percentile) at one residence]; 9 years [national median time (50th percentile) at one residence]

BW: 70 kg (adult, average); Age-specific values

AT: pathway-specific period of exposure for noncarcinogenic effects (*i.e.*, ED × 365 days/year), and 70-year lifetime for carcinogenic effects (*i.e.*, 70 years × 365 days/year), averaging time.

14.2.4.2 Noncancer risks

Noncancer risks are expressed in terms of a hazard quotient (HQ) for a single substance, or hazard index (HI) for multiple substances and/or exposure pathways.

Hazard quotient (HQ) =

$$\frac{\text{Average daily dose during exposure period (mg kg}^{-1}\text{ day}^{-1})}{\text{RfD (mg kg}^{-1}\text{ day}^{-1})}$$ (Eq. 14.8)

EXAMPLE CALCULATION 14.1

Estimation of an Oral Chronic Daily Intake

The mean concentration of 1,2-dichlorobenzene in a water supply is 1.7 μg/L. Determine the chronic daily intake for a 70-kg adult. Assume that 2 L of water are consumed per day.

Solution

The chronic daily intake (CDI) may be calculated using Equation 14.7.

$$CDI = \frac{C \times CR \times EF \times ED}{BW \times AT}$$

where:

CDI = chronic daily intake (mg/kg-day)
C = 1.7 μg/L = 0.0017 mg/L
CR = 2 L/day
EF = 365 days/year
ED = 30 years (standard exposure duration for an adult exposed to a noncarcinogenic)
BW = 70 kg
AT = 365 days/year × 30 years = 10,950 days

Substituting values into the equation yields the chronic daily intake.

$$CDI = \frac{0.0017 \times 2 \times 365 \times 30}{70 \times 10,950} = 4.86 \times 10^{-5} \text{ mg/kg-day}$$

Unlike a carcinogen, the toxicity is important only during the time of exposure, which may be one day, a few days, or years. The HQ has been defined so that if it is less than 1.0, there should be no significant risk or systemic toxicity. Ratios above 1.0 could represent a potential risk, but there is no way to establish that risk with any certainty.

When exposure involves more than one chemical, the sum of the individual hazard quotients for each chemical is used as a measure of the potential for harm. This sum is called the hazard index (HI):

$$HI = \text{Sum of hazard quotients} \qquad \text{(Eq. 14.9)}$$

14.2.4.3 Uncertainty analysis

Uncertainty is inherent in every step of the risk assessment process. Thus, before we can begin to characterize any risk, we need some idea of the nature and magnitude of uncertainty in the risk estimate. Sources of uncertainty include:

- Extrapolation from high to low doses
- Extrapolation from animal to human responses
- Extrapolation from one route of exposure to another
- Limitations of analytical methods
- Estimates of exposure

Although the uncertainties are generally much larger in estimates of exposure and the relationships between dose and response (*e.g.*, the percent mortality), it is important to include the uncertainties originating from all steps in a risk assessment in risk characterization.

EXAMPLE CALCULATION 14.2

Application of Hazard Index and Incremental Carcinogenic Risk Associated with Chemical Exposure

A drinking water supply is found to contain 0.1 mg L^{-1} of acetone and 0.1 mg L^{-1} of chloroform. A 70-kg adult drinks 2 L per day of this water for 5 years. What would be the hazard index and the carcinogenic risk from drinking this water?

First, we need to determine the average daily doses (ADDs) for each of the chemicals and then their individual hazard quotients.

For Acetone

$$ADD = \frac{(0.1 \text{ mg L}^{-1})(2L \text{ day}^{-1})}{70 \text{ kg}}$$

$$= 2.9 \times 10^{-3} \text{ mg kg}^{-1} \text{ day}^{-1}$$

From Table 14.5, the RfD for acetone is 0.1 mg kg^{-1} day^{-1}

$$\text{Hazard quotient (HQ)} = \frac{2.9 \times 10^{-3} \text{ mg kg}^{-1} \text{ day}^{-1}}{0.1}$$

$$= 0.029$$

For Chloroform

$$ADD = \frac{(0.1 \text{ mg L}^{-1})(2L \text{ day}^{-1})}{70 \text{ kg}}$$

$$= 2.9 \times 10^{-3} \text{ mg kg}^{-1} \text{ day}^{-1}$$

From Table 14.5, the RfD value for chloroform is 0.01 mg kg^{-1} day^{-1}

$$HQ = \frac{2.9 \times 10^{-3} \text{ mg kg}^{-1} \text{ day}^{-1}}{0.01}$$

$$= 0.029$$

Thus,

$$\text{Hazard index} = 0.029 + 0.29 = 0.319$$

Since the hazard index is less than 1.0, the water is safe. Notice that we did not need to take into consideration that the person drank the water for 5 years.

The incremental carcinogenic risk associated with chloroform is determined as follows.

$$\text{Risk} = (\text{CDI})(\text{Potency factor})$$

$$CDI = \frac{(0.1 \text{ mg L}^{-1})(2 \text{ L day}^{-1})(365 \text{ days yr}^{-1})(5 \text{ yrs})}{(70 \text{ kg})(365 \text{ days yr}^{-1})(70 \text{ yrs})}$$

$$= 4.19 \times 10^{-5} \text{ mg kg}^{-1} \text{ day}^{-1}$$

From Table 14.6, the potency factor for chloroform is 6.1×10^{-3}

$$\text{Risk} = (\text{CDI})(\text{Potency factor})$$

$$\text{Risk} = (4.19 \times 10^{-5} \text{ mg kg}^{-1} \text{ day}^{-1})$$

$$(6.1 \times 10^{-3} \text{ mg kg}^{-1} \text{ day}^{-1}) = 2.55 \times 2.55 \times 10^{-7}$$

From a cancer risk standpoint, the risk over this period of exposure is less than the 10^{-6} goal.

Two approaches commonly used to characterize uncertainty are sensitivity analyses and Monte Carlo simulations. In **sensitivity analyses**, we simply vary the uncertain quantities of each parameter (*e.g.*, average values, high and low estimates), usually one at a time, to find out how changes in these quantities affect the final risk estimate. This procedure gives us a range of possible values for the overall risk and tells us which parameters are most crucial in determining the size of the risk. In a **Monte Carlo simulation**, however, we assume that all parameters are random or uncertain.

Thus, instead of varying one parameter at a time, we use a computer program to select parameter distributions randomly every time the model equations are solved, the procedure being repeated many times. The resulting output can be used to identify values of exposure or risk corresponding to a specified probability, say, the 50th percentile or 95th percentile.

14.2.4.4 Risk projections and management

The final phase of the risk assessment process is risk characterization. In this phase, exposure and dose–response assessments are integrated to yield probabilities of effects occurring in humans under specific exposure conditions. Quantitative risks are calculated for appropriate media and pathways. For example, the risks of lead in water are estimated over a lifetime, assuming (1) that the exposure is 2 liters of water ingested per day over a 70-year lifetime and (2) that different concentrations of lead occur in the drinking water. This information can then be used by risk managers to develop standards or guidelines for specific toxic chemicals or infectious microorganisms in different media, such as the drinking water or food supply.

14.2.4.5 Hazardous waste risk assessment

Hazardous waste risk assessments are a key part of the Comprehensive Environmental Response, Compensation, and Liability Act (CERCLA). Risk assessments are performed to assess health and ecological risks at Superfund sites and to evaluate the effectiveness of remedial alternatives in attaining a **record of decision (ROD)**. Since specific cleanup requirements have not been established for most contaminants under CERCLA, each site is assessed on an individual basis and cleaned up to a predetermined level of risk, such as 1 cancer case per 1,000,000 people. Risks may be different from one site to the next, depending on characteristics of the site and the potential for exposure.

For example, at one site, a high level of contaminants may be present (10,000 mg per kg of soil), but there is no nearby population, there is a large distance to groundwater, and the soils are of low permeability. Based on a risk assessment, the best remedial action for the site may be to leave the contaminated soil in place, where natural attenuation processes will eventually result in its degradation. Removing the contaminated soil with disposal in a landfill or *in situ* treat-

ment may result in a high risk due to release of wind-blown dusts that may expose workers at the site (Watts, 1998). In contrast, a soil contaminated with 10 mg of hazardous material per kg of soil may be considered a greater risk if the site has sandy soil, shallow groundwater, and nearby drinking water wells, and is located near a school. Cleanup to low levels would be necessary in this case to protect human health (Figure 14.7).

14.3 ECOLOGICAL RISK ASSESSMENT

Ecological risk assessment is a process that evaluates the probability that adverse ecological effects will occur as the result of exposure to one or more stressors. A **stressor** (or agent) is a substance, circumstance, or energy field that has the inherent ability to impose adverse effects upon a biological system. The environment is subject to many different stressors, including chemicals, genetically engineered microorganisms, ionizing radiation, and rapid changes in temperatures. Ecolog-

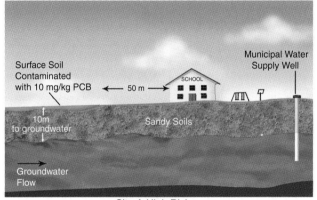

Site A-High Risk

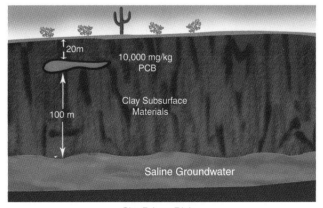

Site B-Low Risk

Figure 14.7 Two extremes of potential risk from contaminated sites. Site A is a high-risk site with potential for migration from the source to nearby receptors. Site B, although characterized by a higher source concentration, has minimal potential for contaminant migration and risk. Modified from Watts, 1998.

ical risk assessment may evaluate one or more stressors and ecological components (*e.g.*, specific organisms, populations, communities, or ecosystems). Ecological risks may be expressed as true probabilistic estimates of adverse effects (as is done with carcinogens in human health risk assessment), or they may be expressed in a more qualitative manner.

In the United States, the Comprehensive Environmental Response, Compensation, and Liability Act (CERCLA) (otherwise known as the Superfund), the Resource Conservation and Recovery Act (RCRA), and other regulations require an ecological assessment as part of all remedial investigation and feasibility studies (see also Section 14.2.4.5). Pesticide registration, which is required under the Federal Insecticide, Fungicide, and Rodenticide Act (FIFRA), must also include an ecological assessment (see Section 14.3). In the CERCLA/RCRA context, a typical objective is to determine and document actual or potential effects of contaminants on ecological receptors and habitats as a basis for evaluating remedial alternatives in a scientifically defensible manner.

The four major phases or steps in ecological assessment (Figure 14.8) are as follows:

• Problem formulation and hazard identification

• Exposure assessment

• Ecological effects/toxicity assessment

• Risk characterization

An ecological risk assessment may be initiated under many circumstances—the manufacturing of a new chemical, evaluation of cleanup options for a contaminated site, or the planned filling of a marsh, among others. The problem-formulation

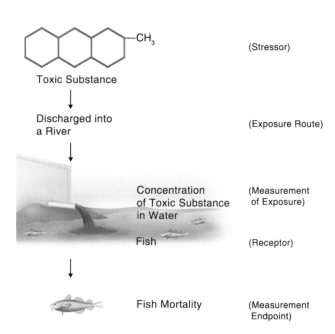

Figure 14.9 Ecological risk assessment.

process begins with an evaluation of the stressor characteristics, the ecosystem at risk, and the likely ecological effects. An endpoint is then selected. An **endpoint** (Figure 14.9) is a characteristic of an ecological component, *e.g.*, the mortality of fish) that may be affected by a stressor. Two types of endpoints are generally used: assessment endpoints and measurement endpoints. **Assessment endpoints** are particular environmental values to be protected. Such endpoints, which are recognized and valued by the public, drive the decisions made by official risk managers. **Measurement endpoints** are qualitatively or quantitatively measurable factors. Suppose, for example, a community that values the quality of sports fishing in the area is worried about the effluent from a nearby paper mill. In this case, a decline in the trout population might serve as the assessment endpoint, while the increased mortality of minnows, as evaluated by laboratory studies, might be the measurement endpoint. Thus, risk managers would use the quantitative data gathered on the surrogate minnow population to develop management strategies designed to protect the trout population.

Exposure assessment is a determination of the environmental concentration range of a particular stressor and the actual dose received by the *biota* (all the plants and animals) in a given area. The most common approach to exposure analysis is to measure actual concentrations of a stressor and combine these measurements with assumptions about contact and uptake by the biota. For example, the exposure of simple aquatic organisms to chemicals can often be measured simply as the concentration of that chemical in the water because the physiologic systems of these organisms are assumed to be in equilibrium with the surrounding water. Stressor measurements can also be combined with quantitative parameters describing the frequency and magnitude of contact. For example, concentrations of chemicals or microorganisms in food items can be

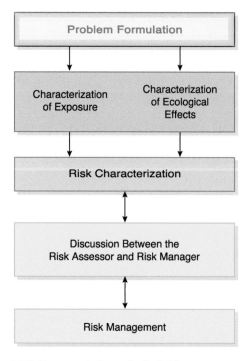

Figure 14.8 Framework for ecological risk assessment. Adapted from U.S. EPA, 1992a.

INFORMATION BOX 14.3

Examples of a Management Goal, Assessment End-point, and Measures

Goal: Viable, self-sustaining coho salmon population that supports a subsistence and sport fishery.

Assessment Endpoint: Coho salmon breeding success, fry survival, and adult return rates.

Measures of Effects

- Egg and fry response to low dissolved oxygen
- Adult behavior in response to obstacles
- Spawning behavior and egg survival with changes in sedimentation

Measures of Ecosystem and Receptor Characteristics

- Water temperature, water velocity, and physical obstructions
- Abundance and distribution of suitable breeding substrate
- Abundance and distribution of suitable food sources for fry
- Feeding, resting, and breeding behavior
- Natural reproduction, growth, and mortality rates

Measures of Exposure

- Number of hydroelectric dams and associated ease of fish passage
- Toxic chemical concentrations in water, sediment, and fish tissue
- Nutrient and dissolve oxygen levels in ambient waters
- Riparian cover, sediment loading, and water temperature

combined with ingestion rates to estimate dietary exposure. Exposure assignment is, however, rarely straightforward. Bio-transformations may occur, especially for heavy metals such as mercury (see Section 13.5.5). Such transformations may result in the formation of even more toxic forms of the stressor. Researchers must therefore use mathematical models to predict the fate and resultant exposure to a stressor and to determine the outcome of a variety of scenarios.

The purpose of evaluating ecological effects is to identify and quantify the adverse effects elicited by a stressor and, to the extent possible, to determine cause-and-effect relationships. During this phase, toxicity data are usually compiled and compared.

Generally, there are acute and chronic data for the stressor acting on one or several species. Field observations can provide additional data, and so can controlled-microcosm and large-scale tests.

The process of developing a stressor–response profile is complex because it inevitably requires models, assumptions, and extrapolations. For example, the relationship between measurement and assessment endpoint is an assumption. It is often expressly stated in the model used, but when it is not specifically stated, it is left to professional judgment. In addition, the stressor–response profile is analogous to a dose–response curve in the sense that it involves extrapolations; in this case, though, a single-species toxicity test is ex-

trapolated to the community and ecosystem level. One of the difficulties in the quantification of the stressor–response profile is that many of the quantitative extrapolations are drawn from information that is qualitative in nature. For example, when we use **phylogenic extrapolation** to transfer toxicity data from one species to another species—or even to a whole class of organisms—we are assuming a degree of similarity based on qualitative characteristics. Thus, when we use green algal toxicity test data to represent all photosynthetic eukaryotes (which we often do), we must remember that all photosynthetic eukaryotes are not, in fact, green algae. Because many of the responses are extrapolations based on models ranging from the molecular to the ecosystem level, it is critically important that uncertainties and assumptions be clearly delineated.

Risk assessment consists of comparing the exposure and stressor–response profiles to estimate the probability of effects, given the distribution of the stressor within the system. As you might expect, this process is extraordinarily difficult to accomplish. In fact, our efforts at predicting adverse effects have been likened to the weather forecaster's prediction of rain (Landis and Ho-Yu, 1995). Thus, the predictive process in ecological risk assessment is still very much an art form, largely dependent on professional judgment.

Conceptual model diagrams can be used to better visualize potential impacts (Figure 14.10). They may be based on theory and logic, empirical data, mathematical models, or probability models. These diagrams are useful tools for communicating important pathways in a clear and concise way. They can be used to ask new questions about relationships that help generate plausible risk hypothesis.

14.4 MICROBIAL RISK ASSESSMENT

Outbreaks of waterborne disease caused by microorganisms usually occur when the water supply has been obviously and significantly contaminated. In such high-level cases, the exposure is manifest, and cause and effect are relatively easy to determine. However, exposure to low-level microbial contamination is difficult to determine epidemiologically. We know, for example, that long-term exposure to microbes can have a significant impact on the health of individuals within a community, but we need a way to measure that impact.

For some time, methods have been available to detect the presence of low levels (1 organism per 1000 liters) of pathogenic organisms in water, including enteric viruses, bacteria, and protozoan parasites. The trouble is that the risks posed to the community by these low levels of pathogens in a water supply over time are not like those posed by low levels of chemical toxins or carcinogens. For example, it takes just one amoeba in the wrong place at the wrong time to infect one individual, whereas that same individual would have to consume some quantity of a toxic chemical to be comparably harmed. Microbial risk assessment is therefore a process that allows us to estimate responses in terms of the *risk of infection* in a quantitative fashion. Microbial risk gen-

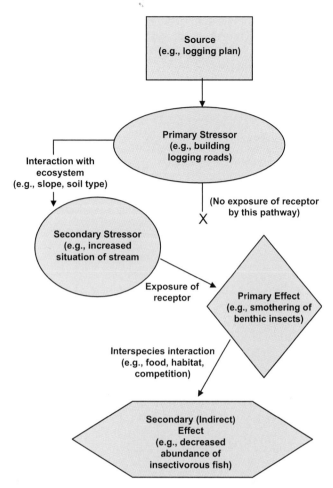

Figure 14.10 Conceptual model for logging. Source: www.epa.gov.

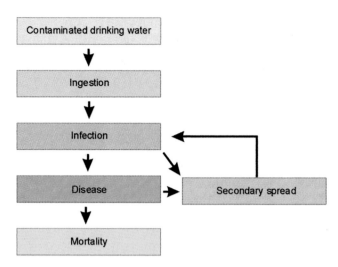

Figure 14.11 Outcomes of enteric viral exposure. From *Pollution Science* © 1996, Academic Press, San Diego, CA.

others—without ever getting sick themselves. The ratio of clinical to subclinical infection varies from pathogen to pathogen, especially in viruses, as shown in Table 14.10. Poliovirus infections, for instance, seldom result in obvious clinical symptoms; in fact, the proportion of individuals developing clinical illness may be less than 1%. However, other enteroviruses, such as the coxsackie viruses, may exhibit a greater proportion. In many cases, as in that of rotaviruses, the probability of developing clinical illness appears to be completely unrelated to the dose an individual receives via ingestion. Rather, the likelihood of developing

TABLE 14.10 Ratio of clinical to subclinical infections with enteric viruses.

VIRUS	FREQUENCY OF CLINICAL ILLNESS[a] (%)
Poliovirus 1	0.1–1
Coxsackie	
A16	50
B2	11–50
B3	29–96
B4	30–70
B5	5–40
Echovirus	
Overall	50
9	15–60
18	Rare–20
20	33
25	30
30	50
Hepatitis A (adults)	75
Rotavirus	
(Adults)	56–60
(Children)	28
Astrovirus (adults)	12–50

[a]The percentage of the individuals infected who develop clinical illness.

From Gerba and Rose (1993). From *Pollution Science* © 1996, Academic Press, San Diego, CA.

erally follows the steps used in other health-based risk assessments—hazard identification, exposure assessment, dose–response, and risk characterization. The differences are in the specific assumptions, models, and extrapolation methods used.

Hazard identification in the case of pathogens is complicated because several outcomes—from asymptomatic infection to death (see Figure 14.11)—are possible, and these outcomes depend upon the complex interaction between the pathogenic agent (the "infector") and the host (the "infectee"). This interaction, in turn, depends on the characteristics of the host as well as the nature of the pathogen. Host factors, for example, include preexisting immunity, age, nutrition, ability to mount an immune response, and other nonspecific host factors. Agent factors include type and strain of the organism as well as its capacity to elicit an immune response.

Among the various outcomes of infection is the possibility of **subclinical infection.** Subclinical (asymptomatic) infections are those in which the infection (growth of the microorganism within the human body) results in no obvious illness such as fever, headache, or diarrhea. That is, individuals can host a pathogen microorganism—and transmit it to

clinical illness depends upon the type and strain of the virus as well as host age, nonspecific host factors, and possibly preexisting immunity. The incidence of clinical infection can also vary from year to year for the same virus, depending on the emergence of new strains.

Another outcome of infection is the development of clinical illness. Several host factors play a major role in this outcome. The age of the host is often a determining factor. In the case of hepatitis A, for example, clinical illness can vary from about 5% in children less than 5 years of age to 75% in adults. Similarly, children are more likely to develop rotaviral gastroenteritis than are adults. Immunity is also an important factor, albeit a variable one. That is, immunity may or may not provide long-term protection from reinfection, depending on the enteric pathogen. It does not, for example, provide long-term protection against the development of clinical illness in the case of the Norwalk virus or *Giardia*. However, for most enteroviruses and for the hepatitis A virus, immunity from reinfection is believed to be lifelong. Other undefined host factors may also control the odds of developing illness. For example, in experiments with the Norwalk virus (norovirus), human volunteers who did not become infected upon an initial exposure to the virus also did not respond to a second exposure. In contrast, those volunteers who developed gastroenteritis upon the first exposure also developed illness after the second exposure.

The ultimate outcome of infection—mortality—can be caused by nearly all enteric organisms. The factors that control the prospect of mortality are largely the same factors that control the development of clinical illness. Host age, for example, is significant. Thus, mortality for hepatitis A and poliovirus is greater in adults than in children. In general, however, one can say that the very young, the elderly, and the immunocompromised are at the greatest risk of a fatal outcome of most illnesses (Gerba et al., 1996). For example, the case-fatality rate (%) for *Salmonella* in the general population is 0.1%, but it has been observed to be as high as 3.8% in nursing homes (Table 14.11). In North America and Europe, the reported case-fatality rates (*i.e.*, the ratio of cases to fatalities reported as a percentage of persons who die) for enterovirus infections range from less than 0.1 to 0.94%, as shown in Table 14.12. The case-

TABLE 14.12 Case-fatality rates for enteric viruses and bacteria.

ORGANISM	CASE-FATALITY RATE (%)
Viruses	
Poliovirus 1	0.90
Coxsackie	
A2	0.50
A4	0.50
A9	0.26
A15	0.12
Cosxsackie B	0.59–0.94
Echovirus	
6	0.29
9	0.27
Hepatitis A	0.30
Rotavirus	
(Total)	0.01
(Hospitalized)	0.12
Norwalk	0.0001
Astrovirus	0.01
Bacteria	
Shigella	0.2
Salmonella	0.1
Escherichia coli 0157:H7	0.2
Campylobacter jejuni	0.1

From Gerba and Rose (1993) and Gerba et al. (1996). From *Pollution Science* © 1996, Academic Press, San Diego, CA.

fatality rate for common enteric bacteria ranges from 0.1 to 0.2% in the general population. Enteric bacterial diseases can be treated with antibiotics, but no treatment is available for enteric viruses.

Recognizing that microbial risk involves a myriad of pathogenic organisms capable of producing a variety of outcomes that depend on a number of factors—many of which are undefined—one must now face the problem of exposure assessment, which has complications of its own. Unlike chemical-contaminated water, microorganism-contaminated water does not have to be consumed to cause harm. That is, individuals who do not actually drink, or even touch, contaminated water also risk infection because pathogens—particularly viruses—may be spread by person-to-person contact or subsequent contact with contaminated inanimate objects (such as toys). This phenomenon is described as the **secondary attack rate**, which is reported as a percentage. For example, one person infected with poliovirus can transmit it to 90% of the persons with whom he or she associates. This secondary spread of viruses has been well documented for waterborne outbreaks of several diseases, including that caused by Norwalk virus, whose secondary attack rate is about 30%.

The question of dose is another problem in exposure assessment. How does one define "dose" in this context? To answer this question, researchers have conducted a number of studies to determine the infectious dose of enteric microorganisms in human volunteers. Such human

TABLE 14.11 Case fatality observed for enteric pathogens in nursing homes versus general population.

ORGANISM	CASE FATALITY (%) IN GENERAL POPULATION	CASE FATALITY (%) IN NURSING HOMES
Campylobacter jejuni	0.1	1.1
Escherichia coli 0157:H7	0.2	11.8
Salmonella	0.1	3.8
Rotavirus	0.01	1.0

Modified from Gerba et al. (1996).

experimentation is necessary because determination of the infectious dose in animals and extrapolation to humans is often impossible. In some cases, for example, humans are the primary or only known host. In other cases, such as that of *Shigella* or norovirus, infection can be induced in laboratory-held primates, but it is not known whether the infectious dose data can be extrapolated to humans. Much of the existing data on infectious doses of viruses has been obtained with attenuated vaccine viruses or with avirulent laboratory-grown strains, so that the likelihood of serious illness is minimized. An example of a dose–response curve for a human feeding study with rotavirus is shown in Figure 14.12.

In the microbiological literature, the term **minimum infectious dose** is used frequently, implying that a threshold dose exists for microorganisms. In reality, the term used usually refers to the ID_{50} dose at which 50% of the animals or humans exposed became infected or exhibit any symptoms of an illness. Existing infectious dose data are compatible with nonthreshold responses, and the term "infectivity" is probably more appropriate when referring to differences in the likelihood of an organism causing an infection. For example, the probability of a given number of ingested rotaviruses causing diarrhea is greater than that for *Salmonella*. Thus, the infectivity of rotavirus is greater than that of *Salmonella*.

Next, one must choose a dose–response model, whose abscissa is the dose and whose ordinate is the risk of infection (see Figure 14.12). The choice of model is critical so that risks are not greatly overestimated or underestimated. A modified exponential (beta-Poisson distribution) or a log-probit (simple lognormal, or exponential, distribution)

model may be used to describe the probability of infection in human subjects for many enteric microorganisms (Haas, 1983). These models have been found to best fit experimental data. For the beta-Poisson model, the probability of infection from a single exposure, *P*, can be described as follows:

$$P = 1 - (1 + N/\beta)^{-\alpha} \qquad \text{(Eq. 14.10)}$$

where N is the number of organisms ingested per exposure and α and β represent parameters characterizing the host-virus interaction (dose–response curve). Some values for α and β for several enteric waterborne pathogens are shown in Table 14.13; these values were determined from human studies. For some microorganisms, an **exponential model** may better represent the probability of infection.

$$P = 1 - \exp(-rN) \qquad \text{(Eq. 14.11)}$$

In this equation, r is the fraction of the ingested microorganisms that survive to initiate infections (host-microorganism interaction probability). Table 14.13 shows examples of results of both models for several organisms.

These models define the probability of the microorganisms overcoming the host defenses (including stomach pH, finding a susceptible cell, nonspecific immunity, and so on) to establish an infection in the host. When one uses these models, one estimates the probability of becoming infected after ingestion of various concentrations of pathogens. For example, Example 14.4 shows how to calculate the risk of acquiring a viral infection from consumption of contaminated drinking water containing echovirus 12 using Equation 14.10.

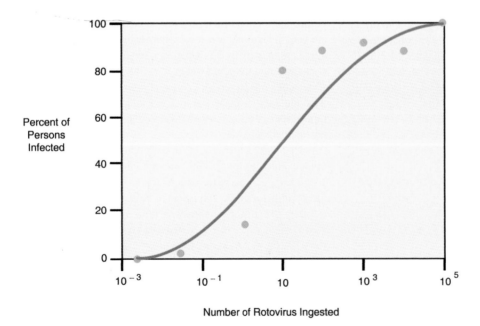

Figure 14.12 Dose–response for human rotavirus by oral ingestion.

TABLE 14.13 Best-fit dose-response parameters for enteric pathogen ingestion studies.

MICROORGANISM	BEST MODEL	MODEL PARAMETERS
Echovirus 12	Beta-Poisson	$\alpha = 0.374$
		$\beta = 186{,}69$
Rotavirus	Beta-Poisson	$\alpha = 0.26$
		$\beta = 0.42$
Poliovirus 1	Exponential	$r = 0.009102$
Poliovirus 1	Beta-Poisson	$\alpha = 0.1097$
		$\beta = 1524$
Poliovirus 3	Beta-Poisson	$\alpha = 0.409$
		$\beta = 0.788$
Cryptosporidium	Exponential	$r = 0.004191$
Giardia lamblia	Exponential	$r = 0.02$
Salmonella	Exponential	$r = 0.00752$
Escherichia coli	Beta-Poisson	$\alpha = 0.1705$
		$\beta = 1.61 \times 10^6$

Modified from Regli et al. (1991). From *Pollution Science* © 1996, Academic Press, San Diego, CA.

EXAMPLE CALCULATION 14.3

Application of a Virus Risk Model to Characterize Risks from Consuming Shellfish

It is well known that infectious hepatitis and viral gastroenteritis are caused by consumption of raw or, in some cases, cooked clams and oysters. The concentration of echovirus 12 was found to be 8 plaque-forming units per 100 g in oysters collected from coastal New England waters. What are the risks of becoming infected and ill from echovirus 12 if the oysters are consumed? Assume that a person usually consumes 60 g of oyster meat in a single serving:

$$\frac{8\ \text{PFU}}{100\ \text{g}} = \frac{N}{60\ \text{g}}\ N = 4.8\ \text{PFU consumed}$$

From Table 14.13, $\alpha = 0.374$, $\beta = 186.64$. The probability of infection from Equation 14.10 is then

$$P = 1 - \left(1 + \frac{4.8}{186.69}\right)^{-0.374} = 9.4 \times 10^{-3}$$

If the percent of infections that result in risk of clinical illness is 50%, then from Equation 14.14 one can calculate the risk of clinical illness:

Risk of clinical illness $= (9.4 \times 10^{-3})\,(0.50) = 4.7 \times 10^{-3}$

If the case-fatality rate is 0.001%, then from Equation 14.15

Risk of mortality $= (9.4 \times 10^{-3})\,(0.50)\,(0.001)$
$$= 4.7 \times 10^{-6}$$

If a person consumes oysters 10 times a year with 4.8 PFU per serving, then one can calculate the risk of infection in one year from Equation 14.12

Annual risk $= P_A = 1 - (1 - 9.4 \times 10^{-3})^{365} = 9.7 \times 10^{-1}$

EXAMPLE CALCULATION 14.4

Risk Assessment for Rotavirus in Drinking Water

Pathogen identified → Rotavirus

↓

Dose Response Model (based on human ingestion studies) → best fit for data is the Beta Poisson Model
$P = (1 + N/\beta)^{-\alpha}$
$\alpha = 0.2631$
$\beta = 0.42$

↓

Exposure (field studies on concentration in drinking water) → 4 rotavirus/1,000 liters

↓

Risk Characterization → Risk of Infection
Assumes: 2 liters/day of drinking water ingested.
Thus,
$N = 0.008$/day
Risk of Infection/day $= 1{:}200$
Risk of Infection/year
$P_A = 1 - (1 - P)^{365}$
$P_A = 1{:}2$

Annual and lifetime risks can also be determined, again assuming a Poisson distribution of the virus in the water consumed (assuming daily exposure to a constant concentration of viral contamination), as follows:

$$P_A = 1 - (1 - P)^{365} \qquad \text{(Eq. 14.12)}$$

where P_A is the annual risk (365 days) of contracting one or more infections, and

$$P_L = 1 - (1 - P)^{25{,}550} \qquad \text{(Eq. 14.13)}$$

where P_L is the lifetime risk (assuming a lifetime of 70 years = 25,550 days) of contracting one or more infections.

Risks of clinical illness and mortality can then be determined by incorporating terms for the percentage of clinical illness and mortality associated with each particular virus:

$$\text{Risk of clinical illness} = PI \qquad \text{(Eq. 14.14)}$$
$$\text{Risk of mortality} = PIM \qquad \text{(Eq. 14.15)}$$

where I is the percentage of infections that result in clinical illness and M is the percentage of clinical cases that result in mortality.

Application of this model allows estimation of the risks of infection, development of clinical illness, and mortality for different levels of exposure. As shown in Table 14.14, for example, the estimated risk of infection from 1 rotavirus in 100 liters of drinking water (assuming ingestion of 2 liters per day) is 1.2×10^{-3}, or almost 1 in 1,000 for a single-day exposure. This risk would increase to 3.6×10^{-1}, or approximately one in three, on an annual basis. As can be seen

TABLE 14.14 Risk of infection, disease, and mortality for rotavirus.

VIRUS CONCENTRATION PER 100 LITERS	RISK DAILY	ANNUAL
Infection		
100	9.6×10^{-2}	1.0
1	1.2×10^{-3}	3.6×10^{-1}
0.1	1.2×10^{-4}	4.4×10^{-2}
Disease		
100	5.3×10^{-2}	5.3×10^{-1}
1	6.6×10^{-4}	2.0×10^{-1}
0.1	6.6×10^{-5}	2.5×10^{-2}
Mortality		
100	5.3×10^{-6}	5.3×10^{-5}
1	6.6×10^{-8}	2.0×10^{-5}
0.1	6.6×10^{-9}	2.5×10^{-6}

Modified from Gerba and Rose (1992). From *Pollution Science* © 1996, Academic Press, San Diego, CA.

from this table, the risk of developing a clinical illness also appears to be significant for exposure to low levels of rotavirus in drinking water.

TABLE 14.15 Comparison of outbreak data to model predictions for assessment of risks associated with exposure to *Salmonella*.

FOOD	DOSE CFU	AMOUNT CONSUMED	ATTACK RATE (%)	PREDICTED P (%)
Water	17	1 L	12	12
Pancretin	200	7 doses	100	77
Ice cream	102	1 portion	52	54
Cheese	100–500	28 g	28–36	53–98
Cheese	10^5	100 g	100	>99.99
Ham	10^6	50–100 g	100	>99.99

Source: Rose et al., 1995. From *Environmental Microbiology* © 2000, Academic Press, San Diego, CA.

The EPA recommends that any drinking water treatment process should be designed to ensure than human populations are not subjected to risk of infection greater than 1:10,000 for a yearly exposure. To achieve this goal, it would appear from the data shown in Table 14.10 that the virus concentration in drinking water would have to be less than 1 per 1,000 liters. Thus, if the average concentration of enteric viruses in untreated water is 1,400/1,000 liters, treatment plants should be designed to remove at least 99.99% of the virus present in the raw water. A further application of this approach is to define the required treatment of a water

---

INFORMATION BOX 14.4

How Do We Set Standards for Pathogens in Drinking Water?

In 1974, the U.S. Congress passed the Safe Drinking Water Act, giving the U.S. Environmental Protection Agency (EPA) the authority to establish standards for contaminants in drinking water. Through a risk analysis approach, standards have been set for many chemical contaminants in drinking water. Setting standards for microbial contaminants proved more difficult because (1) methods for the detection of many pathogens are not available, (2) days to weeks are sometimes required to obtain results, and (3) costly and time-consuming methods are required. To overcome these difficulties, coliform bacteria had been used historically to assess the microbial quality of drinking water. However, by the 1980s it had become quite clear that coliform bacteria did not indicate the presence of pathogenic waterborne *Giardia* or enteric viruses. Numerous outbreaks had occurred in which coliform standards were met, because of the greater resistance of viruses and *Giardia* to disinfection. A new approach was needed to ensure the microbial safety of drinking water.

To achieve this goal a new treatment approach was developed called the Surface Treatment Rule (STR). As part of the STR, all water utilities that use surface waters as their source of potable water would be required to provide filtration to remove *Giardia* and enough disinfection to kill viruses. The problem facing the EPA was how much removal should be required. To deal with this issue, the EPA for the first time used a microbial risk assessment approach. The STR established that the goal of treatment was to ensure that microbial illness from *Giardia lamblia* infection should not be any greater than 1 per 10,000 exposed persons annually (10^{24} per year). This value is close to the annual risk of infection from waterborne disease outbreaks in the U. S. ($4 \ 3 \ 10^{23}$). Based on the estimated concentration of *Giardia* and enteric viruses in surface waters in the United States from the data available at the time, it was required that all drinking water treatment plants be capable of removing 99.9% of the *Giardia* and 99.99% of the viruses. In this manner it was hoped that the risk of infection of 10^{24} per year would be achieved. The STR went into effect in 1991.

To better assess whether the degree of treatment required is adequate, the EPA developed the Information Collection Rule, which required major drinking water utilities that use surface waters to analyze these surface water for the presence of *Giardia*, *Cryptosporidium*, and enteric viruses for a period of almost 2 years. From this information, the EPA set treatment control requirements to ensure that the 10^{24} yearly risk is met. Utilities that have heavily contaminated source water are required to achieve greater levels of treatment (see Figure 14.13).

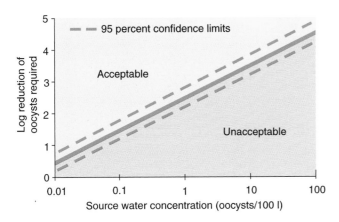

Figure 14.13 Relationship of influent *Cryptosporidium* concentration and log reduction by treatment necessary to produce acceptable water. From Haas et al., 1996. Reprinted from *Journal AWWA*, Vol. 88, No. 9 (September 1996), by permission. Copyright © 1996, American Water Works Association.

source in terms of the concentration of a disease-causing organism in that supply. Thus, the more contaminated the raw water source, the more treatment is required to reduce the risk to an acceptable level. An example of this application is shown in Figure 14.13. The plausibility of validation of microbial risk assessment models has been examined by using data from foodborne outbreaks in which information has been available on exposure and outcomes (Rose et al., 1995; Crockett et al., 1996). These studies suggest that microbial risk assessment can give reasonable estimates of illness from exposure to contaminated foods (Table 14.15).

In summary, risk assessment is a major tool for decision making in the regulatory arena. This approach is used to explain chemical and microbial risks, as well as ecosystem impacts. The results of such assessments can be used to inform risk managers of the probability and extent of environmental impacts resulting from exposure to different levels of stress (contaminants). Moreover, this process, which allows the quantification and comparison of diverse risks, lets risk managers utilize the maximum amount of complex information in the decision-making process. This information can also be used to weigh the cost and benefits of control options and to develop standards or treatment options (see Information Box 14.4).

QUESTIONS AND PROBLEMS

1. List the four steps in a formal risk assessment.
2. Why do we use safety factors in risk assessment?
3. What is the most conservative dose–response curve? What does it mean?
4. What is the difference between risk assessment and risk management?
5. What are some of the differences between the risks posed by chemicals and those posed by microorganisms?
6. Suppose a 50-kg individual drinks 2 L day^{-1} of chloroform and 0.1 mg L^{-1} phenol. What is the hazard index? Is there cause for concern?
7. Estimate the cancer risk for a 70-kg individual consuming 1.5 liters of water containing trichloroethylene (TCE) per day for 70 days.
8. Calculate the risk of infection from rotavirus during swimming in polluted water. Assume 30 ml of water is ingested during swimming and the concentration of rotavirus was 1 per 100 liters. What would the risk be in a year if a person went swimming 5 times and 10 times in the same water with the same concentration of virus?
9. What is a NOAEL and how does it differ from a LOAEL?
10. If 10 oocysts of *Cryptosporidium* are detected in 100 L of surface water, how much reduction (in log$_{10}$) by a water treatment plant is required to achieve a 1:10,000 annual risk of infection?
11. Give an example of a nonthreshold response for a chemical toxin.
12. What is the difference between a stressor and a receptor? Give an example of a chemical stressor and a receptor in an aquatic system. What endpoint would you use?
13. Draw an exposure pathway for pathogens for the disposal of raw sewage into the ocean. Consider likely routes of ingestions and inhalation. As a risk manager, what options may you have to reduce the risks of exposure?
14. Using the U.S. Environmental Protection Agency IRIS database (www.epa.gov/iris), find the critical effect, uncertainty factor, and NOAEL, LOAEL, and RfD for mercury, chromium, and chloroform. In drinking water, which one would be the most toxic?

REFERENCES AND ADDITIONAL READING

Cockerham L.G. and Shane B.S. (1994) *Basic Environmental Toxicology.* CRC Press, Boca Raton, Florida.

Covello V., von Winterfieldt D. and Slovic P. (1986) Risk Communication: A review of the literature. *Risk Analysis* **3,** 171–182.

Crockett C.S., Haas C.N., Fazil A., Rose J.B. and Gerba C.P. (1996) Prevalence of shigellosis: Consistency with dose-response information. *Int. J. Food Prot.* **30,** 87–99.

Gerba C.P. and Rose J.B. (1992) Estimating viral disease risk from drinking water. In *Comparative Environmental Risks,* C.R. Cothern (ed.) Lewis Publishers, Boca Raton, Florida.

Gerba C.P., Rose J.B. and Haas C.N. (1996) Sensitive populations: Who is at the greatest risk? *Int. J. Food Microbiol.* **301,** 113–123.

Gerba C.P., Rose J.B., Haas C.N. and Crabtree K.D. (1997) Waterborne rotavirus: A risk assessment. *Water Res.* **12,** 2929–2940.

Haas C.N. (1983) Estimation of risk due to low levels of microorganisms: A comparison of alternative methodologies. *Am. J. Epidemiology*. **118**, 573–582.

Haas C.N., Crockett C.S., Rose J.B., Gerba C.P. and Fazil A.M. (1996) Assessing the risk posed by oocysts in drinking water. *J. Am. Water Works Assoc*. **88** (9), 131–136.

Haas C.N., Rose J.B. and Gerba C.P. (1999) *Quantitative Microbial Risk Assessment*. John Wiley & Sons, New York.

ILSI Risk Science Institute Pathogen Risk Assessment Working Group. (1996) A conceptual framework to assess the risks of human disease following exposure to pathogens. *Risk Anal*. **16**, 841–848.

Kammen D.M. and Hassenzahl D.M. *Should We Risk It?* Princeton University Press, Princeton, New Jersey.

Kolluru R.V. (1993) *Environmental Strategies Handbook*. McGraw-Hill, New York.

Landis W.G. and Yu M.H. (1995) *Introduction to Environmental Toxicology*. Lewis Publishers, Boca Raton, Florida.

National Research Council. (1983) *Risk Assessment in the Federal Government: Managing the Process*. National Academy Press. Washington, D.C.

National Research Council. (1989) *Improving Risk Communication*. National Academy Press. Washington, D.C.

National Research Council. (1991) *Frontiers in Assessing Human Exposure*. National Academy Press, Washington, D.C.

Regli S., Rose J.B., Haas C.N. and Gerba C.P. (1991) *Modeling the risk from Giardia and viruses in drinking water*. *J. Am. Water Works Assoc*. **83**, 76–84.

Rodricks J.V. (1992) *Calculated Risks. Understanding the Toxicity and Human Health Risks of Chemicals in Our Environment*. Cambridge University Press, Cambridge, England.

Rose J.B., Haas C.N. and Gerba C.P. (1995) Linking microbiological criteria for foods with quantitative risk assessment. *J. Food Safety* **15**, 111–132.

Roseberry A.M. and Burmaster D.E. (1992) Lognormal Distributions for Water Intake by Children and Adults. *Risk Anal*. **12**, 99–104.

Straub T.M., Pepper I.L. and Gerba C.P. (1993) Hazards from pathogenic microorganisms in land-disposed sewage sludge. *Rev. Environ. Contamination Toxicology* **132**, 55–91.

Tseng W.P. (1977) Effects and dose response relationships of skin cancer and blackfoot disease with arsenic. *Environ. Hlth. Perspect*. **19**, 109–119.

Tseng W.P., Chu H.M., How S.W., Fong J.M., Lin C.S. and Yeh S. (1968) Prevalence of skin cancer in an endemic area of chronic arsenic in Taiwan. *J. Natl. Cancer Inst*. **40**, 453–463.

United States Environmental Protection Agency (U.S. EPA) (1990) *Risk Assessment, Management and Communication of Drinking Water Contamination*. EPA/625/4–89/024, Washington, D.C.

United States Environmental Protection Agency (U.S. EPA) (1992a) *Framework for Ecological Risk Assessment*. EPA 1630/R–92/001, Washington, D.C.

United States Environmental Protection Agency (U.S. EPA) (1992b) *Dermal Exposure Assessment: Principles and Applications*. EPA 600/8–91/011B, Washington, D.C.

United States Environmental Protection Agency. (U.S. EPA) (1986) *Guidelines for Carcinogen Risk Assessment*. Federal Register, September 24, 1986.

Ward R.L., Berstein D.I. and Young E.C. (1986) Human rotavirus studies in volunteers of infectious dose and serological response to infection. *J. Infect. Dis*. **154**, 871–877.

Watts, R.J. (1998) *Hazardous Wastes: Sources, Pathways, Receptors*. John Wiley & Sons, New York.

Wilson R. and Crouch E.A.C. (1987) Risk assessment and comparisons: An introduction. *Science* **236**, 267–270.

CHAPTER 15

ENVIRONMENTAL LAWS AND REGULATIONS

C.P. Gerba and C. Straub

The U.S. Congress frequently imposes federal laws and regulations that affect the environment.

15.1 REGULATORY OVERVIEW

In the United States, most environmental legislation consists of federal legislation that has been enacted within the last 30 years (Table 15.1). In addition, there are many state environmental laws and regulations that have been patterned after—and work together with—the federal programs. Federal legislation is, however, the basis for the development of regulations designed to protect our air, water, and food supply, and to control pollutant discharge. Enactment of new legislation empowers the executive branch to develop environment-specific regulations and to implement their enforcement. Among other things, regulations may involve the development of standards for waste discharge, requirements for the cleanup of polluted sites, or guidelines for waste disposal. The Environmental Protection Agency (EPA) is the federal agency currently responsible for development of environmental regulations and enforcement.

When a new environmental law is approved by Congress and the president, it is called an **Act** (Figure 15.1), and the text of the Act is known as a public statute. Some of the better known laws related to the environment are the Clean Air Act, the Clean Water Act, and the Safe Drinking Water Act. Once the Act is passed, the House of Representatives standardizes the text of the law and publishes it in the United States Code. The U.S. Code is the official record of all federal laws. After the law becomes official, how is it put into practice? Laws often do not include all the details. The U.S. Code would not tell you, for example, what the speed limit is in front of your house. In order to make the laws work on a day-to-day level, Congress authorizes certain government agencies—including the Environmental Protection Agency (EPA)—to create regulations. Regulations set specific rules about what is legal and what isn't. For example, a regulation issued by EPA to implement the Clear Air Act might state what levels of a pollutant—such as sulfur dioxide—are safe. It would tell industries how much sulfur dioxide they can legally emit into the air, and what the penalty would be if they emit too much. Once the law is in effect, it is the responsibility of the EPA to enforce it.

During the development of regulations, the agency conducts research and then publishes a proposed regulation in the Federal Register so that members of the public can consider it and send their comments to the agency. The agency considers all the comments, revises the regulation, and issues a final **Rule**. At each stage in the process, the agency publishes a notice in the Federal Register. Once a regulation is completed and has been printed in the Federal Register as a final rule, it is "codified" by being published in the **Code of Federal Regulations (CFR)** (Figure 15.1). The CFR is the official record of all regulations created by the federal government. It is divided into 50 volumes, called titles, each of which focuses on a particular area. Almost all environmental regulations appear in Title 40.

15.2 THE SAFE DRINKING WATER ACT

In 1974, the passage of the **Safe Drinking Water Act (SDWA)** gave the federal government overall authority for the protection of drinking water. Previous to that time, the individual states had primary authority for development and enforcement of standards. Under this authority, specific standards were promulgated for contaminant concentrations and minimum water treatment. **Maximum contaminant levels (MCL)** or **maximum contaminant level goals (MCLG)** were set for specific contaminants in drinking water (see Chapter 10). Whereas an MCL is an achievable, required level, an MCLG is a desired goal, which may or may not be achievable. For example, the MCLG for enteric viruses in drinking water is zero, since even one ingested virus may cause illness. We can't always reach zero, but we are obliged to try. Within the provisions of the SDWA, there are a number of specific rules such as the **Surface Treatment Rule**, which requires all water utilities in the United States to provide filtration and disinfection to control waterborne disease from *Giardia* and enteric viruses. These rules are developed through a process that allows input from the regulated community, special interest groups, the general public, and the scientific community. This process is outlined in Figure 15.2. To aid utilities, the EPA has published guidance manuals on the types of treatment that have been developed (Figure 15.3). Other key sections of the SDWA provide for the establishment of state programs to enforce the regulations. Individual states are responsible for developing their own regulatory programs, which are then submitted to the EPA for approval. The state programs must set drinking water standards equal to, or more stringent than, the federal standards. Such programs must also issue permits to facilities that treat drinking water supplies and develop wellhead protection areas for groundwater drinking supplies.

15.3 THE CLEAN WATER ACT

The **Clean Water Act** began with the Water Pollution Control Act of 1948 and was the first law to deal with comprehensive water pollution control. Designed to control the discharge of effluents into surface waters, this legislation focuses largely on point sources of pollution. The law was designed to protect the recreational and fishing waters of the United States (Figure 15.4).

Under the terms of this act, effluent standards are set and permits are issued on the basis of these standards for existing and new sources of water pollution. These are source-specific limitations. Also, the act lists categories of such point sources as sewage treatment plants, for which the EPA must issue standards of performance. The EPA may provide a list of toxic pollutants and set effluent limitations based on the best available technology economically achievable for designated

TABLE 15.1 Scope of federal regulations governing environmental pollution.

FEDERAL REGULATIONS	PURPOSE/SCOPE
Policy	
National Environmental Policy Act (NEPA). Enacted 1970.	This act declares a national policy to promote efforts to prevent or eliminate damage to the environment. It requires federal agencies to assess environmental impacts of implementing major programs and actions early in the planning stage.
Pollution Prevention Act of 1990.	The basic objective is to prevent or reduce pollution at the source instead of an end-of-pipe control approach.
Water	
Clean Water Act. Enacted 1948.	Eliminates discharge of pollutants into navigable waters. It is the prime authority for water pollution control programs.
1977 Amendment of the Clean Water Act.	Covers regulation of sludge application by federal and state government.
Safe Drinking Water Act (SDWA). Enacted 1974.	Protects sources of drinking water and regulates using proper water-treatment techniques using drinking water standards based on maximum contaminant levels (MCLs).
Clean Air	
Clean Air Act. Enacted 1970.	This act, which amended the Air Quality Act of 1967, is intended to protect and enhance the quality of air sources. Sets a goal for compliance with ambient air quality standards.
Clean Air Amendments of 1977.	Purpose is to define issues to prevent industries from benefiting economically from noncompliance.
Clear Air Amendments of 1990.	Basic objectives of Clean Air Act of 1990 are to address acid precipitation and power plant emissions.
Hazardous Substances	
Comprehensive Environmental Response, Compensation, and Liability (CERCLA). Enacted 1970. Amended in 1980 to include Superfund.	The act, known as Superfund, provides an enforcement agency the authority to respond to releases of hazardous wastes. The act amends the Solid Waste Disposal Act. Note: also affects water resources.
Superfund Amendments and Reauthorization Act (SARA). Enacted 1986.	This act revises and extends CERCLA by the addition of new authorities known as the Emergency Planning and Community Right-to-Know Act of 1986. Involves toxic chemical recall reporting. Note: also affects water resources.
Toxic Substances Control Act. Enacted in 1976.	This act sets up the toxic substances program administered by the EPA. The act also regulates labeling and disposal of polychlorinated biphenyls (PCBs).
Amendment of 1986.	Addresses issues of inspection and removal of asbestos.
Resource Conservation and Recovery Act (RCRA). Enacted in 1976. Amended in 1984.	This completely revised the Solid Waste Disposal Act. As it exists now, it is a culmination of legislation dating back to the passage of the Solid Waste Disposal Act of 1965. Defines hazardous wastes. It requires tracking of hazardous waste. Regulates facilities that burn wastes and oils in boilers and industrial furnaces. Requires inventory of hazardous waste sites.
Federal Insecticide, Fungicide, and Rodenticide Act (FIFRA). Enacted 1947.	Regulates the use and safety of pesticide products.
FIFRA Amendments of 1972.	Intended to ensure that the environmental harm does not outweigh the benefits

point sources. In addition, the EPA may issue pretreatment standards (*i.e.*, treatment preceding discharge of wastes into sewers) for toxic pollutants.

In addition, to direct discharge (*e.g.*, sewage treatment plants), EPA has established regulations and a permit program to regulate storm water discharges. Industries and municipalities are required to obtain storm water permits that incorporate storm water management plans and **Best Management Practices (BMPs)**.

The chief enforcement tool of the Clean Water Act is the permit. Anyone—public or private—engaged in construction or operation of a facility that may discharge anything—

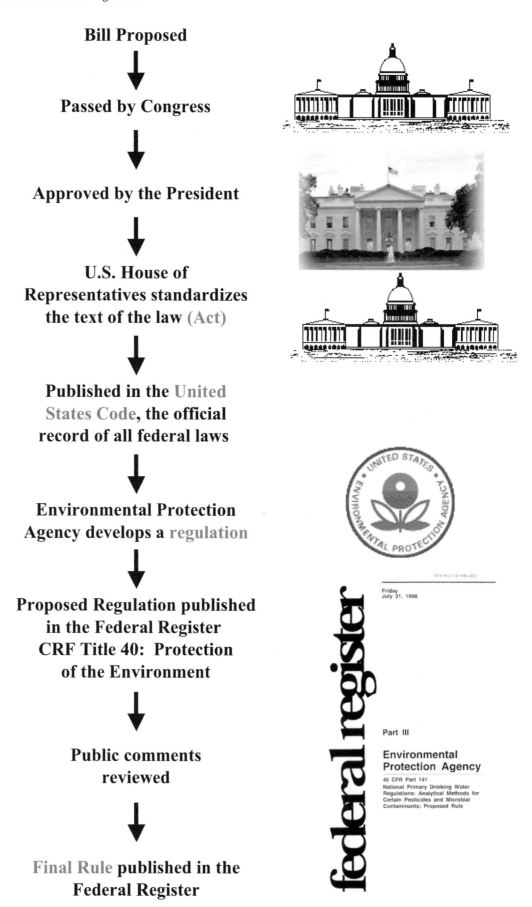

Bill Proposed

↓

Passed by Congress

↓

Approved by the President

↓

U.S. House of Representatives standardizes the text of the law (Act)

↓

Published in the United States Code, the official record of all federal laws

↓

Environmental Protection Agency develops a regulation

↓

Proposed Regulation published in the Federal Register CRF Title 40: Protection of the Environment

↓

Public comments reviewed

↓

Final Rule published in the Federal Register

Figure 15.1 The process for the development of federal regulations.

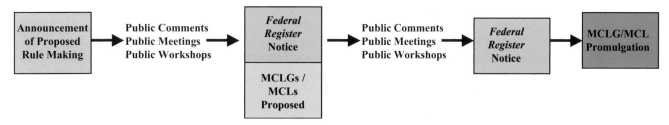

Figure 15.2 Regulatory development processes under the Safe Drinking Water Act. A series of steps is followed in the development of new regulations for contaminants in drinking water or processes for their control. These steps usually involve a series of public notices in the publication the *Federal Register* and meetings to allow for comment from the public, environmental groups, and the regulated industry before the final regulation is developed.

animal, vegetable, or mineral (or even energetic, as in the case of heat) —into navigable fresh waters must first obtain a permit. Permit applications must include a certification that the discharge meets applicable provisions of the act under the **National Pollution Discharge and Elimination Standards**

(NPDES). (Permits for a discharge into ocean water are issued under separate guidelines from the EPA.)

As with the SDWA, states are responsible for the enforcement of the Clean Water Act, that is, they must develop and submit to the EPA a procedure for applying and enforcing these standards. Finally, the Clean Water Act makes provision for direct grants to states to help them in administering pollution-control programs. It also provides grants to assist in the development and implementation of waste-treatment management programs, including the construction of waste-treatment facilities.

15.4 COMPREHENSIVE ENVIRONMENTAL RESPONSE, COMPENSATION, AND LIABILITY ACT

In the late 1970s, when the now-infamous Love Canal landfill in upstate New York was revealed as a major

Figure 15.3 Example of a guidance manual designed to aid water utilities to meet treatment rules under the Safe Drinking Water Act. From EPA.

Figure 15.4 The Clean Water Act was designed to help protect the recreational uses of surface waters from pollutants. Courtesy Wisconsin Department of Natural Resources (http://infotrek.er.usgs.gov/doc/beach/closed_04.gif.)

environmental catastrophe, the attendant publicity spurred Congress to pass the **Comprehensive Environmental Response, Compensation, and Liability Act (CERCLA)** of 1980 (see Figure 15.6). This act makes owners and operators of hazardous waste disposal sites liable for cleanup costs and property damage. Transporters and producers must also bear some of the financial burden. This legislation also established a multi-billion dollar cleanup fund (known as Superfund) to be raised over a 5-year period. Of that sum, 90% was to come from taxes levied on the production of oil and chemicals by U.S. industries and the rest from taxpayers. This tax has since expired and has not yet been renewed by congress. Until it is, all funds will originate from taxpayers.

CERCLA establishes *strict liability* (liability without proof of fault) for the cleanup of facilities on "responsible parties." The courts have found under CERCLA that these parties are "jointly and severally" liable; that is, each and every ascertainable party is liable for the full cost of removal or remediation, regardless of the level of "guilt" a party may have in creating a particular polluted site.

The Superfund is exclusively dedicated to clean up sites that pose substantial threats to human health and habitation, that is, *imminent* hazards. Moreover, the Superfund provides only for the cleanup of contaminated areas and compensation for damage to property. It cannot be used to reimburse or compensate victims of illegal dumping of hazardous wastes for personal injury or death. Victims must take their complaints to the courts.

15.5 FEDERAL INSECTICIDE AND RODENTICIDE ACT

Because of the potentially harmful effects of pesticides on wildlife and humans, Congress enacted the **Federal Insecticide, Fungicide, and Rodenticide Act (FIFRA)** in 1947. This act, which has been periodically amended, requires that all commercial pesticides (including disinfectants against microorganisms) be approved and registered by the EPA (Figure 15.5).

FIFRA include several key provisions: (1) studies by the manufacturer of the risks posed by pesticides (requiring registration); (2) classification and certification of pesticides by specific use (as a way to control exposure); (3) restriction (or suspension) of the use of pesticides that are harmful to health or the environment; and (4) enforcement of the above requirements through inspections, labeling, notices, and state regulation. For example, every pesticide container must have a label indicating that its contents have been registered with the EPA.

15.6 CLEAN AIR ACT

The Clean Air Act, originally passed in 1970, required the EPA to establish **National Ambient Air Quality Standards (NAAQS)** for several outdoor pollutants, such

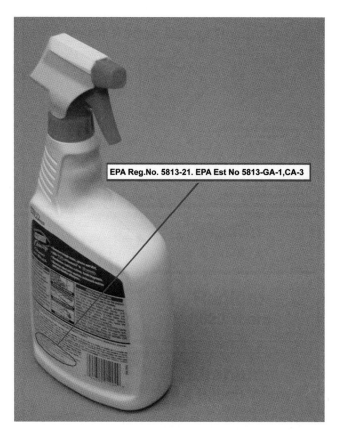

EPA Reg.No. 5813-21. EPA Est No 5813-GA-1,CA-3

Figure 15.5 All pesticides, including disinfectants must be registered with the EPA under FIFRA. Photo courtesy K.L. Josephson.

as suspended particulate matter, sulfur dioxide, ozone, carbon monoxide, nitrogen oxides, hydrocarbons, and lead. (Of these, ozone is the most pervasive and least likely to meet the standard; see also Chapter 23.) Each standard specifies the maximum allowable level, averaged over a specific period of time. These primary ambient air quality standards are designed to protect human health. Secondary ambient air quality standards are designed to maintain visibility and to protect crops, buildings, and water supplies

Each state must monitor the quality of its ambient air for purposes of determining attainment with the NAAQS. Each state is then divided into Air Quality Regions based on compliance with NAAQS. These regions are designated to be **Attainment Areas**, **Unclassifiable Areas**, and **Nonattainment Areas** for each of the six criteria pollutants. For example, an Air Quality Region can be designated as an Attainment Area for Ozone, but Non-Attainment for PM_{10}. Non-Attainment Areas can be further classified from Moderate to Extreme.

This act, which sets air pollution control requirements for various geographic areas of the United States, also deals with the control of tailpipe emissions for motor vehicles. Requirements compel automobile manufacturers to improve design standards to limit carbon monoxide, hydrocarbon, and nitrogen oxide emissions. For cities or areas where the ozone and carbon monoxide concentrations are high, reformulated and oxygenated gasolines are required. This act also addresses power-plant emissions of sulfur dioxide and nitrogen oxide, which can generate acid rain (see Chapter 4).

15.7 RESOURCE CONSERVATION AND RECOVERY ACT (RCRA)

The Federal Solid Waste Disposal Act, as amended by the **Resource Conservation and Recovery Act (RCRA)**, was adopted in 1976 to combat the problems associated with the unregulated land disposal of hazardous wastes. Its primary goals are to protect human health and the environment from hazards posed by waste disposal; conserve energy and natural resources through waste recycling and recovery; reduce or eliminate as expeditiously as possible the generation of hazardous waste; and ensure that wastes are managed in a manner that is protective of human health and the environment.

The most important feature of RCRA provides "cradle-to-grave" management of "hazardous waste" from the point of generation through transportation and its treatment, storage, and disposal. RCRA requires generators to characterize their wastes, properly manage their wastes on site, and then to manifest all wastes sent off site for disposal (Figure 15.6). Transporters of hazardous waste are required to fill in their portion of the manifest, comply with RCRA and federal Department of Transportation placarding requirements, and must deliver the hazardous waste to an approved "Treatment, storage and disposal facility." Those facilities must meet strict performance standards and must control their air emission. They must also monitor and prevent discharges to groundwater. RCRA also prohibits land disposal of certain untreated hazardous wastes.

15.8 THE POLLUTION PREVENTION ACT

The **Pollution Prevention Act of 1990** established as national policy the following waste-management hierarchy designed to prevent pollution and encourage recycling:

1. Prevention—to eliminate or reduce pollution at the source whenever feasible.
2. Recycling—to recycle unpreventable wastes in an environmentally safe manner whenever feasible.
3. Treatment—to treat unpreventable, unrecyclable wastes to applicable standards prior to release or transfer.
4. Disposal—to safely dispose of wastes that cannot be prevented, recycled, or treated.

The EPA, which is integrating pollution prevention into all its programs and activities, has developed unique voluntary reduction programs with public and private sectors.

15.9 OTHER REGULATORY AGENCIES AND ACCORDS

Many nations and international organizations have developed environmental laws and guidelines. For example the

Figure 15.6 CERCLA provides authority to identify and control release of toxic substances. From the Office of Response and Restoration, National Ocean Service, National Oceanic and Atmospheric Administration. (http://response.restoration.noaa.gov/cpr/phototour/strandley).

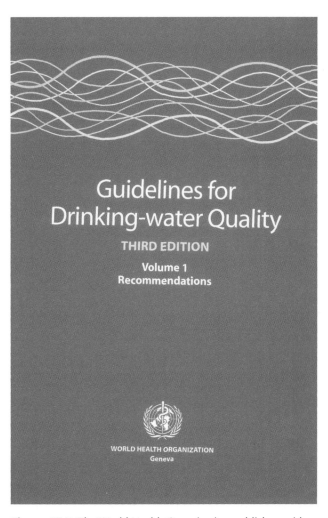

Figure 15.7 The World Health Organization publishes guidelines for safe levels of contamination in water and air.

European Economic Community (EEC) has developed a series of recommendations for standards and guidelines for control of air and waterborne pollutants. The World Health Organization has also published guidelines for the levels of contaminants in air and water, and provisions for their control (World Health Organization, 2004) (Figure 15.7). The United Nations, through the development of international agreements such as the Kyoto Accords, attempted to develop processes by which global warming could be controlled.

QUESTIONS AND PROBLEMS

1. What are the fundamental differences between an act, a regulation, and a rule?
2. Which regulations potentially affect the safety of drinking water?
3. Which regulations potentially affect contaminants in air?

REFERENCES AND ADDITIONAL READING

Pontius F.W. (2003) *Drinking Water Regulation and Health.* John Wiley & Sons, New York.

Outwater A.B. (1994) *Reuse of Sludge and Minor Wastewater Residuals.* Lewis Publishers, Ann Arbor, Michigan.

World Health Organization. (2004) *Guidelines for Drinking Water Quality*, 3rd ed., volume 1. World Health Organization, Geneva, Switzerland.

PART **3**

LAND AND WATER POLLUTION MITIGATION

CHAPTER 16

SOIL AND LAND POLLUTION

J.F. Artiola, J.L. Walworth, S.A. Musil and M.A. Crimmins

Irrigation wheels in Southwest U.S. *Photo courtesy J.F. Artiola.*

16.1 INTRODUCTION

Mining, agriculture, and deforestation are important energy-intensive activities that impact economies and at the same time directly and indirectly cause soil and land pollution. Mining produces vast quantities of almost sterile and structureless geologic materials, such as crushed rock that often contain significant amounts of toxic metals, such as lead and cadmium, and salts. Mine overburden and tailings are often stockpiled next to large open pit excavations. Modern agricultural production requires the use of large quantities of commercial fertilizers and pesticides, and produces animal wastes, all of which can pollute land and water. Land deforestation indirectly affects the quality of land and water by increasing the rates of soil erosion and sediment transport, and accelerating the loss of the nutrient-rich soil surface. Invasive exotic plant species also have a significant impact on the quality of our lands by creating soil conditions that may be toxic to other plants and by increasing fire hazards. All of these activities in turn can affect soil salinity and acidity of surrounding land areas by releasing or concentrating unwanted metals, salts, and acid or acid-forming minerals. These materials can also be released into air or water sources.

Because these activities have played a traditional key role in the growth and development of our modern society, to date their impacts on the environment have not been closely monitored or regulated. It is important to recognize the impacts that these activities have on our environment and there is a need to achieve a balance between their social benefits and the need for the preservation of our environment. The restoration of land adversely affected by these activities is discussed in Chapter 20.

16.2 SURFACE MINING

Mining of coal and metal ores was one of the earliest contributors to the industrial revolution. When transformed, these ores became both the fuel and the building blocks of industrialization. Although carbon-based plastic materials, together with such organic chemicals as pesticides and solvents, have dominated industrial production since the early 1950s, metal-based goods remain fundamental to modern industry. Numerous modern goods—from cars to paints—require the use of such common metals as iron, aluminum, and copper. In addition to these three metals, other less common metals and metalloids, such as lead, cadmium, nickel, mercury, arsenic, and selenium, are essential for the manufacture of these and other goods. Metallic elements are therefore commonly found in industrial wastes, where they have complex and still poorly understood effects on the environment. What is known is that uncontrolled and concentrated releases of metals into the environment present both short- and long-term hazards to human health and adversely affect the environment.

Industries that mine and process ores, drill for oil and gas, and/or burn coal also generate large volumes of salt-containing wastes. For these industries, the predominant chemical species include sodium (Na^+), calcium (Ca^{++}), sulfate ($SO_4^=$), and chloride (Cl^-), and carbonate ($CO_3^=$) ions, which are also very abundant in the natural environment. Because these wastes are not intrinsically hazardous or acutely toxic, they do not pose an immediate risk to health and the environment. Nonetheless, the volumes of these wastes that are generated each year are massive enough to be of concern.

16.2.1 Mine Tailings

Mining activities and, in particular, strip mining of metal ores produce vast quantities of residues called **mine spoils** and **mine tailings** that may contain significant concentrations of metals (Figure 16.1). Mine spoils or **overburden** consist of surface materials that do not contain the metal of interest and that are therefore stockpiled at the surface, often resembling large "mesas." Mine tailings, in contrast, are the crushed mineral rock that has been processed to release the metal of interest. These are often pumped as a slurry in "lifts" of 3-m dimension into valleys or depressions. Mine tailings can be 35

Figure 16.1 Copper mine tailings and spoils in Southwest U.S. Photo courtesy J.F. Artiola.

Figure 16.2 Acid mine drainage from gold mine in South Dakota. From U.S. Fish and Wildlife Service.

m deep due to successive depositions of lifts (see also Chapter 20). Thus, these residues, which are usually composed of unweathered primary minerals, can alter the environment physically and chemically. Strip mining for copper, for example, produces large quantities of tailings that often contain concentrations of 100–10,000 mg kg^{-1} of such metals as cadmium and lead. Similarly, iron pyrites (FeS_2), which are often associated with copper, silver, and lead ores, can have a devastating impact on the aquatic environment because their oxidation releases sulfuric acid into the environment (Figure 16.2). The overall reaction is described as follows:

$$FeS_2 \text{ (pyrite mineral)} + 3.75O_2 + 3.5H_2O \rightarrow$$
$$Fe(OH)_3 \text{ (solid)} + 2SO_4^{2-} + 4H^+ \quad \text{(Eq. 16.1)}$$

In an acid stream (pH <3), fresh pyrite can react in a cascading effect with soluble ferric iron (Fe^{3+}), creating even more acidity (Stumm and Morgan, 1996). The reaction rate is controlled by the oxidation of Fe^{2+} to Fe^{3+} in the presence of O_2, and results in lowering the pH of the environment. This process can also occur biologically via autotrophic bacteria which thrive at pH 2–3) (see Chapter 5).

Mining operations that treat or leach ores and/or store acid chemicals for the extraction of metals can generate large volumes of acidic metal-containing wastewaters and/or leachates. For example, low-grade Cu ore is often extracted by means of sulfuric acid heap leaching. In this process, crushed Cu ore is continuously leached with sulfuric acid until most of the Cu is solubilized due to both the high acidity and formation of Cu-sulfate complexes. Spent acid solutions, usually contaminated with other metals, must be neutralized and stored in lagoons or impoundments. Gold mining also produces vast quantities of spent ores and liquid process streams that usually contain residual levels of cyanide (CN^-) complexes. Metal-cyanide complexes are usually either stable in the soil environment or biologically degraded into nontoxic forms of N (see Chapter 5). However, when released into aquatic systems, these residues can be extremely toxic to fish if free cyanide is released into the water.

16.2.2 Air Emissions

Metal smelting and refining processes generate wastes that may contain multiple hazardous metals, such as lead, zinc, nickel, copper, cadmium, chromium, mercury, selenium, arsenic, and cobalt. These elements may be found in the ores used or they may be added as mixed metals into the melts to produce metal alloys. Thus, metal-containing smelter wastes have to be treated and disposed of as hazardous wastes. Smelting and refining require very high temperatures to reduce the metal ores (such as pyrite and bauxite for iron and aluminum production) into pure metal and to refine metals and alloys. For example, iron melts at 1536°C, copper melts at 1083°C, and aluminum melts at 660°C. At these temperatures, many other metals and metal compounds volatilize; for example, the boiling points of mercury, cadmium, zinc, and arsenic are 357°C, 765°C, 906°C, and 613°C, respectively. Therefore, smelter and metal refining stacks that do not have gas scrubbers can release significant amounts of relatively volatile toxic metals into the atmosphere that eventually deposit onto the land.

16.3 DEFORESTATION

Deforestation is simply the conversion of forested tracts to barren lands. This is usually done by clear-cutting trees and removing the wood (Figure 16.3). Forested areas are typically cleared to make room for agricultural operations or to harvest wood as a fuel source or for lumber products. Much of the deforestation occurring globally is due to slash-and-burn operations that make room for agricultural operations.

The process of deforestation results in many undesirable environmental impacts at multiple scales. Local impacts include decreasing soil stability, increasing erosion and sediment transport into streams, reduction in biodiversity through loss of habitat (see Chapter 24), and alterations to microclimates that typically increase local temperatures because of loss of vegetation and increased numbers of heat islands (see also Chapter 21). Degradation of air quality is often at the regional scale if deforestation is being driven by

Figure 16.3 Example of deforested area with downed slash in foreground. United Nations Environmental Program.

burning downed slash. This promotes episodes of high levels of atmospheric particulate matter and carbon monoxide gas that are harmful to the health of both humans and wildlife (see also Chapter 24). Deforestation can also produce impacts on a global scale. Research over the past decade has shown that the cutting and burning of large forest tracts is quickly liberating large amounts of carbon and increasing levels of the greenhouse gas, carbon dioxide, in the atmosphere (Zhang et al., 2003). Removing forest vegetation further disrupts the global carbon balance by eliminating the living trees that served as a sink for carbon dioxide. Photosynthesis in trees converts atmospheric carbon dioxide into plant cellulose, drawing carbon out of the atmosphere and storing it as biomass (see also Chapter 24).

16.3.1 Local Land Pollution Impacts of Deforestation

Removal of forest vegetation increases the potential of soils to become eroded by wind and/or rainfall. Runoff during precipitation events can promote both the erosion of soils and the transport of sediments into river systems. These sediments will degrade water quality by increasing turbidity and levels of dissolved nutrients (*e.g.*, phosphorus and nitrates) (see also Chapter 18). Experiments to document the effects of deforestation on watershed dynamics and stream water quality have been conducted in several experimental watersheds throughout the U.S. Results from the Hubbard Brook Experimental Forest in New Hampshire show large increases in dissolved nutrient levels and sediment loads in-stream for a deforested watershed area, as compared to a control forested area watershed (Bormann and Likens, 1979; Bonan, 2002).

A more serious form of land-based pollution has been tied to deforestation in areas of South America. High-levels of mercury have been found in the blood of people in many rural communities in Brazil where fish is a staple food. The high mercury levels were initially attributed to gold mining operations found throughout these areas. Further study has shown that naturally occurring pools of mercury found in soil and organic matter were being readily transported into streams through runoff following deforestation (Veiga et al., 1994). The removal of forest vegetation allowed mercury that was initially stabilized in soil organic matter to become mobile and be transported into streams.

16.3.2 Regional Air Quality Impacts of Deforestation

Deforestation is often accompanied by the burning of biomass. Sometimes the burning is done to clear slashed vegetation, and at other times harvested forest vegetation is burned as a fuel source for heating and cooking. In either case, the burning of forest biomass on large scales can cause serious air quality problems. Particulate matter, ozone, and carbon monoxide are all produced when forest biomass is burned, and all pose health risks to humans (see also Chapters 3 and 23). Figure 16.4 shows the impact of large-scale forest

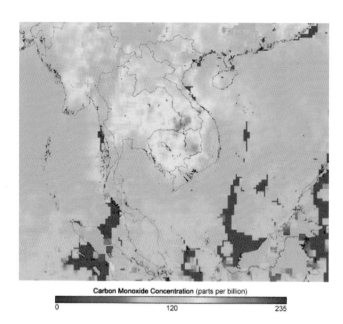

Carbon Monoxide Concentration (parts per billion)

0   120   235

Figure 16.4 Atmospheric carbon monoxide concentrations over Asia estimated using data collected by the NASA Terra satellite. From NASA.

biomass burning for heating and cooking on atmospheric carbon dioxide levels across Asia (Heald et al., 2003).

16.4 SOIL ACIDITY—SALINITY

16.4.1 Acid Soils

Acid soils occur naturally or develop as the result of continuous additions of acid-forming fertilizers. Natural acid soils are usually found in the tropics, the result of thousands of years of excessive weathering of soil minerals. High rainfall and year-around high temperatures leaches all basic cations (such as Na, Ca, Mg, and K) and pH buffering minerals (such as carbonates). Also, this climate promotes the transformation and subsequent leaching of Si from Si-based minerals, leaving acidic iron and aluminum oxides minerals. For example, soluble aluminum can release protons into the soil environment by the following general reaction:

$$Al^{3+} + H_2O \rightarrow Al(OH)_3 \text{ (solid)} + 3H^+ \quad \text{(Eq. 16.2)}$$

Similarly, the presence of pyrite minerals in some soils can lead to the formation of acidic soil conditions, in a reaction similar to Equation 16.2 above.

Agricultural soils can also become acidic due to the continuous additions of large amounts of acid-forming fertilizers such as ammonia and urea. For example, one mole-equivalent weight of ammonium (NH_4^+) can produce two mole-equivalent of H^+ after it is fully oxidized to nitrate (NO_3^-) in the soil environment.

Other sources of acid-forming chemicals that impact the soil environment include coal-burning air emissions (SO_2, NO_x) that are hydrolyzed and scrubbed out of the atmosphere by rain (see also Chapter 20).

TABLE 16.1 Soil salinity rankings.

PARAMETER (mS cm^{-1})	NON-SALINE	SLIGHTLY SALINE	MODERATELY SALINE	SALINE
EC*	<4	4–8	8–16	>16

*Measured on a water-saturated soil paste extract (USDA, 1954).

16.4.2 Salinity

Soil salinity is a measure of the minerals and salts that can be dissolved in water. In most cases, the following mineral ions are found in soil-water extract listed in order of importance:

$$Na^+, Cl^-, Ca^{++}, SO_4^=, HCO_3^-, K^+, Mg^{++}, NO_3^-$$

Increased soil salinity has progressive and often profound effects on the structure, microbial diversity, and plant activity of soils. Soil salinity is measured by using electrical conductivity (EC) measurements of a water-saturated soil paste extract (Table 16.1).

An excessive concentration of Na ions in soils produces an imbalance in the ratio of monovalent cations to divalent cations. This is measured by the exchangeable sodium percent (ESP). Salt-affected soils are thus also classified by their ESP as shown Table 16.2 (see also Chapter 20).

There are numerous sources of soil salinity. Natural soil salinity occurs in hot arid and semi-arid climates with ≤ 27 cm of annual rainfall. Soils and lands that have shallow water tables can develop saline soils due to excessive water evaporation and the concentration of salts. Poor water quality and irrigation practices also contribute to the salinization of thousands of acres of farmland each year around the world. Salt-affected soils occupy, on a global basis, 952.2 million ha of land. These soils constitute nearly 7% of the total land area or nearly 33% of the potential agricultural land area of the world (Gupta and Abrol, 1990).

16.5 SOIL EROSION

Soil particles can act as carriers of other contaminants that are sorbed to particulate surfaces. For example, phosphorus is often associated with soil particles and sediments. When soil particles are eroded and discharged into an aquatic environment, the resulting sediments increase the P nutrient levels of the water and can cause excessive growth of algae and other aquatic plants. This process, together the concomitant reduction in oxygen, is known as eutrophication (see also Chapter 18).

Organic chemicals, including herbicides, insecticides, fuels, solvents, preservatives, and other industrial and agricultural chemicals, can be similarly adsorbed and desorbed from waterborne soil particulates.

Naturally occurring particulate contaminants come from many sources, including agricultural operations, logging, construction-related activities, mining and quarrying, and unpaved roads, and from wind erosion (see Chapter 9). Soil erosion is a natural process that occurs continuously, but is often accelerated by human activities. Several factors are required for soil material to become dislodged and transported into air or water. Soil must be susceptible to erosional processes, which generally requires that the soil be exposed to erosional forces. In the first phase of soil erosion, soil particles become dislodged. Energy inputs must be adequate to dislodge particles. In the second phase, the particles are transported. Various soil properties, which we will examine, determine the susceptibility of soil particles to dislodgement.

16.5.1 Soil Water Erosion and Control

We will first consider particle movement caused by water. As indicated in Chapter 2, soil particles are usually formed into aggregates, which vary considerably in size, shape, and stability. Organic and inorganic materials and certain soil cations are the primary aggregating and inter-particle cementing agents. In the detachment phase, individual particles are dispersed or separated from aggregates or cemented particles. The source of the energy responsible for detaching soil particles is either raindrop impact or the flow of runoff water. When raindrops, which travel at approximately 900 cm s^{-1}, hit bare soil, the kinetic energy of the raindrops is transferred to the soil particles, breaking apart aggregates and dislodging particles. Dislodged particles can be moved over 1 m in the splash from raindrop impact. They are moved larger distances by runoff water, which can dislodge additional particles through scouring action. Smaller particles are transported more easily than large ones, and faster flowing water can carry a heavier particulate load than slow-moving water. When uniform shallow layers of soil are eroded off areas of land, this is called **sheet erosion**. Directed water flow cuts channels into the soil. Small channels are called **rills**; large channels are **gullies** (Figure 16.5).

The process of water erosion has been described by the **United States Department of Agriculture (USDA) Universal Soil Loss Equation (USLE)**, later modified to the **Revised Universal Soil Loss Equation (RUSLE)**.

$$A = 2.24R \times K \times LS \times C \times P \qquad \text{(Eq. 16.3)}$$

where:

A = the estimated average annual soil loss (metric tons/hectare)

TABLE 16.2 Soil exchangeable sodium percentage rankings.

PARAMETER	ADEQUATE	BORDER-LINE	INADEQUATE**	COMMENTS
ESP*	<10	10–15	>15	Sandy soils may tolerate ESP values up to 15

*ESP calculated using a mol-equivalent units, using the concentrations of Na$^+$, Ca^{++}, Mg^{++}, and K$^+$ soil water extractable or exchangeable ions.
**Considered a sodic soil (USDA, 1954).

Figure 16.5 Soil erosion in abandoned farmland in southern Arizona that results in a gully. Photo courtesy J.F. Artiola.

P = the factor for supporting practices. This factor takes into account specific erosion control measures. Erosion control practices reduce the P factor.

In the past few decades, farmers have tried to reduce tillage that leaves soil bare and to minimize the amount of time that the soil surface is exposed to raindrops. These new agricultural practices are collectively known as **Reduced Tillage** or **Minimum Tillage** systems (see Figure 16.6). On highly erodible lands, specific erosion control practices include contour planting strip cropping or terracing, all of which can effectively reduce erosion.

Bare soil at construction sites and along road cuts is often covered with synthetic fabrics called geotextiles. These coarse woven materials provide immediate protection and may be used in conjunction with seeding of cover corps that can provide long-term cover. Grass or other plants can be seeded with a hydroseeder that sprays a mixture of seed, fertilizer, mulch, and polymers that cement the mixture into a cohesive soil covering that gives temporary protection to the soil surface until the seeds can geminate and plants cover the soil.

R = the rainfall and runoff erosivity index. This describes intensity and duration of rainfall in a given geographical area. It is the product of the kinetic energy of raindrops and the maximum 30-minute intensity.

K = the soil erodibility factor. K is related to soil physical and chemical properties that determine how easily soil particles can be dislodged. It is related to soil texture, aggregate stability, and soil permeability or ability to absorb water. It ranges from 1 (very easily eroded) to 0.01 (very stable soil).

LS = a dimensionless topography factor determined by length and steepness of a slope. The LS factor is related to the velocity of runoff water. Water moves faster on a steep slope than a more level one, and it picks up speed as it moves down a slope. Therefore, the steeper and longer the slope, the faster runoff water will flow. The faster water flows, the more kinetic energy it can impart to the soil surface (kinetic energy = mass $\times$ velocity^2).

C = the cover and management factor. Cover of any kind can help protect the soil surface from raindrop impact and can force runoff water to take a longer, more tortuous path as it moves downslope, slowing the water and reducing its kinetic energy.

Figure 16.6 No-till cotton planted into crop residue to protect the soil surface. From USDA, image number K8550–8. From *Environmental Monitoring and Characterization* © 2004, Elsevier Academic Press, San Diego, CA.

Figure 16.7 A rock-filled dam erosion control structure. From *Environmental Monitoring and Characterization* © 2004, Elsevier Academic Press, San Diego, CA.

Permeable barriers made of straw bales or woven fabrics can be used to slow water and reduce its ability to carry sediments. Runoff water can be trapped in settling ponds in which water velocity is eliminated or greatly reduced, allowing suspended particles to settle out, and reducing sediment loads before overflow water is released. If suspended colloids are in a dispersed condition, flocculating agents may be added to aggregate particles into larger assemblages that rapidly settle out of suspension. **Gabions** or wire mesh containers filled with rocks can also be used to control water erosion (Figure 16.7).

16.5.2 Soil Wind Erosion and Control

Like water erosion, wind erosion has two phases: detachment and movement. As the wind blows, soil particles are dislodged and begin to roll or bounce along the soil surface in a process called **saltation**. Larger soil particles can move relatively short distances in this way, but, more importantly, as the large particles bounce and strike smaller particles and aggregates, they provide the energy necessary to break aggregates apart and suspend smaller particles in the air. Smaller particles remain suspended in air for longer periods of time and are therefore more likely to travel much longer distances. As in the case of water erosion, models that examine the factors important in wind erosion are useful in predicting wind erosion (Saxton et al., 2000).

It should be noted that finer textured soils (those with more silt and clay sized particles) are less erodible than sandy soils. This reflects the ability of soil aggregates to hold the soil in place during high wind events. On the other hand, PM_{10} consists largely of silt-sized particles, and $PM_{2.5}$ is mainly clay. Therefore, soil wind erosion and particulate matter production are not directly related (see also Chapter 6).

Wind velocity, a major factor in soil wind erosion, can be decreased with windbreaks. These may be living windbreaks of planted trees, shrubs, or grasses, or they can be constructed material such as fences or screens. Windbreaks are most effective when placed perpendicular to the direction of the prevailing wind. Effects of windbreaks extend to as much as 40 to 50 times the height of the windbreak; however, the area adequately protected by the windbreak is usually smaller. Effective control is usually considered to extend to about 10 times the height of the windbreak. Wind erosion is reduced by a rough soil surface.

Surface roughness can be controlled by creating ridges or a rough surface with tillage implements. Ridges 5 to 10 cm in height are most effective for controlling wind erosion. Soil surface can also be protected by providing vegetative or other surface cover, such as straw, hay, animal manure, or biosolids. The soil water erosion control measures discussed earlier also provide effective wind erosion control.

Various amendments that bind soil particles together, including calcium chloride ($CaCl_2$), soybean feedstock processing byproducts, calcium lignosulfate, polyvinyl acrylic polymer emulsion, polyacrylamide, and emulsified petroleum resin are applied to unpaved roads to reduce particulate emissions. Unpaved roads can also be covered in gravel or similar nonerodible surfacing materials. However, most of these treatments generally offer only temporary dust control and must be periodically repeated.

16.6 AGRICULTURAL ACTIVITIES

16.6.1 Fertilizers

Plants need numerous chemicals in order to complete their life cycles. There are at least 16 **essential elements** required for the growth of all plants: C, H, O, N, P, K, Ca, Mg, S, Fe, Mn, Zn, Cu, Mo, B, and Cl in various ionic forms. Interestingly, soil microbes require these same elements. In undisturbed ecosystems, plants obtain these nutrients from the soil solution via mineral weathering, atmospheric inputs, inputs from stream deposition, and nutrient recycling due to death and decomposition of vegetation. The availability of the nutrients depends on abiotic soil factors (Chapter 2) and chemical and biological properties (Chapters 5 and 7). Agricultural crop production has always relied on soil components for nutrient sources. However, excessive cropping and in particular dense monoculture practices deplete soil plant nutrients, especially N, P, K, and Ca. Thus, over years of continuous crop production, large amounts of nutrients are removed, with a concomitant decline in productivity. Therefore, N, P, K, and other plant nutrients must be periodically augmented by the use of fertilizers, including animal or human wastes. Fertilizers may contain any of the essential nutrients, but the majority of fertilizers applied to agricultural soils contain nitrogen (N), phosphorus (P), potassium (K), or some combination thereof. These are the so-called **macronutrients** because plants take them up in larger amounts than the other essential nutrients.

Figure 16.8 illustrates trends for macronutrient use in the United States. Fertilizer use dramatically increased around the time of World War II, as improved crop varieties and management practices, together with increased mechanization,

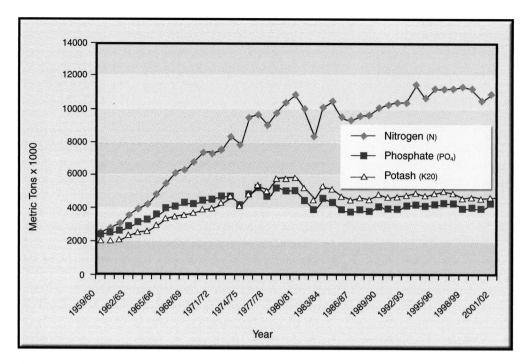

Figure 16.8 Historical trends for fertilizer nutrient use in the United States. Data source: The Fertilizer Institute (TFI, 2005).

made fertilizer use both practicable and profitable. In the 1980s, however, fertilizer use began to level off, reflecting both lower agricultural profitability and increased environmental concerns related to fertilizer use. Nonetheless, the combined annual per capita use in the U.S. of NPK fertilizers is about more than 150 lbs (~70 kg). The aforementioned concerns associated with the use of NPK fertilizers and wastes include excessive surface and groundwater pollution by water-soluble nitrates and colloid-bound phosphates due to poor agricultural fertilizer and waste management practices (see also Chapters 17 and 18).

16.6.2 Pesticides

Extensive use of synthetic pesticides began in the 1940s with DDT used to control mosquitoes. This was quickly followed by the adoption of pesticides in large-scale monocultural agricultural production. Initially pesticide use was credited with significant increases in food production. However, the negative aspects of their indiscriminate use also became evident. For example, extensive use of insecticides and herbicides has created new generations of pesticide resistant insects and plants.

In 1962, Rachel Carson's book *Silent Spring* brought public attention to the fact that chlorinated pesticides were very persistent in the environment. These chemicals can accumulate in animal fatty tissue and produce fish kills when released into waterways. DDT, associated with the rapid decline of some birds of pray, was banned for agricultural use in the United States in 1973. Other chlorinated pesticides were also banned, but have been replaced by much less persistent, but more acutely toxic, pesticides. In addition, in recent years new links have been discovered between some

types of cancer and low-level exposure to some pesticides like 2,4-dichlorophenoxyacetic acid (2,4-D), 2,4,5-trichlorophenoxyacetic acid (2,4,5-T), and other pesticides.

Less persistent pesticides are usually much more soluble in water than chlorinated hydrocarbons. Unfortunately, these new pesticides are more like to leach to groundwater or be found in the agricultural runoff if they are not degraded fast enough in the soil environment (see section 16.6.2.3). Today, pesticides continue to be used extensively in modern farming, urban lawns, parks, and golf courses primarily to control weeds, fungi, and insect infestations. Unfortunately, even less persistent pesticides have their problems. In 2003, the U.S. EPA concluded that atrazine, the second most widely used pesticide (herbicide) in the U.S., could cause sexual abnormalities in frogs. In addition, atrazines, the most common family of herbicide chemicals found in groundwater are also potential endocrine disruptors. Other common pesticides have been linked to or are being studied as possible endocrine disruptors (see Chapter 31).

16.6.2.1 Types of pesticides

The technical definition, stated in the amended **Federal Insecticide, Fungicide, and Rodenticide Act (FIFRA)**, is that a pesticide is any substance or mixture of substances intended for destroying, preventing, or mitigating insects, rodents, nematodes, fungi, weeds, or any other undesirable pests. This also includes plant or insect growth regulators as well as defoliants that are used to cause leaves to drop from plants to facilitate harvest, and desiccants that dry up unwanted plant tissue. Under this definition, many chemicals, both newly developed and familiar, may be considered as pesticides and be regulated as such. For example, insect pheromones (sex attractants) may be used to attract certain insect populations, to confuse mating patterns, and thereby control insect population. In addition, ordinary dish detergent may be used to kill whiteflies or bees. Common table salt (sodium chloride) is used to control weeds in beet fields in humid regions.

Insecticides are formulated to control particular insects. Two common insecticides are chlorpyrifos and malathion. Herbicides are formulated to control weeds. Glyphosate and atrazine are the two most common herbicides, accounting for 70–90% of the total herbicide use in the U.S. (US EPA, 2004). Fungicides are formulated to control fungi including molds and mushrooms. Chloropicrin, metam-sodium and 1,3-dichloropropene are the three commonly used fumigants applied to soil to control nematodes and soil fungi in the U.S. (USEPA, 2004).

Pesticides may also be classified according to their mode of entry into the target pest. Contact pesticides enter the target pest upon direct application, while systemic pesticides must pass through a host organism before they enter their targets. For example, a contact insecticide, or its residue, kills target plants or insects on direct application, while a systemic insecticide kills insects only after moving through the system of the plant hosting the target insect. Thus, if a particular insect does not feed on the plant, it will not be harmed.

Finally, pesticides can be classified by the forms in which they are used. Fumigants, for example, are pesticides applied as gases. Fumigants may be used selectively to control drywood termites in houses or to control the pest population in stored products such as fruits, vegetables, and grains. They may also be released over large areas to remove many pests from soil.

16.6.2.2 Extent of pesticide use

Pesticides are sold or distributed by intra- or interstate commerce in the United States, and they must be registered by the U.S. EPA. The EPA has compiled substantial lists of pesticide ingredients whose applications must be reported. The EPA is also authorized, by the **Federal Food, Drug, and Cosmetic Act (FFDCS)**, to establish tolerances for pesticide residues in raw and processed foods. The **Food and Drug Administration (FDA)** of the Department of Health and Human Services monitors and enforces the established tolerances (see also Chapter 15).

In addition, many individual states in the United States have established other regulatory agencies to control pesticide applications in order to protect wildlife and water supplies. For example, Arizona has compiled a list of chemicals—the Groundwater Protection List—whose use must be reported. Similar requirements exist for the sales of these pesticides, so that significant underreporting of applications cannot occur without alerting the regulatory agency.

According to the U.S. EPA, in 2001 the use of conventional pesticides in the U.S. was estimated to be about 1.2 billion lbs (545 million kg), reflecting a slightly declining trend in use since the mid-1980s (Figure 16.9). These figures places the annual per capita use of pesticides at about 4 lbs (~1.9 kg). The U.S. EPA estimates that in 2001 about 78% of these products were used in agricultural production; 12% in home and garden settings; and the remaining 10% in forestry, industry, and government programs. Therefore, most of these chemicals were applied directly onto plants and animals on agricultural lands and water systems. In addition, industry and water utilities also use chemicals with pesticide-like properties. For example, according to U.S. EPA estimates in 2001, about 790 million lbs (360 million kg) of wood preservative chemicals and 2.6 billion lbs (1.19 billion kg) of chlorine and hypochlorite chemicals were used in the U.S. These highly toxic chemicals include creosote, pentachlorophenol, and CCA (chromate copper arsenate). Presently, pesticide product labeling must list their active ingredients and the EPA's registration number, as well as safe use instructions to minimize personal exposure and damage to soil and water environments.

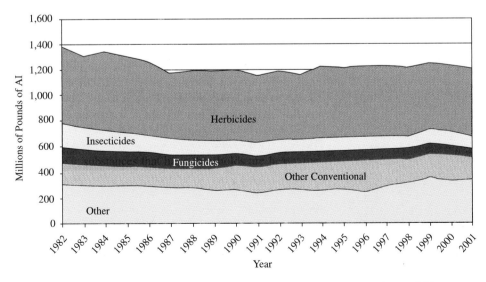

Figure 16.9 Annual amount of pesticide active ingredient used in the U.S. by pesticide type, 1982–2001 estimates. All market sectors. From U.S. EPA, 2004.

It is interesting to note that despite the public awareness about numerous links between pesticide residues, their adverse health and environmental effects, and the increasing public demands for pesticide-free food, the largest growth sector for pesticides is in home and garden applications. Since 1995, the use of pesticides by the private sector (home and garden) has nearly doubled (from 7% to 12% in 2001) (U.S. EPA, 2004).

16.6.2.3 Fate of pesticides

Depending upon their physicochemical properties, patterns of use, and local conditions, some pesticides may leach through the crop root zone and eventually contaminate groundwater at certain locations (see Chapter 17). The two most important properties of a chemical that determine whether a pesticide represents a threat to groundwater are its *persistence* and *mobility* in soil, as discussed in Chapter 6. During the registration of new pesticides, computer programs are used to estimate the potential for groundwater to be contaminated by the specific use of a particular chemical at various locations in the United States. Several states, including Arizona and California, consider the capacity of a compound to leach through the soil into groundwater as a criterion for inclusion in their lists of controlled chemicals.

After a pesticide is applied to a field, it may meet a variety of fates, as shown in Figure 16.10. Some may be lost to the atmosphere through volatilization, carried away to surface waters by runoff and erosion, or photodegraded by sunlight. Pesticides that have entered into soil may be taken up by plants (and subsequently removed), degraded into other chemical forms, or leached downward with water below the crop root zone. The amount of any particular chemical that ends up volatilized, leached, degraded, or in surface runoff depends upon site conditions, weather conditions, management practices, soil properties, and pesticide properties (as described in several previous chapters).

In evaluating the contamination potential of a particular pesticide, it is essential to consider its sorption (retardation) and transformation half-life behavior jointly, as discussed in Chapter 6. For example, a pesticide with low retardation and a long half-life (*e.g.*, more than 100 days) poses a considerable threat to groundwater through leaching, particularly in soils having low organic matter. Conversely, a pesticide with large retardation and a long half-life is more likely to remain on or near the surface of soils with moderate levels of organic carbon content, thereby increasing its chances of being carried to a lake or stream in runoff water. In terms of water-quality protection, pesticides with intermediate retardation and short half-lives may be considered the "safest." Although they are not readily leached, they move into the soil with water, thereby reducing their potential for loss from erosion, and they degrade fairly rapidly, thereby reducing the chance for losses below the root zone. Figure 16.11 provides a schematic representation of the depth of movement of a strongly sorbed (glyphosate), a moderately sorbed (atrazine), and a weakly sorbed (aldicarb) chemical. It was assumed that the rainfall and irrigation amounts exceeded the crop water use by twice the amount of water contained in the root zone at an optimum water content that moved the chemicals downward.

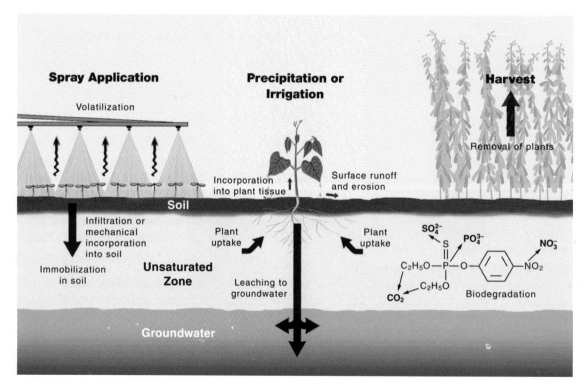

Figure 16.10 Fate of pesticides in soil. From *Pollution Science* © 1996, Academic Press, San Diego, CA.

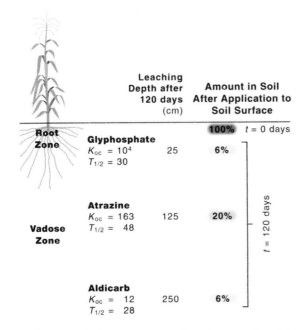

Figure 16.11 Relative movement and persistence of pesticides in soil. The transport of a particular pesticide is strongly influenced by retardation. Here, the degree of retardation increases in the order aldicarb < atrazine < glyphosate. The amount of pesticide remaining in the soil is illustrated by the intensity of color 120 days after the application. From *Pollution Science* © 1996, Academic Press, San Diego, CA.

As shown in Figure 16.11, glyphosate would be concentrated in the root zone to a depth of about 25 cm, atrazine would be concentrated near the bottom of the root zone (about 125 cm), and aldicarb would be concentrated at a depth of about 250 cm. A slightly higher percentage of the applied atrazine would exist in the system, compared with the other two pesticides, because it has a slightly larger half-life. For a growing season of about 120 days, about 6% of the applied aldicarb and glyphosate would remain, while about 20% of the atrazine would remain. This example does not account for numerous differences in management practices that would influence the persistence and soil distribution of these pesticides.

16.7 ANIMAL WASTES

Animal wastes contain several types of land pollutants that are of increasing concern both to the public and regulators. Besides traditional pollutants, discussed below, increasing evidence suggests that excessive use of animals waste on land releases measurable amounts of antibiotics, growth hormones, and pesticides containing toxic metals like arsenic. Animal agricultural wastes can be divided by two production types: range and pasture production, and confined or concentrated animal production.

In range and pasture systems, the concentration of wastes is generally much more diffuse or dispersed than it is when large numbers of animals are confined to relatively small areas. Range and pasture systems have two principal measurable effects on surface water quality: (1) increased turbidity through the movement of soil particles into streams, rivers, and lakes; and (2) increased fecal coliform counts in areas of heavy animal use. Although we know that grazing systems may adversely affect some measures of water quality, we will focus here on the highly concentrated animal production units and the methods of preventing and controlling pollution from these concentrated units. Concentrated animal production is very common and is occurring in increasingly controlled environments to raise productivity and diminish climatic, feeding, and mortality variables. Larger numbers of animals are being raised in **concentrated animal feeding operations** or **CAFOs**—principally, feedlots, dairies, swine operations, poultry houses, and intensive aquaculture.

Following World War II, manure was displaced as the primary fertilizer by fossil-fuel-based fertilizers as farms became increasingly specialized.

With the breakdown of the traditional cycle of reincorporation of wastes back in to the land, what was once an essential source of nutrients has now become a potential pollutant. Thus, the production of large numbers of animals on a small land base has resulted in the stockpiling of wastes, the construction of large waste-storage ponds, and, oftentimes, waste applications to land in excess of agronomic crop needs. To date, few states regulate the land application of animal wastes to the degree that biosolids are regulated (see Chapter 27).

16.7.1 Nonpoint Versus Point Source Pollution

The term "nonpoint pollution" is misleading and is often misused in the context of animal wastes. In animal agricultural systems, true **nonpoint sources** are those in which potential contaminants are not concentrated during production and do not pass through a single or small number of conduits for disposal. These nonpoint sources include corrals, feedlots, and extensive and intensive pasture systems.

Point sources are those facilities that concentrate pollutants or contaminants to a significant degree and pass these contaminants through a pipe, ditch, or canal for disposal. The most common point sources are milksheds and barns, dairy and other food-processing plants, intensive indoor swine facilities, anaerobic and aerobic lagoons, and evaporative storage ponds. In addition, certain types of intensive aquaculture may also be point sources of contaminants, with return flows highly nutrient-laden with fish excreta.

According to EPA regulations, however, some concentrated animal feeding operations may be designated as point sources requiring an individual National Pollution Discharge Elimination System (NPDES) permit. In this case, a concentrated animal feeding operation is defined as a lot or facility without vegetation where animals are confined for 45 or more days per year. The number of animals needed to meet this definition as a CAFO depends on several factors; the key determinant is whether or not the facility discharges into navigable waters, as determined by the method of discharge. The method of discharge is judged by the 25-year, 24-hour storm event, which is the required event that a facility must be designed to meet.

Nonpoint sources, such as nondischarging concentrated animal feeding operations, require a different approach to prevention and mitigation of pollutants than do point source emissions from a pipe or conduit. At present, the nonpoint source approach to mitigation employs **Best Management Practices (BMPs)**, as defined by the 1987 Amendments to the Federal Water Pollution Control Act. In contrast, point source methods employ methods termed **Best Available Demonstrated Control Technology** or **Best Available Control Technology**. In 2003, the U.S. EPA (2003) published a Final Rule on CAFOs that is now used to permit animal feeding operations by establishing requirements that are more protective of the environment. Large amounts of animal wastes are land applied (see Chapter 27).

16.7.2 Specific Pollutants

Concentrated animal agriculture produces specific pollutants in the wastes resulting from animal metabolic activity (Information Box 16.1).

INFORMATION BOX 16.1

Specific Pollutants in Animal Wastes and Common Environmental Effects

Nitrate-nitrogen:	Excessive plant/microbial growth
Phosphates:	Surface water eutrophication
Salts:	Soil salinity
Pathogens such as fecal coliform bacteria, *Salmonella*, *Giardia*, *Campylobacter*, and *Cryptosporidium parvum*:	Water contamination
Pesticides, antibiotics, and growth hormones:	Endocrine disruptors
Carbon dioxide (CO_2):	Global warming
Hydrogen sulfide (H_2S):	Odor
Ammonia (NH_3):	Odor
Methane (CH_4):	Global warming
Arsenic:	Water contamination

16.8 INDUSTRIAL WASTES WITH HIGH SALTS AND ORGANICS

16.8.1 Oil Drilling

The process of drilling for crude oil requires powerful drill rigs that use large quantities of drilling fluids. These fluids contain high-density weighing agents such as barium sulfate (barite). Other drilling fluids are composed of sodium chloride solutions, which are used to force crude oil up to the surface. These fluids must be disposed of once they are "spent," or no longer useful. Prior to 1985, these spend fluids were stored in ponds near the drill sites and often simply bulldozed over when the well was completed. Consequently, many older oilfields have large tracts of land contaminated with spent drilling wastes. These wastes are not considered hazardous because they do not contain significant amounts of metals. Although free barium is very toxic, the mineral barite ($BaSO_4$) is quite inert in the environment. On the other hand, NaCl is very soluble in water and can increase the salinity of surface waters, rendering them nonpotable.

16.8.2 Coal-Burning Electric Power Plants

Electric power plants produce millions of tons of **fly ash** and **flue gas desulfurization wastes** every year. Because these residues are not considered hazardous, they may be stored either in ponds or landfills or, in the case of fly ash, they may be used as fill material. Fly ash is recovered from electrostatic precipitators that scrub out silt-size particulate matter from the flue gases generated from coal combustion. These particles generally arise from the incombustible silt and clay found in coal deposits. Upon exposure to high temperatures, silt and clay (which consists mostly of silica and alumina) combine to yield amorphous Si–Al-based spheres onto which other elements may condense. Typically, fly ash spheres also include Ca, Na, Fe, Mg, K, and Ti, with small amounts of other elements sorbed onto them, such as As, B, Ba, Cd, Cr, Cu, F, MO, NI, Pb, S, and Zn. The concentrations of these elements in fly ash vary widely, depending on the source of the coal. A typical empirical composition of fly ash is

$$10\,Si + 5\,Al + 0.5\,Ca + 0.5\,Na + 0.4\,Fe + 0.2\,Mg + 0.2\,Mg + 0.2\,K + 0.1\,Ti + 0.05\,S + \text{trace amounts of more than 15 other elements}$$

The removal (scrubbing) of sulfur dioxide (SO_2) gas from flue gases produces large quantities of flue gas desulfurization wastes, which consist largely of calcium carbonates, sulfates, and sulfites. These wastes may also contain trace quantities of some of the elements in fly ash, but the concentration of these elements depends on the source of the coal and the type of scrubbing systems used. Because flue gas desulfurization products are usually more than 70% water, these wastes are disposed of in drying ponds and are often treated along with power-plant wastewaters. This waste mixing may add significant amounts of soluble salts (*e.g.*, NaCl) that increase the salinity of sludges (see Table 24.2).

Despite new gas scrubbing technology and stricter emission standards, sulfur dioxide and nitrous oxide have been reduced but not completely eliminated. For example, current emissions of these two acid-forming gases still exceed 10 million tons/year. However, significant reductions (>60%) are mandated by the U.S. EPA in the next 16 years via a new **Clean Air Interstate Rule (CAIR)**, which affects 28 eastern U.S. states.

Mercury metal emissions from coal-burning electric power plants have also been a controversial issue. For the first time, reductions in Hg emissions from coal-burning power plants are being mandated under the new March 2005 **Clean Air Mercury Rule**. Under this rule, by 2020, reductions of 70% Hg emissions are expected from coal-burning power plants.

16.8.3 Industrial Wastes High in Organic Chemicals

Most industrial wastes contain varying amounts of organic chemicals. With few exceptions, carbon-based chemicals, reagents, solvents, feedstocks, and raw materials are extensively used in most phases of industrial processing. Exceptions to this rule may include mine tailings and metal-plating wastes. Wastes high in organic chemicals include those originating from oil refineries, as well as petrochemical, chemical, pharmaceutical, and food-processing industries, and paper mills. However, in recent years, these industries have reduced their polluting waste streams by applying aggressive pollution prevention strategies such as wastewater treatment processes before discharge, the implementation of waste reduction techniques that include recycling, and changes in industrial process with emphasis on waste minimization processes (see also Chapter 21).

16.9 INVASIVE SPECIES

Invasive species are an environmental problem of growing concern worldwide. Invasive species are organisms that have been introduced to a new ecosystem and that have a severe, often irreversible effect on agriculture and natural ecosystems (see Information Box 16.2). Any organism can become an invasive species, including microorganisms, invertebrates, insects, fish, plants, and animals. Invasive organisms often find few enemies (predators and diseases) in their new location, allowing them, at least initially, to grow and reproduce relatively easily.

It is important to note that a key component of this issue is that humans typically introduce the invasive species. Species can gradually spread into new areas as a natural process; this process is usually slow and involves adjustments by all members of the ecosystem. Conversely, human introductions are usually relatively fast, often resulting in large

---

> **INFORMATION BOX 16.2**
>
> **Invasive Species and Associated Problems**
>
> | | |
> |---|---|
> | **Invasive species:** | Non-native species, introduced to a new area primarily by human activity, that can reproduce and spread independently and that cause, or is likely to cause, economic or environmental damage. |
> | **Invasive species problems:** | Competition with and replacement of native species, damage to wildlife habitat, altered fire ecology, potential to spread disease. |

disruptions of the ecosystem. As noted in Information Box 16.2, these disturbances can be obvious or subtle.

In most cases, invasive species take advantage of opportunities in ecosystems that are disturbed by human activity (see Information Box 16.3). Disturbances can be flow control of rivers, disturbed soil along roadways and agricultural areas, human structures (*e.g.*, pigeons and sparrows are better adapted to cities than many native birds), water temperature change due to power plant outflows, and so on.

Some invasive species were introduced to provide erosion control, such as kudzu and salt cedar. Others, such as Lehman lovegrass and red brome, were repeatedly introduced over large areas of the western U.S. as forage grasses for cattle and sheep. These grasses have radically altered the fire ecology of western ranges, which in turn has changed plant populations, wildlife distribution, and nutrient cycling. Other invasive species escaped from gardens. A widely known example is purple loosestrife, which is a colorful perennial plant that is prized in gardens. Purple loosestrife has spread widely in the eastern and northern U.S., choking waterways and supplanting native riparian species. Animals released as hunting stock have caused problems. Examples include opossums in the northwestern U.S., rabbits in Australia, and red deer in New Zealand. Invasive Zebra mussels in the Great Lakes have caused millions of dollars worth

---

> **INFORMATION BOX 16.3**
>
> **Adaptive Traits of Invasive Species**
>
> · High reproductive rates
> · High dispersal rates
> · High genetic variability (allows them to adapt more quickly to a broad range of environmental conditions)
> · Broad range of native habitat, *i.e.*, adapted to a variety of soil and climatic conditions
> · Moved by humans to new locations, whether deliberately (kudzu) or accidentally (tumbleweed)

A

B

Figure 16.12 Kudzu is an invasive perennial vine that grows over everything in its path, including trees and buildings. (A) A cabin in early spring, with kudzu visibly starting to climb on trees and the abandoned cabin. (B) The same location in late summer, showing kudzu that has covered the cabin and most of the bushes and trees in the foreground. Kudzu blocks light from reaching the plants underneath, greatly weakening them, and can become heavy enough to break branches. Photo courtesy Jack Anthony (www.jjanthony.com/kudzu).

of damage to water intake systems, while simultaneously decreasing lake biodiversity. Fire ants (*Solenopsis invicta*) decrease biodiversity while causing major economic damage.

16.9.1 Kudzu

Kudzu (*Pueraria montana* var. *lobata*) is a well-known, highly visible example of an invasive plant species. Kudzu is a broad leaved, fast growing perennial vine (Figure 16.12) from Japan that was deliberately introduced in the early 1900s to control erosion and provide forage in the southern U.S. Kudzu grows rapidly and stabilizes loose soil with large, fleshy roots. In the U.S., kudzu has few serious checks on its growth by insects or disease, which allows it to grow as much as 20 m (60 feet) per season. This growth tends to completely cover existing vegetation, and can break branches and block sunlight from the native plants, eventually greatly weakening or killing them. Kudzu can also cover cars and entire buildings. Once it is well established, kudzu is difficult to remove. In addition, there is some evidence that kudzu is becoming more cold tolerant, extending its range to the north. It is estimated that kudzu covers about 25,000 square kilometers (10,000 square miles) in the U.S. and that it costs somewhere between $100 million to $500 million dollars a year in lost cropland and control costs.

16.9.2 Salt Cedar

Kudzu is quite obvious, even to the untrained eye, and causes visible damage as it smothers other plants and buildings with its extremely rapid growth. Many invasive species do not cause such obvious problems. Salt cedar (*Tamarix* spp.) has rapidly colonized riparian areas throughout the western U.S., causing major changes in this habitat. These changes are not obvious to the casual observer and yet are causing profound changes in riparian ecosystems.

Multiple species of salt cedar were originally introduced more than 100 years ago for erosion control and ornamental use (Figure 16.13). Salt cedar is tolerant of drought and saline and alkaline soils, grows quickly if there is sufficient water, has high seed production, and re-sprouts easily after fire. In addition, salt cedar has been implicated in lowering water tables at the expense of native species and also of salinizing soil. It is generally thought that these characteristics have enabled them to supplant native stands of cottonwood (*Populus* spp.) and willow (*Salix* spp.) that provide wildlife habitat, while producing little useful habitat of their own. As a result, land managers consider salt cedar a prime example of a detrimental invasive species.

Salt cedar is now found throughout the western U.S., thriving in response to human disturbances related to dams and diversion projects along large drainages. Research indicates that salt cedar changes the species composition of riparian communities and reduces their biodiversity (see Chapter 20). Thickets of salt cedar tend to replace native shrubs and trees such as willow and cottonwood, generally without replacing their usefulness as nesting sites and food sources. This may be due in part to the control of floods by dams, as willows and cottonwoods tend to establish new seedlings after flood events. In addition, regulation of river flow in general is causing many riparian areas to become drier and more saline. Recent research indicates that salt cedar may be better adapted to these new growing conditions, which allows them to outcompete the native species.

Salt cedar is also widely regarded as a cause of salinization of soil, which is thought to prevent or diminish the growth of many native species. Salt cedar is very tolerant of salinity and is able to use low-quality surface and ground-water sources that many natives cannot. The plants store excess salts in salt glands in the leaves and also excrete salt onto the surface of the leaves themselves. Since

Figure 16.13 Salt cedar thickets growing along the edge of a seasonally dry retention pond with high salinity in the southwestern U.S. Dark shrubs on the hills above the salt cedar are juniper. The insert shows a flowering salt cedar. Photo courtesy J.F. Artiola.

salt cedar is deciduous, the leaves eventually fall to the soil surface and build up a salt-rich litter. In areas with floods or sufficient rain, this salt is moved out of the root zone, but salt cedar usually grows in arid areas with low rainfall and along regulated waterways (preventing regular floods), thus allowing some salt buildup. Recent research indicates that there may not be salt buildup over a period of years, as rains and occasional floods may leach the salt out of the root zone. Other research indicates that the salt buildup is small enough that it does not affect some of the more salt-tolerant native species.

There is no doubt that salt cedar is an invasive species. Land managers are finding it difficult to remove salt cedar and re-establish native populations, and have blamed salt cedar for drier conditions and more saline soil. However, salt cedar may instead be able to adapt more readily than some native species to human-caused ecosystem disturbances.

Invasive species have caused major ecosystem changes throughout the world, resulting in billions of dollars of damage to agriculture, forestry, power plants, and the like each year. They are almost impossible to eradicate and difficult to control once established (see Chapter 20).

QUESTIONS AND PROBLEMS

1. Explain why mine tailings can be a source of pollutants in the soil and air environments. Give two examples of pollutants associated with mine tailings.

2. Give two examples of land pollution associated with deforestation.

3. Soils can become acidic when (a) basic cations are leached, (b) Al-oxides accumulate, (c) carbonate and bicarbonate minerals disappear, or (d) all of the above. Explain your answer.

4. What is the difference between a saline and a sodic soil? Explain the difference, and the EC and ESP values that a soil would have to have to be called saline-sodic.

5. Which factor would you add to or replace in the water erosion model RUSLE equation presented in Section 16.5.1 to make it a wind erosion model?

6. Explain why aggregated fine particles are less affected by wind erosion than loose coarser soil particulates.

7. Give two examples (each) of chlorinated and nonchlorinated pesticides and their use.

8. Why are chlorinated pesticides more persistent in the environment than nonchlorinated pesticides?

9. Explain how invasive plant species can change soil salinity and water quality.

10. Explain how deforestation affects water quality.

REFERENCES AND ADDITIONAL READING

Anthony J. (2005) Kudzu, The vine. www.jjanthony.com/kudzu.

Blake J., Donald J. and Magette W., eds. (1992) National livestock, poultry and aquaculture waste management. Proceedings of the National Workshop, July 1991. American Society of Agricultural Engineers, St. Joseph, Michigan.

Bonan G. (2002) *Ecological Climatology*. Cambridge University Press, Cambridge, England. TFL. 2005. U.S. fertilizer use statistics. http://www.tfi.org/Statistics/Usfertuse2.asp.

Borman F.H. and Likens G.E. (1979). *Pattern and Process in a Forested ecosystem*. Springer-Verlag, New York.

Bucklin R., ed. (1994) Dairy systems for the 21st century. Proceedings of the 3rd International Dairy Housing Conference, 1994. American Society of Agricultural Engineers, St. Joseph, Michigan.

Glenn E.P. and Nagler P.L. (2005) Comparative ecophysiology of *Tamarix ramosissima* and native trees in western U.S. riparian zones. *J. Arid Environ* **61**, 419–446.

Guptka R.K. and Abrol I.P. (1990) Salt-affected soils: Their reclamation and management for crop production. In R. Lal and B.A. Stewart (1990) *Soil Degradation: Advances in Soil Sci.* **11**, 223–288. Springer-Verlag, New York.

Heald C.L., Jacob D.J., Fiore A.M., Emmons L.K., Gille J.C., et al. (2003) Asian outflow and trans-pacific transport of carbon monoxide and ozone pollution: An integrated satellite, aircraft, and model perspective. *J. Geophys. Res.* **108**(D24), 4804, doi:10.1029/2003JD003507.

Hegg R.O. (2005) Trends in animal manure management research. CRIS database. Symposium State of the Science Animal Manure and Waste Management, January 5–7, 2005. San Antonio, Texas.

Luken J.O. and Thieret J.W. (1997) *Assessment and Management of Plant Invasions*. Springer-Verlag, New York.

Merkel J.A. (1981) *Managing Livestock Wastes*. Avi Publishing Company, Westport, Connecticut.

Saxton K., Chandler D., Stetler L., Lamb B., Clairborn C. and Lee B.H. (2000) Wind erosion and fugitive dust fluxes on agricultural lands in the Pacific Northwest. Trans. *ASAE* **43**, 623–630.

Stumm W. and Morgan J.J. (1996) *Aquatic Chemistry*, 3rd Edition. A Wiley-Interscience Series of Texts and Monographs. John Wiley & Sons, New York.

Texas Environmental Profiles. http://www.texasep.org

USDA. (1954) Saline and sodic soils. Agriculture Handbook No. 60. United States Department of Agriculture.

USDA-CREES. http://www.usawaterquality.org

U.S. EPA. (2004) Pesticide Industry Sales and Usage—2000–2001 Market Estimates. Biological and economic analysis division— Office of Pesticide Programs. Office of Prevention, Pesticides, and Toxic Substances. U.S. Environmental Protection Agency. Washington, DC 20460. May 2004.

U.S. EPA-CAFO Final Rule (2003) http://www.usawaterquality.org/themes/animal/default.html.

Veiga M.M., Meech J.A. and Oñate N. (1994) Mercury pollution from deforestation. *Nature*. **368**, 816–817.

Zhang H., Henderson-Sellers A. and McGuffie K. (2003) The compounds effects of tropical deforestation and greenhouse warming on climate. *Climatic Change*. **49**(3), 309–338. 10.1023/A:1010662425950.

SUBSURFACE POLLUTION

M. L. Brusseau and G. R. Tick

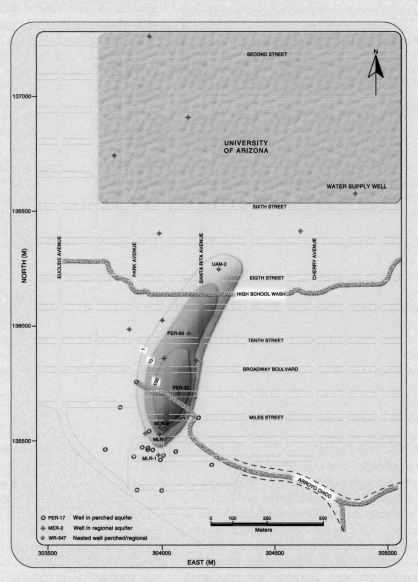

Groundwater contaminant plume (tetrachloroethene) contour map for a site in Tucson, Arizona.
Source: data provided by Arizona Department of Environmental Quality; drawn by Concepción Carreón Diazconti.

INFORMATION BOX 17.1

7,000-Year-Old Hand-Dug Wells in New Mexico and Texas

After the last Ice Age (about 11,000 years ago), the region encompassing western Texas and eastern New Mexico was covered by grasslands with numerous playa lakes. This region experienced a large-scale, long-term drought beginning about 7,500 years ago, resulting in the loss of the playas. It appears that the humans occupying the region adapted to this change in climate by digging wells to obtain water. Evidence of wells dating to approximately 7,000 years ago has been found at sites in New Mexico and Texas. These wells are 1–2 meters in depth, with openings of approximately 1 m in diameter. Artifacts found at the sites indicate that sticks and stones were used to excavate the wells.

Source: http://www.mnsu.edu/emuseum/archaeology/ sites/northamerica/blackwaterdraw.html; and http:// archaeology.about.com/od/mesolithicarchaic/a/ archaicwells.htm.

17.1 GROUNDWATER AS A RESOURCE

Freshwater comprises approximately 3% of all water on Earth (see also Chapter 3). Approximately 95% of this small fraction occurs as water in the subsurface, *i.e.*, groundwater. Thus, groundwater is a critical resource throughout the world.

Groundwater is a major source of potable water, supports food and crop production, and is used for myriad industrial activities. As such, the availability, quality, and sustainability of groundwater resources are issues of great significance. We will briefly explore these issues in this chapter.

Groundwater has long been used by humankind. This likely occurred initially through the use of natural springs and later via hand-dug wells. There are numerous references to groundwater in ancient texts, such as the works of Plato and Aristotle (Fetter, 2001). Chinese archaeologists have found wells in the Hunan Province dating back to 2000 BC (Xinhua News Agency, 2002). Evidence of hand-dug wells has been found at archaeological sites thousands of years old (see Information Box 17.1).

Groundwater use has increased greatly in the past several decades. Groundwater use worldwide has tripled in the past 50 years (Figure 17.1). This is also true for the U.S. (Figure 17.2). These increases are a direct result of increases in population and economic development. A primary use of groundwater is to supply potable water for drinking and other domestic uses. Groundwater serves as a significant source of potable water throughout the world, ranging from 15% in Australia to 75% in Europe (Table 17.1). In the United States, groundwater provides approximately half of the total potable water supply. The percentage of the U.S. population relying on groundwater as their primary source of potable water varies greatly by state (Figure 17.3).

In addition to supplying potable water, groundwater is also used for many other purposes. Agricultural applications, primarily as irrigation for crops, constitute the single largest use of groundwater in the U.S. (Figure 17.4). The majority of

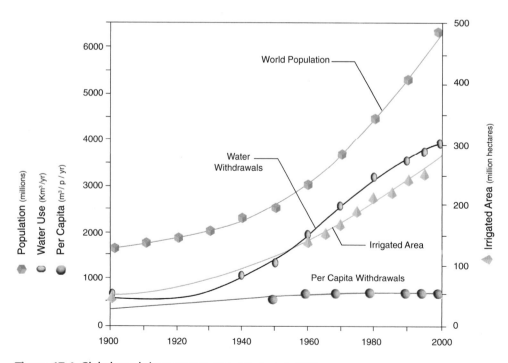

Figure 17.1 Global trends in water use. From Morris et al., 2003.

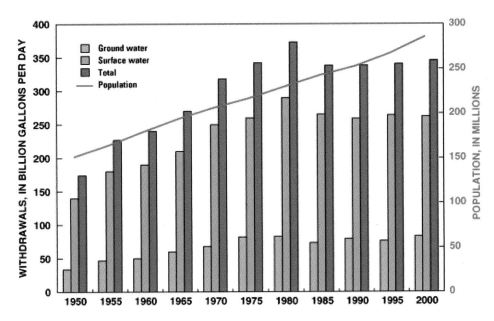

Figure 17.2 Groundwater withdrawals in the USA, 1950–2000. From http://pubs.water.usgs.gov/circ1268/htdocs/figure13.html.

irrigation use occurs in the semi-arid and arid regions of the U.S., as one would expect (Figure 17.5). Other uses of groundwater include industrial, mining, and power generation.

17.2 GROUNDWATER POLLUTION

Hand in hand with the use of groundwater by humans is the pollution of groundwater by humans. This pollution can occur in many forms, including hazardous industrial organic compounds, fuel components, heavy metals, agrochemicals, pathogenic microorganisms, and salinity. Groundwater can be contaminated through numerous means, as illustrated in Figure 17.6. Major sources of groundwater contamination in the U.S. are reported in Figure 17.7. There are two general categories of groundwater contamination: those produced from point sources and those that develop from diffuse or nonpoint sources. Specific examples of major groundwater pollution issues will be presented in later sections.

The growth of population centers (urbanization), with the attendant increase in population densities and industrial/commercial development, has had an enormous impact on groundwater use and quality. Water-resource development during the evolution of an urban center follows a typical pattern, as illustrated in Figure 17.8. Generally, increasing demand for potable water that occurs as population increases results in over-pumping of groundwater from the original well field located in the city center. The amount of water being extracted is greater than the amount recharged, causing a decline in water levels. This eventually requires the city to supplement their potable water supply from other sources. An ancillary effect of the over-pumping and falling water levels is subsidence of the land surface. Another significant effect of over-pumping in coastal areas is seawater intrusion. This will be discussed in detail in a forthcoming section.

The development of high-intensity agriculture and the widespread use of fertilizers and pesticides have lead to major groundwater pollution issues for rural areas. This is of particular concern because groundwater generally is the predominant source of potable water for rural areas. For example, more than 95% of the rural population of the U.S. uses groundwater as their potable water supply. Additionally, most of the potable water supply is obtained from individual or small community wells. Groundwater from these wells does not typically undergo the extensive treatment and monitoring that is prevalent for centralized urban water supply systems (see Chapter 28).

A comprehensive monitoring program is instrumental in managing and protecting groundwater resources. The U.S. Geological Survey's National Water-Quality Assessment (NAWQA) Program is a nation-wide program that is the principal source of information on groundwater quality in the United States. Under this program, the U.S. Geological Survey collects water quality data in 60 special study regions of the country, conducts retrospective analyses of existing data (such as state data), and prepares national-scale syntheses of

TABLE 17.1 Estimated percentage of potable water supply obtained from groundwater (Morris et al., 2003).

REGION	PERCENT	POPULATION SERVED (MILLIONS)
Asia–Pacific	32	1000–2000
Europe	75	200–500
Central and South America	29	150
USA	51	135
Australia	15	3
Africa	NA	NA
World	—	1500–2750

Figure 17.3 Percentage of population relying on groundwater as a drinking source. From EPA, 1999.

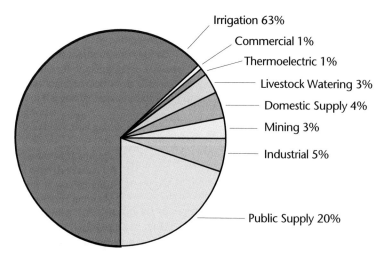

National Ground Water Use

Irrigation 63%

Commercial 1%

Thermoelectric 1%

Livestock Watering 3%

Domestic Supply 4%

Mining 3%

Industrial 5%

Public Supply 20%

Source: *Estimated Use of Water in the United States in 1995.*
U.S. Geological Survey Circular 1200, 1998.

Figure 17.4 Groundwater use in the USA for 1995. From USGS, 1998.

Volume of Ground Water Used for Irrigation in 1995

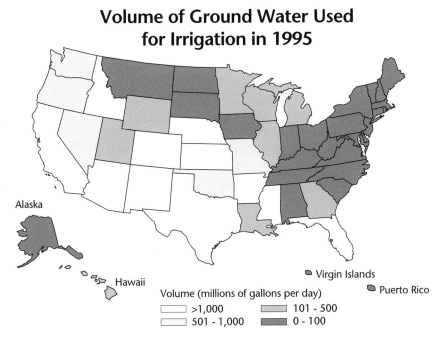

Source: *Estimated Use of Water in the United States in 1995.*
U.S. Geological Survey Circular 1200, 1998.

Figure 17.5 Volume of groundwater used for irrigation in the USA for 1995. From USGS, 1998.

Sources of Ground Water Contamination

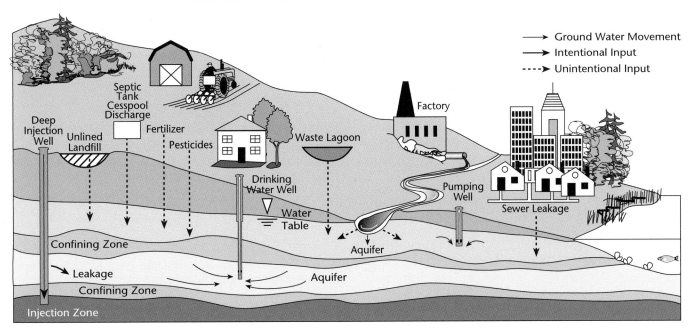

Figure 17.6 Sources of groundwater contamination. From EPA, 1998.

Major Sources of Ground Water Contamination

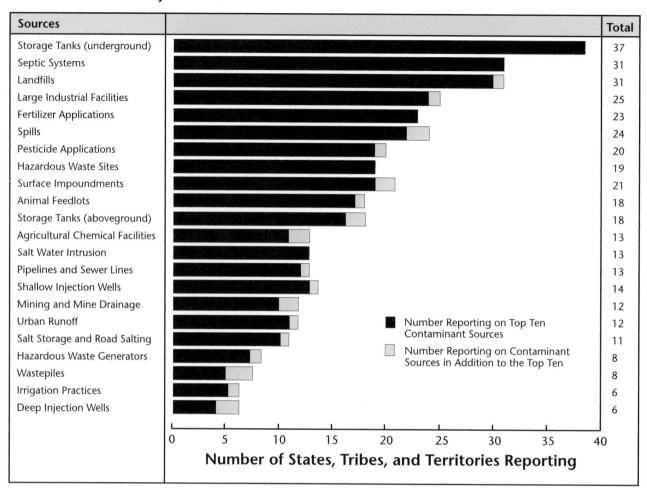

Sources		Total
Storage Tanks (underground)		37
Septic Systems		31
Landfills		31
Large Industrial Facilities		25
Fertilizer Applications		23
Spills		24
Pesticide Applications		20
Hazardous Waste Sites		19
Surface Impoundments		21
Animal Feedlots		18
Storage Tanks (aboveground)		18
Agricultural Chemical Facilities		13
Salt Water Intrusion		13
Pipelines and Sewer Lines		13
Shallow Injection Wells		14
Mining and Mine Drainage		12
Urban Runoff		12
Salt Storage and Road Salting		11
Hazardous Waste Generators		8
Wastepiles		8
Irrigation Practices		6
Deep Injection Wells		6

■ Number Reporting on Top Ten Contaminant Sources

▢ Number Reporting on Contaminant Sources in Addition to the Top Ten

Number of States, Tribes, and Territories Reporting

Figure 17.7 Major sources of groundwater contamination in the U.S. From EPA, 1998.

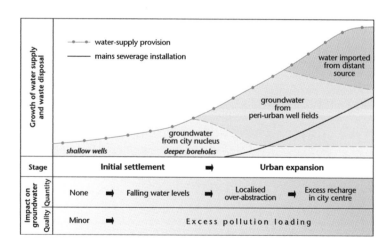

Figure 17.8 Stages in the development of water resources during evolution of an urban center. From Morris et al., 2003.

the results. In addition, many state and local governments in the U.S. carry out groundwater monitoring programs. Specialized monitoring programs are conducted in association with characterization and remediation of contaminated sites (see Chapter 12).

17.3 GROUNDWATER POLLUTION RISK ASSESSMENT

Physical properties of the subsurface result in significant differences in the behavior of groundwater compared to that of surface water. For example, residence times for groundwater range from years to hundreds of years or more; these times are much greater than those for streams. Dilution effects, either in water or the atmosphere, are much less significant for groundwater compared to surface-water systems. In addition, the absence of light eliminates the possibility of photochemical reactions, which are a major route of transformation for aboveground systems. The net result is that once groundwater and the subsurface are contaminated, it is very difficult to decontaminate (the remediation of subsurface contamination is discussed in Chapter 19). Thus, pollution prevention is critical to maintaining sustainable groundwater resources. Groundwater pollution risk assessment is a key aspect of pollution prevention.

Preventing contamination from entering the environment is the only sure way of preventing pollution. Laws and regulations promulgated during the past few decades have helped to reduce the overall contaminant load to the environment (see Chapter 15). Obviously, however, preventing all contamination from entering the environment is not possible, hence the need for groundwater pollution risk assessment, the goal of which is to evaluate the risk posed by a given activity or event to groundwater resources. Implementing risk assessments enhances the effective management of groundwater resources and helps to minimize potential contamination (see also Chapter 14).

Groundwater pollution risk is composed of two components: groundwater vulnerability and contaminant load. Groundwater vulnerability is the intrinsic susceptibility of the specific aquifer in question to contamination. Several factors affect groundwater vulnerability, as shown in Table 17.2. An aquifer that is close to ground surface, overlain by sandy soil, and located in an area with high precipitation rates would clearly be more vulnerable to contamination

than an aquifer that is hundreds of meters below ground surface and located in an area of low precipitation.

Factors involved in the contaminant load are the type of contaminant, the amount of contaminant released, the timescale of release, and the mode of release. The pollution potential of a contaminant is controlled by its transport and fate behavior, as discussed in Chapter 6. Contaminants that are transported readily (*e.g.*, those with high aqueous solubility and low sorption) and that are not transformed to any great extent (*i.e.*, are persistent) generally have greater potential to pollute groundwater. For most contamination events, the contamination enters the environment in close proximity to land surface (*e.g.*, surface spills, leaking storage tanks, and landfills). Thus, transport of contaminants from the source zone to groundwater necessitates travel through the soil and vadose zone. Attenuation processes such as sorption and biodegradation (see Chapters 6, 7, and 8) can act to reduce and limit the transport of contaminants to groundwater. For this reason, the soil and vadose zone is often referred to as a "living filter." The degree to which contaminants will be attenuated is a function of the type of contaminant and the nature of the subsurface. General transport and attenuation properties of common subsurface contaminants are presented in Table 17.3. Generally, the greater the amount of contamination released, the greater the pollution potential. The timescale and mode of release can also affect pollution potential. For example, releases from buried storage tanks may be more prone to cause groundwater contamination than releases from tanks stored aboveground on concrete pads.

The risk of groundwater pollution results from the combination of intrinsic vulnerability and contaminant load factors. Thus, an aquifer that is very vulnerable may have little to no risk of pollution if the contaminant load remains negligible. Conversely, an aquifer that has a relatively low degree of vulnerability may have a significant pollution risk if the contaminant load factor is very high. The greatest pollution risk will be associated with locations where the aquifer has a high vulnerability and the contaminant loading is high.

In general, not much can be done to modify or change the inherent vulnerability of an aquifer. That is why it is critical to focus on controlling the contaminant load for managing and preventing groundwater pollution. This can be done through implementing land-use and facility-operation regulations. For example, aquifer vulnerability assessments can be used in the siting of new facilities that involve production, storage, or disposal of hazardous materials. Including

TABLE 17.2 Factors affecting groundwater vulnerability to contamination.

FACTOR	INCREASED VULNERABILITY	DECREASED VULNERABILITY
Depth to groundwater	Shallow	Deep
Soil type	Well drained (sandy)	Poorly drained (high clay, organic matter content)
Vadose zone physical properties	Preferential flow channels	Horizontal low-permeability layers
Recharge	High precipitation, high infiltration	Low precipitation, low infiltration
Subsurface attenuation processes	Minimal attenuation	Significant attenuation

TABLE 17.3 Transport and attenuation properties of major subsurface contaminants.

CONTAMINANT	SOURCE	ATTENUATION MECHANISM				PERMITTED DRINKING WATER CONCENTRATION	MOBILITY	PERSISTENCE	POTENTIAL TO DEVELOP EXTENSIVE GROUNDWATER PLUME
		Biochemical degradation	Sorption	Filtration	Precipitation				
Pathogens	Sewage	**	**	***	X	Very low (<1 per 100 ml)	Low	Low–moderate	Low
Nitrogen (N)	Agriculture, sewage	*	*	X	X	Moderate (10–20 mg N/L)	Very high	Very high	High
Chloride (Cl)	Sewage, industry, road deicer	X	X	X	X	High	Very high	Very high	High
Sulfate (SO$_4$)	Road-runoff, industry	* 2	*	. X	*	High	High	High	High
Dissolved organic carbon (DOC)	Sewage, industry (esp. food processing, textiles)	**	**	*	X	Not controlled	Moderate	Low–Moderate	Moderate
Heavy metals	Industry	X	***	* 3	**	Low (variable)	Generally low unless pH low (except CR [VI])	High	Low
Halogenated solvents	Industry, commercial	*	*	X	X	Low (5–30 µg/L)	High	High	High
Fuels, lubricants, oils, other hydrocarbons (LNAPLs)	Fuel station, industry	***	**	X	X	Low (10–700 µg/L BTEX 4	Moderate	Low–Moderate	Moderate
Other synthetic organics	Industry, sewage	Variable	Variable	X	X	Low (variable)	Variable	Variable	Variable

KEY = *** highly attenuated, ** significant attenuation, * some attenuation, X no attenuation
1 Ammonia is absorbed, 2 Can be reduced, 3 When it occurs as organic complexes, 4 Aromatic compounds with health guideline limits
Adapted from Morris et al., 2003.

aquifer vulnerability as a siting factor can prevent building such facilities in locations where groundwater is most susceptible to pollution. For existing facilities, operation procedures can be implemented to minimize the production and disposal of wastes (see Chapter 25).

Special procedures and tools have been developed to assess groundwater pollution risk. This is usually done by employing a geographic information system. The first step typically involves constructing an aquifer vulnerability map (Figure 17.9). This map incorporates one or more factors that influence aquifer vulnerability, such as those listed in Table 17.2. A commonly used aquifer vulnerability tool is DRASTIC (See Information Box 17.2). An example of an aquifer vulnerability map is shown in Figure 17.10. Once the vulnerability map has been created, land-use factors can be superimposed to evaluate risk. An example of a groundwater risk map is shown in Figure 17.11, which shows the nitrate contamination risk for shallow aquifers in the U.S. The map was developed using soil drainage as the aquifer vulnerability factor, and fraction of cropland acreage, population density, and nitrogen loading as the loading (land use) factors.

17.4 POINT-SOURCE CONTAMINATION

Point-source systems are characterized by very localized contamination releases. Primary examples include surface spills, leaking storage tanks, disposal pits, and waste-injection wells. The distribution of contamination at sites associated with point-source releases follows a general pattern as illustrated in Figure 17.12. The region of the subsurface where the majority of the original contamination is present is referred to as the **source zone**. It is usually in close proximity to the

---

INFORMATION BOX 17.2

Aquifer Vulnerability is DRASTIC

DRASTIC was developed by the EPA as a standardized system for evaluating aquifer vulnerability to pollution. The method is based on assumptions that the contaminant is released at the ground surface and that it enters the subsurface via infiltrating water. Contaminant attenuation is not considered. Thus, it may be considered to provide a conservative estimate of vulnerability for those situations where attenuation may be significant.

The acronym DRASTIC is derived from the seven hydrogeologic factors considered:

1. **D**epth to groundwater
2. net **R**echarge
3. **A**quifer media
4. **S**oil media
5. **T**opography (slope)
6. **I**mpact of the vadose zone media
7. hydraulic **C**onductivity of the aquifer.

Each of the hydrogeologic factors is assigned a rating from 1 to 10, based on the properties of the specific site. For example, a rating of 10 is given for "depth to groundwater" if the water table is located from 0–5 feet from ground surface. A rating of 1 is given if it is more than 100 feet from ground surface. The ratings are then multiplied by a weighting coefficient ranging from 1 to 5. The most significant factors have a weight of 5; the least significant have a weight of 1. The weight factors are: D = 5, R = 4, A = 3, S = 2, T = 1, I = 5, and C = 3. The products of the rating and weighting coefficient for each factor are summed to produce the final DRASTIC score. The smallest possible DRASTIC score is 23 and the largest is 226.

The DRASTIC score represents a relative measure of aquifer vulnerability. The higher the DRASTIC score, the greater the vulnerability of the aquifer to contamination.

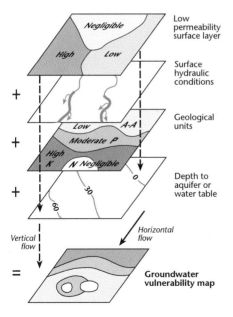

Figure 17.9 Combining factors to create a groundwater vulnerability map. From Morris et al., 2003.

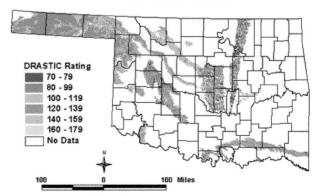

Figure 17.10 DRASTIC-based vulnerability assessment of major aquifers in Oklahoma. From Osborn et al., 1998.

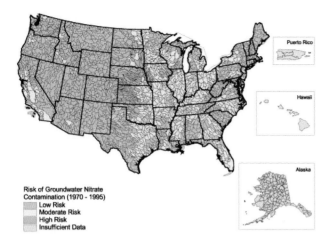

Figure 17.11 Risk of groundwater nitrate contamination for shallow aquifers in the US (< 100 feet deep). Image courtesy of USEPA; http://www.epa.gov/iwi/1999sept/iv21_usmap.html.

location of the contaminant release. The source zone generally encompasses a relatively small area and contains the majority of the contaminant mass. Conversely, the region where groundwater is contaminated by dissolved compounds originating from the source zone is referred to as a **groundwater contaminant plume**. The plume is often much larger than the source zone. However, the amount of contaminant mass associated with the plume may represent a small fraction compared to the mass in the source zone.

The configuration or "architecture" of the source zone (*e.g.*, porous-medium heterogeneity, total contaminant mass, contaminant distribution) and source-zone "dynamics" (*e.g.*, mass-transfer processes, transformation processes) is central to the pollution risk posed by the site. For example, the magnitude (size and concentration level) of the groundwater contaminant plume generated from the source zone is clearly dependent on the magnitude and rate of contaminant mass transfer from the source zone to surrounding regions (*i.e.*, the source-zone mass flux). This mass transfer will be influenced by groundwater flow patterns (which are mediated by the physical properties of the porous media), and by the type, amount, and distribution of contaminant. The groundwater contaminant plume is generally the primary source of human-health risk posed by these types of contaminated sites, as use of contaminated groundwater is usually the major route of potential exposure to subsurface contamination. Thus, it is critical to characterize source-zone and contaminant plume properties at a given site (See Chapter 19).

17.4.1 Hazardous Organic Chemicals

The contamination of groundwater by hazardous organic chemicals and the associated risk to human health and the environment is one of the primary groundwater pollution issues facing the U.S. and other industrialized countries. Major types of organic compounds that are prevalent groundwater contaminants are listed in Table 17.4. It is estimated that there are tens of thousands of sites across the U.S. where groundwater is contaminated with one or more of the

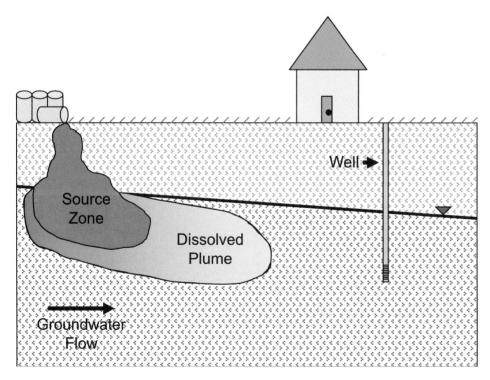

Figure 17.12 Distribution of contamination at a point-source groundwater contamination site. From C.M. McColl.

TABLE 17.4 Major classes of hazardous organic chemicals of significance for groundwater pollution.

CLASS	EXAMPLE COMPOUNDS	SOURCES/USES	PROPERTIES[1]
Chlorinated solvents	Trichloroethene Tetrachloroethene	Manufacturing Degreasing Dry cleaning	DNAPL
Coal tar, Creosote	Polynuclear Aromatic Hydrocardons Phenols	Coal gasification Wood treatment	DNAPL
Hydrocarbon fuels	Benzene	Fuel	LNAPL
Polychlorinated Biphenyls	Aroclor	Transformer fluid	DNAPL

[1]DNAPL = denser than water; LNAPL = less dense than water.

chemical classes listed in Table 17.4 (see Information Box 17.3). This contamination results from the release of the chemicals during their transport, storage, use, and disposal. For example, thousands of underground fuel storage tank releases have been confirmed for every state in the U.S. (Figure 17.13). For another example, it has been reported that up to 75% of the 36,000 active dry-cleaning facilities in the U.S. have experienced releases of tetrachloroethene or other dry-cleaning solvents (EPA, 2003a). Many of these compounds are of concern with respect to human health (*e.g.*, carcinogenic), as discussed in Chapter 13.

One of the most critical issues associated with hazardous waste sites is the potential presence of immiscible-liquid contamination in the subsurface. The chemicals listed in Table 17.4 are liquids under natural conditions, and they are immiscible with water, meaning they do not mix. Immiscible

INFORMATION BOX 17.3

Chlorinated-Solvent Contamination of Groundwater in Arizona

Chlorinated solvents, such as tetrachloroethene, trichloroethene, dichloroethene, carbon tetrachloride, and vinyl chloride, are among the most common groundwater contaminants in the U.S. due to their widespread use as dry-cleaning solvents and as degreasing and cleaning agents for military, industrial, and commercial applications. In Arizona, chlorinated solvents are the primary contaminants at an overwhelming majority of both the state Superfund sites (30 out of 34) and the federal Superfund sites (13 out of 14). Hundreds of millions of dollars have been spent to date on characterization and remediation of these sites. Furthermore, additional hundreds of millions of dollars have been spent to settle toxic-tort lawsuits. In aggregate, these sites encompass hundreds of km^2 in land area and contain billions of liters of contaminated groundwater. As such, chlorinated solvents are a major source of groundwater contamination in Arizona and pose an immediate, significant, and continuing threat to the sustainability of the state's potable water supplies.

liquids released into the subsurface become trapped in pore spaces due to capillary forces. Once entrapped, the immiscible liquid is very difficult to remove. Hence, immiscible liquids serve as long-term sources of subsurface contamination as they dissolve in groundwater or soil-pore water. The presence of immiscible liquids in the subsurface of a site can greatly impact the costs and time required for site remediation. For example, for sites contaminated by dense nonaqueous phase liquids (DNAPLs) (see Chapter 6), it is estimated that upwards of hundreds of years may be necessary to achieve health-based groundwater cleanup objectives using standard pump-and-treat systems (ITRC, 2002). This clearly illustrates the critical importance of addressing immiscible-liquid contamination when it is present at a site. Unfortunately, as is widely acknowledged, cleaning up sites contaminated by immiscible liquids is one of the greatest challenges in the field of environmental remediation. In fact, according to several reviews conducted by expert panels convened by the National Research Council, the presence of immiscible liquids is usually the single most important factor limiting the cleanup of organic-chemical contaminated sites. In addition, the presence of immiscible liquids greatly complicates site characterization and risk assessment efforts. The estimated cost to clean up the immiscible-liquid contaminated sites in the U.S. is $100 billion or more (EPA, 2003).

The distribution of immiscible liquids in the subsurface is controlled by the physical and chemical properties of the porous media and by the properties of the chemical. A primary property of concern is the density of the immiscible liquid in comparison to that of water. In fact, this is such an important property that the immiscible liquids are classified based on whether they are more (DNAPL) or less (LNAPL) dense than water (see Table 17.4), where NAPL = nonaqueous phase liquid. Because DNAPLs are denser than water, they can sink below the water table with sufficient volume of release. A conceptual diagram of the distribution of a DNAPL in the subsurface is presented in Chapter 6 (see Figure 6.1). In contrast to DNAPLS, LNAPLS float on the water table because they are less dense than water (Figure 17.14). The distribution and amount of immiscible liquid present in the source zone has a major impact on the nature of the groundwater contaminant plume that forms at the site.

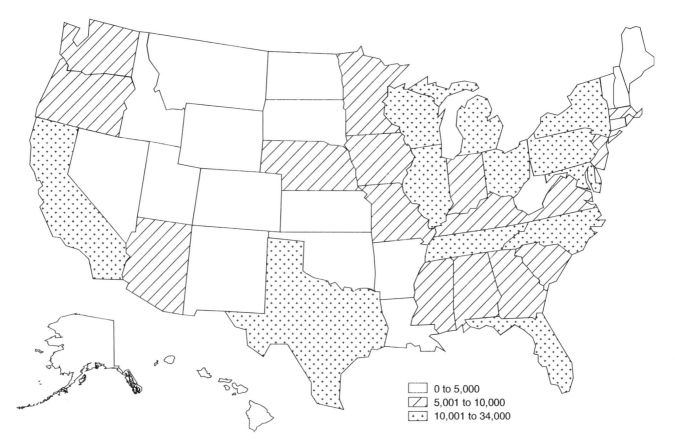

Figure 17.13 Confirmed underground storage tank releases as of February 1999. From EPA, 1999.

In addition to source-zone properties, the size of the dissolved-contaminant plume will be influenced by the nature of the chemicals and by physical, chemical, and biological properties of the subsurface. Plumes are generally relatively small for chemicals that undergo significant attenuation. Conversely, very large contaminant plumes can form for chemicals that undergo minimal attenuation. The general potential of the major classes of subsurface contaminants to

develop extensive groundwater plumes is listed in Table 17.3. For example, groundwater plumes comprised of chlorinated-solvent constituents are typically hundreds to thousands of meters long because these compounds are difficult to biodegrade and have relatively low sorption. Conversely, plumes generated from fuel hydrocarbons such as gasoline are generally much shorter, primarily because the compounds are more amenable to biodegradation.

17.4.2 Landfills

The contamination of soil and groundwater from landfills is another important problem that continues to threaten groundwater resources, posing risks to human health and the environment. Chemicals, both hazardous and nonhazardous, can be leached from the materials that are disposed of in landfills (see Figure 25.7). This landfill-derived contamination or **leachate** can enter the soil and migrate through the vadose zone, eventually contaminating groundwater resources. Landfills are used for the disposal of numerous types of wastes, garbage, and materials. Thus, contamination emanating from landfills can contain mixtures of numerous types of compounds, increasing the complexity of the pollution problem (Table 17.5).

Contaminated water or leachate is produced when water (*e.g.*, precipitation or irrigation) enters the landfill and

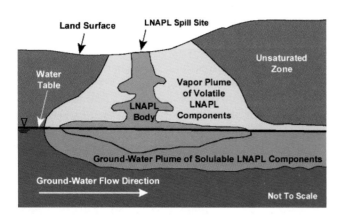

Figure 17.14 Conceptual diagram illustrating the distribution of an LNAPL in the subsurface. Image courtesy of USGS; http://toxics.usgs.gov/definitions/lnapls.html.

TABLE 17.5 Common contaminants and concentrations at landfills.

PARAMETER	"TYPICAL" CONCENTRATION RANGE	AVERAGE
BOD	1,000–30,000	10,500
COD	1,000–50,000	15,000
TOC	700–10,000	3,500
Total volatile acids (as acetic acid)	70–28,000	NA
Total Kjeldahl Nitrogen (as N)	10–500	500
Nitrate (as N)	0.1–10	4
Ammonia (as N)	100–400	300
Total phosphate (PO_4)	0.5–50	30
Orthophosphate (PO_4)	1.0–60	22
Total alkalinity (as $CaCO_3$)	500–10,000	3,600
Total hardness (as $CaCO_3$)	500–10,000	4,200
Total solids	3,000–50,000	16,000
Total dissolved solids	1,000–20,000	11,000
Specific conductance (mhos/cm)	2,000–8,000	6,700
pH	5–7.5	63
Calcium	100–3,000	1,000
Magnesium	30–500	700
Sodium	200–1,500	700
Chloride	100–2,000	980
Sulfate	10–1,000	380
Chromium (total)	0.05–1	0.9
Cadmium	0.001–0.1	0.05
Copper	0.02–1	0.5
Lead	0.1–1	0.5
Nickel	0.1–1	1.2
Iron	10–1,000	430
Zinc	0.5–30	21
Methane gas	60%	
Carbon dioxide	40%	

All values mg/L except as noted.
NA = not available
Source: Lee, G.F. and R.A. Jones (1991b) (http://www.gfredlee.com/lf-conta.htm).

contacts the waste materials (Figure 17.15). The infiltrating water can remove hazardous and nonhazardous chemicals, including metals, minerals, salts, organic chemicals (*e.g.*, chlorinated solvents, petroleum hydrocarbons, and pesticides), and various other toxic compounds. Millions of gallons of leachate can percolate through a landfill, depending on the size of the landfill or disposal facility. A well-designed landfill should contain the disposed materials, isolating potentially harmful leachate from the environment and specifically from groundwater and drinking water resources. However, it is impossible to contain and control all wastes produced at landfill and disposal sites. Leachate can enter groundwater systems as a result of poorly designed or improperly constructed landfills, deterioration of landfill liners, and landfills constructed without liners (*e.g.*, typically older designs).

Given that leachate is likely to leak to some degree from all landfills, it is essential to implement a well-designed monitoring scheme to help manage landfill pollution problems. For example, a series of monitoring wells can be constructed around the perimeter of the landfill. The presence of high salt concentrations (*e.g.*, Cl⁻) indicates the potential threat of contamination to groundwater. Similarly, total dissolved solids (TDS) can be used to indicate potential contamination of

groundwater by landfill leachate. Salts serve as good indicators of leachate contamination because of their high mobility and persistence (see Table 17.3), which means they usually constitute the leading (downgradient) front of the groundwater leachate plume. Monitoring pH is another means to detect potential contamination of groundwater by leachate. Generally, the pH of landfill leachate is lower than that of uncontaminated groundwater. A third method to detect potential leachate contamination of groundwater is by determining the oxidation-reduction potential. Highly reducing conditions typically indicate either low pH or high microbial degradation activity. Waste materials disposed of in landfills are often subject to microbially mediated decay and decomposition. The microorganisms consume oxygen during the degradation process, thus reducing the oxidation-reduction potential. In addition, some microorganisms can degrade waste under anaerobic conditions and in the process release methane and other gases. Thus, the presence of such gases in groundwater is another indicator of landfill-leachate contamination.

As noted above, landfill leachate comprises many compounds. Thus, the transport and fate behavior of leachate is complex and highly variable. Compounds with high mobilities and low degradation potentials (high persistence) will

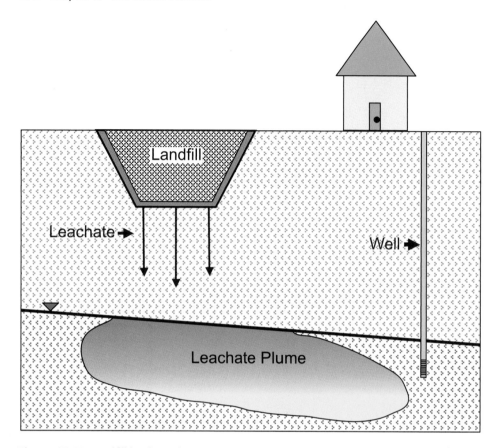

Figure 17.15 Landfill leachate plume. From C.M. McColl.

tend to be transported much further than compounds that sorb to soil and/or that are easily degraded. For this reason, it is difficult to predict the transport and fate behavior of leachate in soil and groundwater.

17.5 DIFFUSE-SOURCE CONTAMINATION

Contamination resulting from nonpoint (diffuse) sources of pollution refers to those inputs that occur over a wide area and are associated with particular land uses. This is in contrast to point-source discharges, which occur from a specific, very localized source such as a leaking fuel tank or pipe. Nonpoint sources generally encompass much larger scales (regional and even global scales) and as a result can create very large zones of pollution compared to point sources. However, the contaminant concentrations associated with nonpoint source pollution are generally lower than those associated with point sources. This section will focus on two major diffuse-source issues: agrochemical contamination and salt-water intrusion.

17.5.1 Agrochemical Pollution of the Subsurface

The advent of intensive agricultural practices during the last century and into the 21st century has greatly increased global food production. However, as discussed in Chapter 16, it has

also greatly increased the use of fertilizers and pesticides, so-called **agrochemicals** (Figure 17.16). The increasing use of these agrochemicals has lead to extensive pollution of groundwater. For example, nitrate, derived from fertilizers, pesticides, and animal wastes, is one of the most widespread and pervasive groundwater contaminants in the United States

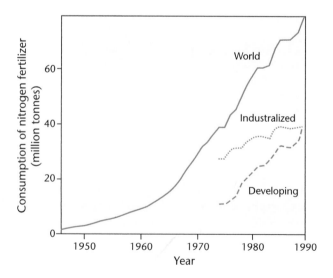

Figure 17.16 Increases in nitrogen fertilizer use in developing countries. Consumption of nitrogen fertilizer from 1946–1989. From Morris et al., 2003.

and the world. In a survey of almost 200,000 water sampling reports, the EPA found that more than 2 million people were using water from public potable-water supply systems for which nitrate standards were exceeded at least once between 1986 and 1995 (EWG, 1996). An additional 3.8 million people were using water from private wells that exceed federal drinking water standards. Researchers predict, due to past and current inputs of nitrates into the environment, that the full effect of overapplication of nitrate fertilizers will not be realized for another 30–40 years (Hallberg and Keeney, 1993). The potential impacts of nitrate contamination on human health are discussed in Chapter 13.

As we might expect, agricultural areas generally have the most significant groundwater nitrate contamination problems. For example, major agricultural regions such as the San Joaquin Valley in California and the Ogallala aquifer system extending from Minnesota to Texas are areas with high vulnerability to nitrate groundwater contamination (see Figure 17.11). In addition, these agricultural regions are also experiencing severe declines in groundwater levels as a result of excessive groundwater extraction for irrigation. It is common in agricultural areas to see the compounding effect of declining groundwater levels coinciding with high levels of nitrate contamination.

Nitrate is generally very mobile in the subsurface. This, in conjunction with the large areal extent of input, results in extensive groundwater plumes of nitrate contamination. It is difficult and expensive to clean up groundwater once it is contaminated by nitrate. For example, the costs associated with managing nitrate contamination problems in California and Iowa are estimated to exceed $200 million per year (EWG, 1996). A standard approach for dealing with shallow groundwater nitrate contamination is to drill deeper wells. However, this can be done only a limited number of times. Another common practice is to blend contaminated water with uncontaminated water. The objective of this technique is to dilute the nitrate to concentrations below the drinking water standard. This approach increases the overall use of water resources. It

is estimated that closing down a well due to nitrate contamination, and drilling another well or blending contaminated water with cleaner supplies, can cost between $200,000–500,000 per well (EWG, 1996). Point-of-use treatment for nitrate is also expensive, requiring methods such as reverse osmosis. A key to solving the problem of nitrate groundwater contamination is to prevent future contamination by using best management practices, as discussed in Chapter 16.

Another major class of agrochemicals, also widely used in agricultural practices, are pesticides (see Chapters 10 and 16). Pesticides are typically applied at the land surface, usually as a chemical spray. Once applied to the ground surface, pesticides can migrate downward through the vadose zone with infiltrating water and contaminate groundwater. Pesticides are used throughout the world, primarily for agriculture, to control weeds, insects, and fungal pests (Table 17.6). A study conducted between 1991 and 2001 by the U. S. Geological Survey found that 42% of wells sampled in agricultural regions across the United States contained the common pesticides atrazine and diethylatrazine. Furthermore, it was reported that about 20% of the wells sampled in major aquifer systems throughout the U.S. contained both atrazine and diethylatrazine. These results illustrate the extent of groundwater pollution by pesticides.

The majority of pesticides are organic compounds. Their transport and fate behavior in the subsurface is a function of their chemical properties (see Chapters 6 and 7). Some pesticides are relatively mobile (*e.g.*, 2,4-D), while others are highly sorbed (*e.g.*, DDT) (see Chapter 16). The classes of pesticides that were initially developed, such as DDT, are very persistent in the environment. Newer pesticides have been designed in part to be less persistent. A complicating factor in the evaluation of pesticide pollution problems is the sheer number of pesticides available for use. It is not customary to analyze for all possible pesticide compounds, their derivatives, and possible degradation products in groundwater monitoring surveys. In addition, analytical limitations have constrained detection capability. However,

TABLE 17.6 Pesticide use and occurrence in groundwater.

REGION	DOMINANT PESTICIDE USE	TYPICAL COMPOUNDS DETECTED
United Kingdom	Pre- and postemergent herbicides on cereals, triazine herbicides on maize and in orchards	Isoproturon, mecoprop, atrazine, simazine
Northern Europe	Cereal herbicides and triazines as above	As above
Southern Europe	Carbamate and chloropropane soil insecticides for soft fruit, triazines for maize	Atrazine, alachlor
Northern U.S.	Triazines on maize and carbamates on vegetables, *e.g.*, potatoes	Atrazine, aldicarb, metolachlor, alachlor and their metabolites
Southern and Western U.S.	Carbamates on citrus and horticulture, and fumigants for fruit and crop storage	Aldicarb, alachlor and their metabolites, ethylene dibromide
Central America and Caribbean	Fungicides for bananas, triazines for sugarcane, insecticides for cotton, and other plantation crops	Atrazine
South Asia	Organophosphorous and organochlorine insecticides in wide range of crops	Carbofuran, aldicarb, lindane
Africa	Insect control in houses and for disease vectors	Little monitoring as yet

From Morris et al., 2003.

recent advancements have resulted in more frequent detection of pesticide compounds in groundwater supplies.

17.5.2 Salt-Water Intrusion

Salinization of freshwater is one of the most serious and widespread groundwater contamination issues throughout the world. Areas along coasts where seas or oceans meet continental landmasses and island systems are most vulnerable to salinization of groundwater resources. In most coastal settings, groundwater beneath land surface generally exists as a lens of "freshwater." The freshwater is separated from the denser seawater by a diffuse interface known as the freshwater/saltwater interface or zone of dispersion (Figure 17.17). The freshwater/saltwater interface typically resides at some point near where the ocean meets the land mass (coastline) and extends vertically in the subsurface, separating freshwater (inland) from high salinity water (seawater). This freshwater lens tends to become thinner as it approaches the shoreline. If recharge into the aquifer equals extraction rates due to pumping, the freshwater/saltwater interface will remain stable (Figure 17.18). However, if groundwater pumping exceeds recharge, the saline water will invade the freshwater aquifer and the freshwater/saltwater interface will progress further inland. As this zone progresses further inland, groundwater supply wells can become contaminated from the invading saltwater. This phenomenon is known as **saltwater intrusion,** and

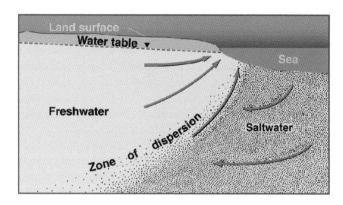

Figure 17.17 Freshwater and saltwater mix in the zone of dispersion. Figure modified from Cooper H. H. (1964). A hypothesis concerning the dynamic balance of fresh water and salt water in a coastal aquifer. U.S. Geological Survey Water-Supply Paper 1613-C, p. 1–12).

it has caused severe degradation and contamination of groundwater.

It does not take very much saltwater to contaminate a fresh groundwater supply. Only 3–4% addition of salinity can make a fresh water supply unsuitable for most uses, including drinking water and even irrigation (Morris et al., 2003). As little as 6% addition of saltwater will render a freshwater source (groundwater) unsuitable for any use except cooling or flushing purposes. Once a freshwater resource is degraded by salt contamination, it will take a very

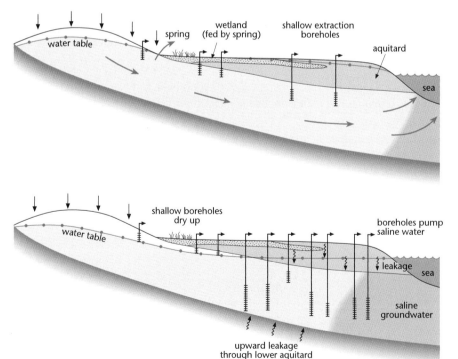

Stage A : Initial condition

- groundwater flows from outcrop to coast
- springs help maintain wetland
- limited extraction from shallow boreholes occurrs

Stage B : Groundwater extraction increases leading to:

- groundwater level decline in aquifer
- springs feeding wetland cease
- shallow wells dry up
- coastal boreholes become saline
- leakage through upper and lower aquitards (if present) begins

Figure 17.18 Saltwater intrusion: invading seawater. Top: Groundwater extraction balanced by recharge. Bottom: Groundwater extraction exceeds recharge. From Morris et al., 2003.

long time for that aquifer to recover, and if positive groundwater recharge conditions are not re-established, it may never recover. Remediation efforts are often cost-prohibitive if not impossible due to the technical constraints associated with removing or decreasing the levels of salt concentration in groundwater. The first response is often abandonment of the contaminated wells, accompanied by drilling of new wells further inland. Effectively, the freshwater resource will have been lost, and supplying new water depends on availability of groundwater supplies further inland. Furthermore, if excessive groundwater pumping continues, the saltwater will continue to invade further inland, again contaminating fresh groundwater supplies. The other option currently available is to construct a de-salinization plant to treat the contaminated groundwater prior to use. Such plants are likely to increase in use in the future, as technology improves and the availability of water supplies dwindles.

Although coastal regions may be most susceptible to saltwater intrusion, many noncoastal areas are also being affected by salinization of groundwater. Many natural geologic systems can lead to the salinization of freshwater resources. Areas once occupied by deep-water oceans or seas are now part of continents. These areas, now deep within the subsurface, contain ancient geologic units that contain high concentrations of salt or brine water. These salt-containing geologic units can contaminate freshwater aquifers when over pumping occurs in the region, causing the water table to decline and encroach into the high-salinity geologic units.

Generally there are no harmful health effects associated with low concentrations of chloride in drinking water. In some cases, salt (chloride) can be harmful to people with heart or kidney conditions. The EPA has set unenforceable secondary drinking water guidelines for chloride at 250 mg/L. However, the contamination of water by saltwater intrusion will increase salt concentrations far beyond what can be tolerated by humans. The primary concern with the contamination of drinking water supplies from saltwater intrusion is the large-scale loss of water resources.

17.6 OTHER GROUNDWATER CONTAMINATION PROBLEMS

Although we have discussed some of the major groups of contaminants threatening groundwater resources, such as hazardous organic chemicals (*e.g.*, chlorinated solvents and fuel-type hydrocarbons), agrochemical pollutants (*e.g.*, nitrates and pesticides), and salinization (*e.g.*, saltwater intrusion and high-salinity groundwater), it is important to note some other contaminants that also present potential threats to the quality of groundwater supplies. We will briefly discuss some of these groundwater contaminants in the following sections.

17.6.1 Pathogen Contamination of Groundwater

Contamination of groundwater by microbial pathogens, including viruses, bacteria, and protozoa, is of significant concern throughout the world. The types of pathogens and their impact on human health were discussed in Chapter 11. The transport behavior of pathogens in the subsurface environment is discussed in Chapter 27.

Potential sources of pathogens for groundwater contamination include land disposal of sewage treatment byproducts (wastewater, biosolids), septic tank systems, and latrines. Risks posed by pathogen-contaminated groundwater are generally believed not to be significant for public supply systems, given the level of treatment applied before use (see Chapter 28). Of much greater concern is potential pathogen contamination of groundwater used for private water supplies, because water from private wells typically undergoes little or no treatment before use. Thus, residential areas with septic systems and private wells are particularly susceptible to potential effects of groundwater contamination by pathogens.

Proper siting and construction of septic and well systems is necessary to minimize potential pollution problems. So-called wellhead protection rules have been developed to prevent the siting or application of pathogen sources too close to water-supply wells. Several recent surveys of groundwater across the U.S. have shown that the incidence of human viruses in groundwater is greater than previously believed and may in part be due to septic tank systems (see Information Box 17.4).

INFORMATION BOX 17.4

Occurrence of Viruses in U.S. Groundwater

Viruses (10–100 nm) are smaller than bacteria (0.5–3 μm) and protozoa (1–15 μm), and thus viruses are generally more mobile in porous media. For example, viruses have been observed to travel more than 100 m in the subsurface. Accordingly, it would be expected that groundwater is more likely to be contaminated by viruses than by other pathogens. To this end, a large-scale study was recently instituted to evaluate the occurrence of viruses in groundwater in the U.S. Information pertaining to physical and geological characteristics of wells and associated subsurface environments, along with various microbial and physico-chemical water-quality parameters, was collected, and possible correlation with the presence of human viruses was investigated. Groundwater samples were collected from 448 sites in 35 states, and assayed for microorganisms and chemical contaminants. Infective viruses, viral nucleic acid, bacteriophages, and bacteria were present in approximately 5, 31, 21, and 15% of the samples, respectively.

Source: Abbaszadegan et al., 2003.

17.6.2 Gasoline Additive: Methyl Tertiary-Butyl Ether (MTBE)

In the mid-1990s it was discovered that methyl tertiary-butyl ether (MTBE), an additive in gasoline, had caused extensive contamination of groundwater throughout the United States (see Information Box 17.5). MTBE had been used since the early 1970s as an oxygenate to promote more efficient combustion of fuel in automobiles. Ironically, while the intended use of MTBE has in fact reduced toxic emissions (*e.g.*, carbon monoxide) released to the atmosphere from automobiles, it has contributed to the widespread contamination of groundwater resources from leaking underground fuel storage tanks, posing serious threats to the quality of drinking water supplies. The contamination of groundwater by MTBE is extensive in the U.S. The EPA reports, citing a study by Chevron, that MTBE concentrations exceeded 1000 μg L^{-} in 47% of 251 California sites surveyed, 63% of 153 Texas sites surveyed, and 81% of 41 Maryland sites surveyed (EPA, 2004). Responding to the rising occurrences of MTBE contamination, California in 1999 became the first state to ban the use of MTBE in gasoline reformulation after 2002. The EPA has now begun regulatory action to phase out the use of MTBE in gasoline.

The primary concern associated with the release of MTBE to the subsurface environment (from spills or leaking underground fuel storage tanks) is its mobility and persistence. For example, at most fuel station sites affected by leaking storage tanks, it is commonly observed that MTBE has migrated much farther than typical gasoline contaminants (*e.g.*, benzene, toluene). Unlike these contaminants, MTBE has a relatively low biodegradation potential. The health effects associated with MTBE through human ingestion of drinking water are not well understood nor well documented. Due to a lack of appropriate data on the health effects of MTBE contamination, the EPA has been unable to define national regulatory standards based on quantitative estimates for health risks. However, the EPA has set provisional drinking water health advisory limits at 20–40 μg L^{-1}.

INFORMATION BOX 17.5

The First Significant Incidence of MTBE Contamination

In 1996, it was discovered that two well fields providing drinking water to the city of Santa Monica, California, were extensively contaminated with MTBE at average levels of 610 μg L^{-1} and 86 μg L^{-1}. The city of Santa Monica lost 50% of its drinking water supply when these two well fields were shut down. Enormous costs were incurred from aquifer decontamination and remediation efforts, which continue to this day. In addition, the city of Santa Monica has had to supplement its drinking water supply by purchasing replacement water from outside resources.

Source: EPA, 2003b.

17.6.3 Solvent Additives: 1,4-Dioxane

In recent years, 1,4-dioxane (dioxane) has gained considerable attention as a primary "emerging" contaminant for groundwater resources. Dioxane, like MTBE, is commonly used as an additive. It is a synthetic organic chemical used as an industrial solvent or solvent "stabilizer" that prevents the breakdown of chlorinated solvents during manufacturing processes. Dioxane is commonly added to chlorinated solvents and other solvents such as tetrachloroethene (PCE), trichloroethene (TCE), 1,1,1-trichloroethane (TCA), and paint thinners. Dioxane is also used as a solvent for the manufacturing of paper, cotton, textiles and various organic products, automotive coolant, shampoos, and cosmetics. It is estimated that TCE and TCA contain approximately 1% and 2–8% 1,4-dioxane, respectively (Mohr, 2001). It was estimated that between 10 and 18 million pounds of 1,4-dioxane were produced in the U.S. in 1990 (Mohr, 2001).

Although 1,4-dioxane has been used as a stabilizer for solvents since the 1940s, the extensive contamination of groundwater by dioxane was not documented until the mid-1990s, when improved analytical methods allowed for the detection of lower concentrations. Dioxane is a very mobile and persistent compound, and is listed as a probable human carcinogen. The EPA has not yet defined a national regulatory standard for dioxane. Action levels have been adopted by some states for 1,4-dioxane. For example, the California Department of Health Services has set an action level of 3 μg L^{-1}.

17.6.4 Perchlorate in Groundwater

Perchlorate, another "emerging" groundwater contaminant, was first detected in drinking water in 1997 and has since been recognized to pose a significant threat to groundwater resources (EPA, 2005). As mentioned previously, it often requires a significant advancement in analytical capability to first observe an emerging contaminant's presence in the environment and in particular groundwater. The development of an analytical method to detect low concentrations of perchlorate allowed for the recognition of its widespread occurrence in groundwater in the U.S. (Information Box 17.6)

INFORMATION BOX 17.6

Perchlorate Contamination in Nevada

The city of Las Vegas gained notoriety as having one of the largest known groundwater perchlorate problems in the U.S. The site of the Kerr-McGee Chemical Corporation near Las Vegas produced some of the highest perchlorate concentrations in groundwater ever reported, with 3.7 million μg L^{-1} in groundwater and 24 μg L^{-1} in drinking water. Strategies for cleanup and remediation are currently in progress and will likely continue well into the future, with incurred costs estimated into the hundreds of millions if not billions of dollars (Struglinski, 2005).

Perchlorate is an inorganic anion and is often present as a salt complex or ammonium salt as ammonium perchlorate. Perchlorate is used for numerous industrial and military purposes. For example, it is a primary constituent in the manufacturing and use of rocket propellants (solid rocket fuel) and other explosives. For this reason, much of the perchlorate contamination of groundwater is derived from military bases and installations. In fact, it is estimated that approximately 90% of perchlorate compounds are produced for use in defense activities and the aerospace industry (EPA, 2005).

Perchlorate, like many salt compounds, is extremely mobile in groundwater. In addition, perchlorate is not readily susceptible to chemical or microbial degradation and is thus extremely persistent in the environment. In studies perchlorate has been shown to interfere with the uptake of iodine by the thyroid. This can disrupt thyroid functioning, including hormone regulation, metabolism regulation, fetus development, and child development. To date, the EPA has not established a maximum contaminant level or enforceable regulatory limit for perchlorate in drinking water. A few states have adopted "action levels" or advisory levels for perchlorate, which generally range from $1 \mu g L^-$ to $6 \mu g L^{-1}$.

17.6.5 Arsenic in Groundwater

Contamination of groundwater by arsenic (As) is another important groundwater contaminant problem throughout the world (see Chapter 10). The high toxicity associated with arsenic is of primary concern for human health. As a result of the high associated toxicity of arsenic, regulatory standards have recently been lowered by the EPA, from $50 \mu g L^{-1}$ to $10 \mu g L^{-1}$. The additional water-treatment costs associated with meeting the revised standard are projected to be in the billions of dollars.

The contamination of groundwater and drinking water can result from natural or human activities. Arsenic is a naturally occurring metallic element that is found in soil, rocks, air, plants, and animals. Arsenic in soil and rocks can act as sources for groundwater contamination. Through processes such as dissolution, weathering, and erosion, arsenic can be released into the environment, resulting in the contamination of groundwater and drinking water supplies. Arsenic sources associated with human activities include agriculture, use as a wood preservative, the burning of fuels and wastes, smelting and mining, paper production, glass manufacturing, and cement manufacturing. In 1997, almost 8 million pounds of arsenic were released to the environment by human activities (EPA, 2000). Extensive adverse health impacts due to arsenic contamination of groundwater have been recently documented in Bangladesh (see Chapter 10).

17.6.6 Acid-Mine Drainage

Another source of pollution that poses threats to groundwater is acid mine drainage. Highly acidic water is produced when rain or groundwater comes into contact with mine tailings or mining wastes. The highly acidic water can change the oxidation-reduction potential in the groundwater and may cause the release of minerals and metals, serving as a source of groundwater contamination. The geology, hydrology, and mining technology employed will determine the nature of the acid mine drainage. Additional issues associated with pollution caused by mining activities are discussed in Chapter 16. Restoration of sites polluted by mining activities is discussed in Chapter 20.

17.7 SUSTAINABILITY OF GROUNDWATER RESOURCES

Clearly, the world's population is greatly dependant on groundwater for many uses. Thus, managing and protecting our groundwater resources is essential to ensuring that sufficient quantities of quality groundwater will be available for future generations. This concept is referred to as **maintaining long-term sustainability of groundwater resources**.

Ensuring groundwater sustainability requires balancing supply and demand. Hydrologic and geologic factors (climate, topography, subsurface properties) exert primary control on the intrinsic supply of groundwater. The potential impact of global climate change on the hydrologic cycle and groundwater supply is of concern. Groundwater pollution affects the fraction of the intrinsic supply that is of sufficient quality for use. Thus, it imposes a constraint on supply. The demand for groundwater is associated with land use and population density. The significant increase in groundwater use observed over the past few decades (see Figure 17.1) is a result of population growth and economic expansion.

The primary means by which to ensure long-term sustainability of groundwater resources is to manage supply and demand. The groundwater supply can be extended through moderating demand and can be supplemented with additional sources of water. Water reuse is one primary method being implemented to enhance sustainability. This includes reusing municipal wastewater directly, either as potable water or for secondary uses such as irrigation, or indirectly (*e.g.*, artificial groundwater recharge). Instituting conservation measures to reduce demand is another method. The use of external supplies is another means of supplementation. For example, the **Central Arizona Project (CAP)** provides surface water from the Colorado River to supplement groundwater resources in Arizona. The CAP canal extends 336 miles from Lake Havasu City to Tucson, and cost $3.6 billion to construct. In Tucson, all water provided by the CAP is recharged into groundwater prior to being pumped to the surface and treated for potable use. A potential problem with using external water supplies, versus water reuse and conservation, to manage supply and demand, is that it imposes a demand on water resources at the point of origin. An example of this is, in fact, the Colorado River, which barely exists as a river close to its entry into Mexico because of its great degree of use upstream. Thus, the demand has just been shifted in part from one location to another. Changes in land use also exert an impact on supply and demand. For example, groundwater use is now shifting from agriculture to residential as urban centers increase in size and population density.

Balancing supply and demand is encapsulated in the concept of **safe yield**. In essence, the principle of safe yield is that the amount of groundwater extracted should not exceed the amount replenished through recharge. This concept is simple in theory. Unfortunately, it is very complex in practice. A primary reason for this is that water resource issues are influenced to a great extent by nonscience factors (political, economic, societal, legal). This causes the management of demand among the competing uses prevalent in large

urban centers to be a complex and difficult task. In addition, planning for future water resource use is difficult and fraught with uncertainty. For example, there is always uncertainty in estimating the future supply of groundwater available in a region. This uncertainty is compounded by uncertainty in demand, which must be estimated based on future population growth and land-use patterns. These difficulties are exacerbated by the "invisibility" of groundwater—as opposed to surface water, it cannot be seen. The prevalent attitude of "out of sight, out of mind" often creates impediments to increasing the awareness of groundwater resource issues.

QUESTIONS AND PROBLEMS

1. What is meant by the term "safe yield" ?
2. Discuss the factors that affect the supply of groundwater available in a specific location.
3. Discuss the factors that affect the demand for groundwater resources in a specific location.
4. What can be done to modify the supply and demand for groundwater resources?
5. Why is nitrate a widespread groundwater contaminant?
6. Comparing heavy metals and solvents, which would you expect to generate larger groundwater contaminant plumes? Why?
7. What is involved in conducting a groundwater pollution risk assessment? What type of information would you need?

REFERENCES AND ADDITIONAL READING

Abbaszadegan M., Lechevallier M. and Gerba C. (2003) Occurrence of viruses in US groundwaters. *J. Amer. Water Works Assoc.* **95**,107–120.

Cooper H.H. (1964) A hypothesis concerning the dynamic balance of fresh water and salt water in coastal aquifer: United States Geological Survey (USGS) Water-Supply Paper 1613-C, pp. 1–12.

Environmental Protection Agency (USEPA). (1998) National Water Quality Inventory Report to Congress.

Environmental Protection Agency (USEPA). (2000) Arsenic Occurrence in Public Drinking Water Supplies. EPA 815-R–00–23.

Environmental Protection Agency. (2003a) The DNAPL Remediation Challenge: Is There a Case for Source Depletion? EPA/600/R-03/143.

Environmental Protection Agency (U.S. EPA). (2003b) Fact Sheet on MTBE. www.epa.gov/mtbe/faq.htm.

Environmental Protection Agency (U.S. EPA). (2004) MTBE Demonstration Project. www.epa.gov/swerustl/mtbe/mtbedemo.htm.

Environmental Protection Agency (U.S. EPA). (2005) Perchlorate Fact Sheet. www.epa.gov/fedfac/documents/perchlorate.htm.

Environmental Working Group (EWG). (1996) Pouring It On: Nitrate Contamination of Drinking Water. www.ewg.org/reports/Nitrate/nitratecontents.html.

Fetter C.W. www.appliedhydrogeology.com.

Hallberg G.R. and Keeney D.R. (1993) Nitrate. In *Regional Ground-Water Quality*, ed. W.M. Alley. Van Nostrand Reinhold. New York.

Interstate Technology and Regulatory Council (ITRC). (2002) *DNAPL Source Reduction: Facing the Challenge.*

Lee G.F. and Jones R.A. (1991b) Groundwater pollution by municipal landfills: Leachate composition, detection and water quality significance. Proc. National Water Well Association's Fifth Outdoor Action Conference on Aquifer Restoration, Ground Water Monitoring, and Geophysical Methods. NWWA, Dublin, OH, pp. 257–271.

Mohr, T. (2001) Solvent Stabilizers White Paper. Santa Clara Valley Water District, San Jose, California.

Morris B.L., Lawrence A.R.L., Chilton P.J.C., Adams B., Calow R.C. and Klinck B.A. (2003). Groundwater and its Susceptibility to Degradation. A global assessment of the problem and options for management. Early Warning and Assessment Report Series. UNEP/DEWA RS. 03–3. United Nations Environment Programme (UNEP), Nairobi, Kenya. 1ISBN: 92–807–2297–2.

Struglinski S. (2005) "Report Perchlorate Cleanup Needs Tracking", Las Vegas Sun, June 2005, Sun Washington Bureau. www.lasvegassun.com/sunbin/stories/lv-gov/2005/jun/23/518950600.html.

Xinhua News Agency. (2002) http://www.china.org.cn/english/24719.htm. Interstate Technology and Regulatory Council, Washington, D.C.

CHAPTER 18

SURFACE WATER POLLUTION

D.B. Walker, D.J. Baumgartner, C.P. Gerba, and K. Fitzsimmons

Water quality sampling from Canyon Lake, Arizona. *Photo courtesy David Walker*

18.1 SURFACE FRESHWATER RESOURCES

Freshwater is a scarce and valuable resource—one that can easily be contaminated. Once contaminated to the extent it can be considered "polluted," freshwater quality is difficult and expensive to restore. Thus, the study of surface water pollution has focused primarily on streams and lakes, and most of the scientific tools developed by such regulatory agencies as the U.S. Environmental Protection Agency have been applied to protecting water quality in this segment of earth's surface waters. The amount and distribution of fresh surface water was illustrated in Chapter 3.

The water stored in reservoirs and lakes, together with the water that flows perennially in streams, is subject to heavy stress, and because it is used for water supplies, agriculture, industry, and recreation, this water can easily be contaminated.

18.2 MARINE WATER RESOURCES

Oceans contain most of the water of the planet. Yet even with the phenomenal volume of water in which contaminants may be dispersed, marine resources can be polluted. Using various biological and physical parameters, we usually classify the ocean environment as three components: the coastal zone, the upper mixed layer, and the abyssal ocean. Several regulatory agencies and international organizations share different responsibilities for those components. The coastal zone, which is most susceptible to the day-to-day kinds of contamination found in freshwater lakes and rivers, is often the province of water quality regulatory agencies established by individual nation states. International organizations have traditionally dealt with pollution concerns of the open ocean and its seabed, which includes the other two components. In addition to these physically described components of the sea, there are legally defined (and disputed) zones, sometimes overlapping, that influence regulatory practices, as indicated in Figure 18.1.

18.2.1 The Coastal Zone

The **coastal zone** extends from the low-tide line to the 200-meter depth contour, tending to match the geophysical demarcation of the continental shelf. The coastal zone can be as wide as 1400 km along some coasts and less than a kilometer along others. The average width of the zone worldwide is about 50 kilometers, comprising about 8% of the surface of the ocean. (The coastal zone of Alaska is larger than that of the rest of the United States.) Within the coastal zone definition, the difference between estuaries and the open coast is important in considering the disposal of wastewater and the potential for pollution problems.

Almost all of the water-carried wastes of a continent enter the coastal zone through an estuary. **Estuaries** are bodies of water with a free connection to the sea whose salinity is measurably diluted with fresh water, as from a river. Because estuaries provide critical and limited habitat for marine organisms to rear and feed their young, water quality is of special concern. Species that inhabit the coastal zone, and especially the estuaries, have to be very resilient to such natural environmental stresses as wide daily variations in salinity, turbidity, temperature, and UV radiation. Owing to this natural resiliency, coastal organisms may be able to tolerate contaminants associated with industrial and municipal wastes better than residents of the continental shelf, where the natural environment is quite stable. The estuarine habitat must be maintained primarily because of its limited extent, as distinguished from the shelf habitat, which is enormous. For this reason, treated wastewater effluents are usually discharged offshore rather than into estuaries in coastal regions. A large pipeline or tunnel, called an **outfall**, is used to transport the effluent to the disposal site.

In disposing of treated wastes offshore, we also need to take the physical features of the coastal zone into account. For example, continental headlands that protrude into the sea can impede both circulation of water and exchange of nearshore water with open ocean water. Outfalls are therefore best located far offshore rather than inside the region of headland influence. Similarly, outfalls should not be located close to shore in the vicinity of estuaries or bays because tidal incursions can carry diluted wastes into the estuary, thereby eroding one of the advantages offered by offshore disposal.

18.2.2 Open Ocean Waters

A variety of the majority of circumstances can contribute to the contamination of the open ocean waters beyond the coastal zone, including atmospheric fallout, oil spills, and dumping of hazardous wastes and sewage sludge as practiced by some countries of the world. Floatable and soluble materials tend to stay in the **upper mixed layer** of the ocean, where they may be decomposed. This upper layer is also the most active photosynthetic zone of the ocean, where the majority of plant—and hence animal—life can thrive. The depth of this layer, which varies between 100 and 1,000 meters, changes with season and geographic location. Although mixing between the upper and deeper layers of the ocean is impeded by strong density gradients, particles formed in the upper mixed layer, or discharged there, may eventually settle so far that they can no longer be resuspended by surface-generated turbulence and thus become part of the detrital sediment load of the deep ocean waters.

Because the quality of the water in the upper mixed layer can significantly affect all life there, it is important to take precautions with waste disposal operations. When ocean disposal of certain materials is justified, we can use technologies to avoid contamination of the upper mixed layer and facilitate transit and long-term retention of the

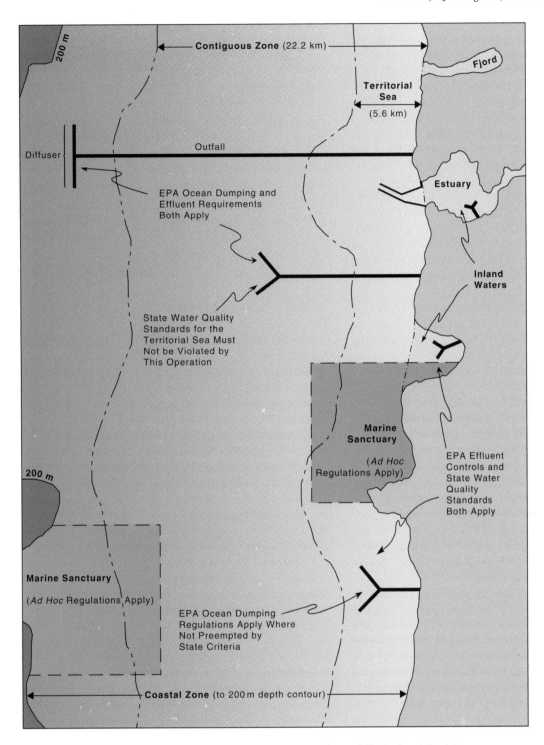

Figure 18.1 Water pollution regulations in the coastal zone. The outfalls depicted (both T- and Y-shaped) all use diffusers. From *Pollution Science* © 1996, Academic Press, San Diego, CA.

material in the deep waters of the open ocean, that is, the **abyssal ocean**. For example, containers have been proposed for disposal of such materials as xenobiotic chemicals or radioactive wastes. Pipelines can also be used to carry liquid carbon dioxide to the seabed, where it can be retained for a long time—conceivably long enough to help reduce the rate of global warming.

18.3 SOURCES OF SURFACE WATER POLLUTION

Water pollution is a qualitative term that describes the situation when the level of contaminants impedes an intended water use. It takes just a small amount of contaminant to pollute a waterbody intended for a drinking water supply. But the same water might not be considered polluted if the water were to be used, for example, for agriculture. Nor is pollution restricted to chemical contaminants. Physical factors of the environment can also contribute to pollution. For example, heated water discharged from a power plant can change the temperature of an aquatic environment. It might not be a problem in a lake or a river during the winter, but it can certainly be a problem in the summertime. Moreover, heated water or water containing some contaminant may not be a problem at any time of the year, provided it is rapidly mixed with the surface water, and the diluted material doesn't accumulate over time. There are also many kinds of contaminants that can usually be accommodated by the natural environment without resulting in pollution, but in many situations, these same contaminants (sometimes in conjunction with other contaminants) can cause pollution even in well-mixed water bodies.

Major sources of surface water contamination are construction, municipalities, agriculture, and industry, however, the water delivered to earth in the form of precipitation is not necessarily pure to begin with. Near the coast, it may contain particulate and dissolved sea salts, and farther inland, it may contain organic compounds and acids scrubbed from contaminants added to the atmosphere both by natural processes and by anthropogenic (human) activities. Gases from plant growth and decay, and gases from geological activity are examples of naturally derived atmospheric contaminants that can be returned to earth via precipitation. The acid rain problem of the New England states is a classic example of anthropogenically derived atmospheric contaminants that contribute to surface water pollution (see Chapter 22).

18.4 SEDIMENTS AS SURFACE WATER CONTAMINANTS

The properties of particulates or sediments in water are described in Chapter 9. Soil water erosion and its control are described in Chapter 16. The ability of rivers to carry sediment over large distances has resulted in the landscape of continents. Certainly, some background level of sediment load in rivers is considered natural and desirable. Problems ensue when anthropogenic activities in a river's watershed increase, or in some cases decrease, sediment load. Running water, wind, and ice are the major factors responsible for the detachment, entrainment, and transport of particulate matter. Geologic erosion is highest in areas with relatively steep gradients such as low- to intermediate-order streams in mountainous areas. Historically, natural erosion has been the largest source of sediment supplied to rivers. As human land use activities increase in watersheds around the globe, anthropogenic effects now are major contributors to both increased sediment supplied to rivers as well as the blocking and impoundment of this sediment behind dams. Both impoundment and increased erosional processes in watersheds

can have profound biological, physical, and chemical impacts on rivers and streams.

Almost any kind of human activity in watersheds can result in an increase of suspended sediment in rivers. A few classic examples of anthropogenic activity known to increase sedimentation are:

Logging, deforestation, wildfire. Specific types of logging activity can increase sediment yield by two orders of magnitude for short periods. Fire suppression and drought can combine to create catastrophic wildfires, which can have devastating impacts on receiving waters from these areas.

Overgrazing by domestic animals. Sedimentation can increase not only due to decreased vegetative interception of precipitation-enhancing erosion, but also through direct trampling of the streambed and channel.

Urbanization and road construction. Road construction commonly results in a 5—20 fold increase in suspended sediment yield. Impervious materials such as pavement, parking lots, or rooftops can increase the velocity of storm water runoff, which will increase erosion once this water comes in contact with soil.

Mining operations. Mines, particularly strip mines, can lead to extraordinarily high levels of erosion and subsequent sedimentation in rivers. An example is the coal strip mines in Kentucky.

On a global scale, rivers discharge roughly 40,000,000 m^3 into the world's oceans annually. For every cubic meter of water reaching the ocean, there is (on average) an accompanying 0.5 kilogram of sediment carried away from the continents.

Suspended sediment is also a major carrier of pollution. While rivers may be transporters of pollution, suspended sediment is the "package" these pollutants are carried in. Heavy metals, organic pollutants, pathogens, and nutrients responsible for eutrophication can all be found attached to sediments in flowing water. The "quality" or overall pollutant load of suspended sediment depends on the degree of pollution in the watershed.

Transport of sediments in water is dependent on many factors, including sediment particle size and water flow rate (see Chapter 9). The quantification and predicted rates of transport of sediment are based upon the assumption that for any given flow and sediment, there is a unique transport rate. Estimates of sediment transport rates are based upon measures of flow (including velocity, depth, shear velocity, viscosity, and fluid density) and both sediment size and density. There are different classification terminologies, based upon the mode of sediment transport in a stream. **Bed load** refers to the sediments moving predominantly in contact with or close to the streambed. In contrast, **suspended load** refers to sediments that move primarily suspended in fluid flow, but that may also interact with bed load. Suspended load has a continual exchange between sediment in fluid flow and on the bed as it is constantly being entrained from the bed and suspended, while heavier particles settle out from the flow to the bed. **Solute load** refers to the total amount of dissolved material (ions) carried in suspension and can only be quantified by laboratory analytical techniques. **Total load** is the total amount of sediment in motion and is the sum of bed load plus suspended load. It is important to remember that these classifications are somewhat artificial. The sediment load carrying capacity of a stream or river constantly changes both spatially and temporally as flow changes. Flow in any river or stream is never homogenous, so the resulting sediment movement in any section of stream or river varies greatly.

Particles that are too heavy to be fully suspended may roll or slide along the bed (**traction load**) or hop as they rebound on impact with the bed. In the latter case, ballistic trajectories occur, and the particle is said to move by **saltation**. **Stream competence** refers to the heaviest particles that a stream can carry. Stream competence depends on stream velocity; the faster the current, the heavier the particle that can be carried. **Stream capacity** refers to the maximum amount of total load (bed and suspended) a stream can carry. It depends on both discharge and velocity, since velocity affects the competence and therefore the range of particle sizes that can be transported. Note that as stream volume and discharge increase, so do competence and capacity. This is not a linear relationship, and doubling the discharge and velocity does not automatically double the competence and capacity. Stream competence varies as approximately the sixth power of velocity. For example, doubling velocity usually results in a 64 times increase in competence. For most streams, capacity varies as a range of squared to cubed values. For example, tripling the discharge usually results in a 9–27 times increase in capacity. Most of the work of streams is accomplished during floods, when stream velocity and discharge (and therefore competence and capacity) are many times their level compared to periods of quiescent flow. This work is in the form of bed scouring (erosion), sediment transport (bed and suspended loads), and sediment deposition.

18.4.1 Suspended Solids and Turbidity

It has been stated that **total suspended solids (TSS)** in water are the most important pollutant. Erosion happens constantly around the planet, and some rivers and streams have naturally high TSS levels without any human intervention. The Yangtze River in China and the Colorado and Mississippi Rivers in the U.S. are examples of rivers that have historically entrained large amounts of sediment due to local topography, geology, and climate (Figure 18.2).

Total suspended solids are defined as all solids suspended in water that will not pass through a 2.0 μm glass-fiber filter (dissolved solids would be the fraction that does pass through the same size filter). The filter is then dried in an oven between 103 and 105°C, and weighed. The increase in weight of the filter represents the amount of TSS.

Problems with TSS arise when excess erosion occurs in a watershed due to human land use practices. Excess levels of TSS can come from either point (municipal and industrial wastewater) or nonpoint (*e.g.*, agriculture, timber harvesting,

Figure 18.2 Flooding in the Santa Cruz River, Arizona. Arid regions are especially prone to increases in suspended sediment concentrations during flood events. Source: http://az. water.usgs.gov/.

mining, and construction) sources. Generally, water with less than 20 mg/L is considered relatively "clear"; levels between 40 and 80 mg/L tend to be "cloudy"; while levels over 150 mg/L would be classified as "dirty" or "muddy." Point sources generally require treatment, usually through settling or flocculation, prior to being released into a river or stream. Nonpoint sources are much more difficult to manage due to several sources acting synergistically. No-till farming, sedimentation basins, and silt fences are common practices to reduce run off from agriculture or construction areas. Stormwater retention ponds and regular street sweeping can reduce the impact of stormwater runoff from urban areas.

Increasing levels of TSS often result in a waterbody being unable to support a diversity of aquatic life. Sedimentation of the stream bed as velocity decreases often results in the suffocation of many aquatic macroinvertebrates and the eggs of fish. Where TSS is deposited results in increased **embeddedness** (the percentage of any piece of substrate covered in sediment) of cobble, rocks, and boulders within the stream. Several species of macroinvertebrates use the bottom of rocks as refuge from predators or from fast-flowing water, and as embeddedness increases, this vital habitat is diminished or completely lost. Additionally, TSS absorbs heat and can increase the temperature of a waterbody. In lakes and reservoirs, this can exacerbate thermal stratification as heat accumulates close to the surface. Suspended solids can also decrease the amount of dissolved oxygen due to consumption of organic matter by respiring bacteria. In lakes or reservoirs with a large algal biomass, sudden inputs of water containing suspended sediments have been known to deplete the oxygen of water and cause massive fish kills. The once-photosynthesizing algae switch to respiration as light for photosynthesis was reduced or eliminated.

Besides the relatively direct effects of suspended solids on water bodies, perhaps the greatest indirect effect are the pollutants that may be attached to suspended sediment.

Examples of pollutants known to sorb to sediment particles are nutrients, metals, organic compounds such as polycyclic aromatic hydrocarbons (PAH) and polychlorinated biphenyls (PCBs), and a wide assortment of herbicides and pesticides. All of these contaminants have differing solubilities and therefore differing fates once they enter a river, stream, or lake. Sediment-associated pesticides in water are an emerging problem in agricultural areas.

Turbidity is related to, but not a proxy for, suspended sediments. Specifically, turbidity is the quantification of the light that is scattered or absorbed rather than transmitted through a water sample. Turbidity is another measure of water clarity but is not a measure of dissolved substances that can add color to water. Particulates are what add turbidity to water and can include such things as silt, clay, organic matter, algae and other microorganisms, and any other particulate matter that can scatter or absorb light. The amount of light scattered or absorbed is proportional to the concentration of particulates in the sample. The exact amount and wavelength of light scattered by a particle is dependent on the particle's shape, size, and refractive index, which makes any correlation between turbidity and suspended solids difficult and impractical. However, turbidity is directly related to the level of particulates and is an excellent general indicator of water quality in its own right. Units of measure for turbidity are in **nephelometric turbidity units (NTUs)** and are measured on a nephelometer (often called a turbidimeter). Turbidimeters operate by shining an intense beam of light up through the bottom of a glass tube containing the sample. Light scattered by particulates in the sample is detected by a sensitive photomultiplier tube at a 90-degree angle from the incident beam of light. The amount of light reaching the photomultiplier tube is proportional to the level of turbidity in the sample. The photomultiplier tube converts the light energy into an electrical signal, which is amplified and displayed on the instrument meter.

18.5 METALS AS SURFACE WATER CONTAMINANTS

Metals that can be toxic to humans and wildlife are often found in industrial, municipal, and urban runoff and in atmospheric deposition from coal-burning plants and smelters and from natural weathering of rocks and soils. Levels of harmful metals in water have risen globally with increasing urbanization and industrialization. Currently, there are over 50 heavy metals that can be toxic to humans. Of these, 17 are considered very toxic and simultaneously readily accessible.

Common heavy metals known to be toxic to humans include arsenic, cadmium, chromium, copper, lead, mercury, and zinc. Interestingly, chromium, copper, and zinc are essential micronutrients required by the human body for growth, and toxicity depends upon enhanced dose.

Heavy metals are also environmentally persistent, which exacerbates any potentially toxic exposure because these metals often accumulate under certain environmental conditions.

18.5.1 Mercury

Mercury in the environment is one of the most widely recognized and publicized pollutants. Under certain environmental conditions, elemental mercury complexes to form methyl-mercury, which is especially mobile in the environment and toxic to humans and wildlife. Use of mercury in the tanning industry and for making hats was the first time that it was widely recognized as a toxic substance affecting the brain; hence the term "mad as a hatter" (see also Chapter 13). Today, release of mercury in smokestack emissions from coal-burning power plants is the primary source of contamination. Bioaccumulation of methyl-mercury in long-lived predatory fish from cold freshwater lakes (pikes and walleyes) and in coldwater marine species (swordfish, sharks, and some tunas) has led to public warnings for pregnant and nursing women to limit consumption of shark and swordfish and for all of the public to limit consumption of fish from certain lakes in parts of the United States and Western Europe.

The cycling of mercury through the environment is complex and depends on several physical, chemical, and, most importantly, biological aspects of the system in question. The manner in which mercury cycles through any given area or ecosystem determines its relative toxicity and subsequent bioaccumulation rate upward through the food chain. The term **bioaccumulation** refers to the net accumulation, over time, of pollutants within an organism from both biotic and abiotic factors. The term **biomagnification** refers to the progressive accumulation of persistent toxicants by successive trophic levels. Biomagnification relates to the concentration ratio in a tissue of predator organisms as compared to that in its prey. Mercury exists in several forms in the environment: elemental mercury (Hg^0), inorganic mercury compounds (Hg^{+1} or Hg^{+2}), and organic mercury compounds ($HgCH_3$ or $Hg(CH_3)_2$), which include both methyl- and dimethyl-mercury (Figure 18.3).

There are numerous pathways by which mercury can make its way into water bodies. Inorganic and methyl-mercury can enter water directly from atmospheric deposition, methyl-mercury and Hg^{+2} can be bound to organic substances in runoff, and surface water flow in upper soil layers can transport Hg^{+2} and methyl-mercury to water bodies. There has been a global increase of mercury released into the atmosphere since the beginning of the Industrial Age, so that atmospheric deposition onto watersheds and surface water often plays as large a role as runoff from natural sources.

Once in an aquatic ecosystem, mercury goes through several complexation and transformation processes. While most forms of mercury are bioavailable, methyl-mercury (MeHg) is the form most readily absorbed and bioaccumulated. The methylation of mercury in aquatic systems not only requires a certain range of physicochemical factors, but also the presence of a group of bacteria known as **sulfate-reducing bacteria (SRBs)**. There are several species of SRBs, but some of the most common include strains of *Desulfovibrio* and *Desulfobacter*. The majority of methylation that occurs in lakes and

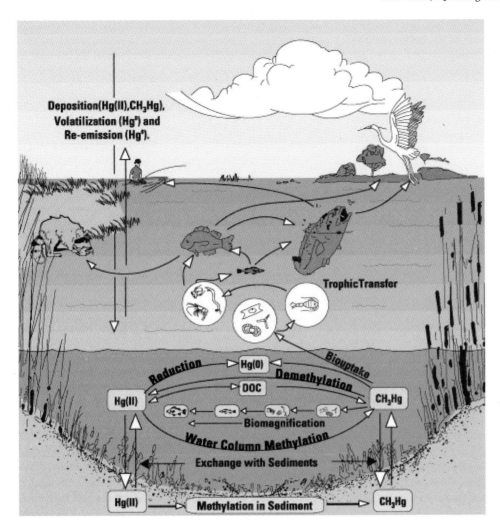

Figure 18.3 Common transformations of mercury in aquatic ecosystems.

TABLE 18.1 Factors influencing the methylation of mercury in aquatic ecosystems.

PHYSICAL OR CHEMICAL CONDITION	INFLUENCE ON METHYLATION
Low dissolved oxygen	Enhanced methylation
Decreased pH	Enhanced methylation within the water column
Decreased pH	Decreased methylation in sediment
Increased dissolved organic carbon	Enhanced methylation within sediment
Increased dissolved organic carbon	Decreased methylation within water column
Increased salinity	Decreased methylation
Increased nutrient concentrations	Enhanced methylation
Increased temperature	Enhanced methylation
Increased sulfate concentrations	Enhanced methylation

reservoirs is within anaerobic sediments. Factors affecting the methylation of mercury are outlined in Table 18.1, while common transformations of mercury are given in Figure 18.4.

Mercury in all forms is a potent toxin that can cause developmental effects in the fetus as well as toxic effects on the liver and kidneys of adults and children. Sublethal effects of mercury toxicity can affect the ability to learn, speak, feel, see, taste and move. Children under the age of 15 are most vulnerable, because their central nervous system is still developing. Mercury is easily passed from pregnant mother to fetus, and even extremely small trace amounts of mercury can have devastating effects to a developing central nervous system. Mercury toxicity can occur through skin contact, in-

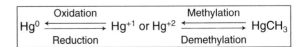

Figure 18.4 Common transformations of mercury.

halation, or ingestion. Due to biomagnification, the most common route of exposure to humans is through consumption of contaminated fish. Approximately 60,000 babies are born in the U.S. each year with some degree of mercury toxicity.

The methylation, biomagnification, bioaccumulation, and toxicity of mercury are often linked to the problem of increasing eutrophication. Increases in most of the factors that cause eutrophication within a waterbody, including increased sulfate content, also increases the rate of methylation and subsequent toxicity to humans and wildlife. Humans have not only increased the availability of mercury deposited into aquatic systems, we have, through cultural eutrophication, also increased the bioavailability and subsequent toxicity of mercury in these systems.

18.5.2 Arsenic

Arsenic is an element widely distributed throughout the earth's crust. As such, it is often introduced into water through the dissolution of minerals and ores and may concentrate in groundwater. Arsenic is also used in industry and agriculture and is a byproduct of copper smelting, mining, and coal burning. One form of arsenic, chromated copper arsenate, is the most common wood preservative in the U.S. and contains 22% arsenic.

Inorganic arsenic occurs in several different forms in the environment, but in natural waters it is most commonly found as trivalent arsenite [As(III)] or pentavalent arsenate [As(V)]. Most of the organic species of arsenic, usually at very high levels in seafood, are less toxic and are readily eliminated by normal body functions.

Symptoms from arsenic exposure include vomiting, esophageal and abdominal pain, and bloody diarrhea. Long-term exposure can cause cancers of the skin, lungs, urinary bladder, and kidney as well as other skin changes, such as pigmentation changes and thickening.

One of the largest mass poisonings in the world recently occurred in Bangladesh, where 53 out of a total of 64 districts had groundwater contaminated with arsenic. The cause of arsenic contamination was related to the onset of intense agriculture in the region where irrigation resulted in large-scale withdrawal of groundwater via wells (see Chapter 10).

18.5.3 Chromium

Chromium is found in natural deposits as ores containing other elements. Additionally, chromium is an important industrial metal, where it is used in alloys such as stainless steel, protective coatings on other metals and magnetic tapes, pigments for paints, cement, paper, rubber, and floor coverings. Chromium has several oxidation states, but the most common are $^{+2, +3}$ and $^{+6}$, with $^{+3}$ being the most stable. Oxidation states of $^{+4}$ and $^{+5}$ are relatively rare.

Toxicity of chromium depends on oxidation state. Chromium(III) is an essential nutrient, while the hexavalent form, chromium(VI), is believed to be carcinogenic in humans. Evidence to date indicates that the carcinogenicity is site-specific, limited to the lung and sinonasal cavity, and dependent on high exposures.

18.5.4 Selenium

Selenium occurs naturally in the environment as selenide and is often combined with sulfide, copper, lead, nickel, or silver. Like chromium, selenium is a micronutrient needed in very small quantities in humans and wildlife to produce the amino acid selenocysteine. However, it can be toxic at higher doses. The relatively narrow range between selenium acting as a beneficial nutrient (50 μg/day) and the initiation of toxicity (400 μg/day) in humans means that it needs to be closely monitored in the environment, especially in areas with alkaline soils, because this is where selenium is often found in its most oxidized and toxic form. As with several other naturally occurring heavy metals, problems may arise due to increased availability of selenium in aquatic systems primarily due to irrigation and farming practices. Symptoms of short-term selenium toxicity include hair and fingernail changes, damage to the peripheral nervous system, and irritability. Long-term symptoms include damage to liver and kidney tissue and nervous and circulatory systems (Figure 18.5).

Selenium is a bioaccumulative pollutant; however, unlike mercury, selenium concentrations do not increase upward through the food chain, *i.e.*, it does not biomagnify. Selenium toxicity can have devastating effects on both terrestrial and aquatic wildlife. Selenium can affect the growth and survival of juvenile fish as well as the offspring of adult fish exposed to sublethal levels. Birds that have eaten fish suffering from selenium toxicity either succumb to the acutely toxic effects of selenium or produce offspring, often stillborn, with gross skeletal deformities. Due to selenium uptake in terrestrial plants, both domestic and wildlife species foraging on these plants can be affected.

18.6 NUTRIENTS AND EUTROPHICATION OF SURFACE WATERS

On a global scale, eutrophication has often been cited as the number one cause of impairment to surface water resources. Eutrophication is the gradual accumulation of nutrients, and organic material subsequently utilizing these nutrients as an energy source, within a body of water. While eutrophication is often cited as an example of anthropogenic pollution of inland waters such as lakes and streams; coastal areas, estuaries, and salt marshes are also commonly affected. Eutrophication often results in increases in algal biomass, and therefore some discussion of what these nutrients are, and what specific ratios cause eutrophication, are in order.

Justus Von Liebig, a German analytical chemist and professor of chemistry at the University of Giessen, made great contributions to the science of plant nutrition and soil fertility in the mid-1800s. Liebig's **Law of the Minimum**

SELENIUM TOXICITY IN KESTERSON RESERVOIR, CALIFORNIA

California's Kesterson Reservoir in the San Joaquin Valley is a classic example of one of the most dramatic cases of heavy metal toxicity known to date. Kesterson Reservoir was built in the late 1960s to address the issue of California's decreasing wetland habitat by using agricultural drainage for the creation of wetlands solely for the purpose of attracting and harboring native wildlife species. Due to its perceived benefit to wildlife, Kesterson was made into a National Wildlife Refuge under the auspices of the U.S. Fish and Wildlife Service.

Mountains forming the western boundary of the San Joaquin Valley consist of shale enriched with selenium. The San Joaquin Valley is an area of poorly drained soils where intensely irrigated agriculture in the otherwise arid valley resulted in selenium becoming highly concentrated in agricultural drainage. By 1981, almost all of the water entering Kesterson Reservoir was agricultural drainage from poorly drained soils.

Mosquitofish collected by the U.S. Fish and Wildlife Service in the early 1980s from Kesterson Reservoir contained levels of selenium approximately 100 times higher than mosquitofish found in neighboring wetlands not receiving agricultural effluent. Several studies were implemented during the 1980s to determine whether selenium or other toxicants were present at levels that could harm wildlife.

Agricultural drainage entering Kesterson Reservoir had an average selenium concentration of 0.3 ppm, seemingly low levels at first glance. The greatest damage, however, rested in the bioaccumulative nature of selenium. Algae had average selenium concentrations of 69 ppm, aquatic plants had 73 ppm, aquatic insects more than 100 ppm, and mosquitofish 170 ppm, which was more than 500 times the concentration of the aquatic habitat in which these mosquitofish lived. All of these levels were much higher than those found in neighboring wetlands not receiving agricultural drainage.

Numerous birds feeding on aquatic organisms in Kesterson Reservoir suffered and died due to selenium toxicity. Symptoms included emaciation, feather loss, degeneration of live tissue, and muscle atrophy. Adult birds that did not immediately succumb to the toxic effects of selenium produced offspring, usually stillborn, that had abnormal or missing eyes, beaks, legs, wings, and feet. The area created to enhance and preserve wildlife, especially migrating and native waterfowl, now appeared to be a death trap to their survival.

Millions of dollars have been spent studying the effects of selenium toxicity at Kesterson Reservoir and millions more have been spent on cleanup efforts. The circumstances that caused the devastating effects to wildlife at Kesterson Reservoir emphasizes the need to find viable solutions to disposing of contaminant-laden waters emanating from agricultural drainage in arid regions.

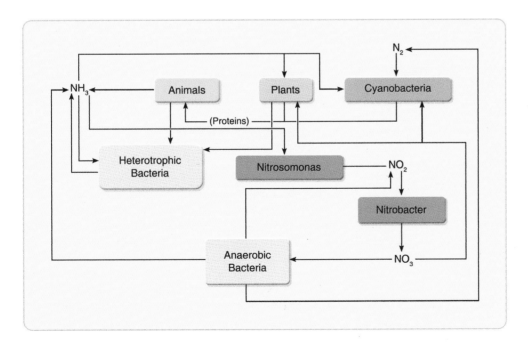

Figure 18.5 The nitrogen cycle in aquatic systems. Source: http://www.marietta.edu/~mcshaffd/ aquatic/sextant/chemistry.htm.

states that yield is proportional to the amount of the most limiting nutrient, whichever nutrient it may be. From this, it may be inferred that if the deficient nutrient is supplied, yields may be improved to the point that some other nutrient is needed in greater quantity than the soil can provide, and the Law of the Minimum would apply in turn to that nutrient. This same law can be applied to aquatic systems. The nutrients that most often limit primary production in aquatic systems are forms of carbon, nitrogen, and phosphorous. The specific ratio of limitation (on a molar basis) is 106C:16N:1P. **Carbon** is ubiquitous in the environment and atmosphere, but can become limiting during intense photosynthesis by algae or aquatic plants. Since carbon dioxide is utilized during photosynthesis, it is possible for this carbon nutrient to be temporarily depleted during daylight hours. This situation would be reversed during the evening, when respiration would exceed photosynthesis, and the carbon dioxide utilized during the day would be released back into the water. Generally, it is uncommon for carbon to be a limiting nutrient.

The idea of nutrient limitation based upon the ratio between C:N:P in aquatic systems only works when one of these essential nutrients is, indeed, "limiting" the growth of primary producers such as algae. In eutrophic or hypereutrophic systems, ratios may indicate that a nutrient is "limiting" in the traditional sense, however, if *all* nutrients are orders of magnitude higher than what it takes to limit primary production than, in this case, ratios can be misleading and nothing is truly limiting growth. Often, in hypereutrophic systems, algal biomass can become so large that the only limiting factor is available light for photosynthesis, as algal cells near the surface shade those at depth.

Nitrogen is an essential plant nutrient used in the synthesis of organic molecules such as amino acids, proteins, and nucleic acids. Nitrogen (mostly as N_2 or "dinitrogen" gas) comprises 78% of the Earth's atmosphere. Most of the abundant nitrogen found in the atmosphere is not yet bioavailable. Nitrogen must be "fixed" into nitrate (NO_3^-), ammonia (NH_3), or ammonium (NH_4) before it can be used by organisms incapable of fixation. Organisms capable of nitrogen fixation include certain species of bacteria, actinomycetes, and cyanobacteria. In aquatic systems, cyanobacteria perform the majority of nitrogen fixation. In aquatic systems, the forms of nitrogen of greatest interest are (in order of decreasing oxidation state):

–Nitrate (NO_3^-)

–Nitrite (NO_2^-)

–Ammonia (NH_3)

–Ammonium (NH_4^+)

–Organic-N (amino groups)

Total oxidized nitrogen is the sum of $NO_3^- + NO_2^-$. Organic nitrogen is the organically bound fraction and includes such natural materials as proteins and peptides, nucleic acids and urea, and numerous synthetic organic materials. Analytically, organic nitrogen and ammonia can be determined together and referred to as "Kjeldahl nitrogen", a term that reflects the technique used in their determination. Total Kjeldahl nitrogen (TKN) is not synonymous with total nitrogen. If TKN and NH_3 are determined individually, "organic nitrogen" can be estimated by the difference.

$$TKN + NO_3^- + NO_2^- = \text{Total Nitrogen}$$

All forms of nitrogen (organic and inorganic) are interconvertible. The nitrogen cycle is an important component of overall biogeochemical cycling in aquatic systems.

Ammonification is an important process in the nitrogen cycle and is, basically, the process of decomposition with production of ammonia or ammonium compounds, especially by the action of bacteria on organic matter. Aquatic animals commonly excrete NH_3 as a waste product of metabolism. The excreted or mineralized NH_3/NH_4 is then available for direct uptake and utilization by other organisms, or it may be converted to more oxidized forms of nitrogen for incorporation into cells. In some nitrogen-poor lakes or reservoirs, the excretory contribution (*e.g.*, ammonification) from zooplankton can provide up to 90% of the nitrogen required by primary producers. Ammonification is difficult to quantify because of the rapid uptake of NH_3 and NH_4 by primary producers. Ammonification is the opposite of assimilation and protein synthesis. Both aerobic and anaerobic bacteria play vital roles in ammonification.

Nitrification is the biological oxidation of NH_4^+ and NH_3 to NO_2^- and then NO_3. Nitrification is important because NH_4^+ and NH_3 are toxic to species of aquatic vertebrates. Nitrification is performed by bacteria that gain energy from oxidizing reduced forms of nitrogen. The aerobic chemoautotrophs involved in nitrification are species of *Nitrosomonas* and *Nitrobacter*. Nitrification consumes and simultaneously requires oxygen, and is a two-part process.

$$NH_4^+ + 1/2\ O_2 \xrightarrow{\substack{\text{Ammonia}\\\text{mono-oxygenase}}} NH_2OH + H^+$$
$$NH_2OH + O_2 \longrightarrow O_2^- + HOH + H^+$$

This process requires 66 Kcal of energy/gram atom of ammonium oxidized.

Under anaerobic conditions:

$$NH_2OH \rightarrow NOH \rightarrow N_2O$$

Ammonium oxidation has important ecological significance in aquatic systems. The microbes that perform nitrification are relatively inefficient autotrophs that use the energy gained from oxidizing ammonia to fix carbon. Thus, these bacteria have a dual ecological role: they are involved in recycling nitrogen and in fixing carbon into organics. The microbes that perform nitrification are fragile. These organisms are acid-sensitive even though they produce acid. If a large source of nitrogen is added into the environment, these organisms can potentially kill themselves by metabolizing it to nitric acid. Since they are also strict aerobes, they can be killed if introduction of wastes leads to excessive growth of other species that deplete oxygen (*i.e.*, eutrophication).

Denitrification is the reduction of nitrate (NO_3) to nitrogen gas or to organic nitrogen compounds and can be a significant pathway for the loss of nitrogen from aquatic systems. There are two types of denitrification, assimilatory and dissimilatory.

Assimilatory nitrate reduction. Many organisms can only acquire nitrogen in the form of nitrate and must reduce nitrate to form the amino groups needed for metabolism.

$$NO_3^- + energy \rightarrow amino\ groups$$

The "energy" in the above equation is usually supplied by enzymatic activity (nitrogenase).

Dissimilatory nitrate reduction. Dissimilatory nitrate reduction is performed by anaerobic bacteria that use nitrate as the terminal electron acceptor in the absence of oxygen. The overall equation is:

$$NO_3^- \rightarrow NO_2^- \rightarrow NO \rightarrow N_2O \rightarrow N_2\ gas$$

The individual steps of dissimilatory nitrate reduction are:

1. Reduction of nitrate to nitrite
   $$2HNO_3^- \rightarrow 2HNO_2^- + 4e$$
   Enzyme: dissimilatory nitrate reductase

2. Reduction of nitrite to nitric oxide
   $$2HNO_2^- \rightarrow 2NO + 2e$$
   Enzyme: dissimilatory nitrite reductase

3. Reduction of nitric oxide to nitrous oxide
   $$2NO \rightarrow N_2O + 2e$$
   Enzyme: dissimilatory nitric oxide reductase

4. Reduction of nitrous oxide to dinitrogen
   $$N_2O \rightarrow N_2 + 2e$$
   Enzyme: dissimilatory nitrous oxide reductase.

Since reductions are energy yielding, 24 ATPs are generated per mole of nitrate reduced.

Although denitrification requires anoxic conditions, it has been observed in aerated lake sediments and can form relatively thin biofilms on rocks in streams. Evidently, denitrification can occur in microzones of anoxia within sediments and biofilms. Oxygen produced through photosynthesis by benthic algae may inhibit denitrification. Denitrification requires an organic carbon source and proceeds faster where more carbon is available in the water and sediments. Denitrification may contribute a significant portion of the oxidative metabolism in water bodies where nitrate levels are high. Within any given waterbody, denitrification can occur simultaneously with nitrification. Denitrification occurs due to microorganisms, usually facultative anaerobes and predominantly two genera: *Pseudomonas* and *Bacillus*. Dissimilatory denitrification is used in sewage treatment and bioremediation where denitrifying bacteria aid in converting organic nitrogen to nitrogen gas that escapes to the atmosphere.

As explained by the previous processes, all forms of nitrogen are interconvertible. While there are losses of nitrogen within any aquatic system, there are simultaneous gains from the atmosphere and from recycling within any given region. Problems with eutrophication arise when humans contribute to loading of nitrogen to a waterbody from either point or nonpoint sources of pollution.

Since the 1940s, the amount of nitrogen available for uptake in aquatic systems at any given time has more than doubled (see also Chapter 16). Human activities now contribute more to the global supply of fixed nitrogen each year than natural processes. Anthropogenic nitrogen totals about 210 million metric tons per year, while natural processes contribute about 140 million metric tons. This influx of extra nitrogen has caused serious distortion of natural nutrient cycling in aquatic systems. Excess nitrogen can wreak havoc with aquatic ecosystem structure affecting the number and kind of species found.

Phosphorous, like nitrogen, is essential to all life. Phosphorous functions in the storage and transfer of a cell's energy and in genetic systems. Cells use adenosine triphosphate (ATP) as an energy carrier that drives a number of biological processes, including photosynthesis, muscle contraction, and the synthesis of proteins. Phosphate groups are also found in nucleotides and therefore nucleic acids. Phosphorous is usually more scarce environmentally than other principle atoms of living organisms including carbon, hydrogen, oxygen, nitrogen, and sulfur.

Phosphorous occurs naturally in rocks and other mineral deposits. During weathering, the rocks gradually release the phosphorus as phosphate ions, which are soluble in water, and the mineralized phosphate compounds breakdown. The phosphorus cycle in aquatic systems is shown in Figure 18.6. Phosphorous exists primarily as phosphates in two forms: orthophosphate and organically bound phosphate. These forms of phosphate occur in living and decaying plant and animal remains as free ions, chemically bonded, or mineralized and chemically bonded in sediments. Analytically, phosphorous in water is usually categorized as being either dissolved or particulate, depending on whether or not it can pass through a 0.45-μm filter. The "dissolved" fraction can have a substantial colloidal component. Within the dissolved fraction, inorganic P (dissolved inorganic phosphorus) occurs as orthophosphate (PO_4). Dissolved inorganic phosphorous is sometimes referred to as **soluble reactive phosphorous (SRP)**. **Total phosphorous (TP)** is determined on a nonfiltered sample by heat and acid digestion, which converts the sample to SRP for measurement.

In unpolluted rivers, SRP averages about 0.01 mg/L on a worldwide basis and total phosphorous averages about 0.025 mg/L (Maybeck, 1982). Agricultural activities may increase SRP levels to 0.05–0.1 mg/L, and municipal effluents may increase SRP concentrations to 1.0 mg/L or much higher. Particulate phosphorous includes P incorporated into mineral structures, adsorbed onto clays, and incorporated into organic matter. Worldwide averages of particulate phosphorous concentrations are about 0.5 mg/L. This level can be much higher depending upon land use and erodibility of the watershed.

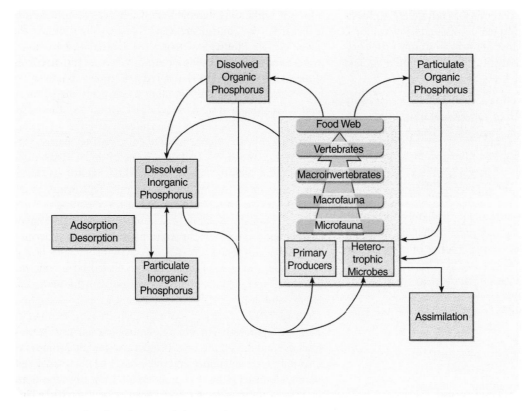

Figure 18.6 The phosphorus cycle in aquatic systems. From Calow P. and Petts G.E. (1992). *The Rivers Handbook.* Vol. 1. Reprinted with permission of Blackwell Publishing.

Phosphorous is often the limiting macronutrient with regard to primary production in aquatic systems. Because of this, and its relative scarcity, it is quickly removed from its dissolved state and incorporated into living biomass. Bacteria and algae are both responsible for turnover rates as fast as 1–8 minutes (Rigler, 1973). Turnover rates usually follow the order of (in order of decreasing turnover times): Bacteria → algae → zooplankton → vertebrates.

It's been estimated that in freshwater lakes, zooplankton excrete about 20% of the phosphorous required by phytoplankton, whereas bacteria can excrete up to 80%. Therefore, food web dynamics play a large role in either the sequestration or the recycling of phosphorous in aquatic systems. The speed at which phosphorous is moved between biotic and abiotic compartments makes interpretation of different forms difficult. It's impossible to distinguish between zooplankton-P, bacterial-P, algae-P, and sometimes even inorganic-P. The best way to quantify phosphorous in a body of water is by analysis of total phosphorous.

Eutrophication, besides increasing algal biomass, often results in depletion of dissolved oxygen, increases in pathogenic bacteria and viruses, increases in potentially toxic species of algae, fish kills, and loss of biodiversity. Remediation efforts of even a small lake are usually cost-prohibitive, and it is very difficult, if not often impossible, to return a lake or reservoir back to an earlier trophic state. The best approach is a proactive, watershed-based one that attempts to protect waterbodies from cultural eutrophication.

This usually requires collaboration among several resource agencies, in addition to municipalities and individual landowners in the watershed.

18.6.1 Harmful Algal Blooms

Planktonic (*i.e.*, free-floating algae) are vitally important components of all marine and freshwater systems on the planet. They form the base of the food chain in all aquatic systems. Of the thousands of known species, a few hundred have the potential to produce a wide variety of toxins under certain environmental conditions. Eutrophication greatly exacerbates the growth and prevalence of potentially toxic species.

18.6.1.1 Harmful algal blooms in marine systems

Most **harmful algal blooms (HABs)** occur in coastal areas where terrestrial runoff of nutrients causes the growth and proliferation of sometimes monospecific blooms of toxic algae. Dinoflagellates (Division Dinoflagellata) are marine phytoplankton often associated with toxic blooms.

Dinoflagellates are microscopic, unicellular, flagellated protists that can be either autotrophic (photosynthetic) or heterotrophic (consuming other organisms). Heterotrophic forms often have life history patterns more akin to an animal than a plant. Additionally, dinoflagellates are routinely

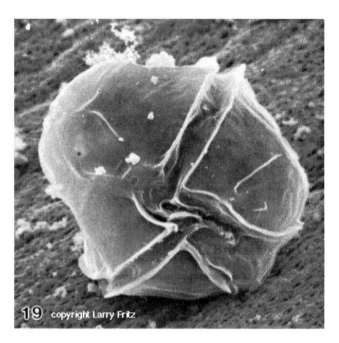

Figure 18.7 Scanning electron micrograph of *Alexandrium tamarense*, a typical dinoflagellate associated with paralytic shellfish poisoning. Source: http://www.whoi.edu/redtide/species/species.html.

found in freshwater and often produce some of the same toxins found in marine systems.

Dinoflagellates have the potential to produce a variety of toxins that can be harmful to humans and wildlife. Some affect humans following the ingestion of shellfish or fish that have consumed toxic dinoflagellate species (Figure 18.7).

Ciguatera poisoning is the most commonly reported disease associated with consumption of seafood. Ciguatera is a lipid-soluble toxin that can affect a variety of fish species and can be very toxic to humans after ingestion of these fish. Tropical and subtropical fish species, including barracuda, grouper, and snapper, are commonly affected. The dinoflagellate species most often associated with ciguatera poisoning is *Gambierdiscus toxicus* but other species including *Prorocentrum mexicanum*, *P. concavum*, *P. lima*, and *Ostreopsis lenticularis* have also been implicated (Figure 18.8). Ciguatera exhibits both gastrointestinal and neurological symptoms, with the time to onset usually less than 24 hours. Gastrointestinal symptoms include diarrhea, abdominal pain, nausea and vomiting. The most common neurological symptoms include abnormal or impaired skin sensations, vertigo, lack of muscle coordination, cold-to-hot sensory reversal, myalgia (muscular pain), and itching. Neurological symptoms may recur intermittently, with gradually diminishing severity for a long as six months. No deaths have been reported from ciguatera in the U.S., although worldwide, the mortality rate is 7–20% of people infected.

Brevetoxin is a large, lipophilic, polyether toxin primarily produced by the dinoflagellate *Karenia brevis* (Figures 18.9 and 18.10). Brevetoxin poisoning occurs with the most frequency in the Gulf of Mexico and has caused sporadic fish kills for decades. These toxins also affect shellfish, which can in turn poison humans who ingest contaminated shellfish. This syndrome is referred to as **neurotoxic shellfish poisoning (NSP)** and produces similar symptoms similar to ciguatera poisoning. There have been no reported fatalities from NSP, although it has been known to kill laboratory mammals.

Under certain environmental conditions, dinoflagellates can rapidly multiply in numbers and form "tides." Usually, tides are identified by the color of the dinoflagellate causing the bloom. "Red" tides (Figure 18.11) are often associated with species of *Alexandrium* and "brown" tides with species of *Aureococcus*.

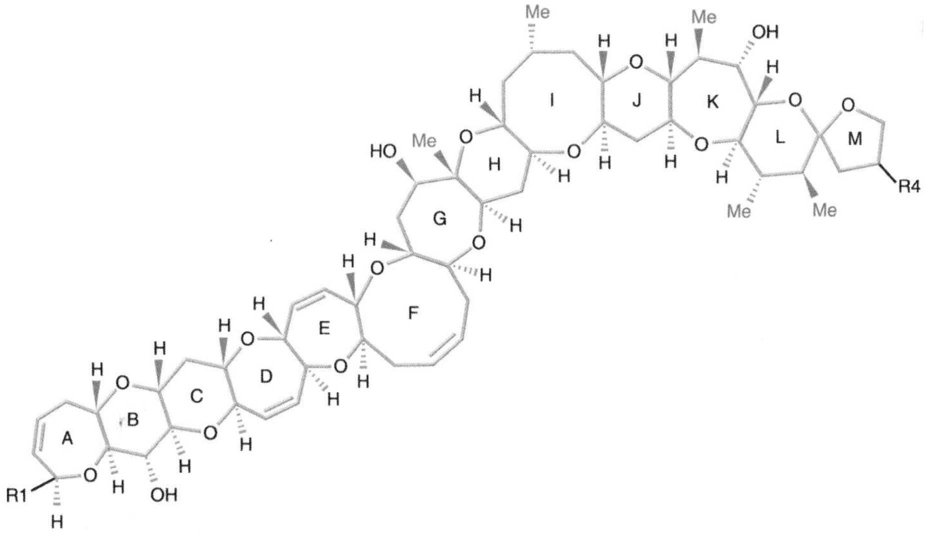

Figure 18.8 Structure of ciguatoxin Type I. Source: http://www. aims.gov.au/arnat/arnat-0004.htm.

Fish and Wildlife Research Institute

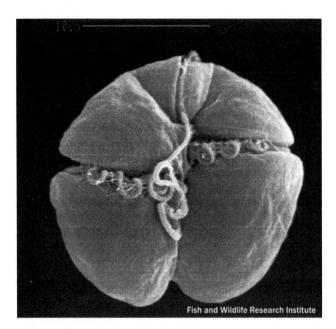

Figure 18.9 Scanning electron micrograph of *Karenia brevis*. Source: Florida Fish and Wildlife Conservation Commission (http://www.csc.noaa.gov/crs/habf/proceedings/intro. html).

How a Toxic Algal Bloom Occurs
The life cycle of one cell
vegetative cells
3: If conditions remain optimal, cells will continue to divide, reproducing exponentially, 2 to 4 to 8. A single cell could produce several hundred cells within a couple of weeks. If other single cells reproduce similarly then toxicity in shellfish may occur
3.
gametes
4.
4 and 5: When nutrients are gone, growth stops and gametes are formed. Two gametes join to form one cell, which develops into a zygote and then into a cyst. This falls to the ocean bottom and is capable of germination the following year.
2: Cysts can germinate only during certain times of the year with warmer temperatures and increased light stimulating germination. The cyst breaks open, and a swimming cell emerges. The cell reproduces by simple division within a few days of "hatching."
2.
zygote
5.
1: Alexandrium resting cysts lay dormant on the ocean floor, buried in sediment. If undisturbed by physical or natural forces, they can stay in this state for years. If oxygen is present, germination may proceed if conditions are right.
1.
cyst

Figure 18.11 Red tide formation. Source: Jack Cook, Woods Hole Oceanographic Institution (http://www.whoi.edu/redtide/whathabs/whathabs.html).

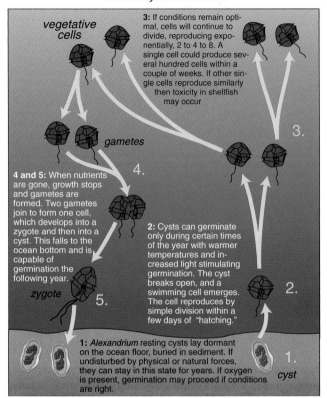

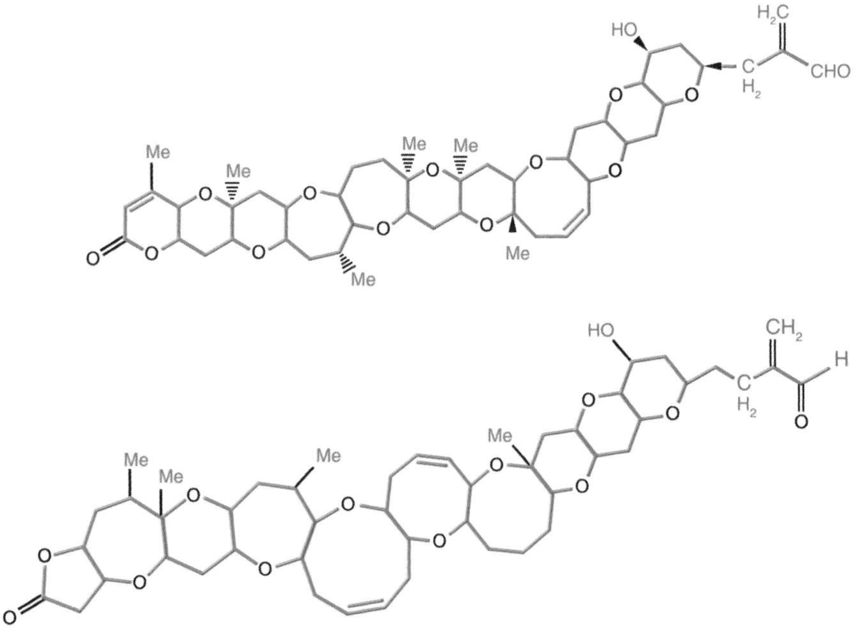

Figure 18.10 Structure of brevetoxin-A and brevetoxin-B respectively. Source: http://www.aims.gov.au/arnat/arnat-0003.htm.

The initiation of either red or brown tides is complex due to the complex life cycles of dinoflagellates but usually involves warm, nutrient-enriched water.

18.6.1.2 Harmful algal blooms in freshwater systems

While freshwaters often contain many of the same dinoflagellate species found in marine systems, and sometimes the same toxins, the majority of freshwater toxins are caused by several different species of cyanobacteria. Cyanobacteria, like dinoflagellates in marine systems, greatly increase in number in eutrophic waters and occur on a global scale (Figure 18.12). Cyanobacterial toxins ("cyanotoxins") can affect both humans and wildlife. Humans are usually affected from ingesting water containing cyanotoxins, and the disease is categorized based upon the type of toxin in the water. Toxins produced by cyanobacteria can be either hepatotoxic or neurotoxic.

One of the ubiquitous cyanotoxins is microcystin, which can be produced by species of *Anabaena* (Figure 18.13), *Nodularia*, *Nostoc*, *Oscillatoria*, and *Microcystis* (Figure 18.14). There are over 50 different analogues of microcystin (Figure 18.15). These toxins mediate toxicity by

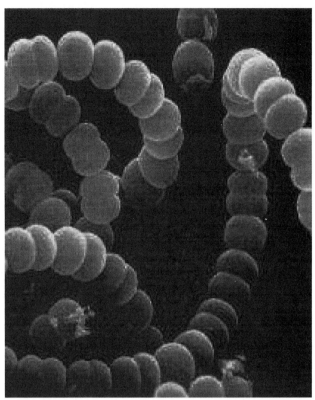

Figure 18.13 Scanning electron micrograph of *Anabaena flos-aquae*. Courtesy Dr. Wayne Carmichael (http://www.nps.gov/romo/resources/plantsandanimals/names/checklists/other_algae/bluegreens/anabaena_f-a.html).

inhibiting liver function (*i.e.*, hepatotoxic) and can often be found at high levels in drinking water reservoirs.

Anatoxin-a is a small, low-molecular-weight neurotoxic alkaloid produced by species of *Anabaena*, *Aphanizomenon*, *Cylindrospermum*, *Microcystis*, and *Oscillatoria*. Anatoxin-a is a powerful, depolarizing, neuromuscular blocking agent that strongly binds to the nicotinic acetylcholine receptor

Figure 18.12 The curtain divides the two halves of the lake. The area of the lake in the lower half of the picture had phosphorus, a limiting nutrient, experimentally added to it and now contains a massive bloom of cyanobacteria. Source: http://www.umanitoba.ca/institutes/fisheries/eutro.html.

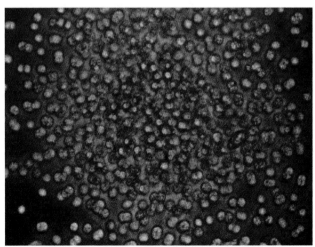

Figure 18.14 Image of *Microcystis sp*. Photo courtesy David Walker.

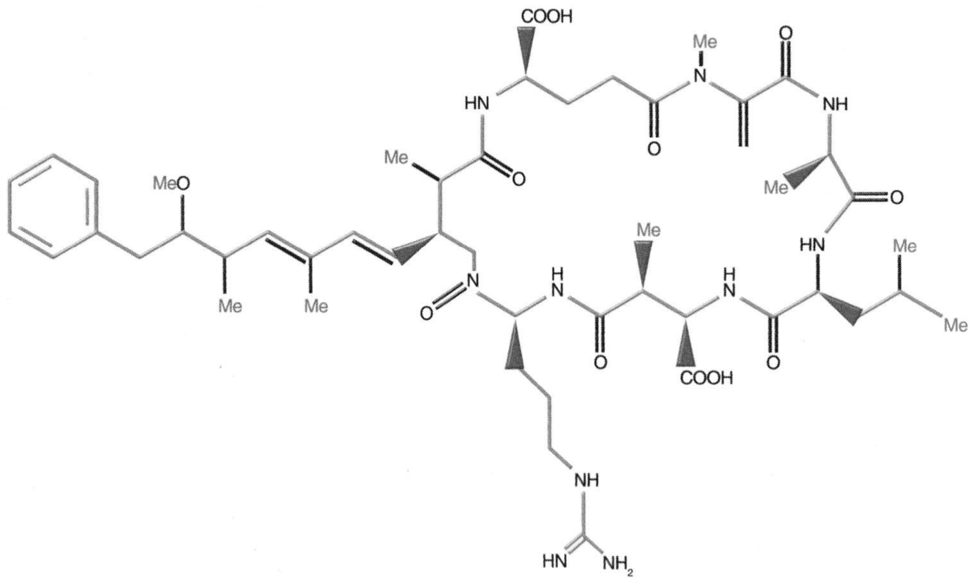

Figure 18.15 Structure of microcystin. Source: http://www.aims.gov.au/arnat/arnat-0002.htm.

(Figure 18.16). This is a potent neurotoxin that can cause rapid death in mammals through respiratory arrest.

Cylindrospermopsin, while having many of the properties of hepatotoxin, also resembles neurotoxin. It is produced primarily by *Cylindrospermopsis raciborskii*, but has also been found in *Umezakia natans* and *Aphanizomenon ovalisporum*. Cylindrospermopsin has poisoned at least 149 people, many of them children requiring hospitalization, in Palm Island, Queensland, Australia. At one time believed to be strictly a tropical to subtropical species, *C. raciborskii* has been found in waters in the north temperate U.S. recently. Like most cyanotoxins, cylindrospermopsin can often be found in drinking water reservoirs.

18.7 ORGANIC COMPOUNDS IN WATER

18.7.1 Persistent, Bioaccumulative Organic Compounds

Certain organic compounds, due to their physicochemical properties (Chapter 7), are very persistent in the environment. These compounds are referred to as persistent, bioaccumulative, toxic (PBTs) contaminants (Chapter 10). These compounds are the most important organic contaminants in aquatic systems. These compounds bioaccumulate and biomagnify within the aquatic ecosystem, and they often accumulate in the sediments of surface water bodies.

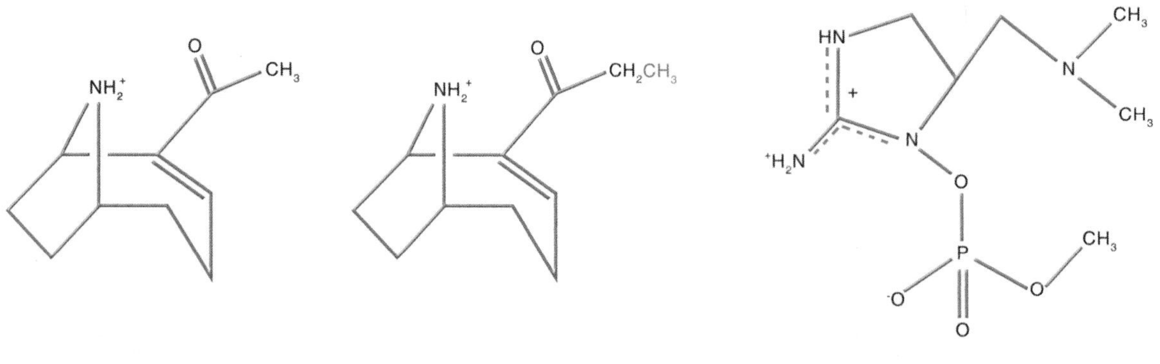

Figure 18.16 Structure of anatoxin-a and the homologue homoanatoxin-a and anatoxin-a(s).
Source: http://www.aims.gov. au/arnat/arnat-0002.htm.

Dichlorodiphenyltrichloroethane (DDT) is an organochlorine pesticide that was in widespread use from the 1940s to the early 1970s, when it was used primarily for agricultural crops or vector-pest control (the control of insects known to carry malaria and typhus).

When DDT is released into the environment, it begins to degrade into several different metabolites. Once bound to sediment particles, DDT and its degradation products can persist for up to 15 years depending upon environmental conditions. Water is the main route of exposure of DDT and its metabolites to humans and wildlife. Ingestion of foodstuffs and in particular consumption of fish is how humans ingest the largest amounts of DDT, primarily due to bioaccumulation. In fish and other wildlife, especially predatory birds feeding on fish, even if acute toxicity and death does not occur, reproductive failure often results.

The use of DDT in the US has been banned since 1972, however, the need to protect agricultural crops and humans from insect-borne vectors of disease still exists. Most organochlorine pesticides, including DDT, have been replaced with less environmentally persistent compounds such as organophosphate, carbamate, and synthetic pyrethyroid pesticides. While these compounds are degraded in the environment at a much faster rate than DDT, they are also more acutely toxic. Even though DDT was banned over 30 years ago, due to its persistence, we still feel its toxic effects in the U.S. Twenty years after the ban of DDT, the U.S. EPA reported that out of 388 sites throughout the nation sampled between 1986 and 1989, total DDT and PCBs (discussed later), were detected at 98 and 90% of all sites respectively. Fish still remain vulnerable to the effects of DDT. A study by Munn and Gruber in 1997 showed total DDT was detected in 94% of whole-fish samples collected in streams of eastern Washington State.

Polychlorinated biphenyls (PCBs) are a group of organic compounds with similar physical structure and chemistry, ranging from oily liquids to waxy solids (Figure 18.17). All PCBs are formed from the addition of chlorine (Cl_2) to biphenyl ($C_{12}H_{10}$), which is a dual-ring structure consisting of two 6-carbon benzene rings linked by a single carbon-carbon bond. The presence of a benzene ring allows a single attachment to each carbon, meaning that there are 10 possible positions for chlorine to replace the hydrogens in the original biphenyl.

Any unique compound in the PCB category is referred to as a "congener" whose individual name is dependent upon the total number and position of each chlorine substitute. There are 209 PCB congeners.

Due to the chemical stability and high boiling point of PCBs, they were used in hundreds of industrial applications, including electrical insulation, hydraulic equipment, and plasticizers in paints, plastics, and rubber products. Prior to their ban in 1977, total production of PCBs in the U.S. was more than 1.5 billion pounds. Because of the vast amount of possible congeners, PCBs were sold as many different trade names but one of the most prevalent was Arachlor®.

Similar to DDT, PCBs are environmentally persistent and adhere strongly to particulates in water, meaning that they can remain intact in sediments of lakes and rivers for extended periods. Because of PCBs' strong adherence to sediments and suspended particles in water, actual contamination of a waterbody may be several times higher than the actual solubility of a particular PCB. Also like DDT, all PCBs are extremely lipophilic, meaning that they can bioaccumulate and biomagnify in aquatic environments. First detected in the 1960s, PCBs were found to be contaminants on a global scale, occupying virtually every component of the environment including air, water, soil, fish, wildlife, and human blood.

Uptake of PCBs by microorganisms is very rapid and extremely high bioconcentration factors are often seen. Uptake by microorganisms is by true absorption into cells rather than adsorption onto the cell.

Fish are especially susceptible to the accumulation and concentration of PCBs, and all life stages of almost every species readily absorb PCBs from the water. PCB congeners with higher chlorination levels are taken up most rapidly by fish. Due to the fact that PCBs are usually at higher levels in sediments, fish such as bottom feeders are most susceptible; however, route of exposure in fish can occur through water, sediment, or prey. Due to rapid uptake of PCBs in fish tissue, birds, especially those that eat fish, are also vulnerable. Egg-laying females can transfer substantial amounts of PCB to eggs with subsequent reproductive failure.

Route of exposure to humans, like DDT, is generally much greater for aqueous environments (through either direct ingestion of water or through eating contaminated fish) than terrestrial. PCBs are probable carcinogens in humans and are known to be carcinogenic to laboratory animals. The risks associated with consuming fish contaminated with PCBs are more than 1,000 times greater than the 1-in-a-million cancer risk used to regulate most hazardous wastes.

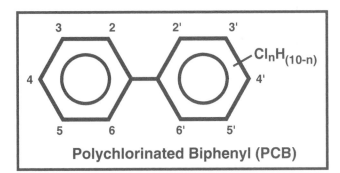

Figure 18.17 Basic structure of a polychlorinated biphenyl. Source: http://www.epa.gov/toxteam/pcbid/defs.htm.

18.8 ENTERIC PATHOGENS AS SURFACE WATER CONTAMINANTS

Almost all animals are capable of excreting disease-causing intestinal microorganisms (enteric pathogens) in their feces (see Chapter 11). Sources of pathogens into surface waters include:

- Urban storm water
- Combined sewer and sanitary sewer overflows
- Animal feeding operations
- Sewage treatment plants
- Septic tanks (onsite systems)

Pathogens can remain infectious for prolonged periods of times in surface waters presenting health risks to recreational users shellfish harvesting, and drinking water treatment plants. While drinking water treatment plants are required to treat water from surface sources, the more pathogens that are present in the raw water, the more treatment is required. Also, after periods of heavy rains, when the amount of suspended matter and pathogens often increases, it is difficult to remove all pathogens. For example, it has been shown that waterborne disease outbreaks in the United States are related to the intensity of rainfall events (Curriero et al., 2001).

Forty percent of rivers and estuaries that fail to meet ambient water quality standards fail because of pathogens,

---

CASE STUDY 18.2

"SILENT SPRING" AND "OUR STOLEN FUTURE"

Rachel Carson (1907–1964) was a scientist, author, and ecologist who spent most of her adult life working for the U.S. Fish and Wildlife Service. She received a letter from a friend in Massachusetts in the summer of 1957 stating that an airplane had flown back and forth over the friends' property spraying DDT to control mosquitoes. The following day, there were numerous dead birds in her yard. She contacted Carson to see what could be done to prevent further spraying. While doing preliminary research on the problem, Carson was shocked at how prevalent the use of pesticides, and DDT in particular, had become. From 1958 to 1962, Rachel Carson performed research for the book that would become *Silent Spring*. To many, this book would become the seminal work exposing the ecological effects of pesticide use in addition to laying the foundation for the modern environmental movement.

Silent Spring opens with a fable of a rural town that suddenly suffers sickness and death, with its citizens realizing they had unwittingly been poisoning themselves with pesticides. Next, Carson explained the science, in easily understandable prose, that this scenario was happening all over the country. She explained how stronger pests survived poisoning so that increasingly stronger pesticides were needed to have the same efficacy over time. Rachel Carson was one of the first to explain about bioaccumulation and biomagnification, and how DDT, at low levels in the water, becomes concentrated as it works its way up the food chain. Her message was that humans could never fully control nature without harmful side effects.

This book was not well received by many, especially companies that manufactured pesticides and were a very powerful and influential group of lobbyists. One company threatened a lawsuit before the book went to press.

The lawsuit never manifested, and the company was later found to be one of the worst offenders with respect to the manufacturing and use of toxic chemicals.

In 1996, Dr. Theodora Colborn wrote *Our Stolen Future*. Dr. Colborn is a senior scientist with the World Wildlife Fund and one of the world's leading experts on endocrine-disrupting chemicals in the environment. This book does not examine the acutely toxic properties of certain synthetic organic compounds, but rather their long-term chronic effects. The endocrine-disrupting effects (see Chapter 31) of a plethora of chemicals are the most insidious and difficult to quantify, but, due to their ubiquity, are perhaps the most devastating on a global scale over the long term. Some endocrine-disrupting chemicals alter sexual development, while others affect intelligence and behavior. Sometimes the effects exerted by these chemicals are not realized until a child reaches puberty or sometime thereafter, even though the effects were brought about in the womb. This book explains that typical dose–response curves do not work with endocrine-disrupting compounds, and that high-dose studies often miss important low-dose biological effects. The author explains that the way we, as a society, use chemicals today is like performing a vast experiment, not in the laboratory, but in the real world. Dr. Colburn was recently awarded the Rachel Carson Award for Integrity in Science by the Center for Science in the Public Interest (CSPI).

What both Rachel Carson and Theodora Colburn epitomize is that it is not enough to elucidate and evaluate scientific problems, but rather that it is of equal importance to communicate the implications of scientific research to nonscientific audiences.

usually measured by fecal coliform bacteria (see Chapter 11) (Smith and Perdek, 2004).

Stormwater can contain a wide variety of pathogens that originate from the feces of wild and domestic animals. Besides pets, other animal sources in urban areas include pigeons, geese, rats, and raccoons. Animal feces accumulate on the ground, and following a storm event are flushed into nearby streams and lakes. This results in a rapid increase in the concentration of enteric organisms, sometimes exceeding that found in raw sewage. In some cities, sewers that collect domestic sewage are combined with stormwater drains or collection systems. These flows are then transported to a sewage treatment plant for treatment. Unfortunately, after periods of heavy rainfall, this combined flow is greater than the sewage plant can treat, requiring the sewage plant to discharge untreated combined sewage and stormwater. These events are referred to as **combined sewer overflows** or **CSOs**. CSOs generally occur in older parts of the country, involving approximately 900 cities. To reduce the impacts of CSOs, cities may blend the untreated wastewater with treated wastewater (Figure 18.18) or may construct large holding reservoirs where the combined flows can be stored until they can be treated later (see Case Study 18.1).

In the United States, there are 238,000 animal feeding operations, or AFOs (sometimes called confined animal feeding operations, or CAFOs) that produce 350 million tons of manure annually. This figure does not include manure from grazing animals. These operations generate approximately 100 times as much manure as municipal wastewater treatment plants produce sewage sludge (biosolids) in the United States. The Clean Water Act (see Chapter 15) requires operations having more than 1,000 animals to have a discharge permit. Runoff from AFOs and farmland can contribute significant levels of pathogens that can infect humans, such as *Cryptosporidium* and *Escherichia coli* O157:H7 (see Chapter 11).

Septic tanks (also referred to as **decentralized** or **onsite wastewater treatment systems**) collect, treat, and release about 4 billion gallons of treated effluent per day from an estimated 26 million homes, business, and recreational facilities in the United States (see Chapter 27). Poorly treated wastewater from improperly operating or overloaded systems can contain enteric pathogens that by transport through the soil can make their way to nearby streams and lakes. Overflows from failing onsite systems can result in the sewage reaching the surface. The discharge of partially treated sewage from malfunctioning onsite systems was identified as a principal or contributing source of degradation in 32% of all harvest-limited shellfish growing areas in the United States. Problems with surface water contamination by onsite systems are most likely to occur in areas with shallow groundwater tables (*e.g.*, within a few feet of the surface).

In the United States it is required that sewage discharges be disinfected to reduce the level of pathogens (see Chapter 26). While very effective in eliminating most enteric bacterial pathogens, significant levels of enteric virus and protozoan parasites may remain.

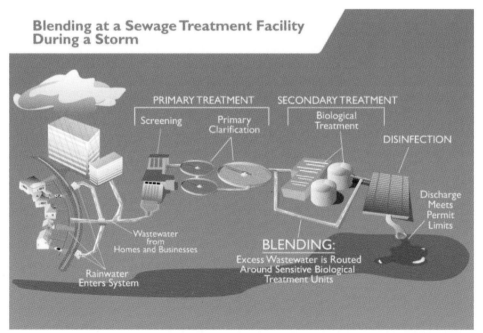

Figure 18.18 Blending at a sewage treatment facility. Source: www.epa.gov.

TUNNEL AND RESERVOIR PLAN (TARP) TO CONTROL EXCESS STORMWATER

Chicago and 51 older municipalities in Cook County, Illinois, have combined sewer systems. This means when rain falls, stormwater runoff drains into a combined sewer, where it mixes with the sewage flow from homes and industry. The net result is one massive quantity of dirty water! A system that was designed to treat 2 billion gallons of wastewater per day may be inundated with more than 5 billion gallons of stormwater runoff (about 1″ of rain) during a single rainstorm.

When the urban area grew and treatment plants were at capacity, there was no alternative but to allow the excess

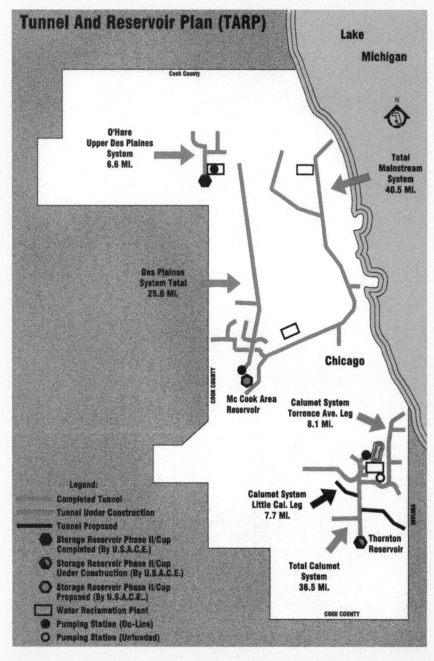

Figure 18.19 Tunnel and reservoir plan (TARP). Photo courtesy C.P. Gerba.

mixture of raw sewage and stormwater to spill directly into the rivers and canals as combined sewer overflow or CSO. This meant that much untreated sewage, diluted with storm runoff, was bypassing treatment plants and polluting lakes, rivers, and streams, and also causing street and basement flooding. A better solution had to be found.

In the 1970s, a team of engineers from the City of Chicago, Cook County, and state agencies considered various plans to solve the problem of flooding and water pollution. The hybrid plan selected as best and most cost-effective was the **Tunnel and Reservoir Plan** or **TARP** (Figure 18.19). Under this plan, 109 miles of huge underground tunnels would be burrowed under the city to intercept combined sewer overflow and convey it to large storage reservoirs. After the storm had subsided, the overflow could then be conveyed to treatment plants for cleaning before going to a waterway.

The Mainstream tunnel is 35 feet in diameter, bored in limestone rock 240 to 350 feet below ground, and holds 1 billion gallons of water (Figure 18.20). Mainstream is one of the largest rock tunnel bores on record. Since tunnel contractors would be working beneath homes, businesses, and streets, excavation by extensive blasting was ruled out. Boring by huge tunnel-boring machines (TBMs) was selected instead, to cause less rock disturbance, noise, and vibration.

Nearly 15 years later, the success of this project is evident by the dramatic improvements in the water quality of the Chicago River, the Calumet River, and other waterways. Game fish have returned, marinas and riverside restaurants abound, and river recreation and tourism has improved. http://www.mwrdgc.dst.il.us/plants/tarp.htm

Figure 18.20 The Mainstream tunnel. Photo courtesy C.P. Gerba.

> **INFORMATION BOX 18.1**
>
> **Microbial Source Tracking**
>
> Fecal contamination of surface waters can result from numerous sources, including human sewage, manure from livestock operations, indigenous wildlife, and urban runoff. Effective watershed management requires identification of, and targeting mitigative action towards, the dominant source of fecal contamination in the watershed. Several **microbiological source tracking (MST)** methods have been developed to fill this need. MST methods are intended to discriminate between human and nonhuman sources of fecal contamination, and some methods are designed to differentiate between fecal contaminations originating from individual animal species.
>
> MST methods involve the isolation of fecal bacteria (*Escherichia coli*, enterococci) or viruses (human or bacterial) from a watershed. Subsequently molecular analyses of DNA or RNA, or patterns of sensitivity to different antibiotics, are performed to "fingerprint" the organisms. Samples of potential sources of the fecal bacteria are then collected in the watershed (*e.g.*, from cattle, ducks, pigeons, dogs, sewage treatment plants, and urban runoff), and a "fingerprint" of these bacteria is obtained. By matching the fingerprints, the major sources of fecal bacteria in a watershed or waterbody can be identified. Identification of human enteric viruses (see Chapter 11) is an indication of human sewage as a source. Certain bacteriophages are only found in humans and others only in animals; hence, this is another approach that can be used.

18.9 TOTAL MAXIMUM DAILY LOADS (TMDL)

A **total maximum daily load (TMDL)** is the maximum amount of pollution that a waterbody can assimilate without violating water quality standards. A TMDL is the sum of the allowable loads of a single pollutant from all contributing point and nonpoint sources, so that a waterbody can meet a designated use, such as swimming or fishing. Under the Clean Water Act of 1972, states are required to identify surface waters not meeting water quality standards and develop a TMDL for each pollutant for each listed waterbody. The processes of TMDL development and implementation are shown in Figure 18.21. Once the impaired body of water is identified, a study is usually conducted to identify the sources and concentration of pollutants. From this, an informational plan is developed to reduce the most significant source(s) so that water quality standards can be met.

TMDL = point sources of waste allocations (*e.g.*, sewage treatment plant discharge) + nonpoint source load allocations (urban runoff) + natural sources (mineral deposits, wild animals) + growth factor (growth of enteric bacteria) + a margin of safety (to compensate for uncertainties about the link between pollutant loads and impairments)

A load allocation is the part of a TMDL/water quality restoration plan that assigns reductions to meet identified water quality targets. For example, in a given watershed it was found that cattle were the major source of fecal coliform bacteria in the streams. To reduce fecal coliform loading, fences were placed to limit direct access of the cattle to the stream.

Fecal coliform bacteria and temperature (largely thermal waters from power plants) are the major contaminants that cause most waterbodies to not meet water quality standards for intended uses in the United States (Figure 18.22).

18.10 QUANTIFICATION OF SURFACE WATER POLLUTION

Even as treatment methods were being developed, methods were simultaneously developed to quantify and assess both the disease threat and the dissolved oxygen problem posed by the discharge of municipal wastes into waterbodies. Quantitative methods are also used to calculate the degree of treatment needed. Such methods are based on an understanding of the physical and biochemical processes controlling the decay of microbes and chemicals over time.

18.10.1 Die-Off of Indicator Organisms

Early in the development of quantitative assessment methods, scientists realized that the many different microorganisms that exist in human wastes could not be effectively cultured and counted. Consequently, they settled on the coliform group of organisms to serve as an indicator of fecal pollution because they could be cultured and counted easily. Known to exist in large numbers in the gut of all warm-blooded animals, the coliform group provides a good indication of fecal pollution; however, it is not very specific, so other indicators such as fecal coliforms and streptococci may also serve as indicators of the sanitary quality of water.

Tests have shown that 99.99% of the indicator bacteria can be removed by wastewater treatment. But the residual 0.01% remains a problem raising concerns about the quality of water for recreational use, for example. The fact is that the number of bacteria in sewage is tremendous (500 million to 2 billion per 100 milliliters), so that, in assessing quality, the percentage of removal is not as useful as is the actual remaining concentration. The microorganism concentrations allowed for various uses of water are relatively small. For drinking purposes, the concentration of fecal coliforms should, of course, be 0, but a concentration of less than 1 per 100 mL may be allowed; for bathing, a concentration of 1000 per 100 mL is frequently accepted. If, for purposes of

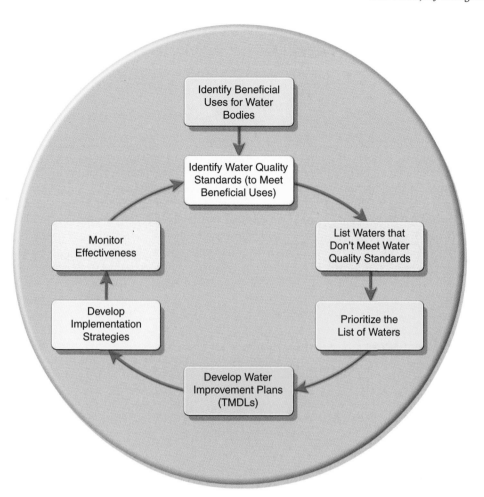

Figure 18.21 TMDL development and implementation.

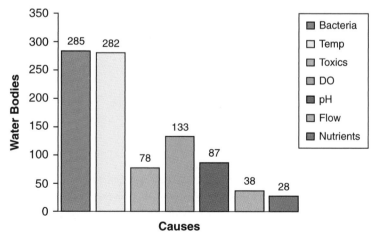

Figure 18.22 Major causes for water bodies that fail to meet water quality standards. Source: www.epa.gov.

CASE STUDY 18.4

TOTAL MAXIMUM DAILY LOAD (TMDL) FOR FECAL COLIFORM BACTERIA IN THE WATERS OF DUCK CREEK IN MENDENHALL VALLEY, ALASKA

Duck Creek is on the list of impaired waters in Alaska because of fecal coliform bacteria. The primary sources of fecal coliform bacteria in the creek were found to be urban runoff and animal waste. As the watershed became more developed, urban runoff and pet populations increased. This increased the level of fecal coliforms entering Duck Creek. Duck Creek is used as a source of drinking water, and the State of Alaska standard for such water bodies is a geometric average of 20 fecal coliforms per 100 mL. This standard was usually exceeded several times per year (Figure 18.23). Based on the water quality standards for fecal coliform bacteria and the hydrologic conditions of Duck Creek, the loading capacity for fecal coliform bacteria was established at $2{,}23 \times 10^{11}$ fecal coliforms per year. To meet these objectives it was recommended that wetlands be constructed to retain stormwater flows, greenbelts be developed to serve as buffers to overland flow, and eroding banks be stabilized. In addition, pet owners were encouraged to clean up and properly dispose of pet waste.

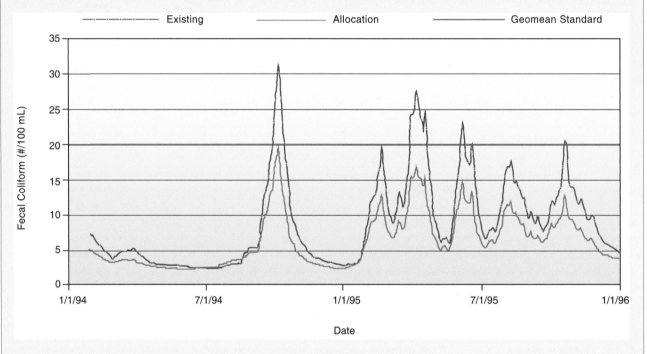

Figure 18.23 Existing fecal coliform concentrations in Duck Creek and allocation to meet standards as a drinking water source water body. Source: www.epa.gov./owow/tmdl.

demonstration, we assume that raw sewage has a population of one billion (10^9) per 100 mL, a removal efficiency of 99.99% would still leave 100,000 per 100 mL. The good news is that the concentration of organisms decreases with time and distance downstream of the discharge point owing to natural processes. Depending on the location and proximity of uses, the point of discharge and method of discharge can be designed to optimize the rate of natural purification, further reducing downstream pollution problems. The bad news is that we have only a partial understanding of all the processes that affect die-off.

The concentration C of bacterial indicators of fecal contamination has been observed to decrease with time t according to a first-order reaction, the equation of which is

$$\frac{dC}{dt} = -KC \qquad \text{(Eq. 18.1)}$$

where K is the **die-off rate constant**. One parameter frequently employed in water pollution analyses obtained by solution of the first-order reaction equation is t_{90}, which is the time required for 90% die-off of the bacteria. This parameter is analogous to the half-life (t_{50}) used in radioactivity studies and is calculated in the same fashion. Thus, given the value of K

$$t_{90} = \frac{2.3}{K}$$ (Eq. 18.2)

The general solution of Equation 18.1 is used to find the concentration at any time C_t, after the initial concentration C_i is determined:

$$C_t = C_i e^{-Kt}$$ (Eq. 18.3)

This information can be used to calculate freshwater or marine die-off. To find the concentration of bacteria after effluent has traveled in the river for, say, 8 hours, it is necessary to determine the value of K for the specific situation being studied. Results of many studies in lakes and streams have shown that K varies widely, depending on the temperature of the water, the amount of sunlight, and the depth at which the plume travels. An average value for fresh water is about $K = 0.038$ per hour; however, values from 0.02 to 0.12 per hour have been measured. Using Equation 18.3 and $K = 0.038$ per hour, the initial concentration $C_0 = 100,000$ per 100 mL would be reduced to about 74,000 per 100 mL in $t = 8$ hours due to die-off alone. In many cases, it is preferable to predict the concentration at a given distance downstream, rather than at time increments. Most U.S. rivers have a remarkably uniform low-flow current of 1.5 to 2 km per hour, which can be used to convert low-flow travel time to distance. In our example, this travel distance would be between 12 and 16 km. Travel times for other flow conditions vary from river to river, so field measurements may be needed to relate bacterial concentrations to specific locations downstream.

In analyzing coliform die-off cases in the marine environment, we often use an average value of 1.2 per hour for K, which is nearly 30 times greater than that of freshwater. This rapid die-off rate is usually attributed to the salinity of the marine environment, although it may also be related to a greater concentration of predatory animals. In addition, the natural flocculation and sedimentation of particles that occurs in estuaries could account for removal of bacteria from the water column. There are, however, other factors that can reduce the die-off rate. For example, when an effluent is discharged at a great depth, the die-off can slow down considerably because sunlight cannot penetrate deeply enough.

Calculating seawater die-off is similar to freshwater die-off. The value of K for marine waters usually ranges from 0.3 to 3.8 per hour. Recently, however, K values as low as 0.02 per hour have been found where an effluent plume is transported in a layer far below the surface, say, 40 meters, suggesting that die-off is reduced because of the low penetration of UV radiation to that depth. What difference would this low K value make in the concentration?

Using the average value of $K = 1.2$ per hour, the original concentration, $C_0 = 100,000$ per 100 mL, would die off to $C_1 = 6.7$ per 100 mL in $t = 8$ hours using Equation 18.3. But using $K = 0.02$ per hour instead of 1.2 per hour, we get 85,000 per 100 mL. Thus, for a 60-fold reduction in K, the concentration is increased by a factor of $85,000/6.7 = 13,000$! It is evident from this example how important it is to have accurate values for K and how widely the results can vary with equally good, but different, estimates for K.

Variations in the rates of indicator bacterial die-off are not the only problem we have to contend with. Noncoliform pathogenic microorganisms may decay at rates different from those of our coliform indicators. Therefore, as water-analysis techniques become more sophisticated, we will need to conduct many field observations to establish values that can be used to predict die-off rates for specific pathogens.

In addition to the decrease of bacterial concentrations due to die-off in either fresh or marine surface waters, bacterial concentrations are decreased as the water is diluted with upstream ambient water at the point of effluent discharge and further diluted as it flows downstream. The effect of dilution may or may not be important in meeting water quality criteria, depending on the initial mixing, the nature of the subsequent flow patterns, and the distance to water use areas. But before discussing the mechanics of the dilution process, observe how the same first-order decay process used for assessing indicator bacteria die-off can be applied to the analysis of the fate of biodegradable organics.

18.10.2 Organic Matter and Dissolved Oxygen

Biodegradable organic compounds are decomposed by bacteria and other organisms that live in surface waters. While some organics are mineralized to carbon dioxide and oxides of nitrogen, others are synthesized into more microbial biomass, most of which is subsequently decomposed as well. All this decomposition consumes dissolved oxygen (DO), upon which many desirable species of fish, other aquatic organisms, and wildlife depend. Thus depressed dissolved oxygen concentrations adversely affect these life forms. For example, some fish can survive at concentrations near 1 mg L^{-1}, most are adversely affected at DO concentrations below 4 mg L^{-1}. The maximum amount of oxygen that pure surface water can hold is a function of salinity, temperature, and atmospheric pressure; compared to the maxima of many other substances, however, it is remarkably low. Note that solubility of O_2 *decreases* with increasing temperature, which is the opposite of the temperature–solubility relationship observed for most substances in water.

Domestic sewage can contain about 300 to 400 mg L^{-1} of organic compounds, 60% of which is readily degradable by bacteria commonly found in nature. Readily degradable implies that most of the material will be decomposed within about a week in a stream or other body of water that is sufficiently large. The change in the concentration of the organic matter with time is conveniently described by the first-order decay equation used to describe bacterial die-off. However, the value of this K depends on the specific organic compounds in the sewage. For domestic sewage, an average value is about $K = 0.4$ per day, ranging from 0.1 to 0.7 per day. As more industrial wastes are contributed to the sewer system, the rate constant may increase or decrease. The amount of organic material discharged to a surface water depends on the population served by the municipal sewer system and treatment technology employed. Each person contributes about 90 grams per day of organics; thus, if the population is 100,000 people, the mass emission rate is 9 metric tons per day. The concentration of organics in the sewage depends on the amount of water added by the individual households and that added by other water uses in the community. If the average water use is 300 liters per person per day, the concentration is 300 mg L^{-1}.

18.10.3 Measurement of Potential Oxygen Demand of Organics in Sewage

18.10.3.1 COD

A parameter frequently used for industrial wastes, particularly where industrial wastes contribute heavily to the sewer system, is the **chemical oxygen demand (COD)**, which is a measure of the amount of oxygen required to oxidize the organic matter—and possibly some inorganic materials—in a sample. Note that the method employed to obtain this parameter, which involves reflux of a sample in a strong acid with an excess of potassium dichromate, does not specifically measure the organic content in the sample, but rather the amount of oxygen required for oxidation. This approach therefore provides a direct measure of the potential impact of oxygen consumption on the oxygen content of the waterbody.

18.10.3.2 BOD

The **biochemical oxygen demand (BOD)**—is the most commonly used parameter in the analysis of oxygen resources in water. The BOD is the amount of oxygen consumed over time, usually 5 to 20 days, as the organic matter is oxidized both microbially and chemically.

18.11 DETERMINING BOD

The laboratory method used to determine BOD has changed very little since it was initiated in the 1930s. We begin by setting up many sample bottles to contain a sample of waste, mixed in water either from the disposal site or from a standard laboratory supply. Then we use standard methods to find the initial concentration of DO, after which the bottles are incubated in a dark water bath at a given temperature, usually 20°C. Every day for five or more days, we open a few of the bottles and measure the remaining DO. The difference between the initial value and the value at each time period, that is, the **demand**, is plotted as the BOD for the series of days.

From the data obtained in the laboratory, we construct a smooth curve that lets us calculate the reaction rate coefficient K by graphical or analytical methods. The curve we construct, whose equation is

$$BOD_t = BOD_L(1 - e^{-Kt}) \qquad \text{(Eq. 18.4)}$$

becomes more and more horizontal as time progresses (Figure 18.24). By extrapolating the curve to horizontal, we can make an estimate of the ultimate value, called the **limiting value of BOD**, or **BOD_L**.

We can use the curve in this example to compute the value of K for the wastewater sample in this laboratory test. First, by extrapolating this curve to horizontal, we see that the estimated value of BOD_L would be about 7.6 mg L^{-1}. Next, by substituting the value read from the curve for day five (6.1 mg L^{-1}), so that $BOD_t = BOD_5 = 6.1$ and $t = 5$, we can find K from Equation 18.4:

$$BOD_5 = BOD_L(1 - e^{-Kt})$$
$$6.1 = 7.6(1 - e^{-5K})$$

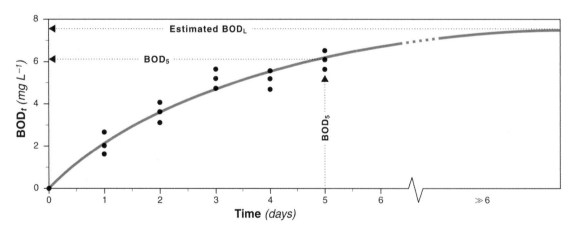

Figure 18.24 Typical biochemical oxygen demand test results illustrating the graphical method for determining BOC$_L$ and BOD$_5$ for use in solving Eq. 18.4 to find K as described in the text. Note that, in actuality, the extrapolated curve is asymptotic to $y = BOD_L$. From *Pollution Science* ©1996, Academic Press, San Diego, CA.

so that

$$1 - \frac{6.1}{7.6} = e^{-5K} \text{ or } \ln 0.197 = -5K$$

Thus

$$K = \frac{-1.62}{-5} = 0.32 \text{ per day}$$

18.11.1 Impact of BOD on Dissolved Oxygen of Receiving Waters

Typically, municipal sewage is treated to some degree to remove organics, and hence to reduce BOD, before discharge to surface waters in developed countries of the world. The amount of BOD remaining after treatment can range from 10 to 70% of the amount originally in the sewage. The impact on receiving waters depends on many environmental and waste characteristics, most of which are briefly mentioned later in this chapter. After the BOD parameters already explained, the next most important considerations are the amount of DO in the waterbody before the sewage is added, called the **initial DO (DO_i)**, and the rate at which additional oxygen is transferred from the atmosphere to the receiving water.

Many rivers and streams have depressed DO concentrations, that is, a **DO deficit**, because of wastewater added by cities upstream. The saturation value of DO is 100% with the quantification in mg/L depending upon temperature, salinity, atmospheric pressure. The local deficit—the difference between the saturation value and the observed initial value at the location of waste discharge—must be included in the computation of the downstream DO deficit caused by a new effluent discharged to the stream. Before presenting an example calculation, let's first examine how nature deals with the DO deficit.

Deficits tend to be redressed by oxygen gas derived from the atmosphere. Such replenishment occurs by a process of gas–liquid mass transfer at the surface and subsequent mixing throughout the depth of water. This overall process can be described in an approximate manner by a first order-equation, the solution of which is

$$D_t = D_i e^{-Rt} \qquad \text{(Eq. 18.5)}$$

where:

D_t is the deficit at time t

D_i is the initial deficit (*i.e.*, $DO_s - DO_t$)

R is the reaeration coefficient

Note that the larger the magnitude of R, the quicker a given deficit is removed. The value of R depends on the degree of vertical mixing in a waterbody, as well as its overall depth. It varies from 0.1 per day in small ponds to above 1 in rapidly moving streams.

Most DO problem situations require that we determine the oxygen deficit resulting from the simultaneous effects of the oxygen demand of a waste and the competing restoration of oxygen from atmospheric reaeration. The combined result is termed the **self-purification capacity** of the waterbody. The deficit shown graphically as a function of time (or distance) is known as the **oxygen sag curve** because of its characteristic spoon shape. The equation for the curve is

$$D_t = \left[\frac{K(BOD_L)}{R-K} \right] (e^{-Kt} - e^{-Rt}) + D_t e^{-} \qquad \text{(Eq. 18.6)}$$

where the terms are as defined above.

An example of the curve produced by the equation is shown in Figure 18.25. The values of K and BOD_L are taken from Figure 18.24; the value for D_i, the initial deficit, was 3 mg L^{-1} due to upstream discharges; and a reaeration coefficient R of 0.5 per day was chosen for a large, slow-flowing stream. This example might represent a case of highly treated sewage, a typical effluent plume after an initial dilution of 4, or poorly treated sewage dispersed from a diffuser that provides an initial dilution factor of 25.

One interesting result from the solution of the oxygen sag equation is that the deficit, irrespective of the saturation value of DO, is the same as long as the initial deficit is the same. Sometimes, water quality standards impose limits both on the amount of deficit *per se* and on the resulting DO value itself. For example, a regulation could require that a waste discharge must neither increase the deficit by more than 10% nor depress resulting DO concentration below 5 mg L^{-1}.

When water systems are heavily used or highly valued water uses are threatened, additional factors require consideration:

1. The diurnal demands and supplies of oxygen from photosynthesizing organisms.

2. The oxygen demand of organic materials deposited in the sediment layer.

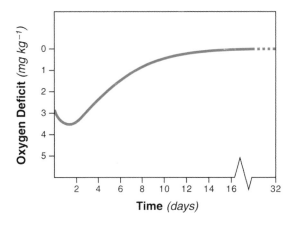

Figure 18.25 Dissolved oxygen sag curve determined using values of BOD_L and K from Figure 18.24. The equation for this curve is given in Eq. 18.6. From *Pollution Science* ©1996, Academic Press, San Diego, CA.

3. The oxygen demand of nitrogen compounds discharged in the effluent.

4. Variations in the reaeration coefficient R with travel time due to flow conditions in the waterbody.

5. The wide range of sewage flows encountered over the lifetime of a river.

Comprehensive computer programs are available to describe the concentration of oxygen in large watersheds consisting of dozens of interconnecting streams and dozens of wastewater inputs; however, data on plume travel, K values, and R values still have to be obtained with time-consuming laboratory studies and physically demanding field studies. Dilution of wastes is one of the most important factors to consider in assessing impacts. The methods used to assess effects on the dissolved oxygen resource of a waterbody and bacterial contamination are applicable to a wide variety of toxic chemical problems.

18.12 DILUTION OF EFFLUENTS

Dilution can be—but is not necessarily—an effective way to prevent pollution of surface waters. Environmental scientists do not categorically embrace the old saw, "Dilution is the solution to pollution." Instead, we recognize through the analytical process of risk assessment that certain principles underlie the utility of dilution in managing waste discharges. The first principle concerns the concentration dependence of the pollutant response mechanism. Here, we want to know if the effect of the contaminant is directly related to its concentration. That is, we ask if the concentration is reduced sufficiently, will the degree of effect be directly reduced?

Further, once the contaminant concentration is reduced, can it subsequently become more concentrated? Reconcentration is a phenomenon often associated with sediments and persistent organic chemicals such as PCBs. Dilution of waste streams containing high concentrations of suspended solids may prevent significant pollution near the discharge site, but such sediments may eventually settle out of the water column and concentrate in depressions in the streambed, where they can cause a variety of problems. Even in the ocean, waste disposal can lead to accumulation of sediments in the seabed. With persistent organic chemicals, whose solubility in water is low and whose affinity for sorption to animal tissues is high, adverse bioaccumulation can occur even from highly diluted mixtures. Thus, dilution may or may not solve a potential concentration-related problem.

18.12.1 Dilution in Streams and Rivers

Aside from concentration–toxicity considerations, social and economic considerations enter into the decision to use dilution. In addition, its use is also dependent on the avail-

ability of a sufficient quantity of dilution water. In the arid southwestern United States, for example, many streams are ephemeral; that is, they contain water only after major rainfall events. In other streams in arid climates, the effluent from municipal treatment plants is the predominant flow for more than 50% of the year; that is, they are effluent-dominated streams. In such situations, dilution is somewhere between small and nil, except during storm runoff. For example, if the stream flow is 10 million liters per day and the effluent flow is 30 million liters per day, the contaminants dissolved in the effluent will be reduced in concentration by just 25%, assuming the concentration of each contaminant upstream is zero (generally, the effective dilution will be even less than that because there is almost always some measurable concentration of contaminants upstream).

The general equation used in determining the concentration after dilution is

$$c_f = \frac{c_e \, v_e + c_a \, v_a}{v_e + v_a} \qquad \text{(Eq. 18.7)}$$

where:

c_f = cross-sectional average final concentration in the stream

c_e = concentration in the effluent

v_e = volume flux of the effluent

c_a = concentration in the ambient dilution water upstream

v_a = volume flux of the ambient dilution water

In large rivers, the amount of dilution achieved depends on the method used to discharge the effluent into the river. To maximize the dilution, it is necessary to employ a diffuser, which consists of a pipeline with many exit orifices across the width of the river. Thus, if the river flow is 120 MLD, an effluent discharge of 30 MLD yields an 80% reduction in concentration of contaminants if the ambient concentration is zero. That is, the final concentration would be one-fifth the effluent concentration. Of course, the ambient concentration is almost always greater than zero, so Equation 18.7 must be used to estimate the final concentration accurately.

In many streams, construction of diffusers may be either inappropriate or prohibited. Consequently, the amount of dilution depends on natural mixing processes that occur during stream flow. But in large streams, the effluent plume may hug the bank for many miles, so it is not actively diluted with the main flow. In cases like this, it is not possible to estimate a range of values for the dilution rate. It is reasonable to assume that without a physical structure, like a diffuser, to mix the effluent into the river, we may consider only the natural die-off process in assessing the impact of bacterial contamination on downstream water uses. Similarly, if there is no dilution of the BOD, we can count only on the decomposition of organics and reaeration to restore or maintain the DO resource of the river.

18.12.2 Dilution in Large Bodies of Water

Large bodies of water, particularly open coastal waters, offer much greater opportunity for effective dilution of waste streams. Effective initial dilution is achieved by a multiport diffuser on the end of the outfall discharge pipe. Because the density of most wastewaters is very close to that of freshwater, the discharge of effluent to deep marine waters creates a strong buoyant force. Thus, the effluent, no matter how deep the discharge, will rise to the surface of the sea if it is not trapped by density gradients below the surface. As it rises toward the surface, the effluent effectively mixes with the surrounding ambient water, resulting in more and more dilution. A good analogy is the increasing width of a smoke plume as it rises in the atmosphere. The dilution is proportional to the square of the plume width.

In the effective placement of ocean outfalls, however, depth is not the only determining factor. All other things being equal, the greater the extent of vertical travel, the greater the amount of initial dilution. But those "other things" must be truly equal. If, for example, a location chosen for its great depth has poor circulation, the net result may be less effective dilution of wastes than that offered by placement in shallower, but more open, water. Such considerations are a major concern in the placement of outfalls in fjords, bays, and, sometimes, estuaries.

Depth does not always provide the same opportunity for greater initial dilution in lakes and reservoirs because the difference in density between wastewater and receiving waters may be very small. In fact, it is not uncommon for industrial wastes to have a density greater than that of lake water, so these wastes tend to settle along the bottom rather than rise to the surface. However, because of their high temperatures, the cooling waters from large thermal-electric power stations have a density less than that of most lake waters. Thus, a deep discharge site can be advantageous for achieving effective reduction of thermal effects.

Many countries still allow the practice of dumping partially treated municipal sewage sludge into the ocean. For many years, before being banned in the United States in the 1980s, such sludge from New York City and Philadelphia was dumped in the Atlantic Ocean. Typically a portion of the sludge is particulate matter possessing a sufficiently high density to settle to the seabed, although currents and turbulence spread the material throughout the disposal zone. Other materials disposed of in the ocean, such as dredged sediments from harbors and waterways and some industrial wastes, behave similarly to sewage sludge.

18.12.2.1 Initial dilution and transport

The term **initial dilution** specifically identifies the amount of dilution achieved in a plume owing to the combined effects of the momentum and buoyancy-induced mixing of the fluid discharged from the orifice. This term is used both in regulatory practice and in plume hydrodynamics. The rate of dilution caused by these forces is quite rapid in the first few minutes after exiting the orifice, and then decreases markedly after the momentum and buoyancy are dissipated. Ambient currents also influence the rate of dilution during the buoyant rise of the plume irrespective of momentum and buoyancy. As current speed increases, so does initial dilution. In many cases, an initial dilution of 100 to 1, commonly sought in design of outfalls, is sufficient to reduce the toxicity of chemical contaminants to an acceptable level. (Note: When bacterial die-off is an important consideration, the distance from a designated use area is usually a more important factor than initial dilution *per se*. In this case, the time it takes microbes to travel a long distance increases the likelihood that they will be inactivated.)

Following initial dilution, waste streams undergo additional dilution, or dispersion, as they are transported by ambient currents and mixed with the surrounding water by turbulence. The process is analogous to the dispersion that takes place when smoke plumes dissipate in the atmosphere after the smoke has risen to an equilibrium level. In some aquatic systems, however, the effluent plume is not as easily observed.

Most modern coastal cities employ multiport ocean outfalls far offshore to protect beaches and nearshore recreational areas from the effects of bacterial contamination. These outfalls are frequently designed to maintain a diluted waste stream below the surface of the sea. Such systems are especially useful during the summer recreational season because they keep the immediate area of discharge free of unsightly messes. Moreover, they reduce landward transport of the diluted waste by onshore wind currents toward peak beach activities. The disadvantage of subsurface trapping lies in the fact that initial dilution is reduced compared with plumes rising to the surface, but this disadvantage is offset by the reduced risk of onshore transport.

18.12.2.2 Measurements and calculations

The dilution achieved in ambient transport of waste stream plumes in large water bodies can be described by physical laws—essentially the same laws used for describing aqueous flow in groundwater and gaseous flow in the atmosphere. However, we cannot solve the equations completely for the general case, because we lack existing data. That is, data obtained from field studies or from reports of previous studies are needed for empirical coefficients in the equations. However, many computer programs are available to obtain *approximate* solutions of the equations for complex cases involving multiple waste inputs and variable current speeds. In addition, we can sometimes use simplifications of the governing equations for many pollution assessment problems to obtain satisfactory estimates of contaminant concentration as a function of travel time or distance.

One simplified equation that has been used successfully over the past thirty years for large bodies of water gives us the maximum concentration at a distance X:

$$C_{\max} = C_{pi} \, erf \sqrt{\frac{Ub^2}{16\varepsilon_o X}} \qquad \text{(Eq. 18.8)}$$

where:

c_{max} = centerline (maximum) concentration at distance X

c_{pi} = plume concentration at the end of initial dilution

erf (#) = standard error function of (#)

U = current speed in the X direction

b = width in the Y direction (orthogonal to X) at the end of initial dilution

ϵ_o = constant horizontal (Y direction) eddy diffusivity

X = travel distance [note that U/X can be replaced by 1/t (time)]

In using this equation, it is important to use values that are expressed in consistent units. For example, the parameters U, b, and ϵ_o also contain a time unit. Another way to appreciate this requirement is to recognize that the argument *arg* of the error function must be dimensionless. The standard error function (erf) serves here as a mathematical representation of the way contaminants are observed to vary laterally (Y-direction) as the plume is transported in the X-direction. It describes the normal distribution curve used in evaluating variance around a mean value. Values can be found from a tabular listing in a handbook or by using the standard error function in a spreadsheet program. The **transport dilution factor** is equal to the reciprocal of the value of erf(*arg*).

In a typical problem involving finding the transport dilution factor, we might estimate the highest concentration that would occur near the beach from an outfall 8 km (*i.e.*, 8 × 10^5 cm) offshore when the onshore current speed U is 15 cm s^{-1}. The width b of the plume after initial dilution, which would be obtained from local observations, would likely be about 1,000 m (*i.e.*, 10^5 cm). We can use a commonly cited value for the eddy diffusivity of $\epsilon_o = 10^4$ cm^2 s^{-1}. Now substituting these values into Equation 18.9), we get:

$$c_{max} = c_{pi}\, erf \sqrt{1.17} = 0.87\, c_{pi} \qquad \text{(Eq. 18.9)}$$

The result is typically surprising: It shows that even with a travel distance of 8 km in a large, open body of water, the concentration is diluted to only 87%, resulting in a transport dilution factor of 1.2. (Contrast this factor of 1.2 to an initial dilution factor of about 100, which is what we expect for an ocean outfall.)

This example shows us that initial dilution is often more important in reducing harmful levels than is dilution due to transport. This is true not only for BOD and DO problems, but also for most contaminants, including metals, ammonia, and toxic organics, that are found in partially treated effluents. However, in the case of indicator bacteria, the value of transport distance is realized, as demonstrated in the following example.

Assume an outfall whose degree of treatment is only 90% effective, so that instead of the 100,000 coliforms per 100 mL used in the example on the freshwater die-off of indicator organisms, the initial count of coliforms is 100,000,000 per 100 mL. The initial dilution of 100:1 would reduce the concentration to 1% of the initial count, or 1,000,000. Then, using the reciprocal of the transport dilution factor found above (0.87), we can calculate that transport to the beach would further reduce the count to C_0 = 870,000 per 100 mL. The 8-km distance would be covered in t = 14.8 hours at a speed of 15 cm s^{-1}. If the die-off rate constant K is found to be above 0.46 per hour, near the usually expected lower range of values, the beach concentration of bacteria (C_t = 14.8), after substituting the values for C_0, K, and t into Equation 18.3, would be about 960 per 100 mL.

Even with some form of secondary treatment, some regulatory agencies require disinfection to reduce bacterial concentrations to bathing water standards. The result in the foregoing example demonstrates that chlorination of the effluent might not be necessary with 90% removal, a level typically achieved in secondary treatment.

Because ocean outfalls often provide greater travel distances, coastal communities may not have to use chlorine disinfection in the sewage treatment plant to reduce microbial contamination of human use areas of the marine environment. This is important because use of chlorine can be hazardous. Also, chlorine may combine with organic materials in the sewage to produce compounds harmful to marine organisms and to the people who consume those organisms. The distance required for indicator organisms to die off to an acceptably low level depends on current speed and direction, as well as on the bacterial concentration at the end of the rapid initial dilution processes.

18.13 DYE TRACING OF PLUMES

Frequently, an oceanographic study is necessary to measure the bacterial concentration in the drifting plume as the current carries the water toward shore. Such studies usually employ a tracer, or dye, which is added to the sewage so that the plume can be followed for several kilometers. The data obtained can then be used to calculate a die-off rate constant, which may be useful for predicting the bacterial concentrations at different distances under a variety of current conditions.

Dye tracing (Figure 18.26a) is a well-known technique commonly used in hydraulic models and prototype field settings, although in deep outfall situations, tracers can be quite costly because of the large volumetric flow rates and large dilutions usually achieved within a short time frame. The rate of dye addition Q_d to the effluent flow V_e needed to provide a dye concentration of C_d following dilution of S_a is

$$Q_d = \frac{V_e\, C_d\, \alpha_a\, S_a}{W \alpha_d} \qquad \text{(Eq. 18.10)}$$

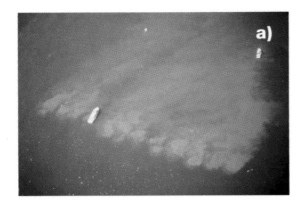

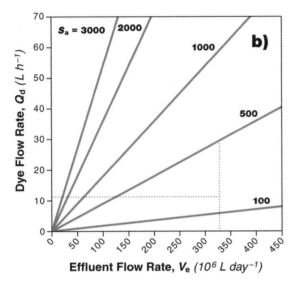

Figure 18.26 (a) A surfacing plume dyed with rhodamine WT of partially treated sewage offshore from San Francisco, California. The dye serves as a tracer for monitoring bacterial counts in drifting sewage plumes. Two monitoring vessels are visible. Photo courtesy W. Smith. b) A graph used to determine the dye required to provide $1\,\mu g\,L^{-1}$ in diluted effluent as used in the study in (a). The dotted lines illustrate using the graph to solve the problem in the text. From *Pollution Science* ©1996, Academic Press, San Diego, CA.

where:

α_a = specific gravity of the diluted plume

α_d = specific gravity of the dye solution

W = weight fraction of dye in stock solution

Figure 18.26b shows the required dye rate in liters per hour for various dilution factors, and effluent flows in million liters per day, to achieve an ambient dye concentration of $1\,\mu g\,L^{-1}$ in seawater. Rhodamine WT, typically used in dye studies, is available as a 20% solution (∞_d = 1.19) in small (57-L) drums. Fluorometers used in field sampling can easily detect this dye at concentrations of 0.5 to $1\,\mu g\,L^{-1}$.

We can use Figure 18.26b to estimate the amount of dye needed to trace an effluent flow in a waterbody or any

similar aquatic mixing question. If the flows marked on the x-axis of the graph and the slanted lines representing the dilution factors do not match exactly with the problems we have, Equation 18.10 may be used to refine the estimate.

Suppose we have an effluent of 330 million liters (86.7 MGD) per day. A regulatory permit requires the effluent to be diluted by a factor of 200 at the end of a mixing zone. Suppose we set up our sampling boat at the mixing zone boundary and hope to measure $1\,\mu g\,L^{-1}$ of dye as the plume passes under our boat. How much dye do we need to add to the effluent? Using the graph, estimate 330 along the abscissa and draw a line up to the dilution factor line $S_a = 500$. Now estimate where $S_a = 200$ would lie on that line. From that point draw a line horizontally to the ordinate and estimate the dye requirement as $12\,L\,h^{-1}$.

For a more precise estimate, we will use Equation 18.10. First convert V_e = 330 million liters per day (*i.e.*, $330 \times 10^6\,L\,day^{-1}$) to 13.7×10^6 liters per hour (L h^{-1}). For C_d, use 10^{-9} g dye per gram of seawater (this is approximately equal to $1\,\mu g\,L^{-1}$, assuming a specific gravity of 1). The specific gravity of sewage effluent diluted 100:1 with seawater is $\infty_a = 1.023$. (If this were a discharge to fresh water, the specific gravity would be 1.0.) Using $S_a = 200$, $W = 0.2$, and $\infty_d = 1.19$ for rhodamine WT, as cited above, and substituting into Equation 18.10, we obtain $Q_d = 12\,L\,h^{-1}$ (rounded from 11.8), verifying our estimate from Figure 18.26b.

18.14 SPATIAL AND TEMPORAL VARIATION OF PLUME CONCENTRATIONS

The concentrations of water quality indicators, such as bacteria, are neither uniform nor steady with respect to the space and time scales involved in regulating the concentrations at the end of the mixing zone. In general, we assume that the concentrations of constituents in the horizontal extent of a plume from an outfall diffuser are uniform. But we can make no such assumption about the vertical direction. Vertical nonuniformity is commonly encountered in design, performance analysis, and compliance monitoring, although in rivers it is not nearly the problem it is in estuaries, coastal water, and some lakes and reservoirs. Generally associated with density stratification in the receiving water, vertical nonuniformity is also associated with transport of a plume in a relatively thin lens as compared to the depth of the water column. For instance, if the plume is traveling on the surface, its constituents will be dispersed downward, and as these constituents disperse into the water column, the concentration of pollutants near the bottom edge of the plume gradually becomes less than that at the surface. (Thus if a permit condition requires that a maximum value be reported, sampling should be done at the surface, not at mid-depth.) Similarly, the dilution water mixed with the effluent being discharged is also vertically variable due to physical processes influencing the advection of ambient water into the region of the discharge. Dissolved oxygen (DO) is an example of one water-quality indicator that exhibits vertical nonuniformity in many riverine impoundments (reservoirs), lake, estuarine, and coastal situations.

Some transport and dispersion models produce estimates in terms of the **centerline concentration**, which is the maximum concentration for the cross section of the plume at a given distance downstream from the orifice. As the plume width expands with increasing distance, the maximum concentration progressively decreases. For example, the centerline (maximum) concentration at a distance of 60 meters from the diffuser may be 100 mg L^{-1}, while at 120 meters from the orifice, the maximum concentration would be closer to 70 mg L^{-1}. Other models calculate an average concentration for the cross section of the plume, and this of course also decreases downstream: the average concentration is always smaller than the maximum concentration. Both values need to be considered in field or lab verification studies, and both values may be useful for regulatory purposes.

18.15 COMPLIANCE MONITORING

Water pollution regulatory practice in the United States is founded on a system of discharge permits—known as the National Pollution Discharge Elimination System (NPDES) permits. Holders of these permits, *e.g.*, municipal sewage treatment authorities and industries, must comply with the restrictions and requirements of their particular permit, such as limits on concentrations and mass emission rates of specific constituents. They also have to meet water quality standards established for the waterbody into which they discharge their effluents. Some permits, especially for coastal water discharges, require elaborate environmental monitoring projects. The permit holder is required to conduct monitoring activities and report the results to demonstrate compliance with permit conditions. On occasion, regulatory agencies conduct studies to verify and revise ongoing programs. Monitoring data reflect the wide variations of conditions found in the natural environment, and dischargers and regulators are often challenged to rationalize monitoring results with predictions used in setting permit conditions.

18.15.1 Mixing Zones

Permit conditions of regulatory agencies usually allow exceptions to one or more of the water quality criteria within a mixing zone adjacent to the point of discharge. A **mixing zone** might be established by purely arbitrary considerations or by use of data and simulations with mathematical models. Many mixing zone determinations are made on the basis of the expected dilution rate that will be provided by efficient diffuser designs intended to optimize initial dilution. For large bodies of water, a common approach is to describe the width of the zone as the depth of water at the disposal site and the length as the length of the diffuser. For large rivers, it is common to restrict a mixing zone so it does not extend completely across the river, thereby leaving a "safe passage" that lets aquatic species avoid high concentrations of wastewater constituents. But sometimes the shape of a mixing zone is entirely arbitrary, say, a rectangular zone downstream from a discharge pipe equal to one-fourth the width of the stream and extending downstream for one kilometer. Frequently, there are two or three mixing zones for different groups of contaminants and degrees of toxicity.

18.15.2 Regulatory Use

Regulatory interest may be appropriately directly toward both discrete and average values of contaminants. For example, the state of California and the U.S. EPA specify maximum allowable instantaneous values for some parameters as well as several temporal average values (*e.g.*, 30-day and 6-month arithmetic means). In some cases, these regulations are based on knowledge of the effects on aquatic organisms. In other cases, these values are specified to acquire statistics on the performance of the wastewater treatment plant.

Criteria that are expressed in terms of temporal averages (daily to semiannual) suggest that plume concentrations be assessed extensively in three dimensions, both at the boundary of the mixing zone and, in some cases, at sensitive biological resource locations down current. Current speed and direction play significant roles when assessing the concentrations at the boundary. By incorporating data on the cyclical variation of effluent composition, density profiles, and current direction, it is possible to construct a running 6-month average (or median) for a number of points on the mixing zone boundary. The 6-month average is expected to be quite variable at these points, and the point with the highest exposure frequency may not have the highest average concentration.

Beyond the mixing zone, there may be regions where current streams of diluted effluent, each leaving the zone at a different time in different direction, would converge over a reef, a kelp forest, or a swimming area. In this case, the frequency and duration of exposure may be more important than the highest observed concentration in assessing the overall impact on these resources.

18.15.3 Verification Sampling

Aside from the question of whether discrete values or cross-sectional averages are used to test compliance with criteria, the way in which field samples are used to verify or compare with model results is an important consideration.

In laboratory or field verification studies of plume performance, the average value is measured or captured in a sample bottle only by chance. Characteristically, the field value measured is from a very small spatial region and represents a signal over a certain time span. Many samples are sought from the same cross section in order to arithmetically compute an average. In the laboratory, using hydraulic models, this is relatively easy to do. But in the field, where multiple plumes are usually involved, sampling is more complicated. We're usually trying to take samples from a moving flow field too deep below the surface to see, using a moving sampler mounted on a moving boat. It is therefore reasonable to assume some uncertainty as to what portion of the cross section the value represents.

For these reasons, field verification studies of submerged plumes in deep rivers, lakes, and coastal waters are best attempted for a cross section as far from the outfall as practical, as long as the region is still within the range where the plume is continuous. Nearer to the outfall, the values are changing more readily and the dimensions of the plume are much smaller, making it much harder to get the sampler in the right place or even in the plume. In addition, it is best to conduct the study when currents are low, so that the plume rises nearest to the surface. Placement of the sampling device may be improved because it may even be possible to see the plume (See Figure 18.34a). Aside from the ease of sampling, samples taken during low currents may be especially useful for verification of regulatory compliance.

QUESTIONS AND PROBLEMS

1. Calculate the time required for the coliform count to diminish to 1000 per 100mL (the bathing water criterion) from a sewage discharge containing 10^7 per 100 mL. The die-off coefficient, K, is expected to be 1.5 per day.

2. Suppose currents in the receiving body of water average 0.8 kilometers per hour. How far away from the bathing area must the discharge be located?

3. The example BOD curve in the chapter shows a BOD value of 6.1 mg L^{-1} at day five. It looks like the ultimate BOD (BOD_L) is about 7.6 mg L^{-1}. We solved the BOD equation using these values to find that the value of K was about 0.32 per day. Suppose the ultimate value was 15. What would be the value at day five?

4. Suppose another waste has an ultimate value of 15, but the waste is more readily degradable. Would the value at day five be lower or higher? Why?

5. Calculate the oxygen concentration in the receiving body of water at 1, 2, 4, and 8 days caused by discharge of partially treated sewage with a BOD of 30 mg L^{-1} and a decay coefficient of 0.1 per day. The reaeration coefficient is 0.3 per day. This is a pristine body of water with no initial deficit in the region of discharge. Consequently, the ambient water is essentially saturated at 10 mg L^{-1} of dissolved oxygen.

REFERENCES AND ADDITIONAL READING

Calow P. and Petts G.E. (1992) *The Rivers Handbook*, vol. 1. Blackwell Scientific Publishers, Oxford, United Kingdom.

Curriero F.C., Patz, J.A., Rose, J.B. and Lele S. (2001) The association between extreme precipitation and waterborne disease outbreaks in the United States, 1948–1994. *American Journal of Public Health.* **91**, 1194–1199.

Maybeck M. (1982) Carbon, nitrogen and phosphorus transport by world rivers. *Amer. J. Sci.* **282**, 401–450.

Rigler F.H. (1973) A dynamic view of the phosphorus cycle in lakes. In E.J. Griffith, A. Beeton, J.M. Spencer, and D.T. Mitchell (eds.), *Environmental Phosphorus Handbook*. John Wiley & Sons, New York.

Scott T.M., Rose J.B., Jenkins T.M., Farrah S.R. and Lukasik J. (2002) Microbial source tracking: Current methods and future directions. *Applied and Environmental Microbiology*. **68**, 5796–5803.

Smith J.E. and Perdek J.M. (2004) Assessment and management of watershed microbial contaminants. *Critical Reviews in Environmental Science and Technology*. **34**, 109–139.

CHAPTER 19

SOIL AND GROUNDWATER REMEDIATION

M. L. Brusseau

Field-scale demonstration project for source-zone remediation at a site in Tucson, Arizona.
Photo courtesy William Blanford.

19.1 INTRODUCTION

Public concern with polluted soil and groundwater has encouraged the development of government programs designed to control and remediate this contamination, as well as to prevent further contamination. In the United States, the first major piece of federal legislation that dealt with remediation of contaminated environments was the **Water Pollution Control Act** of 1972. This act authorized funds for the remediation of hazardous substances released into navigable waters, but did not provide for remediation of contaminated land. Subsequently, in 1976, the first legislation to directly address the problem of contaminated land was passed—the **Resource Conservation and Recovery Act (RCRA)**, which authorized the federal government to order operators to clean up hazardous waste emitted at the site of operation. This act did not, however, cover abandoned or previously contaminated hazardous waste sites.

The **Comprehensive Environmental Response, Compensation, and Liability Act (CERCLA)** of 1980—otherwise known as the **Superfund** program—explicitly addressed the remediation of hazardous waste sites. Together with a later amendment, the **Superfund Amendments and Reauthorization Act (SARA)** of 1986, CERCLA is the major federal act governing activities associated with the remediation of soil, sediment, and groundwater contamination. In addition, the Department of Defense has a separate remediation program (Installation Restoration Program) for military sites, and the Department of Energy has a program designed specifically for remediation of radioactive waste sites associated with the production of materials used for nuclear weapons and nuclear reactor cores. In addition to these federal programs, some states have passed their own versions of the Superfund program.

The institution of Superfund and other regulatory programs mentioned above has generated an entire industry focused on the characterization and remediation of hazardous waste sites. This industry is composed of regulators working for environmental agencies at all levels of government, research scientists and engineers developing and testing new technologies, scientists and engineers working for consulting firms contracted to characterize and remediate specific sites, and attorneys involved in myriad legal activities. In the sections below, basic concepts associated with site characterization and remediation will be presented.

19.2 SUPERFUND PROCESS

We will briefly discuss the major components of Superfund, given its central importance to characterization and remediation activities. Its purpose is twofold: to respond to releases of hazardous substances on land and in navigable waters and to remediate contaminated sites. The former deals with future releases, whereas the latter deals with

sites of existing contamination. There are two types of responses available within Superfund: (1) removal actions, which are responses to immediate threats, such as leaking drums, and (2) remedial actions, which involve cleanup of hazardous-waste sites. The Superfund provisions can be used either when a hazardous substance is actually released or when the possibility of such a release is substantial. They may also be used when the release of a contaminant or possibility thereof poses imminent and substantial endangerment to public health and welfare. The process by which Superfund is applied to a site is illustrated in Figure 19.1.

The first step is to place the potential site in the **Superfund Site Inventory**, which is a list of sites that are candidates for investigation. The site is then subjected to a preliminary assessment and site inspection, which may be performed by a variety of local, state, federal, or even private agencies. The results of this preliminary investigation determine whether or not the site qualifies for the **National Priorities List (NPL)**, which is a list of sites deemed by the U. S. Environmental Protection Agency (EPA) to require remedial action. (Note: Non-NPL sites may also need to be cleaned up, but their remediation is frequently handled by nonfederal agencies, with or without the help of the EPA.)

The two-component **remedial investigation/feasibility study (RI/FS)** is the next step in the process. The purpose of the RI/FS is to characterize the nature and extent of risk posed by the contamination and to evaluate potential remediation options. The investigation and feasibility study components of the RI/FS are performed concurrently, using a "phased" approach that allows feedback between the two components. A diagram of the RI/FS procedure is shown in Figure 19.2.

The selection of the specific remedial action to be used at a particular site is a very complex process. The goals of the remedial action are to protect human health and the environment, to maintain protection over time, and to maximize waste treatment (as opposed to waste containment or removal). Section 121 of CERCLA mandates a set of three categories of criteria to be used for evaluating and selecting the preferred alternative: threshold criteria, primary balancing criteria, and modifying criteria. These criteria are detailed in Information Box 19.1. The **threshold criteria** ensure that the remedy protects human health and the environment and is in compliance with applicable or relevant requirements (ARARs, such as local regulations); the **balancing criteria** ensure that trade-off factors such as cost and feasibility are considered; and the **modifying criteria** ensure that the remedy meets state and community expectations.

After a remedial action has been selected, it is designed and put into action. Some sites may require relatively simple actions, such as removal of waste-storage drums and surrounding soil. However, the sites placed on the NPL generally have complex contamination problems and are therefore much more difficult to clean up. Because of this, it may take a long time to completely clean up a site and to have the site removed from the NPL.

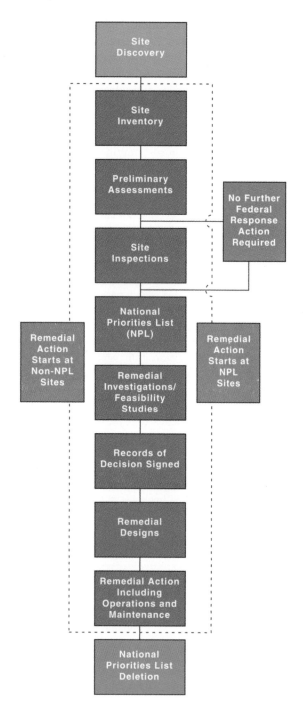

Figure 19.1 The Superfund process for evaluating and remediating a hazardous-waste site. Adapted from U.S. EPA, 1998. From *Environmental Monitoring and Characterization* © 2004, Elsevier Academic Press, San Diego, CA.

An important component of Superfund and all other cleanup programs is the issue of deciding the target level of cleanup. Associated with this issue is the question "How clean is clean?" If, for example, the goal is to lower contamination concentrations, how low does the concentration level have to be before it is "acceptable?" As can be imagined, the stricter the cleanup provision, the greater will be the attendant cleanup costs. It may require tens to hundreds of millions of dollars to return large, complex hazardous waste sites to their original conditions. In fact, it may be impossible to completely clean up many sites. However, a site need not be totally clean to be usable for some purposes. It is very important, therefore, that the technical feasibility of cleanup and the degree of potential risk posed by the contamination be weighed against the economic impact and the future use of the site. Consideration of the risk posed by the contamination and the future use of the site allows scarce resources to be allocated to those sites that pose the greatest current and future risk. The great difficulty of completely remediating contaminated sites highlights the importance of pollution prevention.

19.3 SITE CHARACTERIZATION

Site characterization is a critical component of hazardous waste site remediation. Site characterization provides information that is required for conducting risk assessments and for designing and implementing remediation systems. The primary objectives of site characterization are to identify the nature and extent of contamination. This includes identifying the types of contaminants present, the amount and location of contamination, and the phases in which it is occurring (for example, does it occur only as dissolved contamination in groundwater or is NAPL also present?).

Generally, the first step in identifying the contaminants present at the site involves inspection of material records, if available. Useful information regarding the types of contaminants potentially present at the site can be obtained from chemical-stock purchasing records, delivery records, storage records, and other types of documents. Important information may also be obtained from records of specific chemical spills or leaks, as well as waste disposal records. Unfortunately, such information is incomplete for many sites, especially for older or defunct sites.

After examining available records, site inspections are conducted to identify potential sources of contamination. This includes searching the site for drums, leaking storage tanks, and abandoned disposal pits or injection wells. Once located, actions are taken if necessary to prevent further release of contamination. For example, leaking tanks, along with the surrounding porous media, are excavated and removed.

The next step in site characterization generally involves a "survey" of groundwater contamination. This is accomplished through collecting groundwater samples from wells at and adjacent to the site. This component is often implemented in a phased approach. For example, in the initial characterization phase, when little is known about the type or extent of groundwater contaminants present, it is often necessary to analyze samples for a broad suite of possible contaminants. As noted in Chapter 10, this is typically done using the priority pollutants list associated with the National Primary Drinking Water Standards (see Table 10.2). After

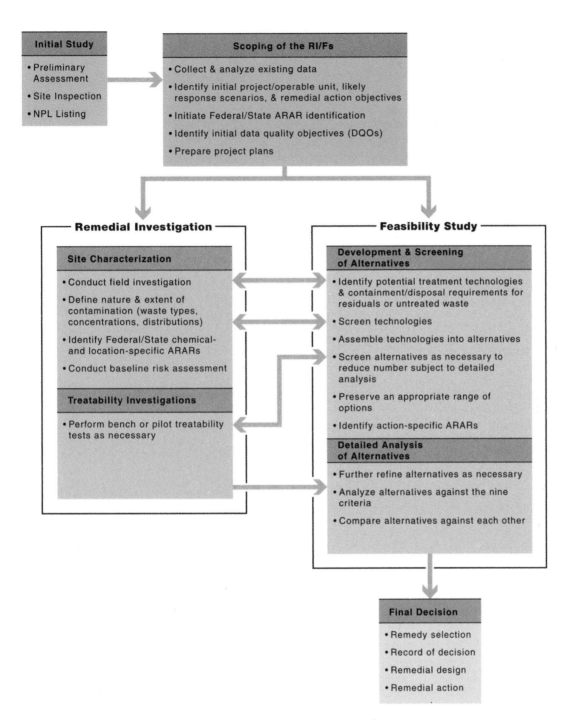

Figure 19.2 The remedial investigation/feasibility study process (RI/FS). Adapted from U.S. EPA, 1998. From *Environmental Monitoring and Characterization* © 2004, Elsevier Academic Press, San Diego, CA.

several rounds of sampling, the list of analytes is often shortened to those that are most commonly found at the site or that pose the greatest risk. In addition, the initial phase typically makes use of existing wells, and a few new wells placed to obtain a complete (but sparse) coverage of the site. Once the specific zones of contamination are identified, additional wells are often drilled in those areas to increase the density of sampling locations. Site characterization programs are generally focused on obtaining information on the areal (x-y

plane) distribution of contamination. However, once this information is available, it is important to characterize the vertical distribution of contamination if possible. Specific methods for developing and implementing a groundwater sampling program are discussed in Chapter 12.

Groundwater sampling provides information about the types and concentrations of contaminants present in groundwater. This is a major focus of characterization at most sites because groundwater contamination is often the primary risk

---

INFORMATION BOX 19.1

Criteria for Evaluating Remedial Action Alternatives

The EPA has developed criteria for evaluating remedial alternatives to ensure that all important considerations are factored into remedy selection decisions. These criteria are derived from the statutory requirements of CERCLA Section 121, particularly the long-term effectiveness and related considerations specified in Section 121 (bX1), as well as other technical and policy considerations that have proved important for selecting among remedial alternatives.

Threshold Criteria

The two most important criteria are statutory requirements that any alternative must meet before it is eligible for selection.
1. *Overall protection of human health and the environment* addresses whether or not a remedy provides adequate protection. It describes how risks posed through each exposure pathway (assuming a reasonable maximum exposure) are eliminated, reduced, or controlled through treatment, engineering controls, or institutional controls.
2. *Compliance with applicable or relevant and appropriate requirements (ARARs)* addresses whether a remedy meets all of the applicable or relevant and appropriate requirements of other federal and state environmental laws or whether a waiver can be justified.

Primary Balancing Criteria

Five primary balancing criteria are used to identify major trade-offs between remedial alternatives. These trade-offs are ultimately balanced to identify the preferred alternative and to select the final remedy.
1. *Long-term effectiveness and permanence* addresses the ability of a remedy to maintain reliable protection of human health and the environment over time once cleanup goals have been met.
2. *Reduction of toxicity, mobility, or volume through treatment* addresses the anticipated performance of the treatment technologies employed by the remedy.
3. *Short-term effectiveness* addresses the period of time needed to achieve protection. It also assesses any adverse impacts on human health and the environment that may be posed during the construction and implementation period, until cleanup goals are achieved.
4. *Implementability* addresses the technical and administrative feasibility of a remedy, including the availability of materials and services needed to implement a particular option.
5. *Cost* addresses the estimated capital and operation and maintenance costs, and net present worth costs.

Modifying Criteria

These criteria may not be considered fully until after the formal public comment period on the Proposed Plan and RI/FS report is complete, although EPA works with the state and community throughout the project.
1. *State acceptance* addresses the support agency's comments. Where states or other federal agencies are the lead agencies, EPA's acceptance of the selected remedy should be addressed under this criterion. State views on compliance with state ARARs are especially important.
2. *Community acceptance* addresses the public's general response to the alternatives described in the Proposed Plan and the RI/FS report.

---

driver–meaning that use of contaminated groundwater represents the primary route of potential exposure. However, sampling programs can also be conducted to characterize contamination distribution for other phases. For example, as discussed in Chapter 12, core material can be collected and analyzed to measure the amount of contamination associated with the porous-media grains. In addition, various methods can be used to collect gas-phase samples in the vadose zone to evaluate vapor-phase contamination.

One component of site characterization that is often critical for sites contaminated by organic chemicals is determining if immiscible liquids are present in the subsurface. As discussed in Chapter 10, many organic contaminants are liquids, such as chlorinated solvents and petroleum derivatives. Their presence in the subsurface serves as a significant source of long-term contamination and greatly complicates

remediation (see Chapter 17). In some cases, it is relatively straightforward to determine the presence of immiscible-liquid contamination at a site. For example, when large quantities of gasoline or other fuels are present, a layer of "free product" may form on top of the water table because the fuels are generally less dense than water. This floating free product can be readily observed by collecting samples from a well intersecting the contamination.

Conversely, in other cases it may be almost impossible to directly observe immiscible-liquid contamination. For example, chlorinated solvents such as trichloroethene are denser than water. Thus, large bodies of floating free product are not formed as they are for fuels (see Figure 6.1). Therefore, evidence of immiscible-liquid contamination must be obtained through the examination of core samples and not from well sampling. Unfortunately, it is very difficult to identify immiscible-liquid contamination with the use of core samples. When these denser-than-water liquids enter the subsurface, localized zones of contamination will usually form due to the presence of subsurface heterogeneities (*e.g.*, permeability variability). Within these zones, the immiscible liquid is trapped within pores and may form small pools above capillary barriers. Given the small diameter of cores (5–20 cm), the chances of a borehole intersecting a localized zone of contamination are relatively small. Thus, it may require a large, cost-prohibitive number of boreholes to characterize immiscible-liquid distribution at a site. Recognizing these constraints, alternative methods for characterizing subsurface immiscible-liquid contamination are being developed and tested. These include methods based on geophysical and tracer-test techniques, some of which provide a larger scale of measurement.

While identifying the nature and extent of contamination is the primary site-characterization objective, other objectives exist. A common objective involves determining the physical properties of the subsurface environment. For example, pumping tests are routinely conducted to determine the hydraulic conductivity distribution for the site (see Chapter 3). This information is useful for determining the location of contaminants, their rates, and their direction of movement, and for evaluating the feasibility and effectiveness of proposed remediation systems. In some cases, core samples will be analyzed to characterize chemical and biological properties and processes pertinent to contaminant transport and remediation. For example, assessing the biodegradation potential of the microbial community associated with the porous medium is critical for evaluating the feasibility of employing bioremediation at the site.

In summary, site characterization is an involved, complex process composed of many components and activities. Generally, the more information available, the better informed are the site evaluations and thus the greater chance of success for the planned remediation system. However, site characterization activities are expensive and often time consuming. Thus, different levels of site characterization may be carried out at a particular site depending on goals and available resources. Examples of three levels of site characterization are given in Information Box 19.2. Clearly, the standard approach is least costly, while the state-of-the-science approach is much more costly. However, the state-of-the-science approach provides significantly more information about the site compared to the standard approach.

The use of mathematical modeling has become an increasingly important component of site characterization and remediation activities. Mathematical models can be used to characterize site-specific contaminant transport and fate processes, to predict the potential spread of contaminants, to help conduct risk assessments, and to assist in the design and evaluation of remediation activities. As such, it is critical that models are developed and applied in such a manner to provide an accurate and site-specific representation of contaminant transport and fate. Employing advanced characterization and modeling efforts that are integrative and

---

INFORMATION BOX 19.2

Various Approaches to Site Characterization

Standard Approach

Activities

· Use existing wells; install several fully screened monitoring wells
· Sample and analyze for priority pollutants
· Construct geologic cross sections by using driller's log and cuttings
· Develop water level contour maps

Advantages

· Rapid screening of site
· Moderate costs involved
· Standardized techniques

Disadvantages

· True nature and extent of problem not identified

State-of-the-Art Approach

Activities

· Conduct standard site-characterization activities
· Install depth-specific monitoring-well clusters
· Refine geologic cross sections using cores
· Conduct pumping tests
· Characterize basic properties of porous media (grain size, organic-matter, clay content)
· Conduct solvent-extraction of core samples to evaluate sorbed- and NAPL-phase contamination
· Conduct soil-gas surveys
· Conduct limited geophysical surveys (resistivity soundings)
· Conduct limited mathematical modeling

Advantages

· Better understanding of the nature/extent of contamination
· Improved conceptual understanding of the problem

Disadvantages

· Greater costs
· Detailed understanding of the problem still limited
· Demand for specialists increased

State-of-the-Science Approach

Activities

· Employ state-of-the-art approach
· Conduct tracer tests and borehole geophysical surveys
· Characterize geochemical properties of porous media (oxides, mineralogy)
· Measure redox potential, pH, dissolved oxygen, etc., of subsurface
· Evaluate sorption-desorption and NAPL dissolution behavior using selected cores
· Assess potential for biotransformation using selected cores
· Conduct advanced mathematical modeling

Advantages

· Information set as complete as generally possible
· Enhanced conceptual understanding of the problem

Disadvantages

· Characterization cost significantly higher
· Field and laboratory techniques not yet standardized
· Demand for specialists greatly increased

iterative in nature enhances the success of remediation projects.

Numerous approaches are available for modeling transport and fate of contaminants at the field scale. These can be grouped into three general types of approaches, differentiated primarily by the level of complexity (see Information Box 19.3). The standard approach is just that, it is the standard approach used almost exclusively in the analysis of Superfund and other hazardous wastes sites. The primary advantage of this approach is the relative simplicity of the model and the relatively minimal data requirements. Unfortunately, the usefulness of such modeling is very limited. Except for the simplest of systems (*e.g.*, a nonreactive, conservative solute), this type of modeling cannot be used to characterize the contribution of specific processes or factors influencing the transport and fate of contaminants. In addition, because there is no real mechanistic basis to the model, its use for generating predictions is severely constrained.

The state-of-the-science approach is based on implementing cutting-edge understanding of transport and fate processes into fully three-dimensional models that incorporate spatial distributions of all pertinent properties. Clearly, this type of modeling, while desirable, is ultimately impractical for all but the most highly characterized research site.

The state-of-the-art approach is an intermediate-level approach that is process based, but also is developed with recognition of the data-availability limitations associated with most sites. Because models based on this approach are process based, this type of modeling can be used to characterize the contribution of specific processes or factors influencing the transport and fate of contaminants. Thus, this modeling approach can be used to effectively evaluate the impact of the key factors influencing the contamination potential of waste sites.

19.4 REMEDIATION TECHNOLOGIES

There are three major categories or types of remedial actions: (1) **containment**, where the contaminant is restricted to a specified domain to prevent further spreading; (2) **removal**, where the contaminant is transferred from an open to a controlled environment; and (3) **treatment**, where the contaminant is transformed into a nonhazardous substance. Since the inherent toxicity of a contaminant is eliminated only by treatment, this is the preferred approach of the three. Containment and removal techniques are very important, however, when it is not feasible to treat the contaminant. Although we will focus on each of the three types of remedial actions in turn, it is important to understand that remedial actions often consist of a combination of containment, removal, and treatment.

19.4.1 Containment Technologies

Containment can be accomplished by controlling the flow of the fluid that carries the contaminant or by directly immobilizing the contaminant. We will briefly discuss the use of

---

INFORMATION BOX 19.3

Various Approaches for Mathematical Modeling

Standard Approach

· "One-layer" model with areal hydraulic-conductivity distribution
· No spatial variability of chemical/biological properties
· Lumped macrodispersion term (lumps all contributions to spreading, mixing, and dilution)
· Dissolution of immiscible liquid simulated using a source term function
· Lumped retardation, with mass transfer processes treated as linear and instantaneous
· Lumped dispersion term for systems with gas-phase transport
· Lumped first-order transformation term

State-of-the-Art Approach

· "Multi-layer" model, with vertical and areal hydraulic conductivity distribution
· Layer/areal distribution of relevant chemical and biological properties
· Layer/areal distribution of immiscible liquid
· Rate-limited immiscible liquid dissolution
· Multiple-component immiscible liquid partitioning (using Raoult's Law)
· Rate-limited, nonlinear sorption/desorption
· Gas-phase transport of volatile organic contaminants (gas-phase advection, diffusion)
· Biodegradation processes coupled to mass-transfer processes
· Geochemical zone-dependent (layer/areal distributed) first-order rate coefficients

State-of-the-Science Approach

· Three-dimensional distributed hydraulic conductivity field
· Three-dimensional distribution of relevant chemical and biological properties
· Three-dimensional distribution of immiscible liquid
· Rate-limited immiscible liquid dissolution
· Immiscible liquid composition effects (multiple-component behavior)
· Rate-limited, nonlinear sorption/desorption
· Gas-phase transport of volatile organic contaminants (gas-phase advection, diffusion)
· Gas-phase retention/mass transfer (air-water mass transfer, adsorption at air-water interface, vapor-phase adsorption, immiscible liquid evaporation)
· Biodegradation coupled to mass-transfer processes
· Microbial dynamics: population growth, death, and cell transport
· Biogeochemical properties: multiple electron acceptors

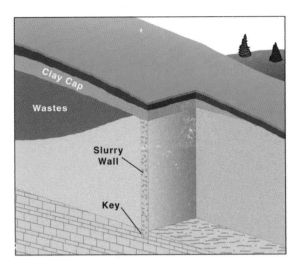

Figure 19.3 Physical containment of a groundwater contaminant plume with the use of a slurry wall. From *Environmental Monitoring and Characterization* © 2004, Elsevier Academic Press, San Diego, CA.

physical and hydraulic barriers for containing contaminated water, which are the two primary containment methods. We will also briefly mention other containment methods.

19.4.1.1 Physical barriers

The purpose of a physical barrier is to control the flow of water to prevent the spread of contamination. Usually, the barrier is installed in front (downgradient) of the contaminated zone (Figure 19.3); however, barriers can also be placed upgradient or both up- and downgradient of the contamination. Physical barriers are primarily used in unconsolidated materials such as soil or sand, but they may also be used in consolidated media such as rock if special techniques are employed. In general, physical barriers may be placed to depths of about 50 meters. The horizontal extent of the barriers can vary widely, depending on the size of the site, from tens to hundreds of meters.

One important consideration in the employment of physical barriers is the presence of a zone of low permeability beneath the site, into which the physical barrier can be seated. Without such a seating into a low permeability zone (a "key"), the contaminated water could flow underneath the barrier. Another criterion for physical barriers is the permeability of the barrier itself. Since the goal of a physical barrier is to minimize fluid flow through the target zone, the permeability of the barrier should be as low as practically possible. Another factor to consider is the potential of the contaminants to interact with the components of the barrier and degrade its performance. The properties of the barrier material should be matched to the properties of the contaminant to minimize failure of the barrier.

There are three major types of physical barriers: slurry walls, grout curtains, and sheet piling. **Slurry walls** are trenches filled with clay or mixtures of clay and soil. **Grout curtains** are hardened matrices formed by cement-like

chemicals that are injected into the ground. **Sheet piling** consists of large sheets of iron that are driven into the ground. Slurry walls are the least expensive and most widely used type of physical barrier, and they are the simplest to install. Grout curtains, which can be fairly expensive, are limited primarily to sites having consolidated subsurface environments. Sheet piles have essentially zero permeability and are generally of low reactivity. They can leak, however, because it is difficult to obtain perfect seals between individual sheets. Moreover, sheet piling is generally more expensive than slurry walls, and it is difficult to drive sheet piles into rocky ground. Thus, they are used primarily for smaller-scale applications at sites composed of unconsolidated materials.

19.4.1.2 Hydraulic barriers

The principle behind hydraulic barriers is similar to that behind physical barriers—to manipulate and control water flow. But unlike physical barriers, which are composed of solid material, hydraulic barriers are based on the manipulation of water pressures. They are generated by the pressure differentials arising from the extraction or injection of water using wells or drains. The key performance factor of this approach is its capacity to capture the **contaminant plume**, that is, to limit the spread of the contamination. Plume capture is a function of the number and placement of the wells or drains, as well as the rate of water flow through the wells or drains. Often, attempts are made to optimize the design and operation of the containment system so that plume capture is maximized while the volume of contaminated water removed is minimized.

The simplest hydraulic barrier is that established by a drain system. Such a system is constructed by installing a perforated pipe horizontally in a trench dug in the subsurface and placed to allow maximum capture of the contaminated water. Water can then be collected and removed by using gravity or active pumping. The use of drain systems is generally limited to relatively small, shallow contaminated zones.

Well-field systems are more complicated—and more versatile—than drain systems. Both extraction and injection wells can be used in a containment system, as illustrated in Figure 19.4. An extraction well removes the water entering the zone of influence of the well, creating a cone of depression. Conversely, an injection well creates a pressure ridge, or mound of water under higher pressure than the surrounding water, which prevents flow past the mound. One major advantage of using wells to control contaminant movement is that this is essentially the only containment technique that can be used for deep (>50 m) systems. In addition, wells can be used on contaminant zones of any size; the number of wells is simply increased to handle larger problems. For example, some large sites have contaminant plumes that are several kilometers long. For these and other reasons, wells are the most widely used method for containment, despite disadvantages that include the cost of long-term operation and maintenance and the need to store, treat, and dispose of the large quantities of contaminated water pumped to the surface.

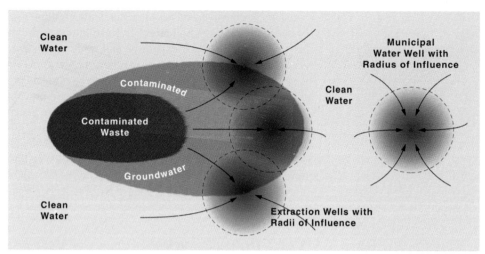

A) Overhead view of the remediation site showing groundwater surfaces, flow directions, and contaminated waste (soil material removed).

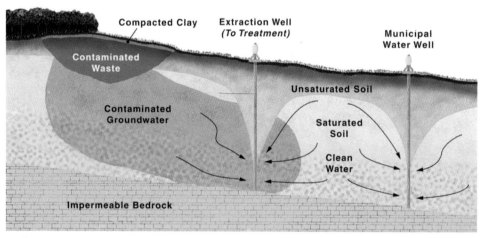

B) Cut-away view of the above site showing lateral movement of leached contaminants and groundwater.

Figure 19.4 Containment of a groundwater contaminant plume with the use of a hydraulic barrier. (A) Overhead view of the site, showing the source zone, the contaminant plume, and the direction of groundwater flow. (B) Cross-section view. Adapted from U.S. EPA, 1985. From *Environmental Monitoring and Characterization* © 2004, Elsevier Academic Press, San Diego, CA.

19.4.1.3 Other containment methods

A variety of techniques have been used to attempt to immobilize subsurface contaminants by fixing them in an impermeable, immobile solid matrix. These techniques are referred to alternatively as solidification, stabilization, encapsulation, and immobilization. They are generally based on injecting a solution containing a compound that will cause immobilization or encapsulation of the contamination. For example, cement or a polymer solution can be added, which converts the contaminated zone into a relatively impermeable mass encapsulating the contaminant.

In another approach, a reagent can be injected to alter the pH or redox conditions of the subsurface, thus causing the target contaminant to "solidify" in situ. For example,

promoting reducing conditions will induce chromium to change its predominant speciation from hexavalent, which is water soluble and thus "bioavailable," to trivalent, which has low solubility and thus precipitates on the porous-media grains (and is therefore no longer readily available).

In general, it is difficult to obtain uniform immobilization due to the natural heterogeneity of the subsurface. Thus, containment may not be completely effective. In addition, a major factor to consider for these techniques is the long-term durability of the solid matrix and the potential for leaching of contaminants from the matrix.

Vitrification is another type of containment, in which the contaminated matrix is heated to high temperatures to "melt" the porous media (*i.e.*, the silica components), which subsequently cools to form a glassy, impermeable block.

Vitrification may be applied either *in situ* or *ex situ*. For treatment of typical hazardous waste sites, potential applications would primarily be *in situ*. Aboveground applications are being investigated for use in dealing with radioactive waste. Vitrification is an energy-intensive, disruptive, and relatively expensive technology, and its use would generally be reserved for smaller-scale sites for which other methods are not viable.

19.4.2 Removal

19.4.2.1 Excavation

A very common, widely used method for removing contaminants is **excavation** of the porous media in which the contaminants reside. This technique has been used at many sites and is highly successful. There are, however, some disadvantages associated with excavation. First, excavation can expose site workers to hazardous compounds. Second, the contaminated media requires treatment and/or disposal, which can be expensive. Third, excavation is usually feasible only for relatively small, shallow areas. Excavation is most often used to remediate shallow, localized, highly contaminated source zones.

19.4.2.2 Pump-and-treat

Pump-and-treat is the most widely used remediation technique for contaminated groundwater. For this method, one or more extraction wells are used to remove contaminated water from the subsurface. Furthermore, clean water brought into the contaminated region by the flow associated with pumping removes, or "flushes," additional contamination by inducing desorption from the porous-media grains and dissolution of NAPL. The contaminated water pumped from the subsurface is directed to some type of treatment operation, which may consist of air stripping, carbon adsorption, or perhaps an aboveground biological treatment system. An illustration of a pump-and-treat system is provided in Figure 19.5. Pump-and-treat and hydraulic control using wells are essentially the same technology; they just have different objectives. For containment, we usually want to minimize water and contaminant extracted, whereas for pump-and-treat, we want to maximize contamination extracted.

Usually discussed in terms of its use for such saturated subsurface systems as aquifers, the pump-and-treat method can also be used to remove contaminants from the vadose zone. In this case, it is generally referred to as *in situ* **soil washing**. For this application, infiltration galleries, in addition to wells, can be used to introduce water to the contaminated zone.

When using water flushing for contaminant removal (as in pump-and-treat), contaminant-plume capture and the effectiveness of contaminant removal are the major performance criteria. Recent studies of operating pump-and-treat systems have shown that the technique is very successful at containing contaminant plumes and, in some cases, shrinking them. However, it appears that pump-and-treat is frequently ineffective for completely removing contaminants from the subsurface. There are many factors that can limit the effectiveness of water flushing for contaminant removal including:

1. **Presence of low-permeability zones.** When low-permeability zones (*e.g.*, silt/clay lenses) are present within a sandy subsurface, they create domains through which advective flow and transport are minimal in comparison to the surrounding sand. The groundwater flows preferentially around the clay/silt lenses, rather than through them. Thus, contaminant located within the clay/silt lenses is released to the flowing water primarily by pore-water diffusion, which is a relatively slow process. This limits the amount of contaminant present in the flushing water, thereby increasing the time required to completely remove the contamination.

2. **Rate-limited, nonlinear desorption.** Research has revealed that adsorption/desorption of many solutes by porous media can be significantly rate limited. When the rate of desorption is slow enough, the concentration of contaminant in the groundwater is lower than the concentrations obtained under conditions of rapid desorption. Thus, less contaminant is removed per volume of water, and removal by flushing will therefore take longer. In addition, many contaminants have nonlinear sorption ($n <$ 1; see Chapter 7). When this occurs, it becomes more difficult to remove contamination as the concentration decreases, because the proportion sorbed increases.

3. **Presence of immiscible liquid.** In many cases, immiscible organic liquid contaminants may be trapped in portions of the contaminated subsurface. Since it is very difficult to displace or push out this trapped contamination with water, the primary means of removal will be dissolution into water. It can take a very long time to completely dissolve immiscible liquid, thus greatly delaying removal. The immiscible liquid, therefore, serves as a long-term source of contaminant.

Because pump-and-treat is a major remedial action technique, methods are being tested to enhance its effectiveness. One way to improve the effectiveness of pump-and-treat is to contain or remove the **contaminant source zone**, that is, the area in which contaminants were disposed or spilled. If the source zone remains untreated or uncontrolled, it will serve as a continual source of contaminant requiring removal. Thus, failure to control or treat the source zone can greatly extend the time required to achieve site cleanup. It is important, therefore, that the source zone at a site be delineated and addressed in the early stages of a remedial action response. This might be done by using a physical or hydraulic barrier to confine the source zone. Other methods of treating the source zone involve enhancing the rate of contaminant removal, as detailed below.

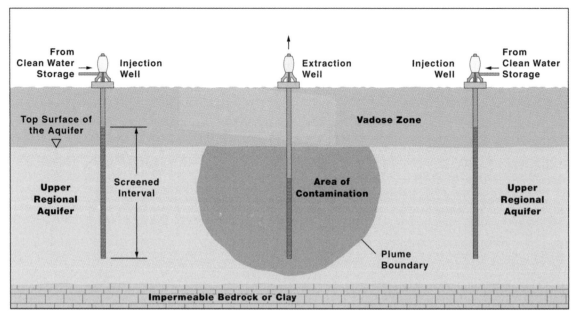

A) Before Treatment

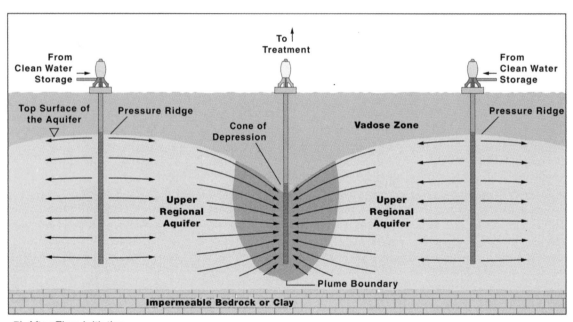

B) After Flow Initiation

Figure 19.5 Remediation of groundwater contamination by the pump-and-treat process. (A) Before treatment. (B) After start of remediation. From *Pollution Science* © 1996, Academic Press, San Diego, CA.

19.4.2.3 Enhanced flushing

Contaminant removal can be difficult because of such factors as low solubility, high degree of sorption, and the presence of immiscible-liquid phases, all of which limit the amount of contaminant that can be flushed by a given volume of water. Approaches are being developed to enhance the removal of low-solubility, high-sorption contaminants. One such approach is to inject a reagent solution into the source zone, such as a **surfactant** (*e.g.*, detergent molecule), that will promote dissolution and desorption of the contaminant, thus enhancing removal effectiveness. Such surfactants work like industrial and household detergents, which are used to remove oily residues from machinery, clothing, or dishes: individual contaminant molecules are "solubilized" inside of surfactant micelles, which are groups of individual surfactant molecules ranging from 5 to 10 nm in diameter. Alternatively, surfactant molecules can coat oil droplets and emulsify them into solution. Other enhanced-removal reagents are also being tested, such as alcohols and sugar compounds.

A key factor controlling the success of this approach in the field is the ability to deliver the reagent solution to the places that contain the contaminant. This would depend, in part, on potential interactions between the reagent and the soil (*e.g.*, sorption) and on properties of water flow in the subsurface. An important factor concerning regulatory and community acceptance of this approach is that the reagent should be of low toxicity. It would clearly be undesirable to replace one contaminant with another. Another important factor, especially with regard to the cost effectiveness of this approach, is the potential for recovery and reuse of the reagent.

19.4.2.4 *Soil vapor extraction*

The principle of **soil vapor extraction**, or **soil venting**, is very similar to that of pump-and-treat: a fluid is pumped through a contaminated domain to enhance contaminant removal. In the case of soil venting, however, the fluid is air rather than water. Because air is much less viscous than water, much less energy is required to pump air. Thus, to remove volatile contaminants from the vadose zone, it is usually cheaper and more effective to use soil venting rather than soil washing. There are two key conditions for using soil venting. First, the subsurface must contain a gas phase through which the contaminated air can travel. This condition generally limits the use of soil venting to the vadose zone. In some cases, groundwater is pumped to lower the water table, thus allowing the use of soil venting for zones that were formerly water saturated. Second, contaminants must be capable of transfer from other phases (solid, water, immiscible liquid) to the gas phase. This requirement limits soil venting to volatile and semi-volatile contaminants. Fortunately, many of the organic contaminants of greatest concern, such as chlorinated solvents (trichloroethene, tetrachloroethene) and certain components of fuels (benzene, toluene), are volatile or semi-volatile. Soil venting is the most widely used method for removing volatile contaminants from contaminated vadose-zone systems.

A blower is generally used to extract contaminated air from the subsurface (Figure 19.6). In some cases, passive or active air injection is used to increase air circulation. Passive air injection simply involves drilling boreholes through which air can then move, as opposed to the use of a blower for actively injecting air. A cap made of plastic or asphalt is often placed on the ground surface to increase venting effectiveness and to prevent water from infiltrating into the subsurface.

Once the contaminant is removed from the vadose zone, it is either released to the atmosphere or placed into a treatment system. The major performance criteria for soil venting are the effectiveness of capturing and removing the contaminant. The effectiveness of contaminant removal by soil venting can be limited by many of the same factors that limit removal by water flushing.

A closely associated technology is that of two-phase or dual-phase extraction. For this technique, both contaminated air and water are extracted simultaneously from the same

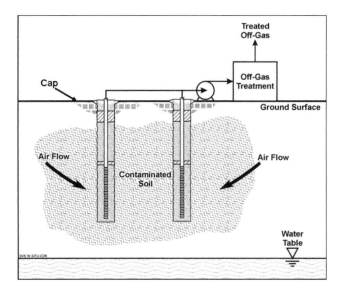

Figure 19.6 Schematic of a soil venting system for remediation of volatile organic contamination in the vadose zone. Image courtesy of U.S. Navy. http://enviro.nfesc.navy.mil/erb/restoration/technologies/remed/phys_chem/phc-26.asp.

well (Figure 19.7). This method is useful for sites at which both groundwater and the vadose zone are contaminated by volatile organic compounds. This approach would be used to target groundwater contamination in the vicinity of the vadose zone (near the water table). Deeper contamination would be targeted with conventional pump-and-treat or other appropriate methods.

19.4.2.5 *Air sparging*

In situ air sparging is a means by which to enhance the rate of mass removal from contaminated saturated-zone systems. Air sparging involves injecting air into the target contaminated zone, with the expectation that volatile and semi-volatile contaminants will undergo mass transfer (volatilization) from the groundwater to the air bubbles (Figure 19.8). Due to buoyancy, the air bubbles generally move upward toward the vadose zone, where a soil-venting system is usually employed to capture the contaminated air.

Recent laboratory and pilot-scale research has shown that the effectiveness of air sparging is often limited by a number of factors in practice. One major constraint is the impact of "channeling" on air movement during sparging. Studies have shown that air injected into water-saturated porous media often moves in discrete channels that constitute only a fraction of the entire cross section of the zone, rather than passing through the entire medium as bubbles (as proposed in theory). This channeling phenomenon greatly reduces the "stripping efficiency" of air sparging.

Another significant limitation to air sparging applications is the presence of low-permeability zones overlying the target zone. As noted above, air-sparging systems are designed to operate in tandem with a soil-venting system, so that the contaminated air can be collected and treated. The

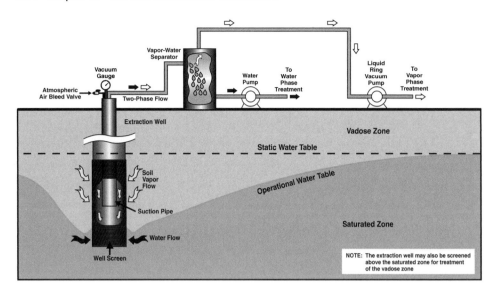

Figure 19.7 Schematic of a two-phase system for extraction of contaminated air from the vadose zone and contaminated water from the saturated zone. Image courtesy of the USEPA. (EPA 540-F-97-004 PB 97-963501 April 1997)

presence of a low-permeability zone overlying the target zone can prevent the air from passing into the vadose zone, preventing capture by the soil-venting system. In such cases, the air-sparging operation may act to spread the contaminant. While limited by such constraints, air sparging may be of potential use for specific conditions, such as for targeting localized zones of contamination.

19.4.2.6 Thermal methods

As noted previously, contaminant removal can be difficult because of such factors as low solubility, large magnitude of sorption, and the presence of immiscible-liquid phases, all of

which limit the amount of contaminant that can be flushed by a given volume of water or air. One means by which to enhance desorption, volatilization, dissolution, and evaporation, and thus improve the rate of contaminant removal is to raise the temperature in the contaminated zone. This can be done in several ways. One method is to inject hot air or water during standard soil-venting or pump-and-treat operations. This approach may increase temperatures by several degrees. Another approach involves injecting steam into the subsurface, which can enhance contaminant removal through a number of complex processes.

A third approach is based on the application of electrical heating methods (*e.g.*, directing electrical current or radio

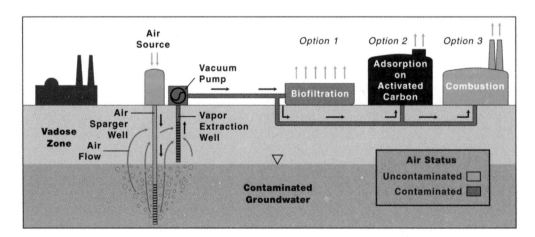

Figure 19.8 Air sparging to remove volatile contaminants from the saturated zone. The contaminated air extracted from the system can be treated above-ground using a number of different methods. Adapted from National Research Council, 1993. From *Environmental Monitoring and Characterization* © 2004, Elsevier Academic Press, San Diego, CA.

waves into the subsurface), to heat the target zone. This last approach can generate higher temperatures than the other methods (*e.g.*, 100–150°C), but is much more energy intensive and thus expensive. However, electrical heating may be particularly useful for remediating low-permeability zones, which are difficult to address with many other methods. The feasibility and effectiveness of these methods are still being tested.

19.4.2.7 Electrokinetic methods

All electrokinetic processes depend on creation of an electrical field in the subsurface. This is done by placement of a pair or series of electrodes around the area to be treated to which direct current (typically between about 50 and 150 V) is applied. Electrokinetic treatment encompasses several different processes that individually or in combination can act to enhance the transport of contaminants, depending on the particular characteristics of the contaminants present.

Electromigration, electrophoresis, and electroosmosis all enhance mobilization of the target contaminant to a location for extraction or treatment (Figure 19.9). The first two mobilization processes act on contaminants that are charged (ionic) or highly polar. Thus, these processes work for metals, radionuclides, and selected ionic or ionizable organic contaminants. Electroosmosis, the movement of water in response to an electrical gradient, can work for charged and uncharged organic contaminants, since the dissolved molecules will be carried along with the moving water. The presence of water is required for these methods to work.

One potentially promising application for electrokinetic methods is for removing contaminants from low-permeability zones. As noted above, such zones are difficult to remediate with existing methods, most of which are based on water or air flow. By placing the electrodes directly into the low-permeability zone, contaminant movement can be generated within that zone. The feasibility and effectiveness of electrokinetic techniques are still being tested.

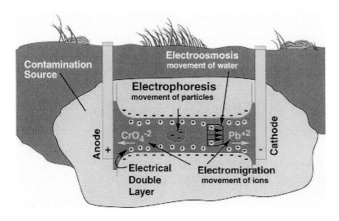

Figure 19.9 Application of the electrokinetic method of remediation. Photo courtesy of Sandia National Laboratories. (http://www.sandia.gov/subsurface/factshts/ert/ek.pdf)

19.4.3 In situ Treatment

In situ treatment technologies are methods that allow in-place cleanup of contaminated field sites. There is great interest in these technologies because they are in some cases cheaper than other methods. Second, *in situ* treatment can, in some cases, eliminate the risk associated with the contaminant by promoting destruction of the contaminant via transformation reactions. The two major types of *in situ* treatment are biological (*in situ* bioremediation) and chemical. The majority of these methods are based on injecting a reagent of some type into the subsurface to promote a transformation reaction (Figure 19.10).

19.4.3.1 Bioremediation

Bioremediation makes use of the activity of naturally occurring microorganisms to clean up contaminated sites–specifically their ability to biodegrade contaminants (see Chapter 8). There are two types of bioremediation: **in situ bioremediation** is the in-place treatment of a contaminated site, and **ex situ bioremediation** is the aboveground treatment of contaminated soil or water that is removed from a contaminated site. Biological methods have been used for many years for aboveground treatment of municipal sewage waste (see Chapter 26) and industrial waste (see Chapter 25). This section will focus on *in situ* bioremediation.

In situ bioremediation is used primarily for organic contaminants. It has become a proven, dependable remediation method for contaminants such as fuels. It also may be successful for other types of contaminants, depending upon specific site conditions. For example, biodegradation is clearly dependent on the structure of the contaminant (see Chapter 8). The status of *in situ* bioremediation feasibility for a range of organic contaminants is presented in Table 19.1.

Bioremediation, while based on a simple concept, incorporates a complex system of biological processes. Thus, it is sometimes difficult to accurately predict the performance of an *in situ* bioremediation system, which complicates planning and design efforts. Also, *in situ* bioremediation is a relatively slow process, especially compared to some of the *in situ* chemical treatment methods discussed in the next section.

Several key factors are critical to successful application of *in situ* bioremediation: environmental conditions, contaminant and nutrient availability, and the presence of microorganisms capable of degrading the contaminant. Thus, when planning a bioremediation effort, it is important to characterize the site—and to identify possible limiting factors. Soil or water samples collected from the site can be tested to determine the presence of microorganisms capable of degrading the contaminant. Monitoring of groundwater at the site can be done to characterize environmental factors that affect biodegradation, such as oxygen concentrations, nutrient concentrations, temperature, and pH.

Not all potentially limiting factors are easy to identify. For example, contaminants are often present as mixtures, and one component of the mixture may inhibit the activity of the

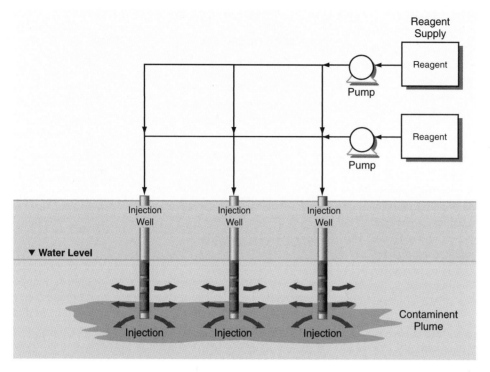

Figure 19.10 *In situ* remediation of a contaminated saturated zone using biological or chemical methods. Image courtesy of the U.S. Navy.

microorganisms. Similarly, low bioavailability, which is another factor that can limit bioremediation, can be very difficult to evaluate in the environment. To be biodegraded, the contaminant must be available to the microorganisms. Contaminants that are present as NAPL, or that are sorbed, are generally not directly available to microbial cells. The rate and magnitude of biodegradation of such contaminants is therefore reduced.

One of the most common limiting factors for *in situ* bioremediation in the saturated zone is the availability of oxygen. Oxygen is required for aerobic biodegradation. However, oxygen has limited solubility in water, and once it is used up, it must be replenished via aqueous diffusion (which is slow). The combination of these factors is the reason that oxygen often limits the use of bioremediation for groundwater systems. Several methods have been developed to supply oxygen to the saturated zone. The simplest method is to inject air directly into groundwater, which is done similarly as the air sparging method discussed above. Other methods involve injecting compounds into the subsurface that release oxygen.

Nutrients, nitrogen and phosphorus in particular, are common additives used to enhance the rate of biodegradation in bioremediation applications. Many contaminated sites contain organic wastes that are rich in carbon but poor in nitrogen and phosphorus. Nutrient solutions are usually injected from an aboveground tank. The goal of nutrient injection is to optimize the carbon/nitrogen/phosphorus ratio (C:N:P) in the subsurface to approximately 100:10:1 (see Chapter 8).

While most of the past *in situ* bioremediation applications were based on aerobic biodegradation, anaerobic-based processes have recently become of interest. There are two potential reasons for considering anaerobic-based bioremediation. First, it is sometimes difficult to establish and maintain aerobic conditions in saturated subsurface systems. Second, some organic contaminants are resistant to aerobic biodegradation. Several alternative electron acceptors have been proposed for use in anaerobic degradation, including nitrate, sulfate, and iron (Fe^{+3}) ions, as well as carbon dioxide. In some cases, labile organic compounds (such as molasses) are added to the subsurface to promote anaerobic conditions—microorganisms aerobically biodegrade the added carbon supply, thereby using available oxygen. Another approach under investigation is the use of sequential anaerobic/aerobic biodegradation processes, wherein biodegradation is first enhanced under anaerobic conditions, followed by aerobic-based bioremediation. This approach may be useful for mixed-waste sites.

Cometabolic biodegradation, in which microorganisms use one compound for energy (its food source), but in the process also biodegrade other compounds (contaminants) present, is another variation of bioremediation (see also Chapter 8). This approach is of particular interest for the bioremediation of chlorinated solvents. For example, methanotrophic organisms produce the enzyme methane monooxygenase to degrade methane, and this enzyme also happens to cometabolically degrade several chlorinated solvents, such as trichloroethene. Thus, methane could be injected into the subsurface to promote biodegradation of the trichloroethene if the appropriate microorganisms were present at the site.

TABLE 19.1 Current feasibility of bioremediation.

CHEMICAL CLASS	FREQUENCY OF OCCURRENCE	STATUS OF BIOREMEDIATION	EVIDENCE OF FUTURE SUCCESS	LIMITATIONS
Hydrocarbons and derivatives				
Gasoline, fuel oil	Very frequent	Established		Forms nonaqueous-phase liquid
Polycyclic aromatic hydrocarbons	Common	Emerging	Aerobically biodegradable under a narrow range of conditions	Sorbs strongly to subsurface solids
Creosote	Infrequent	Emerging	Readily biodegradable under aerobic conditions	Sorbs strongly to subsurface solids; forms nonaqueous-phase liquid
Alcohols, ketones, esters	Common	Established		
Ethers	Common	Emerging	Biodegradable under a narrow range of conditions using aerobic or nitrate-reducing microbes	
Halogenated aliphatics				
Highly chlorinated	Very frequent	Emerging	Cometabolized by anaerobic microbes; cometabolized by aerobes in special cases	Forms nonaqueous-phase liquid
Less chlorinated	Very frequent	Emerging	Aerobically biodegradable under a narrow range of conditions; cometabolized by anaerobic microbes	Forms nonaqueous-phase liquid
Halogenated aromatics				
Highly chlorinated	Common	Emerging	Aerobically biodegradable under a narrow range of conditions; cometabolized by anaerobic microbes	Sorbs strongly to subsurface solids; forms nonaqueous phase—solid or liquid
Less chlorinated	Common	Emerging	Readily biodegradable under aerobic conditions	Forms nonaqueous phase—solid or liquid
Polychlorinated biphenyls				
Highly chlorinated	Infrequent	Emerging	Cometabolized by anaerobic microbes	Sorbs strongly to subsurface solids
Less chlorinated	Infrequent	Emerging	Aerobically biodegradable under a narrow range of conditions	Sorbs strongly to subsurface solids
Nitroaromatics	Common	Emerging	Aerobically biodegradable; converted to innocuous volatile organic acids under anaerobic conditions	
Metals				
Cr, Cu, Ni, Pb, Hg, Cd, Zn, etc.	Common	Possible	Solubility and reactivity can be changed by a variety of microbial processes	Availability highly variable—controlled by solution and solid chemistry

From National Research Council. *In Situ Bioremediation*. National Academy Press, Washington, D.C., 1993.

If microorganisms capable of biodegrading the target contaminants are not present in the subsurface at the site, specific microorganisms can be injected into the subsurface. This process is known as **bioaugmentation**. This process is often difficult to accomplish successfully. First, the introduced microbes often cannot establish a niche in the environment and thus may not survive. Second, microorganisms can be strongly sorbed by solid surfaces and trapped in small pores; so there are difficulties in delivering the introduced organisms to the site of contamination (see also Chapter 29).

Bioventing is a technique used to promote biodegradation in the vadose zone. Bioventing is a combination of soil venting technology and bioremediation. Air flow through the contaminated zone is initiated with a blower or vacuum system, which increases the supply of oxygen throughout the zone and hence the rate of contaminant biodegradation. The rate of air flow is significantly lower for bioventing applications compared to soil venting. In the case of volatile contaminants, remediation will result from a combination of biodegradation and removal in the extracted air (*i.e.*, soil venting).

19.4.3.2 In situ chemical treatment

In situ chemical remediation is a process in which the contaminant is degraded by promoting an abiotic transformation reaction, such as hydrolysis, oxidation, or reduction, within the subsurface. Although this approach has been used much less frequently than *in situ* bioremediation, it has begun to receive increasing attention in the past few years. It can be accomplished by using active methods, such as injecting a reagent into the contaminated zone of the subsurface, or by using passive methods, such as placing a permeable treatment barrier downgradient of the contamination. An advantage of these methods is that they promote *in situ* destruction of the contamination, thus reducing or eliminating the associated hazard potential.

As an alternative to *in situ* bioremediation, *in situ* chemical oxidation (ISCO) has recently become a popular focus of investigation. The objective of this method is to inject into the contaminated zone an oxidizing compound that will react with organic contaminants, oxidizing them to carbon dioxide and other oxidant-specific byproducts. Three oxidizing reagents in particular, permanganate (MnO_4), Fenton's reagent (Fe(II) and H_2O_2), and ozone (O_3), are being tested. Both ozone and permanganate have been used for decades as oxidants for aboveground disinfection and purification of drinking water. As noted previously, chlorinated compounds are in general relatively resistant to biodegradation, which complicates or limits the use of bioremediation methods. There is thus great interest in using *in situ* chemical oxidation to remediate sites contaminated by chlorinated compounds. See the Case Study for more detailed information on the use of ISCO.

All of the oxidizing reagents used for treatment of organic subsurface contamination are relatively nonselective and will thus react with most organics present in the subsurface. Therefore, soils and groundwater containing high natural organic carbon concentrations can exhibit a high oxidant demand, competing with the demand from the actual organic contaminant. Given that this approach is based on injection (flow) of the oxidant solution, the presence of low permeability zones can reduce the overall effectiveness of *in situ* oxidation techniques. Treatment depends on the chemical oxidant being in contact with the contamination; significant heterogeneity will cause preferential channeling of the oxidant solution to the high conductivity zones, with poor transport to the low conductivity zones, thus limiting contact with the contamination.

In situ treatment walls or barriers are of particular interest for chlorinated solvents such as trichloroethene. This approach generally involves digging a trench downgradient of the contaminant plume and filling that trench with a wall of permeable, reactive material that can degrade the contaminant to nontoxic byproducts (Figure 19.11). The wall is permeable so that the water from which the contaminant has been removed can pass through and continue to flow downgradient. For example, iron filings can degrade compounds such as trichloroethene via reduction reactions while permitting water

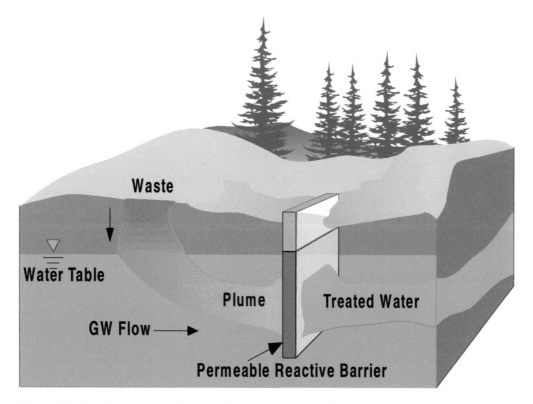

Figure 19.11 *In situ* treatment of a groundwater contaminant plume using a permeable reactive wall. Photo courtesy of US EPA. (EPA/600/R-98/125 September 1998; http://www.clu-in.org/download/rtdf/prb/reactbar.pdf)

to pass through unimpeded. One potential promising application of treatment walls is to use them to prevent the downgradient spread of contamination. For example, the barrier can be emplaced at the boundary of a site property, on the downgradient side of the contaminant zone, with the objective being that the barrier would prevent movement of contamination off the site property. *In situ* treatment walls are used to control spread of a contaminant plume, whereas *in situ* chemical treatment based on injecting a reagent (such as ISCO) is generally targeted to remediate source zones.

19.4.4 Other Technologies: Monitored Natural Attenuation

Monitored natural attenuation (MNA) has recently attracted great interest as a low-cost, low-tech approach for site remediation. Monitored natural attenuation is based upon natural transformation and retention processes reducing contaminant concentrations *in situ*, thereby containing and shrinking groundwater contaminant plumes. The application of MNA requires strict monitoring to ensure that it is working; thus the name "monitored" natural attenuation. Extensive information has been reported concerning the role of natural attenuation processes for remediation of contaminated sites. This information has recently been summarized in a report released by the NRC (2000). Two of the major conclusions reported in this document are that: (1) MNA has the potential to be used successfully at many sites, and (2) MNA should be accepted as a formal remedy only when the attenuation processes are documented to be working and that they are sustainable.

Several processes, such as biodegradation, hydrolysis, sorption, and dilution, may contribute to attenuation of the contaminant plume. However, biodegradation is generally the predominant process involved and usually the key factor for successful use of MNA. MNA has proven to be a success for sites contaminated by petroleum hydrocarbons (*e.g.*, gasoline). Conversely, its use for chlorinated-solvent contaminated sites is more difficult because of the greater resistance of chlorinated compounds to biodegradation (see Chapter 8).

There are two major questions to address when evaluating the feasibility and viability of applying MNA to a particular site. These are: (1) Are natural attenuation processes occurring at the site; and (2) Is the magnitude and rate of natural attenuation sufficient to accomplish the remediation goal (*e.g.*, plume containment, plume reduction), and be protective of human health and the environment? Successful implementation of MNA requires an accurate assessment of these two questions.

As noted above, the first step in evaluating the feasibility and viability of applying MNA to a particular site is to determine whether or not natural attenuation processes are occurring at the site. Evaluating and documenting the potential occurrence of natural attenuation involves the following steps:

1. Developing a general conceptual model of the site that is used to provide a framework for evaluating the predominant transport and transformation processes at the site.

2. Evaluating contaminant and geochemical data for the presence of known "footprints" indicative of active transformation processes. This includes analyzing contaminant concentration histories, the temporal and spatial variability of transformation products and reactants (*e.g.*, O_2, CO_2, and so on), and isotope ratios for relevant products and reactants.

3. Screening groundwater and soil samples for microbial populations capable of degrading the target contaminants.

The second question to address for MNA application is the magnitude and rate of natural attenuation at the site. Characterizing the magnitudes and rates of attenuation at the field-scale is a complex task. For example, the initial mass of contaminant released into the subsurface is not known at most sites. This means it is difficult to accurately quantify the magnitude and rate of attenuation. Thus we can generally obtain estimates of such rates at best. One method is based on characterizing temporal changes in contaminant concentration profiles, electron acceptor concentrations (such as O_2, NO_3^-, SO_4^{2-}, Fe^{3+}), or contaminant transformation products. However, the complexity of typical field sites makes it difficult to accurately determine attenuation rates from temporal changes in these parameters. Another approach often used to characterize potential attenuation processes is to conduct bench-scale studies in the laboratory using core samples collected from the field. However, results obtained from laboratory tests may not accurately represent field-scale behavior. *In situ* microcosms, in which tests are conducted in a small volume of the aquifer, can be used to minimize this problem. Given the complexity of most field sites, including spatial variability of physical and chemical properties and of microbial populations, a prohibitive number of sampling points may often be required to fully characterize a field site using either of these methods. Tracer tests have recently been proposed as an alternative method for field-scale characterization of the *in situ* attenuation potential associated with subsurface environments.

The rates of natural attenuation processes are not sufficient to maintain plume containment at many sites that have large source zones. The flux of contamination emanating from the source zone is greater than the attenuation capacity, so the plume continues to grow. In such cases, there is great interest in combing MNA with an aggressive source-zone remediation action (such as ISCO). The idea is that the source-zone remedial action will remove sufficient contamination to reduce the contaminant flux to levels that are below the attenuation capacity of the system. One possible concern for this approach would be that the source-zone remedial action does not impede the attenuation processes of the system (particularly that it does not harm the microbial community responsible for biodegradation).

19.4.5 Other Technologies: Phytoremediation

A relatively new method being investigated for the remediation of hazardous waste sites is the application of plants, a

USE OF ISCO REMEDIATION AT A SUPERFUND SITE

Pilot-scale tests were performed as part of a collaborative project between academia, industry, consulting firms, and government agencies to assess the efficacy of potassium permanganate for *in situ* chemical oxidation of trichloroethene (TCE) contamination at a Superfund site in Tucson, Arizona. From the early 1950s and 1960s to approximately 1978, industrial solvents including TCE, trichloroethane (TCA), and methylene chloride; machine lubricants and coolants; paint thinners and sludges; and other chemicals were disposed of in open unlined pits and channels at the site. In addition, wastewater containing chromium from plating operations flowed through a drainage ditch that crosses part of the site. A 1981–1982 investigation determined that TCE had moved down through overlying sediments into the regional aquifer. In 1983, the site was added to the national priorities list as a federal Superfund site.

A groundwater pump-and-treat remediation system was started in 1987 and a soil-venting system was initiated in 1998. The pump-and-treat system has successfully controlled and reduced the extent of contamination, while the soil-venting system has removed large quantities of contaminant mass from the vadose zone. Unfortunately, the cleanup target has not yet been attained. The results of advanced characterization studies indicated that contaminant removal is constrained by site heterogeneities, physical and chemical mass transfer limitations, and the presence of immiscible-liquid contamination within the source zones. Remediation of the site is constrained by the difficulty of removing contamination from the source zones.

In situ chemical oxidation (ISCO) is a technique that shows promise for directly treating source-zone contaminants. The simplified proposed reaction mechanism for potassium permanganate ($KMnO_4$) with chlorinated compounds is as follows:

$$C_2Cl_3H \text{ (TCE)} + 2\, MnO_4^- \text{ (Permanganate)}$$
$$\rightarrow 2\, CO_2 + 2\, MnO_2 \text{ (s)} + 3\, Cl^- + H^+$$

with 0.81 kilogram (kg) of chloride (Cl^-) produced and 2.38 kg manganese oxide (MnO_2) reduced per kg of TCE oxidized. While this method promotes destruction of TCE and other contaminants, there are some possible negative impacts. Production of manganese oxide (MnO_2) precipitate has the potential to clog the aquifer, and production of hydrochloric acid (HCl) has the potential to dissolve aquifer material, especially carbonates. Additionally, interaction with co-contaminants is possible. For example, chromium is a fairly common contaminant in the industrial waste sites for which ISCO is a potential remediation strategy. When oxidized, chromium converts to the hexavalent form, which is more toxic as well as more mobile, and is therefore a health concern. *In situ* oxidation for remediation of chlorinated compounds temporarily creates an oxidizing environment in the subsurface. Thus, the potential for the mobilization of chromium and other metals, and the subsequent potential for reduction and stabilization (natural attenuation), is an important consideration for this technology.

ISCO pilot tests with potassium permanganate were conducted at two source zones at the site. These tests were designed to affect a relatively small area of the subsurface and provide useful information for the design of a full-scale application. The ISCO pilot-scale test was performed with the collaboration of Raytheon Systems Environmental

Division, the University of Arizona, Errol L. Montgomery and Associates, Inc., and IT Corporation.

Groundwater samples from all wells in which permanganate was observed showed a decrease in TCE concentrations coincident with the arrival of the potassium permanganate in concentrations above approximately 50 mg/L. During the operational period, when permanganate solution was active in the treatment area, the average TCE concentration reduction was 84% for Site 3 and 64% for Site 2. As permanganate left the treatment area, TCE concentrations in all wells increased to some extent. Possible causes of rebounding include movement of contaminated groundwater into the treatment zone from upgradient, and mass transfer of untreated TCE from the aquifer material by desorption from aquifer material, diffusion from areas of lower permeability, and dissolution from non-aqueous phase liquid (NAPL) contamination. Rebounding to an average of 98% of pre-test concentrations was observed in water-flushing studies conducted previously at the site. However, in the potassium permanganate study, rebound concentrations averaged only 50% of the pre-test concentrations. The significant difference in rebound likely reflects a decrease in mass of TCE associated with nonaqueous phases within the source zones due to permanganate oxidation. These results indicate that *in situ* oxidation using potassium permanganate was successful in removing TCE associated with the two source zones. On this basis, a full-scale application of permanganate-based ISCO is being implemented at the site.

The presence of chromium as a co-contaminant at the site was an important aspect of this pilot project. Chromium concentrations increased as expected. However, concentrations quickly returned to low levels. Further attenuation is expected with time based on the history of the site and review of other ISCO projects. The observed increase in soluble chromium was anticipated and may be due more to the presence of chromium as an impurity in the injection solution than the mobilization of chromium associated with the aquifer solids.

Potassium permanganate is fairly inexpensive; for example, the cost for the industrial grade potassium permanganate purchased for the field project was approximately $4.40/kg. Since extraction of groundwater is not required, disposal costs are minimized. Mobilization, demobilization, and costs associated with the delivery system are the primary costs for ISCO treatment. Overall, ISCO is a relatively inexpensive technology compared to many other methods. Since oxidation occurs rapidly with adequate delivery, the time frame of ISCO projects may be significantly shorter in comparison to other methods such as pump-and-treat and bioremediation.

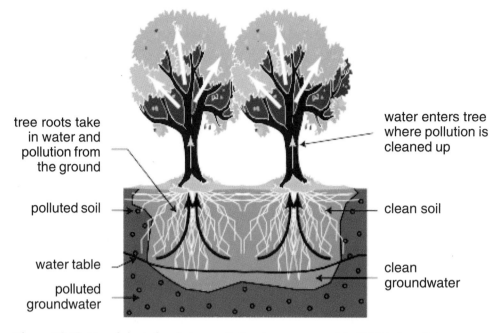

tree roots take in water and pollution from the ground

water enters tree where pollution is cleaned up

polluted soil

clean soil

water table

clean groundwater

polluted groundwater

Figure 19.12 Use of plants for phytoremediation. Image courtesy of the US EPA. (http://clu-in.org/download/citizens/citphyto.pdf; EPA 542-F-01-002 April 2001)

process called **phytoremediation**. Phytoremediation can be used to treat both organic and inorganic contaminants (Figure 19.12). Plants can interact with contaminants in several ways (see Table 19.2).

Accumulation is the uptake of contaminants into plant tissues. For example, some plants, referred to as **hyperaccumulators**, are extremely efficient at metal accumulation. The advantage of this approach is that contaminants are actually removed from the site by harvesting the plants after the contaminants are accumulated. A potential disadvantage of this approach is that wildlife may be exposed to the contaminants if they forage at the site before the plants are harvested (see also Chapter 29).

Stabilization is a second phytoremediation approach and involves stabilization of contaminants in the root zone through complexation with plant root exudates and the plant root surface. This approach is attractive from two perspectives. First, contaminants are not accumulated in above-ground plant tissues, thus avoiding any potential risk of wildlife exposure. Second, this approach does not require plant harvesting (see also Chapter 29).

A third approach is the use of plants as a form of a hydraulic barrier. For example, poplar trees have been used to remove organic contaminants such as trichloroethene from groundwater. Poplar trees are a type of plant whose root system can extend to shallow water tables. Thus, they

TABLE 19.2 Types of phytoremediation processes.

TREATMENT METHOD	MECHANISM	MEDIA
Rhizofiltration	Uptake of metals in plant roots	Surface water and water pumped through troughs
Phytotransformation	Plant uptake and degradation of organics	Surface water, groundwater
Plant-assisted bioremediation	Enhanced microbial degradation in the rhizosphere	Soils, groundwater within the rhizosphere
Phytoextraction	Uptake and concentration of metals via direct uptake into plant tissue with subsequent removal of the plants	Soils
Phytostabilization	Root exudates cause metals to precipitate and become less bioavailable	Soils, groundwater, mine tailings
Phytovolatilization	Plant evapotranspirates selenium, mercury, and volatile organics	Soils, groundwater
Removal of organics from the air	Leaves take up volatile organics	Air
Vegetative caps	Rainwater is evapotranspirated by plants to prevent leaching contaminants from disposal sites	Soils

From J. Chappell, Phytoremediation of TCE using Populus. Status report for the U.S. EPA Technology Innovation Office, 1997.

can draw water directly from the saturated zone, thereby causing a lowering of the water table. In this regard, they can be used to create a hydraulic barrier to contain a groundwater contaminant plume. The contaminated water is drawn up into the trees and expelled via evapotranspiration. It is also possible that some contaminants such as trichloroethene may undergo transformation reactions induced by the tree, either within the plant tissues or within the rhizosphere.

Phytoremediation is becoming a popular approach, partly because it is relatively inexpensive. However, its use is limited to relatively shallow applications (*i.e.*, the depth of the root system). In addition, it is relatively slow in comparison to some other methods discussed above.

QUESTIONS AND PROBLEMS

1. What are the essential components of a typical hazardous waste site characterization program? Why are these the essential components?

2. How do so-called LNAPLs and DNAPLs behave when spilled into the ground? How do these behaviors influence our ability to find them during site characterization projects?

3. Compare and contrast the use of physical barriers versus well fields to contain a groundwater contaminant plume.

4. What is a major limitation of the pump-and-treat method? What are some of the methods that can be used to enhance the performance of pump-and-treat?

5. For which of the following contaminants would you propose to use air sparging as a possible remediation method, and why/why not: trichloroethene, phenol, nitrate, benzene, pyrene, cadmium?

6. What are some of the potential benefits of using *in situ* bioremediation? What are some of the major potential limitations to its use?

7. How is potassium permanganate injection used for remediation?

8. What are the criteria used to evaluate proposed remedial actions?

9. What are the major advantages and disadvantages of using excavation?

REFERENCES AND ADDITIONAL READING

Boulding J.R. (1995) *Practical Handbook of Soil, Vadose Zone, and Ground-Water Contamination*. Lewis Publishers, Boca Raton, Florida.

Natural Research Council. (1994) *Alternatives for Groundwater Cleanup*. National Academy Press, Washington, D.C.

Natural Research Council. *Groundwater and Soil Cleanup*. (1999) National Academy Press, Washington, D.C.

Natural Research Council. *In Situ Bioremediation, When Does It Work?* (1993) National Academy Press, Washington, DC.

CHAPTER 20

ECOSYSTEM RESTORATION AND LAND RECLAMATION

E. P. Glenn, W. J. Waugh, and I. L. Pepper

The Cienega de Santa Clara, a human-made wetland in the delta of the Colorado River where it enters the Gulf of California in Mexico. This wetland is formed by the discharge of agricultural wastewater onto salt flats (visible in background) in the intertidal zone. It supports tens of thousands of resident and migratory water birds and endangered fish and bird species. *Photo courtesy E.P. Glenn.*

20.1 INTRODUCTION

Ecosystem restoration can be defined as the process of manipulating a disturbed ecosystem to achieve compositional, structural, and functional patterns similar to the predisturbed condition. If possible, this would include soil, vegetation, and wildlife. However, depending on the severity of the disturbance, it may not be possible to achieve predisturbance conditions at all levels. For example, the deposition of mine tailings onto vast areas of soil, to a depth of 35 meters, means that the soil surface has been totally altered. In this case, **land reclamation** may be the best alternative. This involves the process of improving disturbed land to achieve a land capability equivalent of the predisturbed condition. For some disturbed ecosystems, even sustainable revegetation of any kind is a major accomplishment.

Humans have had the capacity to greatly disturb many of the world's natural ecosystems. As such, this is pollution on a vast scale. Activities that result in degraded lands that become candidates for restoration include deforestation, overgrazing, secondary salinization from poor irrigation management, wetland clearing and draining, oil production, mining, and toxic spills. In fragile ecosystems such as the desert and semi-desert regions of the world, overuse of land can lead to the irreversible loss of fertile top soil, vegetation, and nutrient cycling, a process called **desertification.** According to the United Nations Atlas of Desertification (Middleton and Thomas, 1997), over half of the world's arid and semi-arid lands have been affected by desertification. In wet regions of the world, deforestation and other unsustainable land use practices have left large tracts of land with unusable, unfertile soils prone to water and wind erosion. In any ecosystem, there is a threshold for self-repair, but once that threshold has been crossed, severe degradation occurs. It has become apparent that land and, in particular, soil are finite commodities, and we need to preserve and make the most of what we have. Today, for example, nearly all mining activi-

---

INFORMATION BOX 20.1

Terminology Used in Ecosystem Restoration

Rehabilitation:
Repairing some or most of the damage done to land so that it can serve some productive function. For example, salinized farmland, unable to support native plants, can be planted with salt-tolerant plants (**halophytes**) to prevent erosion and provide wildlife habitat.

Revegetation:
Planting or seeding an area that has received minor damage.

Reclamation:
Restoring biotic function and productivity to the most severely degraded land, such as an EPA Superfund site.

From *Environmental Monitoring and Characterization* © 2004, Elsevier Academic Press, San Diego, CA.

INFORMATION BOX 20.2

Flow Chart of the Restoration Process

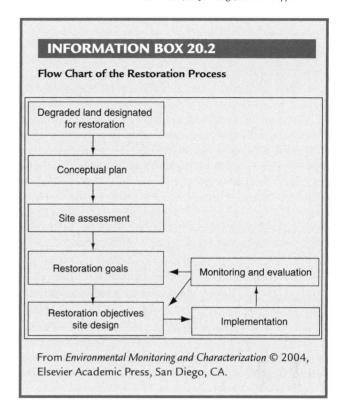

From *Environmental Monitoring and Characterization* © 2004, Elsevier Academic Press, San Diego, CA.

---

ties in developed countries require a **closure plan**, which describes how the land will be restored to a productive state after mining ceases. In return for permits to build new factories and power plants, developers are now often required to provide **environmental offsets**, in which they restore abandoned farmland, create wetlands, or plant trees on logged-over property. The science of **ecological restoration** has been developing rather recently, to evaluate ways of repairing damage to disturbed ecosystems.

Technically, **rehabilitation, revegetation,** and **reclamation** all fall under the umbrella of **restoration** (Information Box 20.1).

In this chapter, we are concerned with methods to repair human-caused damage to natural ecosystems, and like many restoration ecologists we tend to use the terms rehabilitation, revegetation, and reclamation interchangeably. The overall "restoration process" is shown in Information Box 20.2.

20.2 SITE CHARACTERIZATION

20.2.1 Conceptual Plan and Site Assessment

A conceptual plan often begins with a detailed history of the site, where information is collected to determine the anthropogenic changes that have lead to the current state of degradation. This information is cross-referenced against historical and current topographic, geologic, and vegetation maps to determine what changes have occurred spatially and temporally. Information on the physiochemical soil properties and the water quality of the site prior to disturbance can show the restoration potential of the area. Before deciding to proceed with a project, social and cultural values of the neighboring residents need to be assessed to ensure that the proposed restoration objectives are compatible with the local socioeconomic needs of the land.

Once a complete history of the site has been compiled and it has determined that the restoration project is feasible, a thorough site assessment of abiotic and biotic conditions takes place. This step is probably one of the most important in any restoration project, since it not only provides baseline measurements on such parameters as hydrologic features, soil conditions, and biological information, but it also serves as the benchmark upon which to evaluate the project through time. This step involves placing the site in the context of the regional landscape with respect to habitat fragmentation, disconnected surface and subsurface hydrologic flows, water quality issues, physical and chemical properties of the soil, and finally plant and sometimes animal inventories. Depending upon the nature of degradation, additional data may be collected on the presence or absence of toxic chemicals such as organophosphates, heavy metals, and radioactive waste. In assessing a site, it is important to determine which basic ecological functions are damaged or fragmented, since this often sets the restoration priorities and ultimately the success of the project over time. It is worth noting that the soil conditions and water quality in a given area generally dictate the type of vegetation cover and thus the biodiversity of biotic components. Therefore, extra care should be taken in analyzing and describing these two elements both horizontally and vertically across the landscape.

20.2.2 Plant Surveys

Plants provide valuable information about site environmental conditions. The occurrence and relative abundance of certain plant species and their physiological and ecological tolerances provide evidence of environmental conditions that are of importance for understanding the nature of the site, potential human health and ecological risks, and the feasibility of different restoration alternatives. Typically, plant ecology investigations include four types of studies: (1) **plant species survey**; (2) estimates of the **percent cover** and age structure of dominant, perennial plant species; (3) evaluation of the composition, relative abundance, and distribution of **plant associations**; and (4) **vegetation mapping**.

The plant species survey is conducted by traversing a site, usually on foot, and noting each species present.

The percent cover study attempts to quantify the percent of the site that is covered by bare soil or individual plant species utilizing a **line intercept** method. First, the plant community to be described is delineated on a map and then 30-meter transect lines are randomly chosen where actual plant counts will take place. In the field, a 30-meter tape is stretched out and the total distance intercepted by each plant species is recorded and used to calculate the percent cover of each species. For example, a transect might consist of 12% fourwing saltbush, 10% black greasewood, and 78% bare soil. Plant associations and ultimately vegetation mapping are used to delineate land management units for future revegetation efforts.

20.3 SITE RESTORATION

Setting realistic goals for a restoration project is important, and it is important to recognize that ecosystems are dynamic, so that, potentially, restoration projects have a range of short- and long-term outcomes (Hobbs and Harris, 2001). Therefore, the focus of the project should be on the *desired* characteristics of the ecosystem in the future, rather than on the characteristics that were previously there (Pfadenhauer and Grootjans, 1999).

The methods chosen for the restoration of a particular site will be determined by the nature of the site, the level of existing degradation, and the desired outcome over time. The underlying causes of the degradation must be identified as either biotic, abiotic, or a combination of both. The restoration ecologist may ask, for example, whether the degradation of land was caused by simple overgrazing or if the physiochemical soil and hydrological processes were changed to the point where biodiversity had been compromised. In highly polluted sites, it may be necessary to remove contaminants from the soil and/or water prior to beginning restoration work. Mining activities may have contaminated the soil to the extent that it is toxic to humans, animals, and plants, in which case it must be removed and replaced by noncontaminated soil prior to the restoration process. In other cases, minimal work such as managing grazing will be all that is necessary to make the site suitable for plant establishment and growth. The degree of intervention in restoring or remediating a site can be natural restoration through passive to active restoration. The information gathered for the conceptual plan and from the site assessment should present a fairly clear picture of the path a restoration project will follow, and will shape the achievable goals through carefully constructed project objectives, site design, and implementation strategies.

20.4 SITE MONITORING

Up until a few years ago, most projects were monitored in the field using typical agronomic and plant ecology techniques, while controlled greenhouse experiments measured soil, microbial, and plant interactions. While these methods are considered vital in describing the status of plant establishment, water quality, changes in soil chemistry and microbial populations, the current trend is to try to integrate this information into a broader ecological picture. With increased computing power and sophisticated data collecting techniques, real-time data can more easily be obtained without additional personnel, thus giving restoration ecologists larger data sets with which to work.

Most restoration projects rely on short-term monitoring to assess the success of a project.

Contractors are required to post a **performance bond** guaranteeing their work, sponsors want to see the results of their investments, and the public expects immediate results. The reality of the situation is that ecological processes can take from decades to centuries to achieve a level of maturity, a time span that is not economically feasible for continuous monitoring programs. Instead shorter term monitoring programs collect data that is then extrapolated to predict generalized ecological patterns of change against a referenced ecosystem.

20.5 APPROACHES TO ECOSYSTEM RESTORATION

20.5.1 Natural Restoration

Natural restoration is essentially the process of allowing the ecosystem to heal itself without active management or human interference. Essentially this is the same concept as "intrinsic bioremediation" and "monitored natural attenuation" (see Chapter 19). Depending on site-specific characteristics, natural restoration may not be a viable alternative. Natural ecosystems develop over long periods through the process of **ecological succession**. Think of a lava flow, such as those that still occur on the island of Hawaii on the slopes of Mauna Loa, burning through portions of the native rain forest. The lava cools quickly, but lays barren for many years, too hostile to support life. Eventually rain and wind erosion open up tiny fissures in the lava, where life can establish a foothold. Microorganisms, usually bacteria, are often the first forms of life to become established, followed by lichens. In fact, very few disturbed sites are microbiologically sterile. Microorganisms are a prerequisite for plant growth. Lichens and cyanobacteria are the next colonists on the flow. Continual breakdown of the parent lava by acids secreted by the lichens produces a thin layer of soil in which the first higher plants can root. Generally, small ferns appear first, followed by grasses and shrubs as the fissures widen due to the action of the plant roots. Cyanobacteria fix nitrogen, which also supports the plant life. Each stage of succession conditions the lava substrate to favor the next stage, and finally the rain forest is restored. This process may take many hundreds or even thousands of years.

20.5.2 Passive Ecological Restoration

Passive ecological restoration projects primarily apply to land whose system is still functionally intact, but that has lost vegetative cover and biodiversity from such activities as overgrazing or habitat fragmentation. The implicit goals of these projects are to reduce or eliminate the causes of degradation while encouraging the growth of indigenous plants to increase the productivity of the area in a sustainable way. In general, minimal soil preparation is needed, soil amendments and irrigation are not required, and seeds are simply broadcast in a designated area. Restoration under these conditions is usually coupled with land conservation objectives.

In semi-arid and arid areas of the world, where water is the limiting factor for successful restoration, the success of dry seeding methods generally declines as aridity increases. Some consider irrigation essential in areas that receive less than 250 mm of annual precipitation, but the need for irrigation, the amount, and application mode have been debated. However, numerous low-cost techniques can be used to capture and retain precipitation where it falls. The simplest and cheapest methods involve placing logs, rocks, or mulch on bare ground to capture moisture, nutrients, seeds, and soil from the surrounding area. Contour furrows, pits, and small depressions in surface soils play the same role in capturing essential elements for plant es-

Figure 20.1 A naturally vegetated farm furrow on an abandoned field in Maricopa County, Arizona. From *Environmental Monitoring and Characterization* © 2004, Elsevier Academic Press, San Diego, CA.

tablishment. Figure 20.1 illustrates a furrow in an abandoned saline cotton field where plant establishment has occurred naturally.

20.5.3 Active Ecological Restoration

Where the abiotic and biotic functions of an ecosystem have been destroyed, the cost of restoration will rise in proportion to the damage incurred. Active ecological restoration projects can fall between reclaiming abandoned land, to cleaning up and restoring U.S. Superfund sites. For example, the restoration of the uranium site in Monument Valley, Arizona is a complicated, expensive restoration project (see Case Study 20.1). Anthropogenic damage to an area can alter the biogeochemical function of an ecosystem, and in many cases it is necessary to reintegrate the damaged land with the surrounding landscape, particularly with regard to the hydrological cycle. For example, farmers may level an area for new fields or irrigation canals can be constructed. In the process of adding new cultivated fields, the surrounding wildland can become fragmented, and surface water may have been redirected and topographical features flattened. Once this land is abandoned, it becomes a candidate for restoration and it will be necessary to reconnect this land to the surrounding wildland, restore the physiochemical functions to the soils, and seed or plant the area to maximize plant establishment and growth. These projects thus become active and generally high cost because they involve additional procedures in addition to simple seeding. Information Box 20.3 shows some of the methods employed for these higher-cost restoration projects. The restoration goals and objectives developed for a project will specify the degree of work necessary to achieve the specified goals.

Many other specialized restoration methods have been developed for specific habitat types. Constructed wetlands have become a popular means for final treatment of municipal sewage effluent and for providing habitat for waterfowl and other birds. Often, however, the constructed wetlands do not serve as well as natural wetlands in supporting diverse plant and animal communities. In the western United States, considerable attention has been paid to restoring riparian zones by removing the invasive plant, salt cedar, and replacing it with native trees such as cottonwood and willow. Often these projects do not succeed, however, because the hydrological conditions of the river have been so altered that they no longer favor the native species.

20.5.4 Ecological Restoration Using Organic Amendment

Highly disturbed sites often result in surface soils being devoid of organic matter. This can occur from a variety of human activities including strip mining (where the surface top soil is removed) and mine tailing (crushed and processed mineral ores deposited over existing topsoil) or from soil erosion. In all cases, organic matter is sparse or entirely absent, there are extremely low microbial populations, and it

---

INFORMATION BOX 20.3

Potential Procedures for Active Ecological Restoration

Parameter	Procedure
Landscape irrigation	Removal of irrigation canals, roads
	Reconnect natural drainage features
	Construction of berms, swales, and gabions to redirect runoff and control erosion
	Landscaping to naturalize site
Soil	Grading to enhance soil stability, reduce erosion, and enhance water harvesting
	Creation of microcatchments for water harvesting
	Ripping to diminish compaction
	Tilling to incorporate soil amendments
	Mulching to enhance water retention
	Fertilizer additions to encourage plant growth
	Amendments to adjust soil pH
Plant	Mulching to enhance germination of direct seeding
	Transplants to enhance establishment
	Hydroseeding
	Mycorrhizal inoculation
Other	Supplemental irrigation
	Wire cages to protect against wildlife

From *Environmental Monitoring and Characterization* © 2004, Elsevier Academic Press, San Diego, CA.

---

is common for these sites to have extreme pH, low soil permeabilities, and high soluble metal concentrations. These conditions are not suitable for sustainable plant growth, and they generally require some form of organic amendment to jump-start the restoration process. Problems that occur due to low organic materials are shown in Table 20.1, and com-

TABLE 20.1 Problems related to low organic matter surface materials.

PARAMETER	PROBLEM
Poor aggregation of primary particles	Compaction
	Low infiltration rates
	Low water holding capacity
	Limited aeration
Low nutrient status	Infertile soils
	Low microbial populations
Extreme pH	Affect chemical and biological properties
High metal concentrations	Toxicity

From *Environmental Monitoring and Characterization* © 2004, Elsevier Academic Press, San Diego, CA.

INFORMATION BOX 20.4	
Sources of Organic Materials for Ecosystem Restoration	
Human and Animal Wastes	**Industrial Wastes**
Animal manures	Paper mill sludges
Biosolids	Sawdust
Composted wastes	Wood chips

From *Environmental Monitoring and Characterization* © 2004, Elsevier Academic Press, San Diego, CA.

mon sources of organic materials used to enhance ecosystem restoration are illustrated in Information Box 20.4.

The concept of organic amendments to enhance plant growth has been utilized for centuries as in the application of "night soil" (human feces and urine) to agricultural land. The use of raw waste material can spread disease, but in the United States, the solid material (**biosolids**) left after treatment of municipal sewage is further refined to largely eliminate potential pathogens before being applied to agricultural land (see also Chapter 27). Biosolids have been successfully applied to mine tailings or smelters that contain high or even phytotoxic levels of heavy metals. Biosolids that have undergone lime stabilization are particularly useful for such restoration, since the increase in pH reduces the bioavailability of metals to plants. Biosolids and composts have also been used to restore diverse ecosystems such as mountain slopes in the Washington Cascades or stabilize sand dunes in southeastern Colorado.

In all cases of organic amendment added to restore soils, the critical parameter appears to be the magnitude of organic material applied. This is particularly important in desert ecosystems, where high temperatures result in rapid decomposition and mineralization of organic materials. If insufficient organic matter is added to a disturbed site, the beneficial effect is not maintained for a sufficiently long time to allow stable revegetation to occur (see Case Study 20.2).

20.5.5 The Invasive Species Problem

Due to increases in global trade and travel, alien plant and animal species have been introduced into native ecosystems around the world. Most of these species quickly die out in the new environment, while others become permanent, but minor components of the ecosystem. A few (less than 1%) may become invasive, spreading over large areas and disrupting ecosystem functions. They may also become pests in human-managed systems such as farms and rangelands. Examples of invasive plants are purple loosestrife and salt cedar, two introductions that have come to dominate many wetland and riparian ecosystems in North America. An example of an invasive animal is the zebra mussel, which clogs the intakes of power plants and depletes the water of phytoplankton that form the base of the native food chain. Over a hundred invasive species have been recognized as causing damage on an ecosystem-wide scale.

Invasive species are a special problem in restoration work, because human-disturbed ecosystems are especially prone to invasion. For example, abandoned farmland in the western U.S. is often saline and high in nutrients. These conditions favor the establishment of Russian thistle (tumbleweed), an annual plant that completes its life cycle on very little water and releases large amounts of seeds into the soil to germinate in subsequent years. Projects that attempt to reintroduce native plants on abandoned farmland often just stimulate the growth of even more tumbleweed. As another example, western U.S. rivers have been dammed and their flow regulated to prevent the occurrence of normal spring floods. Their floodplains have been invaded by salt cedar, which thrives on the saline soil that develops when the floods are disrupted. By contrast, native cottonwood and willow trees are not salt tolerant, and they have become rare on western rivers.

There is no simple answer to the control of invasive plant species. However, some guiding principles are emerging from restoration studies. First, invasive species tend to be early successional plants, dominating an ecosystem just after it is disturbed. For example, tumbleweed often dominates for five years after a soil is disturbed, but during that time it adds organic matter to the soil, and conditions the soil to eventually become habitable by native species. Hence, patience is necessary in restoration work. Second, eradication of invasive species by physical or chemical means is often ineffective. Unless the physical environment is restored, the invasive species simply returns. A more successful approach is to attempt to restore the environment to one favoring establishment of native species. For example, native cottonwood and willow trees along western U.S. rivers can be restored even in the presence of invasive salt cedar if a pulse flood regime is returned to the river channel. The floods wash salts from the floodplain soil and germinate native trees, which then overtop the salt cedar and shade them out. Third, invasive species may come to play a positive role in the ecosystem. Animals that formerly depended on native species may adapt to use the invasive species as a source of food or shelter. Therefore, a viable restoration goal might be to aim for a mix of native and introduced species that still can fulfill ecosystem functions.

Invasive species can also be attacked by using biological control agents. Typically, these are insects from the native range of the invasive plant that are introduced into its new habitat to control its spread. For example, Asian beetles that feed on salt cedar are being tested as possible control agents for salt cedar on western U.S. rivers. Extensive research is needed before the release of a biocontrol agent to ensure that it is effective and will not become a pest for native plants as well as the introduced species (see also Chapter 16).

20.6 LAND RECLAMATION

Ecosystems can be disturbed dramatically and sometimes very quickly, as in the case of the deposition of mine tailings onto land, which covers the existing soil, completely changing

MONUMENT VALLEY, ARIZONA

This case study involves a former uranium mill site on the Navajo Indian Reservation in Monument Valley, Arizona, that the U.S. Department of Energy had placed in their Uranium Mill Tailings Remedial Action (UMTRA) program. Each step in this restoration process follows a formal procedure and is well documented. Starting in the 1950s, the Atomic Energy Commission encouraged the mining of uranium ore in the southwestern United States to provide fuel for the nuclear power industry and material for weapons. The milling process produced great piles of crude ore and tailings, covering many acres, surrounded by unlined evaporation and leaching ponds, from which "yellowcake," a crude form of uranium, was extracted. In the 1970s, the price of yellowcake collapsed, and most of the mills went bankrupt and were subsequently abandoned. The owners made no effort to clean up the sites. By the end of the 1970s, all that was left of the mill were piles of crushed ore and tailings, each pile covering several acres, but the main problem was that this material was mildly radioactive, and it was suspected that toxic chemicals (heavy metals, nitrates, ammonia, and sulfates) had leached into the soil and ultimately the groundwater. In the 1980s, the Department of Energy was given responsibility for restoring these sites.

The first task in the restoration process was to determine the history of the site. A search was made of company records, former workers were interviewed, and archives of aerial photographs were assembled. A picture of how the mill operated was developed and areas of concern for remediation and restoration were pinpointed. A map of the site was made, showing where the different processes in the milling operation took place. The second task was to determine the current state of contamination. Intensive soil sampling for radioactivity, heavy metals, and other potentially toxic chemicals was undertaken. Bore holes were drilled to determine if the underlying aquifer was contaminated, and

TABLE 20.2 Restoration plan for the UMTRA Program.

PHASE	DESCRIPTION
1a)	Removal of contaminated material
b)	Containment of contaminated material
2a)	Restoration of damaged land
b)	Remediation of contaminated water

vegetation cover across the site was assessed. Maps were produced that detailed the extent of the contamination not only on the site, but also on adjacent land.

Once a good picture of how the site had been used and where the contamination problems were, a baseline risk assessment report was released that evaluated the potential for human and environmental damage if the site was not repaired. This Monument Valley site was given a high priority for remediation, and further studies were conducted to develop a set of restoration goals and objectives, which federal, state, and private stakeholders reviewed in a series of public meetings. Those attending the meetings included representatives from the U.S. Department of Energy, the U.S. Environmental Protection Agency (EPA), the Navajo Nation UMTRA, the Navajo Nation EPA, and the local community. Local residents were vocal in opposing plans that would negatively impact their traditional uses of the land. Stakeholder participation during this planning phase was critical to the ultimate acceptance of the plan by the community. The restoration plan for this UMTRA program is shown in Table 20.2.

Phase I of the restoration plan generated little controversy. It was quickly decided that the ore and tailings had to be removed from the site and that the soil around the site had to be removed down to the level at which there was no radioactivity. Subsequently, a fence was placed around the property to prevent grazing animals from entering. A graded road to the site was constructed across 20 miles of desert, and a fleet of trucks was commissioned to haul away the contaminated material. Local Navajos were trained as truckers and equipment operators for the project. The contaminated material was taken to the nearby town of Mexican Hat, where it was spread out over an impermeable bedrock surface and covered with three layers of material to prevent radon gas and radioactivity from escaping. The first layer was compacted clay from a local site; the second layer was bedding sand; and the third layer was made up of large rocks. The rock layer was thick enough that plants could not easily establish on the surface of the containment cell. The design of the containment cell was such that it is expected to prevent contaminants from leaking for at least 1,000 years.

Phase II, still under development, involves, first, repairing the damage to the land from Phase I and, second, dealing with a plume of contaminated water that is migrating underground away from the site. In removing surface contamination, over 100 acres of the site were denuded of native vegetation. This vegetation must be replaced. The main **chemical of concern (COC)** in the contaminated groundwater plume is nitrate, originating as nitric acid that was used to leach uranium from the ore. Nitrate levels in the groundwater greatly exceed EPA standards for drinking water, and this nitrate must somehow be removed. How to deal with the two problems was analyzed through a process called **value engineering**. All possible alternatives for restoring vegetation and cleaning up groundwater were listed after preliminary analysis by the study team. The list was shortened to those that appeared to be both likely to succeed and be cost effective.

Options for restoring vegetation ranged from relatively low-cost measures, such as application of mulch and seed to the land in a liquid spray (**hydroseeding**), to higher cost measures such as transplanting native shrubs grown in a greenhouse to the site and providing irrigation for several years while they established a root system. In desert ecosystems such as Monument Valley, revegetation success generally goes up in direct proportion to the amount of irrigation provided. In general, direct seeding cannot be relied upon in areas receiving less than 250 mm of rainfall per year (the Monument Valley site receives less than 200 mm per year). Options for remediating the groundwater were even more expensive. Conventional treatment methods required that the water be pumped to the surface and passed through a water treatment plant, using **deionization**, **evaporation**, or **distillation** to separate nitrates from the water. This process is known as pump-and-treat (see Chapter 19).

Further analysis, however, showed that the revegetation problem and the plume remediation problem could be solved together by using the plume water as a source of irrigation water for native plants and forage crop planted over the bare areas of the site. The nitrate in the plume water would serve as a fertilizer for the plants and would be converted into plant nitrogen compounds, which would be harmlessly consumed by the Navajo livestock. Using plants to solve environmental problems is called **phytoremediation** (see also Chapter 19), and this became the preferred alternative at the Monument Valley UMTRA site because it provided a combined solution to two problems and did not require construction and operation of an expensive water treatment plant. As of this writing, the phytoremediation option is undergoing review by stakeholders, and a demonstration phytoremediation plot has been established on site.

The planning process undertaken at the Monument Valley UMTRA site illustrates the numerous checks and balances built into a restoration plan. Many different disciplines are involved, and everyone with a possible stake in the outcome of restoration must be brought into the process. In most cases, a restoration strategy is not adopted until it achieves **consensus** among stakeholders as the best possible choice. Many years of study and planning may precede actual restoration or remediation. This is acceptable because land restoration is expensive, so careful planning may help prevent costly mistakes.

MISSION COPPER MINE, ARIZONA: RECLAMATION AND REVEGETATION OF MINE TAILINGS USING BIOSOLID AMENDMENT

In the United States, mining is a large industry that provides valuable raw material and creates economic benefit for local communities. However, the environmental damage incurred from this industry range from unsightly mine tailings to the leaching of toxic elements into nearby waterways and aquifers. Mine tailings are formed in two ways: by the initial removal of vegetation, soil, and bedrock to expose the valuable mineral veins, and then by the disposal of the crushed rock after the ore has been removed. Typically these tailings are 30–40 meters.

The physiochemical characteristics of mine tailings are totally unlike the displaced topsoil that once supported vegetation in any given area. By removing and crushing bedrock from the mines and placing it on the surface, minerals will oxidize when exposed to the atmosphere. For example, pyrite (FeS_2) common around coal mines oxidizes to sulfuric acid (H_2SO_4) and iron oxide ($Fe(OH)_3$). Acid mine drainage (H_2SO_4), the leachate from tailings, can then contaminate surface and groundwaters in addition to increasing the solubility of toxic metals. Mining tailings are not the ideal medium on which to grow plants. The crushed rock consists of large and small fragments with large void spaces in between them. In addition, there is no organic material; the cation exchange capacity (CEC) is very low; the water holding capacity of the material is poor to nonexistent; and there are few macronutrients (NPK) available for the plants. Soil biota, in the form of bacteria and fungi, are present only in low numbers, and finally, the pH is usually low, which increases the likelihood that toxic metals could be taken up by the plants. The goals of reclaiming mine tailings therefore have to include the application of materials to amend the crushed rock substrate and provide a conducive environment for plant growth. One potential solution is the use of biosolids.

In 1994, the Arizona Mined Land Reclamation Act was passed that required reclamation of all mining disturbances on private land to a predetermined post-mining use, and in 1996 the Arizona Department of Environmental Quality (ADEQ) adopted new rules allowing for the use of biosolids during this reclamation. The Arizona Mining Association (AMA) has estimated that there are 13,360 hectares of active mine sites in Arizona that can be reclaimed with the use of biosolids.

Current ADEQ regulations limit the lifetime loading rate of biosolid applications to mine tailings to 333 metric tons/hectare (dry weight). However, this amount can be applied as one application. Here we describe a case study illustrating the use of biosolids to restore and stabilize mine tailings derived from copper mining.

Overall Objective: To evaluate the efficacy of dried biosolids as a mine tailing amendment to enhance site stabilization and revegetation.

EXPERIMENTAL PLAN

Study Site

A 5-acre copper mine tailing plot located near Mission Mine, south of Tucson, Arizona was designated for this study. Biosolids were applied at a rate of 2″ to 3″ (70100 dry tons/acre) across the site in December 1998. Supplemental irrigation was not used in this experiment.

Soil Microbial Response to Biosolids

Pure mine tailings contain virtually no organic matter and very low bacterial populations of approximately 10^3 per gram. A large population of heterotrophic bacteria is essential for plant growth and revegetation, and therefore, monitoring soil microbial populations gives an insight into the probability of revegetation prior to plant growth. Biosolids routinely contain very high concentrations of organic matter, including the macroelements carbon and nitrogen, which are essential for promoting microbial growth and metabolism. Following biosolid amendment of the mine tailings, heterotrophic bacterial populations increased at the surface to around 10^7 per gram. Bacteria decreased with increased depths from the surface, indicating the influence of the biosolid surface amendment on bacterial growth. Overall, the microbial monitoring data showed the success of biosolid amendment in changing mine tailings into a true soil-like material.

Physical Stabilization

One of the main objectives in reclaiming mine tailings is **erosion control** through wind and water erosion. This is generally best accomplished through a revegetation program, since the root structures of the plants help to hold soil particles in place. In this experiment, the application of biosolids and the subsequent growth of plants promoted site stabilization. In the desert Southwest, high summer temperatures and limited rainfall are normal; however, despite these extreme conditions, grasses have become established on these tailings. Table 20.3 shows the results of vegetation transects surveys conducted on this site 14 months, 21 months, and 33 months after initial seeding.

TABLE 20.3a Vegetation transects[1] at the Mission Mine Project site.

	BASAL COVER(%)	CROWN COVER(%)	TOTAL COVER(%)	ROCK(%)	LITTER(%)	BARE(%)
			02/23/00 (14 months)			
T-1	15	28	43	3	0	54
T-2	14	16	30	3	8	59
T-3	7	1	8	0	1	91
T-4	1	8	9	2	4	85
T-5	0	0	0	0	5	95
Average	**7.4**	**10.6**	**18**	**1.6**	**3.6**	**76.8**
			09/25/00 (21 months)			
T-1	13	39	52	5	6	37
T-2	11	20	31	6	6	57
T-3	6	45	51	2	1	46
T-4	13	48	61	3	2	34
T-5	1	52	53	0	4	43
Average	**8.8**	**40.8**	**49.6**	**3.2**	**3.8**	**43.4**
			09/24/01 (33 months)			
T-1	4	66	70	2	4	24
T-2	6	62	68	0	6	26
T-3	4	73	77	0	4	19
T-4	0	84	84	0	1	15
T-5	0	92	92	0	0	8
Average	**2.8**	**75.4**	**78.2**	**0.4**	**3**	**18.4**

[1]On each date five transects were evaluated, over the five acre site. Each transect consisted of 100 feet, evaluated at one foot increments. From *Environmental Monitoring and Characterization* © 2004, Elsevier Academic Press, San Diego, CA.

Note the variability within each data derived from each individual treatment. The vegetation cover increased from 18% at the 14-month survey to 78.2% after 33 months. At 14 months, the predominant plant species were bermuda grass (*Cynodon dactylon*), and the invasive weed, Russian thistle or tumbleweed (*Salsola tragus*), but by the 33rd month, buffelgrass (*Pennisetum ciliare*), and Lehman lovegrass (*Eragrostis lehmanniana*) had replaced the Russian thistle. Figures 20.2, 20.3, and 20.4 show the progressive increase of vegetation on this site over time. In this case, the use of biosolids for enhanced revegetation and stabilization of mine tailings would be considered a success.

Soil Nitrate and Metal Concentrations

At Site 1, soil nitrate (Table 20.4) and total organic carbon (TOC) (Table 20.5) are very high at the surface, but decrease to the levels found in pure mine tailings at lower depths. The fact that nitrate and TOC concentrations are correlated is important, since it creates substrate and terminal electron acceptor concentrations suitable for denitrification (see also Chapter 5). Data show the nitrate concentrations from June 2000 to July 2001. Nitrate concentrations increased during the monsoon rainy season of 2000, most likely due to enhanced ammonification and subsequent nitrification. However, within the soil profile,

TABLE 20.3b Mission-BS test plot vegetative transects composite. Site 1 averages for years 1999–2003.

	% BASAL COVER	% CROWN COVER	% TOTAL COVER	% ROCK	% LITTER	% BARE
02/23/99	7.4	10.6	18	1.6	3.6	76.8
09/25/00	8.8	40.8	49.6	3.2	3.8	43.4
09/24/01	2.8	75.4	78.2	0.4	3	18.4
09/16/02	2	49.6	51.6	1.6	1.2	45.6
10/30/03	8	70	78	1	0	21

Figure 20.2 Mine tailings prior to biosolid amendment. From *Environmental Monitoring and Characterization* © 2004, Elsevier Academic Press, San Diego, CA.

Figure 20.3 Mine tailings two years after biosolid application. From *Environmental Monitoring and Characterization* © 2004, Elsevier Academic Press, San Diego, CA.

Figure 20.4 Mine tailings three years after biosolid application. From *Environmental Monitoring and Characterization* © 2004, Elsevier Academic Press, San Diego, CA.

TABLE 20.4 Nitrate concentrations at the Mission Mine project site from June 2000 to July 2001.

	DEPTH OF SAMPLE				
SAMPLE DATE	0–1′	1–2′	2–3′	3–4′	4–5′
	mg k^{-1}				
06/26/00	650	250	40	5	ND
07/10/00	1520	120	60	5	ND
07/26/00	1030	200	170	70	ND
02/05/01	480	250	330	150	60
03/26/01	190	40	40	15	5
06/11/01	2350	310	140	260	110
07/13/01	2350	590	220	205	50

From *Environmental Monitoring and Characterization* © 2004, Elsevier Academic Press, San Diego, CA.

TABLE 20.5 Total Organic Carbon (TOC) at the Mission Mine from February 2001 to July 2001.

SAMPLE DATE	DEPTH OF SAMPLE				
	0–1'(%)	1–2'(%)	2–3'(%)	3–4'(%)	4–5'(%)
02/05/01	1.4	0.2	0.3	0.4	0.3
03/26/01	1.0	0.1	0.2	0.2	0.1
06/11/01	1.6	0.2	0.1	2.0	0.1
07/13/01	1.6	0.4	0.1	0.2	0.2

From *Environmental Monitoring and Characterization* © 2004, Elsevier Academic Press, San Diego, CA.

nitrate concentrations decreased with depth. By the winter and spring of 2001, the nitrate concentrations at all soil depths had decreased. There was no evidence of the leaching of nitrate since concentrations at the 3'–4' depth were always minimal. Therefore, the most likely explanation for decreased nitrates within the soil profile is the process of denitrification. Soil nitrate concentrations became extremely high at both sites in the summer of 2001, again most likely due to nitrogen mineralization and seasonal nitrogen cycling. Specifically during the warmer summer months, rainfall events appear to have triggered microbial mineralization of nitrogen as nitrate.

The application of biosolids to a project site brings some concern about the introduction of heavy metals to the environment. However, data from this study showed that metal concentrations were fairly consistent with soil depth, indicating that the tailings were the major source of metals,

not the biosolids. Further evidence of this is shown by the high molybdenum and copper values typical of mine tailings. At this site there was little evidence of metals leaching through the soil profile.

Summary

This study on the application of biosolids to mining tailing at the Mission Mine in Arizona shows that soil stabilization has been encouraged through revegetation techniques and that the leaching of nitrate and heavy metals to important water resources has not been observed. This case study gives an indication of the extensive monitoring that is necessary to understand the restoration process and the necessary duration of the monitoring process. With careful attention paid to subsurface geologic and hydrologic features at other sites, the application of biosolids can be a feasible restoration strategy.

the plant root growth medium. In other cases, soils can be altered in more subtle ways, as in the case of agricultural land, where the pH of the soil may be decreased, or increased, due to the agricultural practices. In yet another scenario, the soil itself within a land area is not changed, but the ecosystem can be changed, as in the case of deforestation. Here, we address land reclamation of acidic soils, saline and alkaline soils, and deforested areas.

20.6.1 Acidic Soils

The optimum pH of soils for plant growth is generally between pH 6 to 8. However, agricultural crop production over a period of several years can result in a decreased soil pH or the production of acidic soils (pH < 6).

In many areas of the world, as for example in the U.S. Midwest and several parts of Europe, agricultural crop production is carried out in environments where rainfall exceeds the amount of water used by the plants. This excess water results in the leaching of basic cations through the soil profile. In essence, these basic cations are replaced from cation exchange sites (see Chapter 2) by H^+ ions that are produced during the degradation of plant residues and organic matter. In addition, applications of ammonium-based fertilizers can also result in increased soil acidity following nitrification (see Chapter 5). The decreased soil pH can adversely influence plant growth in a number of ways (Information Box 20.5).

Although some crops such as pineapples (grown in Hawaii) or azaleas can tolerate and even grow well at soil pH values of 4.5 to 5.0, most crops do not grow well. Therefore, land reclamation must be conducted to allow viable crop production to be maintained.

To correct soil acidity, lime amendments are recommended. The liming materials are frequently **calcitic limestone,** which consists of calcium carbonate, or **dolomitic limestone,** which is a mixture of calcium and magnesium carbonates. There are many factors that influence the amount of lime to be added which include initial soil pH, final target pH, cation exchange capacity, and other soil characteristics that are site specific. Many states in the U.S. have specific recommendations for lime requirements issued through the land grant universities or the United States Department of Agriculture (USDA). As an example, however, approximately 10 tonne ha^{-1} of ground limestone might be necessary to raise the pH of a Midwest agricultural soil from pH 6.2 to pH 7. In contrast, once the soil pH falls below 5.0, the amount of lime required increases dramatically. Hence, the same soil at a pH of 5.6 might need 20 tonne ha^{-1} to bring it back up to 7.0. In general, smaller amounts of lime are added more frequently to agricultural land to maintain the pH between 6.0 and 7.0.

20.6.2 Saline and Sodic Soils

In contrast to soil acidity problems, agricultural crop production can also result in the build up of salts in soil and high pH soils. The original source of salts in soils are derived from the primary minerals found in soils including various proportions of the cations sodium, calcium, and magnesium, and the anions chloride and sulfate. In humid regions with adequate rainfall, these salts are transported or leached downwards into groundwater or streams, and carried to the oceans. Therefore, salt accumulations in soils of the humid regions are rare. In arid areas where rainfall is limited, there is no such mechanism to remove salts, and salt concentrations can increase. However, for salt concentrations to increase to problematic levels, additional sources of salts are necessary. Irrigation water used for crop production in arid regions can be a source of such salts. In hot arid areas, evaporation of applied irrigation water from soils, and evapotranspiration of water from plants, removes water from the land, but retains the salts. Therefore, over time, salt concentrations will increase. This situation can be compounded in low-permeability soils with poor drainage characteristics. In addition, if sodium is the dominant cation, it will replace calcium and magnesium, and result in cation exchange sites dominated by sodium. Therefore salt affected soils can be separated into three groups: saline, saline-alkali, and nonsaline alkalai or sodic soils.

The characterization of these three groups is based on: (1) the electrical conductivity in mmhos (mho is a conductance unit that is the reciprocal of resistance measured in ohms); (2) the percent of cation exchange sites occupied by sodium; and (3) the soil pH (Information Box 20.6).

INFORMATION BOX 20.5

Adverse Affects of Soil Acidity on Plant Growth

· Aluminum toxicity   · Calcium deficiency
· Manganese toxicity   · Phosphorus deficiency
· Heavy metal toxicity   · Molybdenum
                                             deficiency

Biological Effects

· Disturbance and/or decreased microbial activity.

INFORMATION BOX 20.6

Characteristics of Salt Affected Soils

Soil	pH	Electrical Conductivity mmhos cm^{-1}	Exchangeable Sodium Percentage
Saline	<8.5	>4	<15
Saline-Alkali	8.5	>4	>15
Sodic	>8.5	<4	>15

Saline soils are generally flocculated and have reasonable drainage characteristics. These soils can adversely affect plant growth due to high and often toxic amounts of salts. In contrast, **sodic** soils are frequently dispersed and have poor drainage characteristics. Plant growth in sodic soils may be restricted by both high pH and toxic amounts of sodium. **Saline-alkalai** soils can have the adverse characteristics of both saline and sodic soils.

Because of the poor plant growth and decreased yields associated with the three groups of soils, soil reclamation procedures have been developed. Generally, land reclamation of salt affected soils can be accomplished in two ways. First, excessive amounts of irrigation water can be used to leach excess salts through the soil profile and out of the root zone. This is appropriate for reclamation of saline soils. This can only be accomplished if the soil has adequate drainage characteristics. For sodic soils, drainage can in turn be improved by amending the soil with calcium amendments, which replace sodium on the exchange sites and cause flocculation of soil particles with increased pore space (see Chapter 2). The most frequently used amendments are limestone ($CaCO_3$) and gypsum ($CaSO_4 \cdot 2H_2O$). Sodic or saline alkali soils may require up to 15–20 tonne ha^{-1} to reduce the exchangeable sodium content to acceptable levels.

Finally, elemental sulfur is also used as a soil amendment to reduce soil pH when values are initially in excess of 8.5. Elemental sulfur is oxidized by autotrophic bacteria such as *Thiobacillus thiooxidans*, resulting in the production of sulfuric acid and a concomitant soil pH decrease (see also Chapter 5). Sulfur amendments can be in the range of 1 to 3 tonne ha^{-1}. In the desert southwest, commercial nitrogen fertilizers frequently have sulfur included to maintain soil pH values within the neutral zone. Overall, the strategies employed to reclaim saline and/or sodic soils can be summarized as: (1) improving drainage by replacing sodium on exchange sites with calcium; (2) leaching soluble salts via large applications of irrigation water; and (3) decreasing soil pH via sulfur amendments.

20.6.3 Reforestation

Reforestation is a restoration strategy for returning trees to a deforested ecosystem. Trees are removed for a variety of reasons. Tropical rainforests are cleared for agriculture, but often the farms fail due to lack of soil nutrients. Temperate zone forests are often clear-cut for timber. Dryland forests in the subtropics are often denuded due to overgrazing of livestock and harvesting of trees for charcoal and firewood. Forests prevent soil erosion, protect against floods, store carbon, and are sites of biodiversity. Hence, efforts are underway to restore forests on a global scale. A wide variety of techniques have been developed for reforestation. The least expensive method is direct sowing of tree seeds, but the success rate is low. Tree nurseries set up near the site of disturbance can be used to grow juvenile trees for transplanting, a more successful method than direct seeding.

Reforestation can use either native or exotic tree species. Fast-growing Eucalyptus trees from Australia have been planted over vast areas of the world to replace native forests that were cleared by humans. Native trees may be more difficult to grow than exotics. However, there is a trend to use native species where possible, because they usually provide greater ecological benefits than introduced species.

The global importance of reforestation was recognized in 2004, when Wangari Maathai was award the Nobel Peace Prize for creating the Green Belt Movement to plant tens of millions of trees across denuded parts of Africa. Reforestation projects not only benefit the environment, they benefit the people who depend on the land for their livelihood.

QUESTIONS AND PROBLEMS

1. Differentiate among: (a) rehabilitation; (b) revegetation; and (c) reclamation.

2. Based on your microbial expertise gained from Chapter 11, what soil microorganisms would sequentially be activated after land application of biosolids?

3. Identify the abiotic factors that influence the approach to ecosystem restoration on any particular project.

4. What factors primarily determine whether active or passive ecological restoration should be undertaken?

5. What is the influence of organic amendments on soil physical and chemical properties?

REFERENCES AND ADDITIONAL READING

Artiola J.F., Brusseau M.L. and Pepper I.L. (2004) *Environmental Monitoring and Characterization.* Academic Press, San Diego, California.

Hobbs R.J. and Harris J.A. (2001) Restoration ecology: Repairing the earth's ecosystems in the new millennium. *Restoration Ecology* **9**, 239–246.

Middleton N. and Thomas D. (eds.) (1997) *World Atlas of Desertification.* Arnold Press, London, England.

Pfadenhauer J. and Grootjans A. (1999) Wetland restoration in Central Europe: Aims and methods. *Appl. Vegetation Science* **2**, 95–106.

PART 4

ATMOSPHERIC POLLUTION

CHAPTER **21**

SENSORY POLLUTANTS, ELECTROMAGNETIC FIELDS, AND RADIOFREQUENCY RADIATION

J. F. Artiola and C. M. McColl

Lightning, a natural but sporadic form of unwanted light and electricity, in contrast to continuous urban light pollution. *Photo courtesy J.F. Artiola.*

21.1 INTRODUCTION

Sensory pollutants derived from sources of heat, light, noise, and odorous air contaminants, as well as electromagnetic and radio frequency fields (EMF/RFs), are ubiquitous in modern times. These physical forms of pollution can not only impact one or more of our senses in subtle and unexpected ways, but also degrade the natural environment. Furthermore, the health and environmental effects of these forms of pollution are often difficult to quantify. Therefore, unlike other forms of pollution, these sensory and physical pollutants are typically uncontrolled and poorly regulated. More research is needed to quantify the short- and long-term effects of these continuous forms of modern pollution. In the meantime, we should all be aware of the potential effects of these pollutants, and devise strategies to limit their generation and exposures.

21.2 HEAT

21.2.1 Sources of Heat Islands

The term "heat island" is used to describe the increase in urban surface and air temperatures above those observed in rural areas. The United States Environmental Protection Agency (U.S. EPA) has reported that downtown urban and suburban air temperatures may be up to 10°F (5.6°C) warmer than temperatures in surrounding areas with natural land cover (Figure 21.1). Heat islands form when natural land cover such as grasses, shrubs, and trees is replaced with urban infrastructure that includes pavement and buildings. An important surface property that influences heat island formation is **albedo**. Albedo is a measurement of a surface's ability to reflect incoming solar radiation and is reported as the ratio of reflected light to incident light. Thus, albedo is

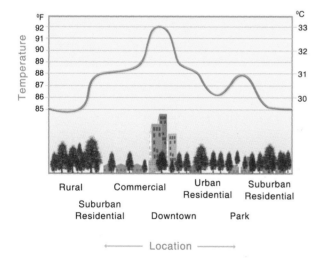

Figure 21.1 Urban heat island late-afternoon temperature profile. From U.S. Environmental Protection Agency. (http://www.epa.gov/heatisland/about/index.html)

measured on a scale from 0 to 1, where a low albedo value indicates high absorbance, and a high albedo indicates high reflectivity of a material (Table 21.1). In general, the lower the albedo value, the higher the temperature, within a given environment. Albedo accounts for the reflectivity of visible, infrared, and ultraviolet wavelengths, and is also called shortwave reflectivity. In general, city structures have a lower albedo than natural cover, yielding an increase in surface and ambient temperatures. Although some flora also have a relatively low albedo, plants provide natural cooling through shading and evapotranspiration. As such, when an area is developed, lower albedo surfaces replace natural land cover, and the natural cooling effect provided by native vegetation is removed. In addition, traffic and urban congestion cause city temperatures to rise as a result of increased waste heat released from vehicles and many other electromechanical devices such as air conditioners. Industrial activity and smokestacks can also increase the magnitude of heat islands.

21.2.2 Effects of Heat Islands

Heat islands can adversely affect human health and the environment. In general, higher rates of heat-related illness, including heat exhaustion and heat stroke, and even death are observed in cities. Populations in cities that are not accustomed to high temperatures are particularly vulnerable. For example, approximately 700 deaths were attributed to a heat wave in Chicago, Illinois, in 1995, and thousands of deaths in France resulted from a massive heat wave in 2003. In 2005, in Phoenix, Arizona, dozens of people died from heat-related causes. In addition, ground-level ozone exposure is of particular concern in and downwind of cities. Although ozone is beneficial in the stratosphere as the "ozone layer," ground-level ozone is a pollutant and a primary cause of smog (see Chapters 23 and 24). In the presence of heat and sunlight, precursor compounds such as nitrogen oxides and volatile organic compounds react to form ground-level ozone. Warmer city temperatures can cause ozone-forming reaction rates to increase, yielding an increase in ground-level ozone concentrations. Exposure to ground-level ozone can induce a variety of health problems, particularly in vulnerable populations such as the young and the elderly. Ozone can irritate and inflame lung tissue reducing lung capacity and may aggravate asthma. In addition, recurring exposure to ground-level ozone may cause permanent lung damage. Ground-level ozone pollution is also a concern for vegetation in and downwind of cities. Specifically, the U.S. EPA (2005) has reported that ground-level ozone interferes with the ability of flora to grow and store photosynthates due to damages to plant foliage. Ground-level ozone can also decrease crop and forest production, leading to an increase in susceptibility to disease, insects, harsh weather, and further pollution.

Heat islands can also benefit society, particularly in cold-climate cities in the winter. Warmer winter temperatures reduce energy demand for heating and help to melt snow and ice on city streets. However, the consequences of

TABLE 21.1 Shortwave reflectivity (albedo) of soils and vegetation canopies.

SURFACE	REFLECTIVITY	SURFACE	REFLECTIVITY
Grass	0.24–0.26	Snow, fresh	0.75–0.95
Wheat	0.16–0.26	Snow, old	0.40–0.70
Maize	0.18–0.22	Soil, wet dark	0.08
Beets	0.18	Soil, dry dark	0.13
Potato	0.19	Soil, wet light	0.10
Deciduous forest	0.10–0.20	Soil, dry light	0.18
Coniferous forest	0.05–0.15	Sand, dry white	0.35
Tundra	0.15–0.20	Road, blacktop	0.14
Steppe	0.20	Urban area (average)	0.15

From Campbell G.S. and Norman J.M. (1998) *Introduction to Environmental Biophysics.* 2nd Ed. Springer-Verlag, New York. Reprinted with kind permission of Springer Science and Business Media.

the heat island effect in the summer are significant, especially in areas with a warmer climate. Higher summer temperatures increase air conditioning and energy demand. They also increase the incidence of air pollutants that include not only greenhouse gases, but also particulates released from power plants in response to increased production to fulfill the higher demand for energy. Increasing greenhouse gases released from power plants also have implications for global warming (see Chapter 24). In addition, the financial cost associated with cooling warmer cities is substantial.

Los Angles, California, is a prime example of an urban heat island. According to the Heat Island group at the Lawrence Berkeley National Laboratory, in 1934, when the Los Angeles basin was primarily comprised of irrigated orchards, the high summer temperature was 97°F. As urbanization of the Los Angeles basin occurred, the high temperature steadily rose to 105°F, and to even higher values in the 21st century. This increase in summer temperatures caused by the heat island effect in Los Angeles has been estimated by Heat Island Group researchers to cost ratepayers approximately $100 million a year.

21.2.3 Controlling Heat Islands

A variety of measures can be taken to mitigate the heat island effect in urban areas, including the application of cool roofs, cool pavements, green roofs, and urban forestry. **Cool roofs** are composed of materials that have a high albedo and therefore high reflectivity. Many cool roof materials also have a high emittance, which refers to the ability of a material to release thermal radiation. Cool roofs can significantly reduce roof temperatures and therefore reduce heat transfer into a building. More specifically, the U.S. EPA (http://www.epa.gov/heatisland) reports that summertime temperatures on a traditional roof may reach as high as 190 °F (88 °C). Cool roofs, by comparison, will only reach 120 °F (49 °C). **Cool pavements** are generally composed of light-colored material with a high permeability. Lighter colored pavements reflect more light and absorb less heat. Pavements with a high permeability allow water to percolate and evaporate, thereby cooling the pavement and surrounding air. **Green roofs** are roofs that have been planted with vegetation to reduce rooftop

temperature and cool the surrounding air. A green roof is comprised of a waterproof membrane, a drainage system, a growing medium, and plants. Green roofs are classified as either intensive or extensive, based on the volume of soil and type of flora planted. Intensive green roofs must have at least 1 foot of soil, usually planted with trees and shrubs. Extensive green roofs have 1 to 5 inches of soil, usually planted with grasses and low-lying plants. Benefits of green roofs are shown in Information Box 21.1.

Urban forestry is the process of incorporating vegetation into an urban area to increase cooling through shading and evapotranspiration. The United States Department of Agriculture Forest Service has reported that urban forestry can decrease mid-day maximum air temperatures by 0.07°F (0.04°C) to 0.36°F (0.2°C) for every 1% increase in canopy cover.

21.2.4 Summary

A heat island occurs when an urban area is warmer than the surrounding area. Urban temperatures may be 10°F (5.6°C) warmer than in rural areas. Heat islands form when natural vegetative cover is removed and replaced with city infrastructure. City infrastructure generally has a lower albedo than natural land cover, and lacks the cooling progenies of the removed flora. Heat islands are of concern for human health and the environment. Heat-related illness and

---

INFORMATION BOX 21.1

Benefits of Green Roofs

1. Reduce roof top temperatures.
2. Reduce temperature of surrounding air.
3. Reduce rainwater runoff to sewer systems.
4. Work as filtration systems for pollutants such as heavy metals and excess nutrients.
5. Provide habitat for birds and other small animals.
6. Aesthetic green space for building residents.

mortality increase in warmer temperatures. Ground-level ozone production also increases with increasing temperature. Ground-level ozone is a health concern for humans and can negatively affect flora. In addition, the energy demand and cost associated with cooling a warmer city are significant. Several measures can be taken to mitigate heat islands including applying cool roofs, cool pavements, green roofs, and urban forestry.

21.3 LIGHT

The part of the electromagnetic radiation spectrum (see also Section 21.6) known as light is of vital importance not only to society, but also to plants and animal life. However, excess light at night induced by human activities can result in light pollution, which has become an issue of environmental concern in recent years. **Light pollution** can be described as the illumination of the night sky by artificial sources of light. **Illuminance**, or **luminous flux density**, is a measure of the total amount of visible light falling on a surface area. Illuminance is measured in either the older unit of foot-candle (Ftc) or the newer SI unit **lux (lx)** (1 Ft = 1 lumen sq ft^{-2} and 1 lux = 1 lumen m^{-2}, where lumen [lm] is a unit of measurement of light). A lumen, also defined as luminous flux, is the SI unit of measurement for the amount of brightness that comes from a light source. For example, a 60-watt bulb may generate 700 lumens and is not as bright as a 100-watt bulb, which can generate 1200 lumens. Light pollution often occurs when night lights are directed upward or outward or fail to deliver all of their light downward.

Key terms used to describe different aspects of light pollution include light trespass, glare, and sky glow. **Light trespass** is defined as intrusive or objectionable light that shines onto neighboring properties. An example of light trespass would be a floodlight shining into a neighbor's bedroom window. **Light glare** is direct light shining into the eye from a bright surface. Glare can come directly from a light fixture or it can be reflected off a surface. Glare can harm vision, cause discomfort, and reduce visibility. **Sky glow** is the yellowish-orange glow seen over many cities and towns and is the composite excess illumination being released from many electrical lighting fixtures. Measuring urban sky glow is not an easy task and depends on a number of variables. Merle Walker (IDA, 1996) measured sky glow for a number of cities in California in the 1970s and developed a formula to estimate sky glow from different distances and populations. The following equation is a simplified form of **Walker's Law**, proposed by the **International Dark-Sky Association (IDA)**, and can be used to estimate the typical level of urban sky glow:

$$I = 0.01 \, Pd^{-2.5} \qquad \text{(Eq. 21.1)}$$

where:

I is the increase in sky glow above natural background light (illuminance) level

P is the population of the city

d is the distance to the city center in km (IDA, 1996).

21.3.1 Sources of Light Pollution

The main source of light pollution is inefficient outdoor lighting. Examples of inefficient outdoor lighting include street lights with inappropriate lamps and fixtures, spot lights pointed toward the sky to mark an event or illuminate an object such as a billboard, yard lights used to illuminate landscaping, and floodlights used for security. Another source of light pollution is light escaping from the inside of buildings at night.

21.3.2 Effects of Light

Light pollution first attracted attention when the general population, and astronomers in particular, were increasingly unable to view the night sky due to excess visible and UV light from cities. In recent years, however, it has become apparent that artificial night lighting is not only a concern for astronomers. Research has shown that artificial night lighting yields negative consequences for a variety of animals including aquatic and terrestrial invertebrates, amphibians, sea turtles and other reptiles, fish, birds, and mammals. For example, light pollution has been shown to affect the reproductive patterns of sea turtles. In Florida, female sea turtles come ashore to dig nests and lay eggs between May 1 and October 31. Bright artificial light on coastlines may deter females from coming ashore to lay their eggs. Further, if female turtles do come to shore, artificial light can disorient them, causing the turtles to wander onto roads, where they may be hit by a vehicle. In addition, sea turtle hatchlings generally hatch at night and instinctively head toward light. Before artificial light was installed along the coastline, the moon reflecting off of the ocean surface was brighter than the inland area. Light pollution can cause hatchlings to wander the shoreline or to head inland instead of into the sea. Therefore, hatchlings are exposed to a higher risk of predation or can be run over by vehicles on nearby roads.

Light pollution can also affect migrating bird navigation. Many bird species use the light from constellations to guide them during their migration. Artificial light can confuse migrating birds, causing them to fly off course. Bird kills are common near floodlit smokestacks and radio transmission towers, where birds become confused by the artificial light and fly into these structures. Light pollution can not only have devastating effects on wildlife by disrupting both navigation and reproductive patterns, some scientists hypothesize that light pollution may affect human health by disrupting hormone regulation.

Presently, it is estimated that about two-thirds of the world's population live in areas where the night sky brightness is considered polluted (Cinzano et al., 2001). In the U.S., the same source estimates that 99% of the population lives in areas polluted with excessive night sky brightness.

Figure 21.2 North American map of artificial night sky brightness. Courtesy Cinzano et al, (2001) The First World Atlas of the Artificial Night Sky Brightness. Monthly Notices of the Royal Astronomical Society, Vol 328 pp.689-707. Reprinted with permission of Blackwell Publishing.

Of those, about 40% can no longer see stars with their eyes adjusted to night vision (Figure 21.2).

Light pollution also contributes to an increase in other environmental pollutants. The excess energy generated to power unnecessary artificial night lighting increases the volume of pollutants released from power plants. For example, power plants release increased volumes of greenhouse and acid rain-causing gases, including carbon dioxide and nitrogen and sulfur oxides, into the atmosphere. In addition, power plant cooling processes contribute to increased thermal pollution of water bodies (see Chapter 25 for additional information).

21.3.3 Controlling Light Pollution

There are currently no federal or state regulations for controlling light pollution. However, several cities in the U.S. have developed guidelines for light pollution. For example, along the Florida coastline, cities have established ordinances that regulate artificial light at night during the nesting season of

sea turtles. In addition, the city of Tucson, Arizona, initiated lighting ordinances to control outdoor night lighting to reduce interference with astronomical observations at the nearby Kitt Peak National Observatory (Figures 21.3a and b).

Light pollution is expensive and contributes to other forms of environmental pollution. The International Dark-Sky Association has estimated that light pollution costs the United States an estimated $2 billion per year in wasted energy. Controlling light pollution conserves energy therefore decreasing energy costs and the pollution associated with energy production.

Light pollution can be controlled by simply turning off unnecessary lights and using appropriate lamps, fixtures/casings, and designs for lighting efficiency. Street lights commonly contain either **high-pressure sodium (HPS) lamps** or **low-pressure sodium (LPS) lamps,** and produce yellowish colored lighting. LPS lamps are more efficient (125 lumens W^{-1}) (W = watt) than the more common street light HPS lamps (83 lumens W^{-1}). However, LPS lamps do not give as true a color to surfaces as HPS lamps do. Other examples of types of lamps include mercury vapor lamps (42 lumens W^{-1}), which are not only less efficient than sodium lamps, but also produce bluish light, glare, and UV light pollution; metal halide lamps, a newer and more efficient but expensive source of white light (59 lumens W^{-1}); the durable compact fluorescent lamps (60–70 lumens W^{-1}); and the most common, less durable and inefficient incandescent (halogen and tungsten) lamps (14–17 lumens W^{-1}) (Calgary Centre 2005).

Full-cut-off (FCO) and **semi-cut-off (SCO)** lamps are important lamp features designed to control light pollution. FCO lamps, when mounted correctly, emit no light above the horizontal. SCO lamps, when mounted correctly, emit little to no light upward. There are also a variety of fixtures and casings that can be used for "shielding" lamps and controlling light pollution. When night lighting is designed and installed correctly, the lamp wattage will be selected to appropriately light the task without over-lighting, and light will only be directed to the area intended to be

Figure 21.3a Car lot lights in Tucson before light pollution control City ordinance. Photo courtesy International Dark-Sky Association(IDA), Tucson, AZ. (http://www.darksky.org)

Figure 21.3b Car lot lights in Tucson after light pollution control City ordinance. Photo courtesy International Dark-Sky Association (IDA), Tucson, AZ. (http://www.darksky.org)

illuminated. In addition, the light should evenly spread across the surface area intended.

21.3.4 Summary

Light pollution is an emerging area of environmental concern. It is caused by inefficient outdoor lighting and indoor lighting that escapes through windows at night. Light pollution is of concern to astronomers, yields negative consequences for wildlife, and contributes to an increase in other environmental pollutants such as greenhouse and acid rain–causing gases. Light pollution can be controlled by turning off unnecessary lights and using appropriate lamps and fixtures and designs for lighting efficiency. Controlling light pollution will protect the night sky, protect light-sensitive wildlife, conserve energy, reduce the volume of environmental pollutants produced as a byproduct of energy production, and save money.

21.4 NOISE POLLUTION

Noise is defined as unwanted or unwelcome sound that produces annoyance or physiological stress. If noise is loud enough and continuous, it can produce temporary and even permanent damage to our hearing system. Loud noises can also place people and animals in danger, because they may prevent the hearing of potential or impending dangers. In the U.S., the **Noise Control Act of 1972** gave the U.S. EPA powers to set noise emission standards for major transportation and industrial sources of noise, to protect public health and welfare. Since the 1970s the Occupational Safety and Health Administration (OSHA) has implemented standards for noise exposure in the workplace for worker protection. However, to date the EPA has set only transportation-related noise regulations. In addition, EPA does not have the funding to enforce or revise these regulations. Thus, there are no explicit national, state, or local laws that protect the public against noise pollution. Existing local noise ordinances vary widely and often they are disregarded or poorly enforced.

This section presents a short summary of the physics of noise, followed by a discussion of the sources of activities that produce noise and the increasing impacts of noise on modern life. Currently, noise is an insidious form of pollution that is increasing in modern day life.

21.4.1 The Physics of Sound

Sound is energy in the form of airborne vibrations or pressure waves that can be sensed or heard through our hearing system. The vibrations of audible sounds range from about 16 to 20,000 Hertz (oscillations per second). All media, including air, liquids, and solids, transmit sound waves, but the speed of sound movement varies dramatically through each medium. For example, sound travels about 346 meters $\sec^{-1}$ through air, and about 4.2 and 14.8 times faster through water and iron metal, respectively.

Since sound is felt as pressure (P_r), its energy is usually given per unit area and in relation to its source or point of origin. Under ideal conditions of a homogeneous medium with no boundary surfaces, the pressure felt from a single point source (P_o) of sound felt at a distance (r) is defined by the following formula:

$$P_r = P_o/r$$

This assumes that under ideal conditions the spherical wave transmits its expanding power to a portion of an area (cm^2) of a sphere. Note that the pressure (P_r) is in dynes cm^{-2}, but the source pressure (P_o) is in dynes cm^{-1} (pressures are the root-mean-squares values) (Liu et al., 1997).

Since the range of sound pressures that the human ear can detect is at least eight orders of magnitude, a logarithmic scale is used to report sound pressure, with a base reference of 0.0002 dynes cm^{-2} being equal to 0 decibels (dB). Note that each order of magnitude in pressure change corresponds to a log scale change of 20 dB.

The sound intensity (I_r) is defined by the following equation:

$$I_r = W/4\pi r^2$$

where:

W = power of the sound per unit area (watts cm^{-2})

r = distance from the source (cm)

The sound pressure (P_o), distance from the source (r), medium density (ρ), and wave velocity through the medium (c) are related as follows:

$$I_r = P_o^2/r^2 \, \rho c$$

Table 21.2 shows a range of sound pressures, levels, and power sources. Note that while the sound pressures vary by about eight orders of magnitude, the power sources and therefore the sound intensities vary by about 18 orders. Table 21.3 shows the scales of common sound sources and their effects on humans and community responses.

Air sound can be transmitted, reflected, absorbed, and distorted by solid objects. For example, a sound wave may pass through a window, transferring some of its energy into the glass in the form of mechanical energy. The remainder of the sound wave is likely to be distorted and much lower in intensity. These and many other complex sound phenomena are studied in detail in the advanced environmental physics field of acoustics (Liu et al., 1997 and Boeker and van Grondelle, 1999).

21.4.2 How We Hear Noise

Noise is felt by our hearing system and converted into electrical impulses that travel to our brain where they are processed. Briefly, sound waves enter our ear canal (outer ear) and travel through our eardrum. On the other side of the eardrum (middle and inner ear), the cochlear organ converts the sound signal into electrical impulses that are sent to the brain. The sensitivity of the human ear to sound depends on the sound frequency and its intensity, which in turn are

TABLE 21.2 Representative sound pressures and power of sources.

SOURCE AND DISTANCE	SOUND PRESSURE (dynes $/\text{cm}^{-2}$)	SOUND LEVEL (dB)
Saturn rocket motor, close by	1,100,000	195
Military rifle, peak level at ear	20,000	160
Jet aircraft takeoff; artillery, 2500'	2,000	140
Planing mill, interior	630	130
Textile mill	63	110
Diesel truck, 60'	6	90
Cooling tower, 60'	2	80
Private business office	.06	50

SOURCE	ACOUSTIC POWER OF SOURCE
Saturn rocket motor	30,000,000 watts
Turbojet engine	10,000 watts
Pipe organ	10 watts
Conversational voice	10 microwatts
Soft whisper	1 millimicrowatt

From Liu H.F., et al (1997). *Environmental Engineer's Handbook* 2nd ed. Lewis Publishers, Boca Raton, FL. Reprinted by CRC Press.

TABLE 21.3 Sound intensity factors, sound levels of common sound sources and their effects.

SOUND INTENSITY FACTOR	SOUND LEVEL (dB)	SOUND SOURCES	EFFECTS — PERCEIVED LOUDNESS	EFFECTS — DAMAGE TO HEARING	EFFECTS — COMMUNITY REACTION TO OUTDOOR NOISE
1×10^{18}	180	Rocket engine		Traumatic injury	
1×10^{17}	170				
1×10^{16}	160				
1×10^{15}	150	Jet plane at takeoff	Painful		
1×10^{14}	140			Injurious range; irreversible damage	
1×10^{13}	130	Maximum recorded rock music			
1×10^{12}	120	Thunderclap, Textile loom		Danger zone; progressive loss of hearing	
		Auto horn, 1 m (3.3 ft) away	Uncomfortably loud		
1×10^{11}	110	Riveter			
		Jet flyover at 300 m (985 ft)			
1×10^{10}	100	Newspaper press			
1×10^{9}	90	Motorcycle, 8 m (26 ft) away, Food blender	Very loud		Vigorous action
1×10^{8}	80	Diesel truck, 0 km/hr (50 mph), 15 m (50 ft) away, Garbage disposal		Damage begins after long exposure	Threats
1×10^{7}	70	Vacuum cleaner	Moderately loud		Widespread complaints
1×10^{6}	60	Ordinary conversation, Air-conditioning unit, 6 m (20 ft) away			Occasional complaints
1×10^{5}	50	Light traffic noise, 30 m (100 ft) away			No action
1×10^{4}	40	Average living room			
1×10^{3}	30	Bedroom, Library	Quiet		
1×10^{2}	20	Soft whisper			
1×10^{1}	10	Broadcasting studio	Very quiet		
1×10^{0}	0	Rustling leaf, Threshold of hearing	Barely audible		

From D.D. Chiras, *Environmental Science, Creating a Sustainable Future*, 6th ed. 2001: Jones and Bartlett Publishers, Sudbury, MA. www.jpub.com. Reprinted with permission.

related to the distance from the source. Humans have different hearing sensitivities to sound. For example, children can hear high frequency sounds (above 150 Hz) much better than low frequency sounds (below 125 Hz, bass) (Boeker & van Grondelle, 1999). However, hearing begins to decline significantly after the age of 30. This change is most pronounced in men and for high-frequency sounds (at or above 4000 Hz) (Liu et al., 1997). Hearing loss, due to repeated exposure to loud noise, can also occur prematurely from injury to the hair cells (sound receptors) found inside the cochlear. There is evidence that exposure to noises from modern society accelerates hearing losses (Liu et al., 1997).

21.4.3 Sources of Noise

Modern life has brought about many types of sources of noise generated from industry, construction, transportation, and community and household activities. With few exceptions, these sources of noise are artificial byproducts of industrialization, mechanization, crowded urban living, mechanized transport, and electronically reproduced and broadcasted sound. Table 21.3 shows the typical noise levels of common noise sources. Noises can be continuous or intermittent and sporadic, depending on the source. In general, industrial noises tend to be continuous-repetitive or momentary. Transportation noises can be continuous, random, or intermittent. In addition, they can have increasing or decreasing sound intensities. Urban environments produce all types of noises that tend to vary significantly by location and time of day. The most disturbing urban noises are sporadic noises produced by sirens, trucks, motorcycles, and car alarms. In suburbia, background noise associated with urban environments is minimal, making sporadic noises associated with motorcycles, car and truck engines, and dogs barking, much more noticeable. Household noises may be internal or include significant outside noise encroachment. Modern homes are better sound insulated, but still have near-constant sources of low-level noise, including air conditioning or heating units, refrigerators, and personal computers. Sporadic sources of noise in household-include telephones, music, dogs, and assorted mechanical noises from doors, chairs, and dishes. Young people are often exposed to excessive sources of noise at rock concerts and also through the use of ear phones to listen to music.

Although the U.S. EPA has noise emission limits for motorcycles of 70–80 dB, these standards (USEPA 40CFR, Part 205) are often exceeded and not enforced. Figure 21.4 shows a motorcycle fitted with loud straight mufflers that can generate noise levels between 100–120 dB at close proximity (see Table 21.3). Increasingly, communities are fighting this form of pollution with their own local enforcement of new noise ordinances.

21.4.4 Effects of Noise

Noise affects us in different ways and to varying degrees, depending on our age, habits, health, and mood. Very loud sustained, repetitive, and even sporadic noises such as gunshots

Figure 21.4 Motorcycle fitted with straight mufflers capable of producing noise levels above 100 dB. Photo courtesy J.F. Artiola.

can damage our hearing progressively or traumatically (see Table 21.3). Progressive loss of hearing starts with sustained exposure to noise around 75 dB or above (Liu et al., 1997). Eardrum rupture can occur when exposed to very loud sharp noises, such as firecrackers or rifle shots. Other factors that affect the degree of hearing loss include the duration of exposure, including constant versus short duration noises. The ear cannot close itself to sound or protect itself well against loud instantaneous noises of frequencies above 2,000 Hz (Liu et al., 1997).

Noises can have varying psychological impacts, depending on the level, duration, location, and time of occurrence, and our mental state. In general, noises can be distracting or annoying. More serious effects include interference with verbal communication, reduced work efficiency, and the production of fatigue. Loud music may be welcomed during a party or in a club, but is not very pleasant when one is reading or is tired or sick. Being awoken from sleep by barking dogs, a passing motorcycle, or car alarm is unpleasant and disruptive of sleep patterns. Repeated sleep disruptions can lead to poor concentration, mood changes, and stress during the day. The documented physiological effects of noise to humans include changes in heartbeat and blood pressure, and increased respiration and pupil dilation (Vijayalakshimi and Phil, 2003).

21.4.5 Summary

Noise, one of the byproducts of modern life, is an insidious and increasing form of pollution. Evidence is mounting that documents the negative physical and mental effects of noise on humans. Yet noise pollution remains one of the most unregulated forms of modern pollution that increasingly degrades our quality of life. The effects of noise can include annoyance, hearing impairment, and adverse physiological effects.

21.5 ODOR AS A SENSORY POLLUTANT

The sensation of odor is due to the stimulation of the olfactory organ in response to an exposure to volatile chemical odorant. Located in the nose, this organ likely evolved as a sensor capable of remotely detecting physical dangers and the presence of other animals at a distance without visual cues. Thus, odors provide an early warning system for natural dangers and modern forms of air pollution. Although low-level exposure to some odorous chemicals may or may not be detrimental to one's health, the sensations they produce can be annoying. Currently, studies are evaluating odors that even at low levels may cause psychosomatic disease (Schiffman et al., 2000). In this section, we will discuss some common odors and the responses associated with them.

21.5.1 Odor Response

The response to a particular odor has a detection threshold that varies significantly from person to person. Thus, odor thresholds are determined by a panel in which 50% of the individuals respond to the odor and 50% do not (Altwicker et al. in Liu et al., 1997). Environmental odors are usually complex mixtures of chemicals whose components are difficult to identify and quantify. In this case, an odor recognition threshold can be determined as described before, but using air/odor dilution ratios. This is commonly done to determine the potential for adverse affects of background or sporadic environmental odors, associated for example, with municipal wastewater treatment plants or industrial air emissions. This odor threshold unit is defined as:

> **Odor Unit** = Volume of sample diluted to threshold response/volume of original sample

See Liu et al. (1997) for an extended description of this and other types of odor thresholds.

21.5.2 Odor Perceptions

The human response to odor types and concentrations varies widely, but has a normal distribution with threshold values that can vary over several orders of magnitude (Altwicker et al. in Liu et al., 1997). The acceptance of an odor is referred to as **hedonic tone**, which determines how pleasant or unpleasant a particular odor is (Table 21.4).

However, repeated or prolonged exposure to a particular odor can become unpleasant or we may become desensitized to it. Tolerance to odors deemed otherwise unpleasant is common in industrial workers and chemists, for example, who work with the same air chemical sources daily. However, nuisance complaints from communities about odors related to industrial and agricultural activities are more common, due in part to the growth of suburbia encroaching into industrial and agricultural environments. Also, there is in-

TABLE 21.4 Common odors and potential powers.

ODOR CHARACTER DESCRIPTOR	POTENTIAL SOURCES
Nail polish	Painting, varnishing, coating
Fishy	Fish operation, rendering, tannin
Asphalt	Asphalt plant
Plastic	Plastics plant
Damp earth	Sewerage
Garbage	Landfill, resource recovery facility
Weed killer	Pesticide, chemical manufacturer
Gasoline	Refinery
Airplane glue	Chemical manufacturer
Household gas	Gas leak
Rotten egg	Sewerage, refinery
Rotten cabbage	Pulp mill, sewage sludge
Cat urine	Vegetation

From Liu H.F., et al (1997). *Environmental Engineer's Handbook* 2nd ed. Lewis Publishers, Boca Raton, FL. Reprinted by CRC Press.

creased public awareness about the potential health effects of some chemicals.

21.5.3 Sources of Odor

Odors associated with raw sewage or municipal wastewater treatment are easily recognized by these two characteristics: the odor of rotten eggs and the odor rotten cabbage. There are numerous chemicals associated with these odors, including sulfur-based compounds like hydrogen sulfide and mercaptans, nitrogenous compounds like amines and scatole, aldehydes like butyraldehyde, and acids butyric acid (Stuetz and Frechen, 2001). Individually and collectively, these chemicals produce odor responses with very low thresholds and are commonly associated with neighborhood nuisance complaints. To date, there is inconclusive evidence that the repeated exposure to these odors produces long-lasting adverse physical effects. However, reported responses include psychological effects, nausea, and stress (Liu et al., 1997). Although chemicals like hydrogen sulfide are very toxic, most odors act as warning signs that often become a nuisance at levels well below what are considered toxic by industrial and ambient air quality standards. For example, in the case of hydrogen sulfide, the human threshold of detection is ~0.15–0.5 ppb–v, but the OSHA has set an occupational air safety exposure of 10 ppm–v. To avoid nuisance complaints, the World Health Organization (WHO) recommends that ambient air hydrogen sulfide concentrations do not exceed 0.5 ppb–v. Ammonia gas is a large component of odors derived from animal wastes and biosolids. Ammonia gas has a very sharp, pungent odor, which is generally only clearly recognized by humans at concentrations above 50 ppm–v (Merck Index, 1996). Common sources of this odor include dairy corrals and animal waste lagoons. Ammonia and other odorous gases emissions are also of concern in the land treatment of biosolids. Figure 21.5 shows the fluxes of

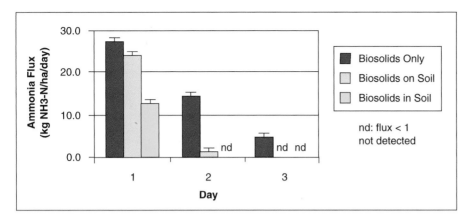

Figure 21.5 Ammonia fluxes were measured using a chamber placed over samples of pure biosolids (~8% biosolids), biosolids applied to the surface of dry soil, and biosolids incorporated within dry soil. The figure illustrates the effectiveness of reducing ammonia emissions by incorporation of the biosolids into the soil. Ammonium N applied within the biosolids was ~364 kg NH_4-N ha^{-1}. From A.D. Matthias and J.F. Artiola, unpublished data.

ammonia from biosolids over a 3-day period under laboratory conditions.

There are numerous types of odorous chemicals associated with industrial and transportation activities. A readily recognized odor is that of methyl methacrylate, which is used in the manufacture of plastics and resins. This synthetic toxic chemical is an irritant that like ammonia affects the mucous membranes and that, like hydrogen sulfide, has a very low odor threshold ~0.1–100 ppb–v. Odors associated with transportation include distinctive gasoline fumes that are composed primarily of chemicals such as benzene, xylenes, and toluene that have relatively high odor thresholds (~0.05–4 ppm–V) and that are known to be carcinogenic and flammable.

21.5.4 Regulations

Odors emanating from farming and animal feed activities are increasingly being regulated under state and local laws. These new regulations attempt to address the nuisance issues that are more common due to the expansion of suburban communities into rural farming areas. However, indirectly, odors associated with toxic chemicals have been regulated in the U.S. under the 1970 Clear Air Act, and the 1990 Amendments. For example, sulfur dioxide, a strong pungent irritant exhaust gas, has a national ambient air quality standard of 0.03 ppm–v (see Chapter 4). Also, since 1974, the OSHA has set worker occupational levels for many individual toxic chemicals like hydrogen sulfide.

21.5.5 Summary

Odors act as early warning signs of dangers that we often cannot see or hear. Odors indicate the presence of common modern and natural pollution sources like fuels and wastes.

Humans can become desensitized to odors, but extended exposures can cause adverse physiological and emotional responses and even death when exposures to some chemicals reach acute toxic levels. Nuisance odors are being regulated locally, and in addition, the U.S. EPA and OSHA have set maximum allowable levels for many individual pollutants found indoors in ambient air and in the work place.

21.6 ELECTROMAGNETIC FIELDS AND RADIOFREQUENCY RADIATION

Electromagnetic fields (EMF) and **radiofrequency (RF) radiation** have been called invisible forms of modern pollution. However, before evidence in support of, or against it, is discussed, we must have some understanding of the nature of these phenomena. Before electromagnetic radiation was fully understood, scientists in the 18th and 19th centuries observed the phenomena of magnetism and electricity from some minerals and metal wires. The scientist Michael Faraday observed and related these two phenomena, describing magnetism as invisible "lines of force" (Stern, 2001). Also related to electricity is an electric field that can be detected in the presence of charge gradient, such as near the opposite ends (+) and (−) of a battery. We are constantly exposed to earth's permanent magnetic field. However, much stronger electric (E) and magnetic (M) fields (F), abbreviated EMF, are present around us, as the result of extensive use of electrical appliances and electrical power lines. One important property of the two forces is that if electricity is shut off, they cease to exist (Figure 21.6). Magnetic (M) field intensities are measured in gauss (G) or milligauss (mG) in the U.S., and in tesla (T) in the rest of the world. The conversion factor between G and T is 10,000, for example:

10,000 G = 1 T

1 G = 100 micro T (μT)

1 microT = 10 milliG (mG)

Electric (E) fields are measured in volts per meter (V m⁻¹).

With the discovery of the association between electricity and magnetism, Ampere and Maxwell determined that light in fact was an electromagnetic (EM) wave composed of electric and magnetic forces. Other scientists like Marconi and Edison later determined that these EM waves could be created and released into space and used to communicate information, such as radiofrequency (RF) waves, and to produce light, for example, the incandescent lightbulb (Stern, 2001). Thus, unlike EMF, electromagnetic waves, which include radiofrequency radiation (RF), exist long after their source is shut off (Figure 21.6). Electromagnetic waves, also described as particles (photons), are defined primarily by their length (wavelength) and frequency of alternation. (See Figure 21.7 for a definition of the electromagnetic spectrum

using these two variables.) This figure shows that there is energy associated with EM waves, defined by the following relationships:

$$E = hv = hc\,\lambda$$

where:

E is the energy in photons

v is the frequency (cycles per unit time) in Hertz (Hz) λ is the wavelength (distance between crests) in meters (m)

h is Planck's constant (6.6262×10^{-34} joules sec)

c is the speed of light in vacuum (3×10^8 m sec⁻¹)

21.6.1 Effects and Levels of EMF

There are concerns about the long-term exposure effects of ever more common electrical and magnetic fields (EMF). Scientists have long known that all living organisms have

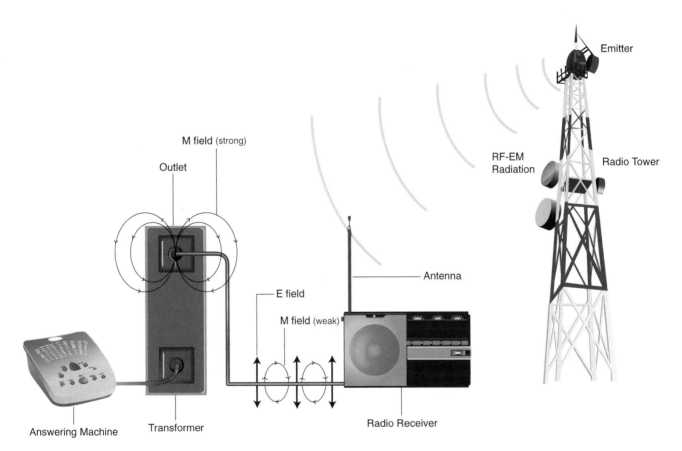

Figure 21.6 Diagram depicting electric (E) and magnetic (M) fields (F) emanating from an electrical wire connected to a household AC outlet. The strongest M fields come from plugged in transformers and running motors, and weak M fields come from paired wires. E fields (arrows) will exist as long as the cable is plugged into a live electrical outlet, even if the radio is turned off, whereas the weak M field (wire circles) will only exist when the radio is on (drawing current). Note the EM fields typically decreased rapidly at short distances (~0.5–1 meter) from the electrical wires or appliances (motors, transformers, etc.) that also emit EM fields. RF radiation can be detected with a common radio with an antenna. The RF is electromagnetic radiation emitted by radio stations (towers).

TABLE 21.5 Guidelines for EMF exposures.

GUIDELINES FOR POWER-FREQUENCY* FIELD EXPOSURE OF THE GENERAL PUBLIC FROM ICNRP**
At 50 Hz: **100 microT (1 G) and 5 kV m^{-1}**
At 60 Hz: **84 microT (0.84 G) and 4.2 kV m^{-1}**
GUIDELINES FOR OCCUPATIONAL EXPOSURE TO POWER-FREQUENCY FIELDS FROM ACGIH***
At 60 Hz: **1,000 microT (10 G)**
GUIDELINES FOR OCCUPATIONAL EXPOSURE TO POWER-FREQUENCY* FIELDS FROM ICNIRP
At 50 Hz: **500 microT (5 G) and 10 kV m^{-1}**
At 60 Hz: **420 microT (4.2 G) and 8.3 kV m^{-1}**

See Moulder (2004) for a more complete list of published EMF exposure guidelines.

*Guidelines of the maximum power of the electric and magnetic fields (EMF) are provided for household alternating electricity (AC), which is delivered at frequencies of 50 Hertz (Hz) in Europe and at 60 Hz in the U.S.

**The International Commission on Non-Ionizing Radiation Protection (ICNIRP).

***The American Conference of Governmental Industrial Hygienists (ACGIH).

electrical fields of their own and that external electrical or magnetic sources can disrupt these fields. We also know that the earth itself is surrounded by a large permanent magnetic field, emanating from its own core, that gives rise to the north and magnetic poles. Epidemiological studies have shown some weak correlations between high occupational EMF exposures and people that live near power lines who seem to have increased cancer risks and many other health problems. However, these claims are frequently anecdotal, and to date no direct connections have been established. Some studies have been inconclusive or contradicting (Moulder, 2004). To date, no adverse health effects have been conclusively demonstrated from low-level exposures to E fields, probably because these fields are much shorter than the concurrent M fields.

With this limited evidence, some agencies and professional organizations groups have published guidelines for exposure to EMF (Moulder, 2004). These are listed in Table 21.5. Until more research helps us to better understand the effects of EMF on human health, experts suggest that exposure to EMF sources should be minimized. At home and at work long-term close proximity (50 cm) to the following electrical devices should be avoided: motors such as hair dryers, fans, or blenders; transformers; and chargers including those wall units extensively used in modern appliances such as telephones, answering machines, computers, and TV screens.

21.6.2 Effects and Levels of RF

Figure 21.7 shows that the energy of short electromagnetic waves like x-rays and UV light have more energy than longer waves like infrared (IR) and radiofrequency (RF) waves. Ultraviolet and x-ray light is called ionizing radiation because there is enough energy in the waves (photons) to break up chemical bonds and force electron transitions in atoms. Thus, we know that exposure to these two types of EM radiation is in general harmful to living organisms.

On the other hand, RF radiation only has enough energy to heat matter and induce currents (Moulder, 2004) (see also Figure 21.7). Thus, RF does not appear to have sufficient energy to damage tissue unless the EM radiation is very strong, as might be the case next to a powerful emitter such as a radio or wireless telephone tower. Therefore, there are safety guidelines for the maximum RF power exposures emitted from wireless telephones and towers, some examples are presented in Table 21.6.

We know that we are constantly bombarded by EM radiation from outer space by both visible (sunlight) and invisible range such as RF, UV, and even x-rays. However, there are also increasing sources of RF emanating from radio and TV, portable telephone transmitters, and other modern wireless communication devices. Scientists are still studying the long-term effects associated with the routine use of devices like cell phones, and the results thus far, taken as a whole, appear to be inconclusive (Moulder, 2004). Nonetheless, experts suggest that the use of portable sources of RF radiation such as cell phones should be limited and kept away from potentially sensitive organs like the brain.

21.6.3 Summary

Numerous studies have thus far failed to show conclusively any adverse health effects associated with the exposures to EMF and RF sources related to modern occupational and general public activities. However, to date science cannot guarantee that there is no risk associated with exposure to these modern and common forms invisible pollution. Here, the classic phrase "more research is needed" is appropriate.

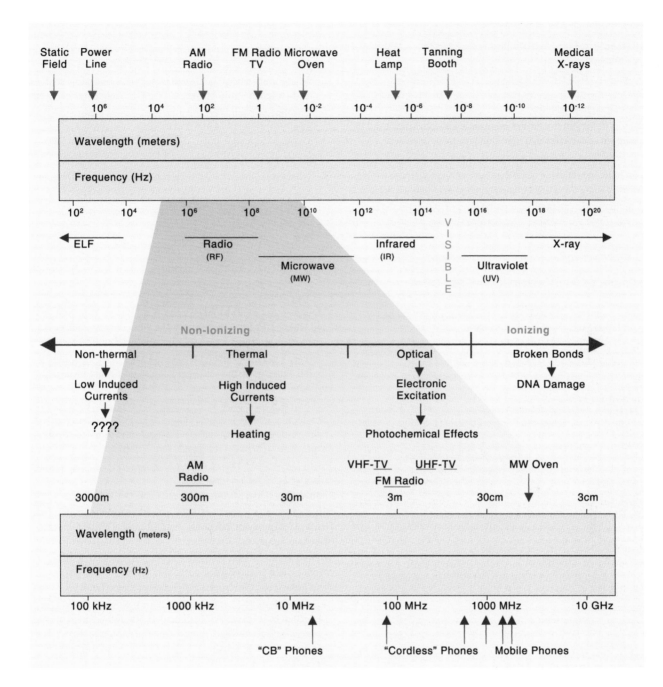

Figure 21.7 The electromagnetic spectrum. From J.E. Moulder, 2003.

TABLE 21.6 Wireless telephone RF power standards.
   Maximum energy in mW cm^{-2}.*

1800–2000 MHz frequency range, 1999 ANSI/IEEE** exposure
   standard for the general public: **1.2 mW cm^{-2}**
900 MHz frequency antennas, ANSI/IEEE exposure standard for
   the general public: **0.57 mW cm^{-2}**

For a complete list of these standards, see Moulder (2004).

*The power density of RF is usually given in mW cm^{-2} = milliwatts
(a unit of energy) per cm^2 (area).

**Electrical and Electronics Engineers and American National Standards Institute (ANSI/IEEE).

QUESTIONS AND PROBLEMS

1. a. Identify and describe two factors that influence heat island formation.

   b. Why are heat islands a concern for human health?

   c. Describe two measures that can be implemented to help mitigate heat islands.

2. Why is light pollution an environmental concern? Include examples of the effects of light pollution in your answer.

3. How can light pollution be reduced or prevented? Explain.

4. Can loud and repeated noise damage our hearing permanently? At what sound levels does damage to our ears begin? Explain.

5. A military rifle can produce a sound level of 160 dB at the barrel. If you are standing 10 meters away, what sound level will you experience? Could repeated exposure to this noise damage your ear? Hint: Use sound pressure to calculate sound level.

6. What common household sources/activities (industrial sources already listed in Table 21.4) could the following odors come from? Gasoline, asphalt, airplane glue, nail polish, damp earth, rotten egg.

   List three specific chemicals associated with one or more of the above sources of odors.

7. What types of EM radiation are known to produce DNA damage? And what types of radiation can heat the skin?

REFERENCES AND ADDITIONAL READING

Boeker E., and van Grondelle R. (1999) Environmental Physics. John Wiley & Sons, New York. Chapter 6.

Calgary Centre. (2005) Light pollution abatement site—Calgary Centre. http://calgary.rasc.ca/lp/greengas.html.

Campbell G.S. and Norman J.M. (1998) *Introduction to Environmental Biophysics.* 2nd Ed. Springer-Verlag, New York.

Chiras, D.D. (2001). *Environmental Science, Creating a Sustainable Future,* 6th ed. Jones and Bartlett Publishers, Sudbury, Massachusetts.

Cinzano P., Falchi F. and Elvidge C.D. (2001) The First World Atlas of the Artificial Night Sky Brightness. *Mon. Not. R. Astron. Soc.* **328**, 689–707.

International Dark-Sky Association. (1996) Estimating the Level of Sky Glow Due to Cities. Information Sheet 11. http://www.darksky.org/infoshts/is011.html.

Lawrence Berkeley National Laboratory. http://eetd.lbl.gov/heatisland.

Liu H.F., Liptak B.G. and Bouis P.A.. (1997) *Environmental Engineer's Handbook.* 2nd ed. Lewis Publishers, Boca Raton, Florida Chapters 5 and 6.

Merck Index. (1996) 12th ed. Merck Research Laboratories Division of Merck & Co., Inc., Whitehouse Station, New Jersey.

Mizon B. (2002) *Light Pollution: Responses and Remedies.* Springer-Verlag London Limited, Singapore.

Moulder J.E. (2004) Electromagnetic Fields and Human Health. http://www.mcw.edu/gcrc/cop/powerlines-cancer-FAQ/toc.html#1.

Moulder J.E. (2004) Mobile Phone (Cell Phone) Base Stations and Human Health. http://www.mcw.edu/gcrc/cop/cell-phone-health-faq/toc.html.

Moulder J.E. (2004) Static Electric and Magnetic Fields and Human Health. http://www.mcw.edu/gcrc/cop/static-fields-cancer-faq/toc.html.

Schiffman S.S., Walker J.M., Dalton P., Lorig T.S., Raymer J.H., Shusterman D. and Williams C.M. (2000) Potential health effects of odor from animal operations, wastewater treatment and recycling of byproducts. *Journal of Agromedicine.* **7**, 7–81.

Stern D.P. (2001) The exploration of the earth's magnetosphere. Chapter 5: Magnetic field lines. http://www-istp.gsfc.nasa.gov/Education/Intro.html.

Stuetz R. and Frechen F.-B. (2001) Odours in Wastewater Treatment; Measurement, Modelling and Control. IWA Publishing, London England.

United States Environmental Protection Agency (U.S. EPA). (2005). Heat Island Effect. http://www.epa.gov/heatisland/about/healthenv.html.

Vijayalakshimi K.S. and Phil M. (2003). Noise Pollution. In M.J. Bunch, V. Madha Suresh, and T. Vasantha Kumaran, eds. *Proceedings of the Third Environmental Conference on Environmental Health.* Chennai, India. 15–17 Dec. 2003, pages 597–603.

CHAPTER 22

INDOOR AIR QUALITY

K. A. Reynolds

Indoor fungal contamination. *Photo courtesy TM Building Damage Restoration, Tucson.*

22.1 FUNDAMENTALS OF INDOOR AIR QUALITY

For more than a decade, the U.S. Environmental Protection Agency (U.S. EPA) has ranked indoor air pollution among the top five risks to public health. Potential hazards that may be associated with indoor air include particulates, microbes, and chemicals. Defining the relative impact of specific indoor air pollutants is difficult because of individual genetic susceptibility and exposure to multiple hazards. Harmful exposure levels are defined for many chemical air pollutants. However, for other indoor pollutants, such as mold, little data is available with respect to acceptable exposure levels. Frequently, the consequences of exposure are evaluated based on retrospective, or even anecdotal, evidence.

In industrialized countries, about 90% of an individual's time is spent indoors, where air quality may be two to five times worse than outdoor air (U.S. EPA, 2001). Approximately one-third of all buildings are expected to have **indoor air quality (IAQ)** problems at some point during their operational lifespan. Health effects of poor IAQ range from mild and acute (cold and flu-like symptoms, headaches, and nausea), to severe and chronic (allergies, asthma, developmental disorders, cancer, and death). While a number of these building-related illnesses have been traced to specific building problems, conditions of complex symptomology related to chemical and/or biological IAQ problems are often vaguely diagnosed as **sick building syndrome (SBS)** and likely involve multiple pollutants acting collectively or synergistically. Excessive complaints, related to IAQ, are generated within 30% of new and remodeled buildings, worldwide. In addition, nearly one in four U.S. workers believe there are air quality problems in their work environments, and most thought that these problems affected their work performance (Kreiss, 1990).

22.2 SOURCES OF INDOOR AIR POLLUTANTS

Human activities and climate control efforts that were originally designed to make our lives more comfortable can exacerbate pollutant levels. For example, the production of energy-efficient homes has resulted in a decrease of indoor-outdoor air exchange, leading to decreased dilution of indoor air pollutants. Elevated humidity levels and temperature throughout buildings or in specific microzones, such as behind bathroom showers, or within cooking areas, also play a role in the increase of certain pollutants. These microzones can increase the growth of molds or result in increased emissions of formaldehyde from manufactured materials. Fuel-burning appliances, humidifiers, pesticides, and construction materials certainly have beneficial applications, but are also associated with toxic emissions and adverse health effects.

Indoor environments harbor a multitude of pollutant contributors, making it difficult to identify the source of adverse health effects or the potential synergistic effects of multiple toxic exposures (Figure 22.1). In the past, IAQ problems or

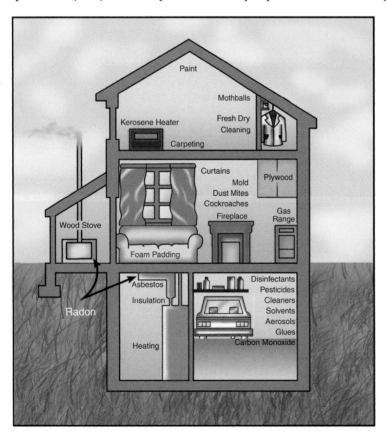

Figure 22.1 Common sources of indoor air pollutants.

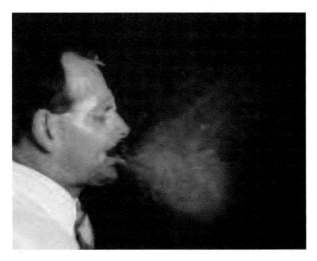

Figure 22.2 A typical sneeze. Found at www.people.virginia. edu/~rjh9u/sneeze.html. From Starr C. and McMillan B. (1996). *Human Biology*, 2nd ed. Wadsworth Publishing. Reprinted by Thomson Learning, Belmont, CA.

SBS were primarily blamed on volatile organic compounds originating from cleaning solutions, paints, or chemicals in building materials and carpets. Biological contaminants (such as mold, dust mites, and cockroach allergens) are now known to be significantly responsible. Other bioaerosols, such as viruses and bacteria, are microscopic and easily spread from person-to-person, where air and inanimate objects serve as intermediary transmission routes of infection (Figure 22.2). For many disease-causing microbes, infection rates are highest in winter months, when people spend more time indoors. Increases in indoor temperature and humidity, and reduced ventilation, have a substantial impact on the exposure levels of many biological and chemical pollutants. In fact, one in three buildings has damp conditions conducive to mold and bacterial growth (U.S. EPA, 2001).

22.2.1 Volatile Organic Compounds

Volatile organic compounds (VOCs) contain carbon and a variety of other elements, such as hydrogen, oxygen, fluorine, chlorine, bromine, sulfur, or nitrogen. They readily vaporize at room temperature, releasing noxious vapors into the air. Common household products, such as those designed for cleaning, disinfecting, degreasing or waxing, often contain high levels of VOCs. Some of these are potent toxins, capable of persisting in the indoor air space for a long time. Even seemingly innocuous materials, like cosmetics, air fresheners, or dry-cleaned fabrics can emit harmful toxins, such as diethanolamine, paradichlorobenzene, and perchloroethylene, respectively. While many organic chemicals have no known health effects, others range from mild irritants to carcinogens (see Chapter 13). In addition, research is lacking on the health effects of exposure to multiple toxins simultaneously. The effects of long-term exposures and individual susceptibilities to different hazards are also unknown.

22.2.1.1 Pesticides

Each year, 75% of all U.S. households report the indoor use of at least one pesticide product. According to the EPA, 80% of a person's exposure to pesticides occurs indoors, and up to 12 different pesticides have been found in single household air samples. Pesticides include insecticides, termiticides, rodenticides, fungicides, and microbial disinfectants, and may be formulated as sprays, foggers, liquids, sticks, or powders. By definition, they aim to kill their designated targets and thus often contain highly toxic compounds. Exposure to pesticides accounts for nearly 80,000 poisonings of children each year, primarily due to ingestion, but inhalation and dermal exposures also occur. Health risks include systematic illness, organ damage, respiratory irritation and disease, neurological disorders, and a host of mild symptomology such as headaches, nausea, and dizziness.

22.2.1.2 Construction materials and furnishings

Materials used in home and building construction and indoor furnishings frequently contain and emit hazardous compounds. Historically, asbestos was used in a variety of construction materials for insulation and as a fire-retardant. Many asbestos products have been removed from buildings and continued use has been banned, but older homes may still contain potentially harmful materials. If in good condition and left undisturbed, asbestos products are generally not a risk; however, aerosolized fibers can be respired, damaging the lungs and abdominal lining and leading to irreversible scarring and cancer (see also Chapter 9).

Carpeting and installation materials, such as adhesives and padding, are known to emit volatile organic compounds. Eye, nose, and throat irritation; headaches; rashes; coughing; fatigue; and shortness of breath have all been reported following new carpet installations. Carpeting may also act as a sink for a multitude of chemical and biological pollutants, including pesticides, dust mites, and molds that may collect in carpet fibers and remain protected from cleaning and vacuuming.

Formaldehyde is a known human carcinogen but is still widely used in the manufacture of household fabrics, paints, and furniture. Pressed wood products, such as particleboard, are generally produced with formaldehyde-containing resins that are released over time into the air. Those with urea-formaldehyde resins are associated with the highest pollutant emissions. Exposure to formaldehyde toxin is generally via the nasal passages or adsorption through the skin. Symptoms of allergies, asthma, throat and nose irritation, headaches, and nausea are well documented at indoor levels above 0.1 ppm, and some states recommend target threshold levels of 0.05 ppm or lower (Liteplo, 2002). Government agencies have set limits on allowable formaldehyde emissions and the use of certain construction materials, such as pressed wood and insulation. Manufactured and mobile homes are of particular concern, because of the use of high-emitting construction materials and small interior air space. Increased ventilation, decreased humidity and temperature, and selection of low-formaldehyde-containing products, such as "exterior-grade" pressed wood materials are several ways that individuals can reduce exposures. In addition, the complete application of water-repellant finishes, such as a polyurethane varnish, can block the release of formaldehyde. Older materials are less of a concern for emissions due to a natural decrease in toxin levels over time.

22.2.2 Combustion Products

Combustion products such as oil, gas, kerosene, wood, and coal are common to indoor environments due to the use of fuel-burning appliances, space heaters, fireplaces, and gas or wood stoves. If not properly vented, harmful pollutants, such as CO, NO_2, and particle or chemical irritants may be released into the air. Improperly installed or maintained chimneys or other ventilation outlets can cause a backdraft of pollutants into the home. **Environmental tobacco smoke (ETS)** is also considered a combustion product, and exposure via second-hand smoke is a major health concern.

22.2.2.1 Carbon monoxide and nitrogen dioxide

Carbon monoxide (CO) and nitrogen dioxide (NO_2) are both colorless, odorless gases that can cause potent health effects even at low levels. At high concentrations, CO inhalation results in rapid illness and death. About 500 Americans die from unintentional CO poisoning each year. Symptoms of exposure include headache, dizziness, weakness, nausea, vomiting, chest pain, and confusion. Nitrogen dioxide is an irritant of the mucous membranes of the eye, nose, and throat. High-level exposure results in respiratory irritation, shortness of breath, and potentially contributes to respiratory infections and lung diseases like emphysema.

22.2.2.2. Environmental tobacco smoke (ETS)

According to the U.S. government's Agency for Health Care Policy and Research (AHCPR), 46 million Americans smoke, exposing themselves and others to hazardous products of combustion. About 75% of the nicotine from a cigarette ends up in the atmosphere. In fact, secondhand smoke contains higher concentrations of toxic and cancer-causing chemicals than smoke inhaled directly, with more than 4,000 chemicals, including 200 known toxins, and 69 known carcinogens. Each year, an estimated 3,000 nonsmoking Americans die of lung cancer, and more than 35,000 die of heart disease from secondhand smoke. Furthermore, an estimated 150,000 to 300,000 children, ≤ 18 months of age experience respiratory tract infections because of secondhand smoke exposures. Young children exposed to secondhand smoke are also more likely to experience increased incidences of pneumonia, ear infections, bronchitis, coughing, wheezing, and asthma. Persons with a history of heart disease are particularly encouraged to avoid second-hand smoke exposures, as one study documented a 40% decrease in heart attacks following the enactment of a smoke-free air ordinance (Sargent, et al., 2004).

22.2.3 Lead and Radon

In the late 20th century, lead was described as the greatest environmental threat to children's health in the U.S. Prior to recognizing the health risks associated with lead, it was used widely in plumbing materials, gasoline, and paints. Atmospheric pollutants from combustion of leaded gasoline, ore smeltering, and the burning of fossil fuels are sources of lead from the outdoor environment that may be tracked indoors via dust, shoes, clothing, or pets. Although soils near roads are contaminated from years of leaded gasoline use, and exposures still occur from hobbies using lead solder, the greatest exposure to lead indoors is from peeling, chipped, or sanded paint. Although banned from use in 1960, older homes may still contain heavily leaded paint.

Lead particulates settle on surfaces of indoor environments and are readily redispersed into the air via air currents common to indoor climates. Contaminated lead particles may be inhaled and ingested. Both result in absorption into the blood, where it is then distributed to soft tissues and bone. As lead accumulates over time, it can eventually affect nearly every system in the body. At levels above 80 μg dL^{-1} of blood, convulsions, coma, and even death can occur. Levels as low as 10 μg dL^{-1} can impair mental and physical development, resulting in lower IQ levels, shortened attention spans, and increased behavioral problems (Lin-Fu, 1992). This is particularly true for actively growing children. Additional exposures can cause problems with the central nervous system, kidneys, and blood cells (see also Chapter 13).

Radon is a radioactive gas released during the natural decay of uranium, a common component of rocks, soil, and water. Radon enters homes through cracks, drains, and even wells. Once inside, the colorless, odorless gas can become trapped in living spaces, increasing in concentration and remaining undetected. Exposure via breathing the radioactive gas into the lungs results in lung cancer and possible death. Radon gas is listed as the second leading cause of lung cancer in the United States. An estimated 21,000 deaths per year could be prevented by addressing radon gas exposures, particularly among smokers, who are known to be at increased risk due to synergistic interactions between radon and smoking (U.S. EPA, 2004).

The only way to be sure that your home, school, or workplace is free of radon is to conduct individual testing. High indoor levels of radon have been documented in every state. Although long-term exposure to any level of radon may be harmful, indoor environments with radon levels ≤ 4 picoCuries per liter (pCi L^{-1}) should be treated. Radon reduction systems are available that can decrease radon levels by 99% in the home.

22.2.4 Biological Pollutants

The term bioaerosol encompasses any biological agent transmitted by the airborne route, *i.e.*, bacteria, viruses, mold, mites, cockroach particles, pollen, and animal dander and saliva (see also Chapters 9 and 27). All of these agents have been associated with adverse health effects, including allergies and asthma, and often coexist in common environments. Allergic diseases have significantly increased worldwide over the last 30 years. More than 50% of all allergic diseases are caused by allergens out of the indoor environment, and nearly one in six persons in the U.S. is effected by hypersensitivity reactions. Indoor molds are a rising concern with ambient air contamination, ranking among the most important allergens of indoor environments. Biological agents can persist in dust particles and animal droppings, or proliferate in humid microzones until they become aerosolized. Natural breezes, air-conditioning systems, humidifiers, and active movement all create eddies that aid in the aerosolization of spores, microbes, and other toxins. In moist environments, mold and bacteria grow in less than 72 hours, colonizing solid surfaces and subsequently releasing toxins, particulates, and allergens into air spaces (Figure 22.3).

22.2.4.1 Mold

The most common indoor molds are *Cladosporium*, *Penicillium*, *Aspergillus*, and *Alternaria*. Although primarily associated with allergic reactions, some molds are responsible for chemical toxigenic responses. Health effects due to molds are caused by direct exposure to spores and cell wall components (exopolysaccharides, glucans) or from the production of metabolites (microbial volatile organic compounds, mycotoxins). Even nonviable mold fragments can initiate an allergic response. Threshold levels of molds are unknown and appear to depend on the type of mold present and individual susceptibilities. Health complaints, however, have been associated with reports of 2,000 CFU m^{-3} of mixed mold populations in air samples.

22.2.4.2 Endotoxin

Bacteria are grouped into two main classes or categories that differ in the chemical composition of their cell walls. One group is called Gram negative, and the other, Gram positive (see Chapter 5). The outer cell wall of the Gram-negative bacteria contains a layer of **lipopolysaccharide** (fat and carbohydrate), or **LPS**. This LPS material, either as intact bacterial cell walls or cell wall fragments from dead cells, can cause adverse effects in the bloodstream of humans or other animals, including fever reactions, clotting of the blood, and alterations in the behavior of the white blood cells. For this reason, it is referred to as **endotoxin**.

LPS is ubiquitous in the environment and commonly detected in household dust associated with decaying bacterial cells. These endotoxins are easily aerosolized, and high concentrations in the respiratory tract (the lungs) can cause adverse effects, including allergic reactions, difficulty in breathing, and other forms of respiratory distress. Most adverse health effects associated with endotoxin have been associated with occupational exposures such as grain houses, cotton dust, or composting plants. The data are conflicting regarding the beneficial or adverse outcomes of endotoxin exposure in individuals at various stages of development, and thus questions remain as to the potential risks associated with routine, low-level exposures in common indoor environments.

Figure 22.3 Bathroom flooring colonized with mold. Photo courtesy K.A. Reynolds.

22.2.4.3 Viruses and bacteria

Infectious disease-causing organisms, such as viruses and bacteria, are easily aerosolized by direct human activity such as sneezing and coughing or via water use such as showering or toilet flushing. In some cases, specific etiological agents are identified, whereas in others, mixed populations of microbes cannot be distinguished. For example, home humidifiers offer a breeding ground for bacteria, increasing indoor humidity and aiding in the overall proliferation of microbes and toxin release from household materials. Improper system maintenance and lack of proper cleaning, and drying encourages the growth of bacteria, protozoa, and fungi. Humidifiers are such a common source of unhealthy bioaerosols that doctors now use the term humidifier fever to describe the associated short-term, flu-like illness.

Infection rates of many microbes increase in the winter months due to an increase in indoor activities. Health agencies estimate that individuals in the U.S. suffer 1 billion colds every year (Figure 22.4). In a given season, up to 50 million Americans contract the flu and an average of 20,000 to 40,000 die each year. Influenza, colds, and illnesses such as the highly contagious chicken pox and tuberculosis are examples of infectious diseases directly transmitted, often via the aerosol route, from one person to another. Conditions of overcrowding and poor air circulation promote the spread of infectious agents. Some microbes become a problem when they grow in building ventilation, cooling, or heating systems. Hot water systems, air-conditioning cooling towers, and evaporative condensers, for example, have all been implicated in outbreaks of Legionnaires' disease and Pontiac fever, primarily caused by the bacterium *Legionella pneumophila* (See also Chapter 11). The disease is found worldwide, with most cases occurring in the summer and fall months. Transmission is thought to be via aerosol inhalation. Infections in hospitals are a concern, with up to 85% of the water supplies testing positive for *Legionella* and 15% of those infected dying of respiratory failure.

Although rare, hantavirus infections associated with exposure to aerosolized rodent droppings of infected deer mice were identified in 1993 and have since occurred in 31 states. The virus causes a severe respiratory syndrome, with a 37% fatality rate. Since activities related to cleaning (*i.e.*, sweeping, vacuuming) are associated with increased aerosolization of the organism, control measures are targeted at reducing rat populations.

22.2.4.4 Animal allergens

Animal dander consists of tiny scales of skin that are constantly shed from animals and easily aerosolized. It is a known allergen, as are animal urine and saliva proteins that are dried in the environment. They are major problems for individuals who suffer from animal allergen sensitization. Pets as well as uninvited animals, such as rats and mice, are primary contributors. Approximately 15% of Americans are allergic to cats or dogs, and 2 million people with cat allergies still have a pet cat.

22.2.4.5 Arthropod allergens

House dust mites are microscopic organisms related to spiders and ticks. They are a major pollutant of indoor air, where allergens from their droppings contribute significantly to nonseasonal allergies and asthma (Figure 22.5). Approximately 20 million Americans are allergic to dust mites. They are probably present in all homes, and nearly half of all homes are thought to have a serious dust mite problem. These dust mites

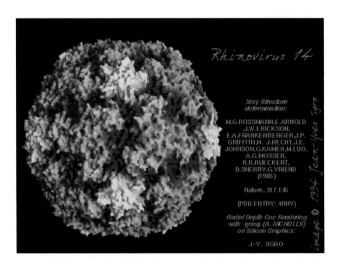

Figure 22.4 Rhinovirus—A leading cause of the common cold. Found at http://rhino.bocklabs.wisc.edu/cgi-bin/virusworld/virustable.pl?virusdata=r14%2C+Human+Rhinovirus+14%2C+4RHV © 2005 Virusworld, Jean-Yves Sgro, Institute for Molecular Virology, University of Wisconsin-Madison.

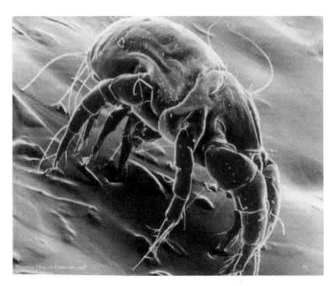

Figure 22.5 A dust mite. Cameron, M. Prevention of Atopic Asthma. *Positive Health Mag* 25: February 1998 - www.positivehealth.com

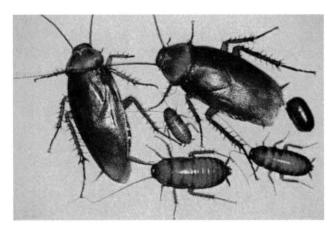

Figure 22.6 American cockroaches. Found at: ianrwww.unl. edu/ianr/pat/roachind.htm. From Pesticide Education Resources Office. The University of Nebraska-Lincoln, Department of Entomology.

live in the house dust, bedding, and carpeting, feeding on dead human skin and thriving in conditions of increased humidity.

Similarly, cockroaches are highly adaptable to extreme environmental conditions and therefore can be found in diverse climates around the world (Figure 22.6). Studies show that up to 98% of urban homes have cockroaches numbering in the hundreds to hundreds of thousands each. Allergies related to cockroach triggers have been recognized since the 1940s. Later studies showed that cockroach allergens contribute to acute asthma attacks. Recently, studies have identified that up to 60% of urbanites are sensitive to cockroach allergens and that frequent hospital admissions of asthmatic children are directly related to cockroach allergen exposures (Rosenstreich et al., 1997).

22.2.4.6 Plant allergens

Allergens from outdoor plants, such as pollen, can enter the indoor environment via pets, shoes, personal objects, or open windows and doors (Figure 22.7). Approximately 35 million

Figure 22.7 Ragweed and pollen. Found at http://www.aaaai. org/media/photos_graphics/pollen.stm © 1996–2005. All rights reserved. American Academy of Allergy Asthma and Immunology.

Americans suffer from pollen allergies, mostly due to ragweed. Although generally seasonal, allergenic triggers from various weeds, grasses, and trees are produced practically year-round.

22.3 FACTORS INFLUENCING EXPOSURE TO INDOOR AIR POLLUTION

Identification of specific hazards, determining exposure levels, and developing specific dose-response characteristics for pollutants are critical steps in assessing an overall health risk (see also Chapter 14). In most instances, minimizing exposure to specific hazards will reduce the risk of adverse health effects. In the case of building-related illnesses, symptoms may be apparent only when a particular environment (work, home, or school) is occupied, or they may be chronic and continue until appropriate treatment is sought. Exposure may be continuous or intermittent. Symptoms may be acute or develop over time, and manifest as large outbreaks or only among individuals that are particularly susceptible.

Indoor pollutant exposures are generally controlled by targeting source prevalence, poorly maintained environments or operated systems, and improper building design, or a combination of factors. An analysis of multiple effects on the prevalence of mucosal irritation and other general adverse symptoms among office workers showed that the concentration of visible floor dust, the type of floor covering, the number of workplaces in the office, the age of the building, and the type of ventilation were all associated with the prevalence of symptoms (Skov et al., 1990). In order to minimize poor IAQ, it is imperative that proactive efforts in building design, construction, operation, and maintenance be implemented.

22.3.1 Source Prevalence

Removing the pollutant source is one way to reduce indoor air quality exposures. However, this may not always be possible. Active educational campaigns on how individuals can reduce indoor exposures to contaminants in addition to efforts to reduce the sources, such as lead and asbestos, have lead to dramatic reductions in reported disease related to these hazards. Similarly, smoking cessation education is important, particularly in the prenatal care medical sector. Following label information for product use regarding exposure precautions and ventilation suggestions, as well as minimizing use of products with known carcinogens, are two primary steps that allow source reduction when use cannot be completely avoided. This is particularly important for common household cleaners and pesticides. Molds and other biological allergens cannot be completely removed, but specific efforts can minimize infestations.

Additional proactive measures of indoor pollutant source reduction include making sure that all fuel-burning appliances are properly installed, maintained, and operated in accordance with manufacturer instructions. Trained service personnel should be utilized annually to inspect combustion appliances, particularly furnaces, water heaters, and gas dryers. Backdrafting of combustion gases can be minimized by making sure that fireplace chimneys and flues are properly vented, regularly cleaned, and inspected yearly.

Hazards of the work environment are not necessarily different than those of the home, but exposures may be much greater. Longer periods of time may be spent in hazardous work environments amidst larger doses of pathogenic molds or mold products. Many molds are inherently present in large numbers in a variety of materials contacted in specialized occupations. Thus, exposure time and the prevalence of molds associated with specific activities increases the individual risk in work environments. Occupational exposures associated with potentially pathogenic molds include the manufacture of bread, cheese, beer, wine, antibiotics, enzymes, and steroids. In addition, farming, commercial composting, and bee-keeping operations subject individuals to numerous species of pathogenic molds. Animal farms and feed houses are frequently associated with increased allergic responses, due in part to mold exposures, where hay, grain, silage, and bedding provide perfect environments for mold proliferation. *Stachybotrys chartarum* has been related to occupational exposures in workers handling decomposable flowerpots, with symptoms of painful inflammation and scaling fingertip skin.

Proliferation of and exposure to mold in the home environment is dependent upon many factors, such as substrate, relative humidity, and temperature. The relative importance of each of these factors in toxin production is poorly understood. Indoor and outdoor microbial environments are complex, dynamic, and transient in nature, and sampling results change with time. Aerosol generation, particle size and concentrations, organism viability, infectivity and virulence, airflow, climate, and the type of environmental sampling and analysis equipment used add to the variability of ambient mold counts. Important microbial ecological factors, such as the presence of competing bacteria, other mold, antifungal metabolites, and insects, greatly influence viability.

Almost every home is subject to some type of moisture problem, either due to leakage (flooding, pipes bursting, or overflows) or routine activities (showering, watering indoor plants, or cooking). According to 1998 U.S. Census data, 10.9 million U.S. homes (16%) had indoor leakage during the last 12 months, and 16.9 million (21%) had leakage from the outside. It is estimated that approximately 30% of all buildings will have indoor air quality problems associated with mold during their useful life span. More than 65% of household occupants in the U.S. report that they live in a damp environment. Dampness has been a significant factor in the prevalence of respiratory and other illnesses in children, and is known to be a major factor in indoor mold proliferation. A study of 4,600 children from six U.S. cities of the Northeast found that home dampness and mold was a strong predictor of several respiratory symptoms and also a number of nonrespiratory symptoms. Brunekreef 1989 et. al. Early exposure of children to molds may predispose them to chronic health problems in the future. The significance of dampness or mold in the home is demonstrated by numerous studies finding indoor moisture and mold to be significant risk factors associated with wheezing, coughing, and asthma in children. Moisture control and routine cleaning can eliminate most mold infestations. Likewise, dust mites are best controlled by eliminating indoor dampness and routine cleaning and maintenance. Covers for mattresses and pillows are available that are designed to reduce dust mites and other home allergens. Bedding should be washed weekly in hot water ($>130°F$).

22.3.2 Building Maintenance and Operation

Indoor environments exchange air with the outdoors via infiltration or natural and mechanical ventilation. If properly maintained, **heating, ventilation, and air-conditioning (HVAC)** systems can reduce the number of pollutants in the air. However, if improperly maintained, they can serve as colonization surfaces and dispersion mechanisms. Indoor air quality problems have become more notable since the application of energy-saving measures. Proper filters in ventilators, correct moisture control, and periodic cleaning and disinfecting (either chemical or ultraviolet) of the entire ventilation system can greatly minimize bioaerosol production problems. Small system maintenance is also important. Water pans associated with various indoor appliance and air conditioning systems must be routinely treated with appropriate disinfectants (*i.e.*, bleach-based). Some systems provide built-in UV lights for continuous disinfection. **Germicidal ultraviolet (GUV)** lights have also been effectively used in central ventilation systems, without adverse effects, to eliminate microbial contamination.

Uninstalled wallboard, available from retailers, can contain a baseline bioburden, including *S. chartarum*, that will colonize surfaces under high humidity conditions. Dry wall (nonsterile gypsum wallboard) is commonly contaminated with pathogenic molds. Under conditions of high relative humidity, the wallboard absorbs moisture, allowing mold spores to propagate. Fungicidal paints have been developed that can be effective in reducing spore counts.

While it is generally recommended that indoor air pollutant management involve increased ventilation with outdoor air infiltration and exchange, such practice increases energy use and associated costs. Therefore, source reduction, via filtration of key pollutants, is another option for improving indoor air quality. A wide range of air purification systems are marketed for the removal of biological contaminants, particulate matter, and specific chemical emissions (Figure 22.8). Mechanical filters provide a physical barrier, often a charged media for attraction or an adhesive media for attachment. These types of filters can be housed in ducts of a building or used in portable devices.

Figure 22.8 Indoor air filtration unit. Photo courtesy K.A. Reynolds.

Electronic air cleaners use an electrical field to attract charged particles that are generally collected on plates or filters. Ion generators charge particles in the air, and these systems usually have a mechanism for collection of the particles. Some systems utilize a combination of techniques to clear the air. Different air cleaning systems have different degrees of effectiveness, which depends largely on design, airflow rates, and pollutant concentration. While some of these are effective at reducing exposures to specific indoor pollutants, others aren't and can even enhance the symptoms of susceptible individuals. For example, the use of electronic air cleaners and ion generators has been controversial, due to the production of ozone, also a pollutant and potential lung irritant. Consumers must carefully evaluate their needs against manufacturer's claims, and conduct periodic maintenance of their purification systems.

Some indoor air quality problems require extensive remediation of the building's structural components. For example, following a mold infestation, general clean-up procedures involve identifying and eliminating the moisture source, cleaning, disinfecting, drying the moldy area, and disposing of any material that has mold residues. Contaminated porous material, such as wallboard, paneling, or carpeting, should be removed from water-damaged areas of the home. Specialized filters can help to remove mold spores from the air, while various disinfectants are effective at removing mold contaminants from nonporous indoor surfaces. In severe situations, respirators and protective clothing may be necessary, since exposure to molds during remediative procedures can be 10–1,000 times higher than background levels. Workers should wear the highest quality (HEPA) respirators while performing clean up. Due to the potential risk associated with re-aerosolization of mold pathogens, and the potential for potent toxins to be released into the ambient air, it is advised that only professionals clean up all but the smallest mold contamination problems.

22.3.3 Building Design

Responding to growing complaints of sick building syndrome or other IAQ-related symptoms, architects and engineers have implemented new approaches in total building design, aimed at balancing a comfortable and healthy indoor environment with practical uses and minimal energy costs. "Sustainable" or "green building" design utilizes new approaches in construction that consider total building operations and systems, paying attention to issues of overcrowding, ventilation, outdoor air intake sources, low-emission or nontoxic construction materials, humidity, temperature, and the overall function of the building. Such designs incorporate planning for changes with season, time, or other factors.

Extensive guidelines are available from a variety of commercial and public health agencies aimed at the reduction of specific contaminants through building design. Radon exposures, for example, are dramatically reduced by utilizing construction designed to minimize soil gas intrusion. Primary prevention techniques include a sub-slab gas permeable system to remove radon from under the building, plastic sheeting barriers to block radon from the home, crack and joint sealing to prevent entry into living spaces, radon reduction vent pipes for passive ventilation into the atmosphere, and installation of an electrical junction box for potential use of active ventilation systems. Approximately one in six homes constructed in the U.S. is now being built with "radon-resistant techniques" utilizing foundation barriers, ventilation piping, and sealing methods.

22.3.4 Sensitive Populations

Everyone is exposed to some level of indoor chemical and biological pollution, but the health effects of those pollutants is highly individualized. Symptom manifestation and severity depends largely on host response/susceptibility, dose, exposure route, the time and frequency of the exposure, and the nature or virulence of the organism (Information Box 22.1). For example, the health effects of many molds and/or mold toxins have not been studied or are not yet known. There is also variation between species within a genus, and between strains within a species. Health effects resulting from mold exposures are dependent upon the health status of an individual, as well as the dosage and route of exposure. As a group, immunocompromised populations are more likely to experience increased morbidity and mortality following exposure to indoor air pollutants. Sensitive populations constitute 20–25% of the population, and include the very young, the elderly, pregnant women, and persons with diminished immunity either due to medical intervention (organ transplants), previous illnesses (cancer, liver disease), preexisting respiratory conditions (allergies, chemical sensitivity, asthma), or infections (AIDS patients). Prolonged use of corticoid steroids may also be a preeminent factor. For these populations, the risk of adverse health effects due to exposure to indoor air pollutants is enhanced. Sometimes popula-

INFORMATION BOX 22.1

Factors Influencing Mold Exposures

Toxicity influencing factors
Exposure factors
Dose, concentration, volume of exposure
Route, rate, site of exposure
Duration and frequency of exposure
Time of exposure (*i.e.*, season)

Host Factors

Immunologic status
Genetic background
Previous illness
Age, gender, body weight
Nutritional status
Hormonal status
CNS status

tions are at increased risk, not due to preexisting health conditions or undeveloped immunity, but due to natural physiological factors. For example, children are more vulnerable to lead exposure than adults, since lead is more easily absorbed into growing bodies.

An estimated 20 million Americans, including 6.3 million children, have asthma. People with asthma have sensitive airways that are more likely to react to various irritants, affecting their breathing. The incidence of asthma has been on a steep incline in recent years, and according to the Asthma and Allergy Foundation of America, every day in America, 40,000 people miss school or work, 5,000 people visit the emergency room, and 12 people die due to asthma. Indoor air pollutants are thought to play a significant role in the incidence of asthma, and questions are particularly focused on highly prevalent biological pollutants.

Conditions of suppressed immunity due to fungal exposures have not been well studied using clinical trials. However, patients with fungal infections or sensitivities appear to be less able to ward off other infections, and are more likely to become afflicted by bacterial and viral diseases. Laboratory studies in animals have shown a clear immunosuppressive quality to certain mycotoxins, in particular, the aflatoxins, ochratoxin, and some trichothecenes. While mold exposure does not always present a health problem indoors, people with preexisting allergies tend to be more sensitive to mold exposures. These people may experience symptoms such as nasal stuffiness, eye irritation, or wheezing when exposed, while others experience more severe respiratory reactions. People with immune suppression or underlying lung disease are more susceptible to fungal infections. Examples of known opportunistic fungal pathogens of the severely immunocompromised include *Aspergillus fumigatus, A. niger, A*

nidulans, A. flavus, A. terreus, Pseudallescheria boydii, Fusarium solani, F. oxysporum, F. moniliforme, and *F. proliferatum.* Many questions remain as to the long-term effects of chronic exposures to low levels of mold contaminants in residential settings in both healthy and sensitive populations.

22.4 MONITORING IAQ

Overall, most indoor air quality problems are due to inadequate ventilation. Therefore, the **American Society of Heating, Refrigerating, and Air-Conditioning Engineers (ASHRAE)** established recommended ventilation rates for indoor air environments in 1973 and currently recommend a standard of 20 cubic feet per minute of outdoor air per person for general office space.

In the U.S., recommended values are given for ventilation rates, particulates, and several chemicals of the indoor air environment. Select indoor air pollutants may be monitored using a variety of methods ranging from low-cost, do-it-yourself kits to high-level, technical laboratory analysis. Radon kits, for example are available at most hardware stores for around $25, and many homes now routinely contain carbon monoxide meters for continuous monitoring. Grab samples or screening samples with direct reading equipment, such as colorimetric detector tubes and particulate monitors may be used for rapid, on-site monitoring of specific contaminants, such as carbon monoxide and nitrogen dioxide. Additionally, portable infrared spectrometers and dosimeter badges are available for a number of volatile organic compounds.

In the occupational arena, airborne, contaminant standards are set for specific hazards (Information Box 22.2). The OSHA PEL (permissible exposure limit) for carbon monoxide is 50 parts per million (ppm) averaged over an 8-hour workday. Indoor ambient levels of CO below 35 ppm and as low as 10 ppm are considered harmful to some, including children and the elderly. Chronic exposure to nitrogen dioxide at even lower levels (0.4–2.7 ppm) has been associated with the development of bronchitis and other cardiovascular problems. Employers must assure that employees are not exposed to greater than 50 μg lead m^3 air averaged over an 8-hour period. The Centers for Disease Control recommends that all children, especially those under 6 years of age, be screened for lead poisoning yearly. Eliminating blood lead levels (BLLs) ≥ 10 μ $=$g dL^{-1} in children is one of the CDC's national health objectives.

Monitoring for biological pollutants can be more problematic, since molds initiate allergic reactions in both the viable and nonviable state. Thus, a variety of monitoring methods must be used, including cultural analysis for viable spore counts and microscopic methods for inactive allergenic particles. Additionally, chemical monitoring methods of mold metabolites, such as mycotoxins, are necessary to fully evaluate the contaminated environment.

INFORMATION BOX 22.2

Exposure Standards for Airborne Contaminants.

Airborne contaminant standards are set, for individual hazards, by various public health, occupational health, and regulatory agencies to guard against exposure levels that could cause adverse health effects.

Exposure Standards are guideline exposure values given below for contaminants that produce irritating or hazardous effects. They may consider ceiling values of acceptable acute exposures, or averages of chronic exposures, where contaminants can accumulate over time. The American Conference of Governmental Industrial Hygienists (ACGIH) sets **Threshold Limit Values (TLVs)** for protecting workers from occupational exposures. These are values thought to be safe for all workers to be exposed to daily without harmful effects. Samples are collected in consideration of

8-hr Time-Weighted Averages (TWAs), reflecting the average exposure level over an 8-hour shift, 5 days per week. TWA evaluate long-term chronic exposures.

Short-term Exposure Limits (STELs), evaluating 15- to 30-minute peak exposure levels. STEL should not exceed three times the TWA more than 30 minutes in an 8-hour workday and never exceed 5 times the TWA.

Permissible Exposure Limits (PELs) are enforceable standards set by Occupational Safety and Health Agency (OSHA) based on 8-hour TWAs.

Currently, there are no EPA regulations or standards for airborne mold contaminants; therefore, sampling of airborne microorganisms is often inconclusive. In addition, there are no absolute monitoring standards or **Threshold Limit Values (TLVs)** for airborne concentrations of mold or mold spores. Therefore, decisions on whether a building has a mold problem are often made arbitrarily. Different methods of collection can give divergent results and adverse outcomes may be heavily subject to interpretation.

In general, in buildings where mold growth is not visible but mold-related SBS is suspected, comparison of indoor versus outdoor mold counts and species diversity may identify, or eliminate, molds as a cause of poor IAQ. High correlations between indoor fungus levels and outdoor levels are observed for most predominant genera. If indoor counts exceed outdoor levels, the source of mold proliferation should be investigated. Previous studies have found concentrations of spores in domestic environments as high as 4.5×10^5 CFU m^{-3}. In a survey of random homes from several countries, moldy odors were associated with a geometric mean of 2.5×10^5 CFU g^{-1} total viable fungi, compared to 1.5×10^5 CFU g^{-1} when moldy odors were not reported. When analyzed for adverse health effects associated with home dampness and mold growth, water damage was associated with a 50% increase in molds.

The Russian Federation is the only governmental agency that has binding quantitative regulations for bioaerosols. Recommended guidelines have been proposed or sponsored by North American and European governmental agencies and private professional organizations. A considerable number of frequently cited guidelines have been proposed by individuals, based either on baseline data or on personal experience. Quantitative standards/guidelines range from less than 100 CFU m^{-3} to greater than 1000 CFU m^{-3} (total fungi) as the upper limit for noncontaminated indoor environments. Major issues with existing quantitative standards and guidelines are the lack of correlation to human dose response data, the reliance on short-term grab samples analyzed only by culture, and the absence of standardized protocols for data collection, analysis, and interpretation.

Improving indoor air quality requires a multidisciplinary approach utilizing educational entities, research programs, public health administrators, architects, engineers, industrial hygienists, and remediators. Benefits are realized by pollution prevention approaches as well as treatment measures following exposure. With ample unknowns and growing concerns, the health benefit of improved IAQ is expected to be a rapidly expanding and highly integrated field of study.

QUESTIONS AND PROBLEMS

1. Describe the characteristics of sick building syndrome. Discuss how you would differentiate potential causative sources of a case of SBS in your workplace.
2. Tour your home or workplace and identify the five top priority contaminants of your indoor environment. Discuss likely approaches to minimize your risk of exposure.
3. From 1976–1980, 88% of children had an elevated level of blood lead, compared to 6% from 1988–1994. Discuss the reasons for this decline and determine additional approaches for further reductions of elevated blood lead levels (consider socioeconomical differences in exposed populations).
4. List the primary biological pollutants of concern in indoor environments and the primary routine control efforts recommended.
5. Although researchers have found an association with total mold counts and reported dampness and adverse health symptoms, a definite dose–response level has not been determined for particular molds or for specific populations. Discuss the problematic factors in determining the health risks of indoor mold exposures and how these factors should be considered in the development of environmental standards for mold exposure limits.

REFERENCES AND ADDITIONAL READING

Brunekreef, B., Dockery, D.W., Speizer, F.E., Ware, J.H., Spengler, J.D., and Ferris, B.G. (1989) Home dampness and respiratory morbidity in children. The American Review of Respiratory Disease. 140(5):1363–7.

Kalnova K. (2000) The concentrations of mixed populations of fungi in indoor air: rooms with and without mold problems; rooms with and without health complaints. *Central European Journal Public Health.* **8**, 59–61.

Kreiss K. (1990) The sick building syndrome: Where is the epidemiologic basis? *American Journal of Public Health.* **80**, 1172–73.

Lin-Fu J.S. (1992) Modern history of lead poisoning: A century of discovery and rediscovery. In H.L. Needleman (ed.), *Human Lead Exposure* (pp. 23–43). CRC Press, Inc, Boca Raton, Florida.

Liteplo R.G., Beauchamp R. and Meek M.E. (2002) Formaldehyde. Concise International Chemical Assessment Document 40. World Health Organization. Geneva, Switzerland.

Rosenstreich D.L., Eggleston, P., Kattan, M., Baker, D., Slavin, R.G., Gergen, P., Mitchell, H., McNiff-Mortimer, K., Lynn, H., Ownby, D., and Malveaux, F. (1997) The role of cockroach allergy and exposure to cockroach allergen in causing morbidity among inner-city children with asthma. *New Eng. J. Med.* **336**, 1156–1163.

Sargent R.P., Shepard R.M. and Glantz S.A. 2004. Reduced incidence of admissions for myocardial infarction associated with public smoking ban: before and after study. *BMJ.* **328**, 977–980.

Skov P., Valbjoern O. and Pedersen B.V. (1990) Influence of indoor climate on the sick building syndrome in an office environment. *Scandinavian Journal of Work, Environment & Health* **16**, 363–371.

Stout J.E. and Yu V.L. 2003. Hospital-acquired Legionnaires' disease: New developments. *Current Opinion in Infectious Diseases.* **16**, 337–341.

United States Environmental Protection Agency (U.S. EPA). (2001). Building Radon Out. United States Environmental Protection Agency. Office of Air and Radiation. EPA/402–K–01–002. April 2001.

United States Environmental Protection Agency (USEPA). (2001) Healthy buildings, healthy people: A vision for the 21st century. Office of Air and Radiation. 402–K–01–003.

United States Environmental Protection Agency (USEPA). (2004). A citizen's guide to radon: The guide to protecting yourself and your family from radon. Indoor Environments Division (6609J), Washington, DC 20460. U.S. EPA 402–K–02–006.

CHAPTER **23**

ATMOSPHERIC POLLUTION

A. D. Matthias, A. C. Comrie, and S. A. Musil

Smelter-produced air pollutants trapped by an inversion layer. *Photo courtesy J.F. Artiola.*

23.1 AIR POLLUTION CONCEPTS

In this chapter, we discuss air pollutants, including their sources and effects on human activity, as well as their transport to, and fate in, the atmosphere. We also describe the role of air pollution in such major environmental issues as global warming and stratospheric ozone depletion.

An **air pollutant** is any gas or particulate that, at high enough concentration, may be harmful to life, the environment, and/or property. A pollutant may originate from natural or anthropogenic sources, or both. Pollutants occur throughout much of the troposphere (see Chapter 4); however, pollution close to the earth's surface within the boundary layer is of most concern because of the relatively high concentrations resulting from sources at the surface (see Section 4.2.5).

Atmospheric pollutant concentrations depend mainly on the total mass of pollution emitted into the atmosphere, together with the atmospheric conditions that affect its fate and transport. Obviously, air pollution has many and varied sources, including cars, smokestacks, and other industrial inputs into the atmosphere as well as wind erosion of soil. Large emissions from both anthropogenic and natural sources over long periods enhance concentrations, as do the chemical and physical properties of these pollutants. For example, when nitrogen oxides and hydrocarbons in car exhaust are emitted into warm, sunlit air, they readily form ozone molecules (O_3). Similarly, the solubility of a pollutant affects how efficiently it is removed by rainfall.

Atmospheric conditions have a major effect upon pollutants once these pollutants are emitted into (*e.g.*, nitrogen oxides from car exhaust) or formed within (*e.g.*, O_3) the atmosphere. Pollution dispersal is controlled by atmospheric motion, which is affected by wind, stability, and the vertical temperature variation within the boundary layer. Stability, in turn, influences both air turbulence and the depth at which mixing of polluted air takes place.

Wind determines the horizontal movement of pollution in the atmosphere. Pollution emitted from a point source, such as a smokestack, is generally dispersed downwind in the form of a **plume**. Wind speed establishes the rate at which the plume contents are transported. Strong winds flowing over rough land surfaces enhance mixing of air by producing shear stress (mechanical mixing) much like that created when an electric fan circulates air in a room. Also, wind direction establishes the path followed by the pollution.

Once present in the atmospheric boundary layer, a pollutant may undergo a series of complex transformations leading to new pollutants, such as O_3. Also, the removal of pollutants from air by rain and snow, by gravity, or by surface deposition is influenced by boundary-layer conditions. These removal processes, in turn, are also affected by the type and roughness of the underlying ground surface.

Even when emissions are relatively constant, pollutant concentrations can quickly change, owing to variations in atmospheric conditions. When atmospheric conditions are stable, relatively low emissions can cause a buildup of pollution to hazardous levels. This situation can occur during radiation inversions at night (see Section 4.2.2). In contrast, unstable conditions may effectively dilute pollution to relatively "safe" concentrations despite a fairly high rate of emissions.

Air pollution, which is of major public concern, is currently the object of extensive scientific research. Its effects on life, including human health, productivity, and property, are not yet fully understood, even though exposure to high levels of pollution is a daily experience for many people. The cost of such pollution, whether expressed in terms of direct biological consequences or in terms of economic impact, is enormous. Worldwide, urban air pollution affects nearly a billion people, exposing them to possible health hazards. In the United States alone, billions of dollars are spent annually to prevent, control, and clean up air pollution; other developed nations are incurring similar costs. The United Nations considers air pollution to be a major global problem.

Most commonly, air pollution poses a health risk that can and does harm life. It harms the human respiratory and pulmonary systems. Emphysema, asthma, and other respiratory illnesses may result from or be aggravated by chronic exposure to certain pollutants, such as O_3 or particulates (see also Chapter 9). Research conducted in Southern California indicates that breathing polluted air slows lung development in children as much as having a parent who smokes tobacco. Teenagers who are chronically exposed to polluted air are five times as likely to have reduced lung function as are teens who breathe clean air. Children and adults with chronic exposure to elevated ozone levels are more likely to develop asthma than individuals in cleaner air (see Information Box 23.1). Breathing polluted air can also thicken human artery walls, which is an important risk factor for heart failure and strokes. In 1991, researchers at the U.S. Environmental Protection Agency

---

INFORMATION BOX 23.1

Asthma and Air Pollution

There is much concern about the possible adverse health effects of air pollution on respiratory health, especially of children within inner-city areas. The U.S. EPA is currently conducting studies of the potential role that several air pollutants may have in inducing and/or exacerbating asthma or asthma symptoms.

The incidence of asthma has increased dramatically in the U.S. during recent decades, the causes of which are not well understood. Increased air pollution in the form of complex mixtures of particulate matter, metals, and tobacco smoke is thought to play an important role in the increase. Indoor air quality is a particular concern (see Chapter 22). Another class of hazardous chemicals that may have an important role is the carbonyl compounds, such as aldehydes and ketones. They can originate from industrial sources and from the burning of diesel fuel. They also form within the troposphere during complex photochemical processes involving organic compounds.

(EPA) concluded that about 60,000 U.S. residents die each year due to heart attacks and respiratory illness caused by breathing particulates (dust) at concentrations within the federal PM_{10} (see Section 23.2.1.3) air quality standard.

Air pollution also poses an ecological risk (see Section 4.1). Vegetation can also be harmed by uptake of pollutants through the leaf stomata or by deposition of pollutants on the leaf surfaces. Sufficiently high concentrations of sulfur dioxide or ozone, for example, may cause leaf lesions in susceptible plants. Chronic exposure to relatively low levels of pollution can harm plants by reducing their resistance to disease and insect predators. Crop yields can be lowered by air pollution. The presence of certain air pollutants, such as ozone (formed from nitrogen oxides and hydrocarbons), enhances the earth's natural greenhouse effect within the troposphere. This enhancement warms the earth and may change rainfall patterns, which could markedly alter the distribution of life on earth (see Chapter 24).

Air pollution can also damage property. It can erode the exterior surfaces of buildings, particularly those constructed of limestone materials that react with acids in precipitation (see Figure 4.1 and Information Box 4.1). Further evidence

of the deleterious effects of air pollution on property can be seen in the damaged paint finishes on cars regularly parked downwind from ore smelters.

23.2 SOURCES, TYPES, AND EFFECTS OF AIR POLLUTION

Air pollution is not a new problem. Lead in Swedish lake sediments indicates that air pollution produced from lead mining and silver production in ancient Greece and Rome affected air quality throughout Europe. Early written accounts of air pollution refer mainly to smoke from burning wood and coal. For example, in the 13th century, King Edward I of England prohibited the use of sea coal, the burning of which produced large amounts of soot and sulfur dioxide (SO_2) in the atmosphere over London (see Information Box 23.2 and Figure 23.1 for a contemporary example from China). The Industrial Revolution increased pollution so markedly that air quality deteriorated significantly in Europe and North America. By the mid-19th century, many cities in the U. S. and Europe were experiencing the consequences of air pollution. By the early

INFORMATION BOX 23.2

Air Pollution in China

China is the second largest energy user in the world, due in part to its rapid industrialization. In 2002, about two thirds of China's energy came from burning coal, which tends to release large amounts of soot and SO_2, depending on the quality of the coal. This is particularly troublesome, since much of the coal emissions are from older industrial plants with poor pollution controls, as well as from domestic use (individuals using coal to cook and heat homes). The high levels of domestic use creates numerous small point sources, low to the ground, that are difficult to control or abate (see Section 23.3 for more information on air pollution dispersion). As a result, some cities are establishing "coal-free zones" and are attempting to increase the use of cleaner burning natural gas.

The World Health Organization noted in a 1998 report that seven of the ten most polluted cities in the world are in China. The government has begun to implement pollution reduction programs, but they face enormous problems, including acid rain, which is estimated to fall on about 30% of China's territory. The government is trying to improve enforcement of existing laws, as well as instituting fines on polluters, designing systems for emissions trading (already in use in North America and Europe), and focusing on new technology to reduce energy use and pollution.

Figure 23.1 A smokestack contributing to smog in Tianjin, China. Photo courtesy J. Walworth.

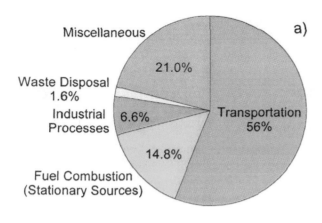

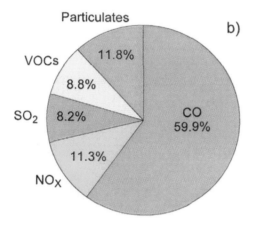

Figure 23.2 (a) Source contributions of primary air pollutants in the U.S. in 2002 and (b) the percentage of total primary air pollutants by type. Data: U.S. Environmental Protection Agency.

20th century, the term "smog" was coined to describe the adverse combination of smoke and fog in London. In Los Angeles, photochemical smog alerts became common by the mid-1940s. The first major air pollution disaster in the United States occurred in 1948, when approximately 20 lives were lost as a result of industrial pollutants trapped in very stable air over Donora, Pennsylvania, in the Monongahela River Valley. During one week in December 1952, stagnant air and coal burning caused severe smog conditions in London that ultimately took the lives of nearly 12,000 people over a three-month period.

Virtually all metropolitan areas are affected by air pollution, especially those situated in valleys surrounded by mountains (*e.g.*, Mexico City) or along coastal mountain ranges (*e.g.*, Los Angeles). But even unpopulated areas far from cities may be affected by long-range transport of pollution, either from urban areas or from such rural sources as ore smelters or coal-burning power plants. For example, pollution from a coal-burning power plant in northern Arizona may be reducing visibility in Grand Canyon National Park, located 400 kilometers west of the plant.

Most of the air we breathe is elemental oxygen (O_2) and nitrogen (N_2) (see Table 4.1 in Chapter 4). About 1% is com-

posed of naturally occurring trace constituents such as carbon dioxide (CO_2) and water vapor. A small part of this 1% may, however, be air pollutants, including gases and **particulate matter** suspended as aerosols. Anthropogenic air pollution enters the atmosphere from both fixed and mobile sources. Fixed sources include factories, electrical power plants, ore smelters, and farms, while mobile sources include all forms of transportation that burn fossil fuels. Mobile sources account for 56% of the pollutants emitted to the atmosphere in the United States (Figure 23.2a). Fuel combustion from stationary sources accounts for nearly 15%, and industrial processes account for about 7% of emissions in the United States. Natural sources of air pollution include winds eroding dust from cultivated farm fields, smoke from forest fires (Figure 23.3), and volcanic ash that is emitted into the troposphere and stratosphere (see Information Box 4.2).

There are many types of air pollutants. Some gases, such as CO_2, although produced by burning fossil fuels, are generally not considered pollutants because they are essential to plant life. Many pollutants, such as dust particles, exist naturally in the atmosphere and become hazardous only when their concentrations exceed air-quality standards set by such regulatory agencies as the U.S. EPA. The EPA classifies air pollutants according to two broad categories: primary and secondary air pollutants.

23.2.1 Primary Pollutants

Primary air pollutants enter the atmosphere directly from various sources. The EPA designates six types of primary air pollutants for regulatory purposes:

- carbon monoxide
- hydrocarbons

Figure 23.3 A natural source of atmospheric pollution. The "Aspen Fire" burned for weeks in the Catalina Mountains near Tucson, Arizona, releasing large amounts of particulates and other air pollutants. Photo courtesy J.F. Artiola.

- particulate matter
- sulfur dioxide
- nitrogen oxides
- lead

23.2.1.1 Carbon monoxide

Carbon monoxide (CO), which is the major pollutant in urban air, is a product of incomplete combustion of fossil fuels. Carbon monoxide has relatively few natural sources. It is a part of cigarette smoke, but the internal combustion engine is the major source, with about 50% of all CO emissions in the United States originating from cars and trucks. Emissions, therefore, are highest along heavily traveled highways and streets (Figure 23.4). Of the EPA-designated primary pollutants in the United States, CO emissions currently contribute about 60% of the total emissions (see Figure 23.2b). Fortunately, CO concentrations are decreasing in the United States, because newer cars have higher fuel efficiencies.

Carbon monoxide is highly poisonous to most animals. The EPA standards currently limit human exposure to a 24-hour average of 9 nL L^{-1} or a 1-hour average of 35 nL L^{-1}. (Note: The alternative unit of parts per billion (ppb) is also commonly used. See footnote to section 4.1 (in Chapter 4) for more information on units for describing gas concentrations.) When inhaled, CO reduces the ability of blood hemoglobin to attach oxygen. Although relatively stable, it is short-lived in the atmosphere because it is quickly oxidized to CO_2 by reaction with hydroxide radicals. Some atmospheric CO may be removed by soil microbes. In order to increase the oxidation of CO to CO_2 during fuel combustion, some cities require the use of oxygenated gasoline containing ethanol or other additives during winter months.

Figure 23.4 Emissions of CO and nitrogen oxides are highest along heavily traveled highways and streets. Photo courtesy S.A Musil.

23.2.1.2 Hydrocarbons

Hydrocarbons (HCs), or **volatile organic carbons (VOCs)**, are compounds composed of hydrogen and carbon. **Methane** (CH_4), the most abundant hydrocarbon in the atmosphere, is an active greenhouse gas. Volatile organics include the **nonmethane hydrocarbons (NMHCs)**, such as benzene, and their derivatives, such as formaldehyde. Some of these compounds (*e.g.*, benzene) are carcinogenic, and some relatively reactive HCs contribute to ozone production in photochemical smog.

Hydrocarbons are produced naturally from decomposition of organic matter and by certain types of plants (*e.g.*, pine trees, creosote bushes). In fact, HCs emitted from vegetation may be a major factor in smog formation in some cities, particularly those near forested areas of the southeastern United States. A large proportion of HCs and NMHCs are generated by human activity. Some NMHCs, including formaldehyde, are readily emitted from indoor sources, such as newly manufactured carpeting. Hydrocarbons are also emitted into the atmosphere by fossil fuel combustion and by evaporation of gasoline during fueling of cars. To mitigate this latter source, some municipalities require that service-station gasoline pumps be fitted with a special trap to collect HC vapors emitted during fueling of vehicles. Because transportation is the primary source of HCs, concentrations tend to be highest near heavily traveled roadways.

23.2.1.3 Particulate matter

The category of particulate matter comprises solid particles or liquid droplets (aerosols) small enough to remain suspended in air. Such particles have no general chemical composition and may, in fact, be very complex. Examples include soot, smoke, dust, asbestos fibers, and pesticides, as well as some metals (including Hg, Fe, Cu, and Pb). We can characterize particulate matter by size. Particles whose diameters are 10 μm or larger generally settle out of the atmosphere in less than a day, whereas particles whose diameters are 1 μm or less can remain suspended in air for weeks. Smaller particulate matter, whose particles are 10 μm or less, have come to be known as PM_{10}. The very small particles with diameters 2.5 μm or less are known as $PM_{2.5}$ (see Information Box 23.3 and Chapter 9).

The effects of particulates in the air are various. Some particulates, especially those containing sulfur compounds, are emitted by volcanoes. These particulates can reach the stratosphere, where they may significantly alter the radiation and thermal budgets of the atmosphere and thus produce cooler temperatures at the earth's surface (see Chapter 4 Information Box 4.2). Tropospheric particulates may cause or exacerbate human respiratory illnesses. Especially harmful to the human respiratory system is the fraction of mid-sized particles, PM_{10} and $PM_{2.5}$. In large cities, particulates also reduce visibility. In the United States in 2002, about 62% of particulates came from roads and transportation, with another 26% contributed by agriculture,

INFORMATION BOX 23.3

Aerosols and Visibility (See also Chapters 9 and 27)

Aerosols are solid or liquid particles suspended in a gas. In relation to environmental pollution, our major concern is with microscopic aerosol particles produced primarily from combustion or windblown dust, or secondarily from gas-to-particle conversion. Aerosols in the atmosphere typically fall into two distributions by mass or size, with coarse particles in the range 2.5–10 μm and fine particles from about 0.1–2.5 μm. The coarse fraction is usually composed of soil dust including minerals and organic particles. Fine particulate matter < 2.5 μm ($PM_{2.5}$) comprises a range of combustion products such as elemental carbon, sulfates, and nitrates. Aerosols are important constituents of acid rain, and they can also alter the atmospheric radiation balance. Because their small size allows them to penetrate deep into the lungs, aerosols have important health effects, and therefore the EPA has established health-based standards for $PM_{2.5}$ and PM_{10}.

Visibility impairment, especially at the regional scale, is typically caused by fine aerosols, especially < 1 μm, as this size range tends to remain suspended longest in the atmosphere (coarser particles may settle out and ultrafine particles can be removed as condensation nuclei in rain-out) (see Figure 23.5). These fine particles also scatter relatively large amounts of light, leading to greater haziness and decreased visibility. Visibility impairment has become a problem in many areas, not only in the eastern U.S., but also in relatively remote places such as the Grand Canyon. The Western Regional Air Partnership is a coordinated effort by the Western Governors' Association and the National Tribal Environmental Council to develop data, tools, and policies needed by states and tribes to improve visibility in parks and wilderness areas across the western U.S.

Figure 23.5 Photos taken of Mount Trumball in the Grand Canyon National Park on a "clear" day and a "hazy" day at noon. The figure to the right clearly shows the impairment of visibility due to aerosols, some of which originate in California urban areas. The National Park Service has a visibility-monitoring program that systematically collects photos at numerous parks across the U.S. Photos courtesy National Park Service.

forestry, and fires (Figure 23.6). Construction is now considered to make a large contribution to PM_{10} levels, but percentages from the EPA aren't available yet.

23.2.1.4 Sulfur dioxide

About 90% of **sulfur dioxide** (SO_2) emissions come from burning sulfur-containing fossil fuels, such as coal, which may contain up to 6% sulfur. Ore smelters and oil refineries also emit significant amounts of SO_2. At relatively high concentrations, SO_2 causes severe respiratory problems. Sulfur dioxide is also a source of acid rain, which is produced when SO_2 combines with water droplets to form sulfuric acid (H_2SO_4). At sufficiently high concentrations, SO_2 exposure is harmful to susceptible plant tissue. Sulfur dioxide and other tropospheric aerosols containing sulfur

Figure 23.6 Agricultural crops and livestock contributed about 19% of particulate matter in 2002 in the U.S. Photo: College of Agriculture and Life Sciences, The University of Arizona.

are believed to affect the radiation balance of the atmosphere, which may cause cooling in certain regions. See Section 4.1 for information about the contribution of SO_2 to acid rain.

23.2.1.5 Nitrogen oxides (NO_x)

Nitrogen oxides (NO_x stands for an indeterminate mixture of NO and NO_2) are formed mainly from N_2 and O_2 during high-temperature combustion of fuel in cars. Catalytic converters are used to reduce emissions. Nevertheless, NO causes a reddish-brown haze in city air that contributes to heart and lung problems and may be carcinogenic. Nitrogen oxides also contribute to acid rain because they combine with water to produce nitric acid (HNO_3) and other acids. Natural sources of nitrogen oxides include those produced during the metabolism of certain soil bacteria.

23.2.1.6 Lead

Lead is highly toxic, and its effects on humans have been recognized since the Roman Empire era. Lead can produce chronic impairment of the formation of blood and it affects infant neurological development. Concentrations of lead in the environment in the United States are no longer as high as they were prior to the introduction of nonleaded gasoline in the 1970s, but it is still a concern in many localities. Lead from human sources may be present in soil and it is often found in particulate matter in older urban environments. Lead-based paints continue to be a source of concern in situations where children are exposed to paints that have peeled from building surfaces (see Chapter 13). Lead-based paint was banned in the U.S. in 1978. Homes built before 1978 may have lead-based paint that can be a health hazard when sanded, chipped, or removed.

23.2.2 Secondary Pollutants

Secondary air pollutants are formed during chemical reactions between primary air pollutants and other atmospheric constituents, such as water vapor. Generally, these reactions must occur in sunlight; thus, they ultimately produce **photochemical smog** (see Information Box 23.4). Photochemical smog is most common in the urban areas where solar radiation is very intense.

A simplified set of some of the reactions involved in photochemical smog formation is given below:

(Eq. 23.1)

$$
\begin{aligned}
N_2 + O_2 &\rightarrow 2NO &&\text{(inside an engine)}\\
2NO + O_2 &\rightarrow 2NO_2 &&\text{(in the atmosphere)}\\
NO_2 + hv &\rightarrow NO + O\\
O + O_2 &\rightarrow O_3\\
NO + O_3 &\rightarrow NO_2 + O_2\\
HC + NO + O_2 &\rightarrow NO_2 + PAN
\end{aligned}
$$

As indicated by the reactions in Equation 23.1, photochemical smog is composed mainly of O_3, **peroxyacetyl nitrate (PAN)**, and other oxidants. Ozone formation is closely tied to weather conditions. Favorable conditions for O_3 formation include:

- air temperatures exceeding 32°C
- low winds
- intense radiation
- low precipitation

Unfortunately, many major U.S. cities exceed the federal air-quality standard for O_3 (an average O_3 concentration > 80 nL L^{-1} for 1 hour 1 day per year averaged over a 3-year period).

As the reactions indicate, HCs are necessary for ozone buildup in the atmosphere. In the absence of HCs, solar UV breaks down the NO_2 into NO and O. Next, the O atom

---

INFORMATION BOX 23.4

Photochemical Smog in Los Angeles

Photochemical smog can be severe in the Los Angeles basin of the California coast. Commuting in Los Angeles requires many cars, which produce high emissions of NO_x and hydrocarbons. At certain times of the year, particularly spring and fall, weather conditions in this area are dominated by subtropical high pressure with clear, calm air conditions that exacerbate air stagnation. The factors influencing smog formation in the Los Angeles basin can be summarized as follows:

- Numerous sources of primary pollutants.
- Inversions that inhibit turbulent mixing of air.
- Few clouds, which result in higher UV intensity.
- Light winds that are unable to disperse pollutants.
- Complex coastal mountain terrain that slows pollutant dispersal.

Figure 23.7 Hydrocarbons interact with nitrogen oxides under the influence of ultraviolet light, resulting in photochemical smog. In urban centers such as Los Angeles, pictured here, atmospheric pollutants can concentrate and pose severe health hazards. Photo: U.S. Environmental Protection Agency.

combines with O_2 to form O_3, which then combines with the NO to reform NO_2 and O_2. Ozone would not accumulate in the atmosphere if it was not for the fact that HCs disrupt the reaction cycle by reacting with NO to form PAN + NO_2. Hydrocarbons from car emissions and other sources, therefore, play an important part in O_3 formation in urban environments. However, not all O_3 in the lower atmosphere results from human activities; natural sources include lightning and the diffusion of some O_3 downward from the stratosphere.

In most western U.S. cities, photochemical smog is often referred to as **brown cloud** (O_3 + PAN + NO_x). Industrial eastern and midwestern U.S. cities also have photochemical smog, but they generally receive less intense sunlight than western cities; thus, smog in those cities is sometimes referred to as **gray air** because of particles (especially $PM_{2.5}$) and SO_2 emanating from burning fossil fuel.

Ozone may be either hazardous or beneficial, depending largely on where it is. For example, it is hazardous as an oxidant in smog (smog ozone), but in the O_3 layer, it is beneficial because it absorbs UV radiation. Smog ozone reduces the normal functioning of lungs because it inflames the cells that line the respiratory tract. Other health effects include increased incidence of asthma attacks, increased risk of infection, and reduced heart and circulatory functions.

Smog O_3 can also damage plant life. In vegetation, the main damage occurs in foliage, with smaller effects on growth and yield. In the United States, it has been implicated in the loss of conifer trees near Los Angeles and is suspected of doing similar damage to trees in the Appalachian Mountains. Some plant species are also very susceptible to PAN in smog, and this is known to affect plants in the Los Angeles area.

Ozone and NO_x pollution in the troposphere is not limited to urban areas. In the mid-1990s, the EPA reported that O_3 and NO_x concentrations were increasing in rural areas in the southeastern and midwestern United States (see Information Box 23.5). Most of this increase is probably attributable to upwind urban sources; however, in some rural areas, soil bacteria may be a greater NO_x source than is fossil fuel combustion. In fact, some estimates indicate that soil bacteria may emit as much as 40% of the total amount of NO_x emitted into the atmosphere. This percentage is very uncertain, since data on measurements of NO_x fluxes from soils are scanty. Soil NO_x fluxes may also be highly spatially and temporally variable.

Tropospheric NO_x strongly controls the concentrations of oxidants such as OH and O_3, which may affect the health of about one-quarter of the U.S. population. As in urban areas, O_3 in the rural atmosphere is controlled by reactions involving NO_x, HCs, OH, and other tropospheric species. The study of the production and destruction processes of O_3 in the rural troposphere is currently an area of active research by the EPA and other government agencies. Many questions remain unanswered concerning the reasons why O_3 concentrations continue to be high in both urban and rural areas despite recent efforts to curb emissions of the O_3 precursors, NO_x and HCs.

INFORMATION BOX 23.5

Regional Ozone Transport

Ground-level ozone ("smog") problems are not just limited to cities. Several times each summer, the central and eastern U.S. comes under the influence of stagnating high-pressure weather systems. Sunshine, high temperatures, and stagnant nonmixing conditions exacerbate the buildup of ozone precursor pollutants and the production of ozone over areas with high pollutant emissions, such as the industrialized Midwest. The slow flow of the polluted air mass traverses multiple states between the Midwest and the Atlantic Coast, and can raise concentrations of ozone to unhealthy levels in rural as well as urban areas (Figure 23.8).

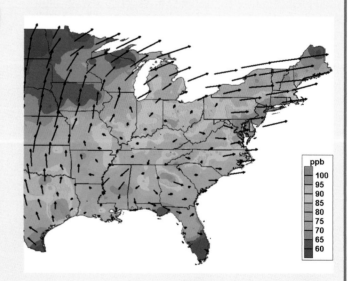

Figure 23.8 Wind can move polluted air across long distances. As a result, rural areas may experience polluted air that originated in metropolitan areas. The black arrows indicate transport winds, where longer arrows indicate faster movement. The color contours indicate ozone concentration. Source: Ozone Transport Assessment Group (1997).

23.2.3 Toxic and Hazardous Air Pollutants

In addition to primary and secondary pollutants, the EPA has identified 188 chemicals (or classes of chemicals) that are considered to be **hazardous air pollutants (HAPs)** or **urban air toxics (UATs)**. Many of these are volatile organic chemicals, such as benzene found in gasoline and used as a solvent, and trichloroethylene, which is used as a solvent/degreaser. Mercury is an example of a hazardous inorganic compound (see Information Box 23.6 and Chapter 13).

23.2.4 Pollutants with Radiative Effects

Some air pollutants greatly influence the interactions between radiation of various wavelengths and the atmosphere. Some radiatively active pollutants contribute strongly to the natural greenhouse effect, while others impact the amount of ozone present within the stratosphere. This section describes these pollutants and their radiative effects.

INFORMATION BOX 23.6

How Does Mercury Get in My Food?

More than 40% of all mercury emissions in the U.S. are from power plants burning coal (about 50 tons Hg yr^{-1} in 2005). The mercury is released as a gas, which eventually is deposited into water or soil. Microbes can then convert inorganic mercury to methyl-mercury in a process called methylation. Small organisms can take up small amounts of methyl-mercury as they feed, storing it in their tissues. As these organisms are eaten by animals higher in the food chain, the methyl-mercury continues to accumulate in body tissues, until they may reach relatively high levels in predators at the top of the food chain, such as swordfish and sharks. The methyl-mercury then becomes part of our food as we consume shellfish or fish from saltwater and freshwater sources.

Mercury tends to accumulate in tissue, so toxic levels can build up slowly. High levels of exposure to methyl-mercury can damage the human brain, resulting in neurological problems, such as increased irritability, shyness, tremors, vision and/or hearing loss, mental retardation, and memory loss. It is also harmful to the kidneys. It is especially hazardous for the fetus, infants, and young children. As many as 4.9 million women of childbearing age in the U.S. may have unsafe levels of mercury, and as many as 60,000 newborns per year are at risk due to dietary exposure.

Fish can be eaten regularly, but some care should be taken in terms of the quantity, frequency and type of fish eaten. Some fish accumulate much higher levels of mercury than others. For instance, studies show that swordfish tend to have much higher levels of mercury than salmon, so it might be wise to eat swordfish only occasionally (see also Chapter 13).

23.2.4.1 *Greenhouse gases*

Carbon dioxide is sometimes not considered to be an air pollutant because it is not hazardous to human health at ambient atmospheric concentrations; moreover, it is essential for carbon fixation by plants. It is, however, an important greenhouse gas and a major byproduct of fossil-fuel burning, which steadily increases the atmospheric concentration of carbon dioxide. This increase is discussed in detail in Chapter 24.

Carbon dioxide is by far the most abundant and important atmospheric trace gas contributing to the natural **greenhouse effect** (see also Sections 4.1 and 4.2.4). It is released to the atmosphere by various processes including deforestation and land clearing, fossil-fuel combustion, and respiration from living organisms. Carbon dioxide is readily absorbed by water, with warm water absorbing more than cold water, so it is removed from the atmosphere by the oceans and other bodies of water. Photosynthesis by land and water plants (phytoplankton) also removes significant amounts of CO_2 from the atmosphere. Removal by plants is particularly apparent during the summer, when average CO_2 concentrations decrease. In addition, large amounts of CO_2 may eventually be fixed as limestone by deposition of the skeletons of some marine invertebrates. The atmosphere currently contains about 750 billion metric tons (BMT) of carbon in the form of CO_2, and a 3-BMT excess enters the atmosphere each year. Research indicates that this excess gives rise to a mean annual increase of about 1.5 $\mu L\ L^{-1}$ in the global concentration of CO_2.

In addition to CO_2, the other main greenhouse gases are CH_4, N_2O, **chlorofluorocarbons (CFCs)**, and O_3. Water vapor is also an important, but variable, greenhouse gas. All of these gases are, as the term "greenhouse" implies, efficient absorbers of longwave radiation. This absorption helps maintain a relatively warm climate on earth. However, because greenhouse gas concentrations continue to increase, it is conceivable that the earth's climate may be altered and become much warmer. Therefore, much scientific research is currently being directed toward improving our understanding of the atmospheric budgets of these trace gases and their role in the greenhouse effect.

Atmospheric CH_4 concentrations have also steadily increased at a rate of about 1% per year in recent decades. This rapid increase is commonly attributed to increased worldwide rice and livestock production. Increased mining of natural gas resources for energy production may also be an important factor.

Synthetic CFCs are also significant contributors to the greenhouse effect. Chlorofluorocarbons are used in refrigerators, air conditioners, foam insulation, and industrial processes. In addition to being very efficient longwave absorbers, CFCs are also involved in depleting stratospheric O_3. Fortunately, because of concerted international effort resulting in the **1987 Montreal Protocol on Substances that Deplete the Ozone Layer**, CFC emissions to the atmosphere have decreased substantially in recent years (see Information Box 23.7 for more details on the Montreal Protocol).

INFORMATION BOX 23.7

Montreal Protocol: International cooperation to reduce an environmental risk from ozone depletion

Research in 1974 at the University of California, Irvine, indicated that synthetic chemicals known as chlorofluorocarbons (CFCs) used as refrigerants and propellants in aerosol cans were slowly diffusing into the stratosphere, breaking down, and releasing chlorine atoms that catalytically destroyed stratospheric ozone. In 1985, observational evidence from the British Antarctic Survey of ozone layer depletion (the ozone hole) over Antarctica spurred several developed (including the U.S.) and developing nations to continue negotiations to significantly reduce the production and use of CFCs. In 1987, the Montreal Protocol on Substances that Deplete the Ozone Layer was adopted. It initially required a 50% reduction of use of CFCs by the year 2000 and a freeze on the use of halons (fire-extinguishing compounds consisting of bromine, fluorine, and carbon) with a 10-year grace period for developing nations. In 1989, the Montreal Protocol entered into force with 13 developed nations announcing a phaseout of CFCs by 1997. Because of mounting scientific evidence of ozone depletion by CFCs, adherents to the Protocol in 1990 agreed to a total ban on CFCs and halons by 2000. Most western industrialized nations have complied. Because of the banning of CFCs and halons, the ozone hole is likely to continue to improve during the coming years.

In the mid-1970s, concern shifted to the possible O_3-depleting effects of manufactured CFCs used as refrigerants, propellants, cleaning compounds, and foam insulation. Intensive study of the effects of CFCs on stratospheric O_3 led to a 1979 U.S. ban on the use of CFC propellants in aerosol spray cans. Growing worldwide use of CFCs, together with evidence of CFC-induced decline of stratospheric O_3 concentrations over Antarctica, convinced 24 industrialized nations to sign the Montreal Protocol (1987 and 1990) (see Information Box 23.7), which called for a complete phase out of the production of CFCs by the year 2000.

Stratospheric O_3 absorbs UV light, decreasing the amount of UV striking living organisms on the earth's surface. Satellite and ground-based measurements have shown that there is a temporary decrease in O_3 concentrations (50 to 75% of total) over Antarctica each year. This has come to be known as the ozone hole, which is defined as the geographic area above the Antarctic where the total ozone is less than 220 Dobson units between 1 October and 30 November (Figure 23.9). Figure 23.10 shows the average area of the Antarctic ozone hole in recent years. Stratospheric O_3 depletion has engendered serious concerns about the causes and possible ecological and human health consequences if this trend continues. In humans, increased UV would probably increase the incidence of skin cancer, including melanoma. Other organisms are also vulnerable to UV; phytoplankton, for example, has declined by 6–12% in areas near Antarctica. The decline in this one-celled organism is thought to be due to in-

Nitrous oxide (N_2O) is an especially good absorber of longwave radiation; 1 molecule of N_2O is equivalent to about 200 CO_2 molecules in terms of its ability to absorb longwave radiation. Currently, atmospheric N_2O accounts for only about 5% of the greenhouse effect, but this percentage is expected to increase in coming years. A 25% N_2O increase in atmospheric concentration may, according to numerical model predictions, increase global mean temperature by about 0.1 K. Worse, N_2O has a very long atmospheric lifetime, estimated to be about 150 years, which is far longer than the atmospheric lifetime of any other nitrogen oxide. Thus, the current buildup of N_2O could affect the earth's climate well beyond the 21st century.

23.2.4.2 Stratospheric pollution

Stratospheric O_3 depletion is another global environmental concern related to pollution. Concern about O_3 first emerged in the early 1970s, when modeling studies indicated that a proposed fleet of supersonic transport (SST) aircraft could emit enough NO_x to damage the O_3 layer. The results from the modeling studies helped put an end to plans to build the fleet, but such considerations remain major factors in plans for aircraft development.

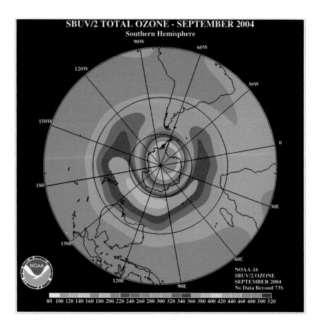

Figure 23.9 Southern Hemisphere map of total ozone for September 2004. The ozone hole, with total ozone lower than 220 Dobson Units, is shown in purple and magenta. Image: U.S. National Oceanic and Atmospheric Administration.

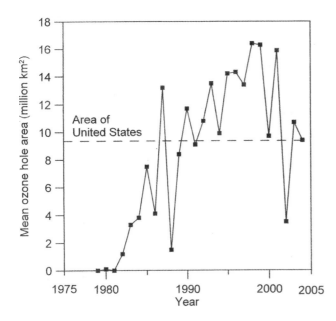

Figure 23.10 Year-to-year variations in the average size of the Antarctic ozone hole between 1 October and 30 November. The size of the ozone hole appears to be decreasing since the phaseout of CFCs. Data: U.S. National Oceanic and Atmospheric Administration.

creased amounts of UV that are reaching surface waters (see also Chapter 24).

The chemical pathways leading to the formation of stratospheric O_3 start with the photodissociation of molecular oxygen (O_2) by solar UV radiation (photons of energy hv, where h is Planck's constant and v is the frequency). The UV photon splits O_2 into two oxygen atoms (O), each of which recombines with undissociated O_2 (in the presence of another chemical species, M) to form two O_3 molecules. These two reactions, which result in a net formation of O_3 are given here:

$$
\begin{aligned}
O_2 + hv &\rightarrow O + O \\
2O_2 + 2O + M &\rightarrow 2O_3 + M
\end{aligned}
$$

$$\text{Net Reaction: } 3O_2 + hv \rightarrow 2O_3 \tag{Eq. 23.2}$$

The two O_3 molecules quickly convert back to molecular oxygen via

$$
\begin{aligned}
O_3 + hv &\rightarrow O + O_2 \\
O_3 + O &\rightarrow 2O
\end{aligned}
$$

$$\text{Net Reaction: } 2O_3 + hv \rightarrow 3O_2 \tag{Eq. 23.3}$$

The process of production and loss of the O_3 molecules by photodissociation is very important because, overall, it helps prevent harmful UV from reaching the earth's surface. These production/destruction schemes indicate that the chemistry of stratospheric O_3 would be straightforward if there were no other reactive chemical species in the stratosphere. However, other chemicals,

TABLE 23.1 Chemical species that are believed to catalyze the destruction of O_3 molecules in the atmosphere.

CYCLE	X	XO
NO_x	NO	NO_2
Water	HO•	HO_2•
CFC	Cl•	ClO•

such as CFCs and NO_x, are present and play an important destructive role. This is indicated by the following general catalytic cycle:

$$
\begin{aligned}
X + O_3 &\rightarrow XO + O_2 \\
O_3 + hv &\rightarrow O + O_2 \\
XO + O &\rightarrow X + O_2
\end{aligned}
$$

$$\text{Net Reaction: } 2O_3 + hv \rightarrow 3O_2 \tag{Eq. 23.4}$$

where X and XO represent the compounds or free radicals that catalyze the destruction of O_3 molecules. Mainly NO_x, water vapor, and CFCs, these species are summarized in Table 23.1.

Of these catalysts, CFCs are entirely anthropogenic, whereas nitrogen oxides come from both natural and synthetic sources. Stratospheric water vapor comes mainly from natural processes. In addition to the three main catalysts, other chemicals may play a role in controlling stratospheric O_3 levels. For example, recent evidence indicates that methyl bromide, which is used as a soil fumigant, may reach the stratosphere, where it can undergo a catalytic reaction sequence with O_3 similar to those of the three main chemical species.

The catalytic reactions do not destroy all the O_3 present in the stratosphere. The reason they don't is that reactions also occur between the catalysts, and these reactions result in chemicals that do not deplete O_3. Some of the chemicals eventually return to the earth's surface (*e.g.*, HNO_3 in rain).

Further aspects of each of these catalytic cycles are described in Graedel and Crutzen (1993) and are briefly discussed in the following paragraphs.

23.2.4.2.1 NO_x/O_3 destruction cycle

The nitrogen oxides in the stratosphere come mainly from photodissociation of nitrous oxide, which originates mostly from microbial processes at the earth's surface. Nitrous oxide is also a major greenhouse gas. It is produced mainly within moist soils by microbial denitrification of nitrate fertilizer, but it can also be biologically produced in oceans. Since the 1970s there has been concern that increased agricultural use of nitrogen fertilizers could increase the amount of nitrous oxide reaching the stratosphere, ultimately depleting O_3. Measurements of atmospheric N_2O indicate that its concentration is increasing by about 0.25% per year. A 25% increase in N_2O by the late 21st century could reduce total stratospheric O_3 by 3–4%, which could increase the incidence of skin cancer by 2–10%.

Nitrous oxide is not known to be lost within the troposphere; however, it is converted to NO in the stratosphere mainly by the following reactions:

$$N_2O + h\nu \rightarrow N_2O \,(^1D)$$
$$N_2O + O\,(^1D) \rightarrow 2NO$$

(Eq. 23.5)

The two NO molecules formed initiate the O_3 destruction reactions described previously. Note that $O(^1D)$ in Equation 23.5 denotes atomic oxygen in an electronically excited state.

23.2.4.2.2 H_2O/O_3 destruction cycle

The stratosphere is generally very dry. However, enough water vapor is present to react with electronically excited atomic oxygen to produce the free radical $HO\bullet$ via

$$H_2O + O\,(^1D) \rightarrow 2HO\bullet$$

(Eq. 23.6)

The catalytic water cycle has less influence upon O_3 concentrations than do the other reaction cycles, but it can be significant when sufficient water vapor is present.

23.2.4.2.3 CFC/O_3 destruction cycle

Chlorofluorocarbons (*e.g.*, $CFCl_3$ and CF_2Cl_2) are relatively stable in the troposphere, but once in the stratosphere, they are photodissociated by UV. This photodissociation produces the catalysts Cl and ClO, both of which deplete O_3. There is evidence that links the O_3 depletion in the Antarctic region to CFCs and other pollutants that carry chlorine and bromine into the stratosphere. Chlorine monoxide (ClO) has been identified as the chief cause of O_3 depletion in polar regions. Weather patterns and volcanic eruptions may also play a part.

Chlorofluorocarbons are also implicated in possible global warming. Because many are extremely efficient absorbers of longwave radiation, they contribute to the earth's greenhouse effect.

23.3 WEATHER AND POLLUTANTS

What happens to pollutants in the atmosphere? The answer depends on several factors. Pollutants are transported by wind and turbulence, and they may undergo chemical transformations before being deposited on the earth's surface. Thus, weather conditions strongly affect the fate of air pollutants.

23.3.1 Stability and Inversions

The stability of boundary-layer air (see Figure 4.7) largely determines how quickly pollutants are moved upward from their ground sources. Stability is primarily a function of the vertical air temperature gradient relative to the adiabatic lapse rate (see Section 4.2.2). Strong instability associated with buoyancy causes efficient air mixing and pollution dispersal over a large mixing depth of the boundary layer (from 100 to 1,000 m). Good mixing often occurs on warm days when the ground is heated by sunlight. In contrast, pollution is poorly dispersed on days or nights when the atmosphere is stable. At those times, turbulent movement of pollution upward is slow or nearly nonexistent.

We know from Section 4.2.2 that temperature inversions influence atmospheric stability; thus, they play an important role in determining the concentrations of air pollutants. The effects of inversions are intensified by limited air drainage out of enclosed valleys, as is the case in Los Angeles (Figure 23.7) and Mexico City. Various processes may generate inversions, including surface cooling caused by loss of infrared radiation or by evaporation, atmospheric subsidence, and topographic effects.

Ground-surface cooling is caused mainly by infrared radiation emission from the surface to the sky. It generally occurs during clear, calm nights, with inversion heights extending about 100 m above the ground. Such inversions commonly occur throughout the western United States during fall, winter, and spring, when the air is relatively dry and skies are clear. Such conditions readily permit cooling by longwave loss of energy from the ground. Tucson, Arizona, for example, often experiences radiation inversions in the cooler months, so that wintertime pollution problems are exacerbated in the area. Cooling of the ground by evaporation of water from soil and plants may also establish inversions. Evaporative cooling can occur during the day or night, particularly over irrigated fields. This type of inversion may be important in relation to certain agricultural activities, such as the aerial application of pesticides over large irrigated fields. The depth of an evaporatively cooled inversion layer is usually just a few meters.

Warming of the atmosphere by subsidence causes inversions over regions that have semi-permanent high pressure (anti-cyclonic flow), such as the southwestern United States. As air subsides (sinks), it encounters higher pressure and thus is warmed (see Section 4.2.2 and Figure 4.7). Within regions of high pressure, an inversion height may be several hundred meters above ground; thus, the air may be very stable over a large depth of the atmosphere. Because subsidence inversions may last from several days to weeks, the result is highly polluted conditions at ground level. For example, subsidence inversions are a major factor in reducing air quality in the Los Angeles basin.

Inversions associated with topography result from adiabatic warming of air as it flows downslope over mountainous terrain. These inversions may exacerbate air pollution problems in populated, mountainous areas such as Denver, Colorado, or Salt Lake City, Utah (see Figure 4.3).

23.3.2 Wind and Turbulence in Relation to Air Pollution

Wind affects turbulence near the ground, thus affecting the dispersion of pollutants released into the air. Turbulence (largely fine scale vertical and horizontal motion of air) is generated in part by air flow over rough ground. The greater the wind speed, the greater the turbulence, and hence the greater the dispersion of pollutants that are near the ground (Figure 23.11).

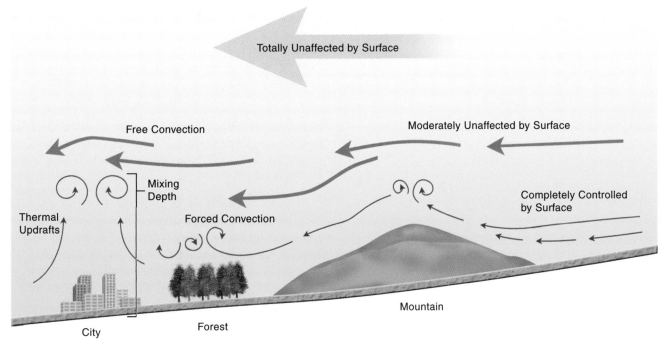

Figure 23.11 Diagrammatic representation of air flow, mixing, and relative velocity over varying terrain as affected by height. From *Pollution Science* © 1996, Academic Press, San Diego, CA.

We can visualize the dispersion of pollutants in air by looking at the familiar cloud or "plume" of pollution emitted continuously by a smokestack (Figure 23.12). As the plume contents are carried away from the stack by the wind, the size of the plume increases owing to dispersion. Because of dispersion, the pollutant concentration within the plume decreases with increasing distance from the source.

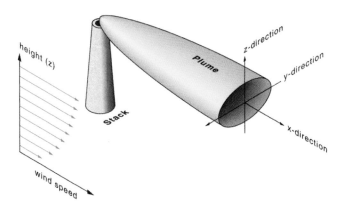

Figure 23.12 Plume pattern (coning) resulting from continuous stack emission into a near-neutral stable boundary layer under moderate winds. From *Pollution Science* © 1996, Academic Press, San Diego, CA.

Dispersion of pollution downwind from a smokestack is affected by the roughness of the ground surface. Because of friction between the atmosphere and the ground, wind speed is slowed markedly near the ground. If the surface is relatively rough, as it is when trees and buildings are present, the air flow tends to be turbulent and the increase of wind speed with increasing height is relatively small. Greater surface roughness increases turbulence, which helps disperse pollutants. Air flow over a smooth surface, such as a large mowed lawn, tends to be less turbulent and the decrease in wind speed near the ground is relatively small (see Figure 23.11).

The cone-shaped plume in Figure 23.12 illustrates the pattern of pollutant dispersal downwind of a point source. Several factors affect the plume, including the effective height (H) of emission, which is a measure of how high the pollutants are emitted into the atmosphere directly above the source. The height is dependent upon source characteristics and atmospheric conditions. Generally, a tall smokestack produces relatively low ground-level pollutant concentrations, because turbulence tends to dilute the pollution before reaching the ground. Driven by buoyancy, fast-moving pollutants are initially transported high up into the atmospheric boundary layer because they are warmer than the surrounding air. But as the pollutants cool and merge with the ambient air, the plume begins to move sideways with the wind. Then turbulence caused by the air flow over the surface and by possible instability governs the diffusion of the plume contents.

Usually, turbulence helps mix plume contents uniformly in such a way that the concentration follows a Gaussian distribution about the plume's central axis. Mathematically, pollutant concentration $\chi_{(x,y,z,H)}$ (kg m^{-3}) at any point in the plume is described by

$$\chi_{(x,y,z,H)} = \frac{Q}{2\pi\sigma_y\sigma_z\bar{u}}$$

$$\cdot \exp\left(-\frac{y^2}{2\sigma_y^2}\right) \qquad \text{(Eq. 23.7)}$$

$$\cdot \left[\exp\left(-\frac{(z-H)^2}{2\sigma_z^2}\right) + \exp\left(-\frac{(z+H)^2}{2\sigma_z^2}\right)\right]$$

where:

Q is the rate of emission of pollution from the source (kg s^{-1})

σ_y and σ_z are the horizontal and vertical standard deviations of the pollutant concentration distributions in the y and z directions

$\bar{u}$ is the mean horizontal wind speed within the plume (m s^{-1}).

This model, which is applicable to continuous sources of gases and particulates less than about 10 µm in diameter (larger particles quickly settle to the ground), can be used to model plume concentrations over horizontal distances of 10^2 to 10^4 m. With this Gaussian plume model, it is assumed that no deposition of plume contents to the ground surface takes place. In fact, it is assumed that plume contents are "reflected" from the ground back to the air. The values of σ in the equation are estimated from any one of several empirical formulas that relate σ to downwind distance (x) and stability conditions. These formulas include the following equations, which were developed by the **Brookhaven National Laboratory (BNL)**.

$$\sigma_y = ax^b \quad \text{and} \quad \sigma_z = cx^d \qquad \text{(Eq. 23.8)}$$

where a, b, c, and d are parameters dependent upon stability. (See Hanna *et al.* [1982] for a summary discussion of BNL equations as well as other approaches.) At ground level, $z = 0$, and along the plume centerline, $y = 0$. Thus, from Equation 23.2, the concentration can be calculated by

$$\chi_{(x,H)} = \frac{Q}{\pi\sigma_y\sigma_z\bar{u}} \exp\left(-\frac{H^2}{2\sigma_z^2}\right)$$

$$= \frac{Q}{\pi ax^b cx^d \bar{u}} \exp\left(-\frac{H^2}{2(cx^d)^2}\right) \qquad \text{(Eq. 23.9)}$$

One type of plume (shown in Figure 23.12) typically occurs under windy conditions with stability conditions at or near neutral. Within such a plume, mixing occurs mainly by frictionally generated turbulence, and pollutant diffusion is nearly equal in all directions (*i.e.,* the σ values are nearly equal and the plume spreads out in the familiar cone pattern, known as **coning**). Coning can occur day or night, and is often seen during cloudy and windy conditions. Depending upon effective source height and atmospheric conditions, the

plume may reach the ground close to the source. Using Equations 23.8 and 23.9, we can estimate the ground level ($z = 0$) concentration of the plume composed of a pollutant, say, SO$_2$, emitted into the atmosphere at a known effective height. Suppose we have the following: $Q = 0.5$ kg s^{-1}, $H = 25$ m; $\bar{u} = 2$ m s^{-1}; near neutral stability, and BNL parameters $a = 0.32$, $b = 0.78$, $c = 0.22$, and $d = 0.78$. Then the ground level SO$_2$ concentration along the plume centerline at an arbitrary distance of $x = 500$ m from the source will be 4.7×10^{-5} kg m^{-3} (or 47 mg m^{-3}).

Plumes may change because of changes in the wind velocity and boundary layer stability. When the atmospheric boundary layer is strongly stable, such as during radiation inversions at night or during subsidence inversions, a **fanning** pattern may be evident in the plume, as illustrated in Figure 23.13a. Under these conditions, there is almost no vertical motion and the BNL parameters are $a = 0.31$, $b = 0.71$, $c = 0.06$, and $d = 0.71$. Lack of vertical motion thus effectively forces the plume into a relatively narrow layer, while changes in wind direction may spread the plume out laterally, resulting in a V- or fan pattern; hence, the term. A constant wind direction, however, forces the plume into a tightly closed fan pattern, which follows a relatively straight and narrow path. Over flat terrain the plume in Figure 23.13a may be unchanged for very long distances. If there is no vertical air movement, ground level concentrations downwind of a tall smokestack can be nearly zero. However, if the source is close to the ground (*i.e.,* H is small), or if changes in topography cause the plume to intercept the ground, the ground level concentrations can be very large.

By midmorning, surface heating by solar radiation typically begins to break down the inversion developed during the previous night, as illustrated in Figure 23.13b. Unstable conditions develop near the ground, resulting in vertical mixing of the air. With moderately unstable conditions, $a = 0.36$, $b = 0.86$, $c = 0.22$, and $d = 0.86$. In this situation, pollution is transported downward toward ground level. Stable conditions above, however, limit dispersion of pollutants upward. Thus the remaining inversion effectively puts a "lid" over the ground level pollution. This situation is known as **fumigation** and generally lasts for periods of an hour or less. Fumigation is highly conducive to enhanced ground level pollutant concentrations.

By early afternoon, lapse conditions (*i.e.,* negative vertical temperature gradient) generally become fully established within the boundary layer due to strong surface heating by the sun. During much of the afternoon, air motion mainly exhibits the large turbulent eddies associated with buoyancy. These eddies are generally larger than the plume diameter and thus transport the plume upward and downward in a sinusoidal path or **looping** pattern, as illustrated in Figure 23.13c. The loops are carried with the overall wind pattern and generally increase in size with increasing distance downwind from the source. The motion may bring the plume contents to ground level quite close to the source. Because of turbulence, however, the plume eventually becomes dispersed at relatively large distances.

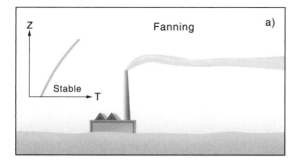

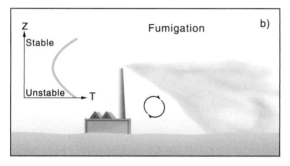

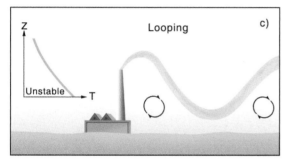

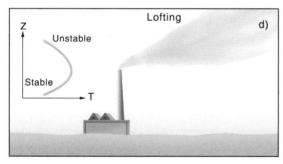

Figure 23.13 Effect of atmospheric stability upon resultant stack plume pattern during (a) inversion (fanning pattern), (b) dissipation of inversion near ground (fumigation pattern), (c) lapse conditions (looping pattern), and (d) lofting pattern.

By early evening, a radiation inversion often rebuilds from ground level upward. Stable conditions near the ground inhibit transport of plume contents downward, but unstable air aloft (above the inversion height) allows dispersal upward. This upward transport, known as **lofting,** is highly favorable for dispersing pollutants, as shown in Figure 23.13d. Lofting is only effective when the source is above the inversion height. Plume contents emitted below the inversion height are essentially trapped in a fan-type plume configuration.

Topography downwind from pollution sources affects air quality, especially in mountainous areas. For example, air drainage into relatively enclosed valleys during winter and/or inversion conditions can cause accumulation of pollutants within the valleys. Thus urban areas in valleys with restricted air flow are particularly prone to high pollution levels. In addition, in coastal areas, air flow from the ocean (sea breezes) can be blocked or channeled by mountain ranges. This situation is common in the Los Angeles basin, which is surrounded by mountains that restrict air flow from the Pacific Ocean. Thus, dispersal of pollutants from sources in the basin is inhibited.

23.3.3 Pollutant Transformation and Removal

As pollutants move with the wind, chemical reactions often occur between the pollutants and other atmospheric chemical species. Although the pathways and rates of many of these chemical reactions are poorly understood, they are an important factor affecting the fates of many air pollutants.

Most pollutants, such as CO, remain in the atmosphere for relatively brief periods, lasting only a few days or weeks. Thus, if emissions were completely curtailed, the lower atmosphere would quickly lose nearly all of its pollutants. However, some pollutants—volcanic ash and sulfur-containing aerosols, for example—emitted high into the stratosphere can remain there for months before settling back to the surface. These long-lasting upper-atmospheric pollutants can alter the earth's climate, as evidenced by lower air temperatures resulting from volcanic eruptions (see Information Box 4.2 in Chapter 4). In addition, synthetic **chlorofluorocarbon** (CFC) compounds can remain in the atmosphere for many years before they break down.

Pollution can be removed from the atmosphere by gravitational settling, dry deposition, condensation, and wet deposition.

Gravitational settling. Gravitational settling removes most particles whose diameters are greater than about 10 μm. Particles less than 10 μm in diameter are often small enough to stay in the atmosphere for long periods. Particles greater than about 10 μm in diameter quickly settle out.

Dry deposition. Dry deposition is a mass-transfer process that results in adsorption of gaseous pollutants by plants and soil. Dry deposition to plants is dependent upon uptake of the pollution through stomatal openings in plant leaves and upon turbulent transport in the air. Dry deposition to bare soil involves not only turbulent transport of pollutants in air above the soil, but also soil microorganisms that take up such pollutants such as CO.

Condensation. Volatile organic compounds can condense on cold surfaces during winter in temperate and polar regions. The process of evaporation, transport, and condensation of toxic compounds, such as dioxins and the pesticide, dichlorodiphenyltrichloroethane (DDT), may be responsible for causing high levels of toxic organic pollutants in the Arctic (see Information Box 23.8).

INFORMATION BOX 23.8

The Grasshopper Effect and Arctic Pollution—Transcontinental Contaminant Transport

There is much public and scientific concern for the health and well being of the wildlife and the 155,000 Inuit inhabitants of the northern Canadian territory of Nunavut. Life there is exposed to some of the highest levels of toxic organic chemicals on earth. These chemicals include carcinogens and compounds that can cause birth defects, such as dioxins, DDT, and PCBs. They can persist within the environment for decades, and they have come to be known collectively as persistent organic pollutants (POPs) (see Chapter 10). Some of the POPs in the Canadian Arctic, such as dioxins and furans, are released locally to the environment by incineration of plastic containers in community garbage dumps. Nearly all of the POPs within the environment, however, are believed to originate from sources (*e.g.*, industry, agriculture) far south of the territory.

How do the POPs reach the Canadian Arctic and what happens to them? To reach the Arctic, it has been recently hypothesized that POPs may undergo multiple cycles of evaporation, transport, and condensation over the earth—with evaporation during warm conditions, transport by global air currents (see Chapter 4), and condensation on surfaces during cool conditions. Because of the "jumping" of POPs due to warming/cooling cycles, the process has been referred to as the "grasshopper effect" or the global "distillation effect." This hypothesized process may be the reason why the very cold Arctic is most susceptible to accumulation of POPs within the environment. The POPs there are essentially trapped because the relatively cold tempera-

tures do not permit them to re-evaporate and move with the winds to other locations.

The POPs condense upon the cold surfaces of the Arctic, including water and vegetation that may be ingested by wildlife. The POPs are soluble in fat and thus tend to collect within the blubber of marine animals, such as beluga, narwhal, and seals. The diet of Inuit people is often supplemented by consumption of blubber, which often has within it large levels of PCBs, DDT, and other POPs. Because of their diet, Inuit women in Nunavut have DDT levels (1,210 ppm) in their breast milk that is on average about seven times larger than what it is in women in southern Canada (170 ppm).

Identifying the sources is controversial but some of the POPs may originate from industrialized countries including the United States. Toxaphene, for example, is still found in the Arctic environment and may have originated at least in part from the southern U.S., where it was used as a cotton crop pesticide. Its use in the U.S. was banned in 1982. Recent research by the U.S. EPA and other agencies indicates, however, that most of the POPs may come from warm regions outside the U.S. The pesticide DDT, for example, is thought to originate from tropical Africa, Asia, and Central America, where it is used to control malaria-carrying mosquitoes. Use of DDT was banned in Canada, the United States, and most developed counties in 1972, but several countries continue to use it for economic and public health reasons.

Wet deposition. Rain is very effective at removing gases and small particulates. Raindrops increase in size as they fall toward the ground, and thus they increasingly capture more pollutants. Raindrops, in effect, "sweep up" pollution as it falls through the air. The ability of the rain to remove pollutants depends upon the rainfall intensity, the size and electrical properties of the drops, and the solubility of the polluting species.

23.4 POLLUTION TRENDS IN THE UNITED STATES

Emissions of nearly all types of primary air pollutants have generally declined or held steady in the United States since about 1970 (Figure 23.14). This decline is mostly attributable to general compliance with the federal air quality regulations set forth in response to the U.S. Clean Air Act of 1970. Although air quality is improving overall, many specific urban areas fail to meet the air quality standards set for some pollutants. Poor air quality is estimated to affect the lives of about 100 million people in the United States alone.

Despite the fact that transportation continues to be a major source of pollution in the United States, the proportion

of its contribution is diminishing. While the number of cars is increasing in most urban areas, fuel efficiencies have increased and pollutant emissions per vehicle are declining owing to improvements in technology such as catalytic converters and other pollution control devices. Evidence of improved air quality is shown by the marked decline of atmospheric lead (Pb) concentrations since 1970. Atmospheric lead comes mainly from the burning of lead-containing gasoline in cars and trucks. Thus, the introduction of unleaded gasoline was a significant factor in this decline. Now required for cars in the United States because of environmental health concerns, unleaded gasoline is also used because leaded fuels deactivate catalytic converters.

It is generally recognized that reducing air pollution through control of emissions at the source is the best approach, which is the goal of the EPA and other regulatory agencies. Total control of pollutant emissions is certainly not feasible for various economic and technological reasons, but efforts at reducing emissions are helping to improve air quality in most locations. There are various physical and chemical precipitators/concentrators/burners that can be used to control emissions. Consult an environmental engineering handbook for more details.

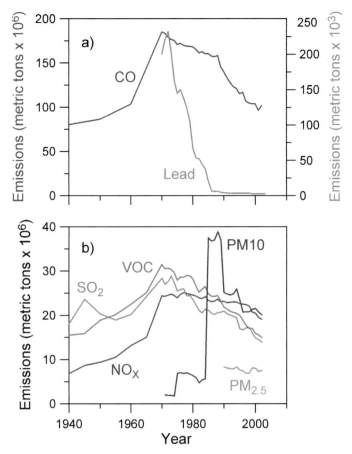

Figure 23.14 Air pollution emissions have generally decreased since the 1970s with the exception of PM_{10}. PM_{10} emissions rose sharply in 1985, largely due to the adoption of reporting PM_{10} emissions in the miscellaneous category, which includes roads, construction, and agriculture. Data: U.S. Environmental Protection Agency.

QUESTIONS AND PROBLEMS

1. Describe how surface air temperature inversions form. Why are air-temperature inversions important relative to air pollution in urban areas?

2. What factors affect atmospheric stability? Explain.

3. Based upon the Brookhaven National Laboratory Equation 23.8, how do numerical values of σ_z values compare at $x = 100$ m for pollution plumes during stable and unstable atmospheric conditions? How do the σ_y values compare? During which condition (stable or unstable) would you expect the plume to intercept the ground closer to the source? Explain.

4. Describe the processes that remove air pollution.

5. What is the difference between EPA-designated primary and secondary air pollution? Give an example of each type of pollutant.

6. What is photochemical smog? Explain how it is formed.

7. Explain how O_3 in the stratosphere is beneficial, whereas O_3 in the troposphere is harmful.

8. Explain how anthropogenic chlorofluorocarbons (CFCs) destroy stratospheric O_3.

REFERENCES AND ADDITIONAL READING

Ahrens C.D. (2003) *Meteorology Today: An Introduction to Weather, Climate and the Environment*, 7th ed. Brooks/Cole-Thomson Learning Inc., Pacific Grove, California.

Albritton D.L., Monastersky R., Eddy J.A., Hall J.M. and Shea E. (1992) Our Ozone Shield: Reports to the Nation on Our Changing Planet. Fall 1992. University Cooperation for Atmospheric Research, Office for Interdisciplinary Studies, Boulder, Colorado.

Dickinson R., Monastersky R., Eddy J., Bryan K. and Matthews S. (1991) The Climate System: Reports to the Nation on Our Changing Planet. Winter 1991. University Cooperation for Atmospheric Research, Office for Interdisciplinary Studies, Boulder, Colorado.

EPA. (2005) Asthma research results highlights. EPA 600/R-04/161. Available from http://www.epa.gov/ord/articles/2005/asthma_fact_sheet.htm.

Graedel T.E. and Crutzen P.J. (1993) *Atmospheric Change: An Earth System Perspective*. W.H. Freeman, New York.

Hanna S.R., Briggs G.A. and Hosker, Jr. R.P. (1982) *Handbook on Atmospheric Diffusion*. U.S. Department of Energy, Washington, D.C.

Mitchell J.F.B. (1989) The "greenhouse" effect and climate change. *Reviews of Geophysics*. **27**, 115–139.

Oke T.R. (1987) *Boundary Layer Climates*. Routledge, New York.

Ozone Transport Assessment Group (1997) Telling the OTAG Ozone Story with Data. Final Report, Vol. I: Executive Summary, OTAG Air Quality Analysis Workgroup. Co-chairs, D. Guinnup and B. Collom. http://capita.wustl.edu/otag/reports/aqafinvol_I/animations/v1_exsumanimb.html.

Schlesinger W.H. (1991) *Biogeochemistry: An Analysis of Global Change*. Academic Press, San Diego, California.

CHAPTER 24

GLOBAL CHANGE

E. P. Glenn and A. D. Matthias

Tree clearing in the Amazon rainforest. This satellite image from the Advanced Spaceborne Thermal Emission and Reflection Radiometer (ASTER) sensors shows deforestation in the state of Rondonia, Brazil, on August 24, 2000. The false-color image combines near-infrared, red, and green light. Tropical rainforest appears bright red, while pale red and brown areas represent cleared land. Black and gray areas have been burned. The Jiparaná River appears as blue. 1 cm = 5 km. *NASA photograph, posted by The Ozone Hole, Inc. (http://www.solcomhouse.com/nasarainforest.htm)*

24.1 INTRODUCTION

In contrast to those human activities that have strictly local effects, **global change** is the term used to describe the effects of human activities on the global environment. Major global change issues are listed in Information Box 24.1. The main change events, discussed in the following sections, include atmospheric warming and climate change, caused by emission of greenhouse gases; deforestation, caused by overharvesting of timber and clearing of forests for agriculture; desertification, caused primarily by overgrazing of arid and semi-arid shrublands and grasslands; depletion of **pelagic** (open ocean) marine species by overfishing; and degradation of coastal ecosystems, caused by discharge of pollutants and overdevelopment of shorelines.

The physical, chemical, and biological factors that produce global change are called **forcing agents**. For example, carbon dioxide is a forcing agent that leads to global warming because it can trap heat in the atmosphere. Global change issues are interrelated in both cause and effect through the **global biogeochemical cycle**, the process by which matter and energy are transformed across the **biosphere** (the living organisms within earth), the **pedosphere** (the soil layer), the atmosphere, and the oceans. For example, the burning of tropical forests to clear the land for agriculture leads to a regional loss of **biodiversity**, the number of species an ecosystem can support. In addition, it also adds carbon dioxide to the atmosphere, which contributes to global warming. Furthermore, forest clearing along tropical coasts leads to soil erosion, which can result in transport of suspended sediment into the sea and resultant smothering of coral reefs. As temperatures rise, the already damaged coral reef ecosystem becomes susceptible to heat stress, leading to further loss of corals. As coral reefs and tropical forests decline, their capacity to fix carbon dioxide into biomass diminishes, increasing the rate at which carbon dioxide accumulates in the atmosphere.

These interactions are called **feedback effects,** and they can be either positive, amplifying global change, as in the above examples, or negative, dampening the effect on global change. An example of a dampening (negative) feedback effect is the **fertilization effect** that carbon dioxide has on plant growth; more carbon dioxide in the air promotes greater plant growth around the globe, thereby removing some of the carbon from the air. Conversely, amplifying (positive) feedback effects can produce a general, downward spiral of ecosystem functions in response to human activities. Change can be gradual, as in the relatively steady increase in carbon dioxide in the atmosphere, or abrupt, as in the collapse of the North Atlantic codfish populations that occurred in the 1990s due to overfishing. These are sometimes called **linear effects** and **nonlinear effects**, respectively.

INFORMATION BOX 24.1

Global change issues that represent challenges to human welfare and natural ecosystems. Many of these issues are interrelated in both cause and effect.

Global Change Issue	Impact
Air Traffic	Jetliners produce contrails at 8–13 km altitude that introduce aerosol pollutants into the upper atmosphere, affecting ozone and greenhouse gas levels.
Coral Reef Destruction	Coral reefs are thought to be home to 24% of all marine life, yet 70% of the earth's coral reefs could be lost over the next few decades.
Deforestation	Tropical rainforests are being cleared for agriculture and settlement around the globe, with negative effects on biodiversity. Forest clearing also increases greenhouse warming and increased erosion.
Desertification	Over 70% of the world's semi-arid zone has been moderately to severely damaged by overgrazing and unsustainable agricultural practices. Desertification interacts with natural drought cycles to produce starvation and refugee crises in Africa.
The Greenhouse Effect and Climate Change	Heat-trapping gases emitted by fossil fuel burning is increasing the earth's surface temperature, with numerous potentially negative impacts on ecosystems and human welfare.
Lower Atmosphere Ozone Enrichment	Ozone produced at ground level from pollutants emitted by cars, power plants, and industrial processes causes lung disease in humans and damages plants.
Species Extinctions	Poaching, habitat destruction, spread of exotic species, pollution, and global warming are driving many species to extinction. Over 19,000 plant species and 5,000 animal species are classified as endangered.
Upper Atmosphere Ozone Depletion	Chlorofluorocarbons (CFCs) injected into the atmosphere from aerosol spray cans and air conditioners have lowered ozone levels, allowing damaging ultraviolet light to penetrate to the earth's surface.

It is important to note that not all global change is due to human activities. The earth has a long history of both gradual and abrupt change in response to forcing events. Volcanoes add gases and particulates to the atmosphere that can perturb climate. Tsunamis and earthquakes can rearrange the landscape in sometimes catastrophic events, and it is suspected that at least one large meteor has struck the earth in past millennia, causing mass extinctions (such as the disappearance of the dinosaurs). Other, less drastic, natural forcing agents act on our climate system. For example, periodic shifts in temperature in tropical Pacific waters produce **El Niño-Southern Oscillation (ENSO)** cycles that affect temperature and precipitation patterns around the world, through a series of connected ocean and atmosphere effects called **teleconnections**. Human activities can interact with natural climate cycles to produce catastrophic effects; for example, in the 1960s, recurrent El Niño cycles combined with overfishing led to the collapse of Peru's anchovy fishery. Hence, it is important to consider natural as well as human-induced forcing agents in understanding global change events.

Humans are not the first organisms to have caused widespread global change. Earth's original atmosphere probably lacked oxygen and was rich in carbon dioxide. However, evolution of the so-called blue-green algae, technically known as cyanobacteria, had the capability to use the energy of sunlight to fix carbon dioxide into organic molecules, at the same time producing gaseous oxygen from water as a byproduct. Over a 2-billion-year period, the atmosphere came to contain 20% oxygen and only a trace amount of carbon dioxide. The main difference between and human-induced global change is in the time scale of events. Natural changes usually take place over thousands or millions of years, giving the earth's life forms time to evolve and adjust to new conditions. Human activities are changing the earth's climate and ecosystems in time scales of decades to hundreds of years, and many life forms may not be able to adjust to such abrupt changes. The interaction between life forms and the biogeochemical cycles on earth might be directed toward preserving conditions favorable for life—a concept that was advanced by James Lovelock as the **Gaia hypothesis** (Gaia is the ancient Greek word for earth goddess). Although the hypothesis is still intensely debated, it is a useful concept in attempting to understand why the diversity of life is an important characteristic of earth system functioning, and why we should be cautious in imposing rapid change on earth's climate system and ecosystems.

The following sections will introduce you to the main types of global change that will pose a challenge to human and natural ecosystems over the next hundred years. We will consider the forcing agents leading to change, the ecosystem responses, the likely effects on human societies, and possible responses to lessen the impacts of global change. Throughout these sections, we will show how the different human activities that are affecting the globe are linked to each other through the biogeochemical cycle.

24.2 GLOBAL WARMING AND THE GREENHOUSE EFFECT

24.2.1 The Greenhouse Effect

Air temperatures near the earth's surface are determined by the balance between incoming and outgoing energy flows through the atmosphere (Figure 24.1). A small amount of the surface energy balance is due to heat flow from earth's molten interior to the surface, but most of the energy affecting the surface comes from incoming solar radiation from the sun. The sun is extremely hot, with a temperature of approximately 6,000°C. It emits relatively high-energy (short wavelength) radiation, ranging from ultraviolet (100 to 400 nm) and visible (400 to 700 nm), to near infrared (700 to 4000 nm) wavelengths. Much of the ultraviolet radiation is absorbed by ozone (O_3) and molecular oxygen (O_2) in the stratosphere layer at the top of the atmosphere (see Chapter 4). A small percentage of the visible and infrared radiation is reflected back to space or absorbed by small particulates or aerosols in the atmosphere. These particles can originate as dust, soot, sea salts, and a variety of chemicals emitted into the air by human activities, plants, soils, volcanoes, and oceanic processes. Clouds typically reflect 40 to 80% of incoming radiation back into space, and they absorb 5–15%. Therefore, only about a half of the incoming solar radiation reaches the earth's surface.

Some of the radiation striking the earth's surface is reflected back into the atmosphere, but most is absorbed. Some of this energy is re-emitted to the atmosphere as long wavelength radiation (4,000 to 50,000 nm), but most is consumed in the evaporation of water from plants, lakes, and oceans. This evaporation releases **latent heat** energy, because the water does not change temperature in passing from the liquid to the vapor stage, but represents potential energy. As this vapor rises in the atmosphere, it eventually recondenses as water or ice crystals, releasing the latent energy of evaporation back into the atmosphere as heat. Radiation absorbed by clouds and water in the atmosphere is also re-emitted as long wavelength radiation. Normally, about two-thirds of the solar energy that enters the atmosphere as ultraviolet, visible, and near infrared energy is re-emitted back to space as reflected radiation or heat energy, and the remainder stays in the atmosphere, warming the earth's surface. Anything that perturbs this delicate balance can affect the surface temperature on earth.

The atmosphere contains trace gases that efficiently absorb outgoing long wavelength, infrared radiation, and that therefore contribute greatly to warming the atmosphere. The most important of these is water vapor, followed by carbon dioxide, methane, and nitrous oxide. These heat-trapping molecules are called **greenhouse gases** and occur naturally (see also Chapter 23). Opposing the action of these gases are trace gases derived from sulfur compounds emitted by soil, plants, volcanoes, and the oceans. These gases form sulfate aerosols in the atmosphere that reflect incoming solar radiation, leading to a po-

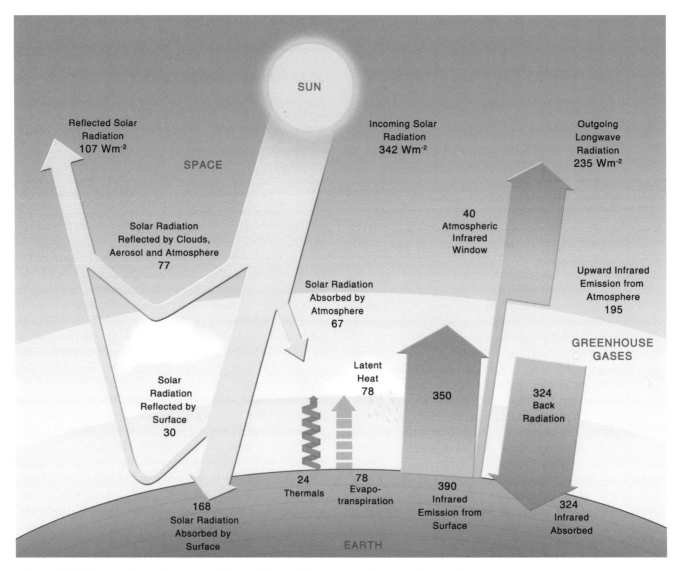

Figure 24.1 The earth's surface energy balance. Solar radiation enters the atmosphere relatively easily, but part of the return infrared radiation emitted by the earth's surface is reabsorbed. Emissions of carbon dioxide, methane, and nitrous oxide from human activities are increasing the amount of heat trapped in the atmosphere, leading to global warming via the so-called "greenhouse effect." From Trenberth et al., 1996. (http://www.atmos.washington.edu/~dennis/Energy_Flow_small.gif)

tential lowering of atmospheric temperature. Except for water, the concentrations of all these trace gases in the atmosphere have been greatly perturbed by human activities over the past century, potentially affecting the atmospheric energy balance and global temperatures.

Svante Arrhenius, a professor of chemistry at Stockholm, Sweden's Hogskola, was better known for developing the electrolytic dissociation theory, for which he received the Nobel Prize in 1903. However, he had a lively interest in all aspects of physics, and he began to wonder how fossil fuel burning, which was increasing exponentially in Europe at the time, might affect global climate. After months of calculations, done by hand, he concluded that a doubling of atmospheric carbon dioxide levels would

increase global temperatures by 5°C, an estimate that is still valid today. This theoretical possibility came to be known as the greenhouse effect, because it superficially resembles the way heat is trapped in a greenhouse during the day.

Little practical attention was paid to the possibility of atmospheric warming until the 1980s, when Charles Keeling began to publish evidence on the rise of atmospheric carbon dioxide based on his measurements made on the top of Mauna Loa volcano in Hawaii. This station in the middle of the Pacific Ocean provides a well-mixed sample of air from the entire Northern Hemisphere. Charles Keeling was a professor of oceanography at the Scripps Institution of Oceanography in California at a time when scientists were unsure whether carbon dioxide emissions from fossil

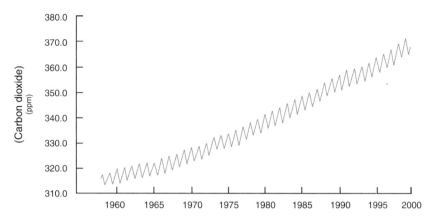

Keeling Curve of Atmospheric Carbon Dioxide from Mauna Loa, Hawaii

Figure 24.2 A graph of Dr. Keeling's curve of increasing CO_2 concentration. The measurements are made at a station on top of Mauna Loa in Hawaii. Data courtesy C.D. Keeling and NOAA (www.noaa.gov). Image from http://earthguide.ucsd.edu/globalchange/keeling_curve/01.html.

fuel burning would actually accumulate in the atmosphere or would be quickly absorbed into vegetation and by the oceans. He worked out that carbon dioxide measurements made on the tops of mountains, far from industrial sources, could be used to track changes in atmospheric levels over time. In 1958, he persuaded the National Science Foundation to set up a monitoring station on the top of Hawaii's Mauna Loa mountain. However, they stopped funding the work in the early 1960s, calling his results "routine." He found other funds to continue, and by the 1980s, it became obvious that there was a steep upward trend in atmospheric carbon dioxide levels, alerting scientists and policy makers to the imminent possibility of an increase in global temperature. President Bush awarded him the National Medal of Science in 2002. The National Science Foundation, which administers the award, declared that the Mauna Loa measurements were "...some of the most important data in the study of global climate change." The type of plot he provided, now called a Keeling curve, showed a typical seasonal pattern of rise and fall of carbon dioxide over each annual cycle, superimposed on a steady interannual increase in carbon dioxide (Figure 24.2). When the Keeling curve for Mauna Loa is extended back in time from measurements of carbon dioxide in gas bubbles trapped in Antarctic and Greenland ice cores, it is seen that carbon dioxide has increased from a concentration of 270 ppb (parts per billion) in the year 1600 to 370 ppb in 2005 (Figure 24.3).

Currently, human activities add about 7–9 gt (gigatons) of carbon to the atmosphere each year in the form of carbon dioxide. About 80% of this is from the burning of fossil fuels, and the remainder is from the clearing of forests for agriculture, especially in the tropics. Of the amount derived from fossil fuel burning, about 60% is from the burning of coal and natural gas to generate electricity, while 40% is from the burning of liquid fuels, mainly to propel cars and trucks. Up until recently, the United States was responsible for over half of all carbon dioxide emissions. However, China, with its booming economy, is rapidly catching up (Table 24.1).

Figure 24.4 shows the increase in carbon dioxide as well as the two other main greenhouse gases, methane and nitrous oxide, in earth's atmosphere over the last 1,000 years. There has been a clear and dramatic increase in these gases since 1800 due to human activities. **Anthropogenic** (human-caused) methane emissions originate from the burning of fossil fuel and are also produced by livestock, landfills, and rice paddies. Nitrous oxide originates from fossil fuel burning as well, but most of it comes from the wide-scale use of chemical fertilizers to boost crop yields around the world. Excess nitrogen, not taken up by crops, is converted to nitrous oxide and dinitrogen gas by soil mi-

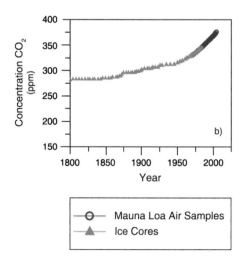

Figure 24.3 Historic trend in atmospheric carbon dioxide levels as measured on bubbles trapped in ice cores and then by direct measurement at the Mauna Loa Observatory.

TABLE 24.1 National inventories of carbon dioxide emissions in 1998 (from Harvey, 2000).

Units are metric tonne carbon dioxide. Note that some nations have negative emissions from land use change—this means their forests are a net sink for carbon dioxide. Other nations, such as Brazil, are net sources of carbon dioxide from forest clearing. Note also the wide disparity in rates of per capita emissions among countries.

COUNTRY	FOSSIL FUELS	CHEMICAL PRODUCTS	LAND USE CHANGES	TOTAL	PER CAPITA
Australia	287	3	130	420	23.5
Brazil	237	13	1200	1,450	9.3
Canada	430	5	–	435	14.8
China	2,970	222	9	3,191	2.7
France	330	11	−37	304	5.3
Germany	815	20	−20	815	10.0
India	874	35	150	1,059	1.2
Indonesia	286	10	455	751	3.8
Iran	256	8	–	264	3.8
Italy	393	17	37	447	6.6
Japan	1,082	45	−90	1,037	8.3
Mexico	346	12	89	447	4.9
N. Korea	256	1	–	257	11.1
Poland	331	7	–	338	8.8
Russia	1,800	18	−587	1,231	8.3
S. Korea	346	1	–	345	8.3
S. Africa	301	5	0	306	7.4
Ukraine	433	6	52	491	7.5
United Kingdom	536	6	−6	536	9.2
United States	5,430	38	−532	4,936	18.5

Data from Harvey, 2000.

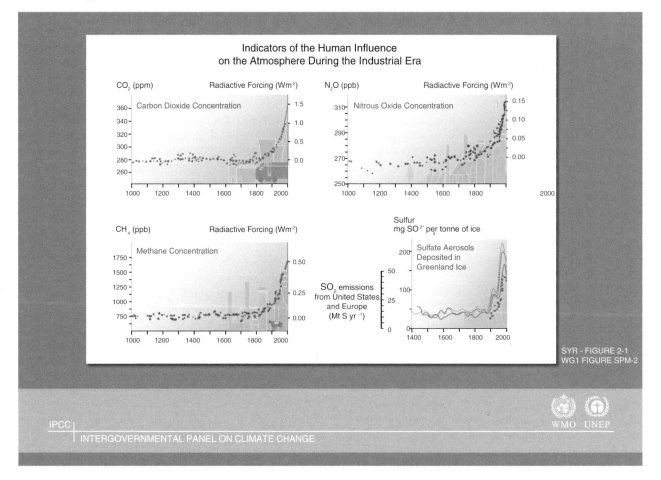

Figure 24.4 Increase in levels of the major greenhouse gases in earth's atmosphere due to human activities. From IPCC, (2001), "Climate change 2001: The scientific basis," Intergovernmental Panel on Climate Change, http://www.grida.no/climate/ipcc_tar/wg1/index.htm (http://www.ipcc.ch/present/graphics.htm).

INFORMATION BOX 24.2

Global Warming from Greenhouse Gases Released by Human Activity

Emission	Global Warming Activity %
Carbon dioxide	70
Methane	23
Nitrous oxide	7

crobes via denitrification (see Chapter 5). The amounts of methane and nitrous oxide emitted from human activities are much lower than carbon dioxide, but they respectively have 26 and 206 times greater heat-trapping ability than carbon dioxide. Currently, 70% of the global warming potential that can be attributed to human activities is due to carbon dioxide emissions, while 23% is due to methane and 7% to nitrous oxide emissions (Information Box 24.2).

Scientists are very certain that the recent rise in carbon dioxide levels is due to human activities. It is relatively easy to know the amount of fossil fuel that is burned each year, because records are kept on the amount of coal and gas that are burned for electricity and on the production of liquid fuels for the transportation sector. It is also relatively easy to measure the increase in carbon dioxide and other greenhouse gases in the atmosphere. In addition to the Mauna Loa observatory, gas measurements are now made around the world, including the Antarctic, and they all show similar increases in greenhouse gases that over time roughly track the amount calculated to have been released by fossil fuel burning. However, only about two-thirds of the amount of carbon dioxide that is expected to enter the atmosphere is actually detected. The other third is called the **missing carbon**. Scientists are greatly interested in solving the missing carbon problem. Do their models on oceanic uptake of carbon dioxide underestimate actual uptake? If so, global warming may not be as severe as expected. Or does the missing carbon represent extra biomass production in forests due to the fertilization effect? If so, it is important to protect these forests from fire and clear-cutting or the carbon will be quickly released back to the atmosphere, making the global warming more severe than expected.

The carbon dioxide emissions entering the atmosphere today will contribute to global warming for the next several centuries. Experts project that at present rates of emissions, atmospheric carbon dioxide levels will double over the next century and will peak at 1,700 ppm in the year 2400 (over five times preindustrial levels). This amount of carbon dioxide is projected to produce a temperature rise of 2 to 4°C at the earth's surface over the next 100 years, and a rise of 4 to 8°C at the peak of carbon dioxide levels. These elevated temperatures, and their effects at sea level, are expected to persist for many centuries after carbon dioxide emissions have stabilized (Figure 24.5). Currently, we are just at the beginning of the rise in greenhouse gas levels in the atmosphere.

24.2.2 Effects of Greenhouse Gas Emissions on the Global Climate

There is great uncertainty about the amount of atmospheric warming, if any, that has already taken place due to greenhouse gas emissions. There is even more uncertainty about the impacts of future warming on global climate systems. The rise in atmospheric carbon dioxide levels has been documented by careful measurements over time, whereas climate projections are based on models and other indirect methods that are subject to error. Nevertheless, there are some logical consequences that follow from a warming of global temperature. These include shifts in regional weather patterns due to unequal heating at the equator and the poles; partial melting of the Antarctic ice shield, resulting in a rise in sea level around the world; and shifts in the distribution of vegetation zones, with impacts on agriculture and natural ecosystems. The evidence for an actual increase in global temperature, and possible consequences over the next 100 years, are discussed briefly below.

The problem of documenting a change in surface temperature is much more difficult than the problem of documenting a rise in atmospheric carbon dioxide levels. The atmosphere at the top of Mauna Loa and other reporting stations is well mixed, and carbon dioxide levels are fairly consistent from station to station. Conversely, surface temperatures vary widely from one location on earth to another, depending on the type of landscape, the time of day and season of year, and method of measurement. For example, the air over a plowed but unplanted field on a summer day might be 10°C warmer than over a fully vegetated adjacent field. The air over cities is warmer than over rural areas, due to the **heat island effect** (absorption of heat by streets and concrete surfaces), and as most historical temperature records were taken near cities, there could be an apparent increase in temperature from the growth of cities quite apart from any effect of greenhouse gases (see also Chapter 21). Ocean temperatures might be a more reliable means to look for global warming; however, over the years, methods for measuring sea temperatures changed from bringing wooden or canvas buckets of water on deck for measurement (which allows the water to cool by evaporation) to making measurements directly in the water. These changing techniques add bias to the temperature record. Indirect methods of measuring surface temperature, such as using the thermal bands on satellites, present technical problems and often do not agree with the results of surface measurements.

Despite the uncertainties, combined annual land-surface and sea-surface temperature databases show a temperature rise of 0.65°C plus or minus 0.15°C over the past 150 years (Figure 24.6). Many scientists believe these data sets confirm the existence of a global warming trend with a certainty of 95% or greater. Other scientists are trying to confirm global warming by using a wide variety of **surrogate measurements** (indirect tests of a warming signal). For example, mountain glaciers around the world retreated in the 20th century at a rate that is consistent with a warming of about 0.6 to

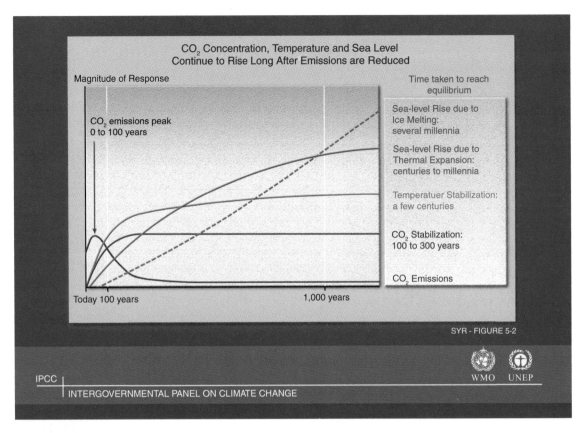

Figure 24.5 Long term projections of the effect of fossil fuel burning on atmospheric carbon dioxide levels, surface temperature, and sea level. Notice that carbon dioxide emissions will decrease after 100 years, but the effects will last for hundreds or thousands of years. From IPCC, 2001, "Climate change 2001: The scientific basis," Intergovernmental Panel on Climate Change, http://www.grida.no/climate/ipcc_tar/wg1/index.htm (http://www.ipcc.ch/present/graphics.htm).

1.0°C. The growing season for wild and cultivated plants in the Northern Hemisphere increased by 12 days during the period 1981–1991, also indicating a warming trend. Furthermore, a study of plant distribution on 30 alpine peaks showed that the distribution of many species has shifted upward in elevation, while the distribution of a number of butterfly species has shifted towards the pole, and mosquito-borne diseases such as dengue fever and malaria are reported at ever-higher elevations. Scientists using different techniques have estimated a rate of sea level rise of about 1.8 cm year^{-1}, expected from the melting of ice caps and the thermal expansion of seawater due to global warming (Figure 24.7). Thus, while we are still at the beginning of the expected global temperature rise due to emission of greenhouse gases, scientists are reasonably certain they have already detected a rise in temperature over the past 150 years.

Projected future effects of greenhouse warming on climate are made using computer models of the atmosphere and oceans, called **Atmosphere-Ocean General Circulation Models (AOGCMs)**. Over a dozen modeling groups around the world are simulating the effects of temperature rise on climate systems, and they periodically compare and check each other's results. Results of AOGCMs are combined with models of the carbon cycle, of atmospheric chemistry and

physics, and of ice sheets to attempt to predict what is in store for the earth under different greenhouse gas emission scenarios. At our current state of knowledge, these models produce results that can at best be regarded as "educated guesses" about the future.

Recognizing the problem of global climate change, the World Meteorological Organization and the United Nations Environment Program established the **Intergovernmental Panel on Climate Change (IPCC)** in 1988. It collects the most recent scientific evidence on all aspects of climate change, including output from AOGCMs, and predicts the potential impacts and options relevant to the risks of human-induced climate change. Their most recent report is listed in the Reference section (Watson and the Core Writing Team, 2001).

Overall, the range of the mean global temperature increase over the next 100 years is expected to be in the range of 2 to 4°C. However, a greater mean annual warming at higher latitudes than near the equator is expected, called a **polar amplification** of the warming. This is in part because, as snow melts, more radiation will be absorbed rather than reflected back to space in the polar regions, leading to greater heating of the land surface. Whereas a rise of less than 1°C is expected at the equator, at 80°N or S latitude a

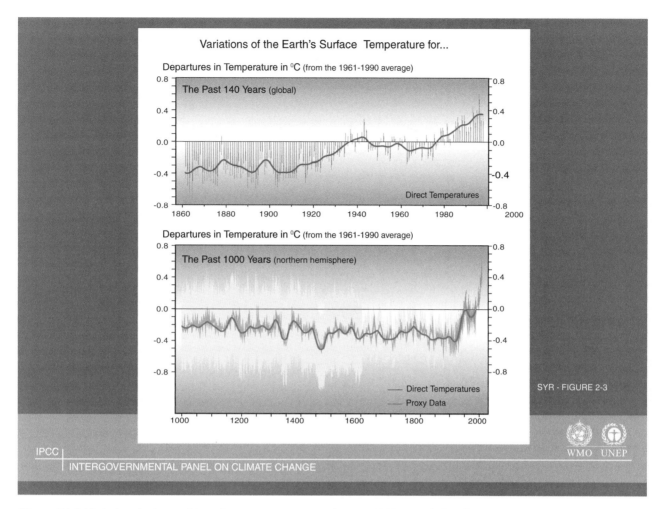

Figure 24.6 Variations in the earth's surface temperature over the past 1000 years, inferred from a number of data sources. From IPCC, 2001, "Climate change 2001: The scientific basis," Intergovernmental Panel on Climate Change, http://www.grida.no/climate/ipcc_tar/wg1/index.htm (http://www.ipcc.ch/present/grahics.htm).

rise of 3°C or more is anticipated. For similar reasons, greater warming in winter than in summer is expected at high latitudes, whereas greater warming in summer is expected in arid and semi-arid regions, where soils become drier in summer. A general tendency for drying of mid-latitude soils is expected and could have implications for agriculture.

The changes in global temperature distribution could have profound effects on the global climate cycles, because climate patterns are driven in part by differences in surface temperature at different latitudes. Although the direction of change is difficult to predict, it can be expected that monsoon rains, tropical cyclones and hurricanes, precipitation patterns over the continents, and the frequency of extreme weather anomalies such as droughts and floods will be affected by global warming. Also, it is expected that sea level will rise around the globe due to melting of the polar ice caps. A mean global temperature rise of 2 to 4°C is expected to raise the sea level by 25 to 75 cm by the year 2100. This would impact large areas of coastal land around the world,

including cities, agricultural areas, and natural coastal ecosystems such as coral reefs, salt marshes, and coastal forests.

24.3 OTHER GLOBAL CHANGES

24.3.1 Acid Rain

Sulfate and nitrogen oxide aerosols are added to the air by the burning of sulfur- and nitrogen-rich fuels. Coal, burned in electric generating plants around the world, is the main source of sulfate aerosols. These aerosols reflect light and may have a small dampening effect on greenhouse warming. Their main effect, however, is to produce acid deposition, called **acid rain**. Wet deposition occurs when the particulates act as condensation nuclei for raindrops. Dry deposition occurs when the aerosols are deposited directly on surfaces. In either case, when they come into contact with water, the

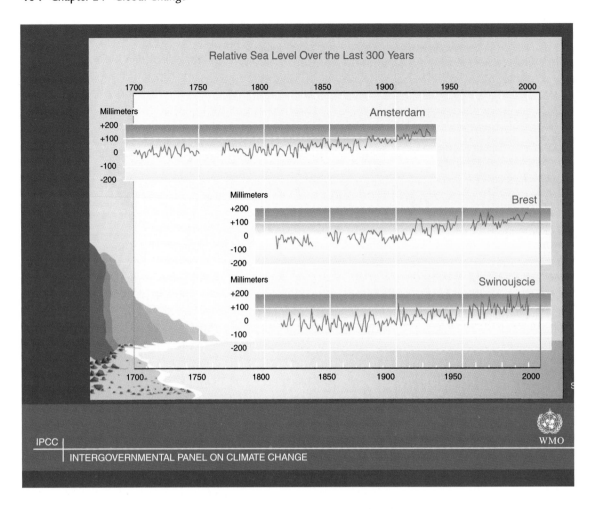

Figure 24.7 Rise in sea level since 1700 at three locations in Europe. Global warming increases sea level through the thermal expansion of water and the melting of polar ice caps. IPCC, 2001, "Climate change 2001: The scientific basis," Intergovernmental Panel on Climate Change, http://www.grida.no/climate/ipcc_tar/wg1/index.htm.

sulfates and nitrogen oxides form sulfuric and nitric acid, respectively. These acids cause acidification of susceptible forest soils, lakes, and streams, and damages trees. They also damage human-made structures such as buildings, statues, and painted surfaces, and can contribute to respiratory diseases (see also Chapters 4 and 23).

24.3.2 Atmospheric Ozone

Ozone is a molecule composed of three oxygen atoms, as opposed to diatomic oxygen, which is composed of two oxygen molecules and makes up most of the oxygen in the atmosphere. Human activities are creating two types of ozone problems: they are depleting the amount of ozone in the upper atmosphere, which screens out harmful ultraviolet radiation (**good ozone**), and they are creating ozone at ground level, which is a major component of urban smog (**bad ozone**).

Good ozone is formed in the stratosphere when incoming ultraviolet radiation is absorbed by ordinary diatomic

oxygen. An ozone molecule can absorb further ultraviolet radiation that regenerates diatomic oxygen. Hence, diatomic oxygen and ozone are interconverted by photochemical processes, but there is no overall loss of ozone when it absorbs ultraviolet light. The **ozone layer** in the stratosphere is vital in protecting life on earth from the harmful effects of ultraviolet radiation (see Chapters 4 and 23).

In the 1970s, scientists from the British Antarctic Service first noticed an apparent decline in the amount of ozone in the stratosphere above the Antarctic in winter. They confirmed their findings with careful, annual measurements from the 1980s to the present (Figure 24.8). The cause of the ozone loss was found to be caused by reactions with chlorine in the upper atmosphere, and the chlorine was found to have originated from human release of **chlorofluorocarbons (CFCs)**, extremely long-lived gases used in air conditioning systems and as propellants for aerosol spray cans. CFCs are also potent greenhouse gases, but in the amounts that have been emitted to the atmosphere so far, their main damage has been to the ozone layer. Today, an **ozone hole** forms over the

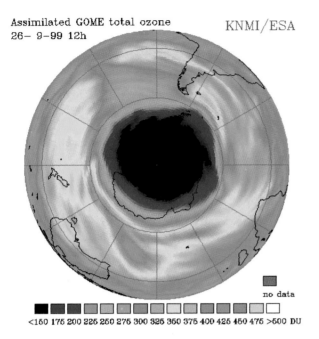

Assimilated GOME total ozone
26- 9-99 12h

KNMI/ESA

no data

<150 175 200 225 250 275 300 325 350 375 400 425 450 475 >500 DU

Figure 24.8 Satellite image of the ozone hole over Antarctica in September 1999. Darker colors denote lower ozone values. Chlorofluorocarbons in the upper atmosphere react with ozone in stratospheric ice clouds over the South Pole to temporarily deplete the atmosphere of ozone each summer. Source: © KNMI/ESA; see the GOME Fast Delivery Service page for details (http://www.xs4all.nl/~josvg/KNMI/hole1999).

South Pole (and more recently the North Pole) each year, due to the **Coriolis effect**, the pattern of global wind circulation caused by the spinning of the earth. In the Southern Hemisphere, these circumpolar winds produce a polar vortex that isolates the air within it during the south pole winter. Then the intense cold leads to the formation of polar stratospheric clouds, and it is within the ice crystals in these clouds that the chemical conditions necessary for ozone destruction by CFCs can take place. The precursors needed to form reactive compounds from CFCs are formed in the ice particles during the dark period, and when sunlight returns, ultraviolet light energizes the destruction of ozone from the byproducts of CFCs. Although the destruction of ozone is local, the depletion affects the total amount of ozone in the atmosphere. Hence, ozone destruction is a global concern.

Bad ozone formed near ground level is a major component of lung-damaging smog within urban environments (see Chapters 4 and 23). High levels of ozone (smog) exacerbate asthma and other respiratory problems, especially in children and the elderly. Ozone can also increase human sensitivity to pollen and other allergens. Ozone is produced photochemically from precursor gases (volatile organic compounds and nitrogen oxides) emitted into the atmosphere by both natural (*e.g.*, vegetation) and human (*e.g.*, fossil fuel combustion) sources. The photochemical reactions that produce ozone are especially sensitive to the temperature of the air in which it forms. As global climate becomes more variable and warmer

due to increasing greenhouse gas concentrations (especially carbon dioxide), urban air quality (especially ozone concentrations) is expected to be affected three ways:

1. Climate models (AOGCMs) indicate increased likelihood of large-scale high-pressure systems, which often produce strong air temperature inversions (see Chapter 23) and very hot weather. These systems are likely to cause stagnant conditions, which enhance the containment of air pollution near the ground surface. The warmer weather may also increase the need for air conditioning to improve human comfort, which could increase emissions of human caused ozone precursors from the burning of fossil fuels for electrical energy generation.

2. Higher air temperatures are expected to directly increase the rates of the reactions producing ozone—thus higher ozone concentrations in the air humans breathe. Increased ozone production is especially expected at temperatures exceeding 32°C.

3. Increased temperature is expected to increase emissions of volatile biogenic organic compounds known as isoprenes from some vegetation, especially woody plants. These compounds are directly involved in the production of urban ozone. In some instances a relatively small 2°C-increase in temperature can increase emissions of isoprenes by 25%, which ultimately increases ozone levels.

Research on the projected consequences of global warming on air pollution (ozone) in urban areas was summarized in 2004 by the **Natural Resources Defense Council (NRDC)**. They reported that by the mid-21st century, 15 selected cities studied in the eastern United States could experience a 60% increase in the number of days (from 12 to almost 20) when ozone levels exceed the health-based 8-hour National Ambient Air Quality Standard established by the Environmental Protection Agency (EPA). Cincinnati, Ohio, for example, could experience a 90% increase (from 14 to 26 days). These findings indicate that more people may be required to restrict outdoor activities when ozone levels are high.

24.3.3 Deforestation

Adding to the expected negative effects of global warming are a series of human land use patterns that have, literally, changed the terrestrial landscape of the earth. **Deforestation** (cutting down of forests) has occurred since prehistoric times. All of Europe was once densely forested, but the trees were cut down to provide wood and to clear the land for agriculture. The United States experienced a wave of deforestation as European settlers began spreading from east to west across the continent in the 1700s.

Today, rapid deforestation is taking place in the tropics, where rainforests in South America, Africa, and Asia are being cleared for agriculture (Figure 24.9). At present there are about 2,000 million ha of tropical forests in the world, but they are being cleared at a rate of about 14

Figure 24.9 Burning of brush to clear land for farming in Tanzania. Image from http://www.rainforests.net/pictures.htm.

to 16 million ha per year, mainly for agriculture. Often, the trees are not utilized, but are burned or left in piles to decompose. Rapid deforestation is also taking place in the subpolar region of the Northern Hemisphere, which covers about one-tenth of the hemisphere's land surface. Hence, boreal forests include spruce, cedar, larch, oak, hemlock, fir, pine, and aspen trees that provide an overstory for rich ecosystems of understory plants and wildlife. Boreal and other northern forests cover 30% of Canada and 45% of Russia. The boreal forests are being clear-cut for lumber, and are also susceptible to fire; NASA scientists estimate that Russia's boreal forests are being destroyed at the same rate as tropical rainforests.

Although they only occupy about 10% of the earth's surface area, forests are estimated to support over half of the estimated species of life on earth (up to 80 million). Many of these species can only live in forest environments, so much of this species biodiversity is lost when forests are cleared. Forests also hold an immense amount of carbon. Typically, each ha of forest stores about 180 tons of carbon, most of which is released into the atmosphere when a forest is logged or burned. Forest clearing is estimated to release 1.6 gt year^{-1} of carbon to the atmosphere (compared to 6.2 gt from fossil fuel burning). Forest clearing also has an impact on regional climate, leading to drier and less productive ecosystems after the forests are gone.

While deforestation of tropical and boreal forests have become a source of carbon dioxide in the atmosphere, temperate-zone forests over much of the globe are regenerating and have become a net **sink** (depository) for carbon. Forests in the U.S. and Europe are experiencing new growth, mainly because they are no longer harvested as a primary fuel source as they were 100 years ago. Some scientists think these regenerating forests represent the missing carbon in the global carbon budget. However, forest fires, disease, and other threats to these forests make them an uncertain form of carbon storage for the future.

24.3.4 Desertification

The world's **drylands** (arid, semi-arid, and subhumid zones where potential evaporation exceeds precipitation) cover about 40% of the earth's land area. In their natural state, these lands vary from bare soil, to grasslands, to mixed grass and shrublands, and to thorn forests, depending on how much rain they receive. However, they have been extensively degraded by human activity, mainly by overgrazing and unsustainable agricultural practices. Land degradation in the drylands is called **desertification**, although this term can be confusing. It conjures up images of desert dunes spreading into fertile areas, which is only part of the problem. More commonly, desertification takes place within a dryland region in a mosaic pattern, rather than at the margins. In fact, the ebb and flow of deserts at their margins can be a natural response to climate variations, rather than an effect of humans on the land. An example of desertification is the conversion of grasslands to shrublands through overgrazing. Too many grazing animals stocked on a range eat the more palatable grasses, which are replaced in time by grazing-resistant shrubs. Grasses tend to hold the soil in place, whereas shrubs tend to be separated by areas of bare soil, which is subject to wind and water erosion. Hence, the landscape is converted from grassland to shrub-desert, even though the climate might remain the same.

Another type of desertification is caused by converting semi-arid land to dryland crop production. The Dust Bowl that arose in the Great Plains region of the United States and in western Canada in the 1930s is an example of this type of land degradation. This region has a variable climate and is only marginally suitable for agriculture. However, in the first part of the 19th century, white settlement converted much land into dryland grain farms (mostly wheat). A wheat crop requires more water than, on average, falls in a year in this region. Therefore, the farmers practiced a type of cultivation called dust mulching. The soil was plowed and pulverized to a fine powder and left fallow for a year. During the fallow period, rain fell on the fine dust surface and soaked into the soil. Then in the second year, a wheat crop was planted and it produced grain, using water stored in the soil from the fallow period plus what fell on the field during the second year.

Unfortunately, when a period of drought came in the early 1930s, the huge areas of cleared fields began to blow and created region-wide dust storms that covered over farms and even farmhouses. One of the worst dust storms occurred on April 14, 1935. Dust rose several thousand feet into the air across the Great Plains, in one large storm involving New Mexico, Texas, Colorado, Kansas, and Oklahoma. Ships entering New York harbor from Europe arrived with a fine coat of soil that was blown out to sea from the Midwest.

Other types of desertification are caused by salinization of land through irrigation with saline water, and destruction of riparian ecosystems through diversion of river water from human use and construction of dams that interfere with the normal hydrological cycle of rivers.

The United Nations Environment Program estimates that 80% of the drylands have been affected by desertification and that 40% have been moderately to severely impacted. Some types of desertification are difficult or even impossible to reverse. For example, replacement of grasslands with shrublands may encourage gully erosion to take place, leading to the formation of deep cuts over the land surface. Like deforestation, desertification is not new. The area of the Middle East known as the Fertile Crescent, now consisting of southern Iraq and part of Iran, where crops were first domesticated and human civilization first arose, has been largely converted to desert, covered by blowing sands.

Since drylands do not store as much biomass on their surface as forests, it has been thought that desertification may not contribute much to the greenhouse effect. However, this might be an erroneous conclusion. Semi-arid and subhumid regions store a large amount of carbon in their soils, and this carbon can be released during desertification, adding carbon dioxide to the atmosphere. Conversely, some forms of desertification, such as replacement of grasses with trees and shrubs, may actually lead to greater storage of carbon on the land and in the soil. A better estimate of the role of drylands in the global carbon budget is needed.

24.3.5 Depletion of Ocean Fish Stocks

One of the first global environmental disasters was the collapse of many of the major open-ocean fisheries, starting in the 1950s with the advent of modern fishing technologies. Until then, fishing was mainly conducted by local fleets of boats using traditional fishing techniques. Starting in the 1950s, large factory ships that could stay at sea for weeks and process and freeze the catch on board were deployed by the top fishing nations (China, Russia, Peru, Chile, the U.S., Norway, and Japan) to roam widely over the oceans. These ships were equipped with increasingly sophisticated fish-finding and capture technologies. Today, open-ocean fishers use GPS-assisted sonar on scout boats or helicopters to locate schools of fish. These are harvested using long-lines (lines of baited hooks several km long), drift nets (large nets that float near the surface), or large-capacity trawl nets that scoop up shrimp and bottom-feeding species. Because modern fishing gear is expensive, fishers are motivated to increase their catch to pay for the added expense of catching fish.

At the same time as the fishing became mechanized, new international markets developed for marine species that were formerly of only local interest. These included many coastal species that reproduced slowly, so the stocks were quickly depleted. Abalone, lobsters, crabs, a wide variety of reef fish, and even sea urchins (harvested for their eggs) became international favorites. Even the lowly seahorse has been harvested to near **extirpation** (local extinction) on many Asian reefs, because it is used as a cure for impotency and asthma in traditional Chinese medicine.

The first large fishery collapse occurred in the 1990s with the loss of the north Atlantic cod fishery (Figure 24.10), which had been in existence since the 1600s. The collapse of

this fishery was preceded by intense negotiations between nations on methods to preserve and regulate the fishery, but each nation saw it in their self-interest to demand their fair share of a declining resource. By the middle of the 1990s, there were insufficient fish left to support a commercial fishing fleet. Fishers around the north Atlantic lost their livelihood. Ten years after this collapse, with a near-total ban on cod fishing in place, stocks have still not recovered sufficiently to support commercial fishing. The cod wars (disputes between nations over the size of the catch) and the subsequent collapse of the fishery showed that modern fishing methods are unsustainable and that the international community lacks a mechanism by which common fishing grounds can be regulated.

The **United Nations' Food and Agriculture Office (FAO)** estimated that, in 2000, 9 to 10% of the world's marine fish stocks had been depleted or were recovering from depletion; 15 to 18% were overexploited and headed for depletion; and 47 to 50% were fully exploited and were at or near their peak of production; while only 25 to 27% were moderately exploited and had some potential for the catch to increase. With 71 to 78% of the world's commercial fish species fully exploited, overexploited, or depleted, FAO concluded that commercial fishing is not a sustainable food resource and is urgently in need of international and national-level management.

Overfishing has environmental consequences beyond the effect on the target species. One problem is the **by-catch**—the marine species that are taken inadvertently with the target species. For example, by-catch in tuna nets is by far the most serious threat to dolphins and other cetaceans. FAO estimates the by-catch of marine species at 20 million tons per year. In some fisheries, such as shrimp trawling, the by-catch is many times the size of the main catch. The by-catch is generally thrown back into the sea dead. In many fisheries, the by-catch includes juveniles of commercial species; for example, in the Gulf of Mexico, by-catch with shrimp has undermined the populations of red snappers. Most of the modern fishing methods are nonselective and produce a by-catch. Longlines catch seabirds, sea turtles, and nontargeted fish, while gillnets catch seabirds and turtles, and continue to kill marine life when the lines are lost or abandoned at sea, a process called "ghost fishing." Bottom trawling not only produces a by-catch, it also disrupts the bottom habitat, sweeping away bottom structure and **benthic** (bottom-dwelling) organisms of all types.

Overfishing can have broader effects on the ecosystem. As stocks of the most desirable species become depleted, fishers have turned to a practice called "fishing down the food chain." Top predators, such as billfish and tuna, tend to be removed first. Then the less desirable fish lower on the food chain are exploited. This practice simplifies the marine **food chain** (the sequence of predator-prey relationships in the ecosystem), affecting future fish and marine mammal populations. The overharvesting of haddock, cod, and mackerel in Alaska, for example, has reduced populations of Steller sea lions, which feed on these fish. Fishing for krill

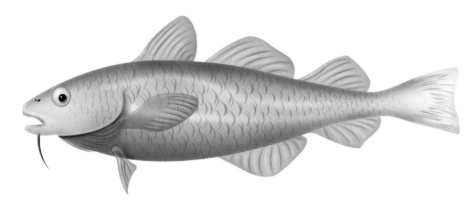

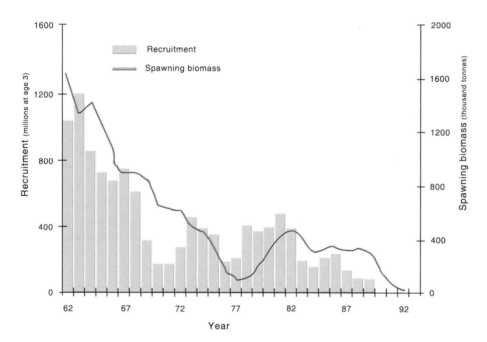

Figure 24.10 Collapse of the cod fishery in the north Atlantic. The fish shown is *Gadus morhua*, or Atlantic cod, which has been fished commercially since the 16th century. Overfishing led to a collapse of the fishery in 1992, and it has not been fished commercially since then. The graph shows the decrease in Canadian cod stocks, but similar decreases occurred throughout the North Atlantic. Image from http://www.erin.utoronto.ca/~w3env100y/env/ENV100/hum/cod.htm.

(small, red marine shrimp about 6 cm long) in the Antarctic has placed stress on the species that depend on them for food, including penguins, whales, squid, and fish. When plant-eating fish are removed from tropical reefs, seaweeds often proliferate, covering over corals.

The size of annual harvest by marine capture fisheries peaked in about 1995, despite increased investment in fishing gear and greater effort per unit catch. Increasingly, **aquaculture** (raising marine species artificially in ponds, pens, or cages) is augmenting the wild catch. When farms are sited in sensitive ecosystems such as mangrove swamps and salt marshes, aquaculture has produced great environmental damage. The **effluent** (wastewater containing feces and uneaten feed) from fish farms can produce **eutrophication** (overfertilization) of coastal waters, leading to harmful algae blooms and loss of oxygen in the water, producing fish kills. Carnivorous aquaculture species such as marine shrimp require large amounts of fish-derived protein in their diets; many fish species that were previously not exploited are now harvested by capture fisheries to provide feed for aquaculture species.

24.3.6 Coastal Degradation

Interacting with overfishing to place stress on marine ecosystems are the problems of pollution and overdevelopment taking place along the world's coastlines. In 1995, 2.2

billion people (39% of the world population) lived within 100 km of a coastline. The trend is towards an even greater concentration of population along coastlines. In the U.S., 54% of Americans live in counties along the Atlantic or Pacific Oceans, the Gulf of Mexico, or the Great Lakes. Populations are expanding rapidly in the coastal states (*e.g.*, Florida and California), so that by 2025, 75% of Americans are expected to live in coastal counties. In China, nearly 700 million people (56% of the population) live in coastal regions, and the trend is expected to continue because the 14 "economic free zones" and 5 "special economic zones," areas targeted by the Chinese government for economic expansion, are in coastal provinces. Similar trends exist around the world.

This high concentration of people along shorelines produces a number of environmental problems. Municipal and industrial effluents add nitrogen, phosphorous, and industrial chemicals to the coastal environment, leading to eutrophication that produces a degradation of the marine food web. Natural ecosystems such as bays, salt marshes, and estuaries are transformed by construction of ports, breakwaters, marinas, and aquaculture facilities. In their natural state, these coastal ecosystems provide protected breeding and nursery grounds for fish and crustaceans, and feeding stations for a wide variety of birds. These functions are lost when they are developed for human use.

Overfishing, pollution, and coastal development are affecting several key coastal ecosystems in the tropics and subtropics. **Seagrass meadows** (beds of marine grasses growing in shallow, sandy areas near coastlines) cover less than 0.2% of the global ocean, but fulfill a critical role in the coastal ecosystem. For example, they are feeding grounds for fish, crustaceans, sea turtles, and marine mammals such as the manatee. Widespread seagrass losses have been caused by human activities including dredging, fishing, anchoring, oil spills, siltation and eutrophication from land runoff, construction projects, and food web alterations. These stresses are expected to be worsened by negative effects of climate change discussed earlier, such as sea level increase, increased storms, and higher levels of ultraviolet radiation.

Coral reef ecosystems are also in decline throughout the tropics. Principle threats are similar to those facing seagrass meadows—overharvesting, pollution, disease, and climate change. The Great Barrier Reef on the north coast of Australia, widely regarded as the most pristine reef in the world, already shows system-wide symptoms of decline in productivity and species richness. Furthermore, coral reefs are susceptible to damage by **coral bleaching**. The hard skeleton of a coral reef is made up of calcium carbonate secreted by millions of small animals called **coral polyps** that filter food particles from the water. Living in symbiosis with the polyps are microscopic, single-celled algae called **zooanthellae**. These are essential to the health of the polyp and the coral reef, since they provide food to the polyps through photosynthesis and help in the deposi-

tion of carbonate. When the coral is put under heat stress, the zooanthellae (which are equipped with flagella) leave the coral, which become white (bleached) in appearance. While not necessarily irreversible, coral bleaching can lead to large-scale die-offs of coral, and it is expected to increase in severity with global warming. Another problem for coral reefs is silt runoff from land, especially when formerly forested land is converted to grazing or is developed for human settlement. The silt disperses in the water, forming a fine sediment called "marine snow" that settles on corals and eventually kills them.

Mangrove swamps are also threatened by human activities. They have traditionally been harvested for wood. In the past 20 years, shrimp farms have been built along many tropical coastlines, and mangrove swamps have been favored sites for development. A rough estimate is that about half of the world's seagrass meadows, coral reefs, and mangrove swamps have already been lost to development. These losses exacerbate the problem of overfishing and loss of marine biodiversity.

24.4 SOLUTIONS TO THE PROBLEMS OF GLOBAL ENVIRONMENTAL CHANGE

Developing responses to these problems is largely in the realm of environmental policy, which is beyond the scope of this chapter. However, scientists can contribute to developing **mitigation strategies** (methods to lessen damaging effects) of global change in two ways. The first is to help develop an understanding of the underlying causes leading to global change, and the second is to develop technical innovations to aid in mitigation. The underlying causes of global change are epitomized by "The Tragedy of the Commons" (Case Study 24.1), which explains why commonly shared resources tend to be overused.

We can now look back at the environmental problems discussed in this chapter and see how they all fit the tragedy of the commons. For example, the atmosphere is shared by everyone. If an electric generating facility voluntarily spends extra money to clean up its pollutants, that entire cost is borne by the company (and its shareholders and customers) and only a very small part of the global problem is solved, so the benefits are marginal. Obviously, individual industries are unlikely to solve pollution problems on their own. Overfishing is another example. As modern fishing methods were developed, fishers invested in new boats, gear, and electronic devices to find fish. Often these were purchased with borrowed money. Acting as individuals, they would be foolish to restrict their catch for the benefit of the greater good. They would simply become the first to go bankrupt.

Hardin described this as a "tragedy" because it was one of those problems facing humans for which there are no strictly scientific or technical solutions. Rather, progress will

CASE STUDY 24.1

THE TRAGEDY OF THE COMMONS.

In 1968, the ecologist Garrett Hardin published an article in the journal *Science* called "**The tragedy of the commons**" that offered a **paradigm** (conceptual model) for understanding global change issues. He originally applied the paradigm to the population problem, but it has since been applied to all of the global change problems discussed in this chapter. Hardin built on an observation published by amateur economist William Lloyd, commenting on the problem of the English commons in 1833. Traditionally, a group of herders shared equal rights to graze sheep on a jointly held pasture called a commons. The system worked well for centuries, until the demand for wool became high due to the invention of new methods for spinning yarn. With wool in high demand, herders began to increase the number of animals placed in the commons until there was not enough grass to support all the sheep.

At this point, it was in each individual herder's self-interest to add more sheep to the commons, to ensure that each got a fair share of the diminishing resource. Consider a commons that is at its maximum carrying capacity. If a herder adds one more sheep to the commons, the positive value to that herder is a return of nearly +1. To the commons as a whole, the negative value is −1, as the commons is now overgrazed by one animal. But the negative value is shared by all the herders, who each see a small diminishing yield in wool from their animals, whereas the positive gain all goes to the single herder who placed the extra sheep on the commons. Hardin then took the process to its logical conclusion:

"Adding together the component partial utilities, the rational herdsman concludes that the only sensible course for him to pursue is to add another animal to his herd. And another.... But this is the conclusion reached by each and every rational herdsman sharing a commons. Therein is the tragedy. Each man is locked into a system that compels him to increase his herd without limit—in a world that is limited. Ruin is the destination toward which all men rush, each pursuing his own best interest in a society that believes in the freedom of the commons. Freedom in a commons brings ruin to all."

require restructuring the way we use and share resources that have been regarded as commons up to now. Hardin thought that the way out of this dilemma was through "mutual coercion mutually agreed upon." This approach has become the basis for international negotiations on the problems discussed in this chapter.

While solutions are in the realm of policy rather than science, technical innovations can greatly accelerate the mitigation process once society accepts that something needs to be done. In the area of global warming, for example, many methods are available for either reducing carbon dioxide emissions through more efficient use of energy or reabsorbing carbon dioxide from the atmosphere once it is emitted. To cite some well-known examples of conservation, **hybrid cars** (combined gasoline-electric engines) get twice the gas mileage as conventional vehicles. Florescent lightbulbs, though more costly than normal incandescent bulbs, last ten times longer and use 60 to 75% less power. Since over half the global carbon dioxide emissions come from electric generating plants, and lighting accounts for about half of global electricity use, replacing incandescent with florescent bulbs could be a major accomplishment in combating global warming.

The world's forests store a great deal of carbon, and many exploratory projects have linked preservation of forests with carbon emissions by utilities. Typically, a power

company will invest in a forest conservation project and will receive **carbon credits** for the amount of carbon in biomass that they **sequester** (store) in the wood in the forest. These demonstration projects may one day be implemented on a large scale if society places a **carbon tax** on emissions or requires polluters to supply **carbon offsets** to balance their emissions. If necessary, it is possible to physically remove carbon dioxide from stack gases at the point of emission. The Idaho National Energy Laboratory and other research groups are developing methods to absorb or precipitate carbon dioxide from stack gases. These products could, in theory, be piped into deep-sea layers, where they would be sequestered for thousands of years. Another carbon sequestration scheme that has received serious attention is to fertilize the southern oceans with iron, a limiting nutrient for algal growth in those iron-poor waters. When the resultant algae blooms died, they would carry excess carbon into the deep ocean. Iron fertilization could also make fisheries more productive by stimulating the growth of algae on which many fish feed. However, many scientists and environmentalists think that the negative side effects of iron fertilization would outweigh the benefits.

It has been argued that solutions to global change problems may be too expensive to implement. This may not be the case. Today, the cost of fossil fuel ranges from about $100 per t carbon for coal to about $300 per t carbon for petroleum. The cost of removing carbon from the air ranges from about $10 per tonne carbon for forest-conservation projects to about $100 per tonne for removal at the point of origin. Hence, controlling carbon dioxide emissions would add from 10 to 30% to the price of fossil fuel energy.

A similar range of technical innovations, from simple to complex, is available for each of the problems in this chapter. As environmental problems worsen, or as society begins to set greater value on **ecosystem goods and services** (benefits provided free to humans from the natural functioning of ecosystems), it can be expected that these innovations will begin to be implemented. However, as long as the atmosphere and oceans are still treated as a commons, mitigation of global environmental problems is unlikely.

QUESTIONS AND PROBLEMS

1. Write a brief essay on any of the global change issues in Table 24.1 not covered in this chapter.
2. Give some additional examples of positive and negative feedback effects related to global change issues.
3. How might a rising sea level affect society in the future? Give some specific examples.
4. Is aquaculture the answer to declining fisheries? Why or why not?
5. Discuss some additional examples of the "Tragedy of the Commons."

REFERENCES AND ADDITIONAL READING

Asner G., Elmore A., Olander L., Martin R. and Harris A. (2004) Grazing systems, ecosystem responses, and global change. *Annual Review of Environment and Resources.* **29**, 261–299.

Bellwood D., Hughes T., Folke C. and Nystrom M. (2004) Confronting the coral reef crisis. *Nature.* **429**, 827–833.

Christianson G.E. (1999) *Greenhouse: The 200-Year Story of Global Warming.* Walker, New York.

Duarte C. (2002) The future of seagrass meadows. *Environmental Conservation.* **29**, 192–206.

Food and Agriculture Organization. (2000) *The State of the World Fisheries and Aquaculture.* United Nations Food and Agriculture Organization, Sofia.

Harvey, L. (2000) *Climate and Global Environmental Change.* Prentice Hall, Harlow, England.

Keeling C., Whorf T., Wahlen M. and van der Plicht J. (1995) Interannual extremes in the rate of rise of atmospheric carbon dioxide since 1980. *Nature.* **375**, 666–670.

Kurlansky M. (1997). *Cod.* Penguin Books, New York.

Middleton N. and Thomas D. (eds.). (1997) *World Atlas of Desertification.* United Nations Environment Program and Oxford University Press, New York.

National Resources Defense Council. (2004). *Heat Advisory—How Global Warming Causes Bad Air Days.* NRDC, New York.

Pauly D., Watson R. and Alder J. (2005) Global trends in world fisheries: Impacts on marine ecosystems and food security. *Philosophical Transactions of the Royal Society of Britain.* **360**, 5–12.

Trenberth K.E., Houghton J.T. and Meira Filho L.G. (1996) The Climate System: An overview. In J.T. Houghton et al. (eds.), *Climate Change 1995: The Science of Climate Change.* Cambridge University Press, Cambridge, England pp. 51–64.

Watson D. and the Core Writing Team (eds.). (2001) *Climate Change 2001: Synthesis Report.* Cambridge University Press, Cambridge, England.

WASTE AND WATER TREATMENT AND MANAGEMENT

CHAPTER 25

INDUSTRIAL AND MUNICIPAL SOLID WASTE TREATMENT AND DISPOSAL

J. F. Artiola

Solid waste incinerator and ash residues that can have hazardous waste characteristics, due to their metal content. *Source: U.S. Environmental Protection Agency*

25.1 INTRODUCTION

Industrial and municipal solid wastes are created by modern societies as unavoidable byproducts of mining, industrial production, and the requirements of today's modern consumers. In the United States, the 19th century was an era of rapid growth during which many new and developing industries were established to systematically process raw materials into finished goods. During the 19th and 20th centuries, and now the 21st century, these industries concentrated on the quality of the final product, while discarding the residues and waste products generated during increasingly complicated manufacturing processes.

At the beginning of the Industrial Revolution, mining of coal and metal ores became both the fuel and the building blocks of Western society's industrialization. With the discovery of petroleum and natural gas deposits, chemical-processing industries quickly followed. From the early 1900s to the late 1960s, hundreds of thousands of petroleum-derived and synthetic chemicals were used in the production of goods.

Although carbon-based plastic materials, together with such organic chemicals as pesticides and solvents, have dominated industrial production since the early 1950s, metal-based goods remain fundamental to modern industry. Numerous modern goods—from cars to paints—require the use of common metals such as iron, aluminum, and copper. In addition, less abundant but more toxic metals like lead, cadmium, nickel, mercury, arsenic, and selenium are also used by industry. Metallic elements are therefore commonly found in industrial wastes, where they

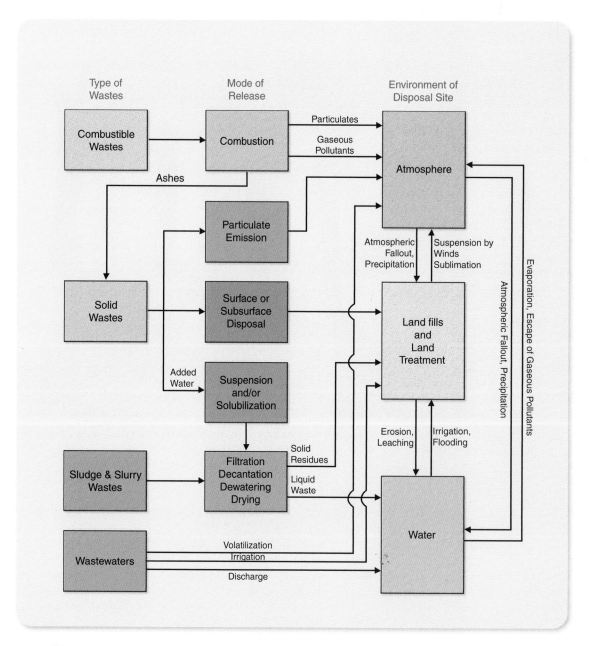

Figure 25.1 Sources and modes of releases of pollutants into the environment. Modified from *Pollution Science* © 1996, Academic Press, San Diego, CA.

have complex and incompletely understood effects on the environment (Figure 25.1). Consumers also discard massive amounts of hazardous wastes. The advent of the development of countries such as China and India has exacerbated the production of industrial wastes. We buy, use, and dispose of increasing quantities of goods that have hazardous characteristics including paints, solvents, pesticides, batteries, lightbulbs and electronic goods.

The disposal of wastes can result in adverse effects not only on the environment, but also on human health. Early in the 20th century, industries, cities, and towns began to use abandoned or natural depressions to dispose of the ever-growing amounts of municipal waste residues. In essence these were the earliest forms of landfills. Until the 1970s in the U.S., it was common practice to dump waste organics on land with little or no regard for the soil and water pollution it produced. Today, there are an increasing number of waste minimization programs recycle and reuse programs, and numerous solid waste disposal programs, which not only regulate hazardous industrial wastes, but also provide guidelines for the safe recycling and disposal of hazardous household wastes.

25.2 RELEVANT REGULATIONS FOR INDUSTRIAL AND MUNICIPAL SOLID WASTES

Before 1970, industrial untreated wastes were disposed of in unlined landfills or lagoons or discharged into surface waters. Other wastes were burned in open pits or in incinerators with no pollution controls. As a result, groundwater and surface water resources, such as lakes and rivers, were polluted. Besides the human toll associated with water and air pollution, animals and plants also experienced adverse effects due to industrial pollution associated with the accumulation of pollutants in the environment.

Since the late 1970s, several laws have been implemented to control the disposal of hazardous wastes to protect our health and the environment (Information Box 25.1) (see also Chapter 15).

INFORMATION BOX 25.1

Laws Implemented to Control the Disposal of Hazardous Waste

The Resource Conservation and Recovery Act (RCRA); the Comprehensive Emergency Response, Compensation and Liability Act; the Safe Drinking Water Act; the Federal Clean Air Act; the Clean Water Act; and the Toxics Control and Safety Act. Industrial wastes are classified as *hazardous* and *nonhazardous* by the **Code of Federal Regulations** (CFR 40, Part 261).

Most of the materials classified as nonhazardous are regulated under such specific industry categories as mining and oilfield wastes. Of all the wastes generated in the United States, more than 95% are classified as nonhazardous municipal. Although industrial, mining, oil- and gasfield wastes are regulated separately, in this chapter we will treat all of them as "industrial wastes."

Also since the late 1970s, industrial wastes classified as hazardous have received considerable attention due to their obvious potential for deleterious impacts on human health and the environment. Consequently, government scrutiny has focused on new regulations for storage and disposal of these wastes (see Information Box 25.1). Nonetheless, less regulated, nonhazardous wastes constitute the bulk of the waste generated by industry. The largest source of nonhazardous wastes is human sewage, which is treated within wastewater treatment plants prior to land application, incineration, or landfilling (see also Chapter 27). Nonhazardous wastes often contain the same potentially polluting components found in hazardous wastes, but at much lower concentrations. To prevent the disposal of household goods with hazardous properties into **municipal solid waste (MSW)** landfills, new guidelines have been implemented that control the disposal of items such as spent crankcase oil, car batteries, tires, paint thinners, or refrigerants. These are implemented through household hazardous waste collection programs (U.S. EPA, 2005).

25.3 MAJOR FORMS OF INDUSTRIAL WASTES

Regardless of how wastes are technically categorized by regulatory agencies, their pollutant-releasing capacities depend on their physical characteristics, which determine their mode of transport into the environment. The four major waste types based on physical characteristics—combustible wastes, solid wastes, sludge and slurry wastes, and wastewaters—are presented in Figure 25.1, which shows how each of these types can eventually release pollutants into the atmospheric, terrestrial, and aquatic environments. **Combustible wastes** yield byproducts that can be released directly into the atmosphere as gases or particulate pollutants when they are not properly filtered. **Solid wastes** can release pollutants into the atmosphere via dust or particular transport, or when these wastes come into contact with water, their soluble constituents can be leached out into the soil surface or below. **Sludge and slurry wastes** can release pollutants into the soil and groundwater from both their solid and liquid phases. **Wastewaters**, owing to their liquid state, are always potential sources of pollution if discharged directly into the aquatic or terrestrial environment without proper treatment. In general, we can see that the pollutants whose impact on the environment is most severe are most usually found in liquid or gas phases. By their fluid nature, these pollutants can be transported large distances and thus affect large segments of the environment. (The various physical methods by which pollutants are translocated within and among environments are discussed in Chapter 6.)

Except for a few isotopic forms, none of the 40 elements that are economically important to modern industrial society can be made synthetically (Table 25.1). Thus, these elements have to be mined, extracted from the natural state, and subsequently purified. Because of such processes, as well as the processes used in manufacturing, these elements—including many precious and strategic metals such as gold (Au), platinum (Pt), cobalt (Co), antimony (Sb), and tungsten (W)—can be found in significant amounts in some industrial wastes. A few elements, such as mercury (Hg) and arsenic (As), can be found in gaseous, liquid, or solid wastes. Yet others, such as lead (Pb) and chromium (Cr), can be discharged into the atmosphere as particulate matter associated with other elements such as sulfur (S) and carbon (C).

However, most of the elements listed in Table 25.1, including metals, metalloids, and salts, can be dissolved in the liquids (usually the aqueous phase) found in wastewaters, sludges, and solid industrial wastes. That is, with few exceptions, most industrial liquid, slurry, and sludge wastes have a water phase that can range from 99% to less than 10% by weight. Therefore, the process of dewatering can be used to reduce much of the mass of these wastes. Moreover, many forms of organic pollutants are also water soluble to various degrees, and are therefore found in the water phase of these wastes. These aqueous phases, once separated from solids, must be processed as wastewaters prior to discharge into the open environment.

TABLE 25.1 Elements of economic importance.

ELEMENT	CHEMICAL SYMBOL	COMMON FORMS IN WASTES
Aluminum	Al	Al^{3+} oxides and hydroxides, Al metal, Al-silicates
Arsenic	As	As^{n+} oxides (arsenate, arsenate)
Barium	Ba	$BaSO_4$ (barite)
Boron	B	B^{3+} hydroxides (borate)
Bromine	Br	Br^-
Cadmium	Cd	Cd^{2+} ion and Cd halides, oxides and hydroxides
Calcium	Ca	$CaCO_3$, $CaSO_4$, Ca oxides and hydroxides
Carbon	C	Inorganic forms: CO_2, CO_3^{2-} and HCO_3^-
		Organic forms include: —C—C—, —C—H, —C—O—, —C—, —C—S—
Chlorine	Cl	Cl^- ion
Chromium	Cr	Cr^{3+} hydroxides, Cr^{6+} oxides, Cr metal
Copper	Cu	Cu^{2+} oxides and hydroxides, Cu metal
Fluorine	F	F^- ion, CaF (apatite)
Iron	Fe	Fe^{n+} oxides and hydroxides, Fe metal
Lead	Pb	Pb^{2+} oxides, hydroxides and carbonates, Pb metal
Magnesium	Mg	$MgCO_3$, $MgSO_4$, Mg oxides and hydroxides, Mg Silicate (asbestos)
Mercury	Hg	Hg^{2+} oxides and halides organo-Hg complexes
Molybdenum	Mo	Mo^{n+}-oxides (molybdate)
Nickel	Ni	Ni^{2+} ion, Ni^{2+} amines, Ni metal
Nitrogen	N	NO_3^- (nitrate) and NH_4^+ (ammonium) ions
		—C—N— (organic chemicals)
Phosphorus	P	PO_4^{3+} ion, Ca phosphates
Potassium	K	K^+ ion
Selenium	Se	Se^{n+} oxides
Silicon	Si	Si^{4+} oxides and hydroxides, Al Silicates
Silver	Ag	Ag^+ as AgCl, Ag metal
Sodium	Na	Na^+ ion
Sulfur	S	SO_4^{2-} (sulfate) ion, SO_2, H_2S, FeS_2 (pyrite)
		—C—S— (organic chemicals)
Titanium	Ti	Ti^{4+} oxides
Uranium	U	U^{n+} oxides and halides, U metal
Vanadium	V	V^{n+}-oxides
Zinc	Zn	Zn^{2+} oxides and hydroxides, Zn metal
Gold	Au	Au^{+3} halides, oxides, $AuCN^-$, Hg-Au amalgam
Platinum	Pt	Pb^{+2} sulfates, carbonates, halides
Cobalt	Co	Co^{+2} as amines and oxides
Tungsten	W	W^{+6} oxides
Antimony	Sb	Sb^{+3}, Sb^{+5} oxides, halides

25.4 TREATMENT AND DISPOSAL OF INDUSTRIAL WASTES

Technologies and practices used for the treatment and disposal of metal- and salt-containing wastes vary widely, but ultimately, the byproducts of these technologies end up being released into the air, land, or water environments (see Figure 25.1). Methods that separate metals from the other waste constituents are driven by the need to reduce waste disposal costs and ultimately by the potential costs associated with liability. Organic wastes that contain low residual concentrations of metals and salts can often be degraded by using thermal or biological destruction processes that completely transform the waste into carbon dioxide and water. Conversely, wastes that contain significant amounts of metals and salts always leave indestructible residues that may or may not be recycled economically. Therefore, these wastes that cannot be eliminated, and must be disposed of in a manner that minimizes their impact on the environment.

25.4.1 Gas and Particulate Emissions

Gases and dust particulates that are generated in the thermal destruction of wastes and smelting can be prevented from escaping into the air using one or more of the following processes: electrostatic precipitators, baghouse and cyclone separators, and wet scrubbers. These technologies are expensive and difficult to operate efficiently. However, without this equipment, smelters and incinerators would discharge large quantities of toxic metals into the atmosphere, thereby contaminating large tracts of land. In the United States, the Clean Air Act requires the control of hazardous emissions by industries. Originally passed in 1970, by 1990 this act was extended to cover 189 industrial chemicals, requiring the installation of air pollution control equipment on all major industrial sources. These regulations also include requirements for the removal of metal-containing particulate matter (PM_{10}) from air emissions (see also Chapter 9).

A very controversial issue is the reduction or trapping of greenhouse gases like carbon dioxide (CO_2) that are released by power-generating plants that use fossil fuels (coal, oil, and natural gas). These gases are known to be linked to global warming (see also Chapter 24). To date, the U.S. government does not have or participate in any program to control CO_2 emissions. While mandated in Europe and other countries around the world that agreed to the **Kyoto Treaty**, only a few U.S. industries have reduced CO_2 emissions. However, future options to aggressively reduce these emissions will have to include federal- and/or state-imposed limits of emissions. In addition, trading programs to buy and sell CO_2 emissions may be implemented, and the expanded use of carbon sequestration techniques is likely. Technologies being evaluated include trapping and liquefying CO_2 gas from smokestacks, followed by storage by injection into depleted deep oil or geologic gas reservoirs.

25.4.2 Chemical Precipitation

Wastewater and slurried sludges can be treated with chemical agents to precipitate and remove metals from the rest of the waste components. For example, solutions containing alkaline materials such as calcium carbonate, sodium hydroxide, aluminum oxide, or sodium sulfide are commonly used to precipitate metals from waste streams. Figure 25.2 shows acidic metal sludges neutralized with lime. These chemicals help form insoluble metal hydroxides, carbonates, and sulfides. Similarly, aluminum and iron oxides sorb metals such as cadmium and metalloids such as arsenic, thus removing them from solution. Precipitation and adsorption of inorganics is discussed in Chapter 7.

General precipitation reactions can be represented by the following:

$$M(free) + CaCO_3(slurry) \rightarrow \overset{\text{(precipitates)}}{MCO_3 + M(OH)_2}$$

$$M(free) + Na_2S \ (pH > 8) \rightarrow \overset{\text{(precipitates)}}{MS}$$

where $M = Cd^{2+}$ or Zn^{2+} or Cu^{2+} or Pb^{2+}.

Figure 25.2 Unlined pond filled with acid and metals (Fe, Zn, Ni, Cr) wastes. The acid liquids and muds (insert) were treated with hydrated lime to neutralize the acidity and precipitate all the metals. Photo courtesy J.F. Artiola.

25.4.3 Flocculation, Coagulation, Dewatering-Filtration-Decanting-Drying

Particulate pollutants in wastewaters can be made to settle out quickly by using chemicals known as flocculants and coagulants. **Flocculants**, which react with dissolved chemicals, facilitate the formation of aggregates or clumps, which can be decanted or filtered out of solution. Conversely, **coagulants** destabilize colloids, thereby permitting suspended particles to form aggregates that can settle out of solution. These chemicals can be useful in removing all kinds of pollutants from wastewaters, including metals and organic constituents. Flocculants and coagulants include iron and copper sulfates and chlorides, as well as complex synthetic organic polymers.

Dewatering and drying is an acceptable option for waste reduction where the liquid components of a liquid waste, such as a slurry or wastewater, have very low levels of pollutants, and these can be treated using conventional wastewater treatment methods (see Chapter 26) or even discharged directly into the environment. In this case, liquids are separated from solids via several dewatering options that include air drying, particle filtration, sedimentation, and decanting. The separated solids are then treated and disposed of as solid wastes. Figure 25.3 shows a pond used to store and dewater flue gas desulfurization waste, a byproduct of sulfur dioxide gas scrubbing generated during coal combustion.

25.4.4 Stabilization (Neutralization) and Solidification

Sludges and slurries containing metals must be chemically and physically stabilized prior to final disposal. Since metals are more soluble in water at low pH, acidic wastes must be neutralized with such basic compounds as calcite ($CaCO_3$) and hydrated lime ($Ca[OH]_2$) to form low water solubility metal-carbonates and metal-hydroxides complexes (see Chapter 7).

The process of waste solidification usually involves trapping or encapsulating the waste into a physically stable matrix. For example, when wet cement is mixed in with sludges, it forms a stable block after a few days of curing (drying). This solidification method encapsulates wastes in a matrix that is relatively low in porosity and cannot easily be deformed or cracked under typical landfill overburden pressures. Consequently, percolating water does not readily infiltrate into the matrix, and metals or salts are less likely to leach out. In some cases, waste products like fly ash, collected from electrostatic precipitators during coal burning, can also be used to neutralize and encapsulate other wastes like metal waste streams, due to their strong cementing (pozzolanic) properties.

25.4.5 Oxidation

Carbon-based waste streams can be oxidized to detoxify and destroy organic pollutants. These are two major types of treatment processes: thermal and chemical. The overall goals are the same in both processes. Thermal oxidation reactions can be described as follows:

$$RCCNOCl + O_2 + heat \rightarrow CO_2 + H_2O + N_2 + NO_x + SO_x + HCl + intense\ heat$$

Thermal oxidation processes include incineration, which use conventional fuel–driven burners. Solid or liquid organic wastes having low water content but high heat values are good candidates for thermal oxidation because they burn hotly enough to sustain the energy these processes require. The major disadvantages of this process include high costs (usually more than $200 a barrel), limited reliability, and a negative public perception of their safety. In addition, large emissions of pollutants can result when the systems are not operated and maintained properly. Incinerators operate most efficiently when designs are tailored to specific waste stream

Figure 25.3 Saline flue gas desulfurization (FGD) waste drying impoundment- playa view of dried FGD wastes with sparse vegetation. Slurried saline FGD waste discharge pipe (insert). Photos courtesy C. Salo and J.F. Artiola.

characteristics. Thus, incinerators are not suitable for the efficient oxidations of mixed waste streams.

Chemical oxidation generally proceeds via the following reaction pathway:

$$RCCNOCl + (O_2, Cl_2, \text{ or } O_3) \rightarrow R'CO(\text{detoxified})$$
$$+ CO_2 + H_2O + N_x + Cl^- + \text{intense heat}$$

where R or R' is the remaining part of the organic molecule. For example, chemical oxidation is used in the destruction of cyanide-containing wastewaters. This process involves chlorination, which can be summarized as follows:

$$2CN^-(\text{liquid}) + 3Cl_2(\text{gas}) + 6NaOH \rightarrow N_2 \text{ gas}$$
$$(pH > 8-9) + 6Cl^- + 2HCO_3^- + 2H_2O$$

Many forms and combinations of chemical oxidation processes utilize chlorine (Cl_2), chlorine dioxide (ClO_2), ozone (O_3), or ultraviolet (UV) radiation to eliminate low concentrations of organics from industrial wastewaters. The chemicals oxidized using these processes include acids, alcohols, and aldehydes (such as oxalic acids and phenols), as well as more stable chlorinated pesticides and some petroleum-derived solvents such as DDT or xylenes.

25.4.6 Landfilling

Industrial wastes that contain significant concentrations of metals that cannot be recycled or recovered economically are usually good candidates for landfill disposal, which is the disposal of waste materials within soil or the vadose zone. However, sludges high in metals must be neutralized and solidified prior to landfill disposal (see Figure 25.2), and only landfills approved for the disposal of solidified hazardous wastes may be used for the disposal of these wastes. Similarly, wastes high in salts may also be buried in special landfills or used as fill materials for road and dam construction. Although such wastes are not considered toxic, they are very soluble in water. Therefore, it is important to minimize water contact to prevent the leaching of salts into the environment.

25.4.7 Landfarming of Refinery Sludges and Oilfield Wastes

Landfarming techniques have been used to treat hazardous wastes, particularly oily sludges, whose major constituents are oil, sediments, and water. In the early 1980s, land treatment of hazardous oily wastes came under intense scrutiny by the U.S. Environmental Protection Agency (EPA). Land disposal restrictions, begun in early 1992, now prohibit land treatment of hazardous oily wastes. These restrictions have compelled the petroleum industry to look at alternative disposal methods, including landfilling and incineration. On the positive side, petroleum refineries have also been spurred to develop waste-minimization strategies. Oilfield wastes generated during well drilling activities are treated in treatment facilities that have to be permitted. These facilities dewater and lower the salinity and oil content of these wastes until they meet state-imposed

TABLE 25.2 Examples of water solubilities of minerals found in wastes.

MINERAL NAMES AND CHEMICAL COMPOSITIONS	WATER SOLUBILITY* (mg L^{-1})
Barite ($BaSO_4$)	2
Calcite ($CaCO_3$)	14
Gypsum ($CaSO_4 \cdot 2H_2O$)	2,400
Salt (NaCl)	370,000

*Varies, depending on temperature and the presence of other minerals in water.

criteria for their safe reuse as fill materials. In addition to residual amounts of oil ($\sim$1–5%), these wastes often contain significant amounts of metals such as zinc, and minerals like barite, and sodium chloride (Table 25.2).

25.4.8 Deep-Well Injection of Liquid Wastes

Deep-well injection of liquid wastes into the subsurface is another waste disposal method. This method greatly reduces the potential hazard posed by wastes through disposing of them in the deep subsurface. According to the U.S. EPA (2001), there are about 470 Class I wells located mostly in the oil-producing states in the central part of the United States and in heavily industrialized states such as California, Michigan, and New York. Examples of deep-well injection of liquid wastes include oilfield wastes (brines) metal-containing wastewaters and slurries and wastewaters with high concentrations of toxic organic chemicals such as chlorinated hydrocarbons, pesticides, and radioactive wastes. According to U.S. EPA (2001), most wells (350 out of 473) inject wastes classified as nonhazardous (mostly oilfield wastes (brines)).

Deep-well injection is most suitable for handling large volumes of liquid or slurry wastes that have a water-like consistency (low viscosity). Watery liquids are usually pumped down into confined, aquifer-like zones composed of highly water-permeable material, such as sandstone or limestone. Similarly, oily wastes can be deep-well injected into high-permeability subsurface zones. The depth of the injection zones, usually hydrologically confined, ranges from about 200 to 4,500 m (600 to 13,000 feet), and most zones are located between 700 and 2500 m (2,000 and 7,000 feet) (Figure 25.4).

Drilling and constructing wells for these depths can be very expensive; therefore, deep-well injection is primarily used by the petroleum industry, which already possesses in-service oilfield wells. Once constructed, these systems are comparatively cheap to operate and maintain.

The practice of deep-well injection of wastes still poses some potential problems. Possible clogging of the injection zone due to solid particles or bacterial growth may occur, and the injected waste may contaminate resident groundwater. This possibility is of major concern if the groundwater is a current or potential potable water source. Waste injection can cause groundwater contamination in several ways: (1) direct injection into the aquifer used for drinking water; (2) leaking wells (waste leaks from the well

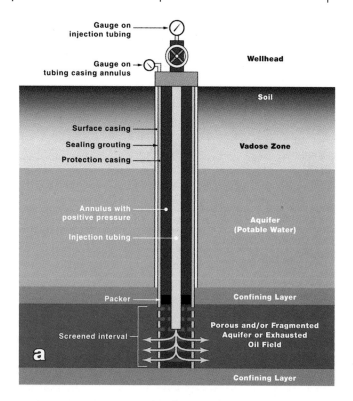

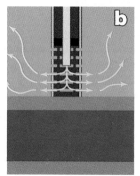

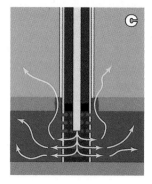

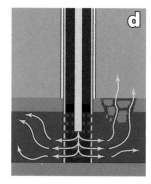

Figure 25.4 Deep-well injection design (a) and potential forms of system failures than can lead to groundwater pollution: (b) direct injection into the aquifer used for drinking water, (c) leaking well bore, and (d) movement of waste to a zone that supplies drinking water. Adapted with permission from *Subsurface Wastewater Injection: The Technology of Injecting Wastewater into Deep Wells for Disposal* by D.L. Warner and J.H. Lehr; copyright © 1981 by Premier Press. All rights reserved.

bore into an aquifer used for drinking water); (3) movement of waste to a zone that supplies drinking water. Despite these concerns, a study by the U.S. EPA (2001) concluded that "Current Class I Well Regulations are protective of Human Health and the Environment."

25.4.9 Incineration

The process of incineration that destroys highly toxic and hazardous organic wastes differs from MSW incineration, where energy is often produced. In general, low-temperature (up to 850°C) and high-temperature (~1200°C) incinerations use energy to oxidize carbon- and water-containing wastes to CO_2, H_2O vapor, and HCl and NOx gases. However, some incinerators can serve as heat-energy sources, especially when oily wastes, such as spent oils, are used as fuel. But

these incinerators must meet stringent emissions levels. Incineration efficiencies are also closely regulated, and destruction of all organic compounds must exceed 99.99%.

Incineration cannot be used with wastes that have high concentrations of water and noncombustible solids, nor can it be used for radioactive materials. Moreover, incinerators are very expensive to build and maintain, and waste incineration expenses often exceed $500 a barrel. Thus, this technology is mostly limited to low-volume wastes such as medical wastes. Finally, besides the gases previously listed, incinerators emit small but significant amounts numerous toxic chemicals including dioxins, furans, polynuclear aromatic hydrocarbons, and metals such as, mercury, beryllium, and cadmium. Often incinerators produce ash residues (bottom and fly ash) that have hazardous characteristics (particularly fly ash) and must be buried in landfills approved to handle such wastes. Thus, incinerators do not totally eliminate the problems of environmental contamination, and this has resulted in the technology being condemned by the public. Grass-roots organizations are currently fighting for the elimination of this technology as a waste treatment option, even as a renewable energy recovery source. Because of the public outcry and health-related concerns raised by the scientific community (NRC, 1999), incineration has undergone a large decline in the U.S. during the last 10 years. The number of municipal solid waste (MSW) incinerators and medical waste incinerators in the U.S. has decreased from a peak of 186 in 1990 to 112 in 2003. Also, the same source indicates that medical waste incinerators decreased from 6200 in 1988 to 115 in 2003 (PR, 2003).

25.4.10 Stockpiling, Tailings, and Muds

Mining activities produce vast quantities of tailings, which are usually stockpiled in the form of terraces on or near the ore-processing mills. The environmental impacts of mining activities, which vary widely from site to site, are usually associated with runoff or percolation of waters contaminated with sediments (see Chapter 16). They are also related to pH, metal solubility, salt concentrations, and the quantities of wind-blown particulates that are contaminated with metals. Oil- and gas-fields produce large quantities of well cuttings (muds). These well cuttings, which contain barite, salts (see Table 25.2), and crude oil, are usually stockpiled or treated until they can be used as fill materials. The potential for offsite releases of pollutants from these sites is normally associated with water and wastewater releases that contain varying concentrations of these chemicals. Mine tailing stabilization and revegetation is discussed in Chapter 20.

25.5 REUSE OF INDUSTRIAL WASTES

Because metal- and salt-containing wastes cannot be destroyed, when improperly disposed they can be a hazard to humans and the environment. At the same time, these wastes have a potential economic value that is becoming more evident as the quality of their natural sources diminish. Precious and strategic metals, such as gold, platinum, cobalt, antimony, and tungsten, are routinely mined out of wastes that contain significant concentrations of these elements (see Table 25.1). However, wastes that contain less valuable metals, such as aluminum and iron, are far less likely to be treated for the removal of these elements. Even more improbable is the extraction of highly reactive metals, nonmetallic elements, and their soluble salts, such as magnesium and calcium sulfates and carbonates, which are found in most industrial wastes.

25.5.1 Metals Recovery

Economically valuable elements such as precious metals can be recovered from waste streams by using complex chemical reactions, including chemical separation or precipitation. Silver, for example, can be recovered from photographic wastes by acidifying the liquid wastes and separating the Ag sludge that precipitates. Subsequently, the supernatant can be neutralized and disposed of safely, while the Ag sludge is sent to a smelter for purification. Metals can also be made to react selectively with synthetic organic chemicals known as chelates (literally, "claws") and synthetic zeolites, which are porous aluminosilicate minerals. In such reactions, metals are grabbed or trapped by the chelates or sequestered in the internal structure of the zeolites, while other materials pass by. Once reacted, the metal of interest is either precipitated or extracted out of the waste solution or is removed with a sorbent. The recovered metal sludge can be smeltered and refined into pure solid metal. Metal-containing residues resulting from the smelting and refining of ores can be further refined by using a combination of chemical and physical separation techniques and high-temperature furnaces.

25.5.2 Energy Recovery

The majority of wastes containing organic carbon forms have large quantities of stored energy. Thus, wastes containing high concentrations of reduced carbon (usually organic carbon), such as organic liquids, woody materials, oils, resins, or asphalts, can be used in incinerators and electrical generators as energy sources. For example, kerosene and natural gas have about 44×10^6 to 49×10^6 Joules of energy Kg^{-1}. Many solvents such as hexane, xylenes, and paraffins, which are found in oils and alcohols, have similar stored energies (42×10^6 to 58×10^6 Joules kg^{-1}). Therefore, many waste mixtures of organic chemicals have energy values approaching those of commercial fuels. However, the limitations of this technology, when applied to wastes, are regulated by very strict air emission standards promulgated under RCRA.

25.5.3 Industrial Waste Solvents

Industrial wastes high in solvents are being successfully recycled using recovery systems that include distillation techniques and chemical and physical fractionation processes. For example, spent solvents used to clean and paint metal parts can be redistilled as a means to separate the heavy impurities from the volatile solvents. Some examples of solvents that can then be reused via solvent recovery stills include chloroform, acetone, benzene, xylene, hexane, and methylene chloride.

25.5.4 Industrial Waste Reuse

Some mine tailings can be used as fills for earthworks. However, economic and potential liability issues related to transportation costs and potential releases of contaminants usually limit their reuse. Consequently, mine tailings are most often stockpiled in place. Oilfield wastes that are high in salts may be leached with water to remove the soluble salts. The salt-free muds are then dried and stockpiled for use as fill material. These treated solid oilfield wastes have been shown to be safe in earthworks and release no residual salts into the surrounding environment. Coal-burning wastes such as fly ash and flue gas desulfurization sludges have been successfully as soil amendments. Coal-burning fly ash has also been used as a fill material for earthworks. These materials can be good sources of gypsum and calcite, which can neutralize acidity in soils and replenish macronutrients such as Ca, Mg, and S, and even trace elements like Zn, Cu, Fe, and Mn. Fly ash additions to acidic agricultural soils have also been shown to improve the soil pit, porosity and overall structure.

25.6 TREATMENT AND DISPOSAL OF MUNICIPAL SOLID WASTE

25.6.1 Municipal Solid Waste

Municipal solid waste (MSW) is the solid waste material commonly called "trash" or "garbage" that is generated by homeowners and businesses. Most state and/or federal rules permit landfills to contain either hazardous or nonhazardous wastes, but more than 95% of landfilling involves MSW. According to the U.S. EPA (2003), municipal solid waste consists primarily of paper, yard waste, food scraps, and plastic (Figure 25.5).

In 2003, EPA estimated that the United States was producing more than 236 million metric tons of MSW each year, or about 2 kg of trash per person per day, or nearly 1.7 times more waste per capita than in 1960 (U.S. EPA 2003). Developed nations such as the United States and Canada lead the world in the production of trash, producing much more discardable material than developing countries. More encouraging is the fact that trends in MSW recycling have been increasing steeply since the 80s, now exceeding 30% in the U.S. Figure 25.6 shows the recycling rates of various MSW materials.

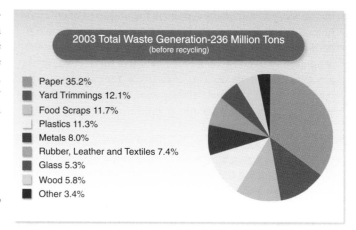

Figure 25.5 Municipal solid waste composition. Source U.S. EPA (2003).

Although it was hoped that much of the landfilled material would biodegrade, in reality, most of the land filled waste from the past 40+ years is still present at landfill sites. Surveys of landfills have shown that conditions are often not favorable for biodegradation. Such conditions include low moisture, low oxygen concentration, and high heterogeneity of materials, many of which are nondegradable or very slow to degrade. Thus, old landfills serve as "chemical repositories," releasing pollutants to the groundwater and the atmosphere (see Chapter 17).

The number of landfills in the United States has been decreasing, from 18,500 in 1979 to less than 8,000 in 1988 and 1,762 in 2002, due mostly to the closure of small landfills (U.S. EPA 2005). However, it appears that the dire predictions of landfill shortages by the turn of the century have not materialized. This is due in part to increases in

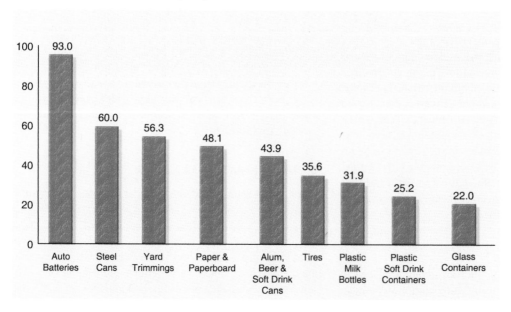

Figure 25.6 Recycling rates of selected MSW materials. Source: U.S. EPA (2003).

recycling and the ongoing construction of much larger modern landfills. This is despite new regulations and landfill designs that have increased construction and permitting costs substantially.

25.6.2 Modern Sanitary Landfills

Until the late 1970s, waste materials were buried or dumped haphazardly within soil or the vadose zone, then simply covered with a layer of soil as a cap. There was no regulation of these landfills nor were protective mechanisms in place to prevent or minimize releases of contaminants from the landfills. Old landfills were usually located in old quarries, mines, natural depressions, or excavated holes in abandoned land (Figure 25.7a). Conditions in these landfills are conducive to pollution, as these sites are often hydrologically connected to surface streams or groundwater sources. Water pollution is caused by the formation and movement of **leachate**, which is produced by the infiltration of water through the waste material. As the water moves through the

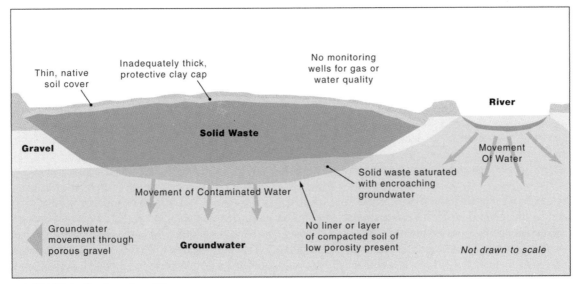

a) *Old-Style Sanitary Landfill*

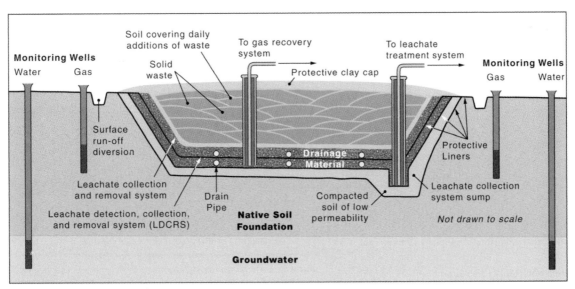

b) *Modern Sanitary Landfill*

Figure 25.7 Landfills: (a) Old-style sanitary landfill where location was chosen more out of convenience or budgetary concerns than out of any environmental considerations. Here, an abandoned gravel pit located near a river exemplifies an all-too-common siting arrangement. (b) Cross section of a modern sanitary landfill showing pollutant monitoring wells, a leachate management system, and leachate barriers. In contrast to the old-style landfill in (a), the modern sanitary landfill emphasizes long-term environmental protection. Additionally, whereas landfills have formerly been abandoned when full, a modern landfill is monitored long after closure. From *Pollution Science* © 1996, Academic Press, San Diego, CA.

waste, it dissolves components of the waste; thus, landfill leachate consists of water containing dissolved chemicals such as salts, heavy metals, and often synthetic organic compounds (see Chapter 17). In addition to contaminating water, landfills release pollutants into the air. Anaerobic microbial processes generate greenhouse gases, such as nitrous oxide (N_2O), methane (CH_4), and carbon dioxide (CO_2).

A modern **sanitary landfill** is designed to meet exacting standards with respect to containment of all materials, including leachates and gases (Figure 25.7b). New design specifications are predicated on minimal impact to the environment, both short- and long-term, with particular emphasis on groundwater protection. Landfill site selection is based on geology and soil type, together with such groundwater considerations as depth of water and use (*i.e.*, aquifer vulnerability, as discussed in Chapter 17). Figure 25.7b shows a schematic drawing of a typical modern landfill. In this design, the new landfill is located in an excavated depression, and fresh garbage is covered daily with a layer of soil. The bottom of the landfill is lined with low-permeability liners made out of high-density plastic or clay. In addition, provisions are made to collect and analyze leachate and gases that emanate from the MSW.

New landfills can cost up to $1 million per hectare to construct. Additionally, they require costly permanent monitoring for potential contaminant releases to the surrounding environment. Thus, many communities are faced with difficult decisions with respect to disposal of MWS—despite the fact that landfill design has improved efficiency with respect to pollution prevention, and increased recycling rates have slowed down MSW production.

Perhaps the most difficult problem associated with construction of new landfills is that of locating or siting a landfill. Few communities want landfills located nearby. Real or perceived local community concerns about potential health effects and property devaluation often delay the location of a landfill for months or years. Thus, the costs of MSW disposal will continue to rise as new landfills are located in increasingly remote areas. However, increasing costs may force communities to look for alternative forms of MSW disposal, such as the development of new, more cost-effective recycling and waste-minimization strategies. Currently, in the United States, 30% of MSW is recovered, recycled or composted; 14% is burned at combustion facilities; and the remaining 56% is disposed of in landfills.

25.6.3 Reduction of MSW

The primary methods used to reduce MSW are recycling or composting (30%) and combustion (14%) (U.S. EPA, 2003). These approaches are widely used in densely populated areas where land scarcity limits the use of landfills. In Europe and Japan, for example, less than 15% of MSW is sent to landfills, in contrast to the U.S., where over 55% is sent to landfills.

Combustion or incineration allows the heat derived from burning to be converted to other forms of energy. **Incineration** usually involves combustion of unprocessed

solid waste; however, in some instances nearly 25% of the waste is processed into pellets, termed **refuse-derived fuel**, prior to burning. A new technique called **mass burning** can burn MSW at temperatures up to 1130°C, trapping the resultant heat to generate steam and electricity. Incineration is a relatively efficient method of reducing MSW, often reducing it as much as 90% by volume and 75% by weight. However, this technique is not without drawbacks. The primary problem associated with incineration is air pollution, due to the release into the air of toxic chemicals like dioxin and metals like mercury. This has led to a growing public resistance to this technology. Additionally, incinerator ash contains concentrated toxic metals and refractory organics that often require their disposal as hazardous waste.

In contrast to incineration, **source reduction** is aimed at prevention; thus, it is a fundamental way of reducing MSW, because it tends to eliminate the need for landfills. Source reduction can be accomplished by a variety of methods, including the use of less material for packaging and the practice of municipal composting. For example, plastic containers may be reduced in size or eliminated altogether; and yard waste may be separated from other sources of MSW and used as compost. Municipalities often sell the composted material for use as a soil amendment or fertilizer.

Recycling, whose trends and growth rate have been previously discussed (see Figure 25.6), involves the collection of certain types of trash, separating them into their components, and subsequently using those components to make new products. Materials that can be recycled include aluminum cans, plastics, glass, paper, cardboard, and metal. Also, as previously indicated, MSW may contain significant amounts of hazardous waste, which to date is not regulated under RCRA. However, in many communities, household hazardous waste programs have been implemented to recycle these wastes and prevent them from entering MSW landfills. Among the hazardous wastes or wastes with hazardous properties that have been banned and thus nearly eliminated from landfills include lead car batteries, tires, and spent motor oil.

Municipal sewage is treated within wastewater treatment plants (see Chapter 26) and over 60% of the biosolids that are produced from treatments, are land applied (see Chapter 27). The remainder of biosolids are either incinerated or landfilled.

25.7 POLLUTION PREVENTION

Industrialized nations like the U.S. and many European countries are controlling sources and releases of pollution into the environment through implementation of the concept of **pollution prevention**. Pollution can be reduced or in some cases eliminated either by reducing the demand for products or by improving the treatment and manufacturing of raw materials and finished goods. Demand-driven pollution can be mitigated by the discovery and use of alternate but equally acceptable less-polluting products. Source reduction implies the modification of chemical processes or

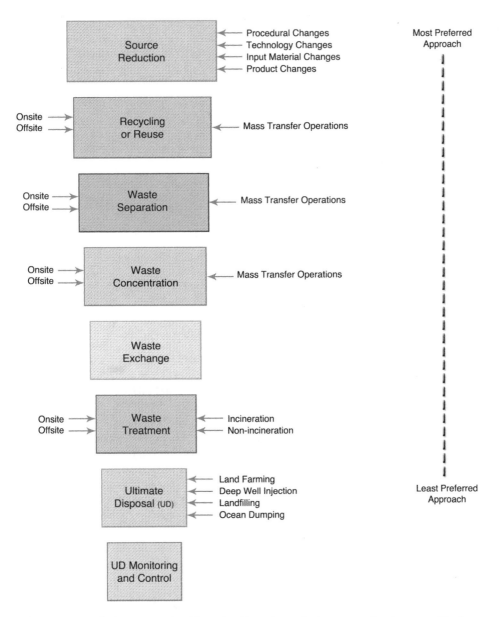

Figure 25.8 Pollution prevention hierarchy. Note that reducing wastes (at the source) is the most desirable form of pollution prevention. From Liu and Liptak, 1997.

the development of alternate processes that ultimately produce less waste or less toxic waste.

In recent years, development of pollution prevention strategies has been driven by stricter environmental regulation and economic factors. For example, new industrial processes that emphasize waste reduction are not only economically desirable, but also increasingly required by government regulations. Thus, pollution prevention strategies are usually process driven. Figure 25.8 shows that, for example, recycling is favored over waste disposal such as landfilling of wastes, which leads to waste minimization and consequently less pollution.

QUESTIONS AND PROBLEMS

1. Using Table 25.1, select the 10 most prevalent elements found in industrial wastes. Give examples of the common mineral compounds of these elements. Example: Oxygen, not listed in Table 25.1, is found in the aqueous phase of wastes as H_2O.

2. What is the most important treatment option difference between organic and inorganic wastes?

3. Two hundred gallons of an industrial wastewater contains the following materials:

   a. 10 lbs of table salt

   b. 5 kg of gypsum

   c. 1 kg of iron rust [$Fe(OH)_3$] (assume solubility $\cong 0.1$ mg L^{-1})

d. 0.5 g of the pesticide DDT (assume solubility $\cong 1$ mg L^{-1}) Using the additional mineral solubilities data from Table 25.2, estimate amounts of each of these four materials that are dissolved in the wastewater, and what amounts of these four materials are in solid forms. (*Hint*: Assume that 1 L of water weighs 1 kg or 2.2 lbs.)

4. Which elements and their chemical forms would be found in solution in the wastewater of Problem 3? List four, in order of decreasing concentrations.

5. Would you consider incineration as a final disposal method for the wastewater in Problem 3? Explain your answer. Can you suggest alternative treatment and disposal methods?

6. Modern landfills have leachate collection systems. Explain why.

7. What concerns are driving the recycling of municipal solid waste in the United States?

REFERENCES AND ADDITIONAL READING

Brown K.W., Carlile B.L., Miller R.H., Rutledge E.M. and Runge E.C.A. (1986) *Utilization, Treatment, and Disposal of Waste on Land*. Soil Science Society of America, Madison, Wisconsin.

Bunce N. (1991) *Environmental Chemistry*. Wuerz Publishing Ltd., Winnipeg, Canada.

Common Dreams Progressive Newswire. (2003 Global protests against incineration signal death knell for deadly technology new report shows dramatic decline in U.S. incinerator industry. http://www.commondreams.org/news2003/0714-06.htm.<

Cope C.B., Fuller W.H. and Willets S.L. (1983) *The Scientific Management of Hazardous Wastes*. Cambridge University Press, Cambridge, England.

Cote P. and Gilliam M. (1989) Environmental aspects of stabilization and solidification of hazardous and radioactive wastes. ASTM STP 1033. American Society of Testing and Materials, Philadelphia, Pennsylvania.

Dohlido J. R. and Best G.A. (1993) *Chemistry of Water and Water Pollution*. Ellis Horwood, New York.

ERI (1984) Land treatability of Appendix VIII constituents present in petroleum industry wastes. Environmental Research and Technology, Inc., Houston, TX. Document B–974–220.

Fuller H.W. and Warrick A.W. (1985) *Soils in Waste Treatment and Utilization*. vol. 2. CRC Press, Boca Raton, Florida.

Hesketh H.E., Cross F.L. and Tessitore J.L. (1990) *Incineration for Site Cleanup and Destruction of Hazardous Wastes*. Technomic Publishing, Lancaster, Pennsylvania.

Jackman A.P. and Powell R.L. (1991) *Hazardous Waste Treatment Technologies: Biological Treatment, Wet Air Oxidation, Chemical Fixation, Chemical Oxidation*. Noyes Data Corporation, Park Ridge, New Jersey.

LaGrega M.D., Buckingham P.L. and Evans J.C. (1994) *Hazardous Waste Management*. McGraw-Hill, New York.

Liu D.H.F. and Liptak B.G. (1997) *Environmental Engineer's Handbook*, 2nd ed. Lewis Publishers, Boca Raton, Florida.

Miller G.T. (1992) *Environmental Science*, 4th ed. Wadsworth, Belmont, California.

National Research Council (2002) *Biosolids Applied to Land: Advancing Standards and Practices*. National Academy Press. National Research Council. The National Academies Press, Washington, D.C.

National Research Council (1999) *Waste Incineration and Public Health*. National Academy Press. National Research Council. The National Academies Press, Washington, D.C.

Pepper I.L. (1991) Agricultural Sludge Utilization. In Annual Report to Pima County Wastewater Management Division. Department of Soil, Water and Environmental Science, University of Arizona, Tucson, Arizona.

U.S. EPA. (1980) Hazardous Waste Land Treatment. United States Environmental Protection Agency, Office of Water and Waste Management, Washington, D.C.

U.S. EPA. (2005) Website on Mercury Emissions. http://www.epa.gov/mercury/control_emissions/.

U.S. EPA. (2005) Website on Wastes. http://www.epa.gov/osw/.

U.S. EPA. (2003) Website on MSW Basic Facts. http://www.epa.gov/epaoswer/non-hw/muncpl/facts.htm.

U.S. EPA. (2001) Website on MSW Facts and Figures 2001. http://www.epa.gov/epaoswer/non-hw/-muncpl/pubs/msw2001.pdf.

U.S. EPA. (2001) Class I Underground injection control program: Study of the risks associated with Class I underground injection wells. U.S. Environmental Protection Agency. Office of Water, Washington, D.C. EPA 816-R-01-007 March 2001.

U.S. EPA (1983) Land Application of Municipal Sludge: Process Design Manual. United States Environmental Protection Agency. Municipal Environmental Research Laboratory, Cincinnati, Ohio. EPA–625/1–83–016.

Wang L.K. and Wang M.H.S. (1992) *Handbook of Industrial Waste Treatment*. Marcel Dekker, Inc., New York.

Warner D.L. and Lehr J.H. (1981) *Subsurface Wastewater Injection: The Technology of Injecting Wastewater into Deep Wells for Disposal*. Premier Press, Berkeley, California.

CHAPTER 26

MUNICIPAL WASTEWATER TREATMENT

C. P. Gerba and I. L. Pepper

A clarifier at a modern wastewater treatment plant. *Photo courtesy K. L. Josephson.*

26.1 THE NATURE OF WASTEWATER (Sewage)

The *cloaca maxima*, the "biggest sewer" in Rome, had enough capacity to serve a city of 1 million people. This sewer, and others like it, simply collected wastes and discharged them into the nearest lake, river, or ocean. This expedient made cities more habitable, but its success depended on transferring the pollution problem from one place to another. Although this worked reasonably well for the Romans, it doesn't work well today. Current population densities are too high to permit a simple dependence on transference. Thus, modern-day sewage is treated before it is discharged into the environment. In the latter part of the 19th century, the design of sewage systems allowed collection with treatment to lessen the impact on natural waters. Today, more than 15,000 wastewater treatment plants treat approximately 150 billion liters of wastewater per day in the United States alone. In addition, septic tanks, which were also introduced at the end of the 19th century, serve approximately 25% of the U.S. population, largely in rural areas.

Domestic wastewater is primarily a combination of human feces, urine, and graywater. **Graywater** results from washing, bathing, and meal preparation. Water from various industries and businesses may also enter the system. People excrete 100–500 grams wet weight of feces and 1–1.3 liters of urine per person per day (Bitton, 1999). Major organic and inorganic constituents of untreated domestic sewage are shown in Table 26.1.

The amount of organic matter in domestic wastes determines the degree of biological treatment required. Three tests are used to assess the amount of organic matter: **total organic carbon (TOC), biochemical oxygen demand (BOD), and chemical oxygen demand (COD).**

The major objective of domestic waste treatment is the reduction of BOD, which may be either in the form of solids (suspended matter) or soluble. BOD is the amount of dissolved oxygen consumed by microorganisms during the biochemical oxidation of organic (carbonaceous BOD) and inorganic (ammonia) matter. The methodology for measuring BOD has changed little since it was developed in the 1930s. The 5-day BOD test (written BOD_5) is a measure of the amount of oxygen consumed by a mixed population of heterotrophic bacteria in the dark at 20°C over a period of 5 days.

Aliquots of wastewater are placed in a 300-ml BOD bottle (Figure 26.1) and diluted in phosphate buffer (pH 7.2) containing other inorganic elements (N, Ca, Mg, and Fe) and saturated with oxygen. Sometimes acclimated microorganisms or dehydrated cultures of microorganisms, sold in capsule form, are added to municipal and industrial wastewaters, which may not have a sufficient microflora to carry out the BOD test.

A nitrification inhibitor is sometimes added to the sample to determine only the carbonaceous BOD. Dissolved oxygen concentration is determined at time 0, and after a 5-day incubation by means of an oxygen electrode, chemical procedures (*e.g.*, Winkler test), or manometric BOD apparatus. The BOD test is carried out on a series of dilutions of the sample, the dilution depending on the sample's source. When dilution water is not seeded, the BOD value is expressed in milligrams per liter, according to the following equation (APHA, 1998).

$$\text{BOD (mg/l)} = \frac{D_1 - D_5}{P} \qquad \text{(Eq. 26.1)}$$

TABLE 26.1 Typical composition of untreated domestic wastewater.

CONTAMINANTS	CONCENTRATION (mg/l)		
	LOW	MODERATE	HIGH
Solids, total	350	720	1200
Dissolved, total	250	500	850
Volatile	105	200	325
Suspended solids	100	220	350
Volatile	80	164	275
Settleable solids	5	10	20
Biochemical oxygen demand[a]	110	220	400
Total organic carbon	80	160	290
Chemical oxygen demand	250	500	1000
Nitrogen (total as N)	20	40	85
Organic	8	15	35
Free ammonia	12	25	50
Nitrites	0	0	0
Nitrates	0	0	0
Phosphorous (total as P)	4	8	15
Organic	1	3	5
Inorganic	3	5	10

[a] 5-day, 20°C (BOD, 20°C).

From *Pollution Science* © 1996, Academic Press, San Diego, CA.

Figure 26.1 BOD bottles used for the determination of BOD. From *Environmental Microbiology* © 2000, Academic Press, San Diego, CA.

where:

D_1 = initial dissolved oxygen (DO), D_5 = DO at day 5
P = decimal volumetric fraction of wastewater utilized

If the dilution water is seeded,

$$\text{BOD (mg/l)} = \frac{(D_1 - D_2) - (B_1 - B_2)f}{P} \quad \text{(Eq. 26.2)}$$

where:

D_1 = initial DO of the sample dilution (mg/l)
D_2 = final DO of the sample dilution (mg/l)
P = decimal volumetric fraction of sample used
B_1 = initial DO of seed control (mg/l)
B_2 = final DO of seed control (mg/l)
f = ratio of seed in sample to seed in control
$\quad$ = (% seed in D_1) / (% seed in B_1)
(See Example Calculation 26.1)

Because of depletion of the carbon source, the carbonaceous BOD reaches a plateau called the ultimate carbonaceous BOD (Figure 26.2). The BOD$_5$ test is commonly used for several reasons:

- To determine the amount of oxygen that will be required for biological treatment of the organic matter present in a wastewater
- To determine the size of the waste treatment facility needed
- To assess the efficiency of treatment processes, and
- To determine compliance with wastewater discharge permits

The typical BOD$_5$ of raw sewage ranges from 110 to 440 mg/l. Conventional sewage treatment will reduce this by 95%.

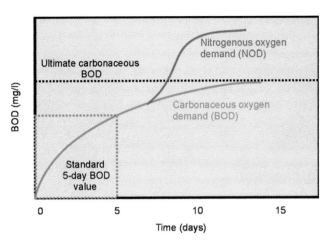

Figure 26.2 Carbonaceous and nitrogenous BOD. From *Environmental Microbiology* © 2000, Academic Press, San Diego, CA.

Chemical oxygen demand (COD) is the amount of oxygen necessary to oxidize all of the organic carbon completely to CO_2 and H_2O. COD is measured by oxidation with potassium dichromate ($K_2Cr_2O_7$) in the presence of sulfuric acid and silver, and is expressed in milligrams per liter. In general, 1 g of carbohydrate or 1 g of protein is approximately equivalent to 1 g of COD. Normally, the ratio BOD/COD is approximately 0.5. When this ratio falls below 0.3, it means that the sample contains large amounts of organic compounds that are not easily biodegraded.

Another method of measuring organic matter in water is the **TOC or total organic carbon test.** TOC is determined by oxidation of the organic matter with heat and oxygen, followed by measurement of the CO_2 liberated with an infrared analyzer. Both TOC and COD represent the concentration of both biodegradable and nonbiodegradable organics in water.

Pathogenic microorganisms are almost always present in domestic wastewater (Table 26.2). This is because large numbers of pathogenic microorganisms may be excreted by infected individuals. Both symptomatic and asymptomatic

---

EXAMPLE CALCULATION 26.1
Calculation of BOD

Determine the 5-day BOD for a wastewater sample when a 15-ml sample of the wastewater is added to a BOD bottle containing 300 ml of dilution water, and the dissolved oxygen is 8 mg/l. Five days later the dissolved oxygen concentration is 2 mg/l.

Using Eq. 26.1,

$$\text{BOD (mg/l)} = \frac{D_1 - D_2}{P}$$

$$D_1 = 8 \text{ mg/L}$$

$$D_2 = 2 \text{ mg/L}$$

$$P = \frac{15 \text{ ml}}{300 \text{ ml}} = 5\% = 0.05$$

$$\text{BOD}_5 \text{ (mg/l)} = \frac{8 - 2}{0.05} = 120$$

TABLE 26.2 Types and numbers of microorganisms typically found in untreated domestic wastewater.

ORGANISM	CONCENTRATION (per ml)
Total coliform	10^5–10^6
Fecal coliform	10^4–10^5
Fecal streptococci	10^3–10^4
Enterococci	10^2–10^3
Shigella	Present
Salmonella	10^0–10^2
Clostridium perfringens	10^1–10^3
Giardia cysts	10^{-1}–10^2
Cryptosporidium cysts	10^{-1}–10^1
Helminth ova	10^{-2}–10^1
Enteric virus	10^1–10^2

From *Pollution Science* © 1996, Academic Press, San Diego, CA.

TABLE 26.3 Incidence and concentration of enteric viruses and protozoa in feces in the United States.

PATHOGEN	INCIDENCE (%)	CONCENTRATION IN STOOL (per gram)
Enteroviruses	10–40	10^3–10^8
Hepatitis A	0.1	10^8
Rotavirus	10–29	10^{10}–10^{12}
Giardia	3.8	10^6
	18–54[a]	10^6
Cryptosporidium	0.6–20	10^6–10^7
	27–50[a]	10^6–10^7

[a]Children in day care centers.

From *Pollution Science* © 1996, Academic Press, San Diego, CA.

TABLE 26.4 Estimated levels of enteric organisms in sewage and polluted surface water in the United States.

ORGANISM	CONCENTRATION (# per 100 ml)	
	RAW SEWAGE	POLLUTED STREAM WATER
Coliforms	10^9	10^5
Enteric viruses	10^2	1–10
Giardia	10–10^2	0.1–1
Cryptosporidium	1–10	0.1–10^2

Source: U.S. EPA (1988).

individuals may excrete pathogens. For example, the concentration of rotavirus may be as high as 10^{10} virions per gram of stool, or 10^{12} in 100 g of stool (Table 26.3). Infected individuals may excrete enteric pathogens for several weeks to months. The concentration of enteric pathogens in raw wastewater varies depending on:

- The incidence of the infection in the community
- The socioeconomic status of the population
- The time of year
- The per-capita water consumption

The peak incidence of many enteric infections is seasonal in temperate climates. Thus, the highest incidence of enterovirus infection is during the late summer and early fall. Rotavirus infections tend to peak in the early winter, and *Cryptosporidium* infections peak in the early spring and fall. The reason for the seasonality of enteric infections is not completely understood, but several factors may play a role. It may be associated with the survival of different agents in the environment during the different seasons. *Giardia*, for example, can survive winter temperatures very well. Alternatively, excretion differences among animal reservoirs may be involved, as is the case with *Cryptosporidium*. Finally, it may well be that greater exposure to contaminated water, as in swimming, is the explanation for increased incidence in the summer months.

Concentrations of enteric pathogens are much greater in sewage in the developing world than the industrialized world. For example, the average concentration of enteric viruses in sewage in the United States has been estimated to be 10^3 per liter (Table 26.4), whereas concentration as high as 10^5 per liter have been observed in Africa and Asia.

26.2 MODERN WASTEWATER TREATMENT

The primary goal of wastewater treatment is the removal and degradation of organic matter under controlled conditions. Complete sewage treatment comprises three major steps, as shown in Information Box 26.1.

26.2.1 Primary Treatment

Primary treatment is the first step in municipal sewage treatment. It physically separates large solids from the waste stream. As raw sewage enters the treatment plant, it passes through a metal grating that removes large debris, such as branches, tires, and the like (Figure 26.4). A moving screen then filters out smaller items such as diapers and bottles (Figure 26.5), after which a brief residence in a grit tank allows sand and gravel to settle out. The waste stream is then pumped into the primary settling tank (also known as a sedimentation tank or clarifier), where about half the suspended organic solids settle to the bottom as **sludge** (Figure 26.6). The resulting sludge is referred to as **primary sludge**. Microbial pathogens are not effectively removed from the effluent in the primary process, although some removal occurs.

26.2.2 Secondary Treatment

Secondary treatment consists of biological degradation, in which the remaining suspended solids are decomposed by microorganisms and the number of pathogens is reduced. In this stage, the effluent from primary treatment may be pumped into a trickling filter bed (Figure 26.7), an aeration tank (Figure 26.8), or a sewage lagoon. A disinfection step is generally included at the end of the treatment.

26.2.2.1 Trickling filters

In modern wastewater treatment plants, the **trickling filter** is composed of plastic units (see Figure 26.7). In older plants, or developing countries, the filter is simply a bed of stones or corrugated plastic sheets through which water drips (Figure 26.9). It is one of the earliest systems introduced for biological waste treatment. The effluent is pumped through an overhead sprayer onto this bed, where bacteria and other microorganisms reside. These microorganisms intercept the organic material as it trickles past and decompose it aerobically.

The media used in trickling filters may be stones, ceramic material, hard coal, or plastic media. Plastic media of polyvinyl chloride (PVC) or polypropylene are used today in high-rate trickling filters. Because of the light weight of

INFORMATION BOX 26.1

The Process of Sewage Treatment (Figure 26.3)

· **Primary treatment** is a physical process that involves the separation of large debris, followed by sedimentation.

· **Secondary treatment** is a biological oxidation process that is carried out by microorganisms.

· **Tertiary treatment** is usually a physicochemical process that removes turbidity nutrients (*e.g.*, nitrogen), dissolved organic matter, metals, and pathogens.

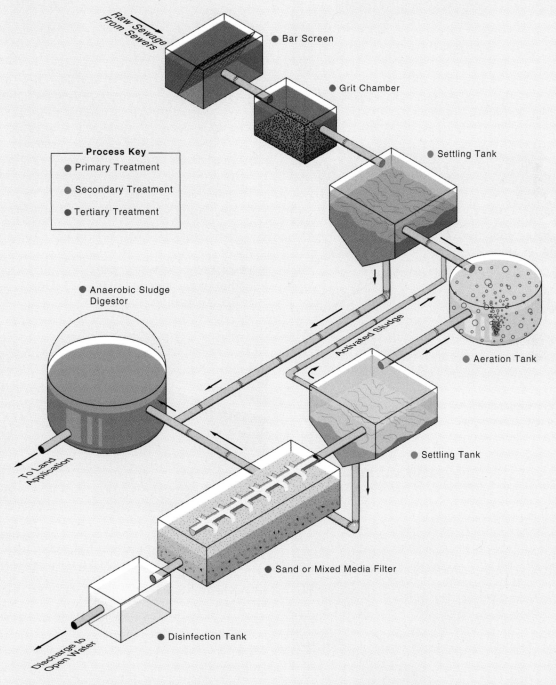

Figure 26.3 Schematic of the treatment processes typical of modern wastewater treatment.
From *Pollution Science* © 1996, Academic Press, San Diego, CA.

Figure 26.4 Removal of large debris from sewage via a "bar screen." Photo courtesy K. L. Josephson.

Figure 26.5 Removal of small debris via a "moving screen." Photo courtesy K. L. Josephson.

Figure 26.6 Primary treatment of sewage. Here, suspended organic solids settle out as primary sludge. Photo courtesy K. L. Josephson.

A

B

Figure 26.7 (a) A unit of plastic material used to create a biofilter. The diameter of each hole is approximately 5 cm. (b) A trickling biofilter or biotower. This is composed of many plastic units stacked upon each other. Dimensions of the biofilter may be 20 m diameter by 10–30 m depth. Photos courtesy K. L. Josephson.

Figure 26.8 Secondary treatment: an aeration basin. Photo courtesy K. L. Josephson.

the plastic, they can be stacked in towers referred to as biotowers (see Figure 26.7). As the organic matter passes through the trickling filter, it is converted to microbial biomass as a biofilm on the filter medium. The biofilm that forms on the surface of the filter medium is called the zooleal film. It is composed of bacteria, fungi, algae, and protozoa. Over time, the increase in biofilm thickness leads to limited oxygen diffusion to the deeper layers of the biofilm, creating an anaerobic environment near the filter medium surface. As a result, the organisms eventually slough from the surface and a new biofilm is formed. BOD removal by trickling filters is approximately 85% for low-rate filters (U.S. EPA, 1977). Effluent from the trickling filter usually passes into a final clarifier to further separate solids from effluent (Figure 26.10).

26.2.2.2 Conventional activated sludge

Aeration-tank digestion is also known as the activated sludge process. In the United States, wastewater is most commonly treated by this process. Effluent from primary treatment is pumped into a tank and mixed with a bacteria-rich slurry known as activated sludge. Air or pure oxygen pumped through the mixture encourages bacterial growth and decomposition of the organic material. It then goes to a secondary settling tank, where water is siphoned off the top of the tank and sludge is removed from the bottom. Some of the sludge is used as an inoculum for primary effluent. The remainder of the sludge, known as secondary sludge, is removed. This secondary sludge is added to primary sludge from primary treatment, and subsequently indigenous anaerobic digestion (see Figure 26.10). The end product of this process is known as biosolids (see Chapter 27). The concentration of pathogens is reduced in the activated sludge process by antagonistic microorganisms as well as adsorption to or incorporation in the secondary sludge.

An important characteristic of the activated sludge process is the recycling of a large proportion of the biomass. This results in a large number of microorganisms that oxidize organic matter in a relatively short time (Bitton, 1999). The detention time in the aeration basin varies from 4 to 8 hours. The content of the aeration tank is referred to as the **mixed-liquor suspended solids (MLSS)**. The organic part of the MLSS is called the **mixed-liquor volatile suspended solids (MLVSS)**, which consists of the nonmicrobial organic matter as well as dead and living microorganisms and cell debris. The activated sludge process must be controlled to maintain a proper ratio of substrate (organic load) to microorganisms or **food-to-microorganism ratio (F/M)** (Bitton, 1999). This is expressed as BOD per kilogram per day. It is expressed as:

$$F/M = \frac{Q \times BOD}{MLSS \times V} \qquad \text{(Eq. 26.3)}$$

Figure 26.9 A trickling filter bed. Here, rocks provide a matrix supporting the growth of a microbial biofilm that actively degrades the organic material in the wastewater under aerobic conditions. From *Pollution Science* © 1996, Academic Press, San Diego, CA.

Figure 26.10 Three clarifiers (foreground—blue) and two anaerobic sludge digesters (background—white). Photo courtesy K. L. Josephson.

where:

$$Q = \text{flow rate of sewage in million gallons per day (MGD)}$$

$$BOD_5 = \text{5-day biochemical oxygen demand}$$

$$MLSS = \text{mixed-liquor suspended solids (mg/L)}$$

$$V = \text{volume of aeration tank (gallons)}$$

F/M is controlled by the rate of activated sludge wasting. The higher the wasting rate, the higher the F/M ratio. For conventional aeration tanks, the F/M ratio is 0.2–0.5 lb BOD_5/day/lb MLSS, but it can be higher (up to 1.5) for activated sludge when high-purity oxygen is used (Hammer, 1986). A low F/M ratio means that the microorganisms in the aeration tank are starved, leading to more efficient wastewater treatment.

The important parameters controlling the operation of an activated sludge process are organic loading rates, oxygen supply, and control and operation of the final settling tank. This tank has two functions: clarification and thickening. For routine operation, sludge settleability is determined by use of the **sludge volume index (SVI)** (Bitton, 1999).

SVI is determined by measuring the sludge volume index, which is given by the following formula:

$$SVI = \frac{V \times 1000}{MLSS} \qquad \text{(Eq. 26.4)}$$

where V = volume of settled sludge after 30 minutes (mL/L).

The microbial biomass produced in the aeration tank must settle properly from suspension so that it may be wasted or returned to the aeration tank. Good settling occurs when the sludge microorganisms are in the endogenous phase, which occurs when carbon and energy sources are limited and the microbial specific growth rate is local (Bitton, 1999). A mean cell residence time of 3–4 days is necessary for effective settling (Metcalf and Eddy, 2003). Poor settling may also be caused by sudden changes in temperature, pH, absence of nutrients, and presence of toxic metals and organics. A common problem in the activated sludge process is filamentous bulking, which consists of slow settling and poor compaction of solids in the clarifier. Filamentous bulking is usually caused by the excessive growth of filamentous microorganisms. The filaments produced by these bacteria interfere with sludge settling and compaction. A high SVI (>150 ml/g) indicates bulking conditions. Filamentous bacteria are able to predominate under conditions of low dissolved oxygen, low F/M, low nutrient, and high sulfide levels. Filamentous bacteria can be controlled by treating the return sludge with chlorine or hydrogen peroxide to kill filamentous microorganisms selectively.

26.2.2.3 Nitrogen removal by the activated sludge process

Activated sludge processes can be modified for nitrogen removal to encourage nitrification followed by denitrification. The establishment of a nitrifying population in acti-

vated sludge depends on the wastage rate of the sludge and therefore on the BOD load, MLSS, and retention time. The growth rate of nitrifying bacteria (μ_n) must be higher than the growth rate (μ_h) of heterotrophs in the system. In reality, the growth rate of nitrifiers is lower than that of heterotrophs in sewage; therefore a long sludge age is necessary for the conversion of ammonia to nitrate. Nitrification is expected at a sludge age greater than 4 days (Bitton, 1999).

Nitrification must be followed by denitrification to remove nitrogen from wastewater. The conventional activated sludge system can be modified to encourage denitrification. Three such processes are:

- **Single sludge system (Fig. 26.11a).** This system comprises a series of aerobic and anaerobic tanks in lieu of a single aeration tank.

- **Multisludge system (Fig. 26.11b).** Carbonaceous oxidation, nitrification, and denitrification are carried out in three separate systems. Methanol or settled sewage serves as the source of carbon for denitrifiers.

- **Bardenpho process (Fig. 26.12).** The process consists of two aerobic and two anoxic tanks followed by a sludge settling tank. Tank 1 is anoxic and is used for denitrification, with wastewater used as a carbon source. Tank 2 is an aerobic tank utilized for both carbonaceous oxidation and nitrification. The mixed liquor from this tank, which contains nitrate, is returned to tank 1. The anoxic tank 3 removes the nitrate remaining in the effluent by denitrification. Finally, tank 4 is an aerobic tank used to strip the nitrogen gas that results from denitrification, thus improving mixed liquor settling.

26.2.2.4 Phosphorus removal by activated sludge process

Phosphorus can also be reduced by the activity of microorganisms in modified activated sludge processes. The process depends on the uptake of phosphorus by the microbes during the aerobic stage and subsequent release during the anaerobic stage. Two of several systems in use are:

1. **A/O (anaerobic/oxic) process.** The A/O process consists of a modified activated sludge system that includes an anaerobic zone (detention time 0.5–1 hour) upstream of the conventional aeration tank (detention time 1–3 hours). Figure 26.13 illustrates the microbiology of the A/O process: During the anaerobic phase, inorganic phosphorus is released from the cells as a result of polyphosphate hydrolysis. The energy liberated is used for the uptake of BOD from wastewater. Removal efficiency is high when the BOD/phosphorus ratio exceeds 10 (Metcalf and Eddy, 2003). During the aerobic phase, soluble phosphorus is taken up by bacteria, which synthesize polyphosphates, using the energy released from BOD oxidation.

2. **Bardenpho process.** This system also removes nitrogen as well as phosphorus by a nitrification–denitrification process (see Figure 26.12 and 26.13).

a

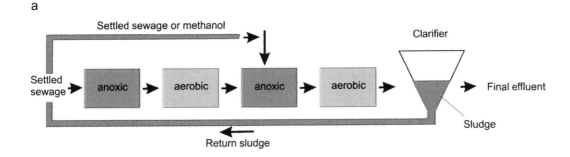

b

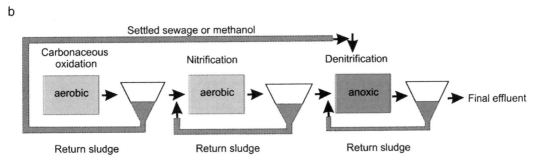

Figure 26.11 Denitrification systems: (a) Single-sludge system. (b) Multisludge system. Modified from Curds and Hawkes, 1983. From *Environmental Microbiology* © 2000, Academic Press, San Diego, CA.

26.2.3 Tertiary Treatment

Tertiary treatment of effluent involves a series of additional steps after secondary treatment to further reduce organics, turbidity, nitrogen, phosphorus, metals, and pathogens. Most processes involve some type of physicochemical treatment such as coagulation, filtration, activated carbon adsorption of organics, reverse osmosis, and additional disinfection. Tertiary treatment of wastewater is practiced for additional protection of wildlife after discharge into rivers or lakes. Even more commonly, it is performed when the wastewater is to be reused for irrigation (*e.g.*, food crops, golf courses), for recreational purposes (*e.g.*, lakes, estuaries), or for drinking water.

26.2.4 Removal of Pathogens by Sewage Treatment Processes

There have been a number of reviews on the removal of pathogenic microorganisms by activated sludge and other wastewater treatment processes (Leong, 1983). This information suggests that significant removal, especially of enteric bacterial pathogens, can be achieved by these processes (Table 26.5). However, disinfection and/or advanced tertiary treatment is necessary for many reuse applications to ensure pathogen reduction. Current issues related to pathogen reduction are treatment plant reliability, removal of new and emerging enteric pathogens of concern, and the ability of new technologies to effect pathogen reduction. Wide vari-

Domestic Wastes and Waste Treatment

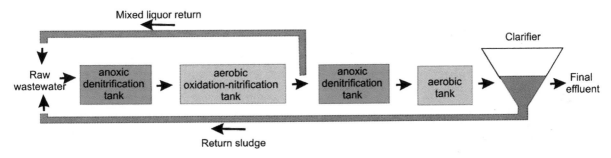

Figure 26.12 Denitrification system: Bardenpho process. From U.S. EPA, 1975. From *Environmental Microbiology* © 2000, Academic Press, San Diego, CA.

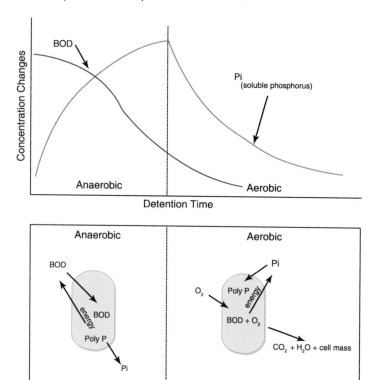

Figure 26.13. Microbiology of the A/O process. Source: Adapted from Deakyne et al., 1984.

ation in pathogen removal can result in significant numbers of pathogens passing through a process for various time periods. The issue of reliability is of major importance if the reclaimed water is intended for recreational or potable reuse, where short-term exposures to high levels of pathogens could result in significant risk to the exposed population.

Compared with other biological treatment methods (i.e., trickling filters), activated sludge is relatively efficient in reducing the numbers of pathogens in raw wastewater. Both sedimentation and aeration play a role in pathogen reduction. Primary sedimentation is more effective for the removal of the larger pathogens such as helminth eggs, but

TABLE 26.5 Pathogen removal during sewage treatment.

	ENTERIC VIRUSES	SALMONELLA	GIARDIA	CRYPTOSPORIDIUM
Concentration in raw sewage (number per liter)	10^5–10^6	5,000–80,000	9,000–200,000	1–3,960
Removal during				
Primary treatment[a]				
% removal	50–98.3	95.8–99.8	27–64	0.7
Number remaining L^{-1}	1,700–500,000	160–3,360	72,000–146,000	
Secondary treatment[b]				
% removal	53–99.92	98.65–99.996	45–96.7	
Number remaining L^{-1}	80–470,000	3–1075	6,480–109,500	
Secondary treatment[c]				
% removal	99.983–99.9999998	99.99–99.999999995	98.5–99.99995	2.7[d]
Number remaining L^{-1}	0.007–170	0.000004–7	0.099–2,951	

[a]Primary sedimentation and disinfection.

[b]Primary sedimentation, trickling filter or activated sludge, and disinfection.

[c]Primary sedimentation, trickling filter or activated sludge, disinfection, coagulation, filtration, and disinfection.

[d]Filtration only.

From Yates (1994), Robertson et al. (1995), Enriquez et al. (1995), Madore et al. (1987). From *Environmental Microbiology* © Academic Press, San Diego, CA.

TABLE 26.6 Average removal of pathogen and indicator microorganisms in a wastewater treatment plant, St. Petersburg, Florida.

	RAW WASTEWATER TO SECONDARY percentage - $\log_{10}$		SECONDARY WASTEWATER TO POSTFILTRATION percentage - $\log_{10}$		POSTFILTRATION TO POSTDISINFECTION percentage - $\log_{10}$		POSTDISINFECTION TO POSTSTORAGE percentage - $\log_{10}$		RAW WASTEWATER TO POSTSTORAGE percentage - $\log_{10}$	
Total coliforms	98.3	1.75	69.3	0.51	99.99	4.23	75.4	0.61	99.999992	7.1
Fecal coliforms	99.1	2.06	10.5	0.05	99.998	4.95	56.8	0.36	99.999996	7.4
Coliphage 15597	82.1	0.75	99.98	3.81	90.05	1.03	90.3	1.03	99.999997	6.6
Enterovirus	98.0	1.71	84.0	0.81	96.5	1.45	90.9	1.04	99.999	5.0
Giardia	93.0	1.19	99.0	2.00	78.0	0.65	49.5	0.30	99.993	4.1
Cryptosporidium	92.8	1.14	97.9	1.68	61.1	0.41	8.5	0.04	99.95	3.2

From Rose and Carnahan (1992). From *Environmental Microbiology* © 2000, Academic Press, San Diego, CA.

solid-associated bacteria and even viruses are also removed. During aeration, pathogens are inactivated by antagonistic microorganisms and by environmental factors such as temperature. The greatest removal probably occurs by adsorption or entrapment of the organisms within the biological floc that forms. The ability of activated sludge to remove viruses is related to the ability to remove solids. This is because viruses tend to be solid associated, and are removed along with the floc. Activated sludge typically removes 90% of the enteric bacteria and 80 to 90–99% of the enteroviruses and rotaviruses (Rao et al., 1986). Ninety percent of *Giardia* and *Cryptosporidium* can also be removed (Rose and Carnahan, 1992), being largely concentrated in the sludge. Because of their large size, helminth eggs are effectively removed by sedimentation and are rarely found in sewage effluent in the United States, although they may be detected in the sludge. However, although the removal of the enteric pathogens may seem large, it is important to remember that initial concentrations are also large (*i.e.*, the concentration of all enteric viruses in 1 liter of raw sewage may be as high as 100,000 in some parts of the world).

Tertiary treatment processes involving physical-chemical processes can be effective in further reducing the concentration of pathogens and enhancing the effectiveness of disinfection processes by the removal of soluble and particulate organic matter (Table 26.6). Filtration is probably the most common tertiary treatment process. Mixed-media filtration is most effective in the reduction of protozoan parasites. Usually, greater removal of *Giardia* cysts occurs than of *Cryptosporidium* oocysts, because of the larger size of the cysts (Rose and Carnahan, 1992). Removal of enteroviruses and indicator bacteria is usually 90% or less. Addition of coagulant can increase the removal of poliovirus to 99% (U.S. EPA, 1992).

Coagulation, particularly with lime, can result in significant reductions of pathogens. The high-pH conditions (pH 11–12) that can be achieved with lime can result in significant inactivation of enteric viruses. To achieve removals of 90% or greater, the pH should be maintained above 11 for at least an hour (Leong, 1983). Inactivation of the viruses occurs by denaturation of the viral protein coat. The use of iron

and aluminum salts for coagulation can also result in 90% or greater reductions in enteric viruses. The degree of effectiveness of these processes, as in other solids separating processes, is highly dependent on the hydraulic design and, in particular, on coagulation and flocculation. The degree of removal observed in bench-scale tests may not approach those seen in full-scale plants, where the process is more dynamic.

Reverse osmosis and ultrafiltration are also believed to result in significant reductions in enteric pathogens, although few studies have been done in full-scale facilities. Removal of enteric viruses in excess of 99.9% can be achieved (Leong, 1983).

26.2.5 Removal of Organics and Inorganics by Sewage Treatment Processes

In addition to microbial pathogens and nutrients such as nitrogen and phosphorus, there are other constituents within sewage that need to be kept at low concentrations. These include inorganics, exemplified by metals, and organic priority pollutants. Metals and organics are normally associated with the solid fraction of sewage and neither are significantly removed by sewage treatment. However, point source control mechanisms to prevent industrial discharges have significantly reduced concentrations of metals and organics within sewage. In particular, over the past 15 years this has resulted in decreased metal concentrations. More recently there has been concern over the presence of pharmaceuticals such as endocrine disruptors in sewage. This topic is discussed in Chapter 31.

26.3 OXIDATION PONDS

Sewage lagoons are often referred to as **oxidation** or **stabilization ponds,** and are the oldest of the wastewater treatment systems. Usually no more than a hectare in area and just a few meters deep, oxidation ponds are natural "stewpots," where wastewater is detained while organic matter is degraded (Figure 26.14). A period of time ranging

Figure 26.14 An oxidation pond. Typically these are only 1–2 meters deep and small in area. Photo courtesy of Taranaki Regional Council (www.trc.govt.nz/region/photos/diary/diary4.htm).

from 1 to 4 weeks (and sometimes longer) is necessary to complete the decomposition of organic matter. Light, heat, and settling of the solids can also effectively reduce the number of pathogens present in the wastewater.

There are four categories of oxidation ponds, which are often used in series: aerobic ponds, anaerobic ponds, facultative ponds, and aerated ponds:

- **Aerobic ponds** (Figure 26.15a), which are naturally mixed, must be shallow (up to 1.5m) because they depend on penetration of light to stimulate algal growth that promotes subsequent oxygen generation. The detention time of wastewater is generally 3 to 5 days.

- **Anaerobic ponds** (Figure 26.15b) may be 1 to 10 m deep and require a relatively long detention time of 20–50 days. These ponds, which do not require expensive mechanical aeration, generate small amounts of sludge. Often, anaero-

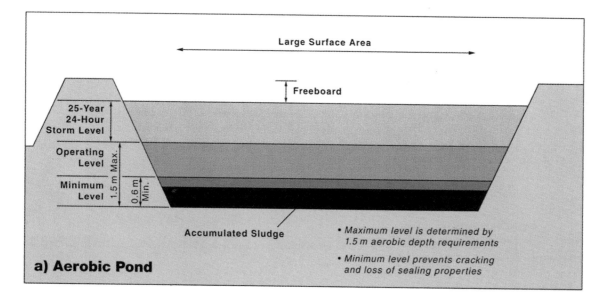

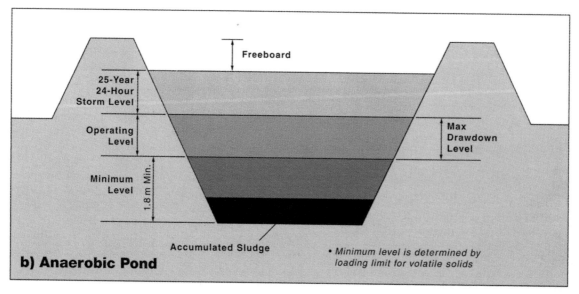

Figure 26.15 Pond profiles: (a) Aerobic waste pond profile, and (b) Anaerobic waste pond profile. From *Pollution Science* © 1996, Academic Press, San Diego, CA.

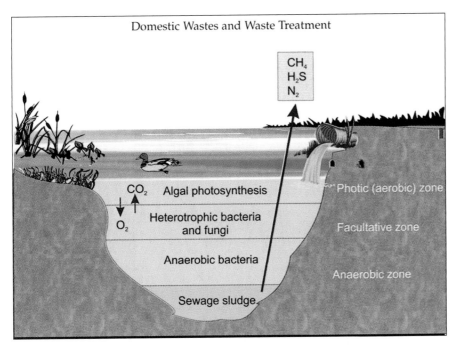

Figure 26.16 Microbiology of facultative ponds. Modified from Bitton, 1980. From *Environmental Microbiology* © 2000, Academic Press, San Diego, CA.

bic ponds serve as a pretreatment step for high-BOD organic wastes rich in protein and fat (*e.g.*, meat wastes) with a heavy concentration of suspended solids.

- **Facultative ponds** (Figure 26.16) are most common for domestic waste treatment. Waste treatment is provided by both aerobic and anaerobic processes. These ponds range in depth from 1 to 2.5 m and are subdivided in three layers: an upper aerated zone, a middle facultative zone, and a lower anaerobic zone. The detention time varies between 5 and 30 days.

- **Aerated lagoons or ponds** (Figure 26.17) which are mechanically aerated, may be 1–2 m deep and have a detention time of less than 10 days. In general, treatment depends on the aeration time and temperature, as well as

the type of wastewater. For example, at 20°C, an aeration period of 5 days results in 85% BOD removal.

Because sewage lagoons require a minimum of technology and are relatively low in cost, they are most common in developing countries and in small communities in the United States, where land is available at reasonable prices. However, biodegradable organic matter and turbidity are not as effectively reduced as during activated sludge treatment.

Given sufficient retention times, oxidation ponds can cause significant reductions in the concentrations of enteric pathogens, especially helminth eggs. For this reason, they have been promoted widely in the developing world as a low-cost method of pathogen reduction for wastewater reuse for irrigation. However, a major drawback of ponds is the potential for short-circuiting because of thermal gradients even in multipond systems designed for long retention times (*i.e.*, 90 days). Even though the amount of short-circuiting may be small, detectable levels of pathogens can often be found in the effluent from oxidation ponds.

Inactivation and/or removal of pathogens in oxidation ponds is controlled by a number of factors including temperature, sunlight, pH, bacteriophage, predation by other microorganisms, and adsorption to or entrapment by settleable solids. Indicator bacteria and pathogenic bacteria may be reduced by 90–99% or more, depending on retention times.

26.4 SEPTIC TANKS

Until the middle of the 20th century in the United States, many rural families and quite a few residents of towns and small cities depended on pit toilets or "outhouses" for waste

Figure 26.17 An aerated lagoon. Photo courtesy C. P. Gerba.

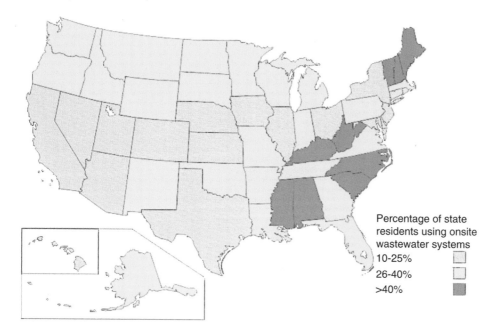

Figure 26.18 Percentage of U.S. residents utilizing septic tanks for onsite wastewater treatment. Source: U.S. Census Bureau, 1990.

disposal. In rural areas of developing countries these are still used. These pit toilets, however, often allowed untreated wastes to seep into the groundwater, allowing pathogens to contaminate drinking water supplies. This risk to public health led to the development of septic tanks and properly constructed drain fields. Primarily, septic tanks serve as repositories where solids are separated from incoming wastewater, and biological digestion of the waste organic matter can take place under anaerobic conditions. In 1997, 25% of the homes in the United States depended on septic tanks. Approximately 33% of all new homes constructed use septic tanks. Most septic tanks are located in the eastern United States (Figure 26.18). In a typical septic tank system

(Figure 26.19), the wastewater and sewage enter a tank made of concrete, metal, or fiberglass. There, grease and oils rise to the top as scum, and solids settle to the bottom. The wastewater and sewage then undergo anaerobic bacterial decomposition, resulting in the production of a sludge. The wastewater usually remains in the septic tank for just 24–72 hours, after which it is channeled out to a drainfield. This drainfield or leachfield is composed of small perforated pipes that are embedded in gravel below the surface of the soil. Periodically, the residual separate in the septic tank known as **septage** is pumped out into a tank truck, and taken to a treatment plant for disposal.

Although the concentration of contaminants in septic tank separate is typically much greater than that found in domestic wastewater (Table 26.7), septic tanks can be an

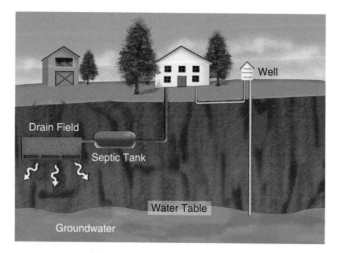

Figure 26.19 Septic tank (On-site treatment system). Source: U.S. EPA, 2002.

TABLE 26.7 Typical characteristics of septage.

	CONCENTRATION (mg/L)	
CONSTITUENT	RANGE	TYPICAL VALUE
Total solids	5,000–100,000	40,000
Suspended solids	4,000–100,000	15,000
Volatile suspended solids	1,200–14,000	7,000
5-day, 20°C BOD	2,000–30,000	6,000
Chemical oxygen demand	5,000–80,000	30,000
Total Kjeldahl nitrogen (as N)	100–1,600	700
Ammonia, NH_3 as N	100–800	400
Total phosphorus as P	50–800	250
Heavy metals[a]	100–1,000	300

[a]Primarily iron (Fe), zinc (Zn), and aluminum (Al).

Modified from Metcalf and Eddy (2003).

effective method of waste disposal where land is available and population densities are not too high. Thus, they are widely used in rural and suburban areas. But as suburban population densities increase, groundwater and surface water pollution may arise, indicating a need to shift to a commercial municipal sewage system. (In fact, private septic systems are sometimes banned in many suburban areas.) Moreover, septic tanks are not appropriate for every area of the country. They do not work well, for example, in cold, rainy climates, where the drainfields may be too wet for proper evaporation, or in areas where the water table is shallow. High densities of septic tanks can also be responsible for nitrate contamination of groundwater. Finally, most of the waterborne disease outbreaks associated with groundwater in the United States are thought to result from contamination by septic tanks.

26.5 LAND APPLICATION OF WASTEWATER

Although treated domestic wastewater is usually discharged into bodies of water, it may also be disposed of via land application for crop irrigation or as a means of additional treatment and disposal. The three basic methods used in the application of sewage effluents to land include low-rate irrigation, overland flow, and high-rate infiltration. Characteristics of each of these are listed in Table 26.8. The choice of a given method depends on the conditions prevailing at the site under consideration (loading rates, methods of irrigation, crops, and expected treatment).

With **low-rate irrigation** (Figure 26.20a), sewage effluents are applied by sprinkling or by surface application at a rate of 1.5 to 10 cm per week. Two-thirds of the water is taken up by crops or lost by evaporation, and the remainder percolates through the soil matrix. The system must be designed to maximize denitrification in order to avoid pollution of groundwater by nitrates. Phosphorus is immobilized within the soil matrix by fixation or precipitation. The irrigation method is used primarily by small communities and requires large areas, generally on the order of 5–6 hectares per 1000 people.

In the **overland flow** method (Figure 26.20b), wastewater effluents are allowed to flow for a distance of 50–100 m along a 2–8% vegetated slope and are collected in a ditch. The loading rate of wastewater ranges from 5 to 14 cm a week. Only about 10% of the water percolates through the soil, compared with 60% that runs off into the ditch. The remainder is lost as evapotranspiration. This system requires clay soils with low permeability and infiltration.

High-rate infiltration treatment is also known as **soil aquifer treatment (SAT)** or **rapid infiltration extraction (RIX)** (Figure 26.20c). The primary objective of SAT is the treatment of wastewater at loading rates exceeding 50 cm per week. The treated water, most of which has percolated through coarse-textured soil, is used for groundwater recharge or may be recovered for irrigation. This system requires less land than irrigation or overland flow methods. Drying periods are often necessary to aerate the soil system and avoid problems due to clogging. The selection of a site for land application is based on many factors including soil types, drainable and depth, distance to groundwater, groundwater movement, slope, underground formations, and degree of isolation of the site from the public.

Inherent in land application of wastewater are the risks of transmission of enteric waterborne pathogens. The degree of risk is associated with the concentration of pathogens in the wastewater and the degree of contact with humans. Land application of wastewater is usually considered an intentional form of reuse and is regulated by most states. Because of limited water resources in the western United States, reuse is considered essential. Usually, stricter treatment and microbial standards must be met before land application. The highest degree of treatment is required when wastewater will be used for food crop irrigation, with lesser treatment for landscape irrigation or fiber

TABLE 26.8 General characteristics of the three methods used for land application of sewage effluent.

| | APPLICATION METHOD | | |
FACTOR	LOW-RATE IRRIGATION	OVERLAND FLOW	HIGH-RATE INFILTRATION
Main objectives	Reuse of nutrients and water Wastewater treatment	Wastewater treatment	Wastewater treatment Groundwater recharge
Soil permeability	Moderate (sandy to clay soils)	Slow (clay soils)	Rapid (sandy soils)
Need for vegetation	Required	Required	Optional
Loading rate	1.3–10 cm/week	5–14 cm/week	>50 cm/week
Application technique	Spray, surface	Usually spray	Surface flooding
Land required for flow of 10^6 l/day	8–66 hectares	5–16 hectares	0.25–7 hectares
Needed depth to groundwater	About 2 cm	Undetermined	5 m or more
BOD and suspended solid removal	90–99%	90–99%	90–99%
N removal	85–90%	70–90%	0–80%
P removal	80–90%	50–60%	75–90%

From *Pollution Science* © 1996, Academic Press, San Diego, CA.

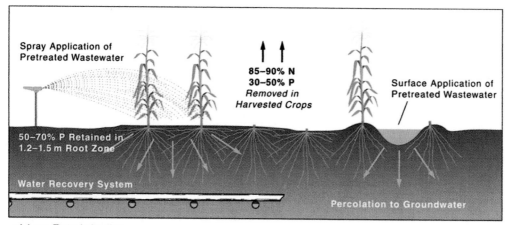

a) Low-Rate Irrigation

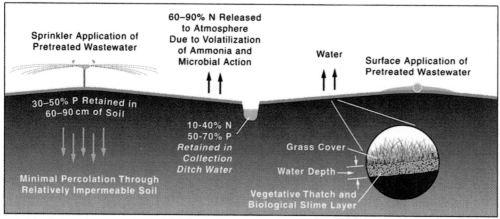

b) Overland Flow

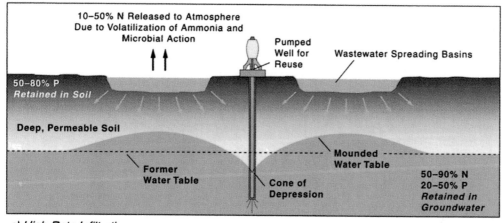

c) High-Rate Infiltration

Figure 26.20 Three basic methods of land application of wastewater. From *Pollution Science* © 1996, Academic Press, San Diego, CA.

crops. For example, the State of California requires no disinfection of wastewater for irrigation and no limits on coliform bacteria. However, if the reclaimed wastewater is used for surface irrigation of food crops and open landscaped areas, chemical coagulation (to precipitate suspended matter), followed by filtration and disinfection to reduce the coliform concentration to 2.2/100 ml is required. In some cities excess effluent is disposed of in river beds that are normally dry. Such disposal can create riparian areas (Figure 26.21).

Because high-rate infiltration may be practiced to recharge aquifers, additional treatments of secondary

Figure 26.21 Effluent outfall of the Roger Road Wastewater Treatment Plant in Tucson, Arizona. Here, extensive growth of vegetation due to the effluent produces a riparian habitat. Photo courtesy K. L. Josephson.

26.6 WETLANDS AND AQUACULTURE SYSTEMS

Wetlands, which are typically less than 1 m in depth, are areas that support aquatic vegetation and foster the growth of emergent plants such as cattails, bulrushes, reeds, sedges, and trees. They also provide important habitat for many animal species. Recently, wetland areas have been receiving increasing attention as a means of additional treatment for secondary effluents. The vegetation provides surfaces for the attachment of bacteria and aids in the filtration and removal of such wastewater contaminants as biological oxygen and excess carbon. Factors involved in the reduction of wastewater contaminants are shown in Table 26.9. Although both natural and constructed wetlands have been used for wastewater treatment, recent work has focused on constructed wetlands because of regulatory requirements. Two types of constructed wetland systems are in general use: (1) **free water surface (FWS)** systems, and (2) **subsurface flow systems (SFSs)**. An FWS wetland is similar to a natural marsh because the water surface is exposed to the atmosphere. Floating and submerged plants, such as those shown in Figure 26.22a, may be present. SFSs consist of channels or trenches with relatively impermeable bottoms filled with sand or rock media to support emergent vegetation.

During wetland treatment, the wastewater is usable. It can, for instance, be used to grow aquatic plants such as water hyacinths (Figures 26.22b) and/or to raise fish for human consumption. The growth of such aquatic plants provides not only additional treatment for the water but also a food source for fish and other animals. Such aquaculture systems, however, tend to require a great deal of land area. Moreover, the health risk associated with the production of

wastewater may be required. However, as some removal of pathogens can be expected, treatment requirements may be less. The degree of treatment needed may be influenced by the amount or time it takes the reclaimed water to travel from the infiltration site to the point of extraction, and the depth of the unsaturated zone. The greatest concern has been with the transport of viruses, which, because of their small size, have the greatest chance of traveling large distances within the subsurface. Factors that influence the transport of viruses are discussed in Chapter 17. Generally, several meters of moderately fine-textured, continuous soil layer are necessary for virus reductions of 99.9% or more (Yates, 1994).

TABLE 26.9 Principal removal and transformation mechanisms in constructed wetlands involved in contaminant reduction.

CONSTITUENT	FREE WATER SYSTEM	SUBSURFACE FLOW	FLOATING AQUATICS
Biodegradable organics	Bioconversion by aerobic, facultative, and anaerobic bacteria on plant and debris surfaces of soluble BOD, adsorption, filtration	Bioconversion by facultative and anaerobic bacteria on plant and debris surfaces	Bioconversion by aerobic facultative, and anaerobic bacteria on plant and debris surfaces
Suspended solids	Sedimentation, filtration	Filtration, sedimentation	Sedimentation, filtration
Nitrogen	Nitrification/denitrification, plant uptake, volatilization	Nitrification/denitrification, plant uptake, volatilization	Nitrification/denitrification plant update, volatilization
Phosphorus	Sedimentation, plant uptake	Filtration, sedimentation, plant uptake	Sedimentation, plant uptake
Heavy metals	Adsorption of plant and debris surfaces	Adsorption of plant roots and debris surfaces, sedimentation	Absorption by plants sedimentation
Trace organics	Volatilization, adsorption, biodegradation	Adsorption, biodegradation	Volatilization, adsorption biodegradation
Pathogens	Natural decay, predation, UV irradiation, sedimentation, excretion of antimicrobials from roots of plants	Natural decay, predation, sedimentation, excretion of antimicrobials from roots of plants	Natural decay, predation sedimentation

Modified from Crites and Tchobanoglous, 1998.

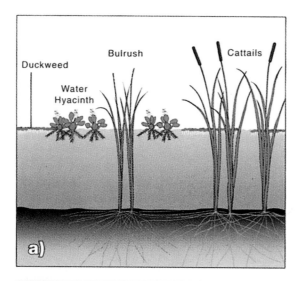

Figure 26.22 (a) Common aquatic plants used in constructed wetlands. (b) An artificial wetland system in San Diego, California, utilizing water hyacinths. Photo courtesy C. P. Gerba. From *Pollution Science* © 1996, Academic Press, San Diego, CA.

Figure 26.23 Sweetwater site, Tucson, Arizona. This is an example of a subsurface flow wetland used to treat secondary treated wastewater. Photo courtesy C.P. Gerba. From *Environmental Microbiology* © 2000, Academic Press, San Diego, CA.

wetlands never reach zero because wetlands are open to wildlife. Reductions in fecal coliforms are generally greater than 99%, but there is a great deal of variation, probably depending on the season, type of wetland, numbers and type of wildlife, and retention time in the wetland. Volume-based and area-based bacterial die-off models have been used to estimate bacterial die-off in surface flow wetlands (Kadlec and Knight, 1996).

In a study of a mixed-species surface flow wetland with a detention time of approximately 4 days, *Cryptosporidium* was reduced by 53%, *Giardia* by 58%, and enteric viruses by 98% (Karpiscak *et al.*, 1996).

26.7 SLUDGE PROCESSING

Primary, secondary, and even tertiary sludges generated during wastewater treatment are usually subjected to a variety of processes.

Raw sludge is sometimes subjected to **screening** to remove coarse materials including grit that cannot be broken down biologically. **Thickening** is usually done to increase the solids content of the sludge. This can be achieved via centrifugation, which increases the solids content to approximately 12%. **Dewatering** can further concentrate the solids content to 20–40%. This is normally achieved via filtration or by the use of drying beds. **Conditioning** enhances the separation of solids from the liquid phase. This is usually accomplished by the addition of inorganic salts such as alum, lime, or ferrous or ferric salts, or synthetic organic polymers known as polyelectrolytes. All of these processes reduce the water content of the sludge, which ultimately reduces transportation costs to the final disposal and/or utilization site.

Finally, **stabilization** technologies are available, reducing both the solids content of the sludge and inactivating pathogenic microbes present in the sludge.

aquatic animals for human consumption in this manner must be better defined.

There has been increasing interest in the use of natural systems for the treatment of municipal wastewater as a form of tertiary treatment (Kadlec and Knight, 1996). Artificial or constructed wetlands have a higher degree of biological activity than most ecosystems; thus, transformation of pollutants into harmless byproducts or essential nutrients for plant growth can take place at a rate that is useful for the treatment of municipal wastewater. Most artificial wetlands in the United States use reeds or bull rushes, although floating aquatic plants such as water hyacinths and duckweed have also been used. To reduce potential problems with flying insects, subsurface flow wetlands have also been built (Figure 26.23). In these types of wetlands, all of the flow of the wastewater is below the surface of a gravel bed containing plants tolerant of water-saturated soils. Most of the existing information on the performance of these wetlands concerns coliform and fecal coliform bacteria. Kadlec and Knight (1996) have summarized the existing literature on this topic. They point out that natural sources of indicators in treatment

CASE STUDY 26.1

RAPID INFILTRATION–EXTRACTION PROJECT IN COLTON, CALIFORNIA.

The cities of San Bernardino and Colton, California, are required to filter and disinfect their secondary effluent prior to discharge to the Santa Ana River, which is a source of drinking water and also used for body contact recreation. The two cities joined under the auspices of the Santa Ana Watershed Project Authority (SAWPA) to seek a regional solution to their wastewater treatment requirements. They hoped to develop a cost-effective alternative to conventional tertiary treatment (chemical coagulation, filtration, and disinfection) that would still result in an effluent that was essentially free of measurable levels of pathogens. SAWPA conducted a 1-year demonstration project to examine the feasibility of a rapid infiltration–extraction (RIX) process to treat unchlorinated secondary effluent prior to discharge to the river and determine whether or not the RIX process was equivalent—in terms of treatment reliability and quality of the water produced—to the conventional tertiary treatment processes specified in the California Department of Health Services Wastewater Reclamation Criteria.

The demonstration was conducted at a site in Colton, California, adjacent to the Santa Ana River bed. Infiltration basins were built at two sites on the property to allow testing of the RIX system under a variety of operating conditions. The soils were coarse sands with clean water infiltration rates of about 15 m/day (50 ft/day). Forty-four monitoring wells were sampled for a number of organic, inorganic, microbiological, and physical measurements.

The study indicted that the optimal filtration rate was 2 m/day (6.6 ft/day) with a wet-to-dry ratio of 1:1 (1 day of flooding to 1 day of drying). Mounding beneath the infiltration basins ranged from 0.6 to 0.9 m (2 to 3 ft), and infiltrated wastewater migrated up to 24.4 m (80 ft) vertically, and over 30.5 m (100 ft) laterally, before it was extracted. Aquifer residence time was 20 to 45 days, depending on the recharge site. Extraction of 110% of the volume of infiltrated effluent by downgradient extraction wells effectively contained the effluent on the RIX site and minimized mixing with regional groundwater outside the project area.

In samples collected approximately 7.6 m (25 ft) below the infiltration basins, the soil–aquifer treatment reduced the concentration of total coliform organisms 99 to 99.9%. Generally, water from the extraction wells prior to disinfection contained less than 2.2 total coliforms per 100 ml. Although viral levels were as high as 316 viruses per 200 L in the unchlorinated secondary effluent applied to the infiltration basins, only one sample of extracted water prior to disinfection was found to contain detectable levels of viruses. In addition, the turbidity of the extracted water generally averaged less than 0.04 NTU.

Although the RIX process greatly reduced microorganism levels in the wastewater, disinfection of the extracted wastewater still proved to be necessary prior to its discharge to the Santa Ana River. Ultraviolet radiation was evaluated as an alternative to chlorination for disinfection. Due in part to the high quality of the extracted water, ultraviolet radiation was shown to be effective for the destruction of both bacteria and viruses and will probably be used instead of chlorine in the full-scale project.

26.7.1 Stabilization Technologies

26.7.1.1 Aerobic digestion

This consists of adding air or oxygen to sludge in a 4 to 8 m L^{-1} feet deep open tank. The oxygen concentration within the tank must be maintained above 1 mg L^{-1} to avoid the production of foul odors. The mean residence time in the tank is 12–60 days, depending on the tank temperature. During this process, microbes aerobically degrade organic substrate, reducing the volatilize solids content of the sludge by 40–50% (U.S. EPA, 1992). Digestion temperatures are frequently moderate or **mesophilic** (30–40°C). By increasing the oxygen content, thermophilic digestion can be induced (>60°C). By increasing the temperature and the retention time, the degree of pathogen inactivation can be enhanced. Pathogen concentrations ultimately determine the treatment level of the product. **Class B biosolids** can contain many human pathogens.

Class A biosolids, which result from more stringent and enhanced treatment, contain very low or nondetectable levels of pathogens. The degree of treatment, Class A versus Class B, has important implications on the reuse potential of the material for land application (see also Chapter 27). Aerobic digestion generally results in the production of Class B biosolids.

26.7.1.2 Anaerobic digestion

This type of microbial digestion occurs under low redox conditions, with low oxygen concentrations. Carbon dioxide is used as a terminal electron acceptor (see Chapter 5), and organic substrate is converted to methane and carbon dioxide. This process reduces the volatile solids by 35–60% (Bitton, 1999), and results in the production of Class B biosolids. The advantages and disadvantages of anaerobic digestion relative to aerobic digestion are shown in Information Box 26.2.

INFORMATION BOX 26.2

Advantages and Disadvantages of Anaerobic Digestion

ADVANTAGES	DISADVANTAGES
· No oxygen requirement, which reduces cost	· Slower than aerobic digestion
· Reduced mass of biosolids due to low energy yields of anaerobic metabolism (see also Chapter 5)	· More sensitive to toxicants
· Methane produced, which can be used to generate electricity	
· Enhanced degradation of xenobiotic compounds	

Source: Adapted from Bitton, 1999.

INFORMATION BOX 26.3

Factors Affecting Efficient Composting

Temperature:	Adequate aeration and moisture must be maintained to ensure temperatures reach 60°C, to inactivate microbial pathogens
Aeration:	Must be provided via blowers or by turning
Moisture:	Must be neither too moist, which promotes anaerobic activity, nor too dry, which limits microbial activity.
C:N ratio:	Should be maintained around 25:1, to ensure adequate but not excessive amounts of nitrogen for the microbes.
Surface area of bulking agent:	Shredded material should be used to increase substrate surface area for microbial metabolism.

26.7.2 Sludge Processing to Produce Class A Biosolids

Class B biosolids that arise following digestion can be further treated to Class A levels prior to land application (see Chapter 27). The three most important technologies here are composting, lime treatment, and heat treatment.

27.7.2.1 Composting

This process consists of mixing sludge with a bulking agent that normally has a high C:N ratio (Figure 26.24). This is necessary because of the low C:N ratio of the sludge. The mixtures are normally kept moist but aerobic. These conditions result in very high microbial activity and the generation of heat that increases the temperature of the composting material. Factors affecting the composting process are shown in Information Box 26.3. There are three main types of composting systems:

1. The **aerated static pile** process typically consists of mixing dewatered digested sludge with wood chips (Figure 26.25). Aeration of the pile is normally provided by blowers during a 21-day composting period. During this active composting period, temperatures increase to the mesophilic range (20–40°C), where microbial degradation occurs via bacteria and fungi. Temperatures subsequently increase to 40–80°C), with microbial populations dominated by thermophilic (heat-tolerant) and spore-forming organisms. These high temperatures inactivate pathogenic microorganisms and frequently result in a Class A biosolid product. Subsequently, the compost is cured for at least 30 days, during which time temperatures within the pile decrease to ambient levels.

2. The **windrow process** is similar to the static pile process except that instead of a pile, the sludge and bulking

Figure 26.24 Scrap timber and wood products are frequently used as a bulking agent in biosolid composting. Photo courtesy I. L. Pepper.

Figure 26.25 The wood bulking agent for composting. The wood is shredded to increase the surface area of the bulking agent for composting. Photo courtesy I. L. Pepper.

Figure 26.26 Biosolid composting via the windrow process. Here, three windrows are illustrated.
Photo courtesy I. L. Pepper

agent are laid out in long rows of the dimension 2 m by 3 m by 80 m (Figure 26.26). Aeration for windrows is provided by turning the windrows several times a week. Once again, if the composting process is efficient, Class A biosolids are produced.

3. In **enclosed systems**, composting is conducted in steel vessels of size 10–15 m high by 3–4 m diameter. For this type of composting, aeration via blowers and temperature of the composting are carefully controlled. This results in a high-quality Class A compost, with little or no odor problems. However, costs of enclosed systems are higher.

26.7.2.2 Lime and heat treatment

Lime stabilization involves the addition of lime as $Ca(OH)_2$ or CaO, such that the pH of digested sludge is equal to or greater than 12 for at least 2 hours. Liming is very effective at inactivating bacterial and viral pathogens, but less so for parasites (Bitton, 1999). Lime stabilization also reduces odors, and can result in a Class A biosolid product.

Heat treatment involves heating sludge under pressure to temperatures up to 260°C for 30 minutes. This process kills microbial pathogens and parasites, and also further dewaters the sludge.

QUESTIONS AND PROBLEMS

1. What are the three major steps in modern wastewater treatment?

2. Why is it important to reduce the amount of biodegradable organic matter and nutrients during sewage treatment?

3. When would tertiary treatment of wastewater be necessary?

4. What are some types of tertiary treatment?

5. What are the processes involved in the removal of heavy metals from wastewater during treatment by artificial wetlands?

6. What are the three types of land application of wastewater? Which one is most likely to contaminate the groundwater with enteric viruses? Why? What factors determine how far viruses will be transported in groundwater? How does nitrogen removal occur? Phosphorus removal?

7. What are the major contaminates in groundwater associated with the use of on-site treatment systems?

8. What factors may determine the concentration of enteric pathogens in domestic raw sewage?

9. Five milliliters of a wastewater sample is added to dilution water in a 300-ml BOD bottle. If the following results are obtained, what is the BOD after 3 days and 5 days?

TIME (DAYS)	DISSOLVED OXYGEN (mg/l)
0	9.55
1	4.57
2	4.00
3	3.20
4	2.60
5	2.40
6	2.10

10. What is the difference between Class A and Class B biosolids? Name three processes that can be used to produce Class A biosolids.

11. List some advantages and disadvantages of the wetland treatment of sewage.

REFERENCES AND ADDITIONAL READING

APHA. (1998) Standard Methods for Water and Wastewater. American Public Health Association, Washington, DC.

Bitton G. (1980) *Introduction to Environmental Virology*. John Wiley & Sons, New York.

Bitton G. (1999) *Wastewater Microbiology*. Wiley-Liss, New York.

Crites R. and Tchobanoglous G. (1998) *Small and Decentralized Wastewater Management Systems*. McGraw-Hill, New York.

Curds C. R. and Hawkes H.A. eds. (1983) *Ecological Aspects of Used-Water Treatment*, vol. 2. Academic Press, London, England.

Deakyne C. W., Patel M.A. and Krichten D.J. (1984) Pilot plant demonstration of biological phosphorus removal. *J. Water Pollut. Control Fed.* **56**, 867–873.

Enriquez V., Rose J.B., Enriquez C.E. and Gerba C.P. (1995) Occurrence of *Cryptosporidium* and *Giardia* in secondary and tertiary wastewater effluents. In Protozoan Parasites and Water, ed. By W.B. Betts, D. Casemore, C. Fricker, H. Smith, and J. Watkins. Royal Society of Chemistry, Cambridge, England, pp. 84–86.

Hammer M. J. (1986) *Water and Wastewater Technology*. John Wiley & Sons, New York.

Kadlec R. H. and Knight R.L. (1996) *Treatment Wetlands*. Lewis Publishers, Boca Raton, Florida.

Karpiscak M.M., Gerba C.P., Watt P.M., Foster K. E. and Falabi J.A. (1996) Multi-species plant systems for wastewater quality improvements and habitat enhancement. *Water Sci. Technol.* **33**, 231–236.

Leong L.Y.C. (1983) Removal and inactivation of viruses by treatment processes for portable water and wastewater—A review. *Water Sci. Technol.* **15**, 91–114.

Madore M.S., Rose J.B., Gerba C.P., Arrowood M.J. and Sterling C.R. (1987) Occurrence of *Cryptosporidium* in sewage effluents and selected surface waters. *J. Parasitol.* **73**, 702–705.

Metcalf and Eddy, Inc. (2003) *Wastewater Engineering*. McGraw-Hill, New York.

Rao V.C., Metcalf T.G. and Melnick J.L. (1986) Removal of pathogens during wastewater treatment. In *Biotechnology*, Vol. 8, ed. H. J. Rehm and G. Reed VCH, Berlin, pp. 531–554.

Rose J.B. and Carnahan R.P. (1992) Pathogen removal by full-scale wastewater treatment. Report to Florida Department of Environmental Regulation, Tallahassee, Florida.

U.S. EPA. (1975) Process design manual for nitrogen control. Office of Technology Transfer, Washington, D.C.

U.S. EPA. (1977) Wastewater treatment facilities for sewered small communities. EPA–62511–77–009, Washington, D.C.

U.S. EPA. (1988) Comparative health risks effects assessment of drinking water. U. S. Environmental Protection Agency, Washington, D.C.

U.S. EPA. (1992) United States Environmental Protection Agency. Guidelines for water reuse. EPA/625/R–92/004, Washington, D.C.

Yates M.V. (1994) Monitoring concerns and procedures for human health effects. In *Wastewater Reuse for Golf Course Irrigation*. CRC Press, Boca Raton, Florida, pp. 143–171.

CHAPTER **27**

LAND APPLICATION OF BIOSOLIDS AND ANIMAL WASTES

I. L. Pepper and C. P. Gerba

One method of land application of biosolids. *Photo courtesy J.P. Brooks.*

Human waste excreted by all of the world's 6.8 billon population must be disposed of by some means. In the United States and many other countries of the world, municipal wastes are treated and subsequently land applied as a method of disposal or recycling. Most methods of municipal sewage treatment produce large amounts of bacteria as the soluble organic matter is converted to bacterial biomass. This material is known as biosolids. Animal wastes result in manures (solids) or liquid effluents that can also be land applied. All of these wastes are routinely applied to soils to provide plant nutrients or as a source of water for plant growth or groundwater recharge.

In the United States, the re-use of treated municipal wastewater effluent provides an opportunity to conserve water resources. Land application of liquid wastes can also provide an alternative to disposal in areas where surface waters have a limited capacity to assimilate elements such as nitrogen or phosphorus, which in excess can result in pollution (see Chapters 10 and 16). The solids that result from municipal wastewater treatment processes contain organic material that, when properly treated and applied to land as "biosolids," can improve the productivity of soils or enhance revegetation of disturbed ecosystems (see Chapter 20). Animal manures can also be land applied beneficially to agricultural land. However, besides the documented benefits of land application, there are also potential hazards, which have caused the public response to the practice to be mixed. In this chapter we review the current science with respect to land application, as well as the current public perceptions and legal aspects of land application.

27.1 BIOSOLIDS AND ANIMAL WASTES: A HISTORICAL PERSPECTIVE AND CURRENT OUTLOOK

Use of animal wastes and manures as a fertilizer source for agricultural crop production has been practiced since the days of the Roman Empire. During the 20th century in both the U.S. and Europe, small agricultural farms frequently consisted of both crop and animal production. Consequently, animal wastes were naturally land applied to enhance crop production. Although fossil-fuel based fertilizers replaced much of the use of manures, following World War II, the practice continues today, particularly in developing countries.

In the U.S., land application of municipal wastewater and biosolids has been practiced for its beneficial effects and for disposal purposes since the advent of modern wastewater treatment about 160 years ago. In Britain in the 1850s, "sewage farms" were established to dispose of untreated sewage. By 1875, about 50 farms were utilizing land treatment in England, as were many others close to other major cities in Europe. In the U.S., sewage farms were established by about 1900. At this same time, primary sedimentation and secondary biological treatment was introduced as a rudimentary form of wastewater treatment, and land application of "sludges" began. It is interesting to note that prior to wastewater treatment, sludge *per se* did not exist. Municipal sludge in Ohio was used as a fertilizer as early as 1907. Early on land application was carried out with little regard to potential pollution.

Since the early 1970s, more emphasis has been placed on applying sludge to cropland at rates to supply adequate nutrients for crop growth (Hinesly et al., 1972). In the 1970s and 80s, many studies were undertaken to investigate the potential benefits and hazards of land application, in both the U.S. and Europe. Ultimately in 1993, Federal regulations were established via the "Part 503 Sludge Rule." This document—"The Standards for the Use and Disposal of Sewage Sludge" (EPA, 1993) —was designed to "adequately protect human health and the environment from any reasonably anticipated adverse effect of pollutants." As part of these regulations, two classes of treatment were defined as "Class A and Class B" biosolids, with different restrictions for land applications, based on the level of treatment.

Land application increased when restrictions were placed on ocean dumping. By the year 2000, 60% of all biosolids were land applied in the U.S. Currently most land application in the U.S. utilizes Class B biosolids. However, due to public concerns over potential hazards, in some areas of the U.S., land application of Class B biosolids has been banned.

27.2 THE NATURE OF WASTEWATER (SEWAGE)

Sewage sludge is defined in the Part 503 rule as the solid, semi-solid, or liquid residue generated during the treatment of domestic sewage in a wastewater treatment plant (see Chapter 26). The term biosolids is not used in the Part 503 rule, but EPA (1995) defines biosolids as "the primarily organic solid product yielded by municipal wastewater treatment processes that can be beneficially recycled" as soil amendments (Information Box 27.1). The term biosolids has been controversial because of the perception that it was created to improve the image of sewage sludge in a public-relations campaign by the sewage industry. For our purposes, the term biosolids implies treatment of sewage sludge to meet the land-application standards in the Part 503 rule (EPA, 1993).

It is estimated that approximately 5.6 million dry tons of sewage sludge were used or disposed of annually in the United States in 2000, of which approximately 60% were used for land application. However, EPA estimates that only

INFORMATION BOX 27.1

Definitions

Sewage sludge: The solid, semi-solid, or liquid residue generated during the treatment of domestic sewage in a treatment works.

Biosolids:

- EPA's definition: The primarily organic solid product yielded by municipal wastewater treatment processes that can be beneficially recycled (whether or not they are currently being recycled).
- National Resource Council (NRC) 2002 Committee's definition: Sewage sludge that has been treated to meet the land-application standards in the Part 503 rule or any other equivalent land-application standards or practices.

approximately 0.1% of available agricultural land in the U.S. is treated with biosolids.

Biosolids are applied to agricultural and nonagricultural lands as a soil amendment because they can improve the chemical and physical properties of soils, and because they contain nutrients for plant growth (see Section 27.5). Land application on agricultural land is utilized to grow food crops such as corn or wheat, and nonfood crops such as cotton. Nonagricultural land application includes forests, rangelands, public parks, golf courses, and cemeteries. Biosolids are also used to revegetate severely disturbed lands such as mine tailings or strip mine areas (see Chapter 20).

INFORMATION BOX 27.2

Part 503 Pathogen Density Limits Adapted from USEPA 2000

Part 503 Pathogen Density Limits

PATHOGEN OR INDICATOR	STANDARD DENSITY LIMITS (DRY WT.)
Class A	
Salmonella	<3 MPN/4 gram total solids *or*
Fecal coliforms	<1000 MPN/gram *and*
Enteric viruses	1 PFU/4 gram total solids *and*
Viable helminth	
Ova	<1/4 gram total solids
Class B	
Fecal coliform density	<2,000,000 MPN/gram total solids

27.3 WASTEWATER (SEWAGE) TREATMENT

Sewage wastewater treatment is covered extensively in Chapter 26. Figure 27.1 provides a simplified schematic of how biosolids are produced.

27.3.1 Class A Versus Class B Biosolids

Biosolids are divided into two classes on the basis of pathogen content: Class A and Class B (Information Box 27.2).

Class A biosolids are treated to reduce the presence of pathogens to below detectable levels and can be used without any pathogen-related restrictions at the application site. Class A biosolids can also be bagged and sold to the public. Class B biosolids are also treated to reduce pathogens, but still contain detectable levels of them. Class B biosolids have site restrictions to minimize the potential for human exposure until environmental factors such as heat, sunlight, or desiccation have further reduced pathogen numbers. Class B biosolids cannot be sold or given away in bags or other containers or used at sites with public use.

27.4 METHODS OF LAND APPLICATION OF BIOSOLIDS

The method of land application of biosolids essentially depends on the percent solids contained within them, which determines whether the biosolids are liquid in nature or a "cake" (Information Box 27.3). Figure 27.2 through 27.7 illustrate all methods of land application and can be summarized as:

- **Injection.** Liquid biosolids are injected to a soil depth of 6–9 inches. Injection vehicles simultaneously disc the field. Injection processes reduce odors and bioaerosols, as well as the risk of runoff to surface waters.
- **Surface Application.** Here, liquid or cake biosolids are surface applied and subsequently tilled into the soil (Figure 27.8).

INFORMATION BOX 27.3

Land Application Methods

% SOLIDS	NATURE OF BIOSOLIDS	METHOD OF APPLICATION
2	Liquid	Sprinkler System (Fig. 27.2)
8	Liquid	Spray Application or Injection (Fig. 27.3)
>20	Cake	Spreaders or Slingers (Figs. 27.4, 27.5, 27.6, 27.7)

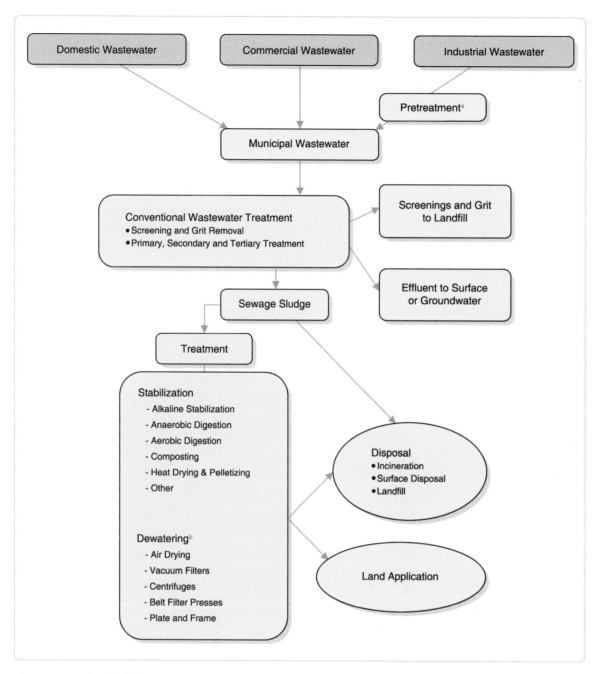

Figure 27.1 Simplified scheme of biosolids production. (From NRC, 2002).

[a]Required by federal and state agencies.

[b]Prior to dewatering, sewage sludge is conditioned and thickened by adding chemicals (*e.g.*, ferric chloride, lime, or polymers).

Figure 27.2 Land application of liquid biosolids via a sprinkler system. Photo courtesy J.P. Brooks.

Figure 27.5 Land application of cake biosolids via a slinger in Solano County California. Photo courtesy J.P. Brooks.

Figure 27.3 Land application of liquid biosolids via a spray applicator. Photo courtesy J.P. Brooks.

Figure 27.6 Loading of biosolids prior to land application. Photo courtesy J.P. Brooks.

Figure 27.4 Land application of cake biosolids via a manure spreader. Photo courtesy J.P. Brooks.

Figure 27.7 Land application of cake biosolids in Eastern Washington. Photo courtesy J.P. Brooks.

Figure 27.8 These photos illustrate a land application of liquid biosolids on an agricultural field in Tucson, Arizona. a) field immediately after land application. b) the biosolids are incorporated via a tractor. From *Environmental Microbiology* © 2000, Academic Press, San Diego, CA.

27.5 BENEFITS OF LAND APPLICATION OF BIOSOLIDS

27.5.1 Biosolids as a Source of Plant Nutrients

Biosolids contain all the elements essential for the growth of higher plants. Nitrogen and phosphorous in particular are abundant, making biosolids attractive as a fertilizer source. Nitrogen and phosphorus are typically present at concentrations of 1–6% on a dry weight basis. Because plants need more nitrogen for growth than phosphorous, when biosolids are applied at a rate to supply sufficient nitrogen, this means that excess phosphorus is applied, which over time can accumulate in the soil.

In addition to nitrogen and phosphorous, biosolids provide other nutrients such as Ca, Fe, Mg, K, Na, and Zn, in amounts adequate for crop needs. A typical analysis of a Class B biosolid is shown in Table 27.1. Finally, note that some liquid biosolids (<8% solids) provide an additional input of water, which can be important in arid regions where crop irrigation is necessary.

27.5.2 Biosolid Impact on Soil Physical and Chemical Properties

Soil organic matter enhances soil structure, through the formation of secondary aggregates (see Chapter 2). This results in increased soil pores, which facilitate air and water

TABLE 27.1 Analysis of anaerobically digested sludge from Tucson, Arizona.

ELEMENT	CONCENTRATION (DRY WEIGHT BASIS)
Metals	—— (mg/kg^{-1}) ——
Copper	520
Nickel	13
Lead	59
Chromium	29
Cadmium	3.5
Zinc	1900
Silver	4.7
Arsenic	ND[a]
Mercury	0.51
Molybdenum	12
Selenium	ND[a]
Other Elements	—— (g 100 g^{-1}) ——
Phosphorus	3.3
Calcium	3.6
Magnesium	0.45
Sodium	0.4
Organic carbon	16.6
Nitrogen	
Total Kjeldahl N	3.4
inorganic N	0.16
Total solids	2.8

[a]ND = Below detection limits.

Source: I.L. Pepper.

movement through the soil. Continuous cropping of soils leads to degradation of both soil organic matter and hence soil structure. Applications of biosolids to soils increases the soil organic content, improving soil structure. Land application improves soil physical properties such as water infiltration and retention, and reduces the soil's susceptibility to erosion. Chemically, land application of biosolids can be beneficial by increasing soil cation exchange capacity (CEC).

27.5.3 Reduced Pollution

In most instances, broad-scale land application of municipal wastes at appropriate loading rates results in far less pollution than if the material had been concentrated by disposal at a single site. When loading rates are controlled, the soil has a chance to transform many waste components into plant-available nutrients. Thus, plants are able to take up these nutrients and complete their natural cycle. These include carbon, nitrogen, sulfur, and phosphorus. Other important sludge components that are recycled include micronutrients such as zinc, iron, and copper. However, when found in high concentrations, these metals can be toxic to plants. Additionally, the soil environment helps stabilize other potential pollutants found in biosolids, such as lead, cadmium, zinc, and arsenic, by trapping them into their solid phases. When this happens, the pollutants do not leach into groundwater and are much less likely to be taken up by plants. However, it is important to note that the metal contaminant itself is still retained in the soil and remains a potential source of pollution.

27.6 HAZARDS OF LAND APPLICATION OF BIOSOLIDS

27.6.1 Site Restrictions

The federal regulations for managing a biosolid land application site include several restrictions.

Some states have much more stringent site criteria than those of the Part 503 rule. Table 27.2 compares Wisconsin criteria with those of the Part 503 rule.

In addition to site restrictions, EPA imposed limitations regarding minimum time durations between applications of Class B biosolids and the harvesting of certain crops, the grouping of animals, and public access to the site.

These limitations are summarized in Table 27.3. If the limitations are followed, EPA concluded that the level of protection from pathogens in Class B biosolids is equal to the level of protection provided by the unregulated use of Class A biosolids.

27.6.2 Nitrates and Phosphates

Biosolids almost always contain large amounts of nitrogen (N) as nitrate (NO_3^-) and ammonium (NH_4^+). In addition, they contain organic forms of nitrogen that are readily transformed to NO_3^- via ammonification and subsequent nitrification. Nitrate ions are very soluble in water and ionic in nature (negatively charged). Therefore excess nitrates leach easily through soils and can reach aquifers (see Chapter 17). Nitrates are of public concern due to the potential for methemoglobinemia. This disease, also known as blue baby syndrome, can occur in young infants due to the drinking of water high in nitrates (>10 ppm). Because of this potential hazard, biosolid land application rates should be matched with crop N requirements. In addition, for irrigated agriculture, irrigation rates should be managed to reduce excess surface runoff or excess sub-surface leaching of NO_3^- (see also Chapter 10).

Biosolids also contain large amounts of phosphate (P) as HPO_4^{2-} or $H_2PO_4^-$. Since plants require more N than P, when biosolid loading rates are based on potential plant N uptake, excess P can accumulate in soil. During many years of continuous land application, soil P concentrations become excessive and can lead to eutrophication of estuaries via surface runoff. Interestingly, in the future, biosolid land application rates may have to be matched to crop P requirements and thus additional N will need to be applied to crops as fertilizer.

27.6.3 Metals and Organics

All sludges have small amounts of essential trace metals needed for plant growth, but they also contain variable amounts of potentially toxic heavy metals. Moreover, even

TABLE 27.2 Wisconsin requirements for biosolids applied to the land in bulk.

SITE CRITERIA	SURFACE	INCORPORATION	INJECTION	PART 503 REQUIREMENTS
Depth to bedrock	3 ft	3 ft	3 ft	
Depth to high groundwater	3 ft	3 ft	3 ft	
Allowable slopes	0–6%	0–12%	0–12%	
Distance to wells				
–Community water supply or school	1000 ft	1000 ft	1000 ft	
–Other[a]	250 ft	250 ft[a]	250 ft[a]	
Minimum distance to residence, business or recreation area	500 ft	200 ft	200 ft	
Minimum distance to residence or business with permission	250 ft	100 ft	100 ft	
Distance to rural schools and health care facilities	1000 ft	1000 ft	500 ft	
Distance to property line	50 ft	25 ft	25 ft	
Minimum distance to streams, lakes, ponds, wetlands, or channelized waterways connected to a stream, lake, or wetland				33 ft
–Slope 0 to <6%	200 ft	150 ft	100 ft	
–Slope 6 to <12%	Not allowed	200 ft	150 ft	
Minimum distance to grass waterways, or dry run with a 50-ft range grass strip				
–Slope 0 to <6%	100 ft	50 ft	25 ft	
–Slope 6 to <12%	Not allowed	100 ft	50 ft	
Soil permeability range (in/h)	0.2–6.0	0–6.0	0–6.0	

[a]Separation distances to nonpotable wells used for irrigation or monitoring may be reduced to 50 ft if the biosolids are incorporated or injected and the department does not determine that a greater distance to the wells is required to protect the groundwater.

Source: Adapted from NRC, 2002.

essential trace elements can be present in such high concentrations that they induce toxicities to plants or microorganisms. Metals of particular concern include Zn, Cu, Cd, Ni, Pb, Hg, Mo, and As. The amount of metal contaminants in a particular sludge depends on the amount of industrial inputs into the municipal sewage system. It is therefore illegal to discharge excessive amounts of metal into municipal wastewater lines, and such wastes must often be treated on site prior to disposal. The soil environment also influences the toxicity of the metals associated with any particular sludge. Sludge-amended soils with high pH have lower plant-available metal concentrations than do sludge-amended low-pH soils. This is because the water solubility of most metals increases as pH decreases. Thus, one management strategy to reduce metal mobility and toxicity towards plants is to lime soils to a neutral or alkaline pH. The organic matter content of soils also affects metal availability. In general, soils with high organic matter content (>5%) exhibit relatively low metal uptake by plants, as metals are sorbed and complexed by the polymer-like organic carbon structure of organic matter. However, when low-molecular-weight organic molecules (usually present in the early stages of plant tissue decay) form complexes with metals, their mobility and plant availability can be dramatically increased in the soil environment. Once metals are introduced into soil, their bioavailability and mobility can be manipulated by changing the valence state of the metal or by altering soil factors that influence their solubil-

TABLE 27.3 Minimum duration between application and harvest/grazing/access for Class B biosolids applied to the land.

CRITERIA	SURFACE	INCORPORATION	INJECTION
Food crops whose harvested part may touch the soil/biosolids mixture (beans, melons, squash, etc.)	14 mo	14 mo	14 mo
Food crops whose harvested parts grow in the soil (potatoes, carrots, etc.)	20/38 mo[a]	38 mo	38 mo
Food, feed, and fiber crops (field corn, hay, sweet corn, etc.)	30 d	30 d	30 d
Grazing of animals	30 d	30 d	30 d
Public access restriction			
High potential[b]	1 yr	1 yr	1 yr
Low potential	30 d	30 d	30 d

[a]The 20-month duration between application and harvesting applies when the biosolids that are surface applied stays on the surface for 4 months or longer prior to incorporation into the soil. The 38-month duration is in effect when the biosolids remain on the surface for less than 4 months prior to incorporation.

[b]This includes application to turf farms that place turf on land with a high potential for public exposure.

Source: Adapted from U.S. EPA, 1993.

TABLE 27.4 EPA pollutants selected for regulation.

INORGANIC CHEMICALS	ORGANIC CHEMICALS
Arsenic	Aldrin and dieldrin
Cadmium	Benzol[*a*]pyrene
Chromium	Chlordane
Copper	DDT, DDD, DDE
Lead	Heptachlor
Mercury	Hexachlorobenzene
Molybdenum	Hexachlorobutadiene
Nickel	Lindane
Selenium	*N*-Nitrosodimethylamine
Zinc	Polychlorinated biphenyls
	Toxaphene
	Trichlorethylene

DDT = 1,1,1-trichloro-2,2-bis(*p*-chlorophenyl)ethane;

DDE = 1,1-dichloro-2,2-bis(*p*-chlorophenyl)ethylene;

DDD = 1,1-dichloro-2,2-bis(*p*-chlorophenyl)ethane.

Source: U.S. EPA 1992.

ities (see Chapter 7). While short-term effects of metal additions to soil can be beneficial, the long-term fate of these metals is more difficult to predict. This is because metal pollutants do not biodegrade and therefore continue to accumulate in the soil environment.

Biosolids can also contain organic compounds that can adversely affect public health. These include trace amounts of pesticides, polyaromatic hydrocarbons (PAHs), plasticizers, volatile organics, and solvents. Many of these compounds are degraded during wastewater treatment. However the more refractory and insoluble compounds such as chlorinated hydrocarbons do not degrade during treatment, particularly if they are sorbed to biosolids. As in the case of metals, it is the bioavailability and mobility of these compounds that determines their ultimate fate and also the potential public health hazard. Biodegradation and mobility are affected by soil type, in particular, soil texture, organic-matter content, pH, and soil moisture content. The complexity of each

site and the interactions of the above factors determine whether or not these organic compounds degrade or accumulate in soil, and whether or not they have the potential to contaminate aquifers (see Chapters 6, 7, 8, and 17).

Chemicals in Table 27.4 were subjected to a formal risk assessment (see Chapter 14) utilizing several exposure pathways. The risk assessment conducted to support the Part 503 rule was then utilized to develop risk-based standards. Table 27.5 illustrates these standards for metals (see also Chapter 14).

Overall, concern over the potential public health hazard with regard to metals has decreased in the United States, for two major reasons. First, a large amount of research has been conducted on crop uptake of metals, including both food and nonfood crops, as well as the fate of metals in soil. Contrary to the dire predictions of some scientists, the "time bomb theory" with respect to metals has not come to pass. Essentially, this theory held that since organic matter in soils is known to complex and accumulate metals, a "flood" of metals would be released when the organic material eventually degraded. In fact, an "aging" effect has been observed for metals, in which bioavailability is observed to decrease with time (Alexander, 2000). The second factor that has reduced fears with respect to metals is enhanced pretreatment technologies that reduce metal inputs into sewage and hence biosolids. Thus biosolid metal contents have decreased dramatically from the 1980s to the present. Despite this, some scientists still have concerns about the effects of metals on the ecosystem, due to the fact that metals do not degrade. This concern is particularly true in Europe.

27.6.4 Emerging Chemicals of Concern

Dioxin-like chemicals

Dioxin and dioxin-like chemicals have been the target of EPA investigation and are considered a group for risk assessment. These compounds are chlorinated hydrocarbons

TABLE 27.5 Pollutant concentration limits and loading rates for land application in the United States, dry weight basis.

CONTAMINANT	CEILING CONCENTRATION LIMIT (mg/kg)	CUMULATIVE POLLUTANT LOADING RATE LIMIT (kg/ha)	POLLUTANT CONCENTRATION LIMIT (mg/kg)	ANNUAL POLLUTANT LOADING RATE (kg/ha-yr)
Arsenic	75	41	41	2.0
Cadmium	85	39	39	1.9
Copper	4,300	1,500	1,500	75
Lead	840	300	300	15
Mercury	57	17	17	0.85
Molybdenum[a]	75	—	—	—
Nickel	420	420	420	21
Selenium	100	100	100	5.0
Zinc	7,500	2,800	2,800	140

[a]Standards for molybdenum were dropped from the original regulation. Currently, only a ceiling concentration limit is available for molybdenum, and a decision about establishing new pollutant limits for this metal has not been made by the U.S. EPA.

Source: U.S. EPA, 1993.

within the family known as polychlorinated biphenyls (PCBs) (see Chapter 10). EPA recently conducted a survey of dioxin concentrations in biosolids, and the results have shown that dioxin is not frequently present in biosolids at levels that would cause concern.

Pharmaceuticals

These compounds are routinely found in biosolids, since they are present in a variety of personal care and skin products, and hence enter wastewater from homes. Some of these compounds are known to be endocrine disruptors. Due to the importance of pharmaceuticals, this topic is covered in detail in Chapter 31.

Flame retardants

Polybrominated diphenyl ethers (PBDEs) are flame retardants and hence enter the food chain through use in fire extinguisher and flame-retardant materials. The use of these compounds is banned in many European countries, and California will become the first U.S. state to ban some PBDEs. The fate and transport of PBDEs is currently unknown.

27.6.5 Pathogens

EPA established two categories of biosolid: Class A biosolids, which have no detectable concentrations of pathogens, and Class B biosolids, which have detectable concentrations of pathogens. In terms of land application, a combination of treatment and site restrictions are intended to result in a reduction of pathogenic and indicator microorganisms to undetectable concentrations prior to potential public contact. Site restrictions for Class B biosolids are shown in Table 27.3.

Principal pathogens of concern in biosolids are identified in Chapter 26. Class B biosolids routinely contain human pathogens. The pathogens found in a particular source of biosolids reflect the incidence of pathogenic disease in the community from which the biosolids are derived.

Also of concern are emerging pathogens such as the SARS virus (severe acute respiratory syndrome). However, regardless of the pathogen of concern, the major routes of potential human exposure to pathogens in biosolids remains the same, specifically via air, soil, and water. Exposure can also occur via vectors, such as flies, and to prevent this, "vector-attraction" reduction requirements are enforced (NRC, 2002). These involve specific biosolid treatment and rapid incorporation of land applied biosolids (< 6 hours).

27.6.5.1 Exposure via air

Human exposure to pathogens via air results from the formation of aerosolized biological particles that are referred to as bioaerosols. Until recently, little was known of the risk of infection from bioaerosols generated during land application of biosolids, and this topic was utilized by environmental activists to challenge the efficacy of land application. However recent national studies across the U.S. have demonstrated that the risk is far lower than previously thought (Brooks et al., 2005a; 2005b).

Characteristics of Biological Aerosols

The term biological aerosol is used to describe biological particles which have been aerosolized (see also Chapter 9). These particles may contain microorganisms (bacteria, fungi, and viruses) or biological remnants such as endotoxin and cell wall constituents such as peptidoglycan. Bioaerosol sizes range typically from 0.5 to 30 μm in diameter and are typically surrounded by a thin layer of water. In other instances, the biological particles can be associated with particulate matter such as soil or biosolids, depending on the place of origin. Bioaerosol particles in the lower spectrum of sizes (0.5 to 5 μm) are typically of most concern, since these particles are more readily inhaled or swallowed.

Bioaerosols generated from the land application of biosolids may be associated with soil or vegetation, depending on the type of land application. For example, if a front-end loader (see Figure 27.7) is used to load a biosolids spreader, it is possible that soil will be in contact with the biosolids and therefore be associated with any aerosol generated by it. In this situation the soil particle or vegetation is known as a "raft" for the biological particles contained with the aerosol. However, for soil particles to be aerosolized, the particles need to be fairly dry, and low soil moisture contents are known to promote microbial inactivation.

27.6.5.2 Exposure via groundwater

In principle, pathogens originally present in biosolids applied to land can contaminate surface or groundwater. However, as discussed in Section 27.6.5.3, most soils limit the movement of microbes to groundwater. Normally, significant migration will only occur in coarse textured soils or karst topography, with a shallow depth to groundwater. Viruses have the greater possibility to migrate through soil; however, they have been found to tightly bind to biosolids, and little leaching appears to occur. (Chetochine et al., 2006) No direct cause and effect has been identified in surface or groundwater near land where biosolids has been applied.

27.6.5.3 Exposure via soil

Soil

Pathogen survival in and transport through soil are considered together in this section. Human pathogens that are routinely found in domestic sewage sludge include viruses, bacteria, protozoan parasites, and helminths. Of those pathogens, viruses are the smallest and least complex, generally have a short survival period in soil, and have the

CASE STUDY: 27.1

POTENTIAL RISK OF INFECTION FROM BIOAEROSOLS GENERATED DURING LAND APPLICATION OF BIOSOLIDS.

In 2002, the National Research Council (NRC) issued a report titled "Biosolids Applied to Land: Advancing Standards and Practices." One of the potential hazards identified in this report was the risk of infection from bioaerosols generated during land application, and it was noted that very little research had been done in this area. In that same year, the National Science Foundation Water Quality Center (WQC) at the University of Arizona initiated a landmark research study on this topic.

The goals of the WQC study were to evaluate the "community" and "occupational" risk of infection from bioaerosols. The scope of the study was larger than any other previously attempted and evaluated the incidence of bioaerosols across the U.S. with differing climatic conditions (Figure 27.9), and various methods of biosolid application (see Figures 27.2, 27.3, and 27.4). Overall, more than 1,000 aerosol samples were collected.

In the study, aerosol samples were collected at finite distances from land application sites, and at 1-meter height intervals to a height of 5 meters to characterize the potential plume of aerosols generated (Figure 27.10).

To evaluate community and occupational risk, the study was undertaken in two parts. First, the emission rate of pathogens generated during loading and land application of biosolids was evaluated. This exposure to workers

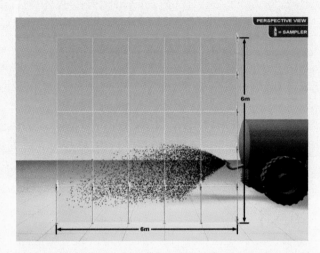

Figure 27.10 Schematic of the plume of aerosols generated during spraying of liquid biosolids, based on data generated during the University of Arizona WQC Study. Drawing courtesy B.D. Tanner.

(occupational risk) was assumed to have no inactivation of pathogens, since the distance to workers was essentially zero. For community risk, fate and transport of pathogens is a factor, since residents live off-site, allowing for natural

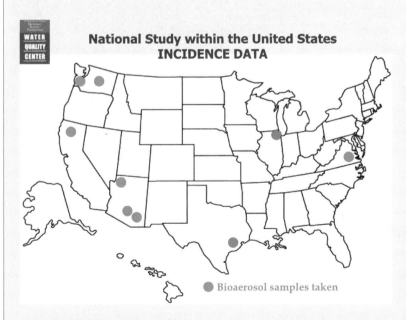

Figure 27.9 Bioaerosol sample sites.

CASE STUDY: (CONTINUED)

attenuation of pathogens due to environmental factors such as desiccation and ultraviolet light. In the following sections, we illustrate the risk assessments and the various assumptions made for these studies.

COMMUNITY RISK OF INFECTION FROM BIOAEROSOLS GENERATED DURING LAND APPLICATION OF BIOSOLIDS

Hazard Identification: Virus Coxsackie Virus A 21

This virus was chosen because it is the only enteric virus likely to be present in biosolids for which dose response via inhalation is known.

Exposure

This was calculated for a distance of 30.5 m ($\cong$100 ft) from a land application site, since this is the minimum distance between residents and a site. Actual exposure to virus was derived from bioaerosol concentrations of virus predicted by an empirically derived transport model. Overall exposure is a function of the virus concentration in the air (#/m³), the duration of exposure, and the human breathing rate. Exposure was based on 1 hour or 8 hours duration, assuming a breathing rate of 0.83 m³ per hour. Annual exposure assumed 6 days of land application.

Dose

$$N = \text{exposure dose in terms of the number of organisms inhaled}$$

Here $N = x \times 0.83 \times t$
where:

$$x = \text{number of organisms m}^{-3}$$
$$0.83 = \text{average breathing rate (m}^3\,\text{h}^{-1})$$
$$t = \text{exposure duration (h)}$$

Dose Response

This was based on the one-hit exponential model

$$P_i = 1 - \exp(-rN)$$

where:

$$r = \text{the constant describing the organisms ability to infect and overcome the host}$$
$$N = \text{number of pathogens inhaled per day}$$

Here, $r = 1/39.4$ or 0.0253
The Annual Risk of Infection is

$$P_{(annual)} = 1 - (1 - P_i)^d$$

where:

$$P_i = \text{one time probability of infection, and}$$
$$d = \text{number of days exposed per year}$$

In order to generate conservative values of risk that were based on worse case scenarios, risks were calculated for a variety of hypothetical sources of biosolids that contained variable concentrations of virus. Table 27.6 shows the annual community risks of infection from exposure to bioaerosols during loading operations of

TABLE 27.6 Annual community risk of infection from Coxsackie virus A21

ANNUAL RISK					
Number of Virus					
10 virus/g biosolid		1 virus/g biosolid		0.1 virus/g biosolid	
Exposure Time					
1h	*8h*	*1h*	*8h*	*1h*	*8h*
4.7×10^{-5}	3.8×10^{-4}	4.7×10^{-6}	$.8 \times 10^{-5}$	$.7 \times 10^{-7}$	3.8×10^{-6}

Assumptions:
1. Distance to land application site $= 30.5$m
2. Loading operations—highest potential for bioaerosol generation

(Brooks et al., 2005).

CASE STUDY: *(CONTINUED)*

biosolids—loading operations were shown in the study to generate the highest concentrations of bioaerosols.

OCCUPATIONAL RISK OF INFECTION FROM BIOAEROSOLS GENERATED DURING LAND APPLICATION OF BIOSOLIDS

Similar risk assessments for occupational risk were calculated (Tanner, 2004).

Assumptions:

1. 8 hours per day, 251 days per year exposure
2. Cubic meters of air breathed per day: 10.0 (light activity)
3. 100 *Salmonella* and 1.0 virus/g in biosolids
4. 1,000,000 coliforms and 10,000 coliphage/g in biosolids
5. 10% of *Salmonella* inhaled are ingested (no dose-response information available for inhalation of *Salmonella*)
6. Percentage of time that the tractor cab will be downwind of the aerosol source: 4 hrs
7. Percentage of aerosols able to enter the tractor cab through the air filter: 4 hrs (ASAE standard 525 specifies 98% reduction)

8. Worker is in the tractor cab loading biosolids for 8 hrs

Annual Risk of Infection from Coxsackievirus A21 = 2×10^{-2}

Annual Risk of Infection from Non-typhi *Salmonella* = 1.3×10^{-4}

SUMMARY

Note that risks for pathogenic bacteria as well as virus were calculated. Based on this study, the following conclusions were drawn:

1. Community risks from land application of biosolids are negligible.
2. Occupational risks during land application are higher than community risks, due to higher exposure, but risks are still low.
3. Risk of disease is ≅one-tenth of the risk of infection.
4. Biosolid workers should utilize sanitary safety precautions, such as hand washing, and should consider wearing face masks.
5. Workers should minimize the amount of time spent downwind of biosolid sources.

greatest potential for transport in soil. Survival of viruses has been shown to be temperature-dependent and decreases as temperature increases. Soil type affects virus survival, with longer survival occurring on clay loam biosolids-amended soils than on sandy loam biosolids-amended soils. Rapid loss of soil moisture also limits virus survival.

Like virus survival, bacteria survival in soil is affected by temperature, pH, and moisture. Soil nutrient availability also plays a role in bacteria survival. Lower temperatures usually increase survival, as do a neutral soil pH and soil at field capacity. Of the pathogenic bacteria, *Salmonella* and *Escherichia coli* can survive for a long time in biosolids-amended soil—up to 16 months for *Salmonella*. In contrast, *Shigella* has a shorter survival time than either *Salmonella* or *E. coli*. Studies on indicator organisms have shown that total and fecal coliforms as well as fecal streptococci can all survive for weeks to several months, depending on soil moisture and temperature conditions (Pepper et al., 1993).

Regrowth is also important when evaluating the survival of pathogenic and indicator bacteria in soil and biosolids compost. *Salmonella, E. coli,* and fecal coliforms are all capable of regrowth. Following land application of biosolids, regrowth of actual pathogens is negligible (Zaleski et al., 2005a). However, regrowth of pathogens can occur in Class A biosolids if they are stored prior to land application and exposed to reinoculation via bird excrement (Zaleski et al., 2005b). Regrowth has also occurred during composting processes. Regrowth of fecal coliforms is more common than pathogens and has been documented even following land application of Class B biosolids (Pepper, et al., 1993).

The protozoan parasites often associated with biosolids include *Giardia* and *Cryptosporidium* spp. However, little research has been conducted on the survival of these parasites in biosolids-amended soil. Helminths are perhaps the most persistent of enteric pathogens. *Ascaris* eggs can survive several years in soils.

The transport of microorganisms through soils or the vadose zone is affected by a complex array of abiotic and biotic factors, including adhesion processes, filtration effects, physiological state of the cells, soil characteristics, water flow rates, predation, and intrinsic mobility of the cells, as well as the presence of biosolids. For viruses, the potential for transport is large, although viruses can adsorb to soil colloidal particles and to the biosolids themselves, thus limiting transport. Virus sorption is controlled by the soil pH. Most viruses are negatively charged (isoelectric point 3–6), so that at a neutral soil pH, soil sorption is reduced, whereas at more acidic soil pH values, the viruses are positively charged, increasing sorption.

The larger size of bacteria means that soil acts as a filter, limiting bacterial transport. Soil would also limit the transport of the even larger protozoa and helminths. However, microorganisms may be transported through soil cracks and macrochannels via preferential flow.

Pathogen survival and transport in soil should be evaluated from a public-health perspective. Pathogens are routinely present in Class B biosolids and are capable of surviving for days, weeks, or even months, depending on the organism and environment. Therefore, site restrictions with durations based on subsequent land use are necessary following land application. For many soils, contamination of aquifers due to vertical migration of pathogens from land-applied biosolids is unlikely because of the sorption of viruses and the soil filtration potential for larger pathogens. However, in coarse textured, sandy soil or high-permeability karst topography, groundwater contamination events are possible.

27.7 SOURCES OF ANIMAL WASTES

Animal wastes predominantly include manures from cows, pigs, and chickens. Animal wastes are pollutants of increasing concern both to the public and to regulatory bodies because they have the potential to contaminate both surface and groundwater. Consequently, animal wastes must now be included as part of the agricultural production cycle and figured into the cost of operating a farm or livestock facility. Animal agricultural wastes can be divided into two production types: range and pasture production, and confined or concentrated animal production.

In range and pasture systems, the concentration of wastes is generally much more diffuse or dispersed than it is when large numbers of animals are confined to relatively small areas. Range and pasture systems have two principal effects on surface water quality: (1) increased turbidity through the movement of soil particles into streams, rivers, and lakes; and (2) increased fecal coliform counts in areas of heavy animal use. Although we know that grazing systems may adversely affect some measures of water quality, we will focus here on the highly concentrated animal production units and the methods of preventing and controlling pollution from these concentrated units.

In the past, animals were concentrated only intermittently. The period of confinement was a transitory phase followed by a return to pasture, after such management activities as milking or shearing. However, animal production is occurring in increasingly controlled environments owing to the success of efforts to raise productivity and diminish climatic, feeding, and mortality variables. Larger numbers of animals are being raised in **concentrated animal feeding operations** or **CAFOs**—principally, feedlots, dairies, swine operations, poultry houses, and intensive aquaculture. The number of CAFO operations more than doubled from 1982 to 1997, increasing from 5,000 to 11,200. Almost every county in the United States has a CAFO with more than 10,000 animals (Figure 27.11). This shift in production methods has changed the age-old method of reincorporation of animal wastes as manure on the farm where it was produced. Specialization has largely divorced animal production from the production of crops: a concentrated animal facility may be located far from crop production, and the same

Figure 27.11 Mounding of manure in dairy corrals within a CAFO facilitates the flow of contaminated runoff water into the appropriate ponds for storage or treatment. In the corral pictured above, a depression in the corral has enabled water to collect in the corral, not only contributing to foot problems in the cows walking in the mud, but also enhancing pollutant leaching and creating an odor problem. From *Pollution Science* © 1996, Academic Press, San Diego, CA.

family (or the same corporation) may not pursue the two types of production.

The production of large numbers of animals on a small land base has resulted in the stockpiling of wastes at specific locations, the construction of large waste-storage ponds, and, oftentimes, waste applications to land in excess of agronomic crop needs.

27.8 NONPOINT VERSUS POINT SOURCE POLLUTION

In discussing sources of animal wastes, it is important to understand the terms point and nonpoint contaminant sources. The term "nonpoint pollution" is misleading and is often misused. In animal agricultural systems, true **nonpoint sources** are those in which potential contaminants are not concentrated during production and do not pass through a single or small number of conduits for disposal. These nonpoint sources include corrals, feedlots, and extensive and intensive pasture systems.

Point sources are those facilities that concentrate pollutants or contaminants to a significant degree and pass these contaminants through a pipe, ditch, or canal for disposal.

According to EPA regulations, however, some concentrated animal feeding operations may be designated as point sources requiring an individual National Pollution Discharge Elimination System (NPDES) permit. In this case, a CAFO is defined as a lot or facility without vegetation, where animals are confined for 45 or more days per year. The number of animals needed to meet this definition as a CAFO depends on several factors; however, the key determinant is whether or not the facility discharges into navigable waters, as determined by the method of discharge. The method of discharge

is judged by the **25-year, 24-hour storm event,** which is the required event that a facility must be designed to meet. This design criterion is a storm of 24 hours' duration whose probability of occurrence is once every 25 years. If a facility does not discharge from storms smaller than the 25-year, 24-hour storm event, it may be regulated as if it were a nonpoint source and thus not require a permit. If a facility does discharge from storms smaller than the 25-year, 24-hour storm event, it may be treated as a point source. This "double standard" recognizes the fact that what appear to be agricultural point sources can safely be treated as nonpoint sources because the pollutants being discharged are either not concentrated enough to warrant specific controls at the pipe or are not laden with hazardous materials.

The 25-year, 24-hour storm criterion does not mean that it is certain that once every 25 years a storm of a certain size and duration will occur. Another, possibly clearer, way of looking at this standard is that within 100 years it is likely that four storms that meet the 25-year, 24-hour storm size and length will occur. However, these four storms may occur back-to-back or be spread out over a period longer than 25 years. Therefore, given the hydrologic record, 25 years is the *average* return period for a storm of this size and duration.

27.9 BENEFITS OF LAND APPLICATION OF ANIMAL WASTES

Land-applied animal wastes provide nutrients for plant growth, similar to land application of biosolids. Land application is one of the best methods for disposal of excess manure or manure slurry. While supplying some of the essential nutrients for crop production, manure is also a beneficial soil amendment for increasing tilth, aeration, and water-holding capacity. Land application does have some disadvantages, however, including the distribution of weed seeds, the possible accumulation of salts, and the application of excess nutrients that may leach and again become pollutants.

Before large quantities of manure and/or slurries can be applied, it is necessary to know: (1) the relative nutrient content; (2) the rate of mineralization (decay rate); (3) the salt load; (4) the concentration and type of toxic elements; (5) the proper time and method of application; and (6) the amount and type of weed seed.

The nutrient content of manure can vary greatly, depending upon the age of the livestock in question, the rate of feed consumption, the feed ration, and the manure handling and storage practices prior to land application. Two facts are especially important when deciding upon the application rate for manure: (1) the moisture content of the manure directly affects the nutrient percentage; and (2) manure loses N as NH_3 when it dries.

Application rates for a given N availability are shown in Table 27.7. The rate of mineralization of manure is the percentage available for plant uptake each year. This rate is

TABLE 27.7 Average manure decay and application rates to maintain 220 kg N ha⁻¹.

Manure Type	DECAY RATE EACH YEAR (% yr⁻¹)				AMOUNT APPLIED EACH YEAR (metric tons HA⁻¹)			
	1st	2nd	3rd	4th	1st	2nd	3rd	4th
Fresh bovine, 3.5% N	75	15	10	7.5	8.5	8.1	7.9	7.9
Corral (dry), 2.5% N	40	25	6.0	3.0	22.5	15.7	14.8	13.9
Corral (dry), 1.5% N	35	15	10	7.5	42.7	30.8	27.4	25.6
Corral (dry), 1.0% N	20	10	7.5	5.0	112.3	67.4	55.0	53.2

From *Pollution Science* © 1996, Academic Press, San Diego, CA.

given as a decay series at three levels of nitrogen at the time of application. The amount applied each year is the quantity of manure that should be added to maintain an annual mineralization rate of 220 kg of nitrogen per hectare per year for a given N% and decay rate. Local climatic conditions such as extreme heat or resistance may cause the decay rate to change considerably, thereby affecting the application rate as well. Note that the amount of manure that needs to be applied decreases each year; this decrease is due to residual N from the previous application. Note also that the decay rate is maximal during the first year of application, but decreases dramatically in subsequent years.

The most effective means of maintaining the nutrient value of manure depends on the form in which it is applied to cropland. If it is being applied as a slurry, it should be injected below the surface; but if the manure is applied as a solid to the surface, the treated soil should be tilled immediately. This reduces nuisance odors, nutrient losses by volatilization, and the potential for groundwater pollution caused by runoff.

27.10 HAZARDS OF LAND APPLICATION OF ANIMAL WASTES

Concentrated animal agriculture produces the following specific pollutants in the wastes resulting from animal metabolic activity:

1. Nitrates and phosphates with the hazards similar to those from biosolids.

2. Pesticides that are used in CAFOs can be a concern for both surface and groundwater contamination. For example, coumaphos, an organophosphate, is commonly used in dips for animals crossing the southern borders of the United States. While relatively immobile, coumaphos is known to be persistent and has been found at some depths when it has not been disposed of properly. Other pesticides, such as toxaphene, are no longer used in animal dips.

3. Biochemical oxygen demand (or BOD) is a measure of the quantity of oxygen (often measured in kilograms of O_2) needed to satisfy biochemical oxidation of organic matter in a waste sample in five days at 20°C. While not a specific pollutant, the BOD is one of the important general indicators of a substance's potential for environmental pollution of surface waters. This measure of pollution capability is important because animal wastes typically contain a high level of BODs, on the order of 0.45–3.6 kg per 454 kg of excreted material.

4. Specific microbial pathogens. Animal manures can be a source of specific microbial pathogens that are not found in biosolids. These include *E. coli* O157:H7, which is an enterohemorrhagic *E. coli* excreted by cows. Cows are also a source of *Cryptosporidium parvum*. Chickens are frequently inhabited with *Campylobacter* spp. Thus, although they do not contain human viruses, animal manures do contain pathogens capable of human infection.

27.11 PUBLIC PERCEPTIONS OF LAND APPLICATION

Public perceptions of land application of organic wastes are highly variable. First, land application of animal manures is and has been considered a "natural country way of life," with few criticisms from the general public. This is despite the fact that cow manures, for example, contain pathogens at least as hazardous as those found in biosolids and despite the fact that such wastes were in part responsible for the Milwaukee *Cryptosporidium* outbreak in 1993. It is also noteworthy that foods raised on land amended with manures are technically classified as "organic," which is not the case when land is amended with biosolids.

Public perception of land application of biosolids in part is a function of land availability and population density. For example, in the desert Southwest, agricultural areas are often located far from urban centers, so that there are fewer surrounding residents who come into contact with the process. In contrast, in the Northeast, the potential impact of land application is much greater, because of less land and larger populations.

QUESTIONS AND PROBLEMS

1. What is the major component of biosolids?
2. Define the differences between sewage, sewage sludge, Class A biosolids, and Class B biosolids.
3. Calculate the annual community risk of infection given the following data.

   - aerosolized virus concentration = 7.16
   - Coxsackie virus per m^3 of air
   - duration of exposure = 8 hours

   - infectivity constant r = 1/39.4
   - breathing rate = 0.83 m^3 per hour

4. How much fresh bovine annual manure needs to be applied each year for 4 successive years to provide 300 kg N/hectare? Give the answer as metric tons/hectare of manure.
5. What are the major hazard differences between land application of Class B biosolids and cow manure?
6. What decreases the potential contamination of groundwater pathogens within biosolids?

REFERENCES AND ADDITIONAL READING

Alexander M. (2000) Aging, bioavailability, and overestimation of risk from environmental pollutants. *Environ. Sci. Technol.* **34**, 4259–4265.

Brooks J.P., Tanner B.D., Gerba C.P., Haas C.N. and Pepper I.L. (2005) Estimation of bioaerosol risk of infection to residents adjacent to a land applied biosolids site using an empirically derived transport model. *J. Appl. Microbiol.* 98:397–405

Gollehon N., Caswell M., Ribaudo M., Kellogg R., Lander C. and Letson D. (2001) Confined Feeding Production and Manure Nutrients. United States Department of Agriculture. Economic Research Service, Washington, D.C.

Hinesly T.D., Jones R.L. and Ziegler E.L. (1972) Effects on corn by application of heated anaerobically digested sludge. *Compost Science.* **13**, 26–30.

NRC (National Research Council). (1996) *Use of Reclaimed Water and Sludge in Food Crop Production.* National Academy Press, Washington, D.C.

NRC (National Research Council). (2002) *Biosolids Applied to Land: Advancing Standards and Practices.* National Academy Press, Washington, D.C.

Pepper I.L., Josephson K.L., Bailey R.L. and Gerba C.P. (1993) Survival of indicator organisms in Sonoran Desert soil amended with sewage sludge. *J. Environ. Sci. Health.* **A28**, 1287–1302.

Rusin P., Maxwell S., Brooks J., Gerba C. and Pepper I. (2003) Evidence for the absence of *Staphylococcus aureus* in land applied biosolids. *Environ. Sci. Technol.* **37**, 4027–4030.

Schekel K.G., Scheinost A.C., Ford R.G. and Sparks D.L. (2000) Stability of layered Ni hydroxide surface precipitates: A dissolution kinetics study. *Geochim. Cosmochim. Acta.* **64**, 2727–2735.

Tanner B.D., J.P. Brooks, C.N. Haas, C.P. Gerba, and I.L. Pepper. (2005) Bioaerosol emission rate and plume characteristics during land application of liquid class B biosolids. *Environ Sci & Technol.* **39**, 1584–1590.

U.S. EPA. (1992) Technical Support Document for Land Application of Sewage Sludge. Vol. I EPA 822/R–93–001A.

U.S. EPA. (1993) The Standards for the Use or Disposal of Sewage Sludge. Final 40 CFR Part 503 Rules. EPA 822/Z–93/001 U.S. Environmental Protection Agency.

U.S. EPA. (1995) A Guide to the Biosolids Risk Assessments for the EPA Part 503 Rule. EPA 832–B–93–005. U.S. Environmental Protection Agency. Available on line: http://www.epa.gov/OW-OWM.html/mtb/biosolids/503rule/index.htm.

U.S. EPA. (1996) Technical Support document for the Round Two Sewage Sludge Pollutants. EPA–822–R–96–003.

U.S. EPA. (1999) Environmental Regulations and Technology: Control of Pathogens and Vector Attraction in Sewage Sludge. EPA/625/R–92/013 Available on line: http://www.epa.gov/ttbn-rmrl/625/R-92/013.htm

U.S. EPA. (2000) A Guide to Field Storage of Biosolids and the Organic By-Products Used in Agriculture and for Soil Resource Management. EPA/832–B–00–007, Washington, DC.

Zaleski K.J., Josephson K.L., Gerba C.P. and Pepper I.L. (2005b) Survival, growth, and regrowth of enteric indicator and pathogenic bacteria in biosolids, compost, soil, and land applied biosolids. *J. Residuals Sci. Technol.* **2**, 49–63.

Zaleski K.J., Josephson K.L., Gerba C.P. and Pepper I.L. (2005b) Potential regrowth and recolonization of *Salmonella* and indicators in biosolids and biosolid amended soil. *Appl. Environ. Microbiol.* 71:3701–3708.

CHAPTER 28

DRINKING WATER TREATMENT AND WATER SECURITY

C. P. Gerba, K. A. Reynolds, and I. L. Pepper

Ozone disinfection chambers. *Photo courtesy of Water-Tech Net. www.water-technology.net*

Rivers, streams, lakes, and aquifers are all potential sources of potable water. In the United States, all water obtained from surface sources must be filtered and disinfected to protect against the threat of microbiological contaminants. Such treatment of surface waters also improves values such as taste, color, and odor. In addition, groundwater under the direct influence of surface waters such as nearby rivers must be treated as if it were a surface water supply. In many cases however, groundwater needs either no treatment or only disinfection before use as drinking water. This is because soil itself acts as a filter to remove pathogenic microorganisms, decreasing their chances of contaminating drinking water supplies.

At first, slow sand filtration was the only means employed for purifying public water supplies. Then, when Louis Pasteur and Robert Koch developed the Germ Theory of Disease in the 1870s, things began to change quickly. In 1881, Koch demonstrated in the laboratory that chlorine could kill bacteria. Following an outbreak of typhoid fever in London, continuous chlorination of a public water supply was used for the first time in 1905 (Montgomery, 1985). The regular use of disinfection in the United States began in Chicago in 1908. The application of modern water treatment processes had a major impact on water-transmitted diseases such as typhoid in the United States (see also Chapter 11). The following sections describe conventional water treat-ment that is practiced in the public sector (*e.g.*, municipal water supplies).

28.1 WATER TREATMENT PROCESSES

Modern water treatment processes provide barriers, or lines of defense, between the consumer and waterborne disease. These barriers, when implemented as a succession of treatment processes, are known collectively as a **treatment process train** (Figure 28.1). The simplest treatment process train, known as **chlorination**, consists of a single treatment process, disinfection by chlorination (Figure 28.1a). The treatment process train known as **filtration**, entails chlorination followed by filtration through sand or coal, which removes particulate matter from the water and reduces turbidity (Figure 28.1b). At the next level of treatment, **in-line filtration**, a coagulant is added prior to filtration (Figure 28.1c). Coagulation alters the physical and chemical state of dissolved and suspended solids and facilitates their removal by filtration. More conservative water treatment plants add a flocculation (stirring) step before filtration, which enhances the agglomeration of particles and further improves the removal efficiency in a treatment process train called **direct filtration** (Figure. 28.1d). In direct filtration, disinfection is

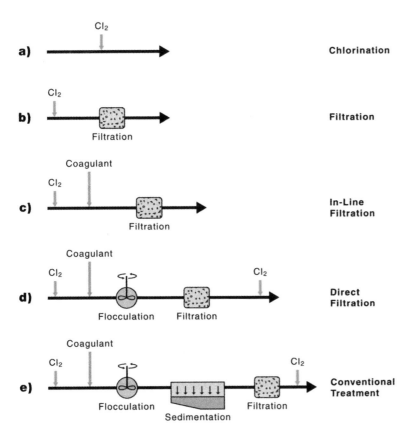

Figure 28.1 Typical water treatment process trains. From *Environmental Microbiology* © 2000, Academic Press, San Diego, CA.

enhanced by adding chlorine (or an alternative disinfectant, such as chlorine dioxide or ozone) at both the beginning and end of the process train. The most common treatment process train for surface water supplies, known as conventional treatment, consists of disinfection, coagulation, flocculation, sedimentation, filtration, and disinfection (Figure 28.1e).

As already mentioned, **coagulation** involves the addition of chemicals to facilitate the removal of dissolved and suspended solids by sedimentation and filtration. The most common primary coagulants are hydrolyzing metal salts, most notably alum [$Al_2(SO_4)_3$ $14H_2O$], ferric sulfate [$Fe_2(SO_4)_3$], and ferric chloride ($FeCl_3$). Additional chemicals that may be added to enhance coagulation are charged organic molecules called polyelectrolytes; these include high-molecular-weight polyacrylamides, dimethyldiallyammonium chloride, polyamines, and starch. These chemicals ensure the aggregation of the suspended solids during the next treatment step, flocculation. Sometimes polyelectrolytes (usually polyacrylamides) are added after flocculation and sedimentation as an aid in the filtration step.

Coagulation can also remove dissolved organic and inorganic compounds. Hydrolyzing metal salts added to the water may react with the organic matter to form a precipitate, or they may form aluminum hydroxide or ferric hydroxide floc particles on which the organic molecules adsorb. The organic substances are then removed by sedimentation and filtration, or filtration alone if direct filtration or in-line filtration is used. Adsorption and precipitation also remove inorganic substances.

Flocculation is a purely physical process in which the treated water is gently stirred to increase interparticle collisions, thus promoting the formation of large particles. After adequate flocculation, most of the aggregates settle out during the 1 to 2 hours of sedimentation. Microorganisms are entrapped or adsorbed to the suspended particles and removed during sedimentation (Figure 28.2).

Sedimentation is another purely physical process, involving the gravitational settling of suspended particles

TABLE 28.1 Factors effecting the removal of pathogens by slow sand filters.

	FACTOR	REMOVAL
↑	Temperature	↑
↓	Sand grain size	↑
↑	Filter depth	↑
↓	Flow rate	↑
↑	Well-developed biofilm layer	↑

From *Environmental Microbiology* © 2000, Academic Press, San Diego, CA.

that are denser than water. The resulting effluent is then subjected to **rapid filtration** to separate out solids that are still suspended in the water. Rapid filters typically consist of 50–75 cm of sand and/or anthracite having a diameter between 0.5 and 1.0 mm (Figure 28.2). Particles are removed as water is filtered through the medium at rates of 4–24 L/min/10 dm². Filters need to be backwashed on a regular basis to remove the buildup of suspended matter. This backwash water may also contain significant concentrations of pathogens removed by the filtration process. Rapid filtration is commonly used in the United States. Another method, **slow sand filtration**, is also used. Employed primarily in the United Kingdom and Europe, this method operates at low filtration rates without the use of coagulation. Slow sand filters contain a layer of sand (60–120 cm deep) supported by a gravel layer (30–50 cm deep). The hydraulic loading rate is between 0.04 and 0.4 m/hr. The buildup of a biologically active layer, called a **schmutzdecke**, occurs during the operation of a slow sand filter. This eventually leads to head loss across the filter, requiring removing or scraping the top layer of sand. Factors that influence pathogen removal by filtration are shown in Table 28.1.

Taken together, coagulation, flocculation, sedimentation, and filtration effectively remove many contaminants as shown in Table 28.2. Equally important, they reduce turbidity, yielding water of good clarity and hence enhanced disinfection efficiency. If not removed by such methods, particles may harbor microorganisms and make final disinfection more difficult. Filtration is an especially important barrier in the removal of the protozoan parasites *Giardia lamblia* and *Cryptosporidium*. The cysts and oocysts of these organisms are very resistant to inactivation by disinfectants, so disinfection alone cannot be relied on to prevent waterborne illness. Because of their larger size, *Giardia* and *Cryptosporidium* are removed effectively by filtration. Conversely, because of their smaller size, viruses and bacteria can pass through the filtration process. Removal of viruses by filtration and coagulation depends on their attachment to particles (adsorption), which is dependent on the surface charge of the virus. This is related to the isoelectric point (the pH at which the virus has no charge) and is both strain and type dependent. The variations in surface properties have been used to explain why

Figure 28.2 Drinking water treatment plant showing sand filter beds in the foreground and tanks containing alum flocculant in the background. Photo courtesy C.P. Gerba.

TABLE 28.2 Coagulation, sedimentation, filtration: typical removal efficiencies and effluent quality.

ORGANISMS (% removal)	COAGULATION AND SEDIMENTATION (% removal)	RAPID FILTRATION (% removal)	SLOW SAND FILTRATION (% removal)
Total coliforms	74–97	50–98	>99.999
Fecal coliforms	76–83	50–98	>99.999
Enteric viruses	88–95	10–99	>99.999
Giardia	58–99	97–99.9	>99
Cryptosporidium	90	99–99	99

Adapted from *Pollution Science* © 1996, Academic Press, San Diego, CA.

different types of viruses are removed with different efficiencies by coagulation and filtration. Thus, disinfection remains the ultimate barrier to these microorganisms.

28.2 DISINFECTION

Disinfection plays a critical role in the removal of pathogenic microorganisms from drinking water. The proper application of disinfectants is critical to kill pathogenic organisms.

Generally, disinfection is accomplished through the addition of an oxidant. Chlorine is by far the most common disinfectant used to treat drinking water, but other oxidants, such as chloramines, chlorine dioxide, and even ozone, are also used (Figure 28.3).

Inactivation of microorganisms is a gradual process that involves a series of physicochemical and biochemical steps. In an effort to predict the outcome of disinfection,

various models have been developed on the basis of experimental data. The principal disinfection theory used today is still the **Chick-Watson Model**, which expresses the rate of inactivation of microorganisms by a first-order chemical reaction.

$$N_t/N_o = e^{-kt} \qquad \text{(Eq. 28.1)}$$

or

$$\ln N_t/N_o = -kt \qquad \text{(Eq. 28.2)}$$

where:

N_o = number of microorganisms at time 0,

N_t = number of microorganisms at time t

k = decay constant (1/time)

t = time

The logarithm of the survival rate (N_t/N_o) plots as a straight line versus time (Figure 28.4). Unfortunately, laboratory and field data often deviate from first-order kinetics. Shoulder

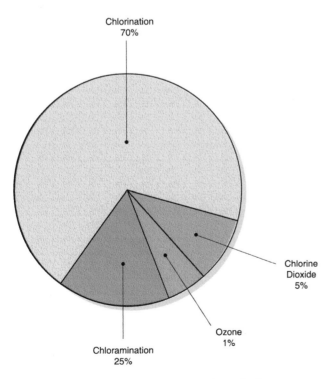

Figure 28.3 Methods of disinfection in the United States, 1993. From Craun, 1993.

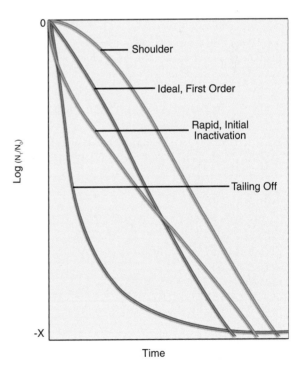

Figure 28.4 Types of inactivation curves observed for microorganisms. From *Environmental Microbiology* © 2000, Academic Press, San Diego, CA.

TABLE 28.3 $C \cdot t$ values for chlorine inactivation of microorganisms in water (99% inactivation)[a]

ORGANISM	°C	pH	$C \cdot t$
Bacteria			
E. coli	5	6.0	0.04
E. coli	23	10.0	0.6
L. pneumophila	20	7.7	1.1
Mycobacterium avium	23	7.0	51–204
Viruses			
Polio 1	5	6.0	1.7
Coxsackie BS	5	8.0	9.5
Protozoa			
G. lamblia cysts	5	6.0	54–87
G. lamblia cysts	5	7.0	83–133
G. lamblia cysts	5	8.0	119–192
Cryptosporidium oocysts	25	7.0	>7200

[a]In buffered distilled water. From Sobsey (1989); Rose et al. (1997); Gerba et al. 2004.

curves may result from clumps of organisms or multiple hits of critical sites before inactivation. Curves of this type are common in disinfection of coliform bacteria by chloramines (Montgomery, 1985). The tailing-off curve, often seen with many disinfectants, may be explained by the survival of a resistant subpopulation as a result of protection by interfering substances (suspended matter in water), clumping, or genetically conferred resistance.

In water applications, disinfectant effectiveness can be expressed as $C \cdot t$, where:

C = disinfectant concentration

t = time required to inactivate a certain percentage of the population under specific conditions (pH and temperature)

Typically, a level of 99% inactivation is used when comparing $C \cdot t$ values. In general, the lower the $C \cdot t$ value, the more effective the disinfectant. The $C \cdot t$ method allows a general comparison of the effectiveness of various disinfectants on different microbial agents (Tables 28.3 through 28.6). It is used by the drinking water industry to determine how much disinfectant must be applied during treatment to achieve a given reduction in pathogenic microorganisms. $C \cdot t$ values for chlorine for a variety of pathogenic microorganisms are shown in Table 28.3. The order of resistance to chlorine and most other disinfectants used to treat water is protozoan cysts > viruses > vegetative bacteria. To obtain the proper $C \cdot t$, contact chambers (Figure 28.5) are used to retain the water in channels before entering the drinking water distribution system or sewage discharge.

28.3 FACTORS AFFECTING DISINFECTANTS

Numerous factors determine the effectiveness and/or rate of kill of a given microorganism. Temperature has a major effect, because it controls the rate of chemical reactions. Thus, as temperature increases, the rate of kill with a chemical disinfectant increases. The pH can affect the ionization of the disinfectant and the viability of the organism. Most waterborne organisms are adversely affected by pH levels below 3 and above 10. In the case of halogens such as chlorine, pH controls the amount of HOCL (hypochlorous acid) and −OCl (hypochlorite) in solution. HOCl is more effective than −OCl in the disinfection of microorganisms. With chlorine, the $C \cdot t$ increases with pH. Attachment of organisms to surfaces or particulate matter in water such as clays and organic detritus aids in the resistance of microorganisms to disinfection. Particulate matter may interfere by either acting chemically to react with the disinfectant, thus neutralizing the action of the disinfectant, or physically shielding the organism from the disinfectant (Stewart and Olson, 1996).

TABLE 28.4 $C \cdot t$ values for chlorine dioxide in water.

MICROBE	ClO$_2$ RESIDUAL (mg/l)	TEMPERATURE (°C)	pH	% reduction	$C \cdot t$
Bacteria					
E. coli	0.3–0.8	5	7.0	99	0.48
Mycobacterium	0.1–0.2	23	7.0	99.9	0.1–11
Viruses					
Polio 1	0.4–14.3	5	7.0	99	0.2–6.7
Rotavirus SA11					
Dispersed	0.5–1.0	5	6.0	99	0.2–0.3
Cell-associated	0.45–1.0	5	6.0	99	1.0–2.1
Hepatitis A	0.14–0.23	5	6.0	99	1.7
Protozoa					
G. muris	0.1–5.55	5	7.0	99	10.7
G. muris	0.26–1.2	25	5.0	99	5.8
G. muris	0.15–0.81	25	9.0	99	2.7
Cryptosporidium	4.03	10	7.0	95.8	7.8

Adapted from Sobsey (1989); Rose et al. (1997). From *Environmental Microbiology* © 2000, Academic Press, San Diego, CA.

TABLE 28.5 $C \bullet t$ values for chloramines in water (99% inactivation)[a]

MICROBE	°C	pH	$C \bullet t$
Bacteria			
E. coli	5	9.0	113
Viruses			
Polio 1	5	9.0	1420
Hepatitis A	5	8.0	592
Coliphage MS2	5	8.0	2100
Rotavirus SA11	5	8.0	4034
Dispersed	5	8.0	6124
Cell-associated			
Protozoa			
G. muris	3	6.5–7.5	430–580
G. muris	5	7.0	1400
Cryptosporidium	25	7.0	>7200

[a]In buffered distilled water.
Adapted from Sobsey (1989); Rose et al. (1997); Gerba et al. (2004). From *Environmental Microbiology* © 2000, Academic Press, San Diego, CA.

Repeated exposure of bacteria and viruses to chlorine appears to result in selection for greater resistance (Bates et al., 1977; Haas and Morrison, 1981). However, the enhanced resistance has not been great enough to overcome concentrations of chlorine applied in practice.

28.4 HALOGENS

28.4.1 Chlorine

Chlorine and its compounds are the most commonly used disinfectants for treating drinking and wastewater (Figure 28.6). Chlorine is a strong oxidizing agent that, when added

TABLE 28.6 $C \bullet t$ values for ozone inactivation of microorganisms in water (99% inactivation).

ORGANISM	°C	pH	$C \bullet t$
Bacteria			
E. coli	23	7.2	0.006–0.02
Mycobacterium	1	7.0	0.10–0.12
Viruses			
Polio 1	5	7.2	0.2
Polio 2	25	7.2	0.72
Rota SA11	4	6.0–8.0	0.019–0.064
Protozoa			
G. lamblia	5	7.0	0.53
Cryptosporidium	7	–	7.0
Cryptosporidium	22	–	3.5

From Sobsey (1989); Rose et al. (1997). From *Environmental Microbiology* © 2000, Academic Press, San Diego, CA.

Figure 28.5 Chlorine contact chambers at a sewage treatment plant. Source: www.mosfet.isu.edu/classes/sato/pocatello WWTP 2003.

as a gas to water, forms a mixture of hypochlorous acid (HOCl) and hydrochloric acids.

$$Cl_2 + H_2O \leftrightarrow HOCl + HCl \qquad \text{(Eq. 28.3)}$$

In dilute solutions, little Cl_2 exists in solution. The disinfectant's action is associated with the HOCl formed. Hypochlorous acid dissociates as follows:

$$HOCl \leftrightarrow H^+ + OCl^- \qquad \text{(Eq. 28.4)}$$

The preparation of hypochlorous acid and OCl^- (hypochorite ion) depends on the pH of the water (Figure 28.7). The amount of HOCl is greater at neutral and lower pH levels, resulting in greater disinfection ability of chlorine at these pH levels. Chlorine as HOCl or OCl^- is defined as **free available chlorine**. HOCl combines with ammonia and organic compounds to form what is referred

Figure 28.6 Chlorine storage tanks at a wastewater treatment plant. Source: www.mosfet.isu.edu/classes/sato/pocatelloWWTP2003.

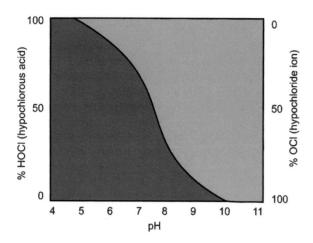

Figure 28.7 Distribution of HOCl and OCl⁻ in water as a function of pH. (From Bitton, 1999). From *Environmental Microbiology* © 2000, Academic Press, San Diego, CA.

to as **combined chlorine**. The reactions of chlorine with ammonia and nitrogen-containing organic substances are of great importance in water disinfection. These reactions result in the formation of monochloramine, dichloramine, trichloramine, etc.

$$NH_3 + HOCl \rightarrow NH_2Cl + H_2O \quad \text{(Eq. 28.5)}$$
$$\text{monochloramine}$$

$$NH_2Cl + HOCl \rightarrow NHCl_2 + H_2O \quad \text{(Eq. 28.6)}$$
$$\text{dichloramine}$$

$$NHCl_2 + HOCl \rightarrow NCl_3 + H_2O \quad \text{(Eq. 28.7)}$$
$$\text{trichloramine}$$

Such products retain some of the disinfecting power of hypochlorous acid, but are much less effective at a given concentration than chlorine.

Free chlorine is quite efficient in inactivating pathogenic microorganisms. In drinking water treatment, 1 mg/1 or less for about 30 minutes is generally sufficient to significantly reduce bacterial numbers. The presence of interfering substances in wastewater reduces the disinfection efficacy of chlorine, and relatively high concentrations of chlorine (20–40 mg/l) are required (Bitton, 1999). Enteric viruses and protozoan parasites are more resistant to chlorine than bacteria and can be found in secondary wastewater effluents after normal disinfection practices. *Cryptosporidium* is extremely resistant to chlorine. A chlorine concentration of 80 mg/l is necessary to cause 90% inactivation following a 90-minute contact time (Korich et al., 1990). Chloramines are much less efficient than free chlorine (about 50 times less efficient) in inactivation of viruses.

Bacterial inactivation by chlorine is primarily caused by impairment of physiological functions associated with the bacterial cell membrane. Chlorine may inactivate viruses by interaction with either the viral capsid proteins or the nucleic acid (Thurman and Gerba, 1988).

28.4.2 Chloramines

Inorganic chloramines are produced by combining chlorine and ammonia (NH_4) for drinking water disinfection. The species of chloramines formed (see Equations 28.5 through 28.7) depend on a number of factors, including the ratio of chlorine to ammonia-nitrogen, chlorine dose, temperature, and pH. Up to a chlorine-to-ammonia mass ratio of 5, the predominant product formed is monochloramine, which demonstrates greater disinfection capability than other forms, *i.e.*, dichloramine and trichloramine. Chloramines are used to disinfect drinking water by some utilities in the United States, but because they are slow acting, they have mainly been used as secondary disinfectants when a residual in the distribution system is desired. For example, when ozone is used to treat drinking water, no residual disinfectant remains. Because bacterial growth may occur after ozonation of tap water, chloramines are added to prevent regrowth in the distribution system. In addition, chloramines have been found to be more effective in controlling biofilm microorganisms on the surfaces of pipes in drinking water distribution systems because they interact poorly with capsular bacterial polysaccharides (LeChevallier et al., 1990).

Because of the occurrence of ammonia in sewage effluents, most of the chlorine added is converted to chloramines. This demand on the chlorine must be met before free chlorine is available for disinfection. As chlorine is added, the residual reaches a peak (formation of mostly monochloramine) and then decreases to a minimum called the **breakpoint** (Figure 28.8). At the breakpoint, the chloramine is oxidized to nitrogen gas in a complex series of reactions summarized in Equation 28.8.

$$2NH_3 + 3HOCl \rightarrow N_2 + 3H_2O + 3HCL \quad \text{(Eq. 28.8)}$$

Addition of chlorine beyond the breakpoint ensures the existence of a free available chlorine residual.

28.4.3 Chlorine Dioxide

Chlorine dioxide is an oxidizing agent that is extremely soluble in water (five times more than chlorine) and, unlike chlorine, does not react with ammonia or organic compounds to form trihalomethane, which is potentially

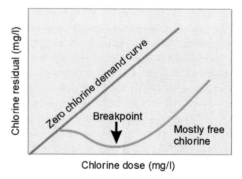

Figure 28.8 Dose-demand curve for chlorine.

carcinogenic. Therefore it has received attention for use as a drinking water disinfectant. Chlorine dioxide must be generated on site because it cannot be stored. It is generated from the reaction of chlorine gas with sodium chlorite:

$$2NaClO_2 + Cl_2 \rightarrow 2\,ClO_2 + 2NaCl \qquad \text{(Eq. 28.9)}$$

Chlorine dioxide does not hydrolyze in water but exists as a dissolved gas.

Studies have demonstrated that chlorine dioxide is as effective as or more effective in inactivating bacteria and viruses in water than chlorine (Table 28.4). As is the case with chlorine, chlorine dioxide inactivates microorganisms by denaturation of the sulfyhydryl groups contained in proteins, inhibition of protein synthesis, denaturation of nucleic acid, and impairment of permeability control (Stewart and Olson, 1996).

28.4.4 Ozone

Ozone (O_3), a powerful oxidizing agent, can be produced by passing an electric discharge through a stream of air or oxygen. Ozone is more expensive than chlorination to apply to drinking water, but it has increased in popularity as a disinfectant because it does not produce trihalomethanes or other chlorinated by products, which are suspected carcinogens. However, aldehydes and bromates may be produced by ozonation and may have adverse health effects. Because ozone does not leave any residual in water, ozone treatment is usually followed by chlorination or addition of chloramines. This is necessary to prevent regrowth of bacteria because ozone breaks down complex organic compounds present in water into simpler ones that serve as substrates for growth in the water distribution system. The effectiveness of ozone as a disinfectant is not influenced by pH and ammonia.

Ozone is a much more powerful oxidant than chlorine (Tables 28.3 and 28.6). Ozone appears to inactivate bacteria by the same mechanisms as chlorine-based disinfection: by disruption of membrane permeability (Stewart and Olson, 1996), impairment of enzyme function and/or protein integrity by oxidation of sulfyhydryl groups, and nucleic acid denaturation. *Cryptosporidium* oocysts can be inactivated by ozone, but a $C \bullet t$ of 1–3 is required. Viral inactivation may proceed by breakup of the capsid proteins into subunits, resulting in release of the RNA, which can subsequently be damaged.

28.4.5 Ultraviolet Light

The use of ultraviolet disinfection of water and wastewater has seen increased popularity because it is not known to produce carcinogenic or toxic byproducts, or taste and odor problems. Also, there is no need to handle or store toxic chemicals. A wavelength of 254 nm is most effective against microorganisms because this is the wavelength absorbed by nucleic acids (Figure 28.9). Unfortunately, it has several disadvantages, including higher costs than halogens, no disinfectant residual, difficulty in determining the UV dose, maintenance and cleaning of UV lamps, and potential photoreactivation of some enteric bacteria (Bitton, 1999) (Figure 28.10). However, advances in UV technology are providing lower cost, more efficient lamps, and more reliable equipment. These advances have aided in the commercial application of UV for water

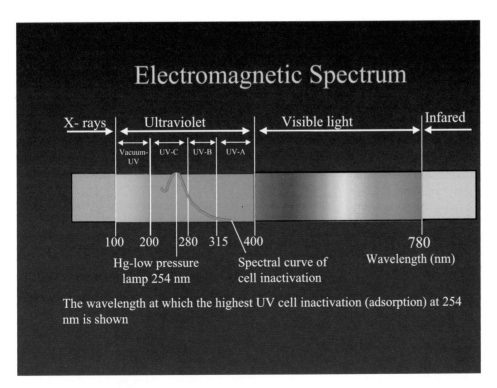

Figure 28.9 The wavelength at which the highest UV cell inactivation (adsorption) at 254 nm is shown.

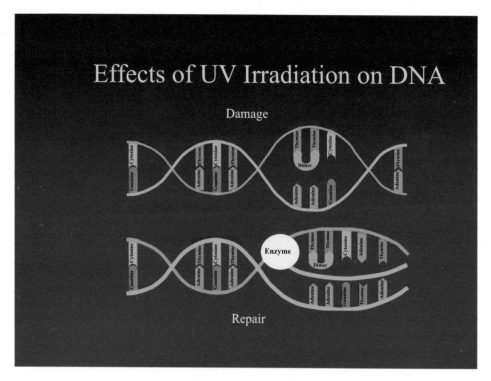

Figure 28.10 UV light damages cells by causing cross-linking of the DNA of bacteria, but some bacteria produce repair enzymes that can remove the cross-linking of the nucleotides in the DNA.

treatment in the pharmaceutical, cosmetic, beverage, and electronic industries in addition to municipal water and wastewater application.

Microbial inactivation is proportional to the UV dose, which is expressed in microwatt-seconds per square centimeter (μW-s/cm^2) or

$$\text{UV dose} = I \bullet t \qquad \text{(Eq. 28.10)}$$

where:

I = μW/cm^2

t = exposure time

In most disinfection studies, it has been observed that the logarithm of the surviving fraction of organisms is nearly linear when it is plotted against the dose, where dose is the product of concentration and time ($C \bullet t$) for chemical disinfectants, or intensity and time ($I \bullet t$) for UV. A further observation is that constant dose yields constant inactivation. This is expressed mathematically in Eq. 28.11.

$$\log \frac{N_s}{N_i} = \text{functional}(It) \qquad \text{(Eq. 28.11)}$$

where:

N_s is the density of surviving organisms (number/cm^3)

N_i is the initial density of organisms before exposure (number/cm^3)

Because of the logarithmic relationship of microbial inactivation versus UV dose, it is common to describe inactivation in terms of log survival, as expressed in Equation 28.12. For example, if one organism in 1000 survived exposure to UV, the result would be a -3 log survival, or a 3 log reduction.

$$\log_{10} \text{survival} = \log_{10} \frac{N_s}{N_i} \qquad \text{(Eq. 28.12)}$$

Determining the UV susceptibility of various indicator and pathogenic waterborne microorganisms is fundamental in quantifying the UV dose required for adequate water disinfection. Factors that may affect UV dose include cell clumping and shadowing, suspended solids, turbidity, and UV absorption. UV susceptibility experiments described in the literature are often based on the exposure of microorganisms under conditions optimized for UV disinfection. Such conditions include filtration of the microorganisms to yield monodispersed, uniform cell suspensions and the use of buffered water with low turbidity and high transmission at a wavelength of 254 nm. Thus, in reality, higher doses are required to achieve the same amount of microbial inactivation in full-scale flow through operating systems.

The effectiveness of UV light is decreased in wastewater effluents by substances that affect UV transmission in water. These include humic substances, phenolic

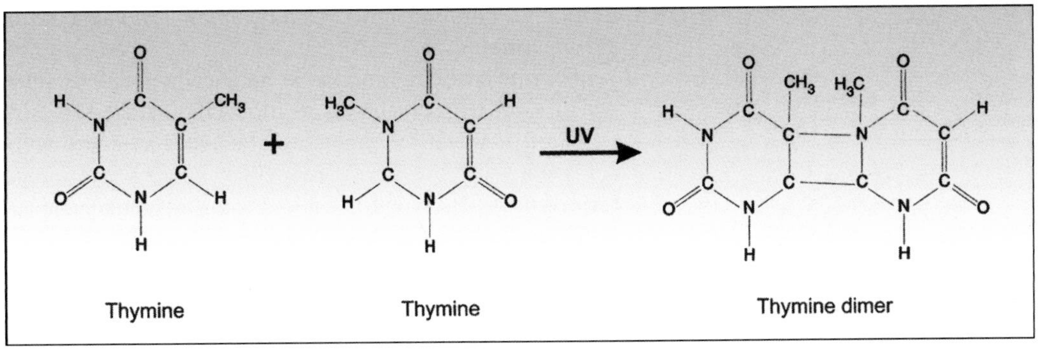

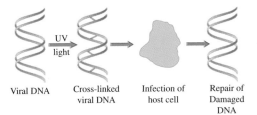

Figure 28.11 Formation of thymine dimers in the DNA. From *Environmental Microbiology* © 2000, Academic Press, San Diego, CA.

compounds, lignin sulfonates, and ferric iron. Suspended matter may protect microorganisms from the action of UV light; thus, filtration of wastewater is usually necessary for effective UV light disinfection.

Ultraviolet radiation damages microbial DNA or RNA at a wavelength of approximately 260 nm. It causes thymine dimerization (Figure 28.11), which blocks nucleic acid replication and effectively inactivates microorganisms. The initial site of UV damage in viruses is the genome, followed by structural damage to the virus protein coat. Viruses with high-molecular- weight double-stranded DNA or RNA are easier to inactivate than those with low-molecular-weight double-stranded genomes. Likewise, viruses with single-stranded nucleic acids of high molecular weight are easier to inactivate than those with single-stranded nucleic acids of low molecular weight. This is presumably because the target density is higher in larger genomes. However, viruses with double-stranded genomes are less susceptible than those with single-stranded genomes because of the ability of the naturally occurring enzymes within the host cell to repair damaged sections of the double-stranded genome, using the nondamaged strand as a template (Roessler and Severin, 1996) (Figure 28.12).

A minimum dose of 16,000 μW s/cm^2 has been recommended for treating drinking water, as this results in a 99.9% reduction in coliforms and is very effective against the pro-

tozoan parasite *Cryptosporidium*. However, this level is not enough to inactivate enteric viruses (Table 28.7). Filtration can be applied before UV light disinfection to improve performance (Figure 28.13).

28.5 DISINFECTION BY-PRODUCTS

All chemical disinfectants produce organic and/or inorganic **disinfection byproducts (DBPs)**, which may be carcinogenic or otherwise deleterious. A summary of the types of disinfectant byproducts is shown in Table 28.8. The most widely recognized chlorination byproducts include chloroform, bromodichloromethane, dibromochloromethane, and bromoform. These compounds are collectively known as the **trihalomethanes (THM)**, and the term **total trihalomethane (TTHM)** refers to their combined concentrations. These compounds are formed by the reaction of chlorine with organic matter—largely humic acids—naturally present in the water.

Haloacetic acids are another group of byproducts produced when chlorine and other disinfectants are used (Figure 28.14). Monochloramine produces lower THM concentra-

TABLE 28.7 UV dose to kill enteric microorganisms.

ORGANISM	ULTRAVIOLET DOSE (μW s/cm^2) REQUIRED FOR 90% REDUCTION
Campylobacter jejuni	1,100
Escherichia coli	1,300–3,000
Legionella pneumophila	920–2,500
Salmonella typhi	2,100–2,500
Shigella dysenteriae	890–2,200
Vibro cholerae	650–3,400
Adenovirus	23,600–50,000
Coxsackievirus	11,900–15,600
Poliovirus	5,000–12,000
Hepatitis A	3,700–7,300
Rotavirus SA11	8,000–9,900
Coliphage MS2	18,600
Cryptosporidium	2,700–6,700

From Roessler and Severin (1996).

Host Cell Repair of Double Stranded DNA Viruses

Viral DNA → UV light → Cross-linked viral DNA → Infection of host cell → Repair of Damaged DNA

Figure 28.12 Viral repair in double-stranded DNA viruses.

Figure 28.13 UV light disinfection of drinking water.
www.mindfully.org

tions than chlorine but produces other DBPs, including cyanogen chloride.

Ozone oxidizes bromide to produce hypohalous acids, which react with precursors to form brominated THMs. A range of other DBPs, including aldehydes and carboxylic acids, may be formed. Of particular concern is bromate, formed by oxidation of bromide. Bromate is mutagenic and is carcinogenic in animals.

Many of these byproducts are classified as possible carcinogens. However, the results of numerous epidemiological studies of populations consuming chlorinated drinking water in the Untied States show that the risks of cancer appear low (Craun, 1993). **Maximum contaminant levels (MCLs)** have been recommended for some of these byproducts, and further regulation is expected in the future. Meanwhile, it is fair to say that the risks of illness and death posed by waterborne microorganisms far outweigh the risk from low levels of potentially toxic chemicals produced during water treatment (Craun, 1993).

The formation of THMs during chlorination can be reduced by removing precursors prior to contact with chlorine—for example, by installing or enhancing coagulation (using high coagulant doses), or reducing the amount of organic matter (pretreatment with activated charcoal). UV irradiation is also an alternative to chemical disinfection, but does not provide any residual disinfection. However, lower doses of chemical disinfectants may be added.

28.6 RESIDENTIAL WATER TREATMENT

Conventional water treatment processes are highly effective at removing contaminants from drinking water sources, and waterborne outbreaks are rare in the United States. However, they

TABLE 28.8 Classes of disinfectant byproducts.

DISINFECTANT	INORGANIC BYPRODUCTS	ORGANIC BYPRODUCTS	
		HALOGENATED	NONHALOGENATED
Chlorine	Chlorate	Trihalomethanes	Aldehydes
		Haloacetates	Carboxylic acids
		Haloacetonitrites	
		Haloaldehydes	
		Halketones	
		Halofuranones	
Chloramine		Cyanogen chloride	
		Others generally	
		thought to be	
		the same as chlorine,	
		but lower in concentration	
Chlorine dioxide	Chlorite	Not well	
	Chlorate	characterized	
Ozone	Bromate	Bromomethanes	Aldehydes
	Hydrogen peroxide	Bromoacetates	Carboxylic acids
		Bromoaldehydes	
		Bromoketones	
		Iodinated analogs	

Source: Craun, 1993.

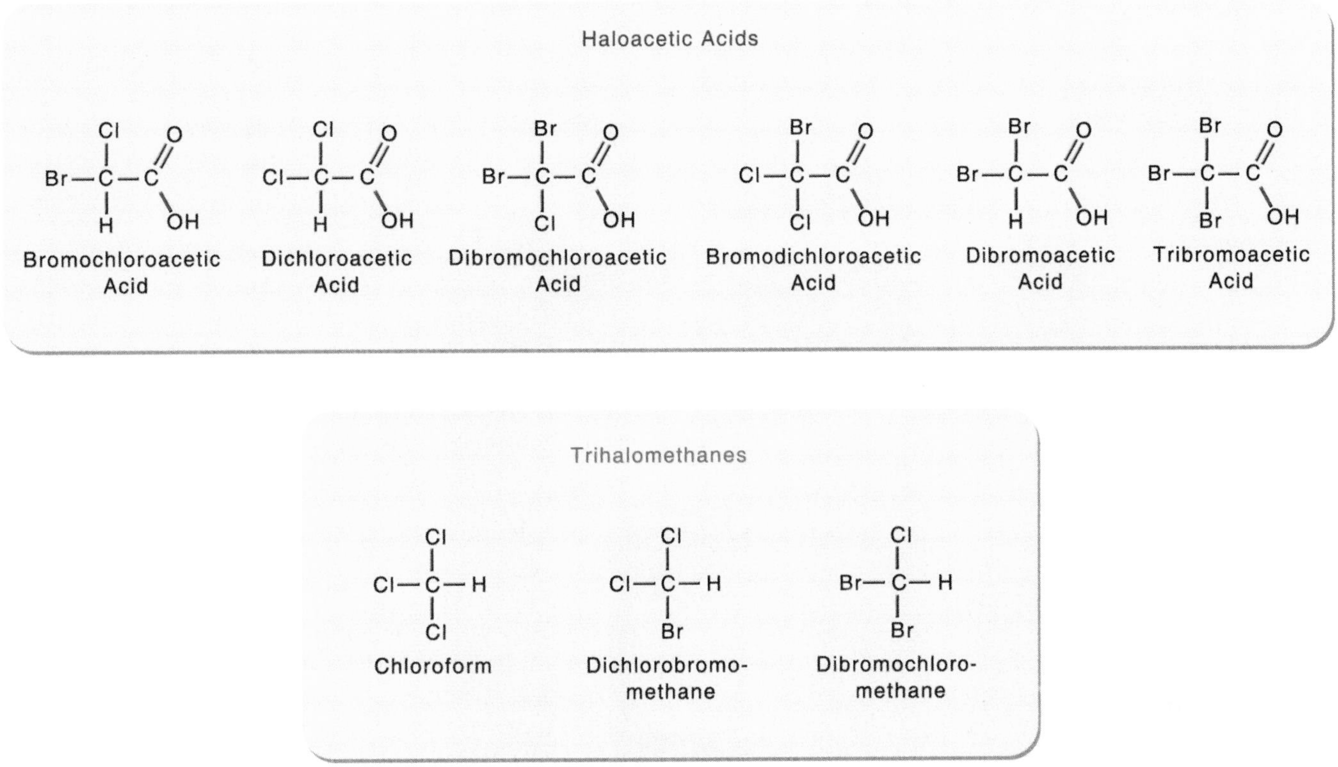

Haloacetic Acids

| Bromochloroacetic Acid | Dichloroacetic Acid | Dibromochloroacetic Acid | Bromodichloroacetic Acid | Dibromoacetic Acid | Tribromoacetic Acid |

Trihalomethanes

Chloroform   Dichlorobromo-methane   Dibromochloro-methane

Figure 28.14 Haloacetic acids and trihalomethanes produced during disinfection with chlorine.

do still occur. Even well-operated, state-of-the-art treatment plants cannot ensure that drinking water delivered at the consumer's tap is entirely free of harmful microbes and chemicals.

Numerous studies reporting the presence of disease-causing microbes and toxins in finished water, designated for human consumption, have led to decreased consumer confidence. This has resulted in recommendations from health agencies that certain populations utilize additional treatment measures for water purification at the point of consumption. **Point-of-use (POU)** residential water purification systems can limit the effects of incidental contamination of drinking water due to source water contamination, treatment plant inadequacies, minor intrusions in the distribution system, or deliberate contamination post-treatment (*i.e.*, a bioterrorism event). Multistage POU water treatment systems are available for the removal of a wide variety of contaminants, such as arsenic, chlorine, microbes, and nitrates, and have the benefit of providing a final barrier for water treatment closest to the point of consumption (Figure 28.15).

28.6.1 Residential Water Contamination

From 1971–2002, 782 outbreaks, 575,457 cases of illness, 2,068 hospitalizations, and 79 deaths were reported from the consumption of drinking water from public and private water systems, involving more than 41 states and three U.S. territories (Blackburn et al., 2004; Calderon, 2004). Protozoa caused the greatest number of outbreaks, followed by bacteria, chemicals, and viruses. No agent was identified in almost

half of the outbreaks; however, the majority of these are suspected to be of microbial origin (Figure 28.16). Scientists recognize that waterborne disease is largely unreported in the general population, and estimates of true illness cases are as high as 7 million, with >1,000 deaths each year in the U.S. (Morris and Levin, 1995).

Although exposure to waterborne contaminants usually results in asymptomatic illness, mild to severe diarrhea or other systemic effects can occur. Health effects are not always immediately realized and can develop slowly over time or manifest later in life. Chronic, long-term illnesses such as ulcers, cancer, myocarditis, arthritis, Guillain-Barré syndrome, diabetes, mental disorders, heart disease, and birth defects have all been linked to contaminants found in drinking water.

28.6.2 Treatment Deficiencies

Drinking water outbreaks occur due to a variety of reasons including deficiency in water treatment, distribution system deficiency, and consumption of untreated or nondisinfected ground water or surface water (Table 28.9). Deficiency in water treatment was the primary cause of outbreaks in community water supplies from 1971–2002 (152/315 or 48%). Predicting drinking water quality is sometimes difficult, because contamination events are not evenly distributed. Contamination can be affected by the initial concentration of toxins in the source water, the age of the distribution system, the quality of the delivered water, or climatic events that can tax treatment plant operations.

Figure 28.15 Example of a plumbed-in, multistage POU water treatment device. Photo courtesy of Kinetico, Inc. and Pall Corp, 2004.

Compliance with current drinking water standards does not ensure safe water at the tap. For example, current national drinking water standards require filtration of all surface water, or groundwater under the direct influence of surface water, to guard against the presence of various microbial pathogens such as *Cryptosporidium*, which are generally resistant to conventional chlorination. Infectious *Cryptosporidium* is, however, frequently detected in finished water samples from U.S. surface water treatment plants (27% prevalence rate in 92 sites) (Aboytes et al., 2004).

Untreated Source Waters. Due to the natural filtration properties of soil, public water utilities with a groundwater source are not currently required to treat their water supply; however, numerous surveys have documented the presence of human pathogens in public and private groundwater sources (Fout et al., 2003; Borchardt et al., 2003, 2004), calling for a re-evaluation of the notion that groundwater is inherently free of pathogens.

Distribution System. Even water that is obtained from adequate sources and properly treated is subject to contamination in the distribution system. From 1971 to 2002, there were 133 documented outbreaks (17%

of all outbreak causes) in the U.S. linked to the distribution system. About half of the waterborne outbreaks caused by distribution system deficiencies were due to cross-connection or backsiphonage. Other causes included contamination of mains due to construction, repair or flushing; inadequate separation of water main and sewer lines; broken or leaking water mains; corrosion; contamination of service lines; contamination of household plumbing; and contamination during storage (Calderon, 2004).

The U.S. EPA requires public systems to maintain a minimum free chlorine residual level of 0.2 mg/L in the distribution system. This chlorine residual is required to minimize bacterial regrowth during distribution, but has proven to be ineffective for eliminating pathogens introduced in the distribution system (Payment et al., 1999; Park et al., 2001). In addition, significant investments in installing, upgrading, or replacing infrastructure for delivering safe drinking water in the U.S. are reportedly needed in 55,000 community water systems to ensure the provision of safe drinking water to their 243,000,000 customers (U.S. EPA, 2001).

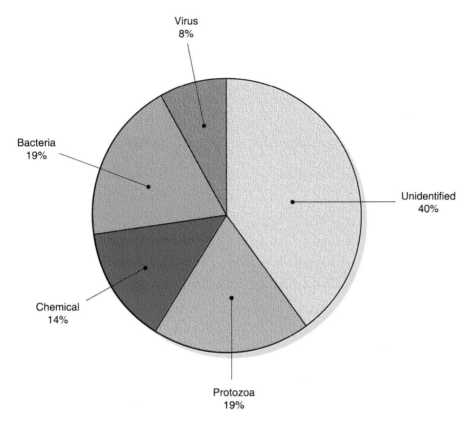

Figure 28.16 Cause of waterborne outbreaks in U.S. 1971–2002. Source: Blackburn et al., 2004; Calderon, 2004.

28.6.3 Point-of-Use Water Treatment Technology

Current treatment technologies are capable of addressing most contaminants of concern in drinking water; however, the site of application is critical. An appropriate barrier at the point-of-use can minimize health risks resultant from treatment failures, untreated source waters, and distribution system contamination, and could be a life-saving choice for susceptible populations.

Choosing the appropriate POU device for individual water quality needs is a difficult task, as water is an ever-changing entity, and delivery of contaminated water can occur randomly and without warning. Many POU treatment devices are designed to improve water aesthetics, such as taste, color, and odor, but do not remove other harmful contaminants, such as *Cryptosporidium* or viruses (Table 28.10).

Everyone is at risk of waterborne disease, but the immunocompromised are generally at increased risk. It is estimated that up to 25 percent of the U.S. population is immunocompromised, including the very young (<5 years), the elderly (>55 years), pregnant women, and persons subject to certain medical interventions (radiation treatment, chemotherapy, transplant therapy). In addition, people with previous illnesses (diabetes, cancer), or prior infections (AIDS patients) are also at higher risk. The U.S. EPA and

TABLE 28.9 Drinking water outbreaks and water supply deficiencies, 1971–2002.

| | OUTBREAKS | | | |
DEFICIENCY	COMMUNITY	NON-COMMUNITY	INDIVIDUAL	TOTAL OUTBREAKS
Deficiency in water treatment	152	133	8	293
Distribution system deficiency	100	25	8	133
Miscellaneous/unknown deficiency	22	33	18	73
Untreated ground water	35	144	60	239
Untreated surface water	6	13	19	38
Total Outbreaks	**315**	**348**	**113**	**776**

From CDC, 2004; Calderon, 2004.

TABLE 28.10 Common POU treatment options.

POU SYSTEM	PRIMARILY REMOVES	ADVANTAGES	LIMITATIONS
Activated carbon	Chlorine, VOCs, pesticides, radon, some metals	Taste and odor improvement, inexpensive	Bacterial regrowth potential
Reverse osmosis	Inorganic chemicals; microbes, nitrates, some organics	Removes wide range of contaminants	Expensive; requires professional maintenance; wastes 2–3x water produced
Distillation	Inorganic chemicals; pathogens, dissolved minerals, trace metals, many organics	Removes wide range of contaminants	Concentrates some organic chemicals, small capacity
Particle filters	Sand, rust	Inexpensive	Not designed for removal of most health-related contaminants
Ion exchange	Softening, iron removal, scale reduction	Improves soap/detergent use	Not designed for removal of most health-related contaminants
Disinfectants (ozone, UV, chlorine, iodine)	Microbes	Simple, inexpensive	Efficacy varies with pathogen and source water quality

Centers for Disease Control and Prevention (CDC) advise severely immunocompromised individuals to purify their water by boiling for one minute, as a safeguard against waterborne exposures to *Cryptosporidium*. As an alternative, the U. S. EPA and CDC recommend POU devices with reverse-osmosis treatment, labeled as absolute one-micrometer filters, or that have been certified by NSF International under standard 53 for "Cyst Removal." A recent publication from the CDC further recommends that homeowners with individual groundwater wells purchase appropriately designed POU devices (Blackburn et al., 2004).

Recent epidemiological studies are conflicted as to the overall health benefit of residential POU water treatment devices for the general population, and thus more research is needed regarding their use (Payment et al., 1991 & 1997; Hellard et al., 2001; Colford et al., 2005). However, a review of the data indicates that some of the documented waterborne disease outbreaks may have been prevented by the additional barrier that an appropriate POU water treatment device offers.

28.7 WATER SECURITY

In the new world politics of the 21st century, terrorism and the potential for terrorist attacks are major issues in the United States. Following the events of September 11, 2001, when the twin towers of the World Trade Center in New York were destroyed, the Department of Homeland Security was initiated with the overall responsibility of protecting the U.S. public from terrorist attacks. One area of concern has been the safety of public water supplies, and public utilities are now mandated by U.S. EPA to incorporate security measures to protect potable water supplies.

28.7.1 Potential Terrorist Attacks

The journey of potable water from its source to the consumer tap is diagrammed in Figure 28.17, and illustrates the potential for terrorist attacks on potable water. Clearly, surface waters—including rivers and dams—could be targets for macro-contamination events. Additionally, infrastructure attacks on dams themselves, such as the Hoover Dam in Nevada, could be attempted. However, this would require massive amounts of chemical or biological contaminants that would clearly be detected during subsequent water treatment by the public utility. Likewise, a direct attack on a dam or the water treatment plant itself would require extensive planning and utilization of terrorist resources, and would have to overcome the intense security placed around such facilities.

Of greater concern to utilities is the potential for a biological or chemical contamination event within the drinking water distribution system. The injection of a contaminant into the distribution system requires less of the contaminant itself and circumvents security measures in place at the water treatment plant. Injection of a contaminant into a utility distribution system is known as a **Water Intrusion Event**. Any homeowner in the community with a bathtub and an 80 lb/inch^2 pump could potentially force a contaminant back into the distribution system.

28.7.2 Real-Time Monitoring

Since no single water-monitoring sensor currently exists that can detect all chemical and biological agents that could be

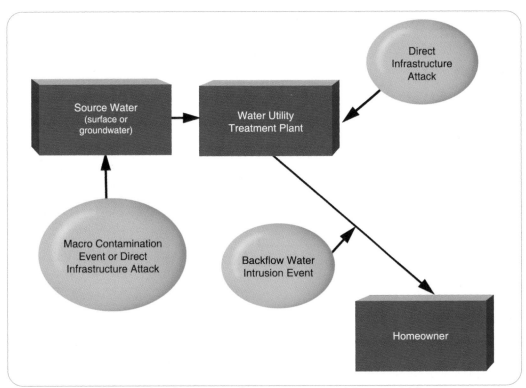

Figure 28.17 Potable water from its source to the consumer tap, and the potential for terrorist attacks.

used to contaminate distribution systems, new approaches are necessary to protect our water supplies.

Currently two fundamental approaches are being considered to monitor water intrusion. One approach involves implementing hydraulic sensors at discrete locations to monitor the water pressure within distribution lines and the amount of water passing through a particular point in the distribution system. These hydraulic sensors coupled to acoustic sensors are capable of detecting leaks in the distribution system and differentiating leaks from backflow events. The second approach involves "real-time" detection of chemical or biological contaminants.

At this time, many cost-effective sensors for real-time monitoring of chemicals exist that allow for a chemical fingerprint of the water to be established. These include measurements such as pH, free chlorine, total organic carbon, and total oxygen. In theory, when a chemical contaminant is injected into the distribution system, the fingerprint changes and the intrusion event is detected real-time. More difficult is the establishment of a biological fingerprint for water, but spectroscopic and molecular (nucleic acid-based) technologies are currently being evaluated. Once the intrusion event is detected, there are many state-of-the-art technologies available to identify specific contaminants, regardless of whether they are chemical or biological. Usually such identification can occur within a window of a few hours.

INFORMATION BOX 28.1

New Real-Time Monitoring Approaches

1. Potentially, water monitoring detection systems need to be placed at key multiple stations throughout the water distribution network.

2. Efficient placement of such stations is necessary to allow discrete sections of the network to be monitored individually, and if need be, shut down.

3. A two-stage detection system is envisioned.

   A. A primary "trigger" mechanism that identifies any addition (chemical or biological) to water within the distribution system. This must be instantaneous or "real-time."

   B. Secondary detection technology that allows for the contaminant to be identified during a period of a few hours, while the discrete segment of the distribution system is shut down.

   C. Depending on the identification of the contaminant, specific water treatment (for chemicals) or disinfection (for biologicals) options are implemented prior to re-opening the distribution network.

CASE STUDY 28.1

THE WATER VILLAGE AT THE UNIVERSITY OF ARIZONA

Figure 28.18 The Water Village at the University of Arizona: Photos courtesy K.L. Josephson.

The University of Arizona is home to the National Science Foundation Water Quality Center (WQC)—the only NSF Center in the U.S. to focus solely on water quality. The mission of the WQC is to maintain and enhance good potable water quality, which is defined as:

"water with acceptable purity, taste, and odor characteristics, which is safe with respect to human health and welfare."

The WQC undertakes many research studies to achieve its mission, and several of them are conducted within a model "Water Village" at The University of Arizona. This Village consists of several dwellings that together are used as an "intermediate field-scale testing facility."

The four houses within the Village are plumbed with unique distribution systems with state-of-the-art access sampling points for water quality monitoring. Studies conducted within the Village include evaluation of point-of-entry (POE) and point-of-use (POU) technology that enhance water safety and water aesthetics. The Village also contains a "Water Intrusion Laboratory" (WIL), where real-time monitoring technologies and strategies are evaluated.

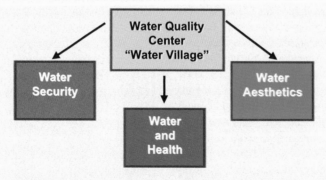

Figure 28.19 Research areas evaluated within the Water Village.

28.8 MONITORING COMMUNITY WATER QUALITY

CASE STUDY 28.2

CITY OF TUCSON WATER EMPACT PROGRAM

Based on a grant received from the **U.S. EPA's Environmental Monitoring for Public Access and Community Tracking (EMPACT) Program**, the City of Tucson implemented a comprehensive water quality monitoring program. The City's EMPACT goals included the following: implementing enhanced monitoring of the utility's potable distribution system, providing the community with near "real-time" water-quality information on Tucson Water's website (www.tucsonaz.gov/water), and creating community partnerships to better inform water consumers about water quality and resource issues. The water quality monitoring and data collection tools provided through EMPACT also enable the utility to track and respond to real-time changes in system water quality. The Tucson Water distribution system consists of one central drinking water distribution system that serves the majority of the customers, and ten isolated drinking water distribution systems. The eleven drinking water distribution systems cover a service area of 375 square miles and serve 680,000 customers in the Tucson metropolitan area. The two types of source water that supply the central distribution system are native groundwater and renewable recharged surface water from the Colorado River. The source water that supplies the ten isolated distribution systems is groundwater.

For the purposes of monitoring, the central distribution system is divided into water quality zones. A water quality zone is defined as an area of the distribution system that has similar water quality and operating system characteristics such as water pressure, source area, and geographical boundaries. Each water quality zone has a set number of dedicated sampling stations and points-of-entry (POEs). The dedicated sampling stations represent the quality of the drinking water in the distribution system before delivery to the customer. POEs are usually individual potable supply wells that represent the water quality of a single well or, in a few cases, combined POE systems that represent the collective blended water quality from a group of wells that directly supply Tucson's drinking water.

In total, there are 262 dedicated sampling stations and approximately 154 active POEs located within the multiple distribution systems. In addition, 22 continuously online water quality monitoring stations (monitoring for chlorine residual, total dissolved solids, pH, and temperature) are located throughout the central distribution system at strategic location such as reservoirs, well sites, and booster stations.

The comprehensive water quality monitoring program encompasses the distribution systems from the water source to the customer's home. Source waters are monitored and sampled according to the Arizona Department of Water Resources and the Arizona Department of Environmental Quality (ADEQ) regulations, while the drinking water is monitoring according to U.S. EPA and ADEQ regulations, and consumer-driven drinking water standards. In addition, nonregulatory monitoring is undertaken to characterize and track changing trends in water quality for both source water and drinking water. Both data sets are utilized to track and monitor changes in water quality compared to baseline water quality operating parameter levels. It is sometimes possible to determine when a contamination event may have occurred by comparing the baseline levels to the contaminating event parameter levels. Based on the water quality measurements collected each month from these sampling location and those at the POEs, the trends in water quality conditions are determined for each water quality zone and for the distribution systems as a whole.

QUESTIONS AND PROBLEMS

1. Which pathogenic microorganisms are the most difficult to remove by conventional water treatment and why?

2. Describe the major steps in the conventional treatment of drinking water.

3. Why are all microorganisms not inactivated according to first-order kinetics?

4. How long would you have to maintain a residual of 1.0 mg/l of free chlorine to obtain a $C \cdot t$ of 18? A $C \cdot t$ of 0.2?

5. Why is chlorine more effective against microorganisms at pH 5.0 than at pH 9.0?

6. What factors interfere with chlorine disinfection? Ultraviolet disinfectant?

7. What is the main site of UV light inactivation in microorganisms? What group of microorganisms are the most resistant to UV light? Why?

8. Why does suspended matter interfere with the disinfection of microorganisms?

9. What are some options for reducing the formation of disinfection byproducts formed during drinking water treatment?

10. How much of a log_{10} reduction and % reduction of *Escherichia. coli* will occur with a UV dose of 16,000 μW s/cm^2?

REFERENCES AND ADDITIONAL READING

Aboytes R. et al. (2004) Detection of infectious Cryptosporidium in filtered drinking water. *Journal AWWA*. **96**,88–98.

Bates R.C., Shaffer P.T.B. and S.M Sutherland. (1977) Development of poliovirus having increased resistance to chlorine inactivation. *Appl. Environ. Microbiol.* **33**, 849–853.

Bitton G. (1999) *Wastewater Microbiology*. Wiley-Liss, New York.

Blackburn B.G., Craun G.F., Yoder, J.S., Hill V., Calderon R.L., Chen N., Lee S.H., Levy D.A. and Beach, M.J. 2004. Surveillance for waterborne-disease outbreaks associated with drinking water—United States, 2001–2002. *MMWR Surveill. Summ.* **53**, 23–45.

Borchardt M.A., Haas N.L. and Hunt R.J. (2004) Vulnerability of drinking-water wells in La Crosse, Wisconsin, to enteric-virus contamination from surface water contributions. *Appl. Environ. Microbiol.* **70**, 5937–5946.

Borchardt M.A., Bertz P.D., Spencer S.K. and Battigelli D.A. (2003) Incidence of enteric viruses in groundwater from household wells in Wisconsin. *Appl. Environ. Microbiol.* **69**, 1172–1180.

Calderon R.L. Measuring benefits of drinking water technology: "Ten" years of drinking water epidemiology. NEWWA Water Quality Symposium, May 20, 2004. Boxborough, MA.

Colford J.M. Jr. et al. (2005) A randomized, controlled trial of in-home drinking water intervention to reduce gastrointestinal illness. *Am. J. Epidemiol.* **161**, 472–82.

Craun G.F. (1993) *Safety of Water Disinfection: Balancing Chemical and Microbial Risks*. ILSI Press, Washington, D.C.

Fout G.S., Martinson B.C., Moyer M.W. and Dahling D.R. (2003) A multiplex reverse transcription-PCR method for detection of human enteric viruses in groundwater. *Appl. Environ. Microbiol.* **69**, 3158–3164.

Gerba C.P., Nwachuko N. and Riley K.R. (2004) Disinfection resistance of waterborne pathogens on the United States Environmental Protection Agency's Contaminant Candidate List (CCL). *J. Water Supply: Research and Technology Aqua.* **52**, 81–94.

Haas C.N. and Morrison E.C. (1981) Repeated exposure of *Escherichia coli* to free chlorine: Production of strains possessing altered sensitivity. *Water Air Soil Pollut.* **16**, 233–242.

Hellard M.E., Sinclair M.I., Forbes A.B. and Fairley C.K. (2001) A randomized, blinded, controlled trial investigating the gastrointestinal health effects of drinking water quality. *Environ Health Perspect.* **109**, 773–778.

Korich D.G., Mead J.R., Madore M.S., Sinclair N.A. and Sterling C.R. (1990) Effect of zone, chlorine dioxide, chlorine, and monochloramine on *Cryptosporidium parvum* oocyst viability. *Appl. Environ. Microbiol.* **56**, 1423–1428.

LeChevallier M.W., Lowry C.H. and Lee R.G. (1990) Disinfecting biofilm in a model distribution system. *J. Am. Water Works Assoc.* **82**, 85–99.

Montgomery J.M., Consulting Engineers Inc. (1985) *Water Treatment Principles and Design*. John Wiley & Sons, New York.

Morris R. and Levin R. (1995) Estimating the incidence of waterborne infectious disease related to drinking water in the United States. In Reichard E.G. and Zapponi G.A., eds. *Assessing and managing health risks from drinking water contamination: Approaches and applications*. International Association of Hydrological Sciences Press. Great Britain.

Park S.R., Mackay W.G. and Reid D.C. (2001) *Helicobacter sp.* recovered from drinking water biofilm sampled from a water distribution system. *Water Research.* **35**, 1624–1626.

Payment P. et al. (1991) A randomized trial to evaluate the risk of gastrointestinal disease due to consumption of drinking water meeting current microbiological standards. *Am. J. Public Health.* **81**, 703–708.

Payment P. et al. (1997) A prospective epidemiological study of gastrointestinal health effects due to the consumption of drinking water. *Int. J. Environ. Health Res.* **7**, 5–31.

Payment P. (1999) Poor efficacy of residual chlorine disinfectant in drinking water to inactivate waterborne pathogens in distribution systems. *Can. J. Microbiol.* **45**, 709–715.

Roessler P.F. and Severein B.F. (1996) Ultraviolet light disinfection of water and wastewater. In Hurst C.J., ed. *Modeling Disease Transmission and Its Prevention by Disinfection*. Cambridge University Press, Cambridge, England, pp. 313–368.

Rose J.B., Lisle J.T. and Lechevallier M. (1997) Waterborne *Cryptosporidiosis*: Incidence, outbreaks, and treatment strategies. In Fayer R., ed. *Cryptosporidium* and *Cryptosporidiosis*. CRC Press, Boca Raton, Florida, pp. 93–109.

Sobsey M.D. (1989) Inactivation of health-related microorganisms in water by disinfection processes. *Water Sci. Technol.* **21**, 179–195.

Stewart M.H. and Olson B.H. (1996) Bacterial resistance to potable water disinfectants. In Hurst C.J., ed. *Modeling Disease Transmission and Its Prevention by Disinfection*. Cambridge University Press, Cambridge, England, pp. 140–192.

Thurman R.B. and Gerba C.P. (1988) Molecular mechanisms of viral inactivation by water disinfectants. *Adv. Appl. Environ. Microbiol.* **33**, 75–105.

U.S. EPA. 2001. Office of Water, Drinking Water Infrastructure Needs Survey, U.S. EPA 816–F–01–001.

EMERGING ISSUES IN POLLUTION SCIENCE

CHAPTER 29

GENETICALLY ENGINEERED CROPS AND MICROBES

C. Rensing, G. Grass, and I. L. Pepper

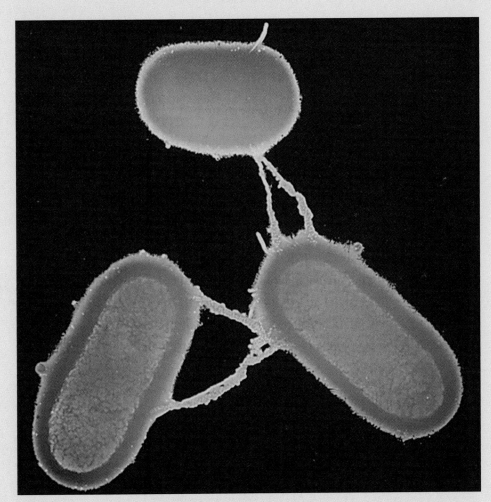

Conjugating cells of *Escherichia coli*. False-color transmission electron micrograph (TEM) of a male *E. coli* bacterium (bottom right) conjugating with two females. This male has attached two F-pili to each of the females. The tiny bodies covering the F-pili are bacteriophage MS2, a virus that attacks only male bacteria and binds specifically to F-pili. Magnification: ×11,250.
Credit: Dr. L. Caro, Photo Researchers, Inc.

29.1 INTRODUCTION TO NUCLEIC ACIDS

The two biologically important forms of nucleic acids are **deoxyribonucleic acid (DNA)** and **ribonucleic acid (RNA)**. DNA carries the genetic blueprint of a cell, and RNA acts as an intermediate to convert the blueprint into protein. Both DNA and RNA are made up of nucleotides, composed of three units: (1) a five-carbon sugar, either ribose (in RNA) or deoxyribose (in DNA); (2) a nitrogen base; and (3) a phosphate molecule. The four nitrogen bases guanine (G), cytosine (C), adenine (A), and thymine (T) are important building blocks of DNA. Guanine and adenine are purine bases, and cytosine and thymine are pyrimidine bases (Figure 29.1a). For convenience these are commonly abbreviated (G, C, A, or T). In RNA thymine is replaced by uracil (U).

In the cell, DNA is present as a double-stranded molecule forming a double helix. A base from one strand always pairs with another base from the other strand and is linked by hydrogen bonds. The base pairing is very specific, as G will only hydrogen bond with C, and A will only bond with T. Because of this specificity, the two strands are said to be complementary, since wherever there is a C in one strand, there will be a G in the other, and wherever there is an A in one strand, the complementary strand will carry a T (Figure 29.1b). In contrast, almost all RNA molecules are single-stranded. However, usually RNA molecules form folded structures in which complementary base pairing can occur within the strand.

29.2 RECOMBINANT DNA TECHNOLOGY

Recombinant DNA technology evolved over time, starting with the discovery by Avery (1944) that DNA is the genetic material. Solving the double-helix structure of DNA by Watson, Crick, and Franklin gave insights into how DNA is replicated and "read" to make RNA and proteins. The real breakthrough came with the discovery and availability of enzymes that recognize and cut specific double-stranded DNA sequences. These enzymes are called **restriction enzymes** or **restriction endonucleases**. Other useful enzymes include ligases that join ends together (see Section 29.3). Other techniques that were developed subsequently, such as the **polymerase chain reaction (PCR)** also proved very helpful in many applications (see Section 29.4.3). These techniques are described in greater detail later in this chapter.

With the molecular biology revolution, people quickly realized the enormous commercial potential of molecular biotechnology. Currently, there is a wide variety of products ranging from medical applications such as production of recombinant insulin and research applications such as synthesis of specialized enzymes (*e.g.*, restriction endonucleases) to agricultural applications such as *Bt*-cotton (the protection of cotton from insects by expressing the *cry1Ab* genes from *Bacillus thuringiensis*) (see Section 29.6.1). Due to this increased prevalence of recombinant material, there has been concern of possible negative consequences.

A basic knowledge of molecular genetics is needed to better understand genetic engineering of plants and microbes. All genetic information stored in DNA is translated into RNA, and ultimately the production of protein molecules responsible for the structural make-up and vital reactions in the cell. A gene is a segment of DNA specifying a protein, a tRNA, or an rRNA.

The processes responsible for the flow of genetic information can be separated into three stages, shown in Information Box 29.1 and Figure 29.2.

The three information transfer steps apply to all cells, but in Figure 29.2, a bacterial cell is shown. In eukaryotic cells, these processes are physically separated between the nucleus and the cytoplasm. In addition, eukaryotic genes encoding proteins are often split into coding and noncoding regions. The coding regions are called **exons** and the intervening noncoding regions are termed **introns**. The gene is first transcribed in its entirety in the nucleus as the pre-mRNA, which is then processed by removal of the noncoding regions. The transcript is then transported into the cytoplasm and translated into protein.

29.3 TRANSFER OF NUCLEIC ACID SEQUENCES FROM ONE ORGANISM TO ANOTHER (Cloning)

In order for the molecular biotechnology revolution to take place, certain tools had to be available. These can be summarized under the broad term **gene cloning** or **recombinant**

INFORMATION BOX 29.1

The Steps of Protein Formation

1. **Replication.** DNA is the genetic storage material. Two long DNA strands form a double helix and each strand is duplicated during replication, producing two double helices.

2. **Transcription.** Information transfer from DNA to RNA is called transcription. RNA that is the intermediate template for protein synthesis is called messenger RNA (mRNA). Some genes do not encode information for protein synthesis but for other types of RNA such as transfer RNA (tRNA) and ribosomal RNA (rRNA).

3. **Translation.** The specific sequence of an mRNA determines the specific sequence of amino acids of a protein. Three bases on the mRNA code for a specific amino acids, therefore a triplet of ribonucleotide bases is called a codon. The actual protein synthesis takes place on the ribosomes.

A

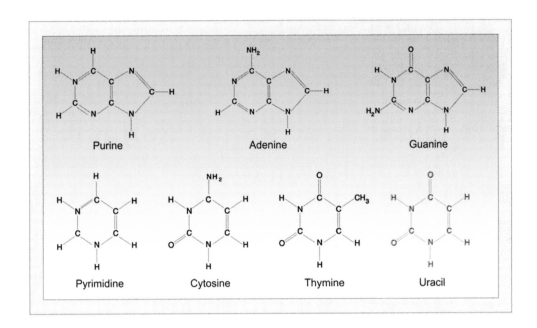

B

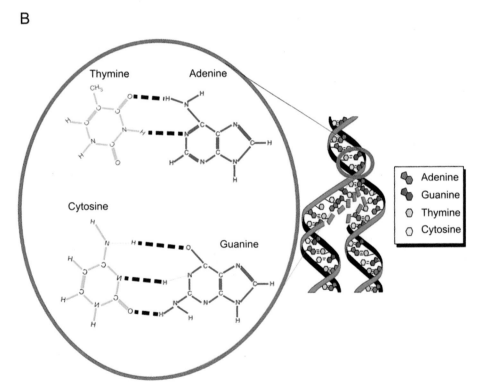

Figure 29.1 (a) Structures of purine and pyrimidine bases used in nucleic acid sequences. (b) Hydrogen bonding between purine and pyrimidine bases. From *Environmental Microbiology* © 2000, Academic Press, San Diego, CA.

DNA technology. It is beyond the scope of this chapter to give an extensive review of these methods. Rather, a typical experiment is described to give a basic understanding of the process.

DNA first has to be extracted from a donor organism. It can then be cleaved by restriction enzymes. Restriction enzymes such as *Eco*RI or *Bam*HI recognize certain DNA sequences that are then cleaved (cut or digested). A DNA double strand that was cleaved in such a way can then be ligated (joined) to another DNA double strand (typically a cloning vector such as a plasmid) that was treated similarly.

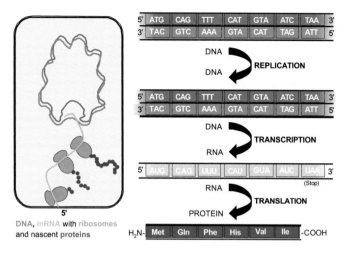

| 5' | ATG | CAG | TTT | CAT | GTA | ATC | TAA | 3' |
| 3' | TAC | GTC | AAA | GTA | CAT | TAG | ATT | 5' |

DNA ⟩ **REPLICATION**
DNA

| 5' | ATG | CAG | TTT | CAT | GTA | ATC | TAA | 3' |
| 3' | TAC | GTC | AAA | GTA | CAT | TAG | ATT | 5' |

DNA ⟩ **TRANSCRIPTION**
RNA

| 5' | AUG | CAG | UUU | CAU | GUA | AUC | UAA | 3' |
| | | | | | | | (Stop) |

RNA ⟩ **TRANSLATION**
PROTEIN

H₂N- | Met | Gln | Phe | His | Val | Ile | -COOH

5'
DNA, mRNA with ribosomes
and nascent proteins

Figure 29.2 The processes of replication, transcription, and translation. Only one of two strands of the DNA double helix is transcribed.

allow selection only of bacteria harboring this construct (Figure 29.3).

29.4 CHEMICAL SYNTHESIS, SEQUENCING AND AMPLIFICATION OF DNA

A variety of new technical developments in molecular biotechnology has made many procedures that were once difficult to perform now much easier to carry out. These techniques are now commonplace and accessible to both large and small research facilities. Synthesizing a DNA molecule, sequencing a DNA strand, and amplifying a specific DNA segment with polymerase chain reaction (PCR) are three techniques, briefly covered here, that today are easy to perform and relatively inexpensive. All of the methods are essential for isolating, characterizing, and expressing cloned genes.

29.4.1 DNA Synthesis

One of the techniques that has greatly expanded the methodologies available in DNA characterization and molecular cloning is the ability to chemically synthesize a strand of DNA with a specific sequence of nucleotides. DNA synthesizers introduce specified nucleotides and the reagents

The resulting construct is a new, recombined DNA molecule. This DNA construct then needs to be introduced into a host bacterium to be propagated. This transfer of DNA into a bacterium is called transformation. The DNA construct usually contains a gene for antibiotic resistance to

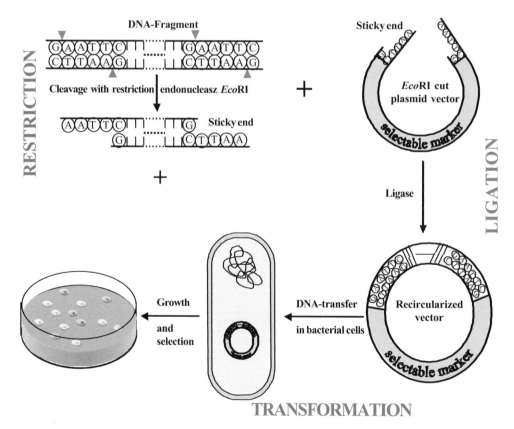

Figure 29.3 Steps in cloning of a DNA fragment. Shown are restriction, ligation, and transformation, as described in the text.

required for the coupling of each consecutive nucleotide to the growing chain. Each incoming nucleotide is coupled to the 5′-end, not to the 3′-end, as in biological synthesis.

29.4.2 DNA Sequencing

The power of current DNA sequencing techniques was recently demonstrated with the completion of the human genome sequence. All current DNA sequencing methods are improvements of the original dideoxy method by Sanger (1977). In this method, a copy of the DNA to be sequenced is made using DNA polymerase, which uses deoxribonucleoside triphosphates (dNTPs) as substrate and adds them to a primer. In the Sanger method, four different incubation mixtures were prepared containing, in addition to dNTPs, small amounts of dideoxy analogs (ddNTPs) of one of the nucleotides in each of the mixtures. The ddNTPs act as specific chain-termination reagents, since the dideoxy sugar lacks the 3′-hydroxyl, preventing further lengthening of the chain. Therefore, fragments of varying lengths are obtained in each of the incubation mixtures, which can be separated by electrophoresis. DNA is labeled using radioactive isotopes or fluorescence, separated on an electrophoresis gel, enabling direct reading of the DNA. This technique has become very affordable and can be performed at most research facilities.

29.4.3 PCR Amplification

In this technique, PCR is used to amplify specific DNA sequences *in vitro*. In principle, PCR involves the repetitive enzymatic synthesis of DNA, using two oligonucleotide primers that hybridize to opposite strands of DNA flanking the target DNA of interest (Figure 29.4). In theory, the number of copies is doubled during each cycle, so that PCR can multiply DNA molecules up to a billionfold in the test tube. Each PCR cycle involves the following:

1. Heat **denaturation** or **separation** of double-stranded target DNA
2. Cooling, to allow **annealing** of specific primers to target DNA
3. Primer **extension** using a heat-stable DNA polymerase

29.5 HETEROLOGOUS GENE EXPRESSION IN PRO- AND EUKARYOTES

The goal of many biotechnological applications is the effective expression of a cloned gene in a selected host organism. Often a high rate of production of the protein encoded by the cloned gene is desired. In order to achieve this, many expression vectors have been created. Often these vectors contain elements that control transcription, translation, protein stability, and secretion of this protein from the host cell. Different cloned genes require different strategies for optimal overexpression, because most cloned genes have distinctive properties. The most commonly manipulated molecular biological features include the promoter and terminator sequences, the number of copies of the cloned gene, the strength of the ribosome-binding site, the efficiency of translation in the host organism, and the stability of the synthesized protein. The level of foreign gene expression also depends on the host organism.

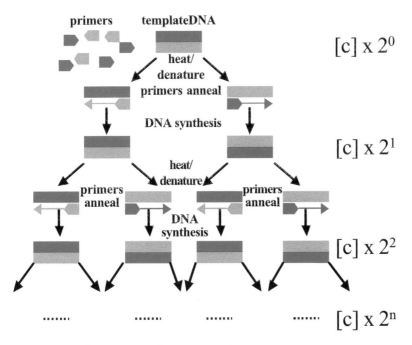

$[c] \times 2^0$

$[c] \times 2^1$

$[c] \times 2^2$

$[c] \times 2^n$

Figure 29.4 The three steps in two PCR cycles.

29.6 GENETICALLY ENGINEERED PLANTS FOR AGRICULTURE

The aim of plant biotechnology is to create new varieties of cultivated plants (cultivars) that have beneficial properties. In addition to strains that give better yields, research has also focused on creating transgenic plants with increased resistance to insects, viruses, herbicides, and environmental stress. In the following sections, examples of engineered plants are described more extensively.

29.6.1 Insect Resistance

The estimated amount spent on chemical insecticides was nearly $5 billion worldwide in 2000. It is therefore desirable to genetically engineer crop plants that produce functional insecticides, since this would lower the amount of chemical pesticides needed. In addition, biological insecticides are usually very specific for a limited number of insect species, minimizing ecological damage.

Bacillus thuringiensis has been used safely for about 50 years in spray formulation for control of insects that damage plants. It is considered safe because specific formulations only harm a narrow range of insects. To be independent of spraying, the gene for the protoxin from *B. thuringiensis* has been introduced and expressed in many plant species including alfalfa, apple, broccoli, cabbage, canola, corn, cotton, cranberry, eggplant, grape, hawthorn, juneberry, peanut, pear, poplar, potato, rice, rutabaga, soybean, spruce, sugar cane, tobacco, tomato, walnut, white clover, and white spruce. There are a number of transgenic plants being commercially used, such as cotton and corn, expressing an insecticidal toxin or protoxin. Currently, more than 25% of the corn and 5% of the cotton worldwide and 26% of corn and 69% of cotton in the United States contain transgenic *B. thuringiensis* (*Bt*) insecticidal protein. Key agricultural pests currently targeted with *Bt* insecticides include stem borers, leafworms, bollworms, and budworms in field crops and grains; the spruce budworm and gypsy moth in forests; the diamond moth and cabbage looper in vegetable crops; and certain beetles.

INFORMATION BOX 29.2

Benefits and Potential Hazards of *Bt* Insecticides

Benefits:

· reduced insect damage
· improved crop yields
· reduced fungal toxins in food supply
· reduction of hazards in the environment and exposure to farm workers due to replacement of toxic chemical pesticides
· potential use in both large and small farms

Potential hazards:

· potential to harm nontarget organisms
· development of resistance
· possible ecological consequences of gene flow from engineered crops to nonengineered crops and wild relatives

---

The insecticidal activity of *B. thuringiensis* is due to intracellular crystal inclusions (ICPs) that are produced during the process of sporulation. One or more related ICPs are encoded by the *cry* (crystal) and *cyt* (cytolytic) genes located mainly on plasmids. *Bt* insecticides do not function on contact, but rather as midgut toxins. Therefore, the insecticides have to be eaten and ingested by the target organism. The toxin disturbs the membrane integrity of the host cells by the formation of spores. This leads to loss of transmembrane potential, cell lysis, leakage of midgut contents, paralysis, and death of the insect. Information Box 29.2 illustrates the benefits and potential hazards of *Bt* insecticides. The concerns highlight the importance of prudent use. In particular, there has been concern about nontarget effects of *Bt* insecticides on the Monarch butterfly (see Case Study 29.1).

CASE STUDY 29.1

THE *Bt* GENE IN CORN AND INFLUENCE ON MONARCH BUTTERFLY

Monarch butterflies are famous not only for their size and beauty but also for their long spring and fall migration. In 1999, a small note published in the journal *Nature* showed that pollen from transgenic *Bt*-corn killed monarch butterfly caterpillars in the laboratory. This raised concerns that beneficial insects and endangered species might be severely affected by transgenic *Bt*-corn. However, subsequent stud-

ies concluded there is no significant risk to monarch butterflies from environmental exposure to *Bt*-corn.

There still is concern that there will be accelerated development of resistance against *Bt* insecticides in insect pests. Once this occurs, farmers and organic growers relying on the natural spray formula will have lost a valuable tool to combat these pests.

29.6.2 Virus Resistance

Plant viruses can cause serious damage and significantly reduce crop yields. At present, the most effective method for conferring some virus protection is the expression of virus-coat protein in transgenic plants. Virus-coat protein is usually the most abundant protein of a virus particle. Although the precise mechanism of this protection is not known, expression of virus-coat protein has been shown to prevent any significant amount of viral synthesis. High levels of virus resistance, but not complete protection, can be achieved by this method.

29.6.3 Herbicide Resistance

Weed infestation causes the loss of approximately 10% of the annual global crop production in spite of application of chemical herbicides at an annual cost of $10 billion. Many herbicides are also not able to discriminate between crop and weed plants, and some persist in the environment. The construction of herbicide-resistant crop plants would be desirable to both improve crop yield and diminish the use of chemical herbicides. Different approaches to create herbicide-resistant plants could be envisioned including:

1. Construction of plants that have the ability to inactivate the herbicide.
2. Changing a herbicide-sensitive target protein so that it binds the herbicide less tightly.
3. Overproduction of a herbicide-sensitive target protein so that amount of inactivated protein is compensated by the increased protein production.
4. Uptake inhibition of a herbicide into a plant.

The first three possibilities have been implemented. A good example is resistance to the herbicide glyphosate. This herbicide is readily degraded to a nontoxic compound in the environment, which is desirable. However, both weeds and crop plants are affected by the glyphosate-inhibiting synthesis of aromatic amino acids. The gene for a bacterial enzyme that is no longer inhibited by glyphosate was used to construct transgenic plants that are now resistant to glyphosate.

29.6.4 Fungus and Bacterial Resistance

Phytopathogenic fungi and bacteria also cause extensive damage. For example, it has been estimated that fungal rice blast costs farmers in Japan, the Philippines, and Southeast Asia more than $5 billion a year. Plants respond to attacks by pathogens with synthesis of **pathogenesis-related (PR) proteins**. These PR proteins include chitinases and β-1,3-glucanases that help lyse fungal cell walls. Expression of these genes makes plant cells more resistant, and this mechanism, called **systemic acquired resistance,** extends to plant tissues not affected by pathogens. Expression of PR proteins is induced by salicylic acid. Therefore overexpression of salicylic acid is one approach to construct more resistant plants.

29.7 GENETICALLY ENGINEERED PLANTS FOR REMEDIATION

One of the many challenges of the 21st century concerns the development of cost-effective remediation techniques for existing contaminated environments (see Chapter 19). Currently, cleanup methods such as physical removal of contaminated soil from a site and disposal in designated landfills are generally too costly, laborious, environmentally destructive, and impractical when applied on the imposing scale that is now required. Recently, the U.S. Environmental Protection Agency (EPA Title 40) set new standards for the determination and disposal of hazardous wastes. For example, soil lead **TCLP (Toxicity Characteristic Leaching Procedure)** concentrations were lowered from 5 mg/L to 0.75 mg/L. The extent of contamination must now be reduced to fall within reasonable limits prior to disposal. Achieving the new lower limits required by the EPA sets even more precedence for innovative remediation strategies. The problem of metal contamination is particularly evident in the western United States due to mining and subsequent contamination (see Chapter 16). Therefore, significant areas of land have to be remediated or at least stabilized to avoid adverse effects on the quality of groundwater, to name just one concern.

Bioremediation can involve the use of organisms to remove or stabilize toxic waste from the environment. Organic compounds that are difficult to degrade include halogenated aromatic compounds, which are the main component of most pesticides and herbicides, crude oil, solvents, and poly-aromatic hydrocarbons (PAHs). Metals are also difficult to remove from soils.

Phytoremediation is a promising, cost-effective, environmentally friendly alternative to most physicochemical treatments typically employed to clean up heavy metal or organic waste contamination from soil, sediments, and ground and surface waters. As a general remediation strategy, phytoremediation—the use of plants to remediate contaminated waters and soils—includes several subsets: phytoextraction, phytostabilization, and phytovolatilization (Figure 29.5). Plants can be used to extract, sequester, and/or detoxify a wide variety of environmental contaminants, including metals and trichloroethylene (TCE).

29.7.1 Metal Remediation via Plants

Elemental pollutants such as heavy or transition metals are essentially immutable by any biological or physical method short of nuclear fission and fusion, and thus their remediation presents special scientific and technical problems. With a few notable exceptions, the best scenarios for the phytoremediation of metals involve plants extracting and translocating a toxic cation to aboveground tissues for later harvest, converting the metal to a less toxic chemical species, or at the very least sequestering the element in roots to prevent leaching from the site. **Phytoextraction** is the uptake of soil-borne metals by plants, and subsequent translocation and

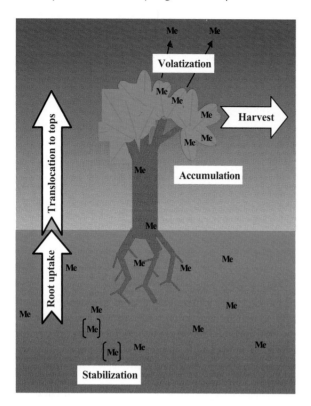

Figure 29.5 Phytoremediation of heavy metals is using plants for stabilization, accumulation, or volatilization of heavy metals.

storage in root and shoot tissue. For effective phytoextraction, **hyperaccumulation** is a desirable trait. Hyperaccumulation is usually defined as the concentration of a metal ion >0.1 to 1% of the dry weight of the plant. At these concentrations, the recovery of metals from the plant tissues is potentially economical. Very few plants are known hyperaccumulators. This field is generating great excitement because phytoremediation may offer the only effective means of restoring the hundreds of thousands of square miles of land and water that have been polluted with metals by certain human activities.

Another area of current research is identifying and characterizing naturally occurring plants capable of hyperaccumulating metals in the aboveground tissue. Natural plant hyperaccumulators such as *Alyssum lesbiacum* and *Thlaspi caerulescens*, when sown in Ni- or Zn-containing medium, can accumulate up to 3% Ni and 5% Zn, respectively, in the aboveground biomass (Cunningham & Ow, 1996). However, recent application of metal hyperaccumulating plants has shown limitations to this approach. For example, plants sown into mine tailing soils heavily laden with a multitude of metal contaminants frequently only accumulate and efficiently translocate a few metals to the aboveground tissue. Metal accumulation by plants is thought to be dependent not only on the bioavailable metal concentration, but also on the transport dynamics of the plant in question. Unfortunately, most hyperaccumulators

identified to date yield low biomass. Therefore, attempts are underway to construct plants yielding high biomass with properties aiding remediation with genes derived from hyperaccumulators.

29.7.2 Organic Remediation via Plants

In contrast to inorganic elements, organic pollutants can be phytoremediated by degradation. Several **phytoremediation** approaches have been defined to aid the degradation of organic pollutants. Organic pollutants can either be completely mineralized to inorganic compounds such as water, carbon dioxide, and chloride, or partially degraded to a stable intermediate. Phytoremediation has been shown to work well for organic pollutants that are mobile in plants, such as trinitrotoluene (TNT) or TCE.

Poplar trees have been studied for their ability to degrade TCE. White-rot fungi have also been studied with a view to enhancing bioremediation of PCBs, PAHs, and TNT (Pointing, 2001). Specifically, the genes for **lignin-modifying enzymes (LMES)** have been utilized to create transgenic plants that produce manganese peroxidase, a specific LME (Iimura et al., 2002). The transgenic plants were able to degrade PCP at higher rates than the nontransgenic control plants. Overall genetic engineering to engineer transgenic plants has great potential for the remediation of many recalcitrant compounds (see also Chapter 19).

29.8 MICROBIAL-ASSISTED REMEDIATION

Microbes are responsible for most of the natural degradative cycles and are vital for the global cycle of nutrients. This is due to their enormous metabolic versatility. This has helped in the degradation of many contaminants in a process called natural attenuation. Pseudomonads are the best-studied degradative bacteria. Various *Pseudomonas* strains have been shown to be able to break down and thereby detoxify more than 100 different organic compounds.

Heavy metals, unlike organic compounds cannot be degraded nor converted to other molecules, but their bioavailability can be reduced by binding agents (Lovley and Lloyd, 2000). Eukaryotic cysteine-rich proteins known as metallothioneins are particularly efficient with respect to their metal binding affinity. Mice metallothionein genes have been introduced into the bacterial *Ralstonia eutropha* genome to enhance the organisms metal binding potential.

Genetic engineering of *Escherichia coli* has been utilized to remove mercury from a contaminated site (Bae et al., 2001). Here, genes encoding for metal binding peptides were introduced in *E. coli* either in the cells cytoplasm (Strain PMC20) or the cell membrane (Strain pLO 20). Both strains enhanced the uptake and bioaccumuatlion of Hg^{2+}.

INFORMATION BOX 29.3

Types of Bioaugmentation

Cell bioaugmentation: Enhanced microbial activity due to the activity of whole cells introduced into the environment.
Gene bioaugmentation: Enhanced microbial activity that occurs subsequent to the transfer of a gene from an introduced donor cell into an indigenous recipient cell in the environment.
Soil bioaugmentation: Here, enhanced microbial activity occurs following the introduction of a foreign soil (that contains specific microbes) into the contaminated soil.

29.8.1 Genetic Engineering of Biodegradative Pathways

Despite the great degradative capability of microorganisms, there are often limits to the biological treatment of some xenobiotic waste materials.

This has led to approaches to aid xenobiotic degradation including:

- Development of genetically engineered microorganisms with increased remediation capabilities. (Case Study 29.2)

- Introduction of plasmids or microorganisms capable of degrading a particular contaminant. This process is known as bioaugmentation (Gentry et al., 2004).

Bioaugmentation is the introduction of specific microorganisms into an environment to enhance the biodegradation of organic contaminants or remediate metal contamination (Information Box 29.3).

29.8.2 Radiation Resistant Degraders

Nuclear weapons programs and production of nuclear energy created huge amounts of radioactive waste. In the United States alone there are about 3,000 of these sites and the estimated costs for cleanup are above $200 billion. In addition to radioactive waste, these sites often contain organic and metal pollutants, making cleanup more difficult. A logical first step would thus be the degradation of organic wastes. However, most organisms are very sensitive to the levels of radiation present at these sites, so the bacteria that are usually used for degradation cannot be used. Therefore, a highly radiation resistant bacterium, *Deinococcus radiodurans*, has been utilized. Since *D. radiodurans* was not able to degrade aromatic contaminants such as toluene and chlorobenzene, the genes encoding toluene dioxygenase were introduced into *D. radiodurans* on a plasmid. This enabled the transformants to degrade these contaminants.

29.9 POTENTIAL PROBLEMS DUE TO GENETICALLY MODIFIED ORGANISMS

Despite the advantages of genetically modified (GM) crops, there have been concerns raised over the environmental safety and the societal impact, especially in the developing world. Genes added to GM crops might escape via pollen to other plants. This could result in a sudden change in the plant environment and disrupt the previous ecological balance. It is generally agreed that some gene transfer will occur among closely related species over the long term. Therefore, each gene construct inserted into transgenic plants should be assessed on how it could disrupt the environment on a case-by-case basis. Examples of concerns raised include:

- Gene constructs that could increase the weediness of wild relatives of the crop plant by conferring a competitive advantage.

- The loss of an effective herbicide if a gene construct conferring herbicide resistance escapes to a weed species.

- The potential for the spread of GM-plants into fields designated "organic," thereby causing the loss of this distinction. This could result in a loss of value and potential

---

CASE STUDY 29.2

GENE BIOAUGMENTATION USING GENETICALLY ENGINEERED *E. COLI*.

Bioaugmentation via genetically manipulated microbes has been demonstrated (Newby et al., 2000). Here, a degradative plasmid pJP4 was introduced into an *E. coli* isolate (D11). Plasmid pJP4 codes for the initial degradation steps of 2,4-dichlorophenoxyacetic acid (2,4-D). However, chromosomal genes are necessary for the complete mineralization of 2,4-D. *E. coli* isolate D11 lacked the needed chromosomal genes for complete mineralization. Therefore, the enhanced degradation of soil contaminated with 2,4-D, when D11 was introduced into the soil, could only be due to gene transfer of the pJP4 plasmid into indigenous soil bacteria that contained the appropriate complementary chromosomal genes.

problems for export into countries with more restricted handling of GM-food.

- Concern over the potential for adverse human health effects from consumption of foodstuffs containing genetically modified crops.

- The public is also very concerned about the application of genetically engineered microbes (GEMs). The National Institutes of Health Biosafety Regulations rigorously control the release of GEMs.

29.10 SUMMARY

Clearly, genetic engineering of plants and microbes has enormous potential to enhance "pollution prevention" and generally favorably impact society. With respect to *bt*-modified crops, there are vast acreages already being utilized in the United States. However, genetically modified crops are not as welcome in other parts of the world, particularly Europe, where they are essentially banned. This, plus the skepticism of large segments of the general public in the U.S., means that the future of GM crops and microbes is still open to debate.

QUESTIONS AND PROBLEMS

1. Describe DNA amplification by PCR.
2. What is phytoremediation and why is it a very attractive method?

3. Name some possible problems associated with genetically modified organisms.

REFERENCES AND ADDITIONAL READING

Avery O.T., MacLeod C.M. and McCarty M. (1944) Studies on the chemical nature of the substance inducing transformation of pneumococcal types; induction of transformation by a desoxyribonucleic acid fraction isolated from *Pneumococcus* type III. *J. Exp. Med.* **79**, 137–158.

Bae W., Mehra R.K., Mulchandani A. and Chen, W. (2001) Genetic engineering of *Escherichia coli* for enhanced uptake and bioaccumulation of mercury. *Applied and Environmental Microbiology.* **67**, 5335–5338.

Cunningham S.D. and Ow D.W. (1996) Promises and prospects of phytoremediation. *Plant Physiol.* **110**, 715–719.

Gentry T.J., Rensing C. and Pepper I.L. (2004) New approaches for bioaugmentation as a remediation technology. *Crit. Rev. Env. Sci. Tec.* **34**, 447–494.

Iimura, Y. et al. (2002) Expression of a gene for Mn-peroxidase from *Coriolus versicolor* in transgenic tobacco generates potential tools for phytoremediation. *Applied Microbiology and Biotechnology.* **59**, 246–251.

Lovley D.R. and Lloyd J.R. (2000) Microbes with a mettle for bioremediation. *Nature Biotechnology.* **18**, 600–601.

Newby D.T., Josephson K. and Pepper I.L. (2000) Detection and characterization of plasmid pJP4 transfer to indigenous soil bacteria. *Appl. Environ. Microbiol.* **66**, 290–296.

Pointing S.B. (2001) Feasibility of bioremediation by white-rot fungi. *Applied Microbiology and Biotechnology.* **57**, 20–33.

Sanger F., Nicklen S. and Coulson A.R. (1977) DNA sequencing with chain-terminating inhibitors. *Proc. Natl. Acad. Sci. USA.* **74**, 5463–5467.

CHAPTER 30

ANTIBIOTIC-RESISTANT BACTERIA AND GENE TRANSFER

C. Rensing and I. L. Pepper

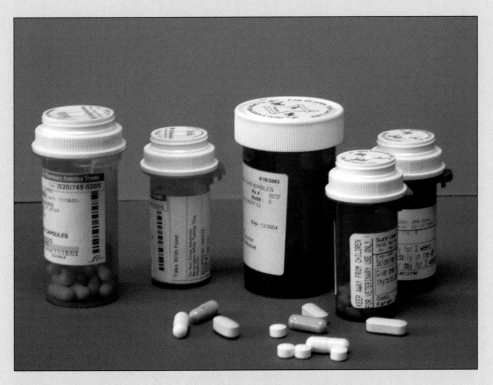

Human consumption of antibiotics is a common medical practice. *Photo courtesy K.L. Josephson.*

30.1 WHY ARE ANTIBIOTICS AN ISSUE?

Antibiotics are compounds produced by microorganisms that kill or inhibit other microorganisms. Thus, antibiotics are a class of chemotherapeutic agents that can be used to control infectious disease. Antibiotics are natural products, which distinguishes them from other agents such as growth factor analogs that are synthetic chemicals (however, antibiotics are also "manufactured," *e.g.*, produced in large quantities and made into "pills.") For example, sulfonamides, first discovered in Germany in the 1930s, is an analog of an essential part of the vitamin folic acid, and prevents microbial growth by inhibiting synthesis of folic acid (see also Chapter 28). The first and perhaps most effective antibiotic discovered was penicillin, isolated by Sir Alexander Fleming in 1929 from the fungus *Penicillium*. *Penicillium* was highly effective in combating staphylococcal and pneumonococcal infections. Later, another potent antibiotic was discovered by Selman Waksman in 1943. The antibiotic, streptomycin, was isolated from the actinomycete *Streptomyces griseus*, a feat for which Waksman was given the Nobel Prize in 1952. However, overall, less than 1% of discovered antibiotics are commercially useful. Despite this, since their inception at the end of World War II, the useful antibiotics have revolutionized our ability to treat infectious diseases, but not without a concomitant cost.

Bacteria are prokaryotic organisms with the ability to metabolize and replicate very quickly. They are also very adaptable genetically. When confronted with an antibiotic, the existence of only one bacterial cell with a genetic or mutational change that confers resistance to that antibiotic is enough to allow for the subsequent proliferation of antibiotic-resistant bacteria. Thus, the more antibiotics are used, the greater the likelihood that antibiotic-resistant strains will develop. The greatest concern with antibiotic resistance is the potential for human pathogenic strains to become resistant to popular antibiotics, which subsequently cannot contain the infectious agent. The widespread, sometimes indiscriminant use of antibiotics has raised the question "Can antibiotic-resistant genes be transferred from nonpathogenic bacteria to human pathogenic strains?" In this chapter we address this issue by examining the incidence of antibiotic-resistant bacteria and the potential for horizontal gene transfer between different groups of bacteria.

30.2 CLASSIFICATION AND FUNCTION OF ANTIBIOTICS

Bacteria are classified as Gram-negative or Gram-positive, based on their Gram stain reaction (see Chapter 5). Essentially, the Gram stain differentiates bacteria on the basis of their cell wall components. Therefore, antibiotics are required that will be effective against either

---

> **INFORMATION BOX 30.1**
>
> **Types of Antibiotic**
>
> | Antibiotic | Effective Against |
> |---|---|
> | Penicillin | Gram-positive bacteria |
> | Polymyxin | Gram-negative bacteria |
> | Chloramphenicol | Gram-negative and Gram-positive bacteria |

Gram-positive or Gram-negative organisms (Information Box 30.1). Some antibiotics are only effective against Gram-positive or Gram-negative organisms. Others are broad spectrum and are effective against either type of organism.

> **INFORMATION BOX 30.2**
>
> **Mode of Action of Widely Used Antibiotics**
>
> | Name | Spectrum | Mode of Action |
> |---|---|---|
> | Chloramphenicol | Broad spectrum | Inhibits protein synthesis by binding to 50S ribosomal subunit |
> | Erythromycin | Mostly gram-positive | Inhibits protein synthesis by binding to 50S ribosomal |
> | Tetracycline | Broad spectrum | Inhibits protein synthesis by binding to 30S ribosomal subunit |
> | Streptomycin | Broad spectrum | Inhibits protein synthesis by binding to 30S ribosomal subunit |
> | Polymyxin | Gram-negative bacteria, especially *Pseudomonas* | Disrupts cell membrane |
> | Nalidixic acid | Gram-negative bacteria | Inhibits DNA synthesis |
> | Novobiocin | Gram-negative bacteria | Inhibits DNA synthesis |
> | Trimethoprim | Broad spectrum | Inhibits purine synthesis |
> | Rifampicin | Gram-positive bacteria | Inhibits RNA synthesis |
> | Penicillin | Mostly Gram-positive bacteria | Inhibits cell wall peptidoglycan synthesis |
>
> From *Environmental Microbiology: A Laboratory Manual, Second Edition* © 2004, Elsevier Academic Press, San Diego, CA.

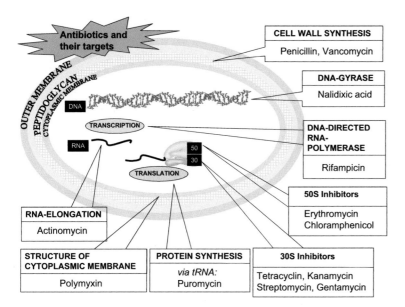

Figure 30.1 Mode of action of major antibiotic drugs.

Some common antibiotics effective against bacteria are shown in Information Box 30.2. Also shown is the mode of action of each antibiotic. The relationship between the antibiotics and their specific mode of action is illustrated in Figure 30.1.

30.3 DEVELOPMENT OF BACTERIAL ANTIBIOTIC RESISTANCE

Antibiotic drug resistance is the acquired ability of an organism to resist the effect of a chemotherapeutic agent to which it is normally susceptible. In general, there are two ways in which a microbe can become resistant:

1. **Mutational change** (spontaneous): One way microbes can become more tolerant of an antibiotic is to alter the target of an agent within the cell. For example, spontaneous mutations in the genes encoding ribosomal RNAs can prevent antibiotics such as tetracycline from binding and blocking gene translation.

2. **Introduction of foreign DNA:** In this case, genes are obtained from other host microorganisms that offer protection against antibiotics. Well-studied examples include efflux of antibiotics such as tetracycline or inactivation of penicillins by β-lactamases. The genes encoding these resistances can be localized on a plasmid or on the chromosome. A gene responsible for tetracycline resistance encodes a membrane transporter responsible for pumping tetracycline out of the cell. The β-lactamases cleave the β-lactam ring of most penicillins, thereby rendering them ineffective.

Because antibiotic-resistant genes can be transferred within and between bacteria, the issue of horizontal gene transfer becomes relevant to the discussion.

30.4 TRANSFER OF GENETIC MATERIAL BY HORIZONTAL GENE TRANSFER

Horizontal Gene Transfer (HGT) is in essence the transfer of genetic material horizontally between bacteria of the same species, or other species or genera. The genetic material can consist of discrete pieces of DNA, whole genes, or multiple genes contained within plasmids or transposons. The mechanisms for gene transfer are shown in Information Box 30.3.

30.5 PREVALENT ENVIRONMENTS FAVORING HGT

The presence of an antibiotic in the immediate environment of a resistant microbe provides selective pressure for the microbe to maintain plasmids that contain antibiotic resistance

---

INFORMATION BOX 30.3

Gene Transfer Mechanisms

Conjugation:
Transfer of genes from one prokaryotic organism to another by a mechanism involving cell-to-cell contact and a plasmid.

Transformation:
Transfer of nucleic acid via uptake of free DNA. It does not require cell-to-cell contact.

Transduction:
Transfer of nucleic acid mediated by a virus.

genes. Initially, it was generally thought that since antibiotics do not persist in the environment, there would be no selective pressure and these resistances would eventually be lost. However, this is not necessarily so. Early studies on the nature of plasmids showed that these mobile elements were often lost from laboratory strains unless selective pressure was exerted. This led researchers to the assumption that later became unexamined dogma: only selective pressure by antimicrobials would keep plasmids in a population; without selective pressure, the bacteria would get rid of this extra burden and the plasmids would be lost from any ecosystem.

However, many studies have demonstrated the persistence of antibiotic-resistant bacteria even after antimicrobial use was discontinued in hospital, community, and agricultural settings. In addition, soils that do not contain antibiotics routinely contain large populations of antibiotic-resistant bacteria. This persistence is based in part on the fact that plasmids, transposons, and integrons on the chromosome usually carry multiple resistances to antimicrobial compounds. Moreover, nonantimicrobial environmental toxins such as heavy metals can also select for multidrug-resistant plasmids. Resistances to metals that occur in the environment, such as copper, arsenic, and cadmium, are found on plasmids of both clinical and environmental Gram-positive and Gram-negative bacteria, often alongside antimicrobial resistance genes. Another reason for the persistence of plasmids in the environment is their selfish nature (Rensing et al., 2002). In its most extreme form, bacterial plasmids carry genes that kill a daughter cell if it fails to get a copy of the plasmid. These genes, referred to as **plasmid addiction** or **postsegregational killing**, are ubiquitous among large transmissible plasmids of Gram-negative bacteria. Therefore, some plasmids can not only control their own acquisition via conjugation or mobilization, but can also prevent their loss. As a result, periodic selection for any plasmid-borne resistance gene allows that plasmid and all the genes it carries to become a fixed component of the population.

One might intuitively infer that the incidence of antibiotic-resistant bacteria would be highly correlated with the use of antibiotics. Therefore, environments that might be conducive to the presence of antibiotic-resistant bacteria include hospitals and animal feedlots, where high concentrations of antibiotics are utilized. In addition, since antibiotic resistance genes are frequently located on plasmids with associated heavy metal resistance genes, then sewage and biosolids could also be a source of acquired resistance.

30.6 ISOLATION AND DETECTION OF ANTIBIOTIC-RESISTANT BACTERIA

Antibiotic-resistant bacteria can easily be isolated by diluting and plating environmental samples, including soils, onto plates of media such as R2A, amended with the antibiotic of interest. In this case, only antibiotic-resistant bacteria will be able to grow and give rise to bacterial colonies. Plating on the same media without the antibiotic allows for the estimation

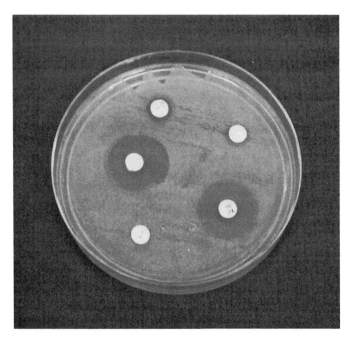

Figure 30.2 Zone of inhibition of bacterial growth on a spread plate. The inhibition is due to diffusion of antibiotics from antibiotic filter disks. Photo courtesy K.L. Josephson. From *Environmental Microbiology: A Laboratory Manual, Second Edition* © 2004, Elsevier Academic Press, San Diego, CA.

of the total culturable population and hence the percentage of antibiotic-resistant bacteria within the environmental sample. Conversely, the tolerance of specific strains of bacteria to multiple antibiotics can be ascertained through the use of antibiotic filter disks. Specifically, a spread plate (see Chapter 5) of a given bacterium can be plated and different antibiotic filter disks placed onto the medium. If the bacterium is resistant to a specific antibiotic, it will grow right up to the disk. If it is sensitive, a zone of inhibition of bacterial growth on the spread plate will occur (Figure 30.2).

30.7 INCIDENCE OF ANTIBIOTIC-RESISTANT BACTERIA IN VARIOUS ENVIRONMENTS

30.7.1 Environmental Samples and Food

When evaluating the incidence of antibiotic-resistant bacteria in environmental samples, it is important to know the total population size within the sample as well as the number of antibiotic-resistant bacteria. For environmental samples with low total bacteria numbers, such as well water, even the presence of a few resistant bacteria can result in a relatively high percentage of antibiotic-resistant microbes. In contrast, soils that contain 10^8 bacteria per gram may have large numbers of resistant bacteria and yet a relatively low percentage of resistant microbes. The relative number of total versus resistant bacteria is also important, since it indicates the potential for subsequent transfer of resistance genes from a host to an initially nonresistant recipient.

CASE STUDY 30.1

INCIDENCE OF ANTIBIOTIC-RESISTANT BACTERIA IN ENVIRONMENTAL SAMPLES AND FOODS*

A recent study at the University of Arizona evaluated the incidence of antibiotic-resistant bacteria in a variety of environmental samples and foods (Tables 30.1 and 30.2).

Tables 30.1 and 30.2 demonstrate the incidence of antibiotic-resistant bacteria. Generally, the highest incidence of resistance was to ampicillin and cephalothin. Cephalothin is a first-generation antibiotic with narrow spectrum effects against Gram-negative bacteria. Ampicillin is often administered with other antibiotics, perhaps explaining the association with cephalothin resistance.

Both environmental samples and foods contained antibiotic-resistant bacteria. Amongst the environmental samples, soils, manures, biosolids, compost, and dust all had high numbers of antibiotic-resistant bacteria. In contrast, tap water, irrigation water, and food contained lower numbers of resistant bacteria. These trends were also true for heterotrophic plate count bacteria (HPC). Interestingly, because HPC counts for water were low, the fraction of resistant bacteria is high when expressed as a percentage. Of the foodstuffs, lettuce appears to be a particularly rich source of antibiotic-resistant bacteria.

TABLE 30.1 Heterotrophic plate count in environmental samples and foods.

SAMPLE	HETEROTROPHIC PLATE COUNT --- CFU/G OR ML ---
Environmental Samples	
Tap water	10^3
Irrigation well water	10^1
Soil	10^7
Biosolids	10^9
Manure	10^8
Compost	10^9
Dust	10^7
Food	
Lettuce	10^7
Tomatoes	10^6
Chicken	10^8
Ground beef	10^6

TABLE 30.2 Antibiotic-resistant bacteria in environmental samples and foods.

SAMPLE	TETRACYCLINE	CIPROFLOXACIN	CEPHALOTHIN	AMPICILLIN
Environmental Samples				
Tap water	10^1	10^1	10^2	10^1
Irrigation well water	10	10	10^1	10^1
Soil	10^5	10^6	10^6	10^6
Biosolids	10^7	10^7	10^8	10^7
Manure	10^5	10^6	10^6	10^5
Compost	10^6	10^6	10^8	10^7
Dust	10^6	10^6	10^6	10^5
Food				
Lettuce	10^5	10^4	10^6	10^6
Tomatoes	10^1	10^1	10^3	10^2
Chicken	10^2	10^3	10^7	10^7
Ground beef	10^4	10^4	10^5	10^5

ANTIBIOTIC NUMBERS CFU/GM OR ML

*Brooks, 2004

30.7.2 Hospitals

Along with bioterrorism and homeland security, antimicrobial resistance has become one of the most important public health issues faced by the industrialized world. It is estimated that the annual cost of infections caused by antibiotic-resistant bacteria in the United States is $ 4–5 billion. The major contributing factor for the selection and propagation of antibiotic-resistant strains is the use of antibiotics. In developed countries, most of the problems of antibiotic resistance are generated within hospitals because of the intensive use of antibiotics (Levy and Marshall, 2004). **Methicillin-resistant *Staphylococcus aureus* (MRSA)** is probably the most important antibiotic-resistant bacterium associated with **hospital-acquired infections (HAI)**, since staphylococci are the most common pathogens causing bacteremia (blood infection). Bacteremia caused by MRSA are much more difficult to treat and causes thousands of deaths every year in the United States.

Measures to manage and prevent the spread of drug resistance in hospitals include:

- isolation of infected patients that are potentially hazardous

- early identification and prompt implementation of control measures

- effective hand washing measures

- new therapeutic approaches

30.7.3 Feedlots

Antibiotics are used as growth promoters to increase feed efficacy and subsequent daily growth of animals. Since many of these drugs are closely related to antibiotics used for treating human diseases, the transfer of these strains and the resistances they carry from animals to humans is of concern.

For example, chickens are fed ciprofloxacin and are also frequently infected with *Salmonella* and/or *Campylobacter* (see also Chapter 11). Therefore, there is the tendency for resistance to ciprofloxacin to increase, since microorganisms develop resistance to this antibiotic fairly easily. If a family consumes chicken that is not cooked properly, they may become infected with an ciprofloxacin-resistant *Campylobacter*. Of additional interest is the fact that ciprofloxacin is one of the primary drugs for the treatment of anthrax.

Cattle are frequently fed a 90 to 100% grain diet prior to slaughter to enhance fat marbling in the meat. However, this diet is unnatural for cows that are used to feeding on low nutrition grasses, since they are ruminant animals. High grain diets can result in acid production, which causes ulcers in the cow's stomachs. Bacteria can then migrate from the ulcers to the liver, where additional abscesses form. Because of this cattle are fed antibiotics to suppress the bacteria. Antibiotics also increase feed conversion efficiency.

However, new studies dispute the ease with which antibiotic resistance develops. A recent study examined 448 *Campylobacter* isolates from U.S. feedlot cattle for resistance to 12 antimicrobials (Englen et al., 2005). These included tetracycline, nalidixic acid, trimethoprim/sulfamethoxazone, and ciprofloxacin. The study demonstrated only low levels of resistance to a broad range of commonly used antibiotics, relative to other recent studies. Therefore, the complete story on this emerging issue has yet to be written. It also raises the question of the ease of transfer of genes between bacteria.

30.8 GENE TRANSFER BETWEEN BACTERIA—HOW PREVALENT IS IT?

Horizontal gene transfer with various outcomes and consequences can take place in all natural environments, aquatic, terrestrial, or biological. The frequency of gene transfer depends on a number of factors including the number and proportion of host and donor cells, natural settings, and the presence of biofilms. For example, high gene transfer frequencies can be observed in sewage plants.

30.8.1 Environmental Samples

The presence of antibiotic resistance genes within nonpathogenic bacteria is not a cause for public concern *per se*. However, if the resistance is passed on to a human pathogenic bacterium, then there is cause for concern. Thus,

---

CASE STUDY 30.2

GENE TRANSFER IN SOIL

The frequency of horizontal gene transfer of a large catabolic plasmid (pJP4) between bacteria was evaluated back in the mid-90s (Neilson et al., 1994). Here, a model gene transfer system was evaluated using an introduced donor organism—*Alcaligenes eutrophus* JMP134, which is now reclassified as *Ralstonia eutropha*. This donor organism contained the plasmid pJP4, and its ability to transfer this plasmid to a model recipient cell—*Variovorax paradoxus*—was evaluated. The plasmid pJP4 is responsible for the initial degradation steps of 2,4-dichlorophenoxyacetic acid (2,4-D).

In pure culture matings on spread plates, gene transfer between the donor and recipient occurred extensively. A transfer frequency of approximately 1 per 10^3 donor and recipient cells was observed on solid agar media. When introduced into sterile soil samples, the transfer frequency decreased to 1 per 10^5 donor and recipient cells. In nonsterile soil, no gene transfer was observed unless 2,4-D was added to the soil as a selective pressure. In this case, transfer frequencies were 1 per 10^6 donor and recipient cells. In soil, we can speculate that the decreased gene transfer was most likely due to special separation and physical restrictions on cell to cell contact.

the potential for horizontal gene transfer in environmental samples warrants discussion.

One theory is that genetic exchange between two organisms is a process that increases an organism's ability to adapt to an environment and thus increase its survival potential (Cohan, 1996). But why in fact do bacteria transfer genes to other bacteria that are likely in fact to be competitors? One answer is the "Selfish DNA Theory" (Rensing et al., 2002), which states that it is the selfish nature of DNA that provides the initial stimulus for adaptation, rather than the host cells themselves. Subsequent to selfish DNA transfer, if useful to host cells, the transferred DNA may become integrated into the host chromosome. Clearly, then, a definitive answer as to why gene transfer occurs is not immediately apparent. In addition, there are genetic and environmental impediments to gene transfer.

Most information on gene transfer in environmental samples has come from conjugation studies. A prerequisite for transfer of plasmids between bacteria is the presence of *tra* genes or *mob* genes. The *tra* region contains genes encoding proteins that function in DNA transfer and replication and others that function in mating pair formation. Conjugative plasmids are called self-transmissible because they carry all the genes that are necessary for a successful conjugative plasmid transfer. In contrast, mobilizable plasmids, carrying their own *mob* (for mobilization) genes, can only be transferred in the presence of a second self-transmissible plasmid in the donor cell. In addition, there are restrictions based on the compatibility of cells and plasmids. For example, some plasmids such as those belonging to the incompatibility group IncP can be maintained in almost all Gram-negative species and are therefore said to have a broad host range. In contrast, plasmids belonging to the incompatibility group ColE1 can only be replicated in *E. coli* and very close relatives, and are called narrow host range plasmids. Finally, in environments such as soils or biosolids, there are physical restrictions that decrease the chances for cell-to-cell contact and resultant conjugation.

30.9 SUMMARY AND CONCLUSIONS

Antibiotic-resistant bacteria appear to be widely dispersed throughout the environment, and yet it is only in specialized environments that their presence appears to adversely affect human health and welfare. In hospitals, antibiotic-resistant bacteria are selected for due to high usage of antibiotics and the fact that sterilized surfaces reduce biotic competition. Therefore, the potential for antibiotic-resistant pathogens is high, as is the potential for nosocomial (hospital aquired) infections. The potential adverse affects from livestock feeding of antibiotics also warrants caution. In contrast, in environments such as soils or biosolids, the potential for gene transfer to human pathogenic bacteria would appear to be lower.

QUESTIONS AND PROBLEMS

1. How can we reduce the incidence of antibiotic-resistant bacteria in the environment?
2. The data below defines bacterial numbers in well water and soil samples. Which sample has the greatest percentage of ampicillin-resistant bacteria, expressed as a percent of (a) the total population and (b) the culturable population?

	WELL WATER	SOIL
Total bacterial count	5000/ml	10^9/g
Culturable count	3000/ml	10^7/g
Culturable ampicillin-resistant Bacteria	50/ml	100/g

REFERENCES AND ADDITIONAL READING

Brooks J.P. (2004) Biological Aerosols Generated from the Land Application of Biosolids: Microbial Risk Assessment: PhD Dissertation: University of Arizona, Tucson, Arizona.

Cohan F.M. (1996) The Role of Genetic Exchange in Bacterial Evolution. ASM News **62**, 631–636.

Englen M.D., Fedorka-Cray P.J., Ladely S.R. and Dargatz D.A. (2005) Antimicrobial Patterns of *Campylobacter* from feedlot cattle. *J. Appl. Microbiol.* In press.

Levy S.B. and Marshall B. (2004) Antibacterial resistance worldwide: Causes, challenges and responses. *Nature Medicine.* **10**, S122–S129.

Neilson J.W. et al. (1994) Frequency of horizontal gene transfer of a large catabolic plasmid (pJP4) in soil. *Appl. Environ. Microbiol.* **60**, 4053–4058.

Rensing C., Newby D.T. and Pepper I.L. (2002) The role of selective pressure and selfish DNA in horizontal gene transfer and soil microbial community adaptation. *Soil Biol. Biochem.* **34**, 285–296.

CHAPTER 31

PHARMACEUTICALS AND ENDOCRINE DISRUPTORS

R. G. Arnold, D. M. Quanrud, C. P. Gerba, and I. L. Pepper

Representative estrogenic compounds.

Pharmaceuticals and personal care products (PPCPs) are bioactive chemicals that include prescription drugs, fragrances, sunscreen agents, and so forth. Pharmaceuticals that are taken orally must, to some extent, resist biochemical transformation within the human body. Thus, excreted metabolites and/or unaltered parent compounds enter sewage treatment facilities, where they can be subjected to further transformations. Bioactive compounds that resist biodegradation are discharged to surface waters that may subsequently be used as potable water sources. The possibilities for exposure of humans and wildlife to PPCPs in wastewater are apparent. However, much remains to be learned about the environmental fate and physiological effects of chronic exposure to these compounds, alone and in combination with others. In recent years, there has been a significant increase in the number of medicines administered for pain relief, control of cholesterol levels, and enhancement of sexual performance. This situation makes the fate and effects of PPCPs critical areas of research. The role of PPCPs as endocrine disruptors is of particular concern in the industrialized world.

31.1 ENDOCRINE DISRUPTORS AND HORMONES

The **endocrine system** is composed of ductless glands that secrete hormones into the bloodstream to act at distant sites. **Hormones** can be proteins, polypeptides, amino acids, or steroids. The best-known hormones are the sex steroids **estrogen**, produced in the ovaries, and **testosterone**, produced in the testes. Estrogen and testosterone are also produced in the adrenal glands of both sexes. Other hormones include thyroxin, produced in the thyroid, and insulin, produced in the pancreas. The pituitary and hypothalamus in the brain release a variety of hormones that affect other organs, including the sex glands. Together with the nervous system, the endocrine system is responsible for the integration of key processes that allow complex biological organisms to function as biochemically coherent units.

From the blood, hormones interact with cells by binding to receptor proteins (Figure 31.1). This results in the synthesis of new proteins that catalyze physiological reactions to stimulate or regulate metabolism, cell development, growth, reproduction, and behavior. In women, for example, estrogen controls the menstrual cycle. In men, testosterone controls sperm production.

Hormones are released into blood in very small amounts. Hormone levels in blood (generally ng/L to μg/L concentrations) are controlled by their respective rates of release and degradation, which normally occurs in the liver or kidneys. Soluble hormone carriers, usually proteins, are sometimes necessary to conduct hydrophobic (low solubility) hormones through the bloodstream. Estrogen carriers are known as **sex hormone binding-globulins (SHBG)**. Timing of hormone release is critical for normal function, particularly during fetal development. Suffice it to say that control of hormone levels in the bloodstream is vital for normal

---

INFORMATION BOX 31.1

Characteristics of EDCs

- EDCs interfere with the synthesis, secretion, transport, binding, action, or elimination of natural hormones in the body that are responsible for the maintenance of homeostasis (normal cell metabolism), reproduction, development, and/or behavior.

- EDCs can be hormone mimics, with hormone-like structures and activities. That is, EDCs sometimes have chemical properties similar to hormones and bind to hormone specific receptors in or on the cells of target organs.

- EDCs frequently have lower potency than the hormones they mimic (*i.e.*, require higher dose to elicit an equivalent response), but may be present in water at high concentrations relative to natural hormones. Furthermore, EDCs may not be subject to normal (internal) regulatory mechanisms.

- For all known EDCs, there is some dose below which there is no observable response.

---

bodily function. In contrast, hormone imbalance can result in disease or disorders. For example, men normally produce small amounts of estrogen, and women produce small amounts of testosterone. Too much estrogen in a male can induce enlarged breasts, whereas too much testosterone in women can result in facial hair. Hormone imbalance can also be life threatening, as in the case of insulin and diabetes.

Endocrine disrupting compounds (EDCs) are chemicals that interfere with endocrine glands, their hormones, or the activities of hormones in tissues where hormones act (Information Box 31.1).

Specifically, EDCs may affect the production or release of hormones by endocrine glands. Alternatively, some EDCs mimic while others block the action of natural hormones at target tissues (see Figure 31.1). In this way, they can accelerate or impede aspects of hormone-dependent metabolism.

31.2 SIGNIFICANCE OF EDCs IN WATER

Environmental exposure to EDCs has produced reproductive and/or developmental failures in wildlife (Information Box 31.2). Endocrine mimics such as DDT have caused spectacular environmental damage, motivating the classic book *Silent Spring* by Rachel Carson, which has been credited with the birth of environmental activism in the U.S. More recently, potential environmental harm due to anthropogenic EDCs was directly analyzed by Theo Colborn and others in *Our Stolen Future* (1996) (see also Chapter 18). In addition, there has been speculation that declining human sperm counts and increased incidence of cancers with endocrine-

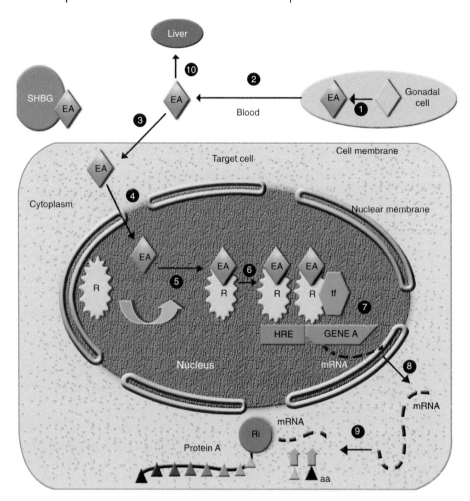

Figure 31.1 Steps necessary for cellular response to sex hormones. The steroidal sex hormones (EA) such as 17β-estradiol, testosterone, and progesterone (1) are produced in the gonads (ovaries and testes). Hormones are carried in the bloodstream (2) to hormone-responsive tissues throughout the body. Due to the limited solubility of steroid hormones, most are bound to sex hormone-binding globulin (SHBG) or other soluble proteins. Since EDCs may be more soluble than the steroid hormones or may dissociate more rapidly from SHBG, they may cross cell membranes more readily than the hormones themselves. The free hormones in the bloodstream cross the cell membrane (3) via diffusion or a more active mechanism involving membrane-bound receptors. The free hormones and EDCs diffuse across the nuclear membrane (4). In the nucleus, they encounter unbound receptor proteins – the estrogen receptor for 17β-estradiol and other estrogenic compounds – forming hormone-receptor complexes (5). Ligand/receptor complexes bind to DNA at regions called hormone receptor elements (HREs) and recruit soluble transcription factors (tf) to form transcriptional complexes (6). Normal transcription/translation follows, in which a messenger RNA (mRNA) is produced (7), crosses the nuclear membrane into the cytoplasm (8) and serves as template for new protein synthesis (9) at the ribosomes (Ri). Amino acids are designated aa. Plasma levels of natural estrogens are regulated by hormone destruction in the liver (10) and excretion. Adapted from WHO, 2002.

related causal factors (*e.g.* testicular and breast cancers) are motivated by environmental exposure to EDCs.

31.3 INCIDENCE OF EDCs IN WATER

In 1999–2000, the United States Geological Survey (USGS) carried out a comprehensive reconnaissance in U. S. streams that are potentially affected by human activities (Kolpin et al.,

2002). Up to 95 trace chemicals frequently present in municipal wastewater were measured at 139 sites across the country (Information Box 31.3; Table 31.1). Results produced a reaction among the nation's environmental community that was almost without precedent (Information Box 31.4). Eighty percent of the waters tested by the USGS contained at least one of the 95 trace contaminants, and 82 of the 95 were present at one or more of the 139 sites. Among the most frequently encountered contaminants were coprostanol, a fecal steroid (86% of samples); cholesterol, a plant and animal steroid (84%); *N*-

diethyltoluamide, an insect repellant (74%); caffeine (71%); 4-nonylphenol, a detergent metabolite and estrogen mimic (51%); and triclosan, a disinfectant (58%). In short, a surprising number of the chemicals that we use in our everyday lives end up in municipal wastewater and subsequently in waters that receive municipal wastewater effluent.

It remains difficult to assign health significance to these measurements. Some perspective is provided by comparing ingestion rates of pharmaceuticals and other organic chemicals in drinking water with their respective medicinal dosages. Table 31.2 provides the highest concentrations encountered in the USGS survey for a few pharmaceuticals and pharmaceutically active compounds. In risk analysis, the volume of water consumed each day by an adult is generally taken as 2 L. Commonplace medicinal doses for pharmaceu-

---

INFORMATION BOX 31.2

Adverse Effects of EDCs on Wildlife

· Eggshell thinning and subsequent reproductive failure of waterfowl

· Reduced populations of Baltic seals

· Reproductive failure in alligators

· Development of male sex organs in female marine animals such as whelks and snails

· Reduced or malformed frog populations

· Disruption of normal sex ratios among exposed populations of fish

---

INFORMATION BOX 31.3

Hormones and Hormone Mimics in U.S. Streams

Citing increasing pressure on the nation's potable waters, the United States Geological Survey (USGS) carried out a nationwide reconnaissance of United States surface water quality during 1999–2000. The survey was unprecedented in terms of the number and type of compounds selected and the sensitivity of the analytical methods employed. In all, USGS sampled 139 streams in 30 states (Figure 31.2). Among the selection criteria was proximity to potential sources of contamination such as highly urbanized areas and livestock production facilities. The compounds measured were classified as **organic wastewater contaminants**

(OWCs). They were selected on the basis of their presence and persistence in municipal wastewater, quantity of use, potential harm to human and environmental health, the degree to which they represent specific classes of contaminants, and feasibility of measurement at concentrations anticipated in U.S. streams. Consequently, results of the survey are representative of the streams most heavily affected by human activity, as opposed to U.S. surface waters in general.

Survey results for compounds that are recognized as hormones and hormone mimics are summarized in Table 31.1.

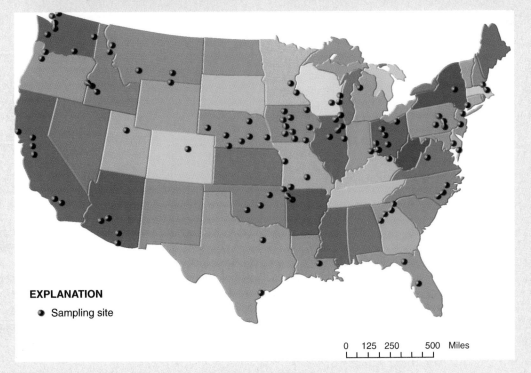

EXPLANATION

· Sampling site

0 125 250 500 Miles

Figure 31.2 Locations of stream sampling sites in the 1999–2000 reconnaissance of United States surface water quality by the USGS. Source: USGS Open File Report 02-94 (Barnes et al., 2002).

TABLE 31.1 Hormones and hormone mimics observed in U.S. surface waters.

COMPOUND	DESCRIPTION	DETECTION LIMIT ($\mu g\,L^{-1}$)	FREQUENCY OF DETECTION (%)	MAX. ($\mu g\,L^{-1}$)	MEDIAN ($\mu g\,L^{-1}$)
Progesterone	Reproductive hormone	0.005	4.1	0.199	0.11
Testosterone	Reproductive hormone	0.005	4.1	0.214	0.017
17β-Estradiol	Reproductive hormone	0.05	9.5	0.093	0.009
17α-Estradiol	Reproductive hormone	0.005	5.4	0.074	0.030
Estriol	Reproductive hormone	0.005	20.3	0.043	0.019
Estrone	Reproductive hormone	0.005	6.8	0.027	0.112
Mestranol	Ovulation inhibitor	0.005	4.3	0.407	0.017
19-Norethisterone	Ovulation inhibitor	0.005	12.2	0.872	0.048
17α-Ethinyl estradiol	Ovulation inhibitor	0.005	5.7	0.273	0.094
cis-Androsterone	Urinary steroid	0.005	13.5	0.214	0.017
4 Nonylphenol	Detergent metabolite	1.0	51.6	40	0.7
4-Nonylphenol monoethoxylate	Detergent metabolite	1.0	45.1	20	1
4-Nonylphenol diethoxylate	Detergent metabolite	1.1	34.1	9	1
4-Octyphenol monoethyoxylate	Detergent metabolite	0.1	41.8	2	0.15
4-Octyphenoldiethoxylate	Detergent	0.2	23.1	1	0.095
Bisphenol A	Plasticizer	0.09	39.6	12	0.13

Adapted from Barnes et al., 2002. Median concentrations were determined on the basis of samples in which the respective chemicals could be measured. That is, negative results were ignored in arriving at a median concentration.

INFORMATION BOX 31.4

Trace organic Measurements in U.S. Surface Waters; the Controversy over Ethinyl Estradiol

The 1999–2000 USGS measurements of organic contaminants in United States surface waters provide the broadest data set of this kind that has ever been produced. The 2-year monitoring program necessary to obtain those data was revolutionary. Its effect on our thinking with respect to water cleanliness and the adequacy of conventional wastewater treatment for protection of surface water quality has been similarly remarkable. For example, relatively few of the 139 chemicals selected for measurement by the USGS are on the EPA Priority Pollutant List. Therefore, this study considered the quality of our water resources in terms of a far broader array of chemicals, including many whose detection in chemically complex waters was not feasible until a few years ago, than did previous surveys.

It frequently is not possible to engender fundamental change without also producing controversy, and some criticism has been directed toward the USGS study. The most instructive of the criticisms surrounds synthetic estrogens. When their work was published, USGS scientists indicated that the highest concentration of 17α-ethinyl estradiol (EE$_2$) among their stream measurements was 831 ng L^{-1}. This figure is less than a part per billion but nonetheless staggering in light of its potential biological impact and relationship to previous measurements of EE$_2$ in U.S. and European surface waters. EE$_2$ at just 1–2 ng L^{-1} leads to vitellogenesis in some continuously exposed male fish. Vitellogenin is a precursor to egg yolk protein and is not normally produced by males. Drinking 2 L of water containing 831 ng L^{-1} EE$_2$ per day would produce a cumulative dose equivalent to that of some birth control pills after just 12 days. The sample that yielded this value was taken at Anoka, Minnesota, on the Mississippi River, seemingly an unlikely location for so striking a result. All other results derived from the Anoka water sample were unremarkable.

The highest EE$_2$ concentrations reported by USGS were questioned based on (1) reasonableness—comparison to expectations based on reported manufacture and use of EE$_2$ in the U.S. population; and (2) previously measured concentrations of EE$_2$ in U.S. and European surface waters (Ericson et al., 2002). It was estimated that the annual use of EE$_2$ in the United States is 170 kg. Dispersed throughout the nation's wastewaters and without attenuation by human metabolism, biodegradation, or dilution upon entry into U.S. receiving waters, the expected concentration of EE$_2$ in all waters would be 3.8 ng L^{-1}, or 245 times lower than that of the Minnesota sample. In earlier measurements of EE$_2$ in surface waters of the United States and Europe, the highest concentration reported was 15 ng L^{-1}.

It was emphasized that procedural differences between the USGS and previous studies could not account for the striking departure of the USGS results from values reported by others. USGS had, however, extracted organics from water samples via continuous liquid-liquid extraction into methylene chloride. Furthermore, samples were extracted without first removing particles via filtration. In these respects, USGS procedures differed from those used by most of their predecessors. Critics concluded that USGS may have reported an interfering signal from natural organics such as humic and fulvic acids in the same sample, citing technical considerations and work by others.

In response, the USGS team admitted that the 831 ng L^{-1} Anoka concentration had been reported in error—that their own chemists had rejected certain measurements, including this one, due to potential interferences. The maximum EE$_2$ concentration measured in the study became 273 ng L^{-1}, which is still almost 20 times higher than previously observed in surface waters of the United States and Europe.

TABLE 31.2 Representative pharmaceuticals measured in the 1999–2000 USGS reconnaissance of U.S. streams (Kolpin et al., 2002). A comparison of drinking water levels with medicinal doses.

CHEMICAL/USE	PERCENTAGE OF SAMPLES WITH COMPOUND	MAXIMUM CONCENTRATION ($\mu g\ L^{-1}$)	MEDICINAL DOSAGE
Caffeine/stimulant	71	6	130 mg[a]
Ibuprofen/anti-inflammatory	9.5	1	400 mg[b]
Cimetidine/antacid	9.5	0.58	800 mg day^{-1c}
17α-Ethinyl estradiol/oral contraceptive	16	0.831	20–35 μg[d]
Testosterone/hormone replacement	2.8	0.214	150–450 mg[e]
Erythromycin/anti-bacterial	21.5	1.7	1000 mg day^{-1f}
Ciprofloxacin/anti-bacterial	2.6	0.03	400–800 day^{-1f}

[a]The mass of caffeine in two Excedrin tablets. There is 135 mg of caffeine in an 8-oz. cup of coffee.

[b]The mass of ibuprofen in two tablets of Advil.

[c]The lowest adult daily dose of cimetidine.

[d]Range of 17α-ethinyl estradiol masses in birth control pills.

[e]Range of testosterone masses provided over 3–6 months when used for hormone replacement.

[f]Recommended adult dosages.

Source of medicinal dosages: *Physicians' Desk Reference* (2002)

ticals and other compounds are also provided. The comparison shows, for example, that daily consumption of 2 L of water containing the highest caffeine level encountered in any U.S. stream (6.0 $\mu g\ L^{-1}$) for 30 years would lead to ingestion of the mass of caffeine in two Excedrin tablets or, for that matter, in a cup of coffee. The amount of ibuprofen, an anti-inflammatory, in two Advil tablets is 400 mg. At the highest concentration measured in the USGS survey (1 $\mu g\ L^{-1}$), drinking 2 L of water per day would provide the same mass of ibuprofen after about 550 years. Clearly, such comparisons are of limited use, since some individuals may not want any ibuprofen in their water. It seems unlikely, however, that ibuprofen or caffeine will produce a health-related response among humans exposed via their drinking water.

Based on these types of crude comparisons, the compounds of primary concern in the USGS survey are pharmaceuticals that play physiological roles at very low concentrations. For example, the mass of 17α-ethinyl estradiol (EE$_2$) in some oral contraceptives is just 20 μg. The highest USGS measurement of EE$_2$ was 0.273 $\mu g\ L^{-1}$. Drinking 2 L of that water for 37 days would provide the mass of EE$_2$ in a birth-control pill. There remains some controversy regarding the accuracy of that EE$_2$ measurement (Information Box 31.4). Nevertheless, contaminants of potential concern due to exposure through water consumption are those that function biologically at exceptionally low concentrations. Hormones and hormone mimics, *i.e.*, EDCs, are in that class.

Among the endocrine disrupters present in wastewater, estrogenic compounds are most frequently acknowledged as a source of environmental problems. The frequency of intersex characteristics among male roach, a freshwater species of fish that inhabits many streams in the United Kingdom, was significantly higher among animals taken downstream from wastewater treatment plant outfalls than from corresponding upstream locations. Similar sets of developmental abnormalities have been observed among fish in at least two locations in the United States, in the South Platte River near

Denver, where sex ratios were heavily skewed toward females and some individuals developed sex tissues of both males and females, and in the Schuylkill River near Philadelphia. The wastewater compounds most frequently cited as responsible for developmental problems in exposed aquatic animals are 17β-estradiol (E$_2$, the primary mammalian estrogen), EE$_2$, and estrone (E$_1$, an estrogen metabolite that is less potent than E$_2$). Less certain are the environment effects of estrogen mimics that are also present in treated wastewater, sometimes at much higher concentrations than the hormones and hormone metabolites. Representative properties of natural estrogens and estrogen mimics are provided in Table 31.3.

31.4 FATE AND TRANSPORT OF ESTROGENIC COMPOUNDS IN MUNICIPAL WASTEWATER

From the foregoing, it is evident that a great many potential EDCs are present at trace quantities in wastewater effluent and persist in water bodies that are influenced by the disposal of treated sewage. Estrogenic compounds are probably the largest contributor to endocrine disruption by domestic wastewater since estrogens are routinely excreted to household sewage. Humans are known to excrete between 10,000 to 100,000 ng 17β-estradiol per day (Tyler et al., 1998).

Synthetic estrogens, such as 17α-ethinyl estradiol, are commonly detected in wastewater effluent and in surface waters affected by effluent discharge. Levels have been detected as high 273 ng L^{-1} (Ericson et al., 2002). Studies have shown that a concentration of 1–2 ng L^{-1} triggers vitellogenin induction in male fish (Denslow et al., 2001). A natural estrogen, 17β-estradiol, has been detected at concentrations as high as 2.5 μg/L (2,500 ng L^{-1}) in surface waters

TABLE 31.3 Physical, chemical and biochemical properties of selected estrogenic chemicals.

CHEMICAL NAME	MOLECULAR WEIGHT	WATER SOLUBILITY (μg L^{-1} AT 20°C)	LOG K$_{OW}$ (L kg^{-1})	RELATIVE ESTROGENIC ACTIVITY*
Nonylphenol (NP)	220	5.43	4.48	1–2×10^{-4}
Nonylphenol monoethoxylate	264	3.02	4.17	$\sim 10^{-5}$
Octylphenol (OP)	206	12.6	4.12	10^{-4}
17β-Estradiol (E$_2$)	272	13	3.94	1.0
17α-Ethinyl estradiol (EE$_2$)	296	4.8	4.15	1.4
Estrone (E$_1$)	270	13	3.43	0.5

*An estrogenic potency, as determined using the yeast-based *in vitro* assay of Routledge and Sumpter (1996) (relative to E$_2$).

impacted by wastewater discharges (Roefer et al., 2000). In the human male, the plasma concentration of 17β-estradiol is just 20 ng L^{-1} (Hulka and Moorman et al., 2002).

Alkylphenolic chemicals are another class of endocrine disruptors present in wastewater effluent. Alkylphenol ethoxylates (APE) are nonionic surfactants that have been used for over 40 years. They are involved in the manufacturing of plastics, elastomers, agricultural chemicals, pulping, and industrial detergent formulations. Nonylphenol ethoxylates (NPE) are the most common, accounting for 82% of production (Tyler et al., 1998). Nonylphenols and octylphenols are weakly estrogenic. They can induce vitellogenin synthesis in male fish at about 10,000 ng L^{-1} and 3,000 ng L^{-1}, respectively. Ethoxylated forms are generally less potent as estrogens.

The highest measured levels of NPE are 1,300,000 ng L^{-1} in wastewater effluent and 13,800 ng L^{-1} in surface water (Field and Reed, 1996). Alkylphenolic chemicals are transformed by biological processes during wastewater treatment. Their breakdown products, however, may be more estrogenic than the original compounds (Panter et al., 1999).

31.5 METHODS FOR MEASURING ESTROGENIC ACTIVITY IN WATER

A great number of individual chemicals will ultimately need to be tested for endocrine-related activity. There is also broad interest in EDCs in waters whose quality is tainted by human activity. These waters may contain a suite of organic contaminants with estrogenic and anti-estrogenic activities. At this time, we have hardly begun to investigate possible sources of estrogenic synergy among multiple contaminants in chemically complex waters. In addition, it will not be possible to subject all such chemicals and environmental samples to animal tests. Furthermore, analytical chemistry alone will not provide answers to questions regarding chemical fate, transport and environmental health effects on humans. Myriad organic chemicals in treated wastewater and other chemically complex waters make complete chemical description of those waters impossible. Compounds with known endocrine activities can affect animal health at very low concentrations, again contributing to the difficulty of generating a relevant chemical description of complex

waters. Although the effects of multiple estrogens simultaneously present in a single sample are generally held to be additive, the possibility of synergistic effects needs consideration. Until more is known about chemical interactions, even a complete chemical description of a water, something that is far beyond the state of the art, will not allow us to anticipate the estrogenic activity of environmental samples.

Consequently, the primary tools for determining the estrogenic or other hormone-related activities of chemically complex water samples are *in vitro* bioassays. Although quite a few such bioassays have been devised, they generally fall into three categories: (1) receptor-binding (competitive) assays; (2) those based on reporter gene expression in recombinant cell lines, and (3) tests based on cell proliferation. The three assay types differ significantly in terms of their sensitivities to known estrogens, the degree of skill or effort necessary to obtain reproducible results, their relevance as predictors of hormonal effects in higher organisms, and cost.

Competitive Assays. Receptor-binding assays are based on the affinity of hormones and hormone mimics for estrogen receptor proteins. Typically, test compounds or chemicals in complex mixtures are asked to displace a compound whose affinity for the hormone receptor is known. Detection of the displacement event can be based on compound radioactivity or other sensitive methods.

Commercial preparations are available for measuring chemical affinities for hER-α and hER-β (human estrogen receptor proteins α and β), androgen receptor protein, the human glucocorticoid hormone receptor protein and other hormone receptors. Assays of this nature can be run in a few hours and do not require extraordinary skills. Compared to alternative bioassays, receptor-binding assays are insensitive although, in environmental situations, test sensitivity can be improved by pre-concentrating sample organics.

The primary disadvantage of this type of bioassay is a consequence of their lack of biophysical relevance. Receptor-dependent response to estrogenic compounds, for example, depends on the ability of estrogens to be carried in blood and to cross both cytoplasmic and nuclear membranes (Figure 31.1). Receptor-binding assays frequently ignore transport requirements, measuring only affinities for receptor proteins. When chemically complex mixtures with relatively high organic content are tested, a degree of nonspecific binding to hormone receptors is expected. Wastewater effluent in particular might produce false positives in receptor-binding tests by displacing the marker ligands with organics that do not bind specifically to receptor active sites or with compounds that cannot be transported to hormone-responsive cells in whole organisms. Measurements of estrogenic activity in environmental samples, including wastewater effluent using both a receptor-binding assay and other *in vitro* bioassays, have been utilized to compare results. Binding assays typically yield much higher values (higher levels of estrogenic activity) than do alternative methods.

Reporter-gene Assays. There are a great number of reporter-gene assays that differ primarily in terms of either the cell type employed or the reporter gene product (β-galactosidase or luciferase). In each case, tests involve a recombinant cell line in which a specific gene, whose protein product is particularly easy to detect, is fused to a hormone response element (region of DNA where hormone/receptor complexes bind to help initiate transcription/translation). Common reporter genes code for the proteins (1) β-galactosidase, the protein that initiates lactose catabolism and that can catalyze a number of color-

producing reactions when specific dyes are substituted for lactose, or (2) luciferase, which produces light in the presence of suitable chemical reagents. Available cell lines for estrogen assays include a strain of yeast, (*Saccharomyces cerevisiae*) and several estrogen-responsive human cell types. Because they are easy to maintain and grow, yeast-based bioassays are very common. However, some reservations exist regarding differences in the transport characteristics of membranes in yeast and human cells. Furthermore, yeast strains used in this regard must be engineered to express one of the estrogen receptor proteins (Figure 31.3). By selecting human cell lines that are naturally estrogen responsive, both of these limitations can be avoided.

It is generally true, however, that *in vitro* bioassays cannot duplicate many aspects of whole-organism metabolism that may affect estrogen activity. The complexity of vertebrate endocrine systems and hormonal response mechanisms assures us of that. When *in vitro* studies provide evidence of estrogenic activity, it is logical to supplement these results with *in vivo* tests.

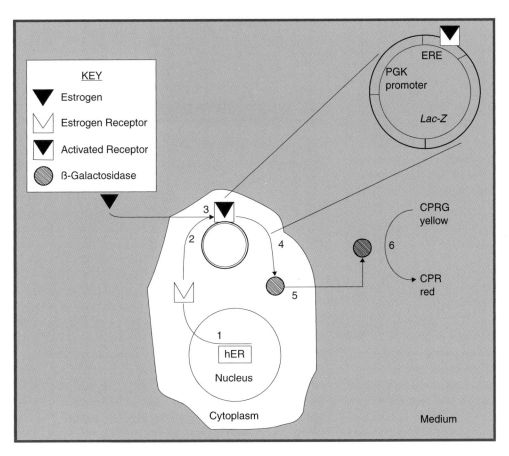

Figure 31.3 The yeast-based reporters gene assay for estrogenic activity in environmental samples (adopted from Routledge and Sumpter, 1996). The activated receptor complex between estrogen and receptor protein binds to an estrogen receptor element on plasmid DNA. In the recombinant yeast strain used, this binding event triggers production of β-galactosidase, which is excreted to the cell surroundings, where it catalyzes a color-altering dye cleavage. Estrogenic activity is assumed to be proportional to the rate of red color development. The human estrogen receptor (hER) is produced from the hER gene locus added to the yeast genome (1). The hER/estrogen complex (2) binds to the estrogen receptor element (ERE) on plasmid DNA (3). β-galactosidase is then produced by normal transcription/translation (4) and passes through the cell membrane (5). β-galactosidase converts a yellow dye to red in the growth medium (6).

TABLE 31.4 Comparion of commonly used *in vitro* bioassays for chemical estrogenic activity with a fish-based (vitellogenin production) *in vivo* assay. Sensitivities represent approximate detection limits for 17β-estradiol.

ASSAY	SENSTIVITY (nM)	TIME REQUIREMENT (DAYS)	REFERENCES
Receptor binding	2.0	1	Bolger et al., 1998
Gene expression	0.05–0.5	1–3	Routledge and Sumpter, 1996
Cell proliferation	0.01	5–7	Soto et al., 1995
Vitellogenin	0.001–0.01	10–30	Harries et al., 2000

Recently, improved *in vitro* assays have been developed that utilize stably transfected reporter-gene constructs in human cell lines (*e.g.*, Wilson et al., 2004). These assays combine the advantages of human cell models with the high-throughput and specificity of response of gene expression assays.

Cell-proliferation Assays. These assays offer great sensitivity and a degree of biochemical reality that is absent from in *vitro* bioassays that do not involve mammalian cell lines. Furthermore, mammalian cells provide possibilities for exploring mechanism-based details that cannot be pursued with recombinant yeast. Disadvantages include cost and skill and time requirements. Basically, an immortal cell line (derived from a carcinoma) that is estrogen responsive, such as T47D and MCF-7 human breast carcinoma cells, is exposed to the test chemical or sample. For the T47D cell line, the estrogen-dependent cell proliferation rate is the test response. The inability to obtain reproducible, exposure-dependent test results without extraordinary care is a limitation of proliferation assays. While test sensitivity is exceptional, the limited range of test results (degree of estrogen-dependent change in cell proliferation rate) necessarily limits test accuracy.

***In vivo* Tests.** The simplest type of *in vivo* test of chemical estrogenic activity is based on production of vitellogenin, a precursor to egg-yolk protein, by exposed males or juvenile organisms, usually fishes. Vitellogenin is assayed in serum samples or in samples developed from tissue homogenates after sacrificing the animals, usually based on an immunoassay procedure. This type of testing is expensive, time-consuming, and demanding in terms of technician skills. The chief advantages of whole-organism tests of this type are their obvious physiological and environmental relevance. The method of exposure to the test compound(s) is identical to the environmental exposure route, and biochemical response necessarily involves all steps leading to normal hormone-dependent response. In Table 31.4, vitellogenin-based assays are compared to the several classes of *in vitro* assays in terms of sensitivity and effort.

It is important to note that EDCs can either act as hormone mimics (agonists) or block the activities of natural hormones (antagonists). Both estrogen agonists and antagonists have been found in chemically complex waters like wastewaters. In general, competition assays measure the activities of both agonists and antagonists, but cannot distinguish between them. Antagonists can interfere with the response of other assays to estrogenic compounds in complex samples— a problem that can be mitigated through sample preparation steps. Such steps usually consist of efforts to separate agonists and antagonists based on hydrophobicity. Modification of either the reporter-gene or cell-proliferation assay procedures permits measurement of hormone antagonism, at least in theory.

Currently there are no federal regulations governing safe EDC levels in the environment. In 1996, amendments to the Safe Drinking Water Act and the Food Quality Protection Act were passed, directing the Environmental Protection Agency (EPA) to develop a screening program to determine which compounds in drinking water act as EDCs. This remains a work in progress.

31.6 WHAT ARE THE RISKS OF EDCs?

Many studies on the potential effects of exposure to EDCs in humans and wildlife have been carried out in the U.S. and in Europe. Similar work has not been carried out in developing countries. There appears to be ample evidence that EDCs in wastewater effluent have, in certain waters, adversely affected fish and other wildlife. There is considerable controversy as to whether human health has also been adversely affected by exposure to endocrine-active chemicals. Generally, studies examining EDC-induced effects in humans have yielded inconsistent and inconclusive results. Evidence showing human susceptibility to environmental EDCs is available from studies at high exposure levels. In contrast, reports of low-dose effects of EDCs are highly controversial and generally inconclusive. Adding to the difficulties of such studies, dose-response relationships are likely to vary for different chemicals and endocrine-disrupting mechanisms, or such relationships may be species dependent. Effects on exposed populations are likely to vary with the age, general health, and so on of the organisms.

In addition to studies on dose-response, there is a pressing need to obtain exposure data for both humans and wildlife. Current data are generally limited to groups that were inadvertently exposed at relatively high chemical doses. Potential sources of exposure include contaminated food or groundwater and contaminants in consumer products. Information on exposure during critical developmental periods is generally lacking. The exposure data that do exist are primarily from chemical levels in various environmental media such as air, food or water, and may not reflect internal concentrations in blood or endocrine-regulated tissues. Exceptions to this are human breast milk and adipose tissue.

Overall, more research is needed. It remains difficult to assess the risk to human and animal health from endocrine disruptors due to the necessity for low-dose risk extrapolations. An additional difficulty in assessing risk from endocrine disruptors is the possible synergistic effect of chemical mixtures. Priority research objectives are shown in Information Box 31.5. Until these data needs are addressed, controversy will continue to surround EDCs and their potential short- and long-term risks to environmental and human health and welfare. For obvious reasons, EDCs remain an "emerging issue."

INFORMATION BOX 31.5

Priority Research Needs for EDCs

- Dose-response effects of EDCs
- Relevant exposure studies for both wildlife and humans
- Enhanced monitoring of "sentinel" wildlife species
- Development of new bioassays for EDCs
- Identification of new EDCs
- Development of global database effects
- Synergistic effects of multiple EDC exposures

QUESTIONS AND PROBLEMS

1. Why is estrogen in drinking water a concern? What are the environmental effects of exposure to anthropogenic estrogens?

2. Why is DDT considered an endocrine disruptor?

3. List four common pharmaceuticals found in wastewater and their sources.

4. How do hormones control body functions?

5. Name three assay methods that are used to measure the activities of EDCs. What are their advantages and limitations?

6. What is a reporter gene?

REFERENCES AND ADDITIONAL READING

Barnes K.K. et al. (2002) Water-quality data for pharmaceuticals, hormones, and other organic wastewater contaminants in U.S. streams, 1999–2000. USGS Open-File Report 02–94.

Bolger R., Wiese T.E., Ervin K., Nestich S. and Checovich W. (1998) Rapid screening of environmental chemicals for estrogen receptor binding capacity. *Environ. Health Perspect.* **106**, 551–557.

Carson R. (2002*). Silent Spring.* Houghton Mifflin Company, Boston, Massachusetts.

Colborn T., Dumanoski D. and Myers J.P. (1996). *Our Stolen Future.* Penguin Books, New York.

Denslow N.D., Bowman C.J., Ferguson R.J., Stehen L.H., Hemmer M.J. and Folmar L.C. (2001). Induction of gene expression in sheepshead minnows (*Cyprinodon variegatus*) treated with 17β-estradiol, diethylstilbestrol, or ethinylestradiol. *Gen. Comp. Endocrinol.* **21**, 250–260.

Ericson J.F., Laenge R. and Sullivan D.E. (2002) Comment on "Pharmaceuticals, hormones, and other organic wastewater contaminants in U.S. streams, 1999–2000: A national reconnaissance." *Environ. Sci. Technol.* **36**, 4005–4006.

Field J.A. and Reed R.L. (1996) Nonylphenol polyethoxycarboxylate metabolites of nonionic surfactants in U.S. paper mill effluents, municipal sewage treatment plants, and river waters. *Environ. Sci. Technol.* **30**, 3544–3550.

Harries J.E. et al. (2000) Development of a reproductive performance test for endocrine disrupting chemicals using pair-breeding fathead minnows (*Pimephales promelas*). *Environ. Sci. Technol.* **34**, 3003–3011.

Hulka B.S. and Moorman P.G. (2002) Breast cancer: Hormones and other risk factors. *Maturitas.* **42**, 105–108.

Kolpin D.W. et al. (2002) Pharmaceuticals, hormones, and other organic wastewater contaminants in U.S. streams, 1999–2000: A national reconnaissance. *Environ. Sci. Technol.* **36**, 1202–1211.

Panter G.H., Thompson R.S., Beresford N. and Sumpter J.P. (1999) Transformation of a non-estrogenic steroid metabolite to an estrogenically active substance by minimal bacterial activity. *Chemosphere.* **38**, 3579–3596.

Physicians' Desk Reference (2002). Medical Economics Company, Montvale, New Jersey.

Roefer P., Snyder S., Zegers R.E., Rexing D.J. and Fronk J.L. (2000) Endocrine disrupting chemicals in a source water. *J. Am. Water Works Assoc.* **92**, 52–58.

Routledge E.J. and Sumpter J.P. (1996) Estrogenic activity in surfactants and some of their degradation products assessed using a recombinant yeast screen. *Environ. Toxicol. Chem.* **15**, 241–258.

Soto A. et al. (1995) The E-Screen assay as a tool to identify estrogens: an update on estrogenic environmental pollutants. *Environ. Health Perspect.* **103**, 113–122.

Tyler C.R., Jobling S. and Sumpter J.P. (1998) Endocrine disruption in wildlife: A critical review of the evidence. *Crit. Rev. Toxicol.* **28**, 319–361.

Wilson V.S., Bobseine K. and Gray L.E. Jr. (2004) Development and characterization of a cell line that stably expresses an estrogen-responsive luciferase reporter for the detection of estrogen receptor agonist and antagonists. *Toxicol. Sci.* **81**, 69–77.

World Health Organization (WHO). (2002). Global Assessment of the State-of-the-Science of Endocrine Disruptors. WHO/PCS/EDC/02.2.

CHAPTER 32

EPILOGUE: IS THE FUTURE OF POLLUTION HISTORY?

I. L. Pepper, C. P. Gerba, and M. L. Brusseau

A pollution free sunrise. *Photo courtesy K. L. Josephson.*

We know that environmental problems exist at all levels: local, state, national, and global. But, as we have seen, new technologies and approaches to mitigate the effects of pollution are available in increasing abundance. In general, we have the knowledge and the technology to solve many pollution problems. What is it, then, that prevents us from making the future of pollution history? The answer is complex, of course, and depends on a number of different factors, including financial, political, and societal parameters. Ultimately, however, we can say that the level of pollution currently around us is a function of how serious we perceive the problem to be, and how much we are willing to pay for clean air, water, and land.

32.1 THE ROLE OF GOVERNMENT IN CONTROLLING POLLUTION

While pollution science can help provide an answer to the question of how clean is clean, and can offer us the technological ability to mitigate the effects of environmental pollution, it cannot resolve all the major questions, such as: What clientele are we trying to satisfy? This question currently revolves around the needs of the general public, corporations, and environmental groups. While none of these groups are well defined, it is generally fair to say that the general public is often caught up in conflict between the other two. On the one hand, corporate entities tend to focus on the economic prosperity of the general public. On the other hand, environmental groups, such as Greenpeace and the Sierra Club, tend to focus on ecological and health-related concerns of the public. Both of these foci are equally legitimate, and each is fundamentally informed by the other.

In addition, the news media—the so-called "fourth estate"—play a critical role in defining both the needs and desires of the general public. With the help of the news media, many disparate environmental groups have succeeded in making the public more environmentally conscious. Moreover, the media have been instrumental in promoting environmentally responsible behavior on the part of corporations and governments. But the news media, which tend to focus on attention-getting headlines, have also exacerbated the conflict between environmental groups and corporate (and sometimes state and national) entities. Arguably, this inevitable tension among environmental activists and corporate groups, filtered through the broad lens of the media, will shape the agenda of politicians who must deal with environmental problems.

Generally, it is the government—the ultimate manifestation of public opinion—that must control the level of pollution in individual countries. This control, however, can take several forms, from outright compulsion and taxes to economic incentives. In the United States, for example, the federal government is currently making tough choices between enhanced or relaxed environmental standards. Among the decision-making tools being used is the concept of cost-benefit ratios, which have been used for many years to frame corporate decisions. As applied to environmental remediation, for example, decision makers attempt to place a dollar value on the cost of a particular cleanup action and likewise a dollar value on its associated benefits. In theory, this method should result in actions whose implementation would yield a positive net benefit. As attractive as this concept is in theory, it is limited in practice because it requires the quantification of qualitative factors. How, for instance, do you put a dollar value on breathing clean air, drinking clean water, and eating uncontaminated food? On the other hand, how can we justify pricing these benefits such that only a very few can afford to enjoy them?

In addition, the advent of new supersensitive analytical methods means that we now have the capability to detect contaminants at very low concentrations in environmental samples. A decade ago, analyses of these same environmental samples would have resulted in zero detection of contamination. Therefore, samples that 10 years ago were noncontaminated would now be found to contain some contaminants. This raises the question of what is an acceptable level of pollution, and to what degree should we clean up pollution?

At times the problem of paying for clean up affects people in the community directly rather than through government. This raises the issue of how are we going to deal with pollution caused by individuals, rather than the private sector. In this case, costs for cleanup become even more prohibitive, as in the case of leaking underground storage tanks located on residential properties (see Case Study 32.1).

32.2 RESEARCH PRIORITIES NECESSARY TO PROTECT HUMAN HEALTH

Human health depends on genetic disposition and environmental exposure. Of these two factors, we can more easily control the latter than the former. Therefore, human health protection can be enhanced by controlling exposure to contaminants via pathways that affect what we breathe, eat, and drink. To this end, there are significant research priorities that need to be addressed in the coming years (Information Boxes 32.1, 32.2, and 32.3).

32.3 POLLUTION PREVENTION OF EARTH, AIR, AND WATER

When examining the contamination caused by a particular activity or at a specific site, we tend to focus on a single medium. For example, when evaluating hazardous waste sites, we often concentrate on groundwater contamination. However contrived, this separation of land, water, and air media is especially pervasive in the regulatory arena. Separate environmental laws exist for controlling pollution

---

INFORMATION BOX 32.1

Research Priorities Affecting What We Breathe

Air pollution is ubiquitous, and frequently long-term exposures occur before adverse health affects are noticed. In particular, studies are needed that evaluate exposure estimates and dose response effects of mixtures of airborne contaminants. These mixtures include both biotic and abiotic entities:

·Dust and *Mycobacterium tuberculosis*
·Ozone and pollen
·Chemicals and pathogens
·Mixtures of pathogens
·Mixtures of organic chemicals
·Mixtures of metals

Adapted from National Research Council, 2006 .

INFORMATION BOX 32.3

Research Priorities Affecting What We Drink

Of all the human resources, water is perhaps the most vital. Ironically, access to good-quality potable water is not available in many parts of the world, and even in the United States, outbreaks of waterborne disease attributable to drinking water still occur quite frequently. Problems that need to be resolved include:

·Evaluation of water quality produced by treating effluent for potable reuse
·Evaluation of the fate of pharmaceuticals in water
·Occurrence of naturally occurring substances such as arsenic in water
·Incidence of emerging pathogens, including opportunistic pathogens, in nondisinfected potable water sources such as well water.

Adapted from National Research Council, 2006.

---

in each medium. In some cases, this compartmentalization of systems, which has facilitated the growth of the existing regulatory structure, has resulted in sets of fractionated, and sometimes overlapping and/or contradictory, rules.

An alternative approach—one that can integrate the several media—is finding increasing acceptance in the scientific and economic community. This integrated approach considers each medium as part of a whole and allows us to evaluate the synergistic and antagonistic interactions that can and do occur in real systems. Such an approach, however, is fundamentally multi- and interdisciplinary in nature, requiring the active collaboration of many individuals trained in such disparate disciplines as science, engineering, economics, and public policy. This will also require the training of scientists with broader areas of expertise who are willing to participate in interdisciplinary research. Although difficult to achieve, this approach will be important for future pollution management and regulation.

Throughout this book we have seen that many human activities in the environment can create pollution problems

---

INFORMATION BOX 32.2

Research Priorities Affecting What We Eat

Many people believe that we are what we eat. However, even now, humans consume food that has been contaminated with pathogens such as *Salmonella* or chemicals such as mercury in fish. Examples of studies that are needed include:

·Evaluation of long-term, low-dose exposure via the food chain to toxic elements such as arsenic
·Evaluation of exposure to multiple metals
·The role of metal speciation in bioavailability in soils for plant uptake and subsequent bioavailability to humans

Adapted from National Research Council, 2006.

---

that are exceedingly difficult to solve and even more difficult to pay for. It should therefore be obvious that the most beneficial and cost-effective approach to pollution control is to stop it before it starts. Thus, pollution prevention has become a widely accepted goal—a goal that is being implemented to an increasing degree. Pollution prevention can take many forms. For example, best management practices have been developed to optimize the amount of fertilizers and pesticides for crop production—that is, to use only the amounts necessary to support growth. Industry, too, is developing optimization procedures to use chemicals efficiently and minimize waste generation.

32.4 IS THE FUTURE OF POLLUTION HISTORY?

Technologically speaking, the answer is yes. But in reality, the answer is no. In the United States, until the late 1960s it was considered acceptable practice to routinely dump chemical wastes into soil or the oceans. Since that time, society has come to understand that such practices are unacceptable, and they are now in fact illegal. However, in many cases we are still paying the price for earlier acts of pollution. In retrospect, it is staggering to think that people actually thought that iron storage tanks could be buried underground without ultimately rusting and leaking. What other acts are we now committing, that 50 years from now will be considered as foolish with respect to pollution prevention? Worldwide there are even greater problems. Enormous amounts of pollution ravage much of Eastern Europe and the Far East. In addition, the developing countries, with their burgeoning economy and quickly growing populations, will place increasing demands on world resources.

CASE STUDY 32.1

A FAMILY WITH A LEAKING UNDERGROUND STORAGE TANK.

As in the case of many properties throughout the U.S., a family-owned farm in Ohio used an underground tank for gasoline storage. The tank was approximately 25 years old, but had not been used for over 2 years. Because of this, the family decided to remove the tank, and following the removal, it was found that the tank had been leaking. The consulting firm involved in the removal was now charged with cleaning up the contamination caused by the leaking fuel. The contaminants found were essentially BTEX components, namely, benzene, toluene, ethyl benzene, and xylene. Now it was necessary to find the extent of contamination, and to do this soil cores were drilled radially around the prior site of the tank, and soil and groundwater samples collected (Figure 32.1).

Data collected showed that significant BTEX components were found within cores 1, 2, and 3, with decreased soil and groundwater contamination at locations 4, 5, and 6. Locations 7 and 8 showed even less contamination. However, location 9 showed the highest level of groundwater contamination (6 ppm benzene), above the regulatory limit of 5 ppb benzene. Based on the hydraulic gradient at the site, contamination was unlikely to have come from the leaking underground storage tank. Further, the location of the contamination suggested that contamination was coming from the neighboring property. However, the neighbors refused to have cores taken on their property. What then can state agencies do in such a situation? The answer is very little, unless a court case ensues. The family with the leaking underground storage tank spent approximately $50,000 on the site characterization and clean up. The extent of contamination has still not been resolved (at the time of press), which precludes the family from selling the residence. Here, then, is a classic case where technology is available to resolve the pollution problem, but economic and social problems prevent the clean up from occurring. Problems like these occur all over the United States, and the extent of contamination of soil and aquifers is in many cases completely unnoticed. Likewise, the incidence of adverse public health effects is unknown.

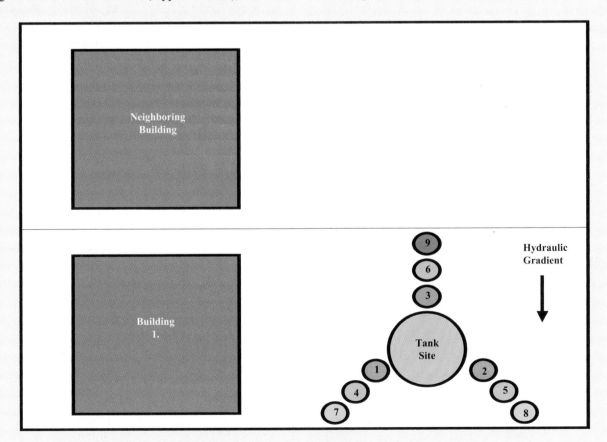

Figure 32.1 Location of soil cores and wells placed around the location of the underground storage tank. Numbers refer to the specific location of a particular well.

Because of these problems, environmental and pollution science will become ever more important. However, it will be society itself that will need to drive not only the application of new technologies for clean up, but also the implementations of pollution prevention and reuse of nonrenewable resources. As the world's population continues to rise, and as natural resources become scarce, this holistic approach will become increasingly important in maintaining a healthy economy and environment, and protecting public health and welfare.

REFERENCES AND ADDITIONAL READING

NRC (National Research Council). (2006) *Research Priorities for Earth Science and Public Health.* National Academy Press, Washington, D.C.

INDEX